## SECOND EDITION

**Harish Kumar Moudgil**
Assistant Professor
Department of Chemistry
Chaudhary Devi Lal University
Sirsa, Haryana

**PHI Learning Private Limited**
Delhi-110092
2015

₹ 595.00

**TEXTBOOK OF PHYSICAL CHEMISTRY, Second Edition**
Harish Kumar Moudgil

© 2015 by PHI Learning Private Limited, Delhi. All rights reserved. No part of this book may be reproduced in any form, by mimeograph or any other means, without permission in writing from the publisher.

**ISBN-978-81-203-5062-5**

The export rights of this book are vested solely with the publisher.

**Second Printing (Second Edition)**    …    …    …      **November, 2014**

Published by Asoke K. Ghosh, PHI Learning Private Limited, Rimjhim House, 111, Patparganj Industrial Estate, Delhi-110092 and Printed by Rajkamal Electric Press, Plot No. 2, Phase IV, HSIDC, Kundli-131028, Sonepat, Haryana.

# Contents

*Foreword* ............................................................................................................ xix
*Preface* ............................................................................................................... xxi
*Acknowledgements* ......................................................................................... xxiii
*Nomenclature* ................................................................................................... xxv

## Part I  THERMODYNAMICS

### 1. Classical Thermodynamics .......................................................... 3–49

    1.1    Some Basic Terms and Concepts   *3*
    1.2    Zeroth Law of Thermodynamics   *9*
    1.3    First Law of Thermodynamics   *9*
    1.4    Types of Work   *10*
    1.5    Application of First Law of Thermodynamics   *10*
    1.6    Limitations of First Law of Thermodynamics   *11*
    1.7    Second Law of Thermodynamics   *12*
    1.8    Entropy(s)   *13*
    1.9    Entropy Change for an Ideal Gas with Temperature and Volume   *14*
    1.10  Entropy Change for an Ideal Gas with Temperature and Pressure   *15*
    1.11  Entropy Change in Reversible Process   *16*
    1.12  Entropy Change in Irreversible Process   *16*
    1.13  Entropy of Phase Transition   *17*
    1.14  Free Energy ($G$) and Work Function ($A$)   *17*
    1.15  Helmholtz Free Energy Function $A = E - TS$   *17*

iii

1.16 Gibbs Free Energy Function ($G$): $G = H - TS$    18
1.17 Variation of Free Energy with Temperature and Pressure    18
1.18 Standard Free Energy Change ($\Delta G^0$) and Equilibrium Constant ($K$)    19
1.19 Criteria for Feasibility of a Process    19
1.20 Chemical Potential ($\mu$)    22
1.21 Gibbs–Duhem Equation    23
1.22 Variation of Chemical Potential with Temperature and Pressure    23
1.23 Gibbs–Helmholtz Equation or Effect of Temperature on Chemical Potential    24
1.24 Clausius–Clapeyron Equation    25
1.25 Integrated Form of Clausius–Clapeyron Equation    25
1.26 Relation between Enthalpy and Chemical Potential    27
1.27 Nernst Heat Theorem    27
1.28 Third Law of Thermodynamics    29
1.29 Concept of Residual Entropy or Unattainability of Absolute Zero    30
1.30 Fugacity    33
1.31 Determination of Fugacity by Graphical Method    34
1.32 Determination of Fugacity from Equation of State    36
1.33 Generalized Method for Determining Fugacity    37
1.34 Activity and Activity Coefficient    38
1.35 Phase Rule    40
1.36 Various Terminology Used in Phase Rule    40
1.37 Types of Equilibrium    41
1.38 Derivation of Phase Rule ($F = C - P + 2$)    41
1.39 Phase Rule for Chemically Reacting Systems    41
1.40 Uses of Phase Rule    42
1.41 Limitations of Phase Rule    42
1.42 Phase Diagram    42
1.43 One-Component ($H_2O$) System    42
1.44 Two-Component System    43
1.45 Systems Forming Solid Compound $A_xB_y$ with Congruent Melting Point    45
1.46 Compounds with Incongruent Melting Point    46
*Exercises*    47
*Suggested Readings*    47

## 2. Irreversible Thermodynamics or Non-equilibrium Thermodynamics ........................................................................ 50–67

2.1 Steady State or Stationary State    50
2.2 Postulates of Non-equilibrium Thermodynamics    51
2.3 Microscopic Reversibility and Onsager Reciprocity Relation (A Proof)    51
2.4 Phenomenological Laws and Onsager's Reciprocol Relations    52
2.5 Entropy    54
2.6 Entropy Production Due to Heat Flow    56
2.7 Entropy Production and Its Rate in Matter (or Mass) Flow    57
2.8 Entropy Production and Its Rate in a Chemical Reaction    59

2.9 Entropy Production Due to Electrochemical Reactions  *61*
2.10 Electrokinetic Effects–Saxen Relation  *62*
2.11 Thermo-molecular Pressure Difference (TPD) and Thermo-mechanical (or Mechano-Calorific) Effect  *64*
*Exercises  66*
*Suggested Readings  66*

## 3. Statistical Thermodynamics .................................................. 68–104

3.1 Basic Terminology Used in Statistical Thermodynamics  *69*
3.2 Partition Function ($Q$)  *70*
    3.2.1 Properties of Partition Function  *71*
    3.2.2 Factorization of Partition Function or Multiplication Theorem of Partition Function  *71*
3.3 Partition Function and Its Thermodynamic Relation with Other Thermodynamic Functions  *71*
    3.3.1 Relation of Partition Function with Energy  *71*
    3.3.2 Relation of Partition Function with Heat Capacity at Constant Volume ($C_V$)  *73*
    3.3.3 Relation of Partition Function with Entropy  *73*
    3.3.4 Another Relation of Entropy, $S$ with Partition Function, $q$  *75*
    3.3.5 Relation of Partition Function with Gibbs Free Energy  *76*
    3.3.6 Relation of Partition Function with Work Function (Helmholtz Free Energy)  *77*
    3.3.7 Relation of Partition Function with Pressure  *77*
    3.3.8 Relation of Partition Function with Enthalpy (Heat Content)  *78*
    3.3.9 Relationship between Equilibrium Constant and Partition Function  *78*
3.4 Translational Partition Function  *80*
    3.4.1 Calculation of Translational Energy from Translational Partition Function  *81*
    3.4.2 Translational Entropy of Monoatomic Ideal Gas or Sackure–Tetrode Equation  *82*
    3.4.3 Translational $C_V$  *84*
    3.4.4 Translational Enthalpy  *84*
    3.4.5 Translational $C_P$  *84*
3.5 Rotational Partition Function  *84*
    3.5.1 Calculation of Rotational Energy  *86*
3.6 Vibrational Partition Function  *87*
    3.6.1 Vibrational Energy  *88*
    3.6.2 Vibrational Entropy  *88*
3.7 The Electronic Partition Function  *89*
3.8 Effect of Change of Zero Point Energy on Partition Function  *91*
3.9 Effect of Change of Zero Point Energy on Energy  *92*
3.10 Effect of Change of Zero Point Energy on Entropy  *93*

3.11  Effect of Change in Zero Point Energy on Gibbs Free Energy  *94*
3.12  Effect of Zero Point Energy on Work Function  *94*
3.13  Separation of Internal Partition Function  *95*
3.14  Polyatomic Molecules  *96*
3.15  Rotational Partition Function for Polyatomic Molecules  *97*
3.16  Free Energy and Equilibrium in Gaseous Systems  *100*
3.17  Partition Function and Equilibrium Constant  *102*
*Exercises  103*
*Suggested Readings  103*

## 4. Statistical Mechanics ............................................................. 105–142

4.1  Ensembles (Collection)  *105*
4.2  Stirling's Formula for $N$  *107*
4.3  Thermodynamic Probability  *107*
4.4  General Statistical Distribution Law  *108*
4.5  Most Probable Distribution  *110*
4.6  Types of Ensemble  *111*
4.7  Fluctuation  *112*
4.8  Distribution Function for Fluctuation: General Theory of Fluctuation  *113*
4.9  Reason for the Fact that $x$ Fluctuates Slightly from $x_0$  *114*
4.10  Energy Fluctuation in Canonical Ensemble  *114*
4.11  Fluctuation in Density  *115*
4.12  Maxwell–Boltzmann Distribution Law  *117*
4.13  Evaluation of Constants $\alpha$ and $\beta$  *120*
4.14  Maxwell–Boltzmann Distribution Law of Velocities  *122*
4.15  Bose–Einstein Statistics  *124*
4.16  Application of Bose–Einstein Statistics for a Photon Gas  *126*
4.17  The Fermi–Dirac Statistics  *129*
　　4.17.1  Applications of Fermi–Dirac Statistics  *130*
4.18  Comparison of Three Statistics  *139*
*Exercises  141*
*Suggested Readings  141*

## Part II  SPECTROSCOPY

## 5. Quantum Mechanics ............................................................. 145–225

5.1  Schrödinger Wave Equation (SWE)  *146*
　　5.1.1  Postulates of Quantum Mechanics  *147*
　　5.1.2  Derivation of SWE Based on de Broglie Dual Nature and Heisenberg Uncertainty Principle  *150*
　　5.1.3  Derivation of SWE Based on Postulates of Quantum Mechanics  *151*
5.2  Solution of Schrödinger Wave Equation to Particle in a Box  *152*
　　5.2.1  Particle in a One-dimensional Box  *152*

- 5.3 Characteristics of the Wavefunction or Quantization of Energy *155*
- 5.4 The Harmonic Oscillator *158*
  - 5.4.1 Linear Harmonic Oscillator *158*
  - 5.4.2 Harmonic Oscillator: Asymptotic Solution *160*
  - 5.4.3 Energy Levels of Harmonic Oscillator *162*
  - 5.4.4 Eigenfunction of Harmonic Oscillator *164*
- 5.5 Zero Point Energy or Residual Energy *164*
- 5.6 The Rigid Rotator *170*
  - 5.6.1 Energy and Moment of Inertia of Rotator *171*
  - 5.6.2 Wave Equation for Rotator *172*
  - 5.6.3 Legendre's Equation for Rigid Rotator *174*
  - 5.6.4 Energy Levels of Rigid Rotator *176*
- 5.7 Hydrogen Atom or Hydrogen Like Atom: A Central Force Field Problem *181*
- 5.8 Separation of Variables *182*
- 5.9 Spherical Eigenfunction *183*
- 5.10 Solution of Radical Wave Function *184*
- 5.11 Atomic Energy Level and Quantum Number *186*
- 5.12 Normalized H-like Eigenfunction *186*
- 5.13 Eigenfunction and Probability Distribution *188*
- 5.14 Approximation Methods *200*
  - 5.14.1 Perturbation Method: An Overview *201*
  - 5.14.2 Perturbation Theory to Solve Eigenfunction and Eigenvalue *202*
  - 5.14.3 Variation Method *206*
  - 5.14.4 Approximation Method for the Ground State of the Helium Atom *208*
- 5.15 Variation Method Applied to Helium Atom *210*
- 5.16 Angular Momentum in Quantum Mechanics *213*
- 5.17 Angular Momentum Operator *215*
- 5.18 The Ladder Operator *217*
- 5.19 Eigenvalues $\hat{J}^2$ and $\hat{J}_z$ *218*
- 5.20 The Pauli Exclusion Principle *220*
- 5.21 Electronic Structure of Atoms: Russell–Saunder Coupling (R–S Term) *221*

*Exercises 224*
*Suggested Readings 224*

# 6. Introduction to Molecular Spectroscopy .......................... 226–246

- 6.1 Introduction *226*
- 6.2 Electromagnetic Spectrum *227*
- 6.3 Types of Spectroscopy *230*
- 6.4 Scattering *230*
- 6.5 Signal-to-Noise Ratio *230*
- 6.6 Resolving Power *231*
- 6.7 Width of the Spectral Lines *231*
  - 6.7.1 Doppler Broadening *232*
  - 6.7.2 Lifetime Broadening *232*

| | | | |
|---|---|---|---|
| | 6.8 | Intensity of Spectral Lines *233* | |
| | 6.9 | Degrees of Freedom of Motion *233* | |
| | 6.10 | Rotational Spectroscopy *234* | |
| | | 6.10.1 Rotational Active Molecules *234* | |
| | | 6.10.2 Energy Level of a Rigid Rotator *235* | |
| | | 6.10.3 Rotational Selection Rules *237* | |
| | | 6.10.4 Rotational Spectra of the Diatomic Molecules *237* | |
| | | 6.10.5 Relative Intensity of Rotational Spectral Lines *239* | |
| | | 6.10.6 Hyperfine Interactions *240* | |
| | 6.11 | Rotational–Vibrational Spectroscopy *241* | |
| | | 6.11.1 Evaluating Vibrational Spectra *241* | |
| | 6.12 | When Does a Molecule Absorb Infrared Light? *243* | |
| | 6.13 | Overtone, Combination Frequencies and Fermi Resonance *244* | |
| | 6.14 | Applications of Vibrational Spectroscopy *244* | |
| | *Exercises 245* | | |
| | *Suggested Readings 246* | | |

## 7. X-Ray Crystallography .................................................................. 247–263

    7.1    Space Lattice and Unit Cell   *247*
    7.2    Hauy's Law of Rationality of Indices   *249*
    7.3    Miller Indices   *250*
    7.4    Deriving Bragg's Law   *251*
    7.5    Experimental Setup of Braggs Method of Structure Determination   *252*
    7.6    Determination of Crystal Structure by Bragg's Method   *253*
    7.7    Separation between Lattice Planes   *255*
    7.8    X-Ray Diffraction Patterns of Cubic Lattices   *257*
    7.9    Experimental Diffraction Patterns   *260*
    7.10   First Discovery of X-ray Diffraction   *261*
    7.11   Detection of Diamonds with the Help of X-Ray Diffraction Study   *261*
    *Exercises 262*
    *Suggested Readings 262*

## 8. Solid State Chemistry .................................................................... 264–334

    8.1    Introduction   *264*
    8.2    Wagner Reaction Mechanism of Solids   *265*
    8.3    Kinetics of Solid State Reactions   *266*
    8.4    Thermal Decomposition Reaction   *266*
    8.5    Free Energy of Nucleation   *267*
    8.6    Laws Governing Nucleation   *268*
    8.7    Growth of Nuclei   *269*
    8.8    Electronic Properties and Band Theory of Metals, Insulators and Semiconductors   *270*
    8.9    Electronic Structure of Solids—Band Theory   *271*

| | | |
|---|---|---|
| 8.10 | Energy Bands in Solids  *273* | |
| 8.11 | Formation of Bands  *273* | |
| 8.12 | Energy Bands in Conductors, Insulators and Semiconductors  *274* | |
| | 8.12.1  Conductors  *274* | |
| | 8.12.2  Insulators  *276* | |
| | 8.12.3  Semiconductor  *276* | |
| 8.13 | Intrinsic Semiconductor  *277* | |
| 8.14 | Extrinsic Semiconductors  *278* | |
| 8.15 | Types of Impurities  *278* | |
| 8.16 | Types of Extrinsic Semiconductors  *278* | |
| | 8.16.1  p-Type or p-Type Semiconductor  *278* | |
| | 8.16.2  n-Type or n-Type Semiconductor  *280* | |
| 8.17 | p–n Junction  *281* | |
| | 8.17.1  Formation of p–n Junction  *281* | |
| | 8.17.2  Formation of p–n Junction Diode  *281* | |
| 8.18 | Explanation and Energy Band Diagram of p–n Junction  *282* | |
| | 8.18.1  Forward Bias  *283* | |
| | 8.18.2  Reversed Bias  *284* | |
| 8.19 | Superconductivity  *285* | |
| | 8.19.1  Some Basic Facts about Superconductivity  *285* | |
| | 8.19.2  Type I and Type II Superconductors  *287* | |
| | 8.19.3  Isotope Effect  *289* | |
| 8.20 | Persistent Current and Flux Quantization  *289* | |
| 8.21 | Specific Heat  *290* | |
| 8.22 | Two Types of Superconductivity  *291* | |
| | 8.22.1  Low Temperature Superconductivity (LTSC)  *291* | |
| | 8.22.2  High Temperature Superconductivity (HTSC)  *291* | |
| 8.23 | Occurrence of Superconductivity  *292* | |
| | 8.23.1  Convential Superconductor  *292* | |
| | 8.23.2  Organic Superconductor  *293* | |
| | 8.23.3  Fullerenes  *294* | |
| 8.24 | Applications of Superconductor  *294* | |
| 8.25 | Organic Solid State Chemistry  *295* | |
| | 8.25.1  Intramolecular Reactions  *295* | |
| | 8.25.2  Intermolecular Reactions  *296* | |
| 8.26 | Electrically Conducting Organic Solids  *296* | |
| | 8.26.1  Conjugated Systems  *296* | |
| 8.27 | Organic Charge Transfer Complexes: New Superconductors  *299* | |
| 8.28 | Defects in Solids: Perfect and Imperfect Crystals  *299* | |
| 8.29 | Types of Defects  *300* | |
| | 8.29.1  Stoichiometric Defects  *300* | |
| | 8.29.2  Nonstoichiometric Defects  *300* | |
| | 8.29.3  Point Defects  *301* | |
| 8.30 | Schottky Defect  *301* | |
| | 8.30.1  Number of Schottky Defects  *302* | |

## Contents

- 8.31 Frenkel Defect  *304*
- 8.32 Thermodynamics of Point Defects  *305*
    - 8.32.1 Thermodynamics of Schottky Defect  *305*
    - 8.32.2 Thermodynamics of Frenkel Defects  *307*
- 8.33 Metal Excess Defects  *307*
- 8.34 Metal Deficiency Defect  *309*
- 8.35 Line Defects or Dislocations  *310*
    - 8.35.1 Edge Dislocations  *310*
    - 8.35.2 Screw Dislocations  *311*
- 8.36 Plane Defects  *311*
- 8.37 Diffraction Methods; Space Lattice  *311*
- 8.38 Unit Cell  *312*
- 8.39 X-Ray Crystallography  *313*
    - 8.39.1 Bragg's X-rays Spectrometric Method  *313*
    - 8.39.2 Determination of Crystal Structure by Bragg's Method  *315*
    - 8.39.3 Powdered Crystal Method  *316*
- 8.40 Structure of Rock Salt (NaCl)  *318*
- 8.41 Structure of Sylvine (KCl)  *320*
- 8.42 Expected Similarity between NaCl and KCl  *320*
- 8.43 Classification of Solids  *321*
- 8.44 Lattice Energy  *322*
- 8.45 Theoretical Treatment of Lattice Energy or Born–Lande's Equation  *325*
- 8.46 Lattice Heat Capacity  *327*
- 8.47 Einstein Theory of Heat Capacity  *327*
- 8.48 Debye Theory of Heat Capacities or Debye T-Cubed Law  *329*

*Exercises  332*
*Suggested Readings  333*

## Part III  INSTRUMENTAL TECHNIQUES

### 9. Reference Electrodes and Potentiometric Methods ........... 337–360

- 9.1 Introduction  *337*
- 9.2 Calomel Electrode  *337*
- 9.3 Silver/Silver Chloride Electrode  *339*
- 9.4 Indicator Electrodes  *340*
    - 9.4.1 Classification of Indicator Electrodes  *341*
    - 9.4.2 Metallic Redox Indicator Electrode  *342*
- 9.5 Membrane Electrode  *342*
- 9.6 The Glass Electrode  *343*
- 9.7 Composition of pH Sensitive Glass Membranes  *344*
- 9.8 Hygroscopicity of Glass Membranes  *344*
- 9.9 Resistance of Glass Membranes  *344*
- 9.10 Theory of Glass–Electrode Potential  *345*
- 9.11 Asymmetric Potential  *346*

| | | |
|---|---|---|
| 9.12 | Glass Electrodes for the Determination of Other Ions | *346* |
| 9.13 | Liquid Membrane Electrodes | *347* |
| 9.14 | Solid State or Precipitate Electrodes | *349* |
| 9.15 | Gas-Sensing Electrodes | *349* |
| 9.16 | Liquid Junction Potential | *351* |
| 9.17 | Ohmic Potential: IR Drop | *352* |
| 9.18 | Polarization | *353* |
| | 9.18.1 Causes of Polarization | *352* |
| | 9.18.2 Polarization Effects | *353* |
| 9.19 | Concentration Polarization | *354* |
| 9.20 | Decomposition Potential | *355* |
| | 9.20.1 Importance and Significance of Decomposition Potential | *355* |
| 9.21 | Overvoltage (Kinetic Polarization) | *356* |
| | 9.21.1 Factors Affecting Overvoltage | *356* |
| | 9.21.2 Importance of Overvoltage | *356* |
| 9.22 | Application of Potentiometric Titrations | *357* |

*Exercises 359*
*Suggested Readings 360*

## 10. Polarography and Voltammetry .................................................. 361–381

| | | |
|---|---|---|
| 10.1 | Theoretical Consideration of Classical Polarography | *361* |
| 10.2 | Polarographic Cell | *361* |
| 10.3 | Polarogram | *362* |
| 10.4 | Interpretation of Polarographic Waves | *363* |
| 10.5 | Effect of Complex Formation on Polarographic Waves | *365* |
| 10.6 | Dropping Mercury Electrode (DME) or Current Variations During the Lifetime of a Drop | *366* |
| | 10.6.1 Advantages of DME | *367* |
| | 10.6.2 Disadvantage of DME | *367* |
| 10.7 | Polarographic Diffusion Currents | *367* |
| 10.8 | Capillary Characteristics and Its Effect on Diffusion Currents | *367* |
| 10.9 | Mixed Anodic–Cathodic Waves | *368* |
| 10.10 | Oxygen Wave | *368* |
| 10.11 | Application of Polarography | *369* |
| | 10.11.1 Determination of Diffusion Currents | *369* |
| | 10.11.2 Analysis of Mixtures | *370* |
| | 10.11.3 Concentration Determination | *370* |
| | 10.11.4 Inorganic Polarographic Analysis | *371* |
| | 10.11.5 Organic Polarographic Analysis | *371* |
| 10.12 | Amperometric Titrations | *372* |
| 10.13 | Amperometric Titration Curves | *372* |
| 10.14 | Apparatus and Techniques of Amperometric Titration | *373* |
| 10.15 | Microelectrodes; The Rotating Platinum Electrode | *373* |
| 10.16 | Application of Amperometric Titrations | *374* |

xii    Contents

        10.17   Modified Voltametric Methods   375
                10.17.1   Pulse Polarography   375
                10.17.2   Current Sampled Polarography or AC Polarography   377
                10.17.3   Stripping Analysis   378
        *Exercises*   379
        *Suggested Readings*   380

## 11. Conductometric Methods .................................................... 382–391

        11.1   Introduction   382
        11.2   Electrolytic Conductivity   382
                11.2.1   Measurement of Electrolytic Conductance   384
                11.2.2   Applications of Conductivity Measurements   385
        *Exercises*   391
        *Suggested Readings*   391

## 12. Coulometric Analysis ........................................................... 392–402

        12.1   Introduction   392
        12.2   Hydrogen–Oxygen Coulometer   393
        12.3   Silver Coulometer   393
        12.4   Iodine Coulometer   393
        12.5   Constant Current Coulometric Analysis   394
        12.6   Controlled Potential Coulometric Analysis   394
        12.7   Characteristics of Coulometric Analysis   394
        12.8   Constant Current Coulometry or Coulometric Titrations   395
        12.9   Cells for Coulometric Titrations   396
        12.10  Apparatus and Method   396
        12.11  External Generation of Reagents   397
        12.12  Application of Coulometric Titrations   398
        12.13  Controlled Potential Coulometric Analysis   499
        12.14  Selection of Experimental Condition in Potentiostatic Coulometry   401
        *Exercises*   401
        *Suggested readings*   402

## Part IV   ELECTROCHEMISTRY

## 13. Electrified Interface .............................................................. 405–435

        13.1   Introduction   405
        13.2   Importance of Electrified Interface   409
        13.3   Polarizable and Nonpolarizable Interface   409
        13.4   Thermodynamics of Electrified Interfaces or Measurement of
               Interfacial Tension as a Function of the Potential Difference
               Across the Interface   410

13.5 Definition of Electrochemical Potential   *412*
    13.5.1 Can the Chemical and Electrical Work be Determined Separately   *413*
    13.5.2 Criterion of Thermodynamic Equilibrium between Two Phases   *413*
    13.5.3 Nonpolarizable Interface and Thermodynamic Equilibrium   *414*
13.6 Some Thermodynamics Thoughts on Electrified Interface   *414*
13.7 Determination of the Charge Density on the Electrode   *417*
13.8 Determination of the Electrical Capacitance of the Interface   *417*
    13.8.1 The Potential at Which an Electrode has a Zero Charge   *419*
13.9 Determination of the Surface Excess   *420*
13.10 The Parallel Plate Condenser Model: Helmholtz–Perrin Model of the Double Layer   *423*
    13.10.1 The Double Layer in Trouble: Neither Perfect Parabola Nor Constant Capacities   *424*
13.11 Ionic Cloud: The Gouy–Chapman Diffuse Charge Model of the Double Layer   *425*
    13.11.1 Ions under Thermal and Electric Forces Near an Electrode or Mathematical Treatment of Diffuse Double Layer   *425*
    13.11.2 A Picture of Potential Drop in the Diffuse Layer   *428*
    13.11.3 An Experimental Test of the Gouy–Chapman Model: Potential Dependence of the Capacitance   *431*
13.12 The Stern Model   *431*
    13.12.1 A Consequence of the Stern Picture: Two Potential Drops Across an Electrified Interface   *432*
    13.12.2 Another Consequence of the Stern Model: An Electrified Interface is Equivalent to two Capacitors in Series   *432*
13.13 The Relative Contributions of the Helmholtz–Perrin and Gouy–Chapman Capacities   *433*
*Exercises*   *434*
*Suggested Readings*   *435*

# 14. Electrodics .................................................................. 436–459

14.1 Introduction   *436*
    14.1.1 Charge Transfer: Its Chemical and Electrical Implications   *436*
    14.1.2 Can an Insolated Electrode-Solution Interface be Used as a Device?   *438*
    14.1.3 Electrochemical System can be Used as Devices   *439*
    14.1.4 An Electrochemical Device: The Substance Producer   *439*
14.2 The Instant of Immersion of a Metal in an Electrolytic Solution   *439*
14.3 The Rate of Charge Transfer Reactions under Zero Field: The Chemical Rate Constant   *440*
    14.3.1 Some Consequences of Electron Transfer at an Interface   *441*
14.4 Rate of an Electron Transfer Reaction under the Influence of an Electric Field   *442*
14.5 The Equilibrium Exchange–Current Density ($i_0$)   *444*

14.6   The Nonequilibrium Drift–Current Density (*i*)   *445*
14.7   The Overpotential ($\eta$)   *445*
14.8   Some General and Special Cases of Butler-Volmer Equation   *446*
   14.8.1   The High-Field Approximation: The Exponential *i* Versus $\eta$ Law   *450*
   14.8.2   The Low Field Approximation: The Linear *i* Versus $\eta$ Law   *451*
14.9   Nonpolarizable and Polarizable Interface   *452*
14.10  Physical Meaning of Symmetry Factor $\beta$   *453*
14.11  Potential–Energy Distance Relations of Particles Undergoing Charge Transfer   *453*
14.12  A Simple Picture of the Symmetry Factor   *455*
14.13  Relationship between $\beta$ and over Potential $\eta$   *458*
*Exercises*   *459*
*Suggested Readings*   *459*

## 15. Contact Adsorption on the Electrode .................................. 460–486

15.1   Introduction   *460*
15.2   Situation Present at Metal Electrode   *461*
15.3   Metal–Water Interactions   *461*
15.4   The Orientation of Water Molecules on Charged Electrodes   *462*
15.5   Closest Approach of Hydrated Ions to a Hydrated Electrode   *462*
15.6   Desolvated Ions Contact Adsorb on the Electrode   *462*
15.7   The Free Energy Change for Contact Adsorption   *463*
15.8   Degree of Contact Adsorption   *463*
15.9   Measurement of Contact Adsorption   *464*
15.10  Contact Adsorption, Specific Adsorption or Super-equivalent Adsorption   *467*
15.11  Contact Adsorption: Its Influence on the Capacity of the Interface   *467*
15.12  The Complete Capacity–Potential Curve   *469*
15.13  The Constant Capacity Region   *469*
15.14  The Position of the Outer Helmholtz Plane and an Interpretation of the Constant Capacity   *471*
15.15  The Capacitance Hump   *472*
15.16  Contact Adsorbed Ions Changes with Electrode Charge   *473*
15.17  Test of the Population Law of Contact Adsorbed Ions   *476*
15.18  Lateral Repulsion Model for Contact Adsorption   *481*
15.19  Flip–Flops Water Molecules on Electrodes   *482*
15.20  The Contribution of Adsorbed Water Dipoles to the Capacity of the Interface   *483*
*Exercises*   *486*
*Suggested Readings*   *486*

## 16. Structure of Semiconductor—Electrolyte Interface ............ 487–503

16.1   Introduction   *487*
16.2   The Band Theory of Crystalline Solids   *487*
16.3   Conductors, Insulators and Semiconductors   *488*

Contents  xv

    16.4    Comparison of Semiconductors and Electrolyte Solution   *492*
    16.5    Impurity Semiconductors, n-Type and p-Type   *492*
    16.6    Current Potential Laws at Other Types of Charged Interfaces: Semiconductor n–p Junctions   *493*
    16.7    The Current Across Biological Membranes   *497*
    16.8    The Hot Emission of Electrons from a Metal into Vacuum   *500*
    16.9    The Cold Emission of Electrons from a Metal into Vacuum   *501*
    *Exercises*   *502*
    *Suggested Readings*   *502*

## 17. Multistep Reactions .................................................................. 504–516

    17.1    Introduction   *504*
    17.2    Queues or Waiting Lines   *505*
    17.3    Relation between Overpotential, $\eta$ and Electron Queue   *505*
    17.4    Relation between the Current Density and Overpotential   *506*
    17.5    Rate-Determining Step   *507*
    17.6    Rate-Determining Steps and Energy Barriers for Multistep Reactions   *511*
    17.7    Determination of Order   *513*
    *Exercises*   *515*
    *Suggested Readings*   *515*

## 18. Energy Conversion .................................................................. 517–538

    18.1    Energy Source   *517*
    18.2    Hydrocarbon Fuel   *517*
    18.3    Atmospheric Pollution from Products of Internal Combustion Engine   *517*
    18.4    Waste of Chemical Energy Available from Burning Hydrocarbons in Air   *518*
    18.5    Direct Energy Conversion   *518*
            18.5.1    Direct Energy Conversion by Electrochemical Means   *519*
    18.6    The Maximum Intrinsic Efficiency   *519*
    18.7    The Actual Efficiency of an Electrochemical Energy Converter   *521*
    18.8    Cold Combustion   *522*
    18.9    Making $V$ near $V_e$ is the Central Problem of Electrochemical Energy Conversion   *522*
    18.10   Condition for Maximum Efficiency   *525*
    18.11   Power Output of Energy Converter   *526*
    18.12   Wrong Step in the Development of Power Sources   *526*
    18.13   Electrochemical Electricity Producers: The Two Types   *527*
    18.14   Fuel Cell   *527*
            18.14.1   The Hydrogen–Oxygen Fuel Cell   *527*
            18.14.2   Hydrogen–Air Fuel Cells   *528*
            18.14.3   Hydrocarbon–Air Cells   *529*
            18.14.4   Natural Gas and CO–Air Cells   *529*
    18.15   Electricity Storage   *530*
            18.15.1   The Important Quantities in Electricity Storage   *532*

xvi    Contents

18.16   Desirable Trend in Storer to Maximize the Energy Density and the Power Output   533
18.17   Lead Acid Storage Battery   534
18.18   A Dry Cell   535
    18.18.1   Two More Electricity Storer   535
Exercises   537
Suggested Readings   538

# 19. Corrosion .................................................................................. 539–577

19.1   What is Corrosion   539
19.2   Classification of Corrosion   539
19.3   Expressions for Corrosion Rate   539
19.4   Electrochemical Principles of Corrosion   540
19.5   Types of Electrochemical Cell Formation   541
19.6   Exchange Current Density ($i_0$)   542
19.7   Polarization of the Electrode   542
19.8   Factors Affecting Activation Polarization   546
19.9   Significance of Activation Polarization   547
19.10  Resistance Polarization   547
19.11  Mixed Potential Theory   548
19.12  Passivity   549
    19.12.1   Electrochemical Behaviour of Active/Passive Metals   550
    19.12.2   Theories of Passivity   552
19.13  Factors Affecting Corrosion Rate   553
19.14  Corrosion Prevention Methods   554
    19.14.1   Cathodic Protection   558
    19.14.2   Anodic Protection Method   561
    19.14.3   Corrosion Protection by the Use Corrosion Inhibitors   562
    19.14.4   Corrosion Protection of the Metal by Protective Coating of the Surface (Surface Treatment)   565
    19.14.5   Volatile Corrosion Inhibitor (VCI Method)   567
    19.14.6   Corrosion Protection by Corrosion Resistant Alloys   568
    19.14.7   Modification of Surface Conditions by Placement of Desiccant Bags   571
19.15  Different Experimental Methods to Find Corrosion Rate   571
    19.15.1   Weight Loss Method   571
    19.15.2   Electrochemical Method to Find Corrosion Rate   572
    19.15.3   Electrochemical Impedance Spectroscopy to Measure Corrosion Rate   573
    19.15.4   Cyclic Votammetry (CV) to Study Corrosion Rate   575
Exercises   576
Suggested Readings   577

## Part V  REACTION DYNAMICS

**20. Chemical Kinetics .................................................. 581–637**

    20.1   Chain Reactions   *581*
    20.2   Formation of HBr from $H_2$ and $Br_2$   *582*
    20.3   Decomposition Acetaldehydes   *583*
    20.4   Decomposition of Ethane   *584*
    20.5   Photochemical Hydrogen and Chlorine Reactions   *585*
    20.6   General Treatment of Chain Reaction   *586*
    20.7   Apparent Activation Energy of Chain Reaction   *589*
    20.8   General Mechanism to Study Decomposition of Organic Compound   *590*
    20.9   Example of Rice Herzfeld Mechanism or Decomposition of Acetaldehyde Molecule   *592*
    20.10  Study of Fast Reactions   *594*
        20.10.1  Flash Photolysis   *594*
        20.10.2  Flash Spectroscopy   *595*
        20.10.3  Kinetic Spectrophotometry   *596*
        20.10.4  Absorption of Photoflash Energy   *596*
        20.10.5  Fast Flow Methods   *597*
        20.10.6  Nuclear Magnetic Resonance (NMR) Methods   *599*
    20.11  Methods of Determining Rate Laws: Arrhenius Law   *600*
    20.12  Collision Theory   *602*
    20.13  Activated Complex Theory or Absolute Reaction Rate Theory: or Transition State Theory   *605*
    20.14  Factors Determining Reaction Rate in Solution   *607*
    20.15  Reactions between Ions   *608*
    20.16  Influence of Solvent: Double Sphere Model   *608*
    20.17  The Single-Sphere Activated Complex Model   *610*
    20.18  Influence of Ionic Strength   *612*
    20.19  Reactions Involving Dipoles: Influence of Solvent   *613*
    20.20  Enzyme Kinetics   *615*
    20.21  Kinetics of Enzymatic Reactions: Michaelis Menten Equation   *617*
    20.22  Kinetic Data Analysis   *620*
    20.23  Competitive Inhibition   *621*
    20.24  Noncompetitive Inhibition   *625*
    20.25  Unimolecular Reaction Rate Theory   *626*
        20.25.1  Lindemann Theory   *626*
        20.25.2  Hinshelwood Treatment for Unimolecular Reaction   *629*
        20.25.3  RRK Theory   *631*
        20.25.4  Rice Rampsberger-Kassel-Marcus (RRKM) Theory   *634*

*Exercises*   *635*
*Suggested Readings*   *636*

## Part VI  ADVANCED CHEMISTRY

**21. Chemistry of Nanomaterials .................................................. 641–669**

    21.1  Introduction  *641*
    21.2  Consequence of Nanoscale  *641*
    21.3  Historical Perspective  *642*
    21.4  Nanoparticle Morphology and Electronic Structure  *642*
    21.5  Properties of Nanomaterials  *643*
        21.5.1  Mechanical Properties  *644*
        21.5.2  Melting Point  *645*
    21.6  Synthesis of Nanomaterials  *646*
        21.6.1  Physical Vapour Deposition  *646*
        21.6.2  Chemical Vapour Deposition  *646*
        21.6.3  Sol–Gel Method  *647*
        21.6.4  Invert Gas Condensation (IGC) Method  *647*
        21.6.5  Microemulsion Technique  *647*
        21.6.6  Biological Method  *648*
    21.7  Characterization Techniques  *648*
        21.7.1  X-ray Diffraction Analysis  *648*
        21.7.2  UV–Vis Spectroscopy Analysis  *652*
        21.7.3  Scanning Electron Microscopy Analysis  *652*
        21.7.4  Transmission Electron Microscope (TEM)  *654*
        21.7.5  Scanning Tunneling Microscope (STM)  *657*
        21.7.6  Atomic Force Microscopy (AFM)  *660*
    21.8  Applications of Nanotechnology  *661*
        21.8.1  Use of Nanotechnology in Everyday Process and Technology  *662*
        21.8.2  Applications of Nanotechnology in Electronics and IT  *664*
        21.8.3  Use of Nanotechnology in Sustainable Energy Applications  *665*
        21.8.4  Environment Remedial Application of Nanotechnology  *665*
        21.8.5  Nanotechnology in Health and Medical Applications  *666*
        21.8.6  Nanotechnology in Future Transportation  *667*
    21.9  Future Applications  *667*
        21.9.1  Environmental Applications  *667*
        21.9.2  Optoelectronics Device  *668*
        21.9.3  Nanoelectronic and Magnetic Devices and New Computing System  *668*
        21.9.4  Microelectromechanical Systems  *668*
    21.10  Safety Issues of Nanotechnology  *668*
    *Exercises*  *668*
    *Suggested Readings*  *669*

**Appendix .................................................................................. 671–673**
**Logarithms Tables ................................................................... 675–678**
**Index ....................................................................................... 679–680**

# Foreword

Undergraduate education in chemistry seems to be in the midst of a major revolution in India. Sophisticated topics are increasingly being introduced into the college chemistry courses. Moreover, there are very few books available in the market which cover almost all the syllabi for B.Sc. (Hons.) and M.Sc. (physical chemistry) students. Dr. H. K. Moudgil has tried to cover almost all the topics of physical chemistry like thermodynamics, quantum chemistry, statistical mechanics, chemical kinetics, irreversible thermodynamics, electrochemistry, solid state chemistry, instrumentation techniques, etc. so that the students of B. Sc. (Hons.), M.Sc. (physical chemistry) and M.Phil. courses are benefited the most.

I have gone through the whole of the text and found it an interesting book. The presentation of the subject is in a lucid style, and simple language speaks of the hard work done by the author. All explanations have well attended with reasons and are supplemented by appropriate data to bring home the point. The inclusion of a large number of worked-out examples and problems gives ample insight into the topics dealt with.

The presentation of the subject matter is systematic and has made the book useful not only to the students, but also to the teachers and researchers engaged in physical chemistry exploration.

I think that a long-felt need for such a book by Indian universities has been fulfilled. I wish the book a great success.

**K. C. Bhardwaj**
Vice Chancellor
Chaudhary Devi Lal University
Sirsa, Haryana

# Preface

There are very few books available on *physical chemistry* which cover almost all the topics of physical chemistry and present the subject matter in a concise manner so that it is interesting and within grasp of undergraduate (B.Sc.) and postgraduate (M.Sc.) students of chemistry.

This book, now in its second edition, is thoroughly revised and updated to cover all essential topics in physical chemistry. It is hoped that it would be useful to physical chemistry students at all levels. A logical and rigorous approach has been adopted, but too much philosophical discussions have been avoided in order to make the material always within the grasp of students. This text gives an introduction to the wide range of topics, i.e. quantum mechanics, statistical thermodynamics, statistical mechanics, classical thermodynamics, solid state, instrumentation techniques, chemical kinetics, electrochemistry, irreversible thermodynamics, corrosion, molecular spectroscopy, X-ray crystallography and nanotechnology that constitute the integral part of physical chemistry.

Traditionally, there are six principal areas of physical chemistry: thermodynamics (which concerns the energetics of chemical reactions), quantum chemistry (which concerns the structure of molecules), chemical kinetics (which concerns the rate of chemical reactions), electrochemistry (which concerns study of electrochemical energy inter-conversion), molecular spectroscopy and nanotechnology. Many physical chemistry courses begin with a study of thermodynamics, then discuss quantum chemistry, and then treat chemical kinetics. This order is a reflection of the historic development of the field.

The book contains twenty-one chapters, each so selected and well-arranged that even the average student would be capable of mastering all the material in it. In general chemistry, one can learn about the three laws of thermodynamics and is introduced to the quantities, enthalpy, entropy, and the Gibbs free energy (formally called the free energy). Besides these, Chapter 1 on classical thermodynamics describes macroscopic chemical systems. Irreversible thermodynamics (Chapter 2) deals with the study of steady states. Statistical thermodynamics and statistical mechanics (Chapters 3 and 4) provide away to describe thermodynamics at a molecular level. Chapter 5 discusses the underlying principles of quantum mechanics and then shows how they can be applied to a number of model systems.

Chapter 6 deals with molecular spectroscopy. Chapter 7 deals with X-ray crystallography. Chapter 8 deals with solid state chemistry. Chapters 9–12 deal with various instrumentation techniques involved in physical chemistry, a basic understanding of which is very necessary to a physical chemist. Chapters 13–17 deal with electrochemical study. Chapter 18 describes various applications of electrochemistry. Chapter 19 deals with the most important part of electrochemistry, i.e. corrosion. Chapter 20 deals with the rate of chemical reactions. The last Chapter 21 gives an introduction to the advanced topic on *nanotechnology*.

Each chapter of the book contains a discussion of the subject matter and important mathematical relations followed by suitable solved examples to illustrate the basic principles. In the text of few chapters, a set of solved problems has been given to help the students become familiar with the concepts and principles presented in the text.

The comments, criticism and suggestions from the readers will be welcomed and appreciated.

**Harish Kumar Moudgil**

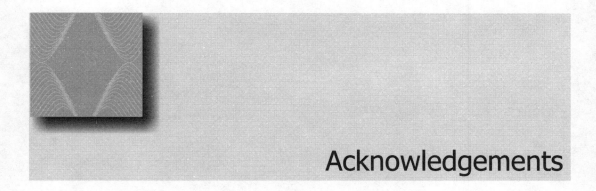

# Acknowledgements

Many people have contributed to the writing of this book. I acknowledge the pivotal role that Prof. R.S. Chaudhary has played in completion of this book. His immensely erudite suggestions and indispensable comments helped me to develop and shape this book in the present form. I also thank my colleagues Prof. Ashok Kumar Malik, Vazid Ali, Gita Rani and my research scholars Renu Rani, Vishal Saini and Dheeraj Kumar, who read and made helpful comments on the entire manuscript.

I also thank to Prof. Parveen Aghamkar, Dean, Faculty of Physical Sciences, for his extremely helpful, supportive and friendly nature that he showed me in writing of this book. I am very much thankful to my parents for their constant support and heartiest blessing.

My deepest appreciation goes to my wife Mrs. Pooja Moudgil who is always a source of strength and love for me.

In the last, I am highly indebted to the almighty for his grace and mercy.

**Harish Kumar Moudgil**

# Nomenclature

**Greek Symbol**

| | |
|---|---|
| $\Lambda_{eq}$ | equivalent conductance |
| $\alpha$ | fluctuation |
| $\alpha$ | Langrenge undetermined multiplier |
| $\alpha$ | volume occupied by real gas |
| $\beta$ | Langrenge undetermined multiplier |
| $\beta$ | symmetry factor |
| $\chi$ | surface potential |
| $\chi^{-1}$ | effective thickness of the ionic cloud |
| $\Delta n_g$ | change in number of moles of gas |
| $\nabla$ | Laplacian operator |
| $\varepsilon$ | dielectric constant of medium |
| $\varepsilon_i$ | energy of $i$ compartment |
| $\varepsilon_{max}$ | maximum efficiency |
| $\varepsilon_p$ | voltage efficiency |
| $\varepsilon^*$ | Fermi energy |
| $\phi$ | Eulerian angle |
| $\phi_0$ | flux quantization |
| $\phi$ | inner potential |
| $\gamma$ | activity coefficient |
| $\gamma$ | inter-atomic spacing |
| $\gamma$ | surface tension |
| $\gamma_\pm$ | mean ionic activity coefficient |
| $\Gamma$ | surface excess |
| $\eta_a$ | activation polarization |
| $\eta_{cc}$ | concentration polarization at cathode |
| $\eta$ | efficiency |
| $\eta$ | overpotential |
| $\lambda_+$ | ionic conductance of cations |
| $\lambda$ | wavelength |
| $\mu$ | chemical potential |
| $\mu$ | electrochemical potential |
| $\mu$ | reduced mass |
| $\pi$ | reduced pressure |
| $\theta_D$ | characteristics Debye temp. |
| $\theta$ | characteristics temperature |
| $\theta$ | Eulerian angle |
| $\theta$ | fraction of surface covered |
| $\sigma$ | conductivity of metal |
| $\tau_{coll}$ | collisional life time |
| $\omega$ | amplitude of any wave |
| $\omega_m$ | mean frequency |
| $\omega$ | vibrational frequency |
| $\xi$ | degree of advancement of reaction |
| $\psi$ | outer potential |
| $\psi$ | wave function |
| $\Psi$ | Eulerian angle |

## Nomenclature

| | | | |
|---|---|---|---|
| $a$ | shape factor | $i_0$ | exchange current density |
| $A$ | work function | $i_d$ | diffusion current |
| $a_i$ | activity of $i$ component | IHP | inner Helmholtz plane |
| $A_r$ | electrochemical affinity | $i_l$ | limiting current |
| $a_\pm$ | mean ionic activity | $J$ | flux |
| $B$ | rotational constant | $J$ | rotational quantum number |
| $\bar{c}$ | average velocity | $J$ | total angular quantum number |
| $C$ | capacitance | $k$ | Boltzmann constant |
| $C$ | component | $K$ | equilibrium constant |
| $C$ | heat capacity | $k$ | force constant |
| $c$ | velocity of light | $k$ | rate constant |
| $C_P$ | heat supplied at constant pressure | $K$ | specific conductance |
| $C_V$ | heat supplied at constant volume | $K_c$ | equilibrium constant |
| $D$ | diffusion coefficient | KE | kinetic Energy |
| $d$ | interplaner distance | $K_m$ | Michelis constant |
| $d\tau$ | $dxdydz$ | $L$ | angular momentum |
| $\bar{E}$ | average energy per molecule | $l$ | Azimuthal quantum number |
| $E$ | internal energy | $L$ | transport coefficient |
| $E$ | total energy | $M$ | Madlung constant |
| $E_a$ | activation energy | $m$ | magnetic quantum number |
| $E_r$ | rotational energy | $M$ | rate of mercury flow |
| $E_t$ | translational energy | $N$ | total number of molecules |
| $E_v$ | vibrational energy | $N_A$ | Avagadro's number |
| $F$ | degree of freedom | $n_{CA}$ | population density of contact adsorbed ions |
| $F$ | Faradaic charge | | |
| $F$ | formality | $n_i$ | number of molecules in $i$ cell |
| $f$ | fugacity | $n_x$ | principal quantum number in $x$ axis |
| $F^0$ | standard free energy per mole | OHP | outer Helmholtz plane |
| $G$ | Gibbs free energy | $P$ | power |
| $G$ | statistical weight factor | $P_C$ | critical pressure |
| $g_{el}$ | electronic degeneracy | $p_x$ | momentum in $X$ direction |
| $g_i$ | degeneracy of $i$ cell | $q$ | heat supplied to the system |
| $H$ | enthalpy | $q$ | partition function |
| $H$ | Hamiltonian operator | $q_{CA}$ | charge density of contact adsorbed ions |
| $h$ | Planck's constant | $q_{el}$ | electronic partition function |
| $H(y)$ | Hermite polynomial | $q_m$ | charge density at metal electrode |
| $h, k, l$ | Miller indices | $q_P$ | heat of reaction at constant pressure |
| $H_c$ | critical magnetic field | $q_r$ | rotational partition function |
| $H_t$ | translational enthalpy | $q_s$ | total charge present in solution side |
| $I$ | intensity of electric current | $q_t$ | translational partition function |
| $\bar{i}$ | electronation current density | $q_V$ | heat of reaction at constant volume |
| $i$ | iota | $q_v$ | vibrational partition function |
| $I$ | moment of inertia | $Q_w$ | fraction of surface covered by water dipole |
| $i$ | nonequilibrium exchange current density | | |

| | | | | |
|---|---|---|---|---|
| $r_b$ | rate of breaking step | $U$ | lattice energy |
| RDS | rate determining step | $V$ | cell potential |
| $r_i$ | rate of initiation step | $V$ | potential energy |
| $r_p$ | rate of propagation step | $V_e$ | thermodynamic equilibrium potential |
| $r_t$ | rate of termination step | $W$ | thermodynamic probability |
| $S$ | entropy | $w$ | work done |
| $S$ | solubility | $X$ | driving force |
| $S_t$ | translational entropy | $\bar{x}$ | mean fluctuation |
| $S_v$ | vibrational entropy | $Y$ | strain energy per unit area |
| $t$ | time interval | $Z_{AA}$ | number of collision between $A$ and $A$ |
| $T_c$ | critical temperature | $Z_i$ | electrovalency of $i$ component |

# Periodic Table

| 1 H 1.0079 | | | | | | | | | | | | | | | | | 2 He 4.0026 |
|---|---|---|---|---|---|---|---|---|---|---|---|---|---|---|---|---|---|
| 3 Li 6.941 | 4 Be 9.012 | | | | | | | | | | | 5 B 10.81 | 6 C 12.011 | 7 N 14.007 | 8 O 15.999 | 9 F 18.998 | 10 Ne 20.179 |
| 11 Na 22.99 | 12 Mg 24.31 | | | | | | | | | | | 13 Al 26.98 | 14 Si 28.09 | 15 P 30.974 | 16 S 32.06 | 17 Cl 35.453 | 18 Ar 39.948 |
| 19 K 39.098 | 20 Ca 40.08 | 21 Sc 44.96 | 22 Ti 47.90 | 23 V 50.94 | 24 Cr 52.00 | 25 Mn 54.94 | 26 Fe 55.85 | 27 Co 58.93 | 28 Ni 58.70 | 29 Cu 63.55 | 30 Zn 65.38 | 31 Ga 69.72 | 32 Ge 72.59 | 33 As 74.92 | 34 Se 78.96 | 35 Br 79.904 | 36 Kr 83.80 |
| 37 Rb 85.47 | 38 Sr 87.62 | 39 Y 88.91 | 40 Zr 91.22 | 41 Nb 92.91 | 42 Mo 95.94 | 43 Tc 98.91 | 44 Ru 101.07 | 45 Rh 102.91 | 46 Pd 106.4 | 47 Ag 107.87 | 48 Cd 112.41 | 49 In 114.82 | 50 Sn 118.69 | 51 Sb 121.75 | 52 Te 127.6 | 53 I 126.9 | 54 Xe 131.3 |
| 55 Cs 132.91 | 56 Ba 137.33 | 71 Lu 174.97 | 72 Hf 178.49 | 73 Ta 180.9 | 74 W 183.85 | 75 Re 186.2 | 76 Os 190.2 | 77 Ir 192.2 | 78 Pt 195.1 | 79 Au 196.97 | 80 Hg 200.59 | 81 Tl 204.37 | 82 Pb 207.2 | 83 Bi 208.98 | 84 Po [210] | 85 At [210] | 86 Rn [222] |
| 87 Fr [223] | 88 Ra 226.025 | 103 Lr [256] | 104 | 105 | 106 | | | | | | | | | | | | |

| 57 La 138.91 | 58 Ce 140.1 | 59 Pr 140.91 | 60 Nd 144.24 | 61 Pm [147] | 62 Sm 150.4 | 63 Eu 151.96 | 64 Gd 157.25 | 65 Tb 158.93 | 66 Dy 162.50 | 67 Ho 164.93 | 68 Er 167.26 | 69 Tm 168.93 | 70 Yb 173.04 |
|---|---|---|---|---|---|---|---|---|---|---|---|---|---|
| 89 Ac [227] | 90 Th 232.03 | 91 Pa 231.036 | 92 U 238.04 | 93 Np 237.05 | 94 Pu [244] | 95 Am [243] | 96 Cm [247] | 97 Bk [247] | 98 Cf [251] | 99 Es [254] | 100 Fm [257] | 101 Md [258] | 102 No [255] |

# Part I
# Thermodynamics

# Classical Thermodynamics

*Thermo* means heat and *dynamics* means flow. Thermodynamics deals with energy changes accompanying all physical and chemical transformations.

*The branch of science which deals with the study of different forms of energy and the quantitative relationship between them is known as thermodynamics.*

Various kinds of energies that a physical system can possess are kinetic, potential, thermal, chemical, mechanical, electrical, radiant, surface, magnetic energy, etc. and under suitable conditions these forms of energy are interconvertible.

Thermodynamics is of great importance in physical chemistry. The main objectives of thermodynamics are:

1. It helps to determine the exact conditions under which a given chemical or physical process is feasible or not. In other words, it helps to establish the criteria for judging the feasibility of given process under given set of conditions of temperature, pressure and concentration.
2. Thermodynamics also helps to determine the extent to which a given process including a chemical reaction would occur before attainment of equilibrium.
3. The important generalizations of physical chemistry such as van't Hoff law of dilute solutions, Raults law of lowering of vapour pressure, distribution law, law of chemical equilibrium, phase rule, etc. can be deduced from the laws of thermodynamics.

## 1.1 SOME BASIC TERMS AND CONCEPTS

*System and surroundings:* The part of universe chosen for thermodynamics consideration or study the effects of temperature, pressure, etc. is called *system*.

The remaining part of universe other than the system is called surroundings.

There are three types of system, i.e. open, closed and isolated.
1. *Open system:* If a system can exchange both matter and energy with surroundings. For example, animals and plants.
2. *Closed system:* If a system can exchange only energy with the surroundings but not matter. For example, water placed in a closed metallic vessel.
3. *Isolated system:* If a system can neither exchange matter nor energy with the surroundings. For example, thermos flask.

*State of a system and state variables:* State of a system means the condition of the system which is described in terms of certain observable properties such as temperature ($T$), pressure ($P$), volume ($V$), etc. of the system and these observable properties of the system like temperature, pressure and volume are called state variables.

*State function:* A physical quantity is said to be state function if the change in its value during the process depends only upon initial and final state of the system and does not depend upon the path by which this change has been brought about.

*Macroscopic system and macroscopic properties:* If a system contains a large number of chemical species, i.e. atoms, ions or molecules, it is called a *macroscopic system*. Properties like temperature, pressure, volume, density, melting point, boiling point, etc. are called *macroscopic properties*. These are further classified as:
1. *Extensive properties:* Those properties which depend upon the quantity of the matter contained in the system. For example, mass, volume, heat capacity, internal energy, enthalpy, entropy, Gibbs free energy, etc.
2. *Intensive properties:* Those properties which depend only on the nature of the substance present in the system. For example, temperature, pressure, refractive index, viscosity, density, surface tension, specific heat, freezing point, boiling point, etc.

*Thermodynamic processes:* There are four types of thermodynamic process, these are:
1. *Isothermal process:* When temperature remains constant throughout the process, it is called *isothermal process*.
2. *Isochoric process:* When volume remains constant throughout the process, it is called *isochoric process*.
3. *Isobaric process:* When pressure remains constant throughout the process, it is called *isobaric process*.
4. *Adiabatic process:* When no heat can flow from the system to surroundings or vice versa, it is called *adiabatic process*.

*Reversible process and irreversible process:* A reversible process is a process which is carried out infinitesimally slow so that all changes occurring in direct process can be exactly reversed and the system remains almost in a state of equilibrium with surrounding at every stage of the process.

On the other hand, a process which does not meet the above requirements is called an *irreversible process*.

*Internal energy* ($E$): The energy thus stored within a substance or a system is called its internal energy and is denoted by $E$.

It is the sum of different types of energies, i.e.
$$E = E_e + E_n + E_c + E_p + E_k$$
Here subscripts $e$ stands for electronic, $n$ for nuclear, $c$ for coulombic, $p$ for potential and $k$ for kinetic.

Absolute value of internal energy is very difficult to determine, but internal energy change can be determined easily and what we are interested is internal energy change and is given by
$$\Delta E = E_2 - E_1 \text{ or}$$
$$\Delta E = E_p - E_r$$
Here subscripts $p$ stands for product and $r$ for reactant species.

Internal energy is a state function quantity and is extensive property. Internal energy of ideal gas is a function of temperature only. Hence in isothermal process, internal energy change is zero. According to latest sign convention, a negative value of $\Delta E$ means energy is evolved and positive value of $\Delta E$ means energy is absorbed. Units of energy are ergs or joules.

*Enthalpy or heat content* ($H$): The thermodynamic quantity $E + PV$ is called the heat content or enthalpy of the system and is represented by the symbol $H$.
$$H = E + PV$$
Enthalpy change of a system is equal to the heat absorbed or evolved by the system at constant pressure.
$$\Delta H = \Delta E + P\Delta V$$
Hence the enthalpy change accompanying a process may also be defined as the sum of the increase in internal energy of the system and the pressure volume work done, i.e. the work of expansion.

*Physical concept of enthalpy or heat content:* The energy stored within the substance or the system that is available for conversion into heat is called the *heat content* or *enthalpy* of the substance or the system.

As we know according to first law of thermodynamics,
$$q = \Delta E + P\Delta V$$
Hence, from above two equations of $q$ and $\Delta H$, we get
$$q = \Delta H$$
But, at constant volume, $\Delta V = 0$, hence $P\Delta V = 0$, so that
$$q = \Delta E$$
*Relationship between heat of reaction at constant pressure and that at constant volume:* As we know,
$$\Delta H = \Delta E + P\Delta V$$
$$\Delta H = \Delta E + P(V_2 - V_1)$$
$$\Delta H = \Delta E + PV_2 - PV_1$$
But for an ideal gas, $PV = nRT$ so that
$$PV_1 = n_2RT \text{ and } PV_2 = n_2RT$$
$$\Delta H = \Delta E + (n_2RT - n_1RT)$$

$$\Delta H = \Delta E + (n_2 - n_1)RT$$
$$\Delta H = \Delta E + \Delta n_g RT$$
$$q_p = q_v + \Delta n_g RT$$

*Heat capacity, specific heat capacity and molar heat capacity:* The heat capacity of a solid is defined as the amount of heat required to raise the temperature of the system through 1°C.

The *specific heat* of a substance is defined as the amount of heat required to raise the temperature of 1 g of the substance through 1°C.

*Molar heat capacity* of a substance is defined as the amount of heat required to raise the temperature of 1 mol the substance through 1°C.

*Heat capacity* at constant volume may also be defined as the ratio of change of internal energy with temperature at constant volume.

$$C = q/\Delta T$$

But according to first law of thermodynamics,

$$q = \Delta E + P\Delta V$$
$$C = \frac{\Delta E + P\Delta V}{\Delta T}$$

At constant volume, $P\Delta V = 0$

$$C_V = \frac{\Delta E}{\Delta T}$$

*Heat capacity* at constant pressure may be defined as the ratio of change of enthalpy with temperature at constant pressure.

$$C = \frac{q}{\Delta T}$$

At constant pressure, $q = \Delta H$

$$C_P = \frac{\Delta H}{\Delta T}$$

**Relationship between $C_P$ and $C_V$:** The difference between the heat capacities of an ideal gas can be obtained by subtracting the two,

$$C_P - C_V = \frac{dH}{dT} - \frac{dE}{dT}$$

But

$$H = \Delta E + P\Delta V$$

or

$$\Delta H = \Delta E + RT$$

Differentiating this equation w.r.t. temperatue, we get

$$\frac{dH}{dT} = \frac{dE}{dT} + R$$

$$\frac{dH}{dT} - \frac{dE}{dT} = R$$

Hence, $\quad C_P - C_V = R$

*Exothermic reaction:* These are those reactions which are accompanied by the evolution of heat.

$$C + O_2 \longrightarrow CO_2 + 393.5 \text{ kJ}$$

$\Delta H$ is negative for exothermic reactions.

*Endothermic reaction:* These are those reactions which are accompanied by the absorption of heat.

$$N_2 + O_2 \longrightarrow 2NO - 180.7 \text{ kJ}$$

$\Delta H$ is positive for endothermic reactions.

*Heat of combustion* is defined as heat change when one mole of substance is completely burnt or oxidized in oxygen.

*Calorific value* of a fuel or food is defined as the amount of heat in calories or joules produced from the complete combustion of one gram of the fuel or the food.

*Heat of formation* of a substance is defined as heat change when one mole of substance is formed from its elements under given conditions of temperature and pressure. It is usually represented by $\Delta H_f$.

*Standard heat of formation* of a substance is defined as heat change when one mole of substance is formed from its elements under standard conditions of temperature and pressure. It is usually represented by $\Delta H_f^0$.

*Heat of neutralization* of an acid by a base is defined as heat change when one gram equivalent of an acid is neutralized by a base.

*Heat of neutralization* of strong acid and strong base is constant and is equal to $-57.1$ kJ/mol. Heat of neutralization of strong acid and weak base or strong base and weak acid is less than 57.1 kJ/mol, because the dissocation process is accompanied by the absorption of energy.

*Heat of solution* is defined as heat change when one mole of the substance is dissolved in such a large volume of the solvent that further addition of the solvent does not produce any more heat change.

*Heat of hydration* is defined as heat change when one mole of the anhydrous salt combines with required number of moles of water so as to change anhydrous salt to hydrated salt.

**Work of expansion or pressure volume work done:** It is the work done when the gas expands or contracts against external pressure. It is a kind of mechanical work. Consider a gas enclosed in a cylinder fitted with a frictionless piston (Figure 1.1).

Area of crosssection of cylinder = $a$ cm². Pressure on the piston = $P$, which is slightly less than internal pressure of the gas. Distance through which gas expands = $dl$ cm.

Pressure is force per unit area, force acting on piston will be $f = P\,a$, therefore, work done

$$w = f \times dl = P \times a \times dl$$

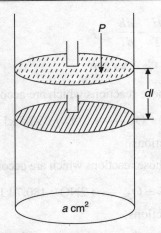

**FIGURE 1.1** Work of expansion.

But, $a \times dl$ is equal to small increase in volume $dV$, therefore,

$$dw = P\, dV$$

If the gas expands from initial volume $V_1$ to the final volume $V_2$, then total work done $w$ will be given by

$$w = \int P\, dV = P(V_2 - V_1) = P\,\Delta V$$

It may be mentioned that $P$ is the external pressure and hence it is sometimes written as $P_{ext}$ so that,

$$w = P_{ext}\,\Delta V$$

According to latest sign convention, work of expansion is negative and work of compression is positive. Therefore,

$$w = -P_{ext}\,\Delta V$$

**Work done in isothermal reversible expansion of an ideal gas:** Total work done when the gas expands from initial volume $V_1$ to final volume $V_2$ will be

$$w = -\int P\, dV$$

But for an ideal gas,

$$PV = nRT$$

i.e.

$$P = nRT/V$$

Hence,

$$w = -\int nRT\, dV/V$$

For isothermal expansion, $T$ = constant, so that,

$$w = -nRT \int dV/V$$

$$= -nRT \ln V_2/V_1$$

or
$$w = -2.303 \, nRT \log V_2/V_1$$

As we know,
$$P_1 V_1 = P_2 V_2 \text{ or } \frac{V_2}{V_1} = \frac{P_1}{P_2}$$

$$w = -2.303 \, nRT \log \frac{P_1}{P_2}$$

## 1.2 ZEROTH LAW OF THERMODYNAMICS

When two bodies are separately in thermal equilibrium with a third body, then two bodies are in thermal equilibrium with each other also. This law can be further explained as follows:

Suppose we have three systems $A$, $B$ and $C$ placed together. $A$ and $C$ are in contact so that the properties of both the systems change and finally the thermal equilibrium is attained and two come at the same temperature; $C$ is in contact with $B$, also then again thermal equilibrium is attained. If this is so, then $A$ and $B$ should also be in thermal equilibrium. That is $A$, $B$ and $C$ should be at the same temperature. Thus, the law provides a rational definition of temperature.

In thermodynamics studies, the temperatures unit is taken as Kelvin (K). The Kelvin is strictly defined as the fraction 1/273.16 of the temperature interval between the absolute zero (−273.16°C) and the triple point of water. In practice, the temperature in Kelvin is obtained by adding 273.15 to the temperature in degree celcius.

## 1.3 FIRST LAW OF THERMODYNAMICS

This law is also called *law of conservation of energy*. *It is impossible to construct a perpetual motion machine which will work continuously without consuming energy.* or
*There is an exact equivalence between heat and work done.* or
*Energy can neither be created nor be destroyed, if one form of energy disappears it must appear in some other equivalent form.*
Mathematical form of first law of thermodynamics can be derived as:

Suppose that a system in its initial state is associated with energy $E_1$ and let heat $q$ be given to it. The heat absorbed by the system may be used to increase its internal energy or perform work or both. If the energy of the system in the final state is $E_2$, then $(E_2 - E_1) = \Delta E$ is the increase in the internal energy. The law of conservation of energy demands that:

$$E_2 = E_1 + q + w$$
$$\Delta E = q + w \quad \text{(first law of thermodynamics)}$$
$$q = \Delta E + P \Delta V \quad \text{(first law of thermodynamics)}$$
$$q_v = \Delta E \quad \text{(at constant volume)}$$
$$q_p = \Delta H \quad \text{(at constant pressure)}$$

The mathematical expression of the law of conservation of energy in the form $\Delta E = dq + dw$ or $\Delta E = q + w$ is called the first law of thermodynamics. The $dq$ and $dw$ are inexact differentials, $q$ and $w$ are path dependent functions.

## 1.4 TYPES OF WORK

| | | |
|---|---|---|
| Mechanical work | $dw = F_{ext}\, dl$ | (force × displacement) |
| Stretching work | $dw = K\, dl$ | (tension × displacement) |
| Gravitational work | $dw = m\, g\, dl$ | |
| Expansion work | $dw = -P_{ext}\, dV$ | |
| Surface work | $dw = \gamma\, dA$ | (surface tension × area) |
| Electrochemical cell work | $dw = \Delta V\, I\, dt$ | (potential difference × current × time) |

## 1.5 APPLICATION OF FIRST LAW OF THERMODYNAMICS

Various possible application of first law of thermodynamics are:

1. *Work done in reversible isothermal expansion of an ideal gas (maximum work):* In isothermal conditions, temperature remains constant, and hence change in internal energy must be zero. Hence,

$$\Delta E = 0$$

According to first law of thermodynamics,

$$q = -w$$

$$w = -\int_{V_1}^{V_2} P\, dV = -nRT \ln \frac{V_2}{V_1}$$

$$w = -\int_{V_1}^{V_2} P\, dV = -2.303\, nRT \log \frac{V_2}{V_1}$$

2. *Work done in isothermal isobaric expansion of an ideal gas:* In isothermal conditions, temperature remains constant, and hence change in internal energy must be zero. Hence,

$$\Delta E = 0$$

According to first law of thermodynamics,

$$q = -w$$

$$w = -P_{ext}\, dV = -P_{ext}\, nRT \left( \frac{1}{P_2} - \frac{1}{P_1} \right)$$

3. *Work done in isothermal isobaric phase change:*

$$q_p = \Delta H$$
$$\Delta E = q + w = 0$$

$$w = -\int_{V_1}^{V_2} P_{ext}\, dV = -nRT \ln \frac{V_2}{V_1}$$

4. *Work done in reversible adiabatic expansion of an ideal gas:*

$$q_p = 0$$

$$\Delta E = w = \int_{T_1}^{T_2} nC_V dT = nC_V(T_2 - T_1)$$

$$\Delta H = \int_{T_1}^{T_2} nC_P dT$$

5. *Work done in isobaric adiabatic expansion of an ideal gas:*

$$q = 0$$

$$\Delta E = \int_{T_1}^{T_2} nC_V dT = nC_V(T_2 - T_1)$$

$$\Delta H = \int_{T_1}^{T_2} nC_P dT$$

Final temperature can be determined by using the following equation:

$$T_2 = T_1 - \frac{P_{ext} dV}{nC_V}$$

6. *In the calculation of the enthalpies of reactions:* Enthalpy of a reaction can be found out from *Hess's law* which states that, *total amount of heat evolved or absorbed in a reaction is same whether the reaction takes place in one step or a number of steps.*

The enthalpies of reactions are usually calculated from enthalpy of formation using the following relationship:

$$\Delta H = \text{(sum of the standard enthalpies of formation of products)} -$$
$$\text{(Sum of the standard enthalpies of formation of reactants)}$$
$$= \Sigma \Delta H_f^0 \text{ (products)} - \Sigma \Delta H_f^0 \text{ (reactants)}$$

In using this formula, the standard enthalpies of formation of elements are taken as zero.

7. *In the calculation of bond energies:* Bond energy is the amount of energy released when one mole of bonds are formed from the isolated atoms in the gaseous state or the amount of energy required to dissociate one mole of bonds present between the atoms in the gaseous molecules.

$$\Delta H = \Sigma \text{ bond energy of reactants} - \Sigma \text{ bond energy of products}$$

## 1.6 LIMITATIONS OF FIRST LAW OF THERMODYNAMICS

1. It does not tell about spontaneity of a process, i.e. whether a process will be spontaneous or not.
2. It does not tell about direction of a reversible reaction.

## 1.7 SECOND LAW OF THERMODYNAMICS

To overcome these limitations of first law of thermodynamics, second law of thermodynamics was introduced. It has following postulates:

1. Heat cannot spontaneously pass from a colder body to a warmer body (R.J.E. Clausius).
2. It is impossible to transfer heat from a colder system to a warmer without other simultaneous changes occuring in the two systems (P.S. Epstein).
3. Every system which is left to itself will on the average, change towards a condition of maximum probability (G.N. Lewis).
4. The state of maximum entropy is the most stable state for an isolated system (E. Fermi).
5. In any irreversible process, the total entropy of all bodies concerned is increased (G.N. Lewis).
6. In an adiabatic process the entropy either increases or remains unchanged (P.S. Epstein).

$$\Delta S \geq 0$$

7. The entropy function of a system of bodies tends to increase in all physical and chemical processes occurring in nature, it will include in the system all such bodies which are affected by the change (Saha).
8. The entropy increases towards a maximum and the energy of the universe is constant (R.J.E. Clausius).
9. Gain of information is loss in entropy (G.N. Lewis).
10. Entropy is times arrow (A. Eddington).
11. Every system left to itself changes rapidly or slowly in such a way to approach a definite final state of rest. No system of its own will change away from the state of equilibrium, except through the influence of external agencies (G.N. Lewis).
12. There is a general tendency in nature for energy to pass from more available to less available forms (J.A.V. Butter).
13. When an actual process occurs, it is impossible to invent a means of restoring every system concerned to its original condition (G.N. Lewis).
14. There exists a characteristic thermodynamic function called entropy. The difference of a system in states (1) and (2) is given by the expression

$$S_2 - S_1 \geq \int_1^2 \frac{dq}{T}.$$

15. All spontaneous processes are thermodynamically irreversible.
16. Without the help of external agency, a spontaneous process cannot be reversed, for example, heat cannot by itself flow from a colder body to a hotter body.
17. The complete conversion of heat into work is impossible without leaving some effect elsewhere.
18. All spontaneous processes are accompanied by a net increase of entropy.
19. Entropy of the universe is continuously increasing.

## 1.8 ENTROPY(S)

The term entropy was introduced by R.J.E. Clausius. *En* means energy and *Trope* means change. Entropy is the measure of change in energy. It is also a measure of the unavailable energy.

Mathematical form of entropy can be derived by taking an example of a reversible Carnot heat engine. A reversible Carnot heat engine works between a higher temperature $T_2$ and a lower temperature $T_1$ with heat $q_2$ absorbed and heat $q_1$ evolved. The efficiency of Carnot's engine is given by

$$\eta = \frac{T_2 - T_1}{T_2} = \frac{q_2 - q_1}{q_2}$$

$\Rightarrow$
$$1 - \frac{q_1}{q_2} = 1 - \frac{T_1}{T_2}$$

$$\frac{q_1}{q_2} = \frac{T_1}{T_2}$$

$$\frac{q_1}{T_1} = \frac{q_2}{T_2}$$

or
$$\frac{q_2}{T_2} - \frac{q_1}{T_1} = 0$$

i.e. $\frac{q}{T}$ = constant (for reversible process). The term $q/T$ is called *reduced heat*. For a reversible Carnot cycle, the algebric sum of the reduced heat is zero. For an irreversible Carnot cycle, we have

$$\frac{q_2 - q_1}{q_2} < \frac{T_2 - T_1}{T_2} \quad \Rightarrow \quad \frac{q_2}{T_2} - \frac{q_1}{T_1} < 0 \quad \text{i.e.} \quad \frac{q}{T} < 0$$

Thus, for an irreversible cycle, the sum of $q/T$ term is less than zero. For an infinitesimally small reversible Carnot cycle,

$$\frac{\partial q_2}{T_2} - \frac{\partial q_1}{T_1} = 0 \quad \text{or} \quad \oint \frac{\partial q}{T} = 0$$

For an infinitesimally small irreversible Carnot cycle, $\oint \frac{\partial q}{T} < 0$

This equation is called *Clausius inequality equation*.

For a such type of cycle, equation $\oint \frac{\partial q}{T} = 0$, modified as (Figure 1.2)

$$\int_A^B \frac{\partial q_1}{T} + \int_B^A \frac{\partial Q_{II}}{T} = 0$$

$$\int_A^B \frac{\partial q_I}{T} = -\int_B^A \frac{\partial q_{II}}{T} \quad \text{or} \quad \int_A^B \frac{\partial q_I}{T} = \int_A^B \frac{\partial q_{II}}{T}$$

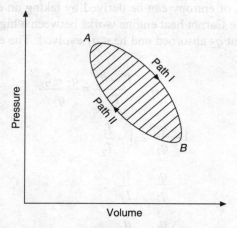

**FIGURE 1.2** Variation of pressure with volume for a cyclic process.

From the above equation, it is clear that the term $\oint \frac{\partial q}{T}$ remain same whether the change occurs through path I or path II. It implies that the term $\oint \frac{\partial q}{T}$ is independent of path, i.e. it is a state function. In general

$$\int_A^B \frac{\partial q_{rev}}{T} = S_B - S_A = \Delta S \quad \text{or} \quad dS = \frac{\partial q_{rev}}{T}$$

## 1.9 ENTROPY CHANGE FOR AN IDEAL GAS WITH TEMPERATURE AND VOLUME

$$dq = dE + PdV \quad \text{(First law of thermodynamics)}$$

Dividing both sides by $T$, we get

$$\frac{dq}{T} = \frac{dE}{T} + \frac{PdV}{T}$$

$$dS = \frac{dE}{T} + \frac{PdV}{T} \quad \text{(Second law of thermodynamics)}$$

But $E = f(T, V)$

$$dE = \left(\frac{dE}{dT}\right)_v dT + \left(\frac{dE}{dV}\right)_T dV$$

On putting value of $dE$ in above equation, we get

$$dS = \frac{1}{T}\left[\left(\frac{dE}{dT}\right)_v dT + \left(\frac{dE}{dV}\right)_T dV\right] + \frac{PdV}{T}$$

By Joule–Thomson law,

$$\left(\frac{dE}{dV}\right)_T = 0$$

Hence,

$$dS = \frac{1}{T}\left[\left(\frac{dE}{dT}\right)_v dT\right] + \frac{PdV}{T}$$

But $\quad PV = RT \quad$ or $\quad P = RT/V \quad$ (For one mole of ideal gas)

$$dS = \frac{1}{T}\left[\left(\frac{dE}{dT}\right)_v dT\right] + \frac{RTdV}{VT}$$

$$\Rightarrow \qquad\qquad = C_V \frac{dT}{T} + R \frac{dV}{V}$$

For a finite change of an ideal gas

$$\int_{S_1}^{S_2} dS = \int_{T_1}^{T_2} \frac{C_V}{T} dT + \int_{V_1}^{V_2} \frac{RdV}{V}$$

$$\Rightarrow \qquad\qquad = C_V \ln \frac{T_2}{T_1} + R \ln \frac{V_2}{V_1}$$

$$\Delta S = 2.303\, C_V \log \frac{T_2}{T_1} + 2.303\, R \log \frac{V_2}{V_1} \qquad (1.1)$$

## 1.10 ENTROPY CHANGE FOR AN IDEAL GAS WITH TEMPERATURE AND PRESSURE

For an ideal gas $\dfrac{P_1 V_1}{T_1} = \dfrac{P_2 V_2}{T_2} \quad$ or $\quad \dfrac{V_2}{V_1} = \dfrac{P_1 T_2}{P_2 T_1}$

Substituting the value of $V_2/V_1$ in Eq. (1.1), for $\Delta S$, we get

$$\Delta S = nC_V \ln \frac{T_2}{T_1} + nR \ln \frac{P_1 T_2}{P_2 T_1}$$

$$\Delta S = nC_V \ln \frac{T_2}{T_1} + nR \left[\ln \frac{P_1}{P_2} + \ln \frac{T_2}{T_1}\right]$$

$$\Delta S = n\left[(C_V + R) \ln \frac{T_2}{T_1} + R \ln \frac{P_1}{P_2}\right]$$

$$\Delta S = nC_P \ln \frac{T_2}{T_1} + nR \ln \frac{P_1}{P_2}$$

$$\Delta S = nC_P \ln \frac{T_2}{T_1} - nR \ln \frac{P_2}{P_1}$$

For an isothermal process

$$\ln \frac{T_2}{T_1} = 0 \qquad \therefore \Delta S_T = R \ln \frac{V_2}{V_1} = -R \ln \frac{P_2}{P_1}$$

For an isobaric process

$$\ln \frac{P_2}{P_1} = 0 \qquad \therefore \Delta S_p = C_P \ln \frac{T_2}{T_1}$$

For isochoric process

$$\ln \frac{V_2}{V_1} = 0 \qquad \therefore \Delta S_V = C_V \ln \frac{T_2}{T_1}$$

## 1.11 ENTROPY CHANGE IN REVERSIBLE PROCESS

If the system absorbs $q_{rev}$ heat from the surrounding at temperature $T$ then heat lost by the surrounding will also be $q_{rev}$.

Entropy change of system $\Delta S_{system} = \dfrac{q_{rev}}{T}$

Entropy change of surrounding $\Delta S_{surrounding} = -\dfrac{q_{rev}}{T}$

Total entropy change = $\Delta S_{system} + \Delta S_{surrounding}$

$$\Delta S = \frac{q_{rev}}{T} - \frac{q_{rev}}{T} = 0$$

There is no net change in entropy for a reversible process.

## 1.12 ENTROPY CHANGE IN IRREVERSIBLE PROCESS

Suppose $q$ amount of heat passes irreversibly from system at a temperature $T_2$, to its surroundings maintained at a temperature $T_1$. Then, decrease in entropy of the system is given by $\Delta S_{system} = -\dfrac{q_{irrev}}{T_2}$ and, increase in the entropy of surroundings

$$\Delta S_{surroundings} = \frac{q_{irrev}}{T_1}$$

∴ Net entropy change

$$\Delta S_T = -\frac{q_{irrev}}{T_2} + \frac{q_{irrev}}{T_1} \Rightarrow q\left[\frac{1}{T_1} - \frac{1}{T_2}\right]$$

$$\Delta S_T = q\left[\frac{T_2 - T_1}{T_1 T_2}\right] \quad \text{or} \quad \Delta S_{\text{irrev}} > 0 \qquad [\because T_2 > T_1]$$

Hence, entropy increases in an irreversible process.

## 1.13 ENTROPY OF PHASE TRANSITION

1. *Entropy of fusion:*

$$\Delta S_{\text{fus}} = \frac{\Delta H_{\text{fus}}}{T_{\text{fus}}} \quad \text{(where } T_{\text{fus}} \text{ is temp. of fusion)}$$

2. *Entropy of vaporization:*

$$\Delta S_{\text{vap}} = \frac{\Delta H_{\text{vap}}}{T_b} \quad \text{(where } T_b \text{ is temp. of boiling)}$$

3. *Entropy of Sublimation:*

$$\Delta S_{\text{sub}} = \frac{\Delta H_{\text{sub}}}{T_s} \quad \text{(where } T_s \text{ is temp. of sublimation)}$$

4. *Entropy change for allotrope:*

$$\Delta S_{\text{allot}} = \frac{\Delta H_t}{T_t} \quad \text{(where } T_t \text{ is transition temp.)}$$

5. *Entropy change in a chemical reaction:*

$$\Delta S = \Sigma S_p - \Sigma S_R \quad \text{(where } S_p \text{ and } S_R \text{ are entropy of product and reactant species respectively.)}$$

## 1.14 FREE ENERGY (*G*) AND WORK FUNCTION (*A*)

Now the question arises: Why there is a need of new function? This is because none of $\Delta E$, $\Delta H$ or $\Delta S$ can alone decide the direction of a spontaneous process. However, $\Delta S_{\text{system}}$ at constant $E$ and $V$, in an isolated system, is maximum. In general, reactions are rarely studied at constant energy ($E$). In actual, reactions are carried at constant temperature and volume or constant temperature and pressure. Hence, it is necessary to introduce new function, which can decide the direction of a spontaneous process.

## 1.15 HELMHOLTZ FREE ENERGY FUNCTION *A* = *E* − *TS*

Helmholtz free energy function (*A*) is sometimes called *work function* or maximum work content of the system. At a constant temperature, the maximum work obtainable from a system is at the expense of decrease in Helmholtz free energy of the system.

$$\Delta A = \Delta E - T\Delta S \quad \text{or} \quad \Delta A = \Delta E - T\frac{q_{\text{rev}}}{T}$$

$$\Delta A = \Delta E - q_{\text{rev}} \quad \text{But} \quad \Delta E = q + w \quad \text{or} \quad w_{\text{rev}} = \Delta E - q_{\text{rev}}$$

In case of work is done by the system, then sign is –ve,
$$-w_{rev} = \Delta E - q_{rev}$$
$$-\Delta A = w_{rev}$$

where $w_{rev}$ represents maximum useful work done during the reversible process. Hence, decrease in Helmholtz free energy gives the maximum work obtained from the system.

## 1.16 GIBBS FREE ENERGY FUNCTION (G): G = H – TS

It is that thermodynamic quantity of a system, the decrease in which value during a process is equal to the useful work done other than work of expansion. Mathematically, it is given by
$$G = H - TS$$
And change in free energy is given by
$$\Delta G = \Delta H - T\Delta S$$

*Physical significance of Gibbs free energy* (Energy available for useful work):

$\Delta G = \Delta H - T\Delta S$, But $\Delta H = \Delta E + P\Delta V$

∴ $\Delta G = \Delta E + P\Delta V - T\Delta S$, but $\Delta A = \Delta E - T\Delta S$

∴ $\Delta G = \Delta A + P\Delta V = -w_{rev} + P\Delta V$

$-\Delta G = w_{rev} - P\Delta V$

Hence, the decrease in free energy function gives the maximum work other than work of expansion that can be obtained from a system at constant temperature and pressure.

At constant volume $P\Delta V = 0$
Therefore $-\Delta G = w_{rev}$

$$\Delta G = -w_{rev} \quad \text{or} \quad \Delta G = \Delta A$$

## 1.17 VARIATION OF FREE ENERGY WITH TEMPERATURE AND PRESSURE

$$G = H - TS$$
But
$$H = E + PV$$
$$G = E + PV - TS$$

On differentiating
$$dG = dE + PdV + VdP - TdS - SdT$$
But
$$TdS = dE + PdV$$
Substituting value of $TdS$ in above equation
$$dG = TdS + VdP - TdS - SdT$$
$$dG = VdP - SdT$$

At constant temperature $dT = 0$
$$(dG)_T = (VdP)_T$$
$$\left(\frac{dG}{dP}\right)_T = V$$

At constant pressure $dP = 0$

$$(dG)_p = -(SdT)_p$$

$$\left(\frac{dG}{dT}\right)_p = -S$$

$$(dG)_T = G_2 - G_1 = \int_{P_1}^{P_2} V dP$$

But for an ideal gas,

$$V = \frac{nRT}{P}$$

$$(dG)_T = G_2 - G_1 = nRT \int_{P_1}^{P_2} \frac{1}{P} dP$$

$$= nRT \ln \frac{P_2}{P_1}$$

For an isothermal process

$$P_1 V_1 = P_2 V_2$$

$$\frac{P_2}{P_1} = \frac{V_1}{V_2}$$

$$(dG)_T = nRT \ln \frac{V_1}{V_2}$$

## 1.18 STANDARD FREE ENERGY CHANGE ($\Delta G^0$) AND EQUILIBRIUM CONSTANT ($K$)

$\Delta G^0$ = [sum of standard free energy of formation of products] − [Sum of the standard free energy of formation of reactants]

$$\Delta G^0 = \sum \Delta G_f^0 \text{ (product)} - \sum \Delta G_f^0 \text{ (reactant)}$$

The standard free energy change is related to the equilibrium constant ($K$) of the reaction as follows.

$$\Delta G^0 = -RT \ln K$$

## 1.19 CRITERIA FOR FEASIBILITY OF A PROCESS

A process is thermodynamically feasible if it is accompanied by net increase in entropy.

$$\Delta S_{\text{system}} + \Delta S_{\text{surroundings}} > 0$$

$$dS = \frac{\partial q_{\text{rev}}}{T} = \frac{dE + PdV}{T} \quad \text{for reversible small change}$$

$$TdS = dE + PdV$$

If small change is brought about irreversibly then heat absorbed will be less because $q_{rev} > q_{irrev}$
That is $TdS > dE + PdV$ for irreversible process.
  By combining the above two processes

$$TdS \geq dE + PdV$$

At constant volume and internal energy, $dV = 0$ and $dE = 0$
Hence, $TdS \geq 0$

$$(dS)_{E,V} \geq 0 \quad \text{or} \quad (\Delta S)_{E,V} \geq 0$$

**Problem 1.1**  0.1 kg $N_2$ at 298 K is held by a piston under 30 atm pressure. The pressure is suddenly released to 10 atm and the gas expands adaibatically to a temperature of 241 K. If $C_V = 20.8$ J K$^{-1}$ mol$^{-1}$. What is the final volume? Calculate $\Delta S$ of system for this expansion. What would be the value of $\Delta S_{surroundings}$.

*Solution:*

$$\frac{P_2 V_2}{T_2} = \frac{P_1 V_1}{T_1}$$

$$V_2 = \frac{30 \times V_1 \times 241}{298 \times 10} = 2.42 V_1$$

$$(\Delta S)_{system} = nC_V \ln \frac{T_2}{T_1} + nR \ln \frac{V_2}{V_1}$$

$$n = \frac{0.1}{0.028} = 3.57$$

$$(\Delta S)_{system} = 10.47 \text{ J K}^{-1}$$

$$(\Delta S)_{surroundings} = \frac{q_{surroundings}}{T} = 0$$

**Problem 1.2**  Evaluate the change in entropy when 3 mol of a gas is heated from 27°C to 727°C at a constant pressure of 1 atm. The molar heat capacity of the gas is 23.7 J K$^{-1}$ mol$^{-1}$.

*Solution:* $(\Delta S) = nC_P \ln \dfrac{T_2}{T_1} \Rightarrow 85.5$ J K$^{-1}$

**Problem 1.3**  The molar entropy of an ideal gas ($C_P = 20.9$ J K$^{-1}$) at 298 K is 146.0 J K$^{-1}$ mol$^{-1}$. Find its value at 500 K.

*Solution:*

$$(\Delta S) = nC_P \ln \frac{T_2}{T_1}; \qquad S_2 - S_1 = nC_P \ln \frac{T_2}{T_1}$$

$$S_{1(298 \text{ K})} = 146.0 \text{ J K}^{-1} \text{ mol}^{-1}$$

$$S_{2(500\ K)} = S_1 + C_P \ln \frac{T_2}{T_1}$$

$$S_{2(500\ K)} = 146.0\ \text{J K}^{-1}\ \text{mol}^{-1} + 20.9 \times 2.303 \times \log \frac{500\ K}{298\ K}$$

$$S_{2(500\ K)} = 156.81\ \text{J K}^{-1}\ \text{mol}^{-1}$$

**Problem 1.4** Calculate the entropy change when $10^{-2}\ m^3$ of an ideal gas ($C_P = 2.5R$) at 27°C and $1.01 \times 10^5\ Nm^{-2}$ pressure, are heated at constant pressure to 127°C.

**Solution:**

$$\Delta S = nC_P \ln \frac{T_2}{T_1} \qquad n = \frac{PV}{RT} \Rightarrow 0.405\ \text{mol}$$

$$\Delta S = 0.405 \times 2.5 \times 8.314 \times 2.303 \log \frac{400\ K}{300\ K}$$

$$\Rightarrow \quad = 2.422\ \text{J K}^{-1}$$

**Problem 1.5** One mole of an ideal gas is allowed to expand isothermally at 27°C. Untill its volume is tripled. Calculate $\Delta S_{system}$ and $\Delta S_{universe}$ under the following conditions
 (a) The expansion is carried out reversibly.
 (b) The expansion is a free expansion.

**Solution:**

(a) $dS = \dfrac{dq_{rev}}{T}$

$$Q = \Delta E - w \qquad \text{because } \Delta E = 0 \quad \therefore \quad q = -w$$

$$-w = 2.303\ RT \log \frac{V_2}{V_1} \Rightarrow 2740.6\ \text{J mol}^{-1}$$

$$\Delta S = \frac{\Delta q_{rev}}{T}$$

$$\Delta S = \frac{-w}{T} = \frac{2740.6}{300} \Rightarrow 9.135\ \text{J K}^{-1}\ \text{mol}^{-1}$$

$$(\Delta S)_{surroundings} = -(\Delta S)_{system}$$

$$(\Delta S)_{univ} = 0$$

(b) In case of free expansion we can use above equation,

$$\Delta S = R \ln \frac{V_2}{V_1} \Rightarrow 9.135\ \text{J K}^{-1}\ \text{mol}^{-1}$$

For surroundings $q = 0$ (isothermal) $(\Delta S)_{surroundings} = 0$

$$(\Delta S)_{univ} = (\Delta S)_{system} = 9.135\ \text{J K}^{-1}\ \text{mol}^{-1}$$

**Problem 1.6** Calculate the difference between $\Delta G$ and $\Delta A$ at 27°C for the reaction

$$H_2(g) + \frac{1}{2} O_2(g) \rightarrow H_2O(l)$$

## Solution:

$$\Delta G = \Delta H - T\Delta S$$
$$\Delta A = \Delta E - T\Delta S$$
$$\Delta G - \Delta A = \Delta H - \Delta E \quad \text{and} \quad \Delta H = \Delta E + P\Delta V$$
$$\Delta H = \Delta E + \Delta V_g RT \quad \text{or} \quad \Delta H - \Delta E = \Delta V_g RT$$
$$\Delta V_g = 0 - 1 - \frac{1}{2} = -\frac{3}{2}$$
$$\Delta H - \Delta E = -\frac{3}{2} \times (8.314 \text{ J K}^{-1} \text{ mol}^{-1}) \times 300 \text{ K}$$
$$= -3741 \text{ J mol}^{-1}$$

## 1.20 CHEMICAL POTENTIAL ($\mu$)

It is partial derivative of free energy w.r.t. amount of species, all other species concentration remain constant at constant temperature. At equilibrium, total sum of chemical potential is zero as free energy is at a minimum.

Also known as partial molar quantity. The thermodynamic properties like $E$, $H$, $S$, $A$, and $G$ are extensive property. In case of an open system, extensive property like $G$ is a function not only of temperature and pressure but also of the number of moles of various components present in the system.

$$G = f(T, P, n_1, n_2, n_3, \ldots, n_i)$$

$$dG = \left(\frac{dG}{dT}\right)_{P,n} dT + \left(\frac{dG}{dP}\right)_{T,n} dP + \left(\frac{dG}{dn_1}\right)_{T,P,n_2,\cdots,n_i} dn_1 + \cdots + \left(\frac{dG}{dn_i}\right)_{T,P,n_1 n_2 \cdots} dn_i$$

The quantity $\left(\frac{dG}{dn_i}\right)_{T,P,n_1 n_2 \cdots}$ is called the *partial molar free energy* and is represented by $\bar{G}_i$ or $\mu_i$ known as chemical potential of the component $i$.

$$\bar{G}_i = \mu_i = \left(\frac{dG}{dn_i}\right)_{T,P,n_1 n_2 \cdots}$$

Hence, chemical potential of a given substance may be defined as the change in free energy of the system when 1 mol of that substance is added at constant temperature and pressure.

Thus, equation for change in free energy of the system may be written as

$$dG = \left(\frac{dG}{dT}\right)_{P,n} dT + \left(\frac{dG}{dP}\right)_{T,n} dP + \mu_1 dn_1 + \mu_2 dn_2 + \cdots + \mu_i dn_i$$

*Physical significance:*

1. It represents the partial change in free energy with respect to its number of moles at constant temperature and pressure.

$$(dG)_{T,P} = \mu_1 dn_1 + \mu_2 dn_2 + \cdots + \mu_i dn_i$$

On integrating, we get

$$G_{T,P,n} = \mu_1 n_1 + \mu_2 n_2 + \cdots + \mu_i n_i$$

2. It is an intensive property and may be regarded as the driving force that compels a system to attain an equilibrium.
3. At equilibrium the chemical potential of a substance has the same value throughout the system, i.e. $\Sigma\, \mu_i dn_i = 0$.
4. For 1 mol of pure substance, the chemical potential is identical to the free energy, $\mu = G$.
5. It has great importance to derive Gibbs–Duhem equation.
6. It is a form of potential energy that can be absorbed or released during a chemical reaction.

## 1.21 GIBBS–DUHEM EQUATION

We know,
$$G_{T,p,n} = \mu_1 n_1 + \mu_2 n_2 + \cdots + \mu_i n_i$$
On differentiating,
$$dG = \mu_1 dn_1 + n_1 d\mu_1 + \mu_2 dn_2 + n_2 d\mu_2 + \cdots + \mu_i dn_i + n_i d\mu_i$$
$$dG = (\mu_1 dn_1 + \mu_2 dn_2 + \cdots + \mu_i dn_i) + (n_1 d\mu_1 + n_2 d\mu_2 + \cdots + n_i d\mu_i)$$
But we know that
$$(dG)_{T,P} = (\mu_1 dn_1 + \mu_2 dn_2 + \cdots + \mu_i dn_i)$$
Therefore, $(n_1 d\mu_1 + n_2 d\mu_2 + \cdots + n_i d\mu_i) = 0$
$$\Sigma n_i d\mu_i = 0$$
This relation is general form of Gibbs–Duhem equation. For a binary solution,
$$n_1 d\mu_1 + n_2 d\mu_2 = 0$$
$$d\mu_1 = -\frac{n_2}{n_1} d\mu_2$$
Gibbs–Duhem equation describes relationship between changes in chemical potential for components in a thermodynamic system.
$$\sum_{i=1}^{I} N_i d\mu_i = -s dT + V dP$$

## 1.22 VARIATION OF CHEMICAL POTENTIAL WITH TEMPERATURE AND PRESSURE

$$dG = \left(\frac{dG}{dT}\right)_{P,n} dT + \left(\frac{dG}{dP}\right)_{T,n} dP + \mu_1 dn_1 + \mu_2 dn_2 + \cdots + \mu_i dn_i$$

$$(dG)_n = \left(\frac{dG}{dT}\right)_{P,n} dT + \left(\frac{dG}{dP}\right)_{T,n} dP$$

$$dG = VdP - SdT$$

By equating the coefficient of $dT$ and $dP$ in the above two equations, we get

$$\left(\frac{dG}{dT}\right)_{P,n} = -S \quad \text{and} \quad \left(\frac{dG}{dP}\right)_{T,n} = V$$

## 1.23 GIBBS–HELMHOLTZ EQUATION OR EFFECT OF TEMPERATURE ON CHEMICAL POTENTIAL

As we know $\left(\dfrac{dG}{dT}\right)_{P,n} = -S$

For initial and final state

$$\left(\dfrac{dG_1}{dT}\right)_{P,n} = -S_1 \quad \text{and} \quad \left(\dfrac{dG_2}{dT}\right)_{P,n} = -S_2$$

Subtracting the two, we get

$$\left(\dfrac{dG_2}{dT}\right)_P - \left(\dfrac{dG_1}{dT}\right)_P = -S_2 - (-S_1) = -dS$$

$$\left(\dfrac{\partial (G_2 - G_1)}{\partial T}\right)_P = -\Delta S$$

$$\left(\dfrac{\partial \Delta G}{\partial T}\right)_P = -\Delta S$$

On putting the value of $\Delta S$ in $\Delta G = \Delta H - T\Delta S$

$$\Delta G = \Delta H + T\left(\dfrac{\partial \Delta G}{\partial T}\right)_P$$

This is Gibbs–Helmholtz equation in terms of free energy.

In terms of work function it is

$$\Delta A = \Delta E + T\left(\dfrac{\partial \Delta A}{\partial T}\right)_V$$

This is Gibbs–Helmholtz equation in terms of work function.

*Alternative form:*

On differentiating $(\Delta G/T)$ with temperature at constant pressure

$$\left[\dfrac{\partial}{\partial T}\left(\dfrac{\Delta G}{T}\right)\right]_P = \Delta G\left(-\dfrac{1}{T^2}\right) + \dfrac{1}{T}\left[\dfrac{\partial \Delta G}{\partial T}\right]_P$$

By multiplying and dividing by $T^2$, we get

$$\Rightarrow \qquad \dfrac{T\left[\dfrac{\partial \Delta G}{\partial T}\right]_P - \Delta G}{T^2}$$

From Gibbs–Helmholtz equation

$$T\left(\dfrac{\partial \Delta G}{\partial T}\right)_P = \Delta G - \Delta H$$

On putting value of $T\left(\dfrac{\partial \Delta G}{\partial T}\right)_P$ in above equation,

$$\left[\dfrac{\partial}{\partial T}\left(\dfrac{\Delta G}{T}\right)\right]_P = \dfrac{\Delta G - \Delta H - \Delta G}{T^2}$$

$$\left[\dfrac{\partial}{\partial T}\left(\dfrac{\Delta G}{T}\right)\right]_P = \dfrac{-\Delta H}{T^2}$$

This equation is also very important form of Gibbs–Helmholtz equation and is applicable to reversible as well as irreversible process.

## 1.24 CLAUSIUS–CLAPEYRON EQUATION

A thermodynamic relation between the change of pressure with change of temperature of a system at equilibrium is called *Clausius–Clapeyron equation*

Phase $A \rightleftharpoons$ Phase $B$

$$dG = VdP - SdT$$

$$(\Delta G)_{T,P} = G_B - G_A = 0$$

$\Rightarrow$ $\qquad G_B = G_A$

If temperature is changed from $T$ to $T + dT$ than pressure of system $P + dP$ will maintain the equilibrium.

Now
$$G_B + dG_B = G_A + dG_A$$

$$dG_B = dG_A, \quad \text{since } (G_B = G_A)$$

$$dG_A = V_A dP - S_A dT \quad \text{and} \quad dG_B = V_B dP - S_B dT$$

$$V_A dP - S_A dT = V_B dP - S_B dT$$

$$(S_B - S_A)dT = (V_B - V_A)dP$$

or
$$\dfrac{dP}{dT} = \dfrac{S_B - S_A}{V_B - V_A} \Rightarrow \dfrac{\Delta S}{\Delta V}$$

But change in entropy $dS = \dfrac{dq_{\text{rev}}}{T}$

$$\dfrac{dP}{dT} = \dfrac{q}{TdV} = \dfrac{q}{T(V_B - V_A)}$$

Above equation is known as the Clausius–Clapreon equation.

## 1.25 INTEGRATED FORM OF CLAUSIUS–CLAPEYRON EQUATION

$\dfrac{dP}{dT} = \dfrac{q}{T(V_g - V_S)} = \dfrac{q}{TV_g}$ because $V_S$ can be neglected in comparison to $V_g$.

For one mol, an ideal gas $V_g = \dfrac{RT}{P}$

$$\dfrac{dP}{dT} = \dfrac{dH}{TV_g} \Rightarrow \dfrac{dHP}{TRT}$$

$$\dfrac{dP}{P} = \dfrac{dH}{R}\left[\dfrac{dT}{T^2}\right]$$

On integrating above equation

$$\int \dfrac{dP}{P} = \dfrac{dH}{R}\int \dfrac{dT}{T^2}$$

$$\ln P = -\dfrac{\Delta H}{R}\left(\dfrac{1}{T}\right) + C'$$

This is equation of straight line $y = mx + C$. When log $P$ is plotted against $1/T$, a straight line with slope $-\dfrac{\Delta H}{2.303\,R}$ is obtained (Figure 1.3).

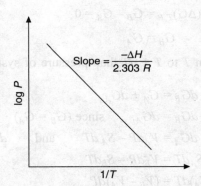

**FIGURE 1.3** Plot of log P against 1/T gives straight line.

$$\ln \dfrac{P_2}{P_1} = -\dfrac{\Delta H}{R}\left(\dfrac{1}{T}\right)_{T_1}^{T_2}$$

$$\ln \dfrac{P_2}{P_1} = -\dfrac{\Delta H}{R}\left(\dfrac{T_2 - T_1}{T_2 T_1}\right)$$

$$\log \dfrac{P_2}{P_1} = -\dfrac{\Delta H}{2.303\,R}\left(\dfrac{T_2 - T_1}{T_2 T_1}\right)$$

This is integrated form of Clausius–Clapeyron equation.

## 1.26 RELATION BETWEEN ENTHALPY AND CHEMICAL POTENTIAL

$$dG = \left(\frac{dG}{dT}\right)_{P,n} dT + \left(\frac{dG}{dP}\right)_{T,n} dP + \mu_1 dn_1 + \mu_2 dn_2 + \cdots + \mu_i dn_i$$

$$dG = VdP - SdT + \Sigma \mu_i dn_i$$

$$G = H - TS \quad \text{or} \quad H = G + TS$$

$$dH = dG + TdS + SdT$$

Substituting the value of $dG$ in above equation

$$dH = VdP - SdT + \Sigma \mu_i dn_i + TdS + SdT$$

$$dH = VdP + TdS + \Sigma \mu_i dn_i$$

If $S$, $P$ and $n_j$ are constant then, $dS = dP = dn_j = 0$

$$\left(\frac{dH}{dn_i}\right)_{S,P,n_j} = \mu_i$$

## 1.27 NERNST HEAT THEOREM

W. Nernst in 1906, studied the variation of enthalpy change ($\Delta H$) and free energy $\Delta G$ with decrease of temperature (Figure 1.4).

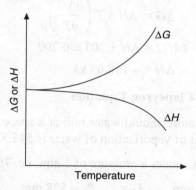

**FIGURE 1.4** Variation of enthalpy or free energy change with temperature.

According to Gibbs–Helmholtz equation,

$$\Delta G = \Delta H + T\left(\frac{\partial \Delta G}{\partial T}\right)_P$$

At absolute zero, i.e. when $T = 0$, $\Delta G = \Delta H$. However, Nernst observed that as the temperature is lowered towards absolute zero, the value of $\partial (\Delta G)/dT$ decreases and then approaches zero asymptotically. This means that $\Delta G$ and $\Delta H$ are not only equal at the absolute zero but the value approach each other asymptotically in the vicinity of this temperature. The result is known as *Nernst heat theorem*.

$$\lim_{T \to 0} \frac{d(\Delta G)}{dT} = \lim_{T \to 0} \frac{d(\Delta H)}{dT} = 0 \qquad (1.2)$$

$\Delta G$ has been shown as greater than $\Delta H$ at temperature away from the absolute zero. However, the reverse is also possible because $d(\Delta G)/\partial T$ can be both positive or negative.

But we know that $\left(\dfrac{d(\Delta G)}{dT}\right) = -\Delta S$ and $\left(\dfrac{d(\Delta H)}{dT}\right)_P = -\Delta C_P$

Combining these results with Eq. (1.2), we get

$$\lim_{T \to 0} \Delta S = 0 \quad \text{and} \quad \lim_{T \to 0} \Delta C_P = 0$$

Since gases do not exist at the absolute zero, this means that the heat theorem is not applicable to gases. Similarly, it has been found to be inapplicable to liquid also.

**Problem 1.7** The free energy change accompanying a given process is $-85.77$ and $-83.68$ kJ at 25°C and 35°C, respectively. Calculate the change in enthalpy for the process at 30°C.

*Solution:*

$$\Delta G \text{ at } 30°C = \frac{-(85.77 + 83.68)}{2} \Rightarrow -84.725 \text{ kJ}$$

$$\left(\frac{d(\Delta G)}{dT}\right)_P = \frac{-83.68 - (-85.77)}{308 - 298} \Rightarrow 0.209 \text{ kJ}$$

We know, 
$$\Delta G = \Delta H + T\left(\frac{\partial \Delta G}{\partial T}\right)_P$$

$$-84.72 = \Delta H + 303 \times 0.209$$

$$\Delta H = -148.05 \text{ kJ}$$

**Problem based on Clausisus–Clapeyron Equation**

**Problem 1.8** At what temperature should water boil at a space station where the atmospheric pressure is 528 mm. Latent heat of vaporization of water is 545.5 cal/g.

*Solution:* Water boils at 100°C under a pressure of 1 atm, i.e. 760 mm of Hg.

$T_1 = 373$ K $\quad P_1 = 760$ mm $\quad T_2 = ? \quad P_2 = 528$ mm $\quad R = 1.987$ cal K$^{-1}$ mol$^{-1}$

$\Delta H_V = 545.5$ cal g$^{-1} = 545.5 \times 18$ cal mol$^{-1} \Rightarrow 9819$ cal mol$^{-1}$

$$\log \frac{P_2}{P_1} = -\frac{\Delta H}{2.303 R}\left(\frac{T_2 - T_1}{T_2 T_1}\right)$$

On solving for $\quad T_2 = 363.02$ K

So, $\quad T_2 = 90.01°C$.

**Problem 1.9** Calculate the change in the freezing point of water at 0°C atm$^{-1}$ change of pressure. Data at 0°C are given as: heat of fusion of ice is 335 Jg$^{-1}$, the density of water is 0.9998 g cm$^{-3}$ and the density of ice is 0.9168 g cm$^{-3}$.

**Solution:**

$$\lambda_f = 335 \text{ J g}^{-1} \qquad T = 273 \text{ K}$$
$$\rho^l = 0.9998 \text{ g cm}^{-3} \qquad \rho^s = 0.9168 \text{ g cm}^{-3}$$
$$V^l = 1.0002 \text{ cm}^3 \text{ g}^{-1} \qquad V^s = 1.0908 \text{ cm}^3 \text{ g}^{-1}$$
$$\Delta V = 1.0002 - 1.0908 = -0.0906 \text{ cm}^3 \text{ g}^{-1}$$

$$\frac{\Delta P}{\Delta T} = \frac{\lambda_{\text{fus}}}{T \Delta V} \quad \Rightarrow \quad -\frac{335 \text{ J g}^{-1}}{273 \times 0.0906} = -13.543 \text{ Jcm}^{-3}\text{K}^{-1}$$

$$\Rightarrow -1.33.7 \text{ atm K}^{-1}$$

$$\frac{\Delta T}{\Delta P} = -0.0075 \text{ K atm}^{-1}$$

Thus, an increase in pressure of 1 atm lowers the freezing point by 0.0075. The negative sign indicates that an increase in pressure causes a decrease in temperature.

**Problem 1.10** When glucose is burnt is a bomb calorimeter, the reaction is found to be exothermic and $\Delta E = -2880$ kJ mole$^{-1}$, $\Delta S = 182.4$ K$^{-1}$ mol$^{-1}$. Calculate free energy change at 298 K for the process. How much of energy change can be extracted as heat, and how much can be extracted as work.

**Solution:**

$$\Delta A = \Delta E - T\Delta S$$
$$= -2880 - 298 \times 182.4 \times 10^{-3} \text{ kJ K}^{-1} \text{ mol}^{-1}$$
$$\Delta A = -2934.3 \text{ kJ mol}^{-1}$$

As we know $\Delta A = W$, hence, the work done during a reversible oxidation of glucose is $-2934.3$ kJ.
Since, oxidation takes place at constant volume $q_v = \Delta E$

$$\Rightarrow \quad -2880 \text{ kJ mol}^{-1}.$$

## 1.28 THIRD LAW OF THERMODYNAMICS

M. Planck (1912) made a suggestion concerning the value of $S_o$, which has become third law of thermodynamics.

*Every substance has a finite positive entropy, but at 0 K the entropy may become zero, and does so become in case of a perfectly crystalline substance.*

$$dS = C_P \frac{dT}{T} \quad \Rightarrow \quad \int_{T_1}^{T_2} dS \quad \Rightarrow \quad \int_{T_1}^{T_2} C_P \frac{dT}{T}$$

$$S(T) = \int_0^T C_P \frac{dT}{T} \qquad (1.3)$$

Entropy value can be calculated, if we integrate RHS of above equation from $T = 0$ to any desired temperature $T$. For this, we must know the values of $C_P$ right from absolute zero to the temperature $T$ concerned. The measurement of $C_P$ up to the absolute zero is inconvenient. For this purpose, the method of extrapolation is employed. The values of $C_P/T$ are plotted against $T$ and the curve is extended up to $T = 0$. The area under the curve, from $T = 0$ to $T = T$ gives the absolute value of entropy of the substance.

In many cases, the value of heat capacity down to 0 K can be obtained by the use of Debye equation as

$$C_V = 3R\left[\frac{4}{5}\pi^4\left(\frac{T}{\theta}\right)^3\right]$$

where, $\theta$ is characteristic temperature of the substance and is calculated as

$$\theta = \frac{hC\omega_m}{k}, \quad \text{where } \omega_m \text{ is mean frequency.}$$

The absolute value of entropy of a gaseous substance at 298 K and 1 atm may be calculated by considering the entropy contribution due to the all the changes, i.e. heat capacity of solid, liquid, gas, entropy of transition (phase), etc.

$$S(T) = \int_0^{T_t} C_{P\alpha}\,d\ln T + \frac{dH_t}{T_t} + \int_{T_t}^{T_m} C_{P\beta}\,d\ln T + \frac{dH_{\text{fus}}}{T_m} + \int_{T_m}^{T_b} C_{P(l)}\,d\ln T + \frac{dH_{\text{vap}}}{T_b} + \int_{T_b}^{298} C_{P(g)}\,d\ln T \quad (1.4)$$

## 1.29 CONCEPT OF RESIDUAL ENTROPY OR UNATTAINABILITY OF ABSOLUTE ZERO

Certain chemical reactions between crystals do not give $\Delta S = 0$ at 0 K, which seems to indicate that exceptions to this law exist. It involves ice, CO, $N_2O$ or hydrogen. In the crystalline state, these four substances appears to retain a finite entropy at 0 K. Ice at zero Kelvin appears to have a residual entropy of 3.3 J K$^{-1}$. Similarly, CO has residual entropy of 5.8, $N_2O$ has 5.8 and hydrogen gas has residual entropy of 6.2 J K$^{-1}$. This suggests that molecules in the crystal of the substances are not in true thermodynamic equilibrium at 0 K.

A number of liquids are readily super cooled below their thermodynamic freezing points and can exist in liquid state even at 0 K. It has been found that liquid at 0 K has a greater entropy than the crystal at 0 K. Thus, the third law is also not applicable to super cooled liquids e.g. glass.

**Problem 1.11** 10 g of ice at 0°C are added to 20 g water at 90°C in a thermally insulated flask of negligible heat capacity. The heat of fusion of ice is 5980 J mol$^{-1}$. What is the final temperature? $\Delta S_{\text{system}}$ and $\Delta S_{\text{surroundings}} = ?$ $C_{P(m)}(\text{water}, l) = 75.42$ J K$^{-1}$ mol$^{-1}$.

*Solution:*

1. solid $\rightarrow$ liquid

    10/18 moles of water (solid, 273 K) $\xrightarrow{\text{melt}}$ $\frac{10}{18}$ moles of water ($l$, 273 K)

    $$q_{(l)} = n\Delta H_f = 5980 \times \frac{10}{18} \text{ J} = 3322.2 \text{ J}$$

2. Water at 273 K is heated to temperature $T_2$.

10/18 moles of water (liquid, 273 K) $\xrightarrow{\text{melt}}$ 10/18 moles of water ($l$, $T_2$)

$$q_{(II)} = nC_{P(m)} (H_2O)(T_2 - T)_1$$
$$= \frac{10}{18} \times 75.42 \times (T_2 - 273)$$
$$= 41.9 \times (T_2 - 273)$$

3. 20/18 moles of water (liquid, 363 K) $\xrightarrow{\text{cool}}$ 20/18 moles of water ($l$, $T_2$)

$$q_{(III)} = \frac{20}{18} \times 75.42 \times (T_2 - 363)$$
$$= 83.4 \times (T_2 - 363)$$

Total heat change $q = q_{(I)} + q_{(II)} + q_{(III)} = 0$ (adiabatic process)

$$= 5980 \times \frac{10}{18} J + \frac{10}{18} \times 75.42 \times (T_2 - 273) + \frac{20}{18} \times 75.42 \times (T_2 - 363) = 0$$

$T_2 = 306$ K

Entropy change $\Delta S_{(I)} = \dfrac{q_1}{273} = \dfrac{5980 \times 10}{273 \times 18} = 12.169$ J K$^{-1}$

$$\Delta S_{(II)} = nC_{P(m)} \ln \frac{T_2}{T_1}$$
$$= \frac{10}{18} \times 75.42 \times \ln \frac{306}{273} = -04.782 \text{ J K}^{-1}$$

$$\Delta S_{(III)} = nC_{P(m)} \ln \frac{T_2}{T_1} = \frac{20}{18} \times 75.42 \times \ln \frac{306}{363} = -14.317 \text{ J K}^{-1}$$

$\Delta S = \Delta S_{(I)} + \Delta S_{(II)} + \Delta S_{(III)} = 2.634$ J K$^{-1}$

$\Delta S_{\text{surroundings}} = 0$, since, process is adiabatic.

**Problem 1.12** One mole of super cooled water at $-10°C$ and 1 atm. pressure turns into ice at $-10°C$. Calculate the entropy change in the surroundings and the system and the net entropy change. Heat capacity of water and ice at 1 atm may be taken as constant at 75.42 J K$^{-1}$ mol$^{-1}$ and 37.2 J K$^{-1}$ mol$^{-1}$, respectively. $\Delta H_{\text{fusion}}$ (273 K) = 6008 J mol$^{-1}$.

*Solution:* The given process is not a reversible one so the entropy change for this can not be calculated directly. But we can consider a series of reversible steps starting from the liquid water at $-10°C$ to solid ice at $-10°C$.

**Step 1:** Heat reversibly the super cooled water at $-10°C$ to $0°C$ keeping the pressure constant.

$$dq_{\text{rev}} = C_P dT = dH$$

$$dS_1 = \int_{263}^{273} \frac{dq_{\text{rev}}}{T} = C_p (H_2O, l) \ln \frac{273}{263} \Rightarrow 2.81 \text{ J K}^{-1}\text{mol}^{-1}$$

**Step 2:** Allow the solidfication of water at 0°C to ice at 0°C.

$$\Delta S_2 = -\frac{\Delta H_{\text{fusion}}}{T} \Rightarrow -\frac{6008 \text{ J mol}^{-1}}{273 \text{ K}} = -22.0 \text{ J K}^{-1} \text{ mol}^{-1}$$

**Step 3:** Cool the ice reversibly from 0°C to –10°C.

$$dq_{\text{rev}} = C_P(\text{H}_2\text{O}, S)dT = dH$$

$$\Delta S_3 = C_P(\text{H}_2\text{O}, S) \ln \frac{263}{273} \Rightarrow -1.38 \text{ J K}^{-1} \text{mol}^{-1}$$

$$\Delta S_{\text{Total}} = \Delta S_1 + \Delta S_2 + \Delta S_3 = -20.57 \text{ J K}^{-1} \text{ mol}^{-1}$$

**Problem 1.13** One mole of ideal gas at 27°C expands isothermally and reversibly from an initial volume of 2 dm$^3$ to a final volume of 20 dm$^3$ against a pressure that is gradually reduced. Calculate $q$, $w$, $\Delta E$, $\Delta A$, $\Delta G$ and $\Delta S$.

*Solution:* Process is carried out reversibly and isothermally. Also given that $n = 1$, $V_1 = 2$ dm$^3$, $V_2 = 20$ dm$^3$, $T = 300$ K.

1. $w = -RT \ln \dfrac{V_2}{V_1} = 2.303 \times (-8.314) \times 300 \text{ K} \times \log \dfrac{20 \text{ dm}^3}{2.0 \text{ dm}^3} \Rightarrow -5744 \text{ J mol}^{-1}$

2. Since gas is ideal and expansion is carried out isothermally ($dt = 0$), $dE = 0$.

3. First law gives $q = \Delta E - w = 0 + 5744$ J mol$^{-1}$ = 5744 J mol$^{-1}$.

4. $\Delta H = \Delta E + \Delta(PV) = \Delta E + \Delta(nRT) = 0 + 0 = 0$

5. $\Delta A = -nRT \ln \dfrac{V_2}{V_1} = w = -5744$ J mol$^{-1}$

6. $\Delta G = nRT \ln \dfrac{P_2}{P_1} = -nRT \ln \dfrac{V_2}{V_1} = -5744$ J mol$^{-1}$

7. $\Delta S = \dfrac{q_{\text{rev}}}{T} = \dfrac{5744 \text{ J mol}^{-1}}{300 \text{ K}} \Rightarrow 19.15 \text{ J K}^{-1} \text{ mol}^{-1}$

**Problem 1.14** One mole is vaporized at its boiling point 111°C. The heat of vaporization at this temperature is 363.3 J g$^{-1}$. Calculate the maximum work done against 1 atm, $q$, $\Delta H$, $\Delta E$, $\Delta G$ and $\Delta S$ for the vaporization of one mole toluene.

$$\text{C}_6\text{H}_5\text{CH}_3 \, (l) \rightarrow \text{C}_6\text{H}_5\text{CH}_3 \, (\text{vapour})$$

*Solution:* The process is isothermal, isobaric and reversible.

$$T = 111 + 273 = 384 \text{ K}$$

$$-w = P\Delta V = P(V_g - V_l) \approx PV_g = R_T [V_g \gg V_l]$$

$$= (8.314)384 = 3193 \text{ J mol}^{-1}$$

$$q_p = \Delta H = 363.3 \times 92$$

$$= 33423.6 \text{ J mol}^{-1}$$

$$\Delta E = \Delta H - P\Delta V = (33424 - 3193) \Rightarrow 30231 \text{ J mol}^{-1}$$

$$\Delta G = \int V dP = nRT \ln \frac{P_2}{P_1} = nRT \ln \frac{1}{1} = 0$$

$$\Delta S = \frac{q_{\text{rev}}}{T} = \frac{\Delta H_{\text{vap}}}{T_b} = \frac{33424}{384} = 87.04 \text{ JK}^{-1}\text{mol}^{-1}$$

**Problem 1.15** Calculate the rate of change of transition temperature with pressure for sulphur. The data given are transition temperature = 95.5°C at 1 atm, enthalpy of transition per gram of sulphur is 13.4 J. Monoclinic sulphur (stable above transition point) has a greater specific volume than that of rhombic sulphur by 0.0126 cm³ g⁻¹.

*Solution:*

$$\lambda_{\text{trans}} = 13.4 \text{ J g}^{-1} \quad T_t = 95.5 + 273 \Rightarrow 368.5 \text{ K}$$

$$\Delta V = 0.0126 \text{ cm}^3 \text{ g}^{-1}$$

$$\frac{dP}{dT} = \frac{\lambda}{T\Delta V} = \frac{13.4}{368.5 \times 0.0126} \Rightarrow 2.886 \text{ J cm}^{-3}\text{K}^{-1}$$

$$\frac{dP}{dT} = 28.5 \text{ atm K}^{-1} \quad \text{or} \quad \frac{dT}{dP} = 0.0351 \text{ K atm}^{-1}$$

## 1.30 FUGACITY

G.N. Lewis in 1901 introduced the concept of fugacity to explain actual behaviour of real gases in chemical equilibrium at high pressures. Fugacity is mainly employed in connection with gas mixture, but the introductory treatment is restricted to pure gases, at a later stage it will be extended to systems consisting of more than one component (gas).

For an infinitesimal, reversible, isothermal expansion of an ideal gas, the work of expansion is given by

$$dw = dF = VdP$$

For 1 mol of an ideal gas $V = \dfrac{RT}{P}$

$$dF = RT \frac{dP}{P} = RT \, d \ln P \tag{1.5}$$

Above equation is valid for ideal gas only.

For real gas, above equation is modified as

$$dF = RT \, d \ln f$$

The function $f$ is known as *fugacity*.

On integrating the above equation, we get

$$F = RT \ln f + C$$

where $F$ is the molar free energy of the gas and $f$ is its fugacity and $C$ is the integration constant, which is dependent upon the temperature and nature of the gas.

On integrating Eq. (1.5) by definite integral of equation

$$F_2 - F_1 = RT \ln \frac{f_2}{f_1} \tag{1.6}$$

where $F_1$ and $F_2$ are molar free energy of the gas in two states, i.e. two pressure at the same temperature and $f_1$ and $f_2$ are the corresponding fugacity. For an ideal gas, the difference of molar free energy in two states at the same temperature is given by

$$F_2 - F_1 = RT \ln \frac{P_2}{P_1} \tag{1.7}$$

Comparison of Eqs. (1.6) and (1.7) reveals that for an ideal gas the fugacity is proportional to the pressure.

$$f \propto \text{pressure} \qquad f = K P$$

It is convenient to take the proportionality constant ($K$) as unity, so that for an ideal gas $\frac{f}{P} = 1$ and fugacity is always equal to the pressure.

For real gases, the fugacity and pressure are not proportional to one another and $\frac{f}{P}$ is not constant. As the pressure of the gas decreased, the behaviour approaches that for an ideal gas and so the gas at very low pressure is chosen as the reference state and it is postulated that the ratio $\frac{f}{P}$ of the fugacity to the pressure than approaches unity, thus

$$\lim_{P \to 0} \frac{f}{P} = 1 \quad \text{or} \quad \frac{f}{P} \to 1 \text{ as } P \to 0$$

## 1.31 DETERMINATION OF FUGACITY BY GRAPHICAL METHOD

For an ideal gas the fugacity is equal to the pressure at all pressures, but for a real gas this is only the case at very low pressure when it behaves ideally. To determine the fugacity of a gas at any pressure where it deviates from ideal behaviour, the following procedure is used.

We know $\quad dF = V dP \qquad dF = RT \dfrac{dP}{P} = RT \, d \ln P$

For real gases $dF = RT \, d \ln f$

Therefore, $\quad V dP = RT \, d \ln f \tag{1.8}$

$$\left( \frac{\partial \ln f}{\partial P} \right)_T = \frac{V}{RT}$$

where $V$ is the actual volume occupied by 1 mol of real gas. For an ideal gas the volume of 1 mol is $\dfrac{RT}{P}$. For a real gas this volume is represented by $\alpha$ and is given by

$$\alpha = \frac{RT}{P} - V \quad \text{so} \quad V = \frac{RT}{P} - \alpha$$

Hence, by putting value of $V$ in Eq. (1.8)

$$RT d\ln f = RT \frac{dP}{P} - \alpha dP$$

Dividing both sides by $RT$, we get

$$d\ln f = \frac{dP}{P} - \alpha \frac{dP}{RT} \quad \Rightarrow \quad d\ln P - \alpha \frac{dP}{RT}$$

$$d\ln \frac{f}{P} = -\frac{\alpha}{RT} dP$$

If this result is integrated between a low, virtually zero, pressure and a given pressure $P$, at constant temperature, the result is

$$\ln \frac{f}{P} = -\frac{1}{RT} \int_0^P \alpha\, dP, \quad \ln \frac{f}{P} \text{ becomes zero at zero pressure}$$

$$\ln f = \ln P - \frac{1}{RT} \int_0^P \alpha\, dP$$

To determine fugacity, $\alpha/RT$ is plotted at various pressures with pressure, the area under the curve between the pressure of zero and $P$ gives the value of the integral in above equation (Figure 1.5).

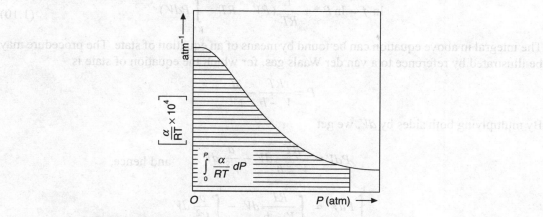

**FIGURE 1.5** Graphical method for the determination of fugacity.

## 1.32 DETERMINATION OF FUGACITY FROM EQUATION OF STATE

If $f$ is the fugacity of a gas at pressure $P$, and $f^*$ is the fugacity value at low pressure $P^*$, then integration of equation

$$RT\, d\ln f = V\, dP \quad \text{or} \quad d\ln f = \frac{V dP}{RT} \quad \text{gives}$$

$$\ln \frac{f}{f^*} = \frac{1}{RT}\int_{P^*}^{P} V dP \tag{1.9}$$

Now integrating $\int_{P^*}^{P} V dP$ between limits $P^* \to P$ and $V^* \to V$ by integrating by parts.

Thus,

$$\int_{P^*}^{P} V dP = \int_{V^*}^{V} PV - \int_{V^*}^{V} P dV$$

$$= PV - P^*V^* - \int_{V^*}^{V} P dV$$

Since the gas then behaves almost ideally, it is permissible to replace $P^*V^*$ by $RT$, and upon substituting the result in Eq. (1.9), we get

$$\ln \frac{f}{f^*} = \frac{1}{RT}\left(PV - RT - \int_{V^*}^{V} P dV\right)$$

At low pressure, $\ln \dfrac{f}{f^*}$ may be replaced by $\ln \dfrac{f}{P^*}$ that is by $\ln f - \ln P^*$; hence,

$$\ln f = \ln P^* + \frac{1}{RT}\left(PV - RT - \int_{V^*}^{V} P dV\right) \tag{1.10}$$

The integral in above equation can be found by means of an equation of state. The procedure may be illustrated by reference to a van der Waals gas, for which the equation of state is

$$P = \frac{RT}{V-b} - \frac{a}{V^2}$$

By multiplying both sides by $dV$, we get

$$P dV = \frac{RT}{V-b} dV - \frac{a}{V^2} dV \quad \text{and hence,}$$

$$\int_{V^*}^{V} P dV = \int_{V^*}^{V} \frac{RT}{V-b} dV - \int_{V^*}^{V} \frac{a}{V^2} dV$$

$$= RT \ln \frac{V-b}{V^*-b} + \frac{a}{V} - \frac{a}{V^*}$$

Since, $V^*$ is very large, $V^*-b$ may be replaced by $V^*$ which is equal to $\dfrac{RT}{P^*}$ and $\dfrac{a}{V^*}$ can be neglected, thus;

$$\int_{V^*}^{V} PdV = RT \ln \frac{V-b}{V^*} + \frac{a}{V}$$

$$\Rightarrow \quad RT \ln \frac{V-b}{RT} + RT \ln P^* + \frac{a}{V} \quad (1.11)$$

van der Waal equation may be written in another form as

$$PV - RT = \frac{RTb}{V-b} - \frac{a}{V} \quad (1.12)$$

On putting the value of $\int_{V^*}^{V} PdV$ and $PV - RT$ from Eqs. (1.11) and (1.12) into Eq. (1.10), we get,

$$\ln f = \ln P^* + \frac{1}{RT}\left[\frac{RTb}{V-b} - \frac{a}{V} - RT \ln \frac{V-b}{RT} - RT \ln P^* + \frac{a}{V}\right]$$

$$\ln f = \frac{b}{V-b} - \frac{a}{RTV} - \ln \frac{V-b}{RT} - \frac{a}{RTV}$$

$$\ln f = \ln \frac{RT}{V-b} - \frac{a}{V-b} - \frac{2a}{RTV}$$

The fugacity of a real gas at any pressure can be calculated at a specified temperature, provided the van der Waal's constant for the given gas are known.

## 1.33 GENERALIZED METHOD FOR DETERMINING FUGACITY

We know, $\alpha = \dfrac{RT}{P} - V \Rightarrow \dfrac{RT}{P}\left(1 - \dfrac{PV}{RT}\right)$

The factor $\dfrac{PV}{RT}$ is the compressibility factor ($K$), so that

$$\Rightarrow \quad \frac{RT}{P}(1-K)$$

This value of $\alpha$ when placed in equation for determination of fugacity by graphical method. The equation is

$$\ln f = \ln P - \frac{1}{RT}\int_0^P \alpha\, dP \Rightarrow \ln P - \frac{1}{RT}\int_0^P \frac{RT}{P}(1-K)dp$$

$$\ln f = \ln P + \int_0^P \frac{K-1}{P} dP$$

On replacing $P$ terms in integral by the corresponding reduced pressure ($\pi$),

$$\ln \frac{f}{P} = \int_0^{\pi^*} \frac{K-1}{\pi} d\pi$$

$\pi$ being equal to $P/P_c$ where $P_c$ is critical pressure of the gas. It was seen, at a given reduced temperature and pressure the compressibility factor ($K$) of all gases are approximately equal; hence, equation is generalized equation which may be plotted graphically on a chart so as to give value of $\ln \frac{f}{P}$ or $\frac{f}{P}$ itself for any gas in terms of the reduced temperature and pressure (Figure 1.6).

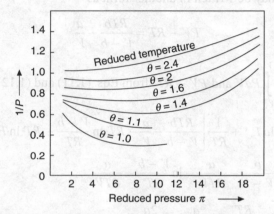

**FIGURE 1.6** Generalized method for the determination of fugacity.

The ratio of fugacity of the gas to its pressure, i.e. $f/P$ is called the *activity coefficient* of the gas at the given pressure.

## 1.34 ACTIVITY AND ACTIVITY COEFFICIENT

The chemical potential value for a real gas $\mu_i$ is given by

$$\mu_i = \mu^0 + RT \ln f_i \tag{1.13}$$

where, $\mu^0$ is chemical potential value in standard state and its value depends upon the nature of the gas and the temperature of the system. The state of unit activity is called the standard state. For a gaseous system, it is advantageous to choose the standard state of unit activity as that in which the fugacity of the gas is unity at a given temperature.

Equation (1.13) for an ideal gas will modify as

$$\mu_i = \mu^0 + RT \ln a_i \tag{1.14}$$

where $a_i$ is the activity of the given gas in the mixture. As already seen, chemical potential are equal for real gas and ideal gas in that particular standard state and so activity $a_i$ of the given gas in the mixture must be equal to its fugacity in that standard state.

In an ideal gas mixture, as should be equal to the partial pressure ($p_i$). Since the partial pressure ($p_i$) is then identical with the fugacity; the ration $a_i/p_i$ or the equivalent quantity $f_i/p_i$ may be taken as a measure of approach to ideality. This dimensionless ratio, which tends to unity as the gas approaches ideal behaviour, i.e. at very low total pressure is called the activity coefficient, and is represented by the symbol $\gamma$. Thus, equation will modify as

$$\mu_i = \mu^0 + RT \ln \gamma p$$

where $p$ is the partial pressure of the given constituent. For the present purpose the partial pressure may be defined by the equation, i.e. $p_i = N_i P$ where $N_i$ is the mole fraction of the gas and $P$ is the total pressure of the mixture.

In some cases, the standard state is chosen as the ideal gas at unit molar concentration, and the chemical potential is expressed as

$$\mu_i = \mu_c^0 + RT \ln a_c(i)$$

where $\mu_c^0$ is the chemical potential of the gas $i$ when behaved ideally at unit concentration. This is equivalent to postulating that the ratio $a_c(i)/C_i$ of the activity to the molar concentration becomes unity in the reference state of low total pressure. It will be observed that whereas the standard state is that in which the defined activity is unity and the corresponding reference state is that in which the defined activity coefficient is unity.

Mean ionic activity ($a_\pm$) in case of ionic compound is found from molality mean ionic coefficient value by using relation

$$a_\pm^V = \left[ m_+^{V+} + m_-^{V-} \right] \gamma_\pm^V$$

where, $m_+$ is the molality of cation and $m_-$ is the molality of anion and $\gamma_\pm$ is the mean ionic activity coefficient.

Ionic strength of a solution or mixture is calculated from molality and valency by the formula.

$$I = \frac{1}{2} \sum C_i Z_i^2$$

According to Debye–Huckel–Gouy–Mann equation (DHG) mean ionic activity coefficient is related to solubility ($S$) by the relation.

$$\ln \gamma_\pm = -\frac{1.177 |Z_+ Z_-| \sqrt{S}}{1 + \sqrt{S}} \tag{1.15}$$

**Problem 1.16** Calculate mean ionic activity ($a_\pm$) for 0.5 molal solution of KCl. Given mean activity coefficient for KCl at 25°C is 0.649.

*Solution:*

$$a_\pm^V = \left[ m_+^{V+} + m_-^{V-} \right] \gamma_\pm^V$$

$$\text{KCl} \rightarrow \text{K}^+ + \text{Cl}^- \quad \text{and} \quad V_+ = 1, \, V_- = 1$$

$$a_\pm^2 = \left[ m^1 + m^1 \right] \gamma_\pm^2$$

$$0.5 \times 0.5 \times (0.649)^2$$
$$a_{\pm} = 0.325$$

**Problem 1.17** Calculate ionic strength of a solution containing 0.1 m $KNO_3$, 0.15 m $K_2SO_4$ and 0.023 m $La(SO_4)_3$.

**Solution:**
$$KNO_3 \rightarrow K^+ + NO_3^-, \qquad K_2SO_4 \rightarrow 2K^+ + SO_4^{2-}$$
$$La_2(SO_4)_3 \rightarrow 2La^{3+} + 3SO_4^{2-}$$

$$I = \frac{1}{2} \sum C_i Z_i^2$$
$$I = \frac{1}{2}\left[0.1 \times (1)^2 + 0.1 \times (1)^2 + 0.15 \times 2 \times (1)^2 + 0.15 \times (2)^2 + 0.046 \times (3)^2 + 0.069 \times (2)^2\right]$$
$$\Rightarrow \quad 0.895$$

## 1.35 PHASE RULE

This rule was given by J.W. Gibbs in 1876 and is popularly known as *Gibbs phase rule*. It was deduced on the basis of principles of the thermodynamics. It is an important tool to predict the conditions necessary to be specified for a heterogeneous system to exhibit equilibrium. The rule is able to predict qualitatively the effect of changing temperature, pressure or concentration on a heterogeneous system in equilibrium.

*For a heterogeneous system in equilibrium at a definite temperature and pressure, the number of degree of freedom is greater than the difference in the number of components and the number of phases by two provided the equilibrium is not influenced by external effects such as gravity, electrical or magnetic forces, surface tension, etc.*

$$F = C - P + 2$$

## 1.36 VARIOUS TERMINOLOGY USED IN PHASE RULE

*Phases (P):* A phase is a homogenous, physically distinct and mechanically separable portion of the heterogeneous system, which is separated from other parts of the system by well defined boundary. For example, water has three phases and air has one phase. Condition is that there must be dynamic equilibrium between different phases.

*Components (C):* Smallest number of independent variables, chemical constituents by means of which the composition of all phases present in the system can be expressed directly or in the form of chemical equation.

For example, $H_2O$ (one), sugar solution (2), $NH_4Cl$ (s) decomposition (1), $CaCO_3$ (s) decomposition (S – R), where S is number of chemical species and R is number of independent chemical equilibria.

$$NaCl, KCl, H_2O \text{ system } [S - (R + 1)] \text{ (5, 4)}$$

*Degree of Freedom (F):* It is the least number of independent variables (temperature, pressure and concentration) which may be changed without changing the number of phases. These variables describe the state of the system. It is also called *variance*.

A gas in cylinder with moveable piston has 2 degree of freedom. Water $\rightleftharpoons$ vapour system has one degree of freedom.

## 1.37 TYPES OF EQUILIBRIUM

*True equilibrium:* If the same state can be attained from either direction. For example, Ice $\rightleftharpoons$ water.

*Metastable equilibrium:* If the same state can be attained from only one direction by a careful change of condition. For example, $H_2O$ liquid ($-2°C$) no ice.

*Unstable equilibrium:* If the same state cannot be attained from either direction.

## 1.38 DERIVATION OF PHASE RULE ($F = C - P + 2$)

To find total number of independent variables:
(1) Temperature, (2) pressure and (3) concentration

*Temperature and pressure are same for all phases.*
Independent concentration variables for one phase in respect to $C$-components $= C - 1$
(The concentration of last component is not independent.)
Independent concentration variable(s) for $P$ phase in respect to $C$-components

$$S = P(C-1)$$

Total number of independent variables $(S) = P(C-1) + 2$. Here 2 stands for temperature and pressure.

To find number of equations of equilibrium $(R)$

For component 1     $\mu_1(a) \rightleftharpoons \mu_1(b) \rightleftharpoons \mu_1(n)$
For component $C$     $\mu_C(a) \rightleftharpoons \mu_C(b) \rightleftharpoons \mu_C(n)$
For each component, the relation for $P$ phases   $= P - 1$
For $C$-components, the relation for $P$ phases   $= C(P-1)$
Total number of independent equation $(R)$   $= C(P-1)$
Degree of freedom $(F)$   $= S - R$
                  $= [P(C-1) + 2] - [C(P-1)]$
                  $F = C - P + 2$

## 1.39 PHASE RULE FOR CHEMICALLY REACTING SYSTEMS

$$F = (C - r) - P + 2$$

where $r$ is the number of chemically reactive equilibria.

## 1.40 USES OF PHASE RULE

1. Provides simple basis for the classification of equilibrium state of any heterogenous systems.
2. Applicable to macroscopic system (no need to know molecular structure).
3. It tells behaviour of system when subjected to change in variables.
4. It is helpful in the study of physical and chemical phase reactions.
5. It tells that under a given set of condition, a number of substances when put together would remain as such in equilibrium or not.

## 1.41 LIMITATIONS OF PHASE RULE

1. It applies to single equilibrium state and does not tell about the number of other equilibria possible in the system.
2. It takes into account the number of phases and not their quantities, even the small quantity of phase, when present counts to the number of phases present in the system, in equilibrium.
3. Attainment of equilibrium state is prerequisite for the application of phase rule. It is thus not applicable for such system which is slow in reaching the equilibrium state.
4. This rule does not give any information regarding the time taken for the system to attain equilibrium.
5. If the number of variables other than concentration variables is different from 2, then this factor of the phase rule has to be adjusted accordingly.

## 1.42 PHASE DIAGRAM

It is a graphical representation giving the conditions of temperature and pressure under which the various phases are capable of stable existence and transform into vapour phase. Hence, the relation between the solid, liquid and the gaseous states of a given substance as a function of the temperature and pressure can be summarized on a single graph known as *phase diagram*.

## 1.43 ONE-COMPONENT ($H_2O$) SYSTEM

Water exists in three forms as solid (ice), Liquid ($H_2O$) and gas (vapour). The application of phase rule reveals that we can have the following types of equilibria for water system (Table 1.1).

TABLE 1.1 Types of equilibria for water system

| P | System | F | Position in phase diagram |
|---|--------|---|---------------------------|
| 1 | Solid, liquid, vapour separately | 2 | Area (3) |
| 2 | Solid–vapour, solid–liquid and liquid–vapour | 1 | Curves (3) |
| 3 | Solid–liquid vapour | 0 | Point (1) |

As the maximum number of degree of freedom is two, water system can be represented by a two-dimensional phase diagram using temperature–pressure as variables. The phase diagram for water system is shown in Figure 1.7. In the phase diagram of water system, curve *AC* is the vapour pressure curve of water. This curve divided the liquid region from the vapour region. This curve

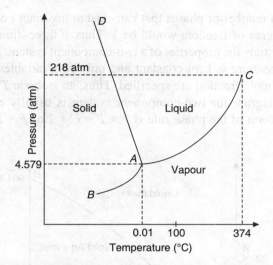

**FIGURE 1.7** Phase diagram of one component system (water).

has one degree of freedom. The slope of curve $AC$ is positive as predicted by Clapeyron equation Curve $AB$ is the sublimation pressure curve of ice. This divides the solid region from the vapour region. This curve gives the pressure of water vapour in equilibrium with solid ice. This curve has a positive slope. Curve $AD$ is the solid–liquid equilibria for water. This curve shows how the melting temperature of ice or the freezing temperature of water varies with pressure. The slope of curve $AD$ is negative. This shows that with increase of pressure the melting point should decrease. This is explained by Clapeyron equation:

$$\frac{dP}{dT} = \frac{S^L - S^S}{(V^L - V^S)} = \frac{+ve}{-ve}$$

$= $ negative, as $V^S > V^L$ for water system.

Point $A$ in the phase diagram of water system is the point where three curves $AB$, $AC$, $AD$ meet at a point $A$ at which solid, liquid and vapour are simultaneously present at equilibrium. This point at 273.16 K and 4.579 mm pressure is called a *triple point*. Since three phases coexist, the system is invariant. The number of degree of freedom is zero.

## 1.44 TWO-COMPONENT SYSTEM

Gibbs phase rule when applied to a two-component system gives

$$F = C - P + 2$$
$$= 2 - P + 2$$

At invariant point, $F = 0$, Therfore $P = 4$.

Therefore, the maximum number of phases that can exist at invariant point ($F = 0$) are four and maximum number of degree of freedom would be 3. Thus, a three-dimensional phase diagram should be constructed to study the properties of a two-component system. In practice, one variable the pressure or the temperature is kept constant and only two variables temperature and mole fraction or pressure and mole fraction are specified. Thus, an isobaric T–x phase diagram or an isothermal P–x phase diagram for two-component system is usually constructed. Under this condition, the modified form of the phase rule is $F + P = C + 1$ or $F + P = 2 + 1 = 3$ for a two-component system.

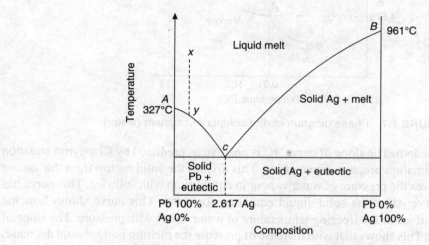

**FIGURE 1.8** Phase diagram of lead silver system.

*e.g. Pb–Ag system.* Both Pb and Ag are completely miscible and form homogenous mixture. Melting point of pure Pb is 327°C and is lowered by addition of Ag. Melting point of pure Ag (961°C) and is lowered by addition of Pb. Point C, where three phase exist together having zero degree of freedom ($F = 0$) is known as *eutectic point*. The temperature and composition at which the three-phase equilibria exist are known as *eutectic temperature* ($T_e$) and *eutectic composition* ($x_e$), respectively (Figure 1.8).

Since, it is a solid–liquid equilibrium system which does not have gaseous phase and is called *condensed system*. So, pressure does not have any effect on this type of equilibrium. Hence, the degree of freedom for such a system will be reduced by one. Hence, the reduced phase rule is $F = C - P + 1$.

Cooling curve for a binary system is drawn in Figure 1.9. Curve XY represents the cooling of liquid solution, point Y is the first break (arrest point) when one solid Pb (present in excess) starts crystallizing, line YC represents the variation of freezing point of the solution. Point C is the second break, when the second component Ag along with the first Pb separates. Line CD is called the *eutectic halt* along which solid Pb, solid Ag and liquid phases are present at equilibrium. Line DE represents cooling of solid mixture.

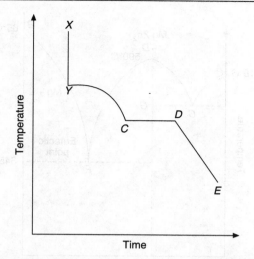

**FIGURE 1.9** Cooling curve for a binary system (Pb + Ag) with eutectic point.

## 1.45 SYSTEMS FORMING SOLID COMPOUND $A_XB_Y$ WITH CONGRUENT MELTING POINT

A compound is said to have congruent melting point if it melts completely at one temperature giving a liquid having the same composition as the solid.

Examples of the binary systems forming solid compounds with congruent melting point are: Zn + Mg, Au + Te, urea + phenol, aniline + phenol, etc. Let us take an example of Zn + Mg system (Figure 1.10). If the compound Zn + Mg is considered as a separate component then this phase diagram may be imagined to be made up of two diagrams of the simple eutectic type *ACD* and *DEB* placed side by side. The region above the curve *ACDEB* represents the existence of only liquid phase. The points *A*, *B* and *D* represent the freezing points of Zn, Mg and $MgZn_2$, respectively. Curve *AC* represents the equilibrium conditions for solid Zn and solution; liquid curve *CDE* describes the equilibrium between the solid compound $MgZn_2$ and the liquid solution. The points on the curve *BE* give the temperature and composition for the equilibrium between the solid Mg and the liquid solution. When Mg is added to Zn the freezing point decreases along *AC* until the first eutectic point *C* is reached where Zn and $MgZn_2$ separate out as eutectic mixture. Similarly, the melting point of Mg changes by the addition of Zn along *BE*. At the second eutectic point *E*, the solid compound $MgZn_2$ and pure Mg separate out. Thus, the liquid having the composition corresponding to points *C*, *E* or *D* will freeze at a constant temperature. Point *D* is known as congruent melting point.

Shape of the curve *CDE* will depend upon the stability of the compound formed. If compound is stable in the liquid phase and solid phase, the bend of the curve will be sharper. Greater the degree of dissociation, more flatter the curve will be.

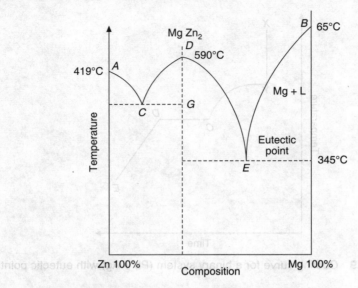

**FIGURE 1.10** Phase diagram of Zn–Mg system.

## 1.46 COMPOUNDS WITH INCONGRUENT MELTING POINT

A system (compound) is said to be possess incongruent melting point, if it decomposes much below its melting point and forms a new solid phase and a solution having different composition from the solid state. It has no any sharp melting point. Figure 1.11 shows phase diagram of K–Na system. The dotted curve is the metastable state of the compound, the summit $E$ is the hypothetical melting point. The compound $Na_2K$ formed undergoes dissociation at $D$ below the melting point and the composition of the liquid is not the same as that of the solid compound. The point $D$ is called *incongruent melting point*. This point represents the limit of the existence of the compound $Na_2K$. It is an invariant point as there are three phases present at point $D$.

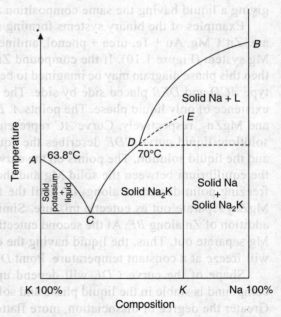

**FIGURE 1.11** Phase diagram of K–Na system.

## EXERCISES

1. Give the mathematical statement of the first law of thermodynamics. Comment on the statement "while $U$ is a definite property, $q$ and $w$ are not definite properties".
2. What do you understand by the term internal energy change $\Delta U$ and $\Delta H$.
3. Derive the expression for work done in reversible isothermal expansion and reversible isothermal compression of an ideal gas. What is meant by maximum work?
4. Distinguish between isothermal and adiabatic process. Derive the relation between temperature and volume and that between temperature and pressure in reversible adiabatic expansion of an ideal gas.
5. State and explain zeroth law of thermodynamics. What is the significance of this law?
6. Calculate $q$, $w$, $\Delta U$ and $\Delta H$ for the reversible adiabatic compression of 0.2 mol of an ideal gas from a volume of 1 dm³ to a volume of 0.25 dm³.

[Ans. 0, –99.8 J, 98.8 J; 138 J]

7. Derive the relation between $\Delta H$ and $\Delta U$.
8. What is cyclic process? Describe in details the Carnot reversible cycle for establishing the maximum convertibility of heat into work.
9. What is meant by change of entropy of a system. Show that $\Delta S = q_{rev}/T$.
10. Discuss entropy change in reversible and irreversible processes. Comment on the statement "entropy of the universe is constantly increasing".
11. Derive Gibbs–Helmholtz equation for a process at constant pressure and at constant volume. Discuss the important application of this equation.
12. What is meant by chemical potential? How does chemical potential vary with temperature and pressure? Derive the Gibbs–Duhem equation.
13. Derive Clasius–Clapeyron equation in the simple and integrated form for liquid ↔ vapour and solid ↔ vapour equilibria. Can we have the integrated equation for solid ↔ liquid equilibrium?
14. Explain the term fugacity and activity? How they are related to chemical potential? What is physical significance of fugacity?
15. Calculate the free energy change accompanying the compression of 1 mol of a gas at 25°C from 20 to 200 atm. The fugacities of the gas be taken as 18 and 120 atm, respectively, at pressure of 18 and 200 atm.
16. Explain the Nernst heat theorem. How does it lead to the enunciation of the third law of thermodynamics?
17. State and explain the third law of thermodynamics. How can it be verified experimentally?

## SUGGESTED READINGS

Aston, J.G. and J.J. Fritz, *Thermodynamics and Statistical Thermodynamics*, John Wiley, New York, 1959.

Atkins, P., J. de Paula, *Atkins Physical Chemistry*, Oxford University Press, New York, 2004.

Atkins, P.W., *The Second Law*, W.H. Freeman, 1984.

Barrow, G.M., *Physical Chemistry*, 5th ed., Tata McGraw-Hill, New Delhi, 2004.

Bauman, R.P., *Equilibrium Thermodynamics*, Prentice Hall, Englewood Cliffs, New Jersey, 1966.

Berry, R.S., S.A. Rice, and J. Ross, *Physical and Chemical Kinetics*, Oxford University Press, 2001.

Berry, R.S., S.A. Rice, and J. Ross, *Physical Chemistry*, John Wiley, New York, 1980.

Caldin, E.F., *An Introduction to Chemical Thermodynamics*, Oxford University Press, New York, 1958.

Callen, Herbert B., *Thermodynamics and Introduction to Thermostatistics*, Wiley, 1985.

Cengel, A. Yunus and Michael A. Boles, *Thermodynamics: An Engineering Approach*, McGraw-Hill, 2005.

Denbigh, K.G., *Principles of Chemical Equilibria*, Cambridge University Press, London, 1971.

Dugdale, J.S., *Entropy and its Physical Meaning*, Taylor & Frances, 1996.

Erying, H., D. Henderson, and W. Jost, *Physical Chemistry: An Advance Treatise*, Vols. 9A and 9B, Academic, New York, 1970.

Fenn, J.B., *Engines, Energy and Entropy*, W.H. Freeman, 1982.

Fermi, E., *Thermodynamics*, Prentice Hall, 1937.

Fong, P., *Foundation of Thermodynamics*, Oxford University Press, New York, 1963.

Gale, J.D. and J.M. Seddon, *Thermodynamics and Statistical Mechanics*, Wiley Interscience, New York, 2002.

Glasstone, S., *Thermodynamics for Chemist*, East-West Press, New Delhi, 2003.

Guggenthum, *Elements of Chemical Thermodynamics*, Monographs for Teachers, 12, Royal Institute of Chemistry, 1968.

Helsdon, R.M., *Introduction to Applied Thermodynamics*, Pergamon Press, New York, 1965.

Hindelwood, C.N., *Structure of Physical Chemistry*, Oxford University Press, 2005.

Klotz, I.M. and W.A. Benjamin, *Chemical Thermodynamics*, New York, 1964.

Larter, Ashley H., *Classical and Statistical Thermodynamics*, Prentice Hall, 2001.

Lewins, J.D., *Teaching Thermodynamics*, Plenum Press, 1986.

Lewis G.N. and M. Randall, *Thermodynamics*, McGraw-Hill, New York, 1961.

Mahan, B.H. and W.A. Benjamin, *Elementary Chemical Thermodynamics*, New York, 1963.

McGlashan, M.L., *Chemical Thermodynamics*, Academic Press, 1978.

McQuarrie, D.A. and J.D. Simon, *Physical Chemistry: A Molecular Approach*, Viva Books, New Delhi, 2004.

Nash, K., *Elements of Chemical Thermodynamics*, Addison-Wesley, Reading Mass., 1962.

Pitzer, K.S., *Activity Coefficient in Electrolytic Solutions*, 2nd ed. CRC Boca Ratan, Florida, 1991.

Prigogine, I. and R. Defay, *Chemical Thermodynamics*, Longmans Green, London, New York, 1954.

Puri, B.R., L.R. Sharma, and M.S. Pathania, *Principles of Physical Chemistry*, Vishal Publishing, Jalandher, 2004.

Rastogi, R.P. and R.R. Mishra, *An Introduction to Chemical Thermodynamics*, 6th revised ed., Vikas Publishing, New Delhi, 2006.

Roy, B.N., *Fundamentals of Classical and Statistical Thermodynamics*, John Wiley, 1982.

Sears, F.W. and G.L. Salinger, *Thermodynamics, Kinetic Theory and Statistical Thermodynamics*, Addison-Wesley, 1975.

Sharma, K.K. and L.K. Sharma, *A Textbook of Physical Chemistry*, Vikas Publishing, New Delhi, 2006.

Silbey, R.J. and R.A. Alberty, *Physical Chemistry*, 3rd ed., John Wiley, 2003.

Smith, E. Brian, *Basic Chemical Thermodynamics*, 5th ed., World Scientific, 2004.

Smith, J.M., H.C. Van Ness and Micheal M. Abbott, *Introduction to Chemical Engineering Thermodynamics*, McGraw-Hill, 2004.

*Thermodynamics of the Steady States*, Methuen, 1951.

Wall, F.T., *Chemical Thermodynamics*, W.H. Freeman, San Francisco, 1965.

Zemansky, M. and R. Dittman, *Heat and Thermodynamics*, McGraw-Hill, 1981.

# 2

# Irreversible Thermodynamics or Non-equilibrium Thermodynamics

Classical thermodynamics deals primarily with the study of systems which are in a state of equilibrium. It adequately deals with the processes which begin and end with equilibrium states, though the intervening states of a given process may be non-equilibrium. All natural processes, on the other hand, are irreversible processes taking place in open systems. As a consequence of several dissipative processes occurring there in, the time invariant of these irreversible processes is a steady state and not an equilibrium state.

## 2.1 STEADY STATE OR STATIONARY STATE

It is a time invariant condition of a system which is open to environment. In contrast, equilibrium state is the limiting condition, when the flux from the environment is zero. As such the steady state is a non-equilibrium state for which the usual equilibrium thermodynamics relations are not valid. A continuously stirred flow reactor as shown in Figure 2.1(a) would reach stationary steady state in course of time.

Steady states can be achieved in phenomenon, where coupled process occur. Such phenomenon is called *cross phenomenon*. One such example is thermo-osmosis. When a fluid placed in two chambers maintained at different temperatures and are separated by a membrane or a capillary, flow of matter can takes place on account of pressure difference (Poiseulle flow) and also on account of temperature difference (thermo-osmotic flow) [Figure 2.1(b)]. If these are in the opposite directions, a state can be reached when these counter balance and this would yield the steady state which is a non-equilibrium state.

In the same manner, in case of electro-osmatic pressure difference as one of the electro-kinetic phenomenon, we have balancing of electro-osmotic flux and hydrodynamic flux. It should be noted that electro-osmotic flux is the mass flux due to potential difference on the two sides of the membrane as shown in Figure 2.1(c). The branch of science dealing with the study of thermodynamic properties of the system, which are not in equilibrium and involve transport processes, which are irreversible, is termed as *non-equilibrium* or *irreversible thermodynamics*. Irreversible thermodynamics is applicable to those systems which are not too far from equilibrium.

A few examples of transport phenomenon include flow (conduction) of heat along a metal bar whose ends are kept at fixed but at different temperatures. The development of heat when an electric current flows through a metallic conductor, diffusion of a solid or a fluid across a concentration gradient, etc.

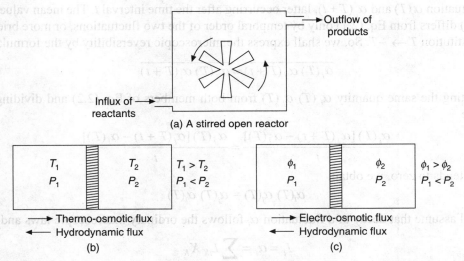

**FIGURE 2.1** (a) A continuously stirred flow reactor, (b) a thermo-osmotic flow, and (c) electroosmotic flow.

## 2.2 POSTULATES OF NON-EQUILIBRIUM THERMODYNAMICS

*Postulate 1: Postulate of local equilibrium:* It states that the parameters such as $T$, $P$, $N$, etc. at any point in the matter of non-equilibrium system can be defined by limiting operations of taking successively smaller and smaller volume of matter around this point, isolating them and determining their values after they have reached an equilibrium state. It is for this reason, the non-equilibrium thermodynamics which describes the rates of flow processes is also sometimes called *irreversible thermodynamics near equilibrium*.

*Postulate 2:* The relation between state functions, defined for non-equilibrium states and expressed in terms of variables $P$, $V$, $T$, etc. as defined by postulate of local equilibrium are identical with the relations between the corresponding state functions that were derived for equilibrium states.

## 2.3 MICROSCOPIC REVERSIBILITY AND ONSAGER RECIPROCITY RELATION (A PROOF)

The value of fluctuation $\alpha_i$ at a time instant $T$ and fluctuation $\alpha_j$ after a time interval $t$ and we form the product of both quantities. The average value of this product during a sufficiently long lapse of time is given by:

$$\overline{\alpha_i(T)\,\alpha_j(T+t)} = \lim_{T \to \infty} \frac{1}{T} \int_0^T \alpha_i(T)\alpha_j(T+t)dT \tag{2.1}$$

Following are the general principles of statistical mechanics, it can be shown that the time average is equivalent to the average taken with the help of probability function ($P$). This is so called *Ergodic theorem*.

We shall next consider the average value of the product $\alpha_j(T)\,\alpha_i(T+t)$ in which we consider the fluctuation $\alpha_j(T)$ and $\alpha_i(T+t)$, latter occurring after the time interval $t$. The mean value $\alpha_j(T)\,\alpha_i(T+t)$ differs from Eq. (2.1) only by temporal order of the two fluctuations, or more briefly by the substitution $T \to -T$. So, we shall express the microscopic reversibility by the formula.

$$\overline{\alpha_i(T)\,\alpha_j(T+t)} = \overline{\alpha_j(T)\,\alpha_i(T+t)} \tag{2.2}$$

Subtracting the same quantity $\alpha_i(T)\,\alpha_j(T)$ from both members of Eq. (2.2) and dividing by $t$, we have

$$\frac{\overline{\alpha_i(T)\,[\alpha_j(T+t) - \alpha_j(T)]}}{t} = \frac{\overline{\alpha_j(T)\,[\alpha_i(T+t) - \alpha_i(T)]}}{t} \tag{2.3}$$

When $t$ tends to zero, we obtain

$$\overline{\alpha_i(T)\,\dot\alpha_j(T)} = \overline{\alpha_j(T)\,\dot\alpha_i(T)} \tag{2.4}$$

We shall assume that decay of a fluctuation $\alpha_i$ follows the ordinary macroscopic laws and write

$$J_i = \dot\alpha_i = \sum_K L_{iK} X_K \tag{2.5}$$

Introducing Eq. (2.5) into Eq. (2.4), we get

$$\sum_K L_{jK}\,\overline{\alpha_i X_K} = \sum_K L_{iK}\,\overline{\alpha_j X_K} \tag{2.6}$$

$$L_{jK}\,\alpha_i = L_{iK}\,\alpha_j \tag{2.7}$$

or

$$L_{ji} = L_{ij} \tag{2.8}$$

Equation (2.8) is popularly known as *Onsager reciprocity relation*. Although, they have been proved, here for small spontaneous fluctuation around thermodynamic equlibria only. We shall accept the validity of Onsager's relation even for systems with systematic deviations from equlibrium as long as the relation between fluxes and affinities remain linear.

## 2.4 PHENOMENOLOGICAL LAWS AND ONSAGER'S RECIPROCOL RELATIONS

The irreversible processes involve the transport of one or more of the quantities such as heat, mass, momentum and electric charge. In all the cases, a quantity called a *flux* is transported as a result of a driving force which is derived from the gradient of some physical property of the system. Thus, the driving force for a heat flux is the temperature gradient, that for a mass flux is the concentration gradient and that for an electric current is the potential gradient. In all the cases, the magnitude of the flux (or flow) is directly proportional to the driving force. In general, the transport phenomenon for one-dimensional system is written as

$$J = LX \qquad (2.9)$$

where $J$ is flux (flow per unit area) of the quantity transported along a given direction, $X$ is the driving force (or the gradient) which causes the flow in that direction and $L$ is the proportionality constant called *transport coefficient*.

For the various transport processes, we can now write the following relations.

1. Heat transfer $J_Q = -K \dfrac{dT}{dx}$ (Fourier law) (2.10)

2. Mass transfer $J_M = -D \dfrac{dc}{dx}$ (Fick's law) (2.11)

3. Momentum transfer $J_M = -M \dfrac{du}{dx}$ (Newton's law) (2.12)

4. Flow of electricity $J_e = -\lambda \dfrac{dE}{dx}$ (Ohm's law) (2.13)

Here $J_i$'s are the corresponding fluxes. The transport coefficients $K$, $D$, $M$ and $\lambda$ depend upon the material properties of the systems. It should be born in mind that Eqs. (2.10), (2.11), (2.12) and (2.13) are not laws in the traditional usage of the term, they are phenomenological laws, introduced to define transport processes. Phenomenological equations describe in a simple way, how the system changes. Fick's law of diffusion,

$$\frac{dm}{dt} = D \frac{dc}{dx}$$

where $\dfrac{dm}{dt}$ is the rate of change of solute across the unit surface area under the influence of concentration gradient $dc/dx$ and $D$ is the diffusion coefficient, may be adequate in a simple system. Suppose, however, that while one solute diffuses under its own concentration gradient, another solute is also diffusing under its concentration gradient. Can we assume that the movement of the second solute has no influence on that of the first? Again suppose that while one solute is diffusing, there also simultaneously exists a temperature gradient in the system resulting in a flow of heat. Will the movement of the solute still be described by the equation given above? The answer to these questions by theory and also by intuition is none.

We can generalize the equation $dm/dt = D(dc/dx)$ for a one-dimensional flow as follows:

$$\frac{dm_1}{dt} = D \frac{dc_1}{dx} + E \frac{dc_2}{dx} \qquad (2.14)$$

where second term incorporates the influence of the gradient of the second solute on the movement of the first solute. Similarly, we can write down another equation for the movement of the second solute.

$$\frac{dm_2}{dt} = F \frac{dc_2}{dx} + G \frac{dc_1}{dx} \qquad (2.15)$$

where $D$, $E$, $F$ and $G$ are the corresponding diffusion coefficients. If there are more solutes and thermal gradients, other terms will have to be written.

To simplify notation, we use a single symbol $J$ for $dm/dt$. Symbols $J_1$, $J_2$,..., are used for the rates of movement of different entities (solute, heat, etc.). Again the multiplicity of coefficients $D$, $E$, $F$ and $G$ can be replaced by a single symbol $L$ with appropriate subscripts. Accordingly, $D$ becomes $L_{11}$, $E$ becomes $L_{12}$. Here, the first subscript refers to the component moving (and is thus the same as the subscript on the corresponding $J$, while the second subscript refers to the component whose gradient is being consideration. The gradient or driving force is designated by the symbol $X$. Thus,

$$X_1 = \frac{dc_1}{dx}, \qquad X_2 = \frac{dc_2}{dx} \text{ and so on.}$$

This notation is both convenient and easy to remember. Thus, Eqs. (2.14) and (2.15) are written as

$$J_1 = L_{11} X_1 + L_{12} X_2 \qquad J_2 = L_{22} X_2 + L_{21} X_1 \tag{2.16}$$

Two simultaneous irreversible flows, provided they are independent of each other, can be described by appropriate phenomenological relations. In practice, however, the simultaneous flows are not independent of each others gradient. Such flows are known as *coupled flows*. Onsager developed irreversible thermodynamic (or non-equilibrium thermodynamic) in 1931 for analyzing coupled irreversible flows.

If the gradients $X_i$'s are not too great the fluxes $J_i$'s are linear functions of the driving forces. Thus, we can write,

$$J_i = L_{i1}X_1 + L_{i2}X_2 + L_{i3} X_3 + \cdots + L_{in} X_n \tag{2.17}$$

The relations depicted in Eq. (2.17) are called *linear phenomenological relations*. The coefficients $L_{ii}$ are called the *primary phenomenological coefficients*, while the coefficients $L_{ij}$ are called *Onsager's phenomenological coefficients*. In the Onsager coefficients, the subscript $i$ denotes the flux and the subscript $j$ denotes the driving force.

The solution to Eq. (2.17) is extremely difficult. Several scientists including Kelvin had attempted its solution in the 19th century. It was, however, Lars Onsager who finally solved it in 1931. The difficulty in solving this equation lies in the fact that the phenomenological coefficients have to be determined experimentally. The coupling coefficients (Onsager's phenomenological coefficients, $L_{ij}$) however, present severe difficulties for their experimental measurement, since they involve the control of many experimental parameters. Onsager theoretically showed that

$$L_{ij} = L_{ji} \tag{2.18}$$

Equation (2.18) gives Onsager's reciprocal relations (also called *reciprocity relations*). Onsager showed that these relations exists for a properly selected pair of flows, called the *conjugate flows*.

## 2.5 ENTROPY

1. Entropy of the system is an extensive property.
2. Change of entropy $dS$ can be split into two parts. Denoting $d_eS$ the flow of entropy, due to interactions with the exterior, and by $d_iS$, the contribution due to changes inside the system, we have

$$dS = d_eS + d_iS \quad (2.19)$$

The entropy increase, $d_iS$, due to change inside the system, is never negative. It is zero, when the system undergoes reversible change only, but it is positive, if the system is subjected to irreversible processes as well.

$$d_iS = 0 \quad \text{(reversible process)} \quad (2.20)$$

$$d_iS > 0 \quad \text{(irreversible process)} \quad (2.21)$$

3. For an isolated system, there is no flow of entropy so that Eqs. (2.19) and (2.21) reduces to

$$dS = d_iS \geq 0 \quad \text{(isolated system)} \quad (2.22)$$

For an isolated system, this relation is equivalent to the classical statement that entropy can never decrease, so that, in this case the behaviour of the entropy function provides a criteria which enables us to detect the presence of irreversible processes.

4. For closed systems at constant temperature and volume, the Helmholtz free energy ($A = E - TS$) decreases, when irreversible changes occur and remains constant otherwise. The only general criteria of irreversibility is given by entropy production.

5. Suppose we enclose a system which we shall denote by I, inside a larger system II, so that global system containing both I and II is isolated. In both parts, I and II, some irreversible process may take place. The classical statements of the second law of thermodynamics would be

$$dS = dS^I + dS^{II} \geq 0$$

Apply now Eqs. (2.20) and (2.21) to each part separately, we shall postulate that $d_iS^I \geq 0$, $d_iS^{II} \geq 0$.

A physical situation such that $d_iS^I > 0$, $d_iS^{II} < 0$, with $d(S^I + S^{II}) > 0$ is excluded. We can therefore say that, absorption of entropy in one part compensated by a sufficient production in another part of the system is prohibited. The term *macroscopic region* refers to any region containing a number of molecules sufficiently large, for microscopic fluctuations to be negligible. Such a formulation may be called a *local formulation* of the second law in contrast to the global formulation of classical thermodynamics.

6. Entropy of one component system (closed), an irreversible process being excluded ($d_iS = 0$) then

$$dS = \frac{dq}{T}$$

where $T$ is a positive quantity, called *absolute temperature* and is intensive property. We know,

$$dq = dE + PdV, \quad \text{(first law of thermodynamics)}$$

$$dS = \frac{dE + PdV}{T} \quad \text{(second law of thermodynamics)} \quad (2.23)$$

7. *Entropy production:* The entropy of an isolated system in equilibrium is maximum. Hence, if such a system is not in equilibrium, the entropy will increase but may not decrease, i.e. equilibrium lies in the direction of increasing entropy. This is termed as

*entropy production*. The concept of entropy production in an irreversible process may be understood in a simple manner as follows:

$$dS \geq \frac{dq}{T} \quad \text{or} \quad dS - \frac{dq}{T} \geq 0 \tag{2.24}$$

The quantity on the left is greater than or equal to zero. So, we may write

$$dS - \frac{dq}{T} = d\sigma \tag{2.25}$$

where $d\sigma$ will be either zero or positive.

If it is assumed that the system is in contact with a reservoir at $T$ and a quantity of heat $dq$ flows into the system, then a quantity $-dq$ flows in the reservoir reversibly and then the entropy change of the reservoir is

$$dS_{\text{rev}} \geq -\frac{dq}{T} \tag{2.26}$$

So that, Eq. (2.25) can be written as

$$dS + dS_{\text{rev}} = d\sigma$$

The quantity $d\sigma$ refers to the entropy increase of the system plus that of the surrounding (the reservoir). The $d\sigma$ is called *entropy production* of the process. For an irreversible process the entropy production is positive while for a reversible process the entropy production is zero.

## 2.6 ENTROPY PRODUCTION DUE TO HEAT FLOW

Consider a system consisting of two closed phases I and II maintained respectively at uniform temperatures $T^{\text{I}}$ and $T^{\text{II}}$. For the whole system, entropy being an extensive variables

$$dS = dS^{\text{I}} + dS^{\text{II}} \tag{2.27}$$

We now split the heat received by each phase into two parts

$$d^{\text{I}}Q = d_i^{\text{I}}Q + d_e^{\text{I}}Q \quad \text{and} \quad d^{\text{II}}Q = d_i^{\text{II}}Q + d_e^{\text{II}}Q \tag{2.28}$$

where $d_i^{\text{I}}$ is the heat received by phase I from phase II, and $d_e^{\text{I}}$ is the heat supplied to phase I from outside.

Rewriting Eq. (2.27), we get

$$dS = \frac{d^{\text{I}}Q}{T^{\text{I}}} + \frac{d^{\text{II}}Q}{T^{\text{II}}} \tag{2.29}$$

By putting the value of $d^{\text{I}}Q$ and $d^{\text{II}}Q$ from Eq. (2.28) into Eq. (2.29)

$$dS = \frac{d_e^{\text{I}}Q}{T^{\text{I}}} + \frac{d_e^{\text{II}}Q}{T^{\text{II}}} + d_i^{\text{I}}Q\left(\frac{1}{T^{\text{I}}} - \frac{1}{T^{\text{II}}}\right) \tag{2.30}$$

The entropy change in Eq. (2.30) consists of two parts. The first

$$d_e S = \frac{d_e^I Q}{T^I} + \frac{d_e^{II} Q}{T^{II}} \tag{2.31}$$

is due to exchange of heat with the exterior while the second part

$$d_i S = d_i^I Q \left( \frac{1}{T^I} - \frac{1}{T^{II}} \right) \tag{2.32}$$

results from the irreversible heat flow from inside the system. In agreement with equation $d_i S > 0$ (for irreversible process), we may postulate that the entropy production is really positive.

In fact
$$d_i^I Q > 0 \quad \text{when} \quad \frac{1}{T^I} - \frac{1}{T^{II}} > 0$$

$$d_i^I Q < 0 \quad \text{when} \quad \frac{1}{T^I} - \frac{1}{T^{II}} < 0$$

The entropy production can only be zero when thermal equilibrium is established, that is when

$$T^I = T^{II} \tag{2.33}$$

Further more, we shall often make use of entropy production per unit time.

$$\frac{d_i S}{dt} = \frac{d_i^I Q}{dt} \left( \frac{1}{T^I} - \frac{1}{T^{II}} \right) > 0 \tag{2.34}$$

This equation is of simple form which is of great importance. It is the product of the rate of the irreversible process $\frac{d_i^I Q}{dt}$ by the function of state $\left( \frac{1}{T^I} - \frac{1}{T^{II}} \right)$. The direction of heat flow is determined by the sign of this function, which, can therefore be considered as its macroscopic cause.

## 2.7 ENTROPY PRODUCTION AND ITS RATE IN MATTER (OR MASS) FLOW

For the purpose of internal matter flow, the system is assumed to be divided into compartments. When a small amount of matter is transformed through the imaginary barrier separating one compartment from the other, the heat content of the substance itself is increased by an amount equal to $dH$ per mole. In addition to this, there is usually a heat effect when the fluid enters the surface of the barrier and passes into the interior. The reverse heat effect occurs when the fluid comes out of the barrier into the other compartment. Thus, heat is transformed with the substance. This quantity is known as *heat of transfer* and is denoted by $Q^*$ per mole.

Now consider the flow of $\delta n$ moles of some components of matter from one region of temperature, say $T_1$, to another region with temperature, say $T_2$. Let $\mu_1$ be the chemical potential of the first region and $\mu_2$ be the chemical potential of the second region. According to Gibbs–Helmholtz equation, the free energy of a system is given by

$$G = H - TS$$

$$S = \frac{H-G}{T} = \frac{H}{T} - \frac{G}{T} \tag{2.35}$$

For one mole of system, we can write

$$\bar{S} = \frac{\bar{H}}{T} - \frac{\bar{G}}{T} = \frac{\bar{H}}{T} - \frac{\mu}{T}$$

where $\bar{S}$ and $\bar{H}$ are called *partial molal entropy* and *partial molal enthalpy* of the system.

With the transfer of matter, there will be heat transfer $Q$ addition to ordinary enthalpy transfer of the substance itself. Also the sum $(\bar{H} + Q^*)$ according to the first law of classical thermodynamics, will be constant whether considered for region 1 or region 2. Therefore, the entropy increase for the transfer of $\delta n$ moles of matter from region 1 to region 2 is given by

$$dS_{\text{irrev}} = (\bar{S}_2 - \bar{S}_1)$$

$$dS_{\text{irrev}} = \delta n \left[ \left( \frac{\bar{H} + Q^*}{T_2} - \frac{\mu_2}{T_2} \right) - \left( \frac{\bar{H} + Q^*}{T_1} - \frac{\mu_1}{T_1} \right) \right]$$

$$dS_{\text{irrev}} = \delta n \left[ \left( \frac{\mu_1}{T_1} - \frac{\mu_2}{T_2} \right) + (\bar{H} + Q^*) d\left( \frac{1}{T} \right) \right]$$

For infinitesimal difference, this equation is rewritten as

$$dS_{\text{irrev}} = \delta n \left[ -d\left( \frac{\mu}{T} \right) + (\bar{H} + Q^*) d\left( \frac{1}{T} \right) \right] \tag{2.36}$$

For the transfer of a mole of single component in a system, where pressure is the only external force, rewriting Eq. (2.35) as

$$\frac{\mu}{T} = \frac{\bar{H}}{T} - \bar{S}$$

$$d\left( \frac{\mu}{T} \right) = d\left( \frac{\bar{H}}{T} \right) - d\bar{S}$$

$$= \bar{H} d\left( \frac{1}{T} \right) + \frac{1}{T}(d\bar{H} - dS T)$$

$$= \bar{H} d\left( \frac{1}{T} \right) + \frac{1}{T} d(E + P\bar{V}) - Td\bar{S}$$

$$= \bar{H} d\left( \frac{1}{T} \right) + \frac{1}{T}(dE + Pd\bar{V} + \bar{V}dP - Td\bar{S})$$

$$= \bar{H} d\left( \frac{1}{T} \right) + \frac{1}{T}(dq - dw + Pd\bar{V} + \bar{V}dP - Td\bar{S})$$

$$= \bar{H}d\left(\frac{1}{T}\right) + \frac{1}{T}(Td\bar{S} - Pd\bar{V} + Pd\bar{V} + \bar{V}dP - Td\bar{S})$$

$$= d\left(\frac{\mu}{T}\right) = \bar{H}d\left(\frac{1}{T}\right) + \frac{\bar{V}}{T}dP \qquad (2.37)$$

So, Eq. (2.36) becomes,

$$dS_{\text{irrev}} = \delta n\left[-\left\{\bar{H}d\left(\frac{1}{T}\right) + \frac{\bar{V}}{T}dP\right\} + (\bar{H} + Q^*)d\left(\frac{1}{T}\right)\right]$$

$$dS_{\text{irrev}} = \delta n_i\left[-\frac{\bar{V}}{T}dP + Q^* d\left(\frac{1}{T}\right)\right] \qquad (2.38)$$

The rate of entropy production due to mass flow is expressed as

$$\left(\frac{dS}{dt}\right) = \frac{\delta n}{dt}\left[-\frac{\bar{V}}{T}dP + Q^* d\left(\frac{1}{T}\right)\right] \qquad (2.39)$$

## 2.8 ENTROPY PRODUCTION AND ITS RATE IN A CHEMICAL REACTION

Consider a closed system in which the amounts of various components, $n_1, n_2, n_3,\ldots$ change due to a chemical reaction. In other words, chemical composition of the system changes. In such cases, the extensive thermodynamic state functions of the system are functions of amounts $n_j$ of various components. The Gibbs function is $G(T, P, n_j)$, the work function is $A(T, V, n_j)$, the enthalpy is $H(S, P, n_j)$ and the internal energy is $E(S, V, n_j)$.

Such a function is applicable to each of the finite number of imaginary cells in the system. The complete differential of the internal energy function is expressed as

$$\partial E = \left(\frac{\partial E}{\partial S}\right)_{V,n_j} \partial S + \left(\frac{\partial E}{\partial V}\right)_{S,n_j} \partial V + \sum_j \left(\frac{\partial E}{\partial n_j}\right)_{V,S,n_j} \partial n_j \qquad (2.40)$$

Now in a system of constant composition

$$dE = dq - dw \qquad \text{(first law of thermodynamics)}$$
$$dE = TdS - PdV$$

$$\left(\frac{\partial E}{\partial S}\right)_{V,n_j} = T \quad \text{and} \quad \left(\frac{\partial E}{\partial V}\right)_{S,n_j} = -P$$

Putting value of $T$ and $P$ in Eq. (2.40), we get

$$dE = TdS - PdV + \sum_{i,j}\left(\frac{dE}{dn_j}\right)_{V,S,n_i} dn_j \qquad (2.41)$$

Now Gibbs chemical potential is defined as

$$\mu_j = \left(\frac{\partial G}{\partial n_j}\right)_{T,P,n_i} = \left(-\frac{\partial A}{\partial n_j}\right)_{T,V,n_i} = \left(\frac{\partial H}{\partial n_j}\right)_{P,S,n_i} = \left(\frac{\partial E}{\partial n_j}\right)_{V,n_i}$$

Hence, Eq. (2.41) reduces to

$$dE = TdS - PdV + \sum_j \mu_j dn_j \qquad (2.42)$$

This Gibbs equation can be applied to each cell in the system and is written as

$$TdS = dE + PdV - \sum_j \mu_j dn_j$$

$$dS = \frac{dE + PdV}{T} - \frac{1}{T}\sum_j \mu_j dn_j \qquad (2.43)$$

In order to maintain the temperature, it is necessary to have heat flow into or out of the system during the reaction. Therefore, we may write that change in the entropy is

$$dS = d_i S + d_e S \qquad (2.44)$$

where, $d_i S$ is the entropy change due to changes with in the system and $d_e S$ is the entropy change due to flow of entropy into the system from the exterior. Further, heat can be transferred reversibly to the isothermal system, so that

$$Td_e S = dq = dE + PdV$$

$$d_e S = \frac{dE + PdV}{T} \qquad (2.45)$$

Combining Eqs. (2.43), (2.44) and (2.45), we have

$$\frac{dE + PdV}{T} - \frac{1}{T}\sum_j \mu_j dn_j = d_i S + \frac{dE + PdV}{T}$$

$$d_i S = -\frac{1}{T}\sum_j \mu_j dn_j \qquad (2.46)$$

Defining the extent of a reaction or degree of advancement of a reaction at a given moment $t$ by $\xi$ as

$$n_j - n_{jo} = V_j \xi \qquad (2.47)$$

where the $V_j$ are the stoichiometric coefficient or number of the reaction, we may write

$$\zeta = \frac{n_j - n_{jo}}{V_j}$$

or

$$n_j - n_{jo} = V_j \zeta \qquad (2.48)$$

Differentiating equation (2.48) $dn_j = 0 + V_j \zeta$

or
$$dn_j = V_j d\zeta$$

Substituting in Eq. (2.46) gives

$$d_i S = -\frac{1}{T} \sum_j \mu_j V_j d\xi \Rightarrow -\frac{1}{T} d\xi \sum_j \mu_j V_j \quad (2.49)$$

Therefore, the rate of entropy production is given by

$$\frac{d_i S}{dt} = -\frac{1}{T} \frac{d\xi}{dt} \sum_j \mu_j V_j \quad (2.50)$$

Since, the chemical affinity of a reaction, $(A) = -\sum_j \mu_j V_j$. So Eq. (2.50) becomes

$$\frac{d_i S}{dt} = \frac{1}{T} \frac{d\xi}{dt} A \quad (2.51)$$

or

$$\frac{1}{V}\left(\frac{d_i S}{dt}\right) = \frac{1}{T}\left(\frac{1}{V}\frac{d\xi}{dt}\right) A \quad (2.52)$$

where, $V$ is the volume of the system.

Since, $\left(\dfrac{1}{V}\dfrac{d\xi}{dt}\right)$ gives the rate of reaction per unit volume, it can be concluded from Eq. (2.52) that rate of entropy production per unit volume in a chemical reaction is directly proportional to the product of the affinity, A (generalized force) and the rate of reaction per unit volume (generalized flux). Such a relation is valid in the neighbourhood of the equilibrium.

## 2.9 ENTROPY PRODUCTION DUE TO ELECTROCHEMICAL REACTIONS

The transport of some electrically charged components from a position, where the electrical potential is $\psi^I$ to a position with potential $\psi^{II}$. For convenience, we shall imagine that our system consists of two parts, each at a well defined electrical potential, while the system as a whole is closed. Introducing the degree of advancement of the phase change, we have

$$-dn_r^I = dn_r^{II} = d\xi_r \quad (2.53)$$

We shall denote by $Z_r$ the electrovalency of the ionic component, which is being transported, and $F$ by the Faraday, i.e., electrical charge associated with one gram ion of a specimen having an electrovalency $I$ ($F = 0.9649 \times 10^5$ C). The intensity of electric current is then related to the degree of advancement $\zeta_r$ and to the rate of phase change $v_r$ by the following relation.

$$I = Z_r F \frac{d\xi_r}{dt} = Z_r F v_r \quad (2.54)$$

The energy equation now contains an additional term expressing the change of electrical energy into internal energy.

$$dE = dQ - PdV + (\psi^I - \psi^{II}) I\, dt \qquad (2.55)$$

We shall assume that Gibbs equation is valid. This is equivalent to the assumption that the entropy can be completely expressed as a function of the energy, volume and composition, even in the presence of electric field. Such a hypothesis is in agreement with the statistical treatment of entropy in an electric field as long as possible variation of polarization of matter are not taken into account. Polarization is mainly associated with orientation of molecules and orientation of molecules in an electrical field is followed by a decrease of entropy.

For our two-phase system, the Gibbs equation can be written as

$$dS = \frac{1}{T} dE + \frac{P}{T} dV - \sum_r \left( \frac{\mu_r^I}{T} dn_r^I + \frac{\mu_r^{II}}{T} dn_r^{II} \right) \qquad (2.56)$$

It is assumed that the temperature is uniform throughout the whole system. We find the entropy balance by inserting Eqs. (2.53), (2.54) and (2.55) into Eq. (2.56) gives

$$dS = \frac{dQ}{T} + \widetilde{A}_r \frac{d\xi_r}{T} \qquad (2.57)$$

With the notation

$$\widetilde{A}_r = A_r + Z_r F (\psi^I - \psi^{II}) - \overline{\mu}_r^I - \overline{\mu}_r^{II} \qquad (2.58)$$

$\widetilde{A}_r$ is the electrochemical affinity corresponding to the transfer of the component $r$ from phase I to phase II. The expression

$$\overline{\mu}_r = \mu_r + Z_r F \psi$$

is called the *electrochemical potential*. It consists of ordinary chemical contribution $\mu_r$ and the electrical part $Z_r F \psi$. The entropy production corresponding to Eq. (2.57) is given by

$$d_i S = \frac{\widetilde{A}_r d\xi_r}{T} \qquad (2.59)$$

and

$$d_e S = \frac{dQ}{T}$$

This rather remarkable result expresses the entropy production due to single irreversible process.

## 2.10 ELECTROKINETIC EFFECTS–SAXEN RELATION

Consider a system consisting of two vessels I and II, which communicate by means of a porous wall or a capillary. The temperature and concentration are supposed to be uniform throughout the entire system and both phases differ only with respect to pressure and electrical potentials. The entropy production, due to the transfer of the constituents from vessel I to vessel II is given by

$$d_i S = \frac{1}{T} \sum_r \widetilde{A}_r d\xi_r = -\frac{1}{T} \sum_r \widetilde{A}_r dn_r^I \qquad (2.60)$$

where, $\widetilde{A_r}$ is the electrochemical affinity given by

$$\widetilde{A_r} = (\mu_r^I - \mu_r^{II}) + Z_r F(\psi^I - \psi^{II}) \tag{2.61}$$

More simply,
$$\widetilde{A_r} = \Delta\mu_r + Z_r F \Delta\psi \tag{2.62}$$

where $\Delta$ denotes the difference in the value of a given variable between vessel I and vessel II. Temperature and composition being the same in both vessels, we have

$$\Delta\mu_r = v_r \Delta P \tag{2.63}$$

where $v_r$ is the specific molar volume of constituent $r$.
Equation (2.62), thus becomes

$$\widetilde{A_r} = v_r \Delta P + Z_r F \Delta\psi \tag{2.64}$$

Putting value of (2.64) in Eq. (2.60), we get

$$d_i S = \frac{1}{T}\sum_r v_r \frac{dn_r^I}{dt}\Delta P - \frac{1}{T}\sum_r Z_r F \frac{dn_r^I}{dt}\Delta\psi \tag{2.65}$$

We introduces the fluxes

$$J = -\sum_r v_r \frac{dn_r^I}{dt} \quad \text{and} \quad I = -\sum_r Z_r F \frac{dn_r^I}{dt} \tag{2.66}$$

where $I$ is the electrical current, due to transfer of charges from I to II, and $J$ the resultant flow of matter. The entropy production now becomes

$$\frac{d_i S}{dt} = \frac{J\Delta P}{T} + \frac{I\Delta\psi}{T} \tag{2.67}$$

and the phenomenological equations are given by

$$I = L_{11}\frac{\Delta\psi}{T} + L_{12}\frac{\Delta P}{T} \quad \text{and} \quad J = L_{21}\frac{\Delta\psi}{T} + L_{22}\frac{\Delta P}{T} \tag{2.68}$$

with the Onsager relation

$$L_{12} = L_{21} \tag{2.69}$$

We have here two irreversible effects, transport of matter under the influence of a difference of pressure and electrical current due to difference of electrical potential. Moreover, we have a cross effect related by the coefficient $L_{12} = L_{21}$ which is due to the interference of the two irreversible processes.

We now turn to the definition of the electrokinetic effects.

In the first place, we have streaming potential defined as the potential difference per unit pressure difference in the state with zero electrical current. From Eq. (2.68), we get

$$\left(\frac{\Delta\psi}{\Delta P}\right)_{I=0} = -\frac{L_{12}}{L_{11}} \quad \text{(streaming potential)} \tag{2.70}$$

The second *electro-kinetic* effect is called *electro-osmosis* and is defined as the flow of matter per unit electrical current in the state with uniform pressure. Using Eq. (2.68), it follows that

$$\left(\frac{J}{I}\right)_{\Delta P=0} = \frac{L_{21}}{L_{11}} \quad \text{(electro-osmosis)} \tag{2.71}$$

The third effect is called *electro-osmotic pressure* and is defined as the pressure difference per unit potential differences when the flow $J$ is zero,

$$\left(\frac{\Delta P}{\Delta \psi}\right)_{J=0} = -\frac{L_{21}}{L_{22}} \quad \text{(electro-osmosis pressure)} \tag{2.72}$$

The fourth effect is the streaming current

$$\left(\frac{I}{J}\right)_{\Delta\psi=0} = \frac{L_{12}}{L_{22}} \quad \text{(streaming current)} \tag{2.73}$$

Between these four effects, which can be studied experiment-independently, the Osnager relation gives the two connections.

$$\left(\frac{\Delta\psi}{\Delta P}\right)_{I=0} = \left(\frac{J}{I}\right)_{\Delta P=0} \tag{2.74}$$

$$\left(\frac{\Delta P}{\Delta\psi}\right)_{J=0} = -\left(\frac{I}{J}\right)_{\Delta\psi=0} \tag{2.75}$$

These two relations both relate an osmotic effect to a streaming effect. Relations (2.74) and (2.75) are known as *Saxen's relation*.

## 2.11 THERMO-MOLECULAR PRESSURE DIFFERENCE (TPD) AND THERMO-MECHANICAL (OR MECHANO-CALORIFIC) EFFECT

We have already calculated rate of entropy production for such a system. Modifying this equation suitably and assuming that no reaction is taking place, we have

$$\frac{d_i S}{dt} = \frac{d^I \phi}{dt}\left(\frac{1}{T^I} - \frac{1}{T^{II}}\right) - \left(\frac{d\mu_r^I}{T^I} - \frac{d\mu_r^{II}}{T^{II}}\right)\frac{d_e^I \mu_r}{dt} \tag{2.76}$$

Temperature gradient

$$X^{th} = \left(\frac{1}{T^I} - \frac{1}{T^{II}}\right) = \frac{T^{II} - T^I}{T^I T^{II}} = -\frac{\Delta T}{T^2}$$

If $\Delta T$ is small $T^I \approx T^{II} = T$

Then, chemical potential gradient

$$X_m = -\frac{(\Delta \mu)_T}{T} = -\frac{(VdP + SdT)_T}{T}, \quad \text{but } dT = 0$$

$$X_m = -\frac{V}{T}\Delta P$$

Here $V$ is specific molar volume and $S$ is specific molar entropy.

The thermal energy flux $J_{th}$ and the mater flux $J_m$ are given by

$$J^{th} = \frac{d^l\phi}{dt} \quad \text{and} \quad J_m = \frac{d_e^l n_r}{dt}$$

The entropy production becomes

$$\frac{d_iS}{dt} = -J^{th}\frac{\Delta T}{T^2} - J_m\frac{V}{T}\Delta P \tag{2.77}$$

And the phenomenological laws now becomes

$$J_{th} = -L_{11}\frac{\Delta T}{T^2} - L_{12}\frac{V\Delta P}{T} \tag{2.78}$$

$$J_m = -L_{21}\frac{\Delta T}{T^2} - L_{22}\frac{V\Delta P}{T} \tag{2.79}$$

We have again the Onsager relation $L_{12} = L_{21}$.

We shall study two phenomena viz. thermo-molecular pressure difference (TPD) and thermo-mechanical effect. The thermo-molecular pressure difference is defined as the difference of pressure, which arises between two phases in the stationary state $J_m = 0$, when a temperature difference is maintained. From Eq. (2.79), this pressure is given by

$$\left(\frac{\Delta P}{\Delta T}\right)_{J_m=0} = -\frac{L_{21}}{L_{22}VT} \tag{2.80}$$

This cross phenomenon is due to interference of the irreversible processes of transport of energy and matter. The TPD is called *Knudsen effect*, if the system consists of a gas and vessels are separated by narrow capillaries or small openings. When the same effect occurs in gases or liquids with a membrane separating the two phases, it is called *thermo-osmosis*.

Let us now study the other phenomenon, we deal with the same system as defined above. If we maintain a pressure difference between the two vessels and a uniform temperature through out the system, matter flows from one vessel to the another and an associated energy flow, proportional to the matter flow, is observed. This energy flow can be measured by determining the heat necessary to maintain a uniform temperature in the system. This effect is known as *thermo-mechanical effect*. It is expressed in terms of phenomenological coefficients Eq. (2.78) and Eq. (2.79).

$$\left(\frac{J_{th}}{J_m}\right)_{\Delta T=0} = \frac{L_{12}}{L_{22}} \quad \text{(thermo-mechanical effect)} \tag{2.81}$$

The quantity $L_{12}/L_{22}$ is called the *heat of transfer* defined as energy transfer per unit transfer of mass. Thus

$$q^* = \frac{L_{12}}{L_{22}} \tag{2.82}$$

From Eqs. (2.80) and (2.81) using the Onsager relation, we obtain the following relationship between TPD and the thermo-mechanical effect

$$\left(\frac{\Delta P}{\Delta T}\right)_{J_m=0} = -\frac{1}{VT}\left(\frac{J_{th}}{J_m}\right)_{\Delta T=0}$$

Thus, both effects will appear in the same system.

## EXERCISES

1. What are coupled fluxes? Give some examples.
2. Differentiate between: (i) steady flow and stationary states and (ii) equilibrium states and metastable state.
3. Write down the equation for entropy production for thermo-osmotic phenomenon and show that the rate of entropy production is the sum of product of fluxes and forces.
4. What do you understand by conjugate pair of fluxes and forces?
5. Write down the linear equation for heat flux and mass flux in the case of thermo-osmosis. What are limitations of such equations?
6. What are Onsager relations? Indicate how will you verify Onsager reciprocity relation in case of electro-osmosis.
7. Show that for steady state close to equilibrium, entropy production is minimum.
8. Discuss the limitations of linear thermodynamics of irreversible phenomenon.
9. Obtain the equation for entropy production for electro-kinetic phenomenon and identify the appropriate thermodynamic fluxes and forces. Hence, write the phenomenological relations.
10. Show that if Onsager relations are valid

$$\left(\frac{\Delta P}{\Delta \psi}\right)_{J=0} = -\left(\frac{I}{J}\right)_{\Delta\psi=0}$$

$$\left(\frac{\Delta \psi}{\Delta P}\right)_{I=0} = \left(\frac{J}{I}\right)_{\Delta P=0}$$

11. Write down postulates of non-equilibrium thermodynamics.
12. Derive the relation for entropy production due to heat flow.
13. Derive the relation for entropy production due to mass flow.
14. Derive the relation for entropy production due to chemical reaction.
15. Derive the relation for entropy production due to electrochemical reaction.
16. Discuss in detail thermo-molecular pressure difference (TDP) and thermo-mechanical effect.

## SUGGESTED READINGS

De Groot, S.R. and P. Major, *Nonequilibrium Thermodynamics*, Academic Press, 1962.

De Groot, S.R., *Thermodynamics of Irreversible Process*, North Holland Publishing Company, Amsterdam, 1952.

Katchalsky, A. and P.F. Currow, *Nonequilibrium Thermodynamics in Biophysics*, Harvard University Press, Cambridge, Massachusetts, 1965.

Muschik, W., *Aspects of Nonequilibrium Thermodynamics*, Amazing Publishing House, 1990.

Ottihger, H.C., *A Systematic Approach to Nonequilibrium Thermodynamics*, Wiley, 2005.

Prigogine, I., *Introduction to Thermodynamics of Irreversible Process*, 2nd rev. ed., John Wiley, New York, 1961.

Rastogi, R.P., R.C. Srivastava, and S.N. Singh, Nonequilibrium Thermodynamic of Electro-kinetic Phenomena, *Chemical Reviews*, USA, 1993.

# 3

# Statistical Thermodynamics

In classical thermodynamics, we deal with macroscopic properties of the system without regarding the contribution of individual particles (molecules, ions, atoms). In quantum mechanics, we deal with microscopic properties like position, velocity, orientation, distribution, energy, etc. of the microscopic systems associated with microscopic particles like protons, electrons, neutrons, etc. In classical thermodynamics, we discussed thermodynamic principles without considering any model or theory on microscopic level. This treatment has both advantages and disadvantages. The advantage includes is that its predictions can be extended to as complex systems as biological processes even though the exact mechanism may not be known. The disadvantage of classical thermodynamics is that we are not certain about the rate and mechanism of the process under investigation. The relevant information can be provided if we know the link between the macroscopic properties and microscopic properties of the system. *Such types of correlation or link between classical thermodynamics and quantum mechanics are established by considering statistical thermodynamics. Statistical thermodynamics acts as a bridge between classical thermodynamics and quantum mechanics.*

*Thus, the subject of the study of the link between thermodynamics and quantum mechanics is called statistical thermodynamics.* Quantum mechanics provides information about the energy of the molecular system, statistical mechanics tells us about the possible arrangement of the energy among various molecules of the system and introduces the concept of probability and partition functions. Statistical thermodynamics deals with the relationship between the probability, partition function and the thermodynamic properties. The methods of statistical thermodynamics were first developed by Boltzmann (Germany) and Gibbs (USA). With the development of the methods of quantum theory some modifications in Boltzmann's original ideas were introduced by Bose (Indian Physicist) and Einstein (Germany) and Fermi (Italian Physicist) and Dirac (English Physicist).

Answer to the question, how the molecules are distributed amongst the various possible energy levels? will be answered by statistical mechanics. In other words, if we want to know that, how many molecules are present in the lowest energy level, how many are present in highest energy level and how many are present in various intermediate energy levels. That is to say, we wish to have information about the best possible arrangement of molecules in various quantum or energy levels. Such a description is given by the term *probability* denoted by $W$ in statistical thermodynamics.

## 3.1 BASIC TERMINOLOGY USED IN STATISTICAL THERMODYNAMICS

*Ensemble:*  A number of $N$ identical entities is called an *assembly*. If the entities are assemblies of particles then we call the number $N$ as assembly of assemblies or an *ensemble*. Thus, an ensemble consists of a large number of replicas of the system under consideration. If each member of the ensemble has same number of molecules $N$, same volume $V$ and same energy $E$, it is called a microcanonical ensemble.

*Probability:*  The probability of a state of a system is defined as the number of configurations leading to that particular state divided by the total number of configurations possibly available to the system. For example, tossing of a coin. It can either show head or tail. Thus, total number of possible configurations of the state of the coin is two, i.e. one head and one tail. The probability of showing head in one out of two configurations, i.e. 1/2, similarly the probability of showing tail is 1/2.

*Thermodynamic probability* ($W$):  The thermodynamic probability of a system is equal to the number of ways of realising the distribution. The symbol of thermodynamic probability is $W$. Here, it is to be noted that thermodynamic probability is not equivalent to probability since former is always equal to or greater than unity whereas the latter is always less than one. Speaking more accurately, $W$ is the number of quantum states in the energy range between $E$ and $E + dE$ for given $V$ and $N$.

*Statistical weight factor* ($G$):  It is the degree of degeneracy of a particular energy level, and is equal to the energy states of an energy level. For example, the energy level of a particle in three-dimensional box is given by $E_i = \dfrac{\left(n_x^2 + n_y^2 + n_z^2\right)h^2}{8\,m\,V}$. The energy of the quantum states 211, 121, 112 is the same, but the three states are distinct. Hence, the degree of degeneracy is 3 and the statistical weight factor $g$ is also 3.

*Partition function* ($q$):  In quantum mechanics, all the information about microscopic system, i.e. position, velocity, momentum, energy, etc. is stored in wave function $\psi$. In statistical mechanics also, there exists a function which contain all information about macroscopic system, that function is known as *partition function*. The partition function is a dimensionless quantity. It summarizes in a convenient mathematical form as to how the energy of a system of molecules is partitioned among the molecules. The value of partition function depends upon the molar mass, the temperature, the molar volume, the internuclear distances, the molecular motion and the internuclear forces. The partition function provides a bridge to link the microscopic properties of individual molecules such as their discrete energy levels, moments of inertia, etc., with the macroscopic properties like entropy, heat capacity, etc., of a system containing a large number of molecules.

## 3.2 PARTITION FUNCTION (Q)

According to Boltzmann distribution law, the fraction of molecules which are in the most probable state at temperature $T$ possessing the energy $E_i$ is given by

$$\frac{n_i}{N} = \frac{g_i\, e^{-\varepsilon_i/kT}}{\sum g_i\, e^{-\varepsilon_i/kT}}$$

The denominator of the above equation which gives the sum of terms $g_i\, e^{-\varepsilon_i/kT}$ for all the energy levels is called the *partition function*. It is represented by $q$. Thus,

$$q = \sum g_i\, e^{-\varepsilon_i/kT}$$

Partition function ($q$) indicates how the particles are distributed among the various energy levels (states).

From Maxwell–Boltzmann law, we have

$$n_i = \frac{g_i}{e^{\alpha+\beta\varepsilon_i}} = \frac{g_i}{e^{\alpha+\varepsilon_i/kT}} = \frac{g_i}{e^{\alpha}e^{\varepsilon_i/kT}} = \frac{g_i e^{-\varepsilon_i/kT}}{e^{\alpha}} \qquad (3.1)$$

Taking summation law, we have

$$\sum_i n_i = \frac{\sum g_i e^{-\varepsilon_i/kT}}{e^{\alpha}}$$

$$N = \frac{\sum g_i e^{-\varepsilon_i/kT}}{e^{\alpha}} = \frac{q}{e^{\alpha}} \qquad (3.2)$$

From Eq. (3.1), we have

$$n_i = \frac{g_i e^{-\varepsilon_i/kT}}{e^{\alpha}}$$

In ground state $n_i = n_0$, $g_i = g_0 = 1$, hence the above equation becomes

$$n_0 = \frac{1}{e^{\alpha}}$$

Putting this value of $\frac{1}{e^{\alpha}}$ in Eq. (3.2), we get

$$N = n_0 q$$

or

$$q = \frac{N}{n_0}$$

*So, partition function may also be defined as the ratio of number of particles in a system to the number of particles in the ground state.*

Or

*It is the measure of extent to which the particles leave the ground state.*

### 3.2.1 Properties of Partition Function

1. It is dimensionless quantity.
2. Its value depends upon the molecular weight, temperature, internuclear forces, etc.
3. It can never be zero but $q \gg 1$, i.e. at absolute zero as $N = N_o$, $q = 1$ and at higher temperature as $N > 1$, therefore, $q > 1$.

### 3.2.2 Factorization of Partition Function or Multiplication Theorem of Partition Function

Partition function may be defined by

$$q = \sum g_i \, e^{-\varepsilon_i/kT} \tag{3.3}$$

The energy $\varepsilon$ of a molecule is the sum of contributions from, the different modes of motion like translational, rotational, vibrational, electronic, etc.

If we assume that energy associated with any one mode of behaviour is independent of all other modes, then we can define energy by

$$E = E_T + E_r + E_v + E_e$$

and

$$g_i = g_T + g_r + g_v + g_e$$

where $E_T$, $E_r$, $E_v$ and $E_e$ are translational, rotational, vibrational and electronic contribution, respectively and $g_T$, $g_r$, $g_v$ and $g_e$ are degeneracy of translational, rotational, vibrational, and electronic energy level, respectively.

So, Eq. (3.3) can be written as

$$q = \sum (g_t g_r g_v g_e) \, e^{-(\varepsilon_t + \varepsilon_r + \varepsilon_v + \varepsilon_e)/kT}$$

$$q = \sum g_t e^{-\varepsilon_t/kT} \sum_i g_r e^{-\varepsilon_r/kT} \sum_i g_v e^{-\varepsilon_v/kT} \sum_i g_e e^{-\varepsilon_e/kT}$$

$$q = q_t \cdot q_r \cdot q_v \cdot q_e \tag{3.4}$$

Equation (3.4) is known as *multiplication theorem* or *factorization of partition function*. This factorization means that we can investigate each contribution separately.

## 3.3 PARTITION FUNCTION AND ITS THERMODYNAMIC RELATION WITH OTHER THERMODYNAMIC FUNCTIONS

### 3.3.1 Relation of Partition Function with Energy

If molecules are non interacting, i.e. gas is ideal, the total energy $E$ of the system is given by

$$E = \sum n_i \varepsilon_i \tag{3.5}$$

According to Maxwell–Boltzmann law, we have

$$n_i = \frac{g_i}{e^{\alpha+\beta\varepsilon_i}} = g_i e^{-\alpha} e^{-\beta\varepsilon_i}$$

$$n_i = g_i e^{-\alpha} e^{-\varepsilon_i/kT}$$

Taking summation on both sides, we get

$$\sum n_i = e^{-\alpha} \sum g_i e^{-\varepsilon_i/kT} \tag{3.6}$$

$$N = e^{-\alpha} \sum g_i e^{-\varepsilon_i/kT}$$

$$N = e^{-\alpha} q$$

$$e^{-\alpha} = \frac{N}{q} \tag{3.7}$$

Putting the value of $\sum n_i$ from Eq. (3.6) into Eq. (3.5), we get

$$E = e^{-\alpha} \sum g_i e^{-\varepsilon_i/kT} \varepsilon_i \tag{3.8}$$

Putting the value of $e^{-\alpha}$ from Eq. (3.7) into Eq. (3.8), we get

$$E = \frac{N}{q} \sum g_i e^{-\varepsilon_i/kT} \varepsilon_i$$

$$\frac{Eq}{N} = \sum g_i e^{-\varepsilon_i/kT} \varepsilon_i \tag{3.9}$$

Now as $q = \sum g_i e^{-\varepsilon_i/kT}$

Differentiating with respect to $T$ at constant $V$, we get

$$\left(\frac{\partial q}{\partial T}\right)_V = \frac{1}{kT^2} \sum g_i e^{-\varepsilon_i/kT} \varepsilon_i$$

$$kT^2 \left(\frac{\partial q}{\partial T}\right)_V = \sum g_i e^{-\varepsilon_i/kT} \varepsilon_i \tag{3.10}$$

From equation (3.9) and (3.10), we get

$$kT^2 \left(\frac{\partial q}{\partial T}\right)_V = \frac{Eq}{N}$$

$$NkT^2 \frac{1}{q} \left(\frac{\partial q}{\partial T}\right)_V = E$$

$$NkT^2 \left(\frac{\partial \ln q}{\partial T}\right)_V = E$$

For one mole of gas $N = N_A$ = Avogadro's number

$$N_A kT^2 \left(\frac{\partial \ln q}{\partial T}\right)_V = E$$

$$E = RT^2 \left(\frac{\partial \ln q}{\partial T}\right)_V \qquad (\because R = kN_A) \qquad (3.11)$$

This gives energy for one mole of particles (molecules). For finding average energy per molecule, we have to divide above equation by $N_A$, thus

$$\bar{E} = \frac{E}{N_A} = \frac{RT^2}{N_A}\left(\frac{\partial \ln q}{\partial T}\right)_V$$

$$\bar{E} = kT^2 \left(\frac{\partial \ln q}{\partial T}\right)_V$$

This can be written as,

$$\bar{E} = k\left(\frac{\partial \ln q}{\partial (1/T)}\right)_V \qquad (3.12)$$

### 3.3.2 Relation of Partition Function with Heat Capacity at Constant Volume ($C_V$)

Heat capacity at constant volume is given by

$$C_V = \left(\frac{\partial E}{\partial T}\right)_V \qquad (3.13)$$

and

$$E = RT^2 \left(\frac{\partial \ln q}{\partial T}\right)_V$$

Putting this value of $E$ in Eq. (3.13), we get

$$C_V = \frac{\partial}{\partial T}\left(RT^2 \left(\frac{\partial \ln q}{\partial T}\right)_V\right)_V$$

$$C_V = R\left[2T\left(\frac{\partial \ln q}{\partial T}\right)_V + T^2\left(\frac{\partial^2 \ln q}{\partial T^2}\right)_V\right]$$

This can also be written as,

$$C_V = \frac{R}{T^2}\left[\left(\frac{\partial^2 \ln q}{\partial (1/T)^2}\right)\right]$$

### 3.3.3 Relation of Partition Function with Entropy

From thermodynamics, we know that for most probable distribution of particles, the entropy of system is given by

$$S = k \ln W \qquad (3.14)$$

From Maxwell–Boltzmann statistics

$$W = N! \prod_i \frac{g_i^{n_i}}{n_i!} \quad \text{(for distinguishable particles)}$$

For indistinguishable particles, we have

$$W = \prod_i \frac{g_i^{n_i}}{n_i!}$$

Taking log on both sides, we get

$$\ln W = \sum (n_i \ln g_i - \ln n_i!)$$

Applying Sterling approximation

$$\ln W = \sum (n_i \ln g_i - n_i \ln n_i + n_i) \tag{3.15}$$

Putting the value of $\ln W$ from Eq. (3.15) in (3.14), we get

$$S = k \sum (n_i \ln g_i - n_i \ln n_i + n_i)$$

$$S = k \sum \left( n_i \ln \frac{g_i}{n_i} + n_i \right) \tag{3.16}$$

But, $n_i = g_i e^{-(\alpha + \varepsilon_i/kT)}$

$$\frac{n_i}{g_i} = e^{-(\alpha + \varepsilon_i/kT)}$$

$$\frac{g_i}{n_i} = e^{(\alpha + \varepsilon_i/kT)}$$

$$\ln \frac{g_i}{n_i} = (\alpha + \varepsilon_i/kT)$$

Putting this value of $\ln \frac{g_i}{n_i}$ in Eq. (3.16), we get

$$S = k \sum (n_i (\alpha + \varepsilon_i/kT) + n_i) \tag{3.17}$$

$$S = k \sum (n_i \alpha + n_i \varepsilon_i/kT + n_i)$$

$$S = k \left( \sum n_i \alpha + \sum \frac{n_i \varepsilon_i}{kT} + \sum n_i \right)$$

$$S = k \left( N\alpha + \frac{E}{kT} + N \right)$$

For one mole of particles $N = N_A$ = Avogadro's number

$$S = k\left(N_A \alpha + \frac{E}{kT} + N_A\right)$$

$$S = \left(kN_A \alpha + \frac{E}{T} + kN_A\right)$$

$$S = R\alpha + \frac{E}{T} + R \tag{3.18}$$

But, $n_i = g_i e^{-(\alpha + \varepsilon_i/kT)}$

$$n_i = g_i e^{-\alpha} e^{-\varepsilon_i/kT}$$

$$\sum n_i = \sum g_i e^{-\alpha} e^{-\varepsilon_i/kT}$$

$$\sum n_i = e^{-\alpha} \sum g_i e^{-\varepsilon_i/kT}$$

$$N = e^{-\alpha} q \quad \left[\because q = \sum g_i e^{-\varepsilon_i/kT}\right]$$

$$e^{-\alpha} = \frac{N}{q}$$

$$e^{\alpha} = \frac{q}{N}$$

or 
$$\alpha = \ln \frac{q}{N} \tag{3.19}$$

Putting this value of $\alpha$ in Eq. (3.18), we get

$$S = R \ln \frac{q}{N} + \frac{E}{T} + R \tag{3.20}$$

### 3.3.4 Another Relation of Entropy, S with Partition Function, q

As we know from Eq. (3.19), $\alpha = \ln \frac{q}{N}$, putting this in Eq. (3.17), we get

$$S = k \sum \left(n_i \left(\ln \frac{q}{N} + \frac{\varepsilon_i}{kT}\right) + n_i\right)$$

$$S = k \sum (n_i \ln q - n_i \ln N + n_i \varepsilon_i/kT + n_i)$$

$$S = k \left[\sum n_i \ln q - \sum n_i \ln N + \sum n_i \varepsilon_i/kT + \sum n_i\right]$$

$$S = k[N \ln q - N \ln N + N + E/kT]$$

$$S = k[\ln q^N - (N \ln N - N) + E/kT]$$

$$S = k[\ln q^N - \ln N! + E/kT]$$

$$S = \left[k \ln \frac{q^N}{N!} + \frac{E}{T}\right] \tag{3.21}$$

$$E = RT^2 \left(\frac{\partial \ln q}{\partial T}\right)_V$$

$$S_t = RT \left(\frac{\partial \ln q_t}{\partial T}\right)_V + k \ln \frac{q_t^N}{N!} \quad \text{(for indistinguishable particles)} \tag{3.22}$$

In driving above relation, we have assumed that particles (molecules) are indistinguishable. It follows, therefore, that particles are moving because as long as they are moving, it is impossible to distinguish them. Thus, above relation between $S$ and $q$ is for translational motion of molecules.

Now, if we consider that molecules in a system have only rotational and/or vibrational energies and so translational energy, then molecules in such a system may be considered localized. Hence, for such a system molecules are distinguishable and we can represent mathematical probability by

$$W = N! \prod_i \frac{g_i^{n_i}}{n_i!}$$

And relation between $S$ and $q$ is given by

$$S = RT \left(\frac{\partial \ln q}{\partial T}\right)_V + k \ln q^N$$

$$S = RT \left(\frac{\partial \ln q}{\partial T}\right)_V + Nk \ln q \tag{3.23}$$

For one mole, $N = N_A =$ Avagadro's number, and we get

$$S_r = RT \left(\frac{\partial \ln q_r}{\partial T}\right)_r + Nk \ln q_r$$

$$S_V = RT \left(\frac{\partial \ln q_V}{\partial T}\right)_V + Nk \ln q_V \tag{3.24}$$

Above equations give rotational entropy and vibrational entropy of a diatomic molecule, respectively.

### 3.3.5 Relation of Partition Function with Gibbs Free Energy

From thermodynamics, we have

$$G = H - TS$$
$$G = E + PV - TS$$

For one mole of ideal gas $\quad G = E + RT - TS$

But
$$S = R\ln\frac{q}{N} + \frac{E}{T} + R$$

$$G = E + RT - T\left[\frac{E}{T} + R\ln\frac{q}{N} + R\right]$$

$$G = -RT\ln\frac{q}{N}$$

or for one mole of ideal gas
$$G = -RT\ln\left(\frac{q_m}{N_A}\right) \qquad (3.25)$$

### 3.3.6 Relation of Partition Function with Work Function (Helmholtz Free Energy)

$$A = E - TS$$

$$S = k\left[\ln\frac{q^N}{N!} + \frac{E}{kT}\right]$$

$$S = k\ln\frac{q^N}{N!} + \frac{E}{T}$$

$$A = E - T\left[k\ln\frac{q^N}{N!} + \frac{E}{T}\right]$$

$$A = E - kT\ln\frac{q^N}{N!} - E$$

$$A = -kT\ln\left(\frac{q^N}{N!}\right)$$

### 3.3.7 Relation of Partition Function with Pressure

$$A = -kT\ln\left(\frac{q^N}{N!}\right) \qquad (3.26)$$

As
$$A = E - TS$$
$$dA = dE - TdS - SdT$$
$$dA = dE - dq - SdT \qquad \left(dS = \frac{dq}{T} \text{ second law of thermodynamics}\right)$$
$$dA = dw - SdT \qquad \text{(First law of thermodynamics)}$$
$$dA = -PdV - SdT$$
$$dA + SdT = -PdV$$

Differentiating with respect to $V$ at constant $T$, we get

$$\left(\frac{\partial A}{\partial V}\right)_T = -P$$

or
$$P = -\left(\frac{\partial A}{\partial V}\right)_T \tag{3.27}$$

Putting the value of $A$ from Eq. (3.26) into Eq. (3.27), we get

$$P = -\left[\frac{\partial}{\partial V}\left(-kT \ln \frac{q^N}{N!}\right)\right]_T$$

$$P = kT\left(\frac{\partial}{\partial V} \ln \frac{q^N}{N!}\right)_T$$

### 3.3.8 Relation of Partition Function with Enthalpy (Heat Content)

Enthalpy is defined as
$$H = E + PV$$

Also
$$P = -\left(\frac{\partial A}{\partial V}\right)_T$$

Using the expression for $A$ as $A = -kT \ln q$

$$P = kT\left(\frac{\partial \ln q}{\partial V}\right)_T$$

But
$$E = kT^2\left(\frac{\partial \ln q}{\partial T}\right)_V$$

Therefore,
$$H = kT^2\left(\frac{\partial \ln q}{\partial T}\right)_V + kT\left(\frac{\partial \ln q}{\partial V}\right)_T V$$

$$H = kT\left(T\frac{\partial \ln q}{\partial T}\right)_V + V\left(\frac{\partial \ln q}{\partial V}\right)_T \tag{3.28}$$

### 3.3.9 Relationship Between Equilibrium Constant and Partition Function

Let us consider a general reaction involving gaseous reactants and products:

$$aA + bB \Leftrightarrow mM + nN$$

Equilibrium constant, $K$ in terms of concentration of the various species is given by

$$K = \frac{[M]^m[N]^n}{[A]^a[B]^b} = \frac{[C_M]^m[C_N]^n}{[C_A]^a[C_B]^b} \tag{3.29}$$

At equilibrium free energy changes are zero, and

$$m\mu_M + n\mu_N = a\mu_A + b\mu_B \qquad (3.30)$$

But, $\mu_i = -RT \ln\left(\dfrac{q_i}{n_i}\right)$

Putting the chemical potential expression in Eq. (3.30)

$$mkT \ln\left(\dfrac{q_M}{n_M}\right) + nkT \ln\left(\dfrac{q_N}{n_N}\right) = akT \ln\left(\dfrac{q_A}{n_A}\right) + bkT \ln\left(\dfrac{q_B}{n_B}\right)$$

Dividing throughout by $kT$, we get

$$m \ln\left(\dfrac{q_M}{n_M}\right) + n \ln\left(\dfrac{q_N}{n_N}\right) = a \ln\left(\dfrac{q_A}{n_A}\right) + b \ln\left(\dfrac{q_B}{n_B}\right)$$

or

$$\ln\left(\dfrac{q_M}{n_M}\right)^m + \ln\left(\dfrac{q_N}{n_N}\right)^n = \ln\left(\dfrac{q_A}{n_A}\right)^a + \ln\left(\dfrac{q_B}{n_B}\right)^b$$

$$\ln \dfrac{[q_M]^m [q_N]^n}{[q_A]^a [q_B]^b} = \ln \dfrac{[n_M]^m [n_N]^n}{[n_A]^a [n_B]^b}$$

$$\dfrac{[q_M]^m [q_N]^n}{[q_A]^a [q_B]^b} = \dfrac{[n_m]^m [n_n]^n}{[n_A]^a [n_B]^b} \qquad (3.31)$$

Let $V$ be the total volume then the concentration ($C_i$) of any species per unit volume will be given as

$$C_i = \dfrac{n_i}{V}$$

or

$$n_i = C_i V$$

Substituting expressions for $n_i$ in Eq. (3.31), we get

$$\dfrac{[q_M]^m [q_N]^n}{[q_A]^a [q_B]^b} = \dfrac{[C_M]^m [C_N]^n}{[C_A]^a [C_B]^b} V^{(m+n)-(a+b)}$$

$$\dfrac{[q_M]^m [q_N]^n}{[q_A]^a [q_B]^b} \times \dfrac{1}{V^{(m+n)-(a+b)}} = \dfrac{[C_M]^m [C_N]^n}{[C_A]^a [C_B]^b} \qquad (3.32)$$

Comparing Eqs. (3.32) and (3.29), we get

$$K = \dfrac{[q_M]^m [q_N]^n}{[q_A]^a [q_B]^b} \times \dfrac{1}{V^{\Delta V}} \qquad (3.33)$$

where $\Delta V = (m+n) - (a+b)$

Equation (3.33) is an important link between the quantum mechanics and chemistry. If we know the energy distribution of the molecules, we can calculate the molecular partition function and then by the use of Eq. (3.33) the equilibrium constant $K$ for the chemical reaction can be calculated.

## 3.4 TRANSLATIONAL PARTITION FUNCTION

By definition, partition function is given by,

$$q = \sum g_i e^{-\varepsilon_i/kT}$$

For translational energy, as the translational energy level are non-degenerate, i.e. $g_i = 1$. Translational partition function can be written as

$$q_t = \sum e^{-\varepsilon_t/kT} \qquad (3.34)$$

Further, we know that the translational energy of a molecule moving in a rectangular box of dimension $a$, $b$ and $c$ is given by

$$\varepsilon_t = \frac{h^2}{8m}\left[\frac{n_x^2}{a^2} + \frac{n_y^2}{b^2} + \frac{n_z^2}{c^2}\right] \qquad (3.35)$$

where, $n_x$, $n_y$ and $n_z$ are principal quantum numbers determining the possible value of translational energy of particle along $x$, $y$ and $z$ directions, respectively.

From Eqs. (3.34) and (3.35), we can write

$$q_t = \sum e^{-\frac{h^2 n_x^2}{8mkTa^2}} e^{-\frac{h^2 n_y^2}{8mkTb^2}} e^{-\frac{h^2 n_z^2}{8mkTc^2}}$$

$$q_t = q_t(x)\, q_t(y)\, q_t(z) \qquad (3.36)$$

If we consider motion in $x$ direction only, then corresponding partition function can be written as:

$$q_t(x) = \sum_{n_x=0}^{n_x=\infty} e^{-\frac{h^2 n_x^2}{8mkTa^2}}$$

Since the energy levels are very close to each other, summation in above equation can be replaced by integration, so we have

$$q_t(x) = \int_0^\infty e^{-\frac{h^2 n_x^2}{8mkTa^2}} dn_x \qquad (3.37)$$

Using standard integrals

$$\int_0^\infty e^{-ax^2} dx = \frac{1}{2}\sqrt{\frac{\pi}{a'}}$$

where $a' = \dfrac{h^2}{8mkTa^2}$ and $x = n_x$

Equation (3.37) reduces to

$$q_t(x) = \frac{1}{2}\pi^{1/2}\left(\frac{8mkTa^2}{h^2}\right)^{1/2}$$

$$q_t(x) = (2\pi mkT)^{1/2}\frac{a}{h} \tag{3.38}$$

Similarly, for motion along $y$ and $z$ directions, we have

$$q_t(y) = (2\pi mkT)^{1/2}\frac{b}{h} \tag{3.39}$$

and

$$q_t(z) = (2\pi mkT)^{1/2}\frac{c}{h} \tag{3.40}$$

From Eqs. (3.36), (3.38), (3.39) and (3.40), we can write

$$q_t = \frac{(2\pi mkT)^{3/2}}{h^3}abc$$

$$q_t = \left(\frac{2\pi mkT}{h^2}\right)^{3/2}V$$

where $V$ is the volume of container.

### 3.4.1 Calculation of Translational Energy from Translational Partition Function

For one mole, relation between $E$ and $q$ is

$$E = RT^2\left(\frac{\partial \ln q}{\partial T}\right)_V \tag{3.41}$$

For translational energy, we can write

$$E_t = RT^2\left(\frac{\partial \ln q_t}{\partial T}\right)_V$$

where $q_t$ is the translational partition function.

Now translational partition function is gives by

$$q_t = \left(\frac{2\pi mkT}{h^2}\right)^{3/2}V$$

$$q_t = \left(\frac{2\pi mk}{h^2}\right)^{3/2}T^{3/2}V$$

Taking log on both sides, we get

$$\ln q_t = \ln\left(\frac{2\pi mk}{h^2}\right)^{3/2} + \ln T^{3/2} + \ln V$$

$$\ln q_t = \frac{3}{2}\ln\left(\frac{2\pi mk}{h^2}\right) + \frac{3}{2}\ln T + \ln V$$

Differentiating it w.r.t. $T$, at constant volume, we get

$$\left(\frac{\partial \ln q_t}{\partial T}\right)_V = \frac{3}{2}\left(\frac{\partial \ln T}{\partial T}\right) = \frac{3}{2T} \tag{3.42}$$

Comparing Eq. (3.41) and (3.42), we can write

$$E = RT^2\,\frac{3}{2T}$$

or

$$E_t = \frac{3}{2}RT \tag{3.43}$$

### 3.4.2 Translational Entropy of Monoatomic Ideal Gas Or Sackure–Tetrode Equation

Translational entropy is given by

$$S_t = \frac{E_t}{T} + k\ln\frac{q_t^N}{N!}$$

$$S_t = \frac{E_t}{T} + k\ln q_t^N - k\ln N!$$

Using the Sterling approximation $S_t = \dfrac{E_t}{T} + Nk\ln q_t - k(N\ln N - N)$

$$S_t = \frac{E_t}{T} + Nk\ln q_t - kN\ln N + Nk$$

$$S_t = \frac{E_t}{T} + Nk\left[\ln q_t - \ln N + 1\right] \tag{3.44}$$

Now, number of molecules ($N$) = number of moles × Avogadro's number

$$N = n\,N_A$$

Hence, molar translational entropy can be written as,

$$S_t(\text{molar}) = \frac{E_t}{T} + nN_A k\left[\ln q_t - \ln N + 1\right]$$

$$S_{t,m} = \frac{E_t}{T} + nR\left[\ln q_t - \ln N + 1\right] \tag{3.45}$$

Now, translational energy is given by

$$E_t = \frac{3}{2}RT \quad \text{(for one mole)}$$

$$E_t = \frac{3}{2}nRT \quad \text{(for } n \text{ mole)}$$

$$S_{t,m} = \frac{3}{2}nR + nR[\ln q_t - \ln N + 1] \tag{3.46}$$

As we know,

$$q_t = \frac{(2\pi mkT)^{3/2}}{h^3} V$$

Hence,

$$S_{t,m} = nR\left[\frac{3}{2} + 1 + \ln\frac{(2\pi mkT)^{3/2}}{h^3} V - \ln N\right]$$

$$S_{t,m} = nR\left[\frac{5}{2} + \ln\frac{(2\pi mkT)^{3/2}}{Nh^3} V\right] \tag{3.47}$$

For ideal gases, $PV = nRT$

or

$$V = \frac{nRT}{P}$$

$$S_{t,m} = nR\left[\frac{5}{2} + \ln\frac{(2\pi mkT)^{3/2}}{Nh^3}\frac{nRT}{NP}\right] \tag{3.48}$$

and

$$S_{t,m} = nR\left[\frac{5}{2} + \ln\frac{(2\pi mkT)^{3/2}}{Nh^3}\frac{RT}{N_A P}\right] \quad \because \frac{N_A}{n} = \bar{N}_A \text{ and } \frac{R}{N_A} = k \tag{3.49}$$

This equation is known as *Secure–Tetrode* equation. It can be written in some other form also, as shown below.

For one mole above equation becomes

$$S_{t,m} = R\left[\frac{5}{2} + \ln\frac{(2\pi mkT)^{3/2}}{h^3}\frac{kT}{P}\right]$$

$$S_{t,m} = R\ln\left[\frac{(2\pi mkT)^{3/2}}{h^3}\frac{kT}{P}\right] + \frac{5}{2}R$$

$$S_{t,m} = R\ln\left[\frac{(2\pi mN_A kT)^{3/2}}{N_A h^3}\frac{kT}{P}\right] + \frac{5}{2}R$$

$$S_{t,m} = R\ln\left[\frac{(2\pi mkT)^{3/2}}{N_A h^3}\frac{kT}{P}\right] + \frac{5}{2}R \tag{3.50}$$

$$S_{t,m} = R\left[\frac{3}{2}\ln m + \frac{5}{2}\ln T - \ln P + \ln\frac{(2\pi)^{3/2}}{N_A h^3} + \frac{5}{2}\right] \tag{3.51}$$

Inserting the values of constants and converting the units so that pressure is in atmosphere, we get

$$S_{t,m} = R\left[\frac{3}{2}\ln m + \frac{5}{2}\ln T - \ln P - 0.5053\right] \qquad (3.52)$$

### 3.4.3 Translational $C_V$

$$C_V = \left(\frac{\partial E_t}{\partial T}\right)_V \Rightarrow \left(\frac{\partial 3/2 RT}{\partial T}\right)_V$$

$$C_V = \frac{3}{2}R \qquad (3.53)$$

### 3.4.4 Translational Enthalpy

$$H_T = E_T + PV$$
$$H_T = E_T + RT$$
$$H_T = \frac{3}{2}RT + RT \qquad (3.54)$$
$$H_T = \frac{5}{2}RT$$

### 3.4.5 Translational $C_P$

$$C_P = \left(\frac{\partial H_t}{\partial T}\right)_P \Rightarrow \left(\frac{\partial 5/2 RT}{\partial T}\right)_P = \frac{5}{2}R \qquad (3.55)$$

## 3.5 ROTATIONAL PARTITION FUNCTION

The rotational energy for a single molecule is then given by equation,

$$E_r = J(J+1)\frac{h^2}{8\pi^2 I} \qquad (3.56)$$

where $I$ is the moment of inertia of the diatomic molecule. The rotational quantum number $J$ can have zero or integral values. The expression for the energy, may be put in the form,

$$E_r = J(J+1)B\,hc$$

where $B$ is relational constant and is equal to $h/8\pi^2 I c$ and $c$ is velocity of light. For every value of $J$ these are $2J+1$ eigenstates corresponding to approximately the same magnitude of the rotational energy, each rotational level thus has a degeneracy of $2J+1$. The expression for rotational partition function of a single molecule is consequently,

## Statistical Thermodynamics

$$q_r = \sum g_r e^{-\varepsilon_r/kT}$$

$$= \sum_{J=0}^{\infty} (2J+1) e^{-J(J+1)h^2/8\pi^2 IkT}$$

$$= \sum_{J=0}^{\infty} (2J+1) e^{-J(J+1)Bhc/kT} \tag{3.57}$$

Ignoring for the present, the effect of the nuclear spin if the symbol $\rho$ is defined in the following manner,

$$\rho = \frac{Bhc}{kT} = \frac{h^2}{8\pi^2 I kT} \tag{3.58}$$

It is seen that the rotational partition function may be expressed as,

$$q_r = \sum_{J=0}^{\infty} (2J+1) e^{-J(J+1)\rho} \tag{3.59}$$

The result of this summation can be obtained with a far degree of accuracy by means of the Euler–Maclaurin formula.

If $\rho$ is less than unity as will be the case at all temperatures with the possible exception of very light molecules, such as hydrogen and deuterium, then

$$q_r = \frac{1}{\rho}\left(1 + \frac{\rho}{3} + \frac{\rho^2}{15} + \frac{4\rho^3}{315} + \cdots \right) \tag{3.60}$$

If $\rho$ is quite small, e.g. less than 0.05 which will be due for nearly all substances at moderately high or high temperature, all terms beyond the first in the parenthesis may be neglected, so that rotational partition function of a rigid diatomic molecule is given by

$$q_r = \frac{1}{\rho} = \frac{8\pi^2 IkT}{h^2} \tag{3.61}$$

The Eq. (3.59), can be written as,

$$q_r = e^{\rho/4} \sum_{J=0}^{\infty} 2\left(J + \frac{1}{2}\right) e^{-(J+1/2)^2 \rho} \tag{3.62}$$

and, if $\rho$ is small, summation can be replaced by integration, so that,

$$q_r = 2e^{\rho/4} \int_{J=0}^{\infty} \left(J + \frac{1}{2}\right) e^{-(J+1/2)^2 \rho} dJ \tag{3.63}$$

Utilizing standard integrals, this becomes

$$q_r = e^{\rho/4} \frac{1}{\rho} \approx \frac{1}{\rho} \tag{3.64}$$

and if $\rho$ is small, $e^{\rho/4}$ is virtually unity, so that the result is same as given by Eq. (3.61), It is of great interest to show that this approximate partition function applicable at high temperature, which represent correspondence principle limits can be obtained by utilizing the classical definition of the partition function. The energy of a rotator with free axis, expressed in Hamiltonian form is

$$H_{(p,q)} = \frac{1}{2I}\left(p_\theta^2 + \frac{p_\phi^2}{\sin^2\theta}\right) \qquad (3.65)$$

and since the diatomic molecule has two rotational degrees of freedom, the classical partition function can be expressed as

$$q_r = \frac{1}{h^2}\iiiint_\infty e^{-(p_\theta^2 + p_\phi^2/\sin^2\theta)2IkT} dp_\theta dp_\phi\, d\theta d\phi \qquad (3.66)$$

$$q_r = \frac{1}{h^2}\int_{-\infty}^{\infty} e^{-p_\theta^2/2IkT} dp_\theta \int_{-\infty}^{\infty} e^{-p_\phi^2/2IkT} \sin^2\theta\, dp_\phi \int_0^\pi d\theta \int_0^{2\pi} d\phi \qquad (3.67)$$

The first two integrals are standard forms, and so, it is readily found that,

$$q_r = \frac{2\pi IkT}{h^2}\int_0^\pi \sin\theta\, d\theta \int_0^{2\pi} d\phi \qquad (3.68)$$

$$q_r = \frac{8\pi^2 IkT}{h^2} \qquad (3.69)$$

Above equation is in agreement with Eq. (3.61).

The contribution of rotational degree of freedom to the total energy and heat capacity of a diatomic molecule may be obtained by combining the foregoing result with Eq. (3.65). Since,

$$\frac{\partial \ln q_r}{\partial T} = \frac{1}{T}$$

it follows that $E_r = RT$ and $C_V = R$ \hfill (3.70)

This is just the contribution for two degrees of freedom to be expected from the classical principle of the equi-partition of energy.

### 3.5.1 Calculation of Rotational Energy

The rotational energy is given by

$$E_r = RT^2\left(\frac{\partial \ln q_r}{\partial T}\right)_V \qquad (3.71)$$

Now $q_r$ is given by $q_r = \dfrac{8\pi^2 IkT}{h^2}$

$$\ln q_r = \ln \frac{8\pi^2 Ik}{h^2} + \ln T$$

$$\left(\frac{\partial \ln q_r}{\partial T}\right)_V = \frac{1}{T} \tag{3.72}$$

Combining Eqs. (3.71) and (3.72), we get

$$E_r = RT \tag{3.73}$$

## 3.6 VIBRATIONAL PARTITION FUNCTION

Vibrational partition function is given by,

$$q_v = \sum g_v e^{-\varepsilon_v/kT} \tag{3.74}$$

Degeneracy of vibrational energy level is unity $g_v = 1$ \hfill (3.75)

Furthermore, energy of vibrational energy level is given by

$$E_v = \left(v + \frac{1}{2}\right)\overline{\omega} \text{ cm}^{-1}$$

$$E_v = \left(v + \frac{1}{2}\right)hc\overline{\omega} \text{ J} \tag{3.76}$$

where $\overline{\omega}$ vibrational frequency and $v$ is vibrational quantum number 0, 1, 2, 3, …

Putting $v = 0$, in Eq. (3.76) gives

$$E_v = \frac{1}{2}hC\overline{\omega} \tag{3.77}$$

Energy given by above equation is zero point energy. Thus, vibrational energy corresponding to lowest energy state is given by

$$E_v = vhc\overline{\omega} \tag{3.78}$$

By combining Eqs. (3.74), (3.75) and (3.78), we get

$$q_v = \sum_{v=0}^{v=\infty} e^{-vhc\overline{w}/kT} \tag{3.79}$$

Putting $x = \dfrac{hc\overline{\omega}}{kT}$, above equation becomes

$$q_v = \sum_{v=0}^{v=\infty} e^{-vx} \Rightarrow 1 + e^{-x} + e^{-2x} + e^{-3x} + \cdots$$

$$q_v = \frac{1}{1 - e^{-hc\overline{\omega}/kT}} \tag{3.80}$$

## 3.6.1 Vibrational Energy

The vibrational energy is given by

$$E_v = RT^2 \left( \frac{\partial \ln q_v}{\partial T} \right)_V \tag{3.81}$$

Now, also

$$q_v = \frac{1}{1 - e^{-hc\bar{\omega}/kT}}$$

$$\ln q_v = \ln \frac{1}{1 - e^{-hc\bar{\omega}/kT}}$$

Taking differentiation w.r.t. $T$, we get

$$\frac{\partial}{\partial T} \ln q_v = \frac{-1}{1 - e^{-hc\bar{\omega}/kT}} \left( -e^{-hc\bar{\omega}/kT} \right) \frac{hc\bar{\omega}}{kT^2}$$

$$\frac{\partial}{\partial T} \ln q_v = \frac{e^{-hc\bar{\omega}/kT}}{1 - e^{-hc\bar{\omega}/kT}} \frac{hc\bar{\omega}}{kT^2} \left( \frac{e^{hc\bar{\omega}/kT}}{e^{hc\bar{\omega}/kT}} \right)$$

$$\frac{\partial}{\partial T} \ln q_v = \frac{hc\bar{\omega}}{kT} \frac{1}{T} \frac{1}{(1 - e^{-hc\bar{\omega}/kT})e^{hc\bar{\omega}/kT}}$$

Putting $x = \dfrac{hc\bar{\omega}}{kT}$ in the above equation, we get

$$\frac{\partial}{\partial T} \ln q_v = \frac{x}{T} \frac{1}{(e^x - 1)}$$

Putting the value in Eq. (3.81)

$$E_v = RT \left( \frac{hc\bar{\omega}}{kT(e^{hc\bar{\omega}/kT} - 1)} \right) \tag{3.82}$$

## 3.6.2 Vibrational Entropy

Relation between vibrational entropy and vibrational partition function is

$$S_v = RT \frac{\partial}{\partial T} \ln q_v + R \ln q_v \tag{3.83}$$

But,

$$q_v = \frac{1}{1 - e^{-hc\bar{\omega}/kT}}$$

$$\ln q_v = \ln \frac{1}{1 - e^{-hc\bar{\omega}/kT}} \tag{3.84}$$

Taking differentiation w.r.t. $T$, we get

$$\frac{\partial}{\partial T} \ln q_v = \frac{hc\bar{\omega}}{kT^2} \frac{1}{(e^{hc\bar{\omega}/kT} - 1)} \tag{3.85}$$

Combining Eqs. (3.83), (3.84) and (3.85), we get

$$S_v = RT \frac{hc\bar{\omega}}{kT^2} \frac{1}{(e^{hc\bar{\omega}/kT} - 1)} + R\left[-\ln(1 - e^{-hc\bar{\omega}/kT})\right]$$

$$S_v = \frac{hc\bar{\omega}}{kT} \frac{R}{(e^{hc\bar{\omega}/kT} - 1)} - R\ln(1 - e^{-hc\bar{\omega}/kT}) \qquad (3.86)$$

## 3.7 THE ELECTRONIC PARTITION FUNCTION

For most of the molecules the excited electronic energy level lies, so far above the ground state compared with $kT$ (a typical state value being greater than $2\,\text{eV} = 3 \times 10^{-19}\,\text{J}$) that all the molecules may be considered to be in the ground state at ordinary temperature. Thus, contributions to the electronic partition function arising from excited electronic states may be neglected. The electronic partition function is given by

$$q_{el} = \sum g_{el,i} e^{-\varepsilon_i/kT}$$

$$q_{el} = g_0 e^0 + g_1 e^{-\varepsilon_1/kT} + \cdots \qquad (3.87)$$

$$q_{el} = g_0 \qquad (3.88)$$

Since the second, third and subsequent terms are considered negligible.

Thus, the electronic partition function is simply the degeneracy $g_o$ of the ground electronic state. In the Russel–Saunders coupling scheme, the degeneracy of an atomic electronic level is given by,

$$g_0 = 2J + 1 \qquad (3.89)$$

where $J$ is the total angular momentum quantum number. The degeneracy of the electronic ground state of atoms are given in Table 3.1.

**TABLE 3.1** Ground state electronic degeneracy

| Atom | H | He | Na | Ca | Cl | Pb |
|---|---|---|---|---|---|---|
| Term/symbol | $^2S_{1/2}$ | $^1S_0$ | $^2S_{1/2}$ | $^1S_0$ | $^4P_{3/2}$ | $^3P_0$ |
| $g_0 = 2J + 1$ | 2 | 1 | 2 | 1 | 4 | 1 |

The term symbol $^{2s+1}L_J$ is a short hand notation for all the angular momenta of an atom, viz. the spin angular momentum $S$, the orbital angular momentum $L$, and the total angular momentum $J$. Mathematically

$$S = \Sigma\, S_i;\ L = \Sigma L_i\ \text{and}\ J = L + S \quad \text{(for atoms with}\ Z \leq 30).$$

where $S_i$ and $L_i$ are the spins and the orbital angular momenta of individual electrons in the atom, respectively.

The ground state of electronic states of free atoms are generally degenerate—for hydrogen atom with electronic configuration $_1S^1$, spin $s = 1/2$ and $L = 0$ so that $J = L + s = 1/2$ and $g_0 = 2 \times 1/2 + 1 = 2$. For helium atom with electronic configuration $_1S^2$, spin, $s = 1/2 - 1/2 = 0$ and $L = 0$ so that $J = 0$ and hence $g_0 = 1$. Thus, for hydrogen atom, $q_{el} = 0$ and for He atom, $q_{el} = 1$.

Similarly, the ground state of chlorine (Cl) atom has a degeneracy of 4 so that $q_{el}=4$. The alkali metal atoms having one electron in their outermost orbital, like the H atom have $q_{el}=2$.

The ground electronic states of most molecules and stable ions are invariably non-degenerate, i.e. are singlets so that $g_o = 1$ and hence $q_{el} = 1$. A very important exception is the oxygen molecule, $O_2$ having a triply degenerate ground state, i.e. $g_{el} = 3$ so that $q_{el} = 3$, for NO molecule $g_o = 2$ so that $q_{el} = 2$.

For atoms like chlorine and molecules like NO, the difference between the ground state energy level and the first excited state energy level is small. Hence, the contribution of the excited state must be taken into account in calculating the electronic partition function.

**Problem 3.1** The first excited state of chlorine atom $^2P_{1/2}$ lies at 0.11 eV above the ground state, $^2P_{3/2}$. Calculate the electronic partition of Cl at 1000 K.

*Solution:* For the ground state $g_0 = 2\left(\dfrac{3}{2}\right)+1 = 4$

For the first excited state $g_i = 2\left(\dfrac{1}{2}\right)+1 = 2$

Hence at 1000 K, $q_{el} = g_0 e^{-\varepsilon_0/kT} + g_i e^{-\varepsilon_i/kT}$

$$\Rightarrow \qquad 4e^0 + 2\exp\left[\dfrac{(0.11\,\text{eV})\,(1.602\times 10^{19}\,\text{J eV}^{-1})}{1.38\times 10^{-23}\,\text{J K}^{-1}\times 1000\,\text{K}}\right]$$

$$= 4 + 2\,(0.28) = 4.56$$

Notice that the first excited state contributes about 13% of $q_{el}$.

**Problem 3.2** The energies of the first three energy levels of Flourine atom, determined from spectroscopy, are as follows (Table 3.2):

| Energy level | Energy (cm$^{-1}$) |
|---|---|
| $^2P_{3/2}$ | 0.0 |
| $^2P_{1/2}$ | 404.0 |
| $^2D_{5/2}$ | 102,406.5 |

Calculate (i) the electronic partition function and (ii) the fractions of flourine atoms in the three energy level at 1000 K.

*Solution:*

(i) According to Eq. (3.89), the degeneracy of the three levels are,

$$g_0 = 2\left(\dfrac{3}{2}\right)+1 = 4$$

$$g_1 = 2\left(\dfrac{1}{2}\right)+1 = 2$$

$$g_2 = 2\left(\frac{5}{2}\right) + 1 = 6$$

From Eq. (1), $q_{el} = g_0 e^{-\varepsilon_0/kT} + g_1 e^{-\varepsilon_1/kT} + g_2 e^{-\varepsilon_2/kT}$

We must first convert energies in cm$^{-1}$ to energies in Joules by multiplying by $hc$. Thus, substituting for $k$, $h$, $c$, and $T$ and after simplifying, we get

$$q_{el} = 4e^0 + 2e^{-0.58123} + 6e^{-147.4}$$

$$q_{el} = 5.118$$

(ii) From Maxwell–Boltzmann distribution law

$$n_i = \frac{N g_i e^{-\varepsilon_i/kT}}{q}$$

Hence for ground state ($i = 0$)

$$\frac{n_0}{N} = \frac{g_0 e^{-\varepsilon_0/kT}}{q_{el}} = \frac{4 \times 1}{5.118} = 0.72$$

Similarly, for the first excited state ($i = 1$)

$$\frac{n_1}{N} = \frac{g_1 e^{-\varepsilon_1/kT}}{q_{el}} = \frac{2 \times e^{-0.5813}}{5.118} = 2.218$$

and for the second excited state ($i = 2$)

$$\frac{n_2}{N} = \frac{g_2 e^{-\varepsilon_2/kT}}{q_{el}} = \frac{6 \times e^{-147.4}}{5.118} = 0$$

Notice that at all the given temperatures, the fractions of molecules of the successive energy levels go on decreasing. The second excited state is empty even at 1000 K.

## 3.8 EFFECT OF CHANGE OF ZERO POINT ENERGY ON PARTITION FUNCTION

Since, the partition function involves the summation of a number of exponential terms containing the energy of all possible levels in the molecule, it is necessary to consider, and it is general practice to take the zero point level of each molecule, i.e. the level for which the translational, vibrational and rotational quantum numbers are all zero, as to energy zero. It is of great interest, however, to see what would be the effect of changing the energy zero to any other arbitrary level. Let $\varepsilon_i$ be the energy of $i$th level, the energy being reckoned from the zero point level of the molecule the partition function is then given in the usual manner by

$$Q = \sum g_i e^{-\varepsilon_i/kT} \tag{3.90}$$

Suppose now that the energy zero is changed by an amount $\varepsilon_0$ the energy value of each value is then changed by an equal amount, so that the new partition function designated $Q_{E\varepsilon}$ will be given by

$$Q_{E\varepsilon} = \sum g_i e^{-(e_0+\varepsilon_i)/kT}$$

$$Q_{E\varepsilon} = e^{-e_0/kT} \sum g_i e^{-\varepsilon_i/kT}$$

$$Q_{E\varepsilon} = e^{-e_0/kT} Q \qquad (3.91)$$

The new partition function is thus equal to the original value multiplied by $e^{-e_0/kT}$ where $\varepsilon_0$ is the change in the energy zero per molecule. If $E_0$ is the energy change per mole, which is equal to $N\varepsilon_0$, where $N$ is the Avogadro number, then Eq. (3.91), may be written as

$$Q_{E_0} = e^{-E_0/RT} Q \qquad (3.92)$$

Upon taking logarithms of Eq. (3.92), it is seen that,

$$\ln Q_{E_0} = -\frac{E_0}{RT} + \ln Q \qquad (3.93)$$

and, hence

$$\frac{\partial \ln Q_{E_0}}{\partial T} = \frac{\partial \ln Q}{\partial T} + \frac{E_0}{RT^2} \qquad (3.94)$$

From these equations and bearing in mind the expression for the various energy functions $E$, $H$, $G$ and $A$, it can readily be seen that the values of these functions based on the new energy zero, will be $E_0$ per mole greater than those based on the original zero level. This is of course, exactly what is to be expected. On the other hand, the entropy and heat capacity are found to be independent of the arbitrarily chosen energy zero.

## 3.9 EFFECT OF CHANGE OF ZERO POINT ENERGY ON ENERGY

As

$$E = RT^2 \frac{\partial}{\partial T} \ln q \qquad (3.95)$$

After making change in zero point energy, the expression for energy is

$$E' = RT^2 \frac{\partial}{\partial T} \ln q_{\varepsilon_0} \qquad (3.96)$$

Now

$$\frac{\partial}{\partial T} \ln q_{\varepsilon_0} = \frac{\partial}{\partial T} \ln q + \frac{E_0}{RT^2}$$

Putting this value in Eq. (3.96), we get

$$E' = RT^2 \left( \frac{\partial}{\partial T} \ln q + \frac{E_0}{RT^2} \right)$$

$$E' = RT^2 \frac{\partial}{\partial T} \ln q + \frac{E_0}{RT^2} RT^2$$

$$E' = E + E_0$$

Hence, new energy increases by $E_0$ from old energy.

## 3.10 EFFECT OF CHANGE OF ZERO POINT ENERGY ON ENTROPY

Expression for entropy is given by

$$S = k \ln \frac{q^N}{N!} + RT \left( \frac{\partial}{\partial T} \ln q \right)_V \qquad (3.97)$$

After changing zero point energy, expression for entropy becomes

$$S' = k \ln \frac{q_\varepsilon^N}{N!} + RT \left( \frac{\partial}{\partial T} \ln q_E \right)_V \qquad (3.98)$$

$$S' = k(\ln q_\varepsilon^N - \ln N!) + RT \left( \frac{\partial}{\partial T} \ln q_E \right)_V$$

$$S' = k \ln q_\varepsilon^N - k \ln N! + RT \left( \frac{\partial}{\partial T} \ln q_E \right)_V$$

$$S' = Nk \ln q_\varepsilon - k \ln N! + RT \left( \frac{\partial}{\partial T} \ln q_E \right)_V$$

For one mole $N$ = Avogadro's number, above equation becomes

$$S' = R \ln q_\varepsilon - k \ln N! + RT \left( \frac{\partial}{\partial T} \ln q_E \right)_V \qquad (3.99)$$

Now also,

$$\ln q_E = \ln q - \frac{E_0}{RT} \quad \text{and} \quad \frac{\partial}{\partial T} \ln q_E = \frac{\partial}{\partial T} \ln q + \frac{E_0}{RT^2}$$

Putting these values in Eq. (3.99), we get

$$S' = R \left( \ln q - \frac{E_0}{RT} \right) - k \ln N! + RT \left( \frac{\partial}{\partial T} \ln q + \frac{E_0}{RT^2} \right)_V$$

$$S' = R \ln q - \frac{E_0}{T} - k \ln N! + RT \left( \frac{\partial}{\partial T} \ln q \right)_V + \frac{E_0}{T}$$

$$S' = R \ln q - k \ln N! + RT \left( \frac{\partial}{\partial T} \ln q \right)_V$$

$$S' = kN \ln q - k \ln N! + RT \left( \frac{\partial}{\partial T} \ln q \right)_V$$

$$S' = k \ln q^N - k \ln N! + RT \left( \frac{\partial}{\partial T} \ln q \right)_V$$

$$S' = k \ln \frac{q^N}{N!} + RT \left( \frac{\partial}{\partial T} \ln q \right)_V \tag{3.100}$$

On comparing Eqs. (3.97) and (3.100), we get

$$S = S'$$

Hence, there is no change in entropy on changing zero point energy.

## 3.11 EFFECT OF CHANGE IN ZERO POINT ENERGY ON GIBBS FREE ENERGY

Equation for Gibbs free energy is given as

$$G = -RT \ln \frac{q}{N} \tag{3.101}$$

After changing zero point energy, expression for free energy

$$G' = -RT \left( \ln \frac{q_\varepsilon}{N} \right) \tag{3.102}$$

Above equation can be written as

$$G' = -RT \ln q_e + RT \ln N \tag{3.103}$$

Also, $\ln q_\varepsilon = \ln q - \dfrac{E_0}{RT}$

Putting this value in Eq. (3.103), we get

$$G' = -RT \left( \ln q - \frac{E_0}{RT} \right) + RT \ln N$$

$$G' = -RT \ln q - E_0 + RT \ln N$$

$$G' = -RT (\ln q - \ln N) + E_0$$

$$G' = -RT \ln \frac{q}{N} + E_0$$

$$G' = G + E_0$$

## 3.12 EFFECT OF ZERO POINT ENERGY ON WORK FUNCTION

Work function is given by

$$A = -kT \ln \frac{q^N}{N!} \tag{3.104}$$

After changing zero point energy, work function is given by

$$A' = -kT \ln \frac{q_\varepsilon^N}{N!} \tag{3.105}$$

$$A' = -kT \ln q_\varepsilon^N + kT \ln N!$$
$$A' = -NkT \ln q_\varepsilon + kT \ln N!$$

For one mole, above equation becomes

$$A' = -RT \ln q_\varepsilon + kT \ln N!$$

But

$$\ln q_E = \ln q - \frac{E_0}{RT}$$

$$A' = -RT\left(\ln q - \frac{E_0}{RT}\right) + kT \ln N!$$
$$A' = -RT \ln q + E_0 + kT \ln N!$$
$$A' = -NkT \ln q + kT \ln N! + E_0$$
$$A' = -kT \ln q^N + kT \ln N! + E_0$$
$$A' = -kT \ln \frac{q^N}{N!} + E_0$$
$$A' = A + E$$

## 3.13 SEPARATION OF INTERNAL PARTITION FUNCTION

Because of the relationship between nuclear spin and rotational level, it is convenient to include the nuclear spin factor with the rotational contribution to the partition function. For the present, therefore, the internal partition function of a stable diatomic molecule, may be regarded as made up solely of the contribution from rotational and vibrational degree of freedom. If the rotational and vibrational energies of a molecules are quite independent, then the energies $\varepsilon_i$ due to internal degrees of freedom would be equal to the sum $\varepsilon_r$, the rotational energy and $\varepsilon_v$, the vibrational energies. It can be seen therefore that internal partition function of a stable diatomic molecule can be separated into two parts, one $q_r$ for rotational and the other $q_v$ for vibrational, degrees of freedom; thus

$$q_i = q_r \, q_v$$

where $q_r = \sum g_r e^{-\varepsilon_r/kT}$ and $q_v = \sum g_v e^{-\varepsilon_v/kT}$

The determination of the partition function for internal degrees of freedom resolves itself into the evaluation of the separate partition function for rotation and vibration. It will be observed that the expressions for all the thermodynamic quantities, viz. entropy, total energy, heat content, heat capacity, free energy and the maximum work function involve either the logarithm of the partition function or a derivative of this logarithm. Since, the total partition function may be resolved into the product of the partition function for various degree of freedom, assuming that the respective energies are independent of each other, it is obvious, that the thermodynamic quantity may be represented as the sum of contribution of the separate degree of freedom. As, it is sometimes convenient to consider the effect of rotational and vibrational energies individually. It is interesting to see how the separation of the thermodynamic quantities can be made.

The expressions for the various parts of the total energy and heat capacity present no difficulty, taking the complete partition function $q_i$ for a single molecule as the product of translational ($q_t$), rotational ($q_r$) and vibrational ($q_v$) contributions, so that

$$q = q_t q_r q_v$$

and
$$\ln q = \ln q_t + \ln q_r + \ln q_v$$

It follows that, general Eq. (3.96) for the total energy per mole can be written as

$$E = RT^2 \left(\frac{\partial \ln q_t}{\partial T}\right)_V + RT^2 \left(\frac{\partial \ln q_r}{\partial T}\right)_V + RT^2 \left(\frac{\partial \ln q_v}{\partial T}\right)_V$$

The first term on the right-hand side is, of course, the translational energy and the second and third term, represents the contribution to the total energy of the one mole of an ideal gas made by the rotational and vibrational degrees of freedom, respectively. Since, the partition functions for rotation and vibration are independent of the volume of the vessel occupied by the gas, the subscript $v$ may be omitted; it is thus possible to write

$$E_r = RT^2 \left(\frac{\partial \ln q_r}{\partial T}\right)_V \quad \text{and} \quad C_r = \frac{\partial}{\partial T}\left(RT^2 \frac{\partial \ln q_r}{\partial T}\right)$$

$$E_v = RT^2 \left(\frac{\partial \ln q_v}{\partial T}\right)_V \quad \text{and} \quad C_V = \frac{\partial}{\partial T}\left(RT^2 \frac{\partial \ln q_v}{\partial T}\right)$$

where $C_r$ and $C_V$ are the rotational and vibrational contributions to the molar heat capacity, respectively. Since, the respective partition function are independent of pressure or volume, these degrees of freedom make the same contribution to the heat capacities at both constant pressure and volume.

## 3.14 POLYATOMIC MOLECULES

A molecule containing $n$ atoms has $3n$ degrees of freedom, i.e. $3n$ coordinates. Using the term, in a general sense, are required to specify the complete configuration of the molecule. Of these coordinates, three are so called external coordinates, which give the position of the centre of gravity of the molecule in space. In general, except for diatomic and linear molecules, three coordinates, to which reference will be made shortly are required to indicate the orientation of the molecule, this is equivalent to the statement that molecules, with the exceptions noted have three rotational degrees of freedom. There are consequently $3n - 6$ coordinates still remaining to be specified to give the relative positions of the atoms and the centres of gravity of internally rotating groups. Assuming, for the present that there are no groups in the molecule that exhibit internal rotation, the $3n - 6$ degrees of freedom represent vibrational modes. The orientation of a linear molecule in space can be represented by two coordinates, so that molecules of this type have two degrees of rotational freedom, they consequently possess $3n - 5$ vibrational modes. If groups are present in the molecule that are capable of undergoing free internal rotation, the number of degrees of vibrational freedom is reduced by the number of such rotations. In many cases, the molecules possess elements of symmetry. In the evaluation of the partition function of a polyatomic molecule, the foregoing facts must be taken into consideration.

After separating the translational degrees of freedom from those due to internal coordinates, in the usual manner, an exact evaluation of the partition functions of certain, relatively simple, polyatomic molecules have been achieved by the summation method. In most cases, however, the molecule has been treated as rigid, and the rotational and vibrational energies have been taken as independent. The combined rotational, vibrational partition function is then obtained as the product of the separate contributions of the rotational and vibrational degrees of freedom. The classification of polyatomic molecules according to their electronic configuration for diatomic molecules is possible for a limited number of substances only, in other cases, it is assumed that the lowest electronic level of the molecule is a singlet state, and that excited states makes no contribution to the total partition function, unless there is evidence to the contrary. Therefore, the electronic factor is unity. Exceptions to this rule are molecules, such as nitrogen dioxide, containing an odd number of electrons.

## 3.15 ROTATIONAL PARTITION FUNCTION FOR POLYATOMIC MOLECULES

In the derivation of the rotational contribution to the partition function of a polyatomic molecule, it is therefore, adequate to use the classical method of calculating the Eulerian angles (to define the orientation in space of the molecule, assumed to be a rigid body) while three Cartesian coordinates are used to determine the position of its centre of gravity. Let $x$, $y$ and $z$ be the Cartesian axes fixed in space and $X$, $Y$ and $Z$, the three principle axes, at right angle, within the molecule itself, both sets of axes are referred to the same origin $O$ (Figure 3.1).

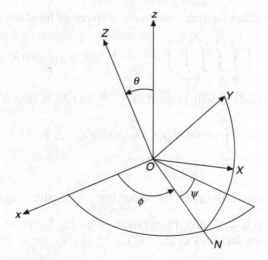

**FIGURE 3.1** Cartesian axes and principle axes.

The planes containing the axes $xy$ and $XY$ meet at the line $ON$ which is called the *nodel line*. The Eulerian angles $\theta$, $\phi$ and $\psi$ are then defined in the following manner. The angle $\theta$ is that between the body axis $Z$ and the space axis $z$, the angle between the model line and the $x$-axis is $\phi$. The third angle $\psi$ is the one between the nodal line and the $X$-axis in the molecule. These three angles define the orientation of the molecule with fixed center of gravity. The angle $\theta$ may vary between zero and $\pi$ while the angle $\phi$ and $\psi$ can vary from zero to $2\pi$. The rotational energy of the molecule is expressed in Hamiltonian form, that is in terms of the Eulerian angles $\theta$, $\phi$, $\psi$ and their conjugate momenta $P_\theta$, $P_\phi$, $P_\psi$ by the equation

$$H_{(p,q)} = \frac{\sin^2\psi}{2A}\left\{P_\theta - \frac{\cos\psi}{\sin\theta\sin\psi}(P_\phi - P_\psi\cos\theta)\right\}^2$$
$$+ \frac{\cos^2\psi}{2B}\left\{P_\theta + \frac{\sin\psi}{\sin\theta\cos\psi}(P_\phi - P_\psi\cos\theta)\right\}^2 + \frac{1}{2C}P_\psi^2 \qquad (3.106)$$

where $A$, $B$ and $C$ are the three principal moments of inertia of the molecule. For ease of integration, this expression is put into the alternative form

$$H_{(p,q)} = \frac{1}{2}F\left\{P_\theta - \frac{\sin\psi\cos\psi}{F\sin\theta}\left(\frac{1}{B} - \frac{1}{A}\right)(P_\phi - P_\psi\cos\theta)\right\}^2$$
$$+ \frac{1}{2AB\sin^2\theta}\frac{1}{F}(P_\phi - P_\psi\cos\theta)^2 + \frac{1}{2C}P_\psi^2 \qquad (3.107)$$

where $F$ is a function of $\psi$ defined by

$$F = \frac{\sin^2\psi}{A} + \frac{\cos^2\psi}{B}$$

The classical partition function for three rotational degrees of freedom is

$$Q_r = \frac{1}{h^3}\int_0^\pi\int_0^{2\pi}\int_0^{2\pi}\int_{-\infty}^{+\infty}\int_{-\infty}^{+\infty}\int_{-\infty}^{+\infty} e^{-H_{(p,q)}/kT}\, d\theta\, d\phi\, d\psi\, dP_\theta\, dP_\phi\, dP_\psi \qquad (3.108)$$

The integration is first carried out with respect to $P_\theta$, $P_\phi$ and $P_\psi$ in turn making use of the fact that

$$\int_{-\infty}^{+\infty} e^{-a(x+b)^2}\, dx = \int_{-\infty}^{+\infty} e^{-ax^2}\, dx = \left(\frac{\pi}{a}\right)^{1/2}$$

Integration first over $P_\theta$, there is obtained the factor $\left(\dfrac{2\pi kT}{F}\right)^{1/2}$ the integration w.r.t. $P_\phi$ then gives $F^{1/2}(2\pi kTAB)^{1/2}\sin\theta$, while finally integration over $P_\psi$ leads to the factor $(2\pi kTC)^{1/2}$. Multiplication of these three factors yield $(2\pi KTC)^{3/2}(ABC)^{1/2}\sin\theta$, so that

$$Q_r = \frac{(2\pi kT)^{3/2}(ABC)^{1/2}}{h^3}\int_0^\pi \sin\theta\, d\theta \int_0^{2\pi}\sin\theta\, d\theta \int_0^{2\pi} d\phi \int_0^{2\pi} d\psi$$

$$Q_r = \frac{8\pi^2(8\pi^3 ABC)^{1/2}(kT)^{3/2}}{h^3} \qquad (3.109)$$

The calculation, so far, has not taken into account molecular symmetry or the nuclear spin statistical weight. The allowance must be course, always be made for the former, while the latter may be excluded, if desired. The complete rotational partition function, including the nuclear spin factor is thus

$$Q_r = g_n \frac{8\pi^2 (8\pi^3 ABC)^{1/2} (kT)^{3/2}}{\sigma h^3} \qquad (3.110)$$

in which, $\sigma$ is the symmetry number and $g_n$ is equal to the product of the $2_i+1$ terms, where $i$ is the nuclear spin for every atom in the molecule. The symmetry numbers of some polyatomic molecules are as follows: $H_2O$ and $SO_2$ (2), $NH_3$ (3), $C_2H_4$ (4), $CH_4$ (12) and $C_6H_6$ (12).

If the moment of inertia, $A$, $B$ and $C$ are known from spectroscopic data. The rotational partition function can be calculated directly from Eq. (3.110). Alternatively, if the molecular dimensions are known, e.g. from electron diffraction measurements, the product $ABC$ may be derived from the determinant.

$$ABC = \begin{vmatrix} I_{xx} & -I_{xy} & -I_{xz} \\ -I_{xy} & I_{yy} & -I_{yz} \\ -I_{xz} & -I_{yz} & I_{zz} \end{vmatrix} \qquad (3.111)$$

where the moments of inertia $I_{xx}$, $I_{yy}$ and $I_{zz}$ are defined by

$$I_{xx} = \sum_i m_i (y_i^2 + z_i^2), \text{ etc.}$$

and the products of inertia $I_{zy}$, $I_{xz}$ and $I_{yz}$ are given by

$$I_{xy} = \sum_i m_i x_i y_i \text{ etc.}$$

where $m_i$ is the mass of the $i$th atom, and $x_i y_i z_i$ are its coordinates in any system of Cartesian coordinates, whose origin coincides with the centre of gravity of the molecule. The summations are in each case carried over the coordinates of all the atoms.

The contribution $E_r$ to the total energy of a polyatomic molecule made by the rotational degrees of freedom may be derived from the partition function, given by Eq. (3.110) by making use of equation $E_r = RT^2 \frac{d \ln Q_r}{dT}$. It is readily seen that for 1 mol

$$E_r = RT^2 \frac{d \ln Q_r}{dT} = \frac{3}{2} RT \qquad (3.112)$$

which is in agreement, as it should be, with the classical value derived from the equi-partition principle for a rotator with three degrees of freedom, i.e. with energy in three square terms. The contribution of the rotator to the molar heat capacity of the polyatomic molecule is obviously $3/2\ R$. The general expression Eq. (3.110) for the rotational partition function of any polyatomic molecule can be put in a simpler form. If $\rho_A$, $\rho_B$ and $\rho_C$ are defined in a manner similar to that employed previously for $\rho$ i.e.

$$\rho = \frac{Bhc}{kT} = \frac{h^2}{8\pi^2 IkT}$$

Therefore, $\rho_A = \dfrac{h^2}{8\pi^2 AkT}$, $\rho_B = \dfrac{h^2}{8\pi^2 BkT}$ and $\rho_C = \dfrac{h^2}{8\pi^2 CkT}$

It is seen that

$$q_r = \dfrac{g_n}{\sigma} \dfrac{\pi^{1/2}}{\rho_A \rho_B \rho_C} \qquad (3.113)$$

This is the equation applicable to all polyatomic molecules and particularly to asymmetrical tops, for which $A$, $B$ and $C$ are different, for planer molecules of this type, e.g. water and benzene

$$A + B = C$$

For a symmetrical top molecules, two of the three moments of inertia, e.g. $A$ and $B$ are equal while the third, i.e. $C$ is different, equation (3.113) then becomes

$$q_r = \dfrac{g_n}{\sigma} \dfrac{\pi^{1/2}}{\rho_A (\rho_C)^{1/2}} \qquad (3.114)$$

If the three principal moments of inertia $A$, $B$ and $C$ are all equal for a spherical molecule, the partition function for rotation is

$$q_r = \dfrac{g_n}{\sigma} \dfrac{\pi^{1/2}}{\rho^{3/2}} \qquad (3.115)$$

As far rotation is concerned a linear molecule behaves like a diatomic molecule that is as a rigid rotator with two identical moments of inertia. The rotational partition function is thus, represented by the expression

$$q_r = g_n \dfrac{8\pi^2 IkT}{\sigma h^2} \qquad (3.116)$$

The molecules HCN, $N_2O$, $CO_2$, COS, $CS_2$ and $C_2H_2$ are all linear and to the above Eq. (3.116) is applicable.

## 3.16 FREE ENERGY AND EQUILIBRIUM IN GASEOUS SYSTEMS

The free energy of a mole of an ideal gas can be related to the partition function by means of equation $F = -RT \ln Q/N$; since, the methods of deriving partition functions for various types of molecules are available, at least in principle, it should be possible to calculate the free energies of the substance taking part in a reaction. The free energy $\Delta F^0$, for reactants and products in their standard states, is related to the equilibrium constant $K$ of the reaction by the equation

$$-\Delta F^0 = RT \ln K \qquad (3.117)$$

Hence, there should be a close connection between the partition function of the substance involved in the reaction and the equilibrium constant of the system.

It was seen that, the choice of the energy zero for the evaluation of the partition function at a given molecule has no effect on the heat capacity and entropy but it introduces an additive term in the expressions for the total energy and free energy. If the energy difference between the arbitrary zero, from which all energies are to be reckoned and the zero point level of a particular molecular species is represented by $E_0$ per mole (Figure 3.2), then

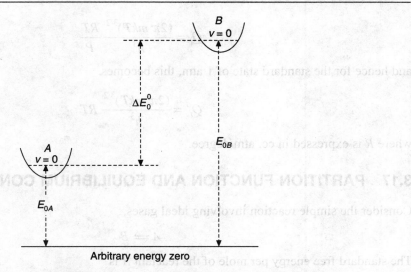

**FIGURE 3.2** Energy difference between arbitrary zero and zero point level.

according to equation

$$Q_{E_0} = e^{-E_0/RT} Q \tag{3.118}$$

where $Q_{E_0}$ is the partition function based on the new energy zero, while $Q$ is the value used in some other levels.

If the new expression for the partition function is inserted into equation $F = -RT \ln Q/N$, the free energy per mole, based on the new energy zero, becomes

$$F = -RT \ln \frac{Q_{E_0}}{N} \tag{3.119}$$

and consequently by Eq. (3.118),

$$F = -RT \ln \frac{Q}{N} + E_0 \tag{3.120}$$

For the evaluation of equilibrium constants, it is necessary to consider the free energy changes when the substances are in their standard states; using the superscript zero, in the usual manner to represent the standard state, it follows that

$$F^o = -RT \ln \frac{Q^0}{N} + E_0^0 \tag{3.121}$$

For gases, and it is only to ideal gas systems that the partition function concept as developed in this chapter is applicable, the standard state is chosen as the ideal gas at 1 atm pressure. This choice does not affect any of the contributions to the partition function associated with internal degree of freedom, since these are all independent of the volume or pressure of the system. The only contribution that has to be considered, is that, due to the translational degree of freedom, this may be written in the form of equation

$$Q_t = \frac{(2\pi\, mkT)^{3/2}}{h^3} \frac{RT}{P}$$

and hence for the standard state of 1 atm, this becomes

$$Q_t^o = \frac{(2\pi\, mkT)^{3/2}}{h^3} RT \qquad (3.122)$$

where $R$ is expressed in cc. atm/degree.

## 3.17 PARTITION FUNCTION AND EQUILIBRIUM CONSTANT

Consider the simple reaction involving ideal gases,

$$A \rightleftharpoons B$$

The standard free energy per mole of the reactant $A$ is

$$F_A^0 = -RT \ln \frac{Q_A^0}{N} + E_{0A}^0$$

While for a mole of product $B$,

$$F_B^0 = -RT \ln \frac{Q_B^0}{N} + E_{0B}^0$$

The standard free energy decrease $-\Delta F^0$ for the reaction is then,

$$-\Delta F^0 = F_A^0 - F_B^0 = -RT \ln \frac{(Q^0/N)_A}{(Q^0/N)_B} - \Delta E_0^0 \qquad (3.123)$$

where $\Delta E_0^0$ is given by

$$\Delta E_0^0 = E_{0B}^0 - E_{0A}^0 \qquad (3.124)$$

As seen in Figure (3.2), $\Delta E_0^0$ is the difference in energy between the zero point levels of the product $A$ and the reactant $B$, on the other hand, $\Delta E_0^0$ is the energy change in the reaction calculated for the absolute zero of temperature, i.e. when all the molecules are in their lowest (zero point) vibrational and rotational states. Making use of Eq. (3.117) in the form

$$-\Delta F^0 = RT \ln K_P$$

Since, the standard state 1 atm, it follow from Eq. (3.124), that

$$RT \ln K_P = RT \ln \frac{(Q^0/N)_B}{(Q^0/N)_A} - \Delta E_0^0$$

$$K_P = \frac{(Q^0/N)_B}{(Q^0/N)_A} e^{-\frac{\Delta E_0^0}{RT}} \qquad (3.125)$$

where the partition function, $Q^0$ refer to 1 atm pressure. It is obvious that equation (3.125) is of fundamental importance, for it gives a direct connection between the equilibrium constant of an ideal gas reaction and the partition function of the reacting substance and products. This expression can be readily extended to cover the case of a reaction of greater complexity, thus for the general reaction

$$aA + bB + \cdots \rightleftharpoons lL + mM + \cdots$$

$$k_p = \frac{(Q^0/N)_L^l \, (Q^0/N)_M^m \cdots}{(Q^0/N)_A^a \, (Q^0/N)_B^b \cdots} e^{-\frac{\Delta E_0^0}{RT}} \quad (3.126)$$

As before, the partition function refers to the standard state of 1 atm pressure and $\Delta E_0^0$ is the standard change in the total energy of the reaction at 0 K.

## EXERCISES

1. Maximizing the thermodynamic probability of a macrostate and invoking Lagrange's undetermined multipliers derive the expression for Maxwell–Boltzmann statistics.
2. Derive an expression for molecular translational partition function of an ideal gas?
3. Derive an expression for molecular rotational partition function of an ideal diatomic gas?
4. Derive an expression for molecular vibrational partition function of an ideal diatomic gas?
5. Calculate the translational partition function for hydrogen atom at 1000 K and 1 atm.
   [**Ans.** $4.94 \times 19^{29}$]
6. Calculate the translational partition function for hydrogen molecule at 1000 K and 1 atm pressure. [**Ans.** $1.396 \times 19^{30}$]
7. Derive relation between partition function and energy.
8. Derive relation between partition function and entropy.
9. Derive relation between partition function and equilibrium constant.
10. Derive Sackure–Tetrode equation.
11. Define electronic partition function.
12. Explain the effect of change of zero point energy on partition function.
13. Derive expression for rotational partition function for polyatomic molecules.
14. Derive relationship between free energy and equilibrium constant in gaseous systems.

## SUGGESTED READINGS

Aston, J.G. and J.J. Fritz, *Thermodynamics and Statistical Thermodynamics*, John Wiley, New York, 1959.

Atkins, P. and J. De Paula, *Atkins Physical Chemistry*, Oxford University Press, New York, 2004.

Berry, S.R., S.A. Rice, and J. Ross, *Matter in Equilibrium: Statistical Mechanics Thermodynamics*, 2nd ed., Oxford University Press, 2001.

Carter, A.H., *Classical and Statistical Thermodynamics*, Prentice Hall, 2001.

Chang, Y., *Statistical Thermodynamics*, World Scientific, 2006.

Couture, L. and R. Zitoum, *Statistical Thermodynamics and Properties of Matter*, Gordon and Breach, 2001.

Dole, M., *Introduction to Statistical Thermodynamics*, Prentice-Hall, New York, 1954.

Flower, R.H., *Statistical Thermodynamics*, Cambridge University Press, 1985.

Garvod, C., *Statistical Mechanics and Thermodynamics*, Oxford University Press, 1995.

Gasser, R.P.H. and W.H. Richards, *Statistical Thermodynamics*, World Scientific, 1995.

Glasstone, S., *Theoretical Chemistry*, Affiliated East-West Press, 1965.

Hill, T.L., *An Introduction to Statistical Thermodynamics*, Dover Publication, 1960.

Jackson, E.A., *Equilibrium Statistical Thermodynamics*, Printice Hall, 1968.

Le Bellae, M., F. Mortessange, and G.G. Batrouni, *Equilibrium and Nonequilibrium Statistical Thermodynamics*, Cambridge University Press, 2004.

Maczek, A., *Statistical Thermodynamics*, Oxford University Press, 1998.

Robertson, H.S., *Statistical Thermodynamics*, Prentice Hall, 1993.

Roy, B.N., *Fundamentals of Classical and Statistical Thermodynamics*, John Wiley, 1982.

Schrödinger, E., *Statistical Thermodynamics*, Dover Publications, 1989.

Sears, F.W. and G.L. Salinger, *Thermodynamics, Kinetic Theory and Statistical Thermodynamics*, Addison-Wesley, 1975.

Stowe, K.S., *Introduction to Statistical Mechanics and Thermodynamics*, John Wiley, 1984.

Tien, C., *Statistical Thermodynamics*, Hemisphere Publications, 1985.

# Statistical Mechanics

Quantum mechanics deals with the individual atoms and molecules of a system by assigning definite values of the wave functions. On the other hand, thermodynamics deals with the properties like internal energy, enthalpy, entropy, free energy, etc. of the matter in the bulk and not with the individual particles. It was therefore, realized that by suitable averaging method, it should be possible to calculate the thermodynamic properties of the bulk matter from those of its individual atoms and molecules.

The discipline which deals with the computation of the macroscopic properties of matter from the data on the microscopic properties of individual atoms (or molecules) is called *statistical mechanics* or *statistical thermodynamics*.

As in quantum mechanics, wave function contains all the information about a particle, similarly in statistical mechanics a thermodynamic function, called *partition function*, which contains all the thermodynamic information about the system. A macroscopic system is made up of the microscopic constituents, it should be possible to consider the properties of the individual atoms or molecules and then find their sum for the whole system and calculate the average value. Thus, it should be possible to calculate theoretically the thermodynamic properties of the system and these calculated values could be compared with the experimental determined values. The method of averaging the behaviour of a large number of individuals is called *statistical method*. For example, kinetic energy of a gas at a particular temperature is a statistical average of the kinetic energies of all the individual molecules.

## 4.1 ENSEMBLES (COLLECTION)

A number of things observed at a whole. The calculation of thermodynamic properties ($E$, $S$, $H$, $G$, etc.) is based upon the states of the individual molecules, which however, are changing with time (because of intermolecular forces) and the calculation of average over time for such a large number of molecules is impossible. To overcome this problem, instead of considering a single system, a large number of systems are considered which are exactly identical with the system under consideration in a number of aspects such as total volume, total number of molecules, total energy, etc.

A collection of such a large number of systems, which are identical with the system under consideration in a number of aspects such as total volume, total number of molecules, etc. is called an ensemble of systems. They may however, differ in some thermodynamic property such as energy, etc.

Consider an ensemble consisting of a large number of independent systems or a gas consisting of a large number of molecules in phase space. The state of an individual system or molecule may be represented in phase space by a point known as *phase* or *representative point*. The phase space may be divided into cells 1, 2, 3, ..., $i$.

Each arrangement of specified system or molecules with their representative points in particular cells is called a *microstate*. In other words, a microstate of the ensemble may be defined by the specification of the individual position of phase points for each system or molecule of the ensemble. Thus, in a microstate we have to state to which cell each system or molecule belongs temporarily.

A macrostate of the ensemble may be defined by the specification of phase point in each cell, i.e. by specifying the numbers only and overlooking the identities of the systems or molecules.

A change of phase points between two cells in phase space shifts the position of unit cell because the microstate changes. But from the point of view of macroscopic properties, this exchange makes no difference, i.e. corresponds to same macrostate. Thus, there may be many different microstates, which may correspond to the same macrostate.

Let us clear it from the following example.

Let there be cell 1, cell 2, cell 3, cell $i$ in phase space. Suppose there are four phase points $a$, $b$, $c$ and $d$ in cell 1, three phase points $e$, $f$ and $g$ in cell 2, one phase point, $h$ in cell 3 and two phase points $j$ and $k$ in cell $i$.

| $n_1 = 4$ | $n_2 = 3$ | $n_3 = 1$ | $n_i = 2$ |
|---|---|---|---|
| a b<br>c d | e f<br>g | h | j k |
| cell 1 | cell 2 | cell 3 | cell $i$ |

The macrostate is specified by merely giving the phase point $n_1 = 4$, $n_2 = 3$, ..., $n_i = 2$ of different cells. This also represents a particular microstate by specifying the positions of phase points. If position of two phase points $a$ and $e$ from different cells are interchanged then microstate are changed, while the macrostate remains the same as the number of phase points in cells remain the same.

The microstate, which is allowed under given restriction, is called *accessible microstate*. One of the most fundamental postulate of the statistical mechanics is that all accessible microstates corresponding to possible macrostate are equally probable.

## 4.2 STIRLING'S FORMULA FOR N

The calculation of $N!$ becomes laborious for large values of $N$. The Stirling's formula gives the approximate values of $\log N!$ when $N$ is very large. According to formula we have,

$$\log N! = N \log N - N \text{ (Stirling's approximation)}$$

By the definition of $N!$ we have

$$N! = 1 \times 2 \times 3 \times \cdots \times (N-2) \times (N-1) \times N$$

Therefore, $\ln N! = \ln 1 + \ln 2 + \ln 3 + \cdots + \ln(N-2) + \ln(N-1) + \ln N$

$$= \sum_{m=1}^{N} \ln m$$

In this summation, except for the first few terms whose value are small, as $m$ increases and attains large values, the increase in the value of $m$ by unity is very small. Hence in the above summation $\ln m$ can be approximately treated as continuous so that it gives the area under the curve from $m = 1$ to $m = N$ obtained by plotting $\ln m$ versus $m$ (Figure 4.1). This is a turn and is equal to the integration of $\ln x \, dx$ between the limits $x = 1$ and $x = N$. Hence, above equation can be approximated to

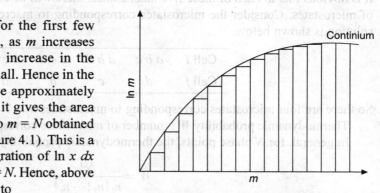

**FIGURE 4.1** A plot of $\ln m$ versus $m$.

$$\log N! = \int_{1}^{N} 1 \log x \, dx$$

Integration by parts

$$\log N! = (x \log x)_{1}^{N} - \int_{1}^{N} \frac{1}{x} x \, dx$$

$$\log N! = N \log N - N + 1$$

Here, we can neglect 1 in comparison with the large quantity $N$, then

$$\log N! = N \log N - N$$

This is simplified as stirling theorem.

## 4.3 THERMODYNAMIC PROBABILITY

$$\text{Probability of an event} = \frac{\text{Number of cases in which event occurs}}{\text{Total number of cases}}$$

Let us consider that an event happens in $x$ ways and fails to happen in $y$ ways, then

$$\text{Probability of happening the event} = \frac{x}{x + y}$$

The thermodynamic probability of a macrostate is defined as the number of microstates corresponding to that macrostate. Consider two cells in phase space represented by $i$ and $j$ and four phase points (molecules) $a$, $b$, $c$, and $d$. Let $n_i$ and $n_j$, respectively, be the five possible macrostates as shown below.

| $n_i$ | 4 | 3 | 2 | 1 | 0 |
|---|---|---|---|---|---|
| $n_j$ | 0 | 1 | 2 | 3 | 4 |

It is obvious that to each of these five macrostates, there will be corresponding different number of microstates. Consider the microstates corresponding to macrostates $n_i = 3$ and $n_j = 1$. The solution is shown below.

| Cell $i$ | $a\,b\,c$ | $a\,b\,d$ | $a\,c\,d$ | $b\,c\,d$ |
|---|---|---|---|---|
| Cell $j$ | $d$ | $c$ | $b$ | $a$ |

So there are four microstates corresponding to macrostates $n_i = 3$ and $n_j = 1$.

Thermodynamic probability $W$ = number of microstates corresponding to that macrostate = 4.

In general, for $N$ phase points, the thermodynamic probability is given by

$$W = \frac{N!}{n_1! n_2! \cdots n_i!}$$

where $n_1, n_2, ..., n_i$ denote the number of phase points in cell 1, 2, ..., $i$ respectively.

Taking logarithm of both sides of above equation, we get

$$\log W = \log N! - \Sigma \log n_i!$$

Applying Stirling's theorem

$$\log W = N \log N - N - \Sigma n_i \log n_i! + \Sigma n_i$$
$$= N \log N - N - \Sigma n_i \log n_i! + N$$
$$= N \log N - \Sigma n_i \log n_i!$$

The condition of maximum probability gives

$$\delta \log W = \delta [N \log N - \Sigma n_i \log n_i] = 0$$
$$-\Sigma (1 + \log n_i) \delta n_i = 0$$

## 4.4 GENERAL STATISTICAL DISTRIBUTION LAW

Consider the case of large flat box having compartments or cells of size $a_1, a_2, a_3, ..., a_i$ as shown in Figure 4.2. Now we wish to find the most probable distribution of particles, say marbles along the compartments. The distribution is defined by stating the number of particles in each cell as $n_1$, $n_2, n_3, \ldots, n$. The total number of particles is fixed, i.e.

$$n_1 + n_2 + \cdots + n_z = N$$

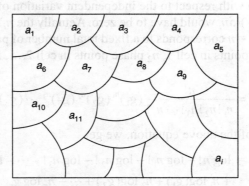

**FIGURE 4.2** Large box having $a_i$ cells of different sizes.

Mathematically, the probability of given distribution $W$ is given by the product of two factors. The first is the number of ways the particles may be rearranged and still leave the distribution same, i.e. the total number of particles in each cell left as before. Suppose, first of all we choose $n_1$ marbles which are to be placed in first cell. This can be done in ${}^nC_{n_1}$ ways, i.e.

$$\frac{N!}{n_1!(N-n_1)!}$$

Now remaining numbers are $(n - n_1)$ and we want $n_2$ marbles in second cell, which can be done ${}^{n-n_1}C_{n_2}$ ways, i.e.

$$\frac{(N-n_1)!}{n_2!(n-n_1-n_2)!}$$

Therefore, the number of ways in which the particles in different cells may be exchanged, leaving total number in each cell fixed, is

$$\frac{N!}{n_1!(n-n_1)!} \times \frac{(N-n_1)!}{n_2!(n-n_1-n_2)!} \times \cdots \times \frac{N!}{n_1!n_2!n_3!\cdots n_z!}$$

The second factor is a priori probability of a particle falling into the given cell. The probability, $g_i$ of any one particle falling in certain cell is given by the ratio of the size of that cell to the size of the box, $g_i = (a_i/a)$, where $a = a_1 + a_2 + \cdots + a_i$. The sum of probabilities of all cells is of course unity. The probability of $n_i$ particles falling in the $i$th cell is $(g_i)^{n_i}$. Therefore, a priori probability of the distribution is

$$(g_1)^{n_1}(g_2)^{n_2}(g_3)^{n_3}\cdots(g_z)^{n_z}$$

$$W = \frac{N!}{n_1!n_2!n_3!\cdots n_z!}(g_1)^{n_1}(g_2)^{n_2}(g_3)^{n_3}\cdots(g_z)^{n_z}$$

## 4.5 MOST PROBABLE DISTRIBUTION

We must select the distribution which has the largest value of $W$. The methods of calculus are used for this purpose. Here, we work with log $W$, which obviously corresponds to a maximum $W$. If log $W$ were to be at maximum with respect to the independent variation of each $n_1$ separately, all its partial derivatives $\delta(\log W)/\delta n_i$ would have to be zero. Actually the $n_i$'s must satisfy an auxiliary condition $n_1 + n_2 + \cdots + n_z = n$ corresponds to a fixed total number of particles. The probability $W$ of a distribution ($n_1$ phase points in cell 1, $n_2$ phase points in cell 2,..., $n_z$ phase points in cell $z$) is given by

$$W = \frac{n!}{n_1! n_2! n_3! \cdots n_z!} (g_1)^{n_1} (g_2)^{n_2} (g_3)^{n_3} \cdots (g_z)^{n_z} \tag{4.1}$$

Taking natural logarithm of the above equation, we get

$$\log W = \log n! - \log n_1! - \log n_2! - \log n_3! - \cdots - \log n_z! \\ + n_1 \log(g_1) + n_2 \log(g_2) + \cdots + n_z \log g_z \tag{4.2}$$

Using Stirling's formula, we have

$$\log W = n \log n - n - n_1 \log n_1 + n_1 - n_2 \log n_2 + n_2 - \cdots - n_z \log n_z \\ + n_z + n_1 \log g_1 + n_2 \log g_2 + \cdots + n_z \log g_z$$

$$= n \log n - n_1 \log n_1 - n_2 \log n_2 - \cdots - n_z \log n_z + n_1 \log g_1 + n_2 \log g_2 + \cdots + n_z \log g_z$$

$$= n \log n - \sum_z n_z \log n_z + \sum_z n_z \log g_z \tag{4.3}$$

For most probable distribution

$$\frac{\partial (\log W_{max})}{\partial n_i} = 0 \tag{4.4}$$

From Eq. (4.3)

$$\delta \log W_{max} = -\sum_z \left[ n_z \frac{1}{n_z} \partial n_z + \partial n_z \log n_z \right] + \sum_z \partial n_z \cdot \log g_z$$

$$= -\sum_z \partial n_z - \sum_z \log n_z \partial n_z + \sum_z \log g_z \cdot \partial n_z$$

$$= -\sum_z \log n_z \partial n_z + \sum_z \log g_z \cdot \partial n_z \quad \left( \because \sum_z \partial n_z = 0 \right)$$

But, $\delta \log W_{max} = 0$, hence

$$= -\sum_z \log n_z \partial n_z + \sum_z \log g_z \cdot \partial n_z = 0 \tag{4.5}$$

Lagranges method of undetermined multipliers is employed in solving problems of this type. So, we multiply Eq. (4.5) by undetermined multiplier $\alpha$ which does not depend upon any of the $n_z$, i.e.

$$\sum_z \alpha \partial n_z = 0 \tag{4.6}$$

Adding Eqs. (4.5) and (4.6), we get,

$$-\sum_z \log n_z \partial n_z + \sum_z \log g_z \cdot \partial n_z + \sum_z \alpha \partial n_z = 0 \tag{4.7}$$

$$\sum (-\log n_z + \log g_z + \alpha) \partial n_z = 0 \qquad (\therefore \partial n_z\text{'s are independent variables})$$

$$n_z = g_z e^\alpha \tag{4.8}$$

$$\sum_z n_z = e^\alpha \sum_z g_z \qquad (\text{because } \alpha \text{ does not depend on } Z)$$

$$\sum_z n_z = e^\alpha \qquad \left(\text{because } \sum_z g_z = 1\right)$$

$$n = e^\alpha \tag{4.9}$$

Substituting the value of $e^\alpha = n$ from Eq. (4.9) into Eq. (4.8), we get

$$n_z = g_z n$$

$$\frac{n_z}{n} = g_z \tag{4.10}$$

Thus, we have finally obtained the result that the most probable number of particles in each cell is proportional to the size of cell, a result which is obviously correct.

## 4.6 TYPES OF ENSEMBLE

Depending upon the thermodynamic variables kept constant, the ensembles are of many different kinds.

1. *Microcanonical ensemble:* If the original system is an isolated one having volume $V$, total number of molecules $N$ and energy $E$ (with an extremely narrow range) then we can have an ensemble of $n$ such systems, each member having the same value of $N$, $V$ and $E$. Microcanonical ensemble is shown in box below.

| NVE | NVE | NVE | NVE | NVE |
|-----|-----|-----|-----|-----|
| NVE | NVE | NVE | NVE | NVE |
| NVE | NVE | NVE | NVE | NVE |
| NVE | NVE | NVE | NVE | NVE |

It can be set up by imaging rigid impermeable walls separating the systems so that neither energy nor material particles can flow from one system to the other. Thus, each system in this ensemble is like the isolated system in the thermodynamic sense. Such an ensemble of systems in which each member has the same value of $N$, $V$ and $E$ is called a *microcanonical ensemble*. In the microcanonical ensemble shown in box there are 20 members. However, in statistical mechanics, we consider a very large value of $n$ and in a number of calculations, we take the limit $n = \infty$.

2. *Canonical ensemble:* If all the members of an ensemble have the same value of $N$, $V$ and $T$, then it is called a *canonical ensemble*. It can be set up by imagining rigid but conducting wall separating the different systems through which energy can pass but not the material particles. Thus, each system in this ensemble is like a closed system in the thermodynamic sense. Such an ensemble is shown in box below.

| NVT | NVT | NVT | NVT | NVT |
|-----|-----|-----|-----|-----|
| NVT | NVT | NVT | NVT | NVT |
| NVT | NVT | NVT | NVT | NVT |
| NVT | NVT | NVT | NVT | NVT |

Due to the conducting walls, when the equilibrium is attained, each member of the ensemble has the same temperature $T$, but may not have the same energy $E$. Canonical ensemble is the most commonly used ensemble in statistical mechanics.

3. *Grand Canonical Ensemble:* A third type of ensemble called grand canonical ensemble used in more advanced calculations was introduced by Gibbs. In this ensemble, each system is an open system. Hence, the matter can flow between the systems and the composition of each one may fluctuate. Consequently, the number of molecules in different systems is not the same. However, $V$, $T$ and $\mu$ for each component is same for each member of the ensemble.

Thus, the main features of the three different types of ensembles may be summed up as given in Table 4.1.

**TABLE 4.1** Main features of the three types of ensembles

| Types of ensemble | Properties same for each system |
|---|---|
| (a) Microcanonical | $N, V, E$ |
| (b) Canonical | $N, V, T$ |
| (c) Grand Canonical | $\mu, V, T$ |

## 4.7 FLUCTUATION

The difference between true (experimental) and observed value is called *fluctuation*. It may be positive or negative. But the mean of fluctuation is always low positive. Fluctuation in some cases, e.g. fluctuation of density caused scattering of light.

1. *Mean square fluctuation:* It is the mean of square of fluctuation and is always positive. If $x$ is a true value and $\bar{x}$ is the mean of various observed values, i.e.

$$\bar{x} = \frac{x_1 + x_2 + x_3 + \cdots + x_n}{x}$$

Then mean square fluctuation

$$(x - \bar{x})^2 > 0$$

2. *Relative fluctuation*: Fluctuation relative to the true value of an observable property expressed as

$$\left(\frac{x-\bar{x}}{\bar{x}}\right)$$

For energy relative fluctuation is

$$\left(\frac{E-U}{U}\right)$$

3. *Mean square relative fluctuation*: It is always a positive quantity, i.e.

$$\overline{\left(\frac{x-\bar{x}}{\bar{x}}\right)^2} > 0$$

We can show that many systems have energy extremely close to a certain mean value and that deviation form this mean are extremely small in comparison with the total energy.

## 4.8 DISTRIBUTION FUNCTION FOR FLUCTUATION: GENERAL THEORY OF FLUCTUATION

Suppose we have a quantity $x$ and $\bar{x}$ is the mean of this quantity. Let forces be a distribution function such that $f(x)\,dx$ represent the fraction of all system of assembly for which $x$ lies in the range between $x$ and $x + dx$. If $x$ is a measurable property and $x$ is mean of that property, then mean square value of function is given by

$$\overline{(x-\bar{x})^2} = \int (x-\bar{x})^2 f(x)\,dx \qquad (4.11)$$

Let us assume that energy level of the system are so close to each atom that they can be treated as continuous distribution of energy. Arranging the energy level in accordance of value of $x$. If $e^{\delta x/v}$ is a density fluctuation. Then $f(x)\,dx$ will give number of energy levels which lie in the range between $x$ and $x + dx$. Thus, fluctuation of system in this range

$$f(x)\,dx = e^{-\frac{[E_{(x)} - TS_{(x)}]}{kT}} \qquad (4.12)$$

where $f(x)$ = energy corresponds to the levels in $dx$. We may evaluate constant, from the condition that the integral of $f(x)$ overall value of $f(x)$ must be unity, so

$$\text{Constant} = \frac{1}{\int e^{-\left[\frac{(E_{(x)} - TS_{(x)})}{kT}\right]} dx}$$

So Eq. (4.12) becomes,

$$f(x) = \frac{e^{-\frac{[E_{(x)} - TS_{(x)}]}{kT}}}{\int e^{-\left[\frac{(E_{(x)} - TS_{(x)})}{kT}\right]} dx} \qquad (4.13)$$

Let us consider that $f(x)$ has very sharp and narrow maxima at a certain value.

Let it be $x_0$ rapidly falling to zero on both sides of the maxima. This leads to fact that $x$ fluctuate only slightly from $x$ in a system of assembly (Figure 4.3).

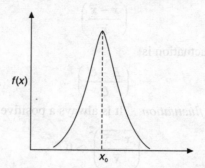

**FIGURE 4.3** Distribution function for fluctuation.

## 4.9 REASON FOR THE FACT THAT $x$ FLUCTUATES SLIGHTLY FROM $x_0$

The fraction $E_{(x)} - TS_{(x)}$ must have a maxima at $x_0$ in order that $f_{(x)}$ may have maxima. There the function $f_{(x)}$ will be then reduced to $1/e$ of maximum value. The $E_{(x)} - TS_{(x)}$ is greater than its minimum by $kT$ only. But where $E$ is the energy of whole system of the order of $mkT$ then we shall find likewise that $TS_{(x)}$ must be of same magnitude. So change in $x$ will increase the function $E_{(x)} - TS_{(x)}$ by $kT$ or more.

Let us expand $E_{(x)} - TS_{(x)}$ in Tayler series about $x_0$ (since $f_{(x)}$ has a minima at $x_0$, so that its first derivative is $\infty$ (infinity).

So, we have

$$E_{(x)} - TS_{(x)} = E_{(x_0)} - TS_{(x_0)} - \frac{1}{2}\left[\frac{d^2E}{dx^2} - \frac{Td^2S}{dx^2}\right](x-x_0)^2 \quad (4.14)$$

Numerator of Eq. (4.13) becomes

$$e^{-\frac{[E_{(x)} - TS_{(x)}]}{KT}} e^{-\alpha(x-x_0)^2} \quad (4.15)$$

where

$$\alpha = \frac{1}{2KT}\left[\frac{d^2E}{dx^2} - \frac{Td^2S}{dx^2}\right] \quad (4.16)$$

The function $e^{-\alpha(x-x_0)^2}$ is called *Gausseous curve* and is equal to unity when $x = x_0$ and falls on both sides of this point symmetrically when $x - x_0 = \pm\frac{1}{\sqrt{\alpha}}$.

## 4.10 ENERGY FLUCTUATION IN CANONICAL ENSEMBLE

Let us consider a system of a perfect gas obeying Boltzmann's distribution. If $E$ be the energy of particular system in canonical assembly and $u$ is average or mean energy. Now we will compute mean square deviation of energy from its mean, i.e. $\overline{(E - \overline{U})^2}$.

Now we have a general relation to measure mean square value by

$$\overline{(x-\overline{x}_0)^2} = \frac{kT}{\frac{d^2E}{dx^2} - \frac{Td^2S}{dx^2}} \qquad (4.17)$$

Putting $x = E$, $x_0 = U$, and $\frac{d^2E}{dx^2} = 0$. Then, mean square fluctuation is $\overline{(E-\overline{U})^2}$.

$$\overline{(E-\overline{U})^2} = \frac{kT}{0 - \frac{Td^2S}{dx^2}} \qquad (4.18)$$

In thermodynamics $\left(\frac{dS}{dE}\right)_V = \frac{1}{T}$

$$\left(\frac{d^2S}{dE^2}\right)_V = -\frac{1}{T^2}\left(\frac{dT}{dE}\right)_V \qquad (4.19)$$

$$= -\frac{1}{T^2}\left(\frac{1}{C_V}\right) \qquad \left(\because C_V = \left(\frac{dE}{dT}\right)_V\right)$$

$$\overline{(E-\overline{U})^2} = -kT^2 C_V \qquad (4.20)$$

For perfect gas $C_V = 3/2\, Nk$, where, $N$ is number of atoms.

$$\overline{(E-\overline{U})^2} = \frac{3}{2} k^2 T^2 N$$

Now mean square relative energy fluctuation

$$\left(\frac{E-U}{U}\right)^2 = \frac{3/2N - k^2T^2}{(3/2N - kT)^2} \quad \text{or} \quad \frac{2}{3N} \qquad (4.21)$$

$$\left(\frac{E-U}{U}\right)^2 = \frac{2}{3N} \qquad (4.22)$$

## 4.11 FLUCTUATION IN DENSITY

Instead of fluctuation of density itself, we find fluctuation of volume occupied by a certain group of molecule. The relative function will be square in either case. Since a given proportional increase in volume will give an equal proportional decrease in density, so using relation.

$$\overline{(x-x_0)^2} = \frac{kT}{\frac{d^2E}{dx^2} - \frac{Td^2S}{dx^2}}$$

Putting $x = V$ and $x_0 = V_0$ average value of volume,

$$\overline{(V - V_0)^2} = \frac{kT}{\dfrac{d^2 E}{dx^2} - \dfrac{T d^2 S}{dx^2}}$$

$$\overline{(V - V_0)^2} = \frac{kT}{\left(\dfrac{d^2 A}{dV^2}\right)_T} \tag{4.23}$$

$$A = E - TS$$

$$dA = -PdV - SdT \tag{4.24}$$

$$\left(\frac{dA}{dV}\right)_T = -P \text{ and } \left(\frac{dA}{dT}\right)_V = -S \tag{4.25}$$

$$\left(\frac{dA}{dV}\right)_T = \left(\frac{dE}{dV}\right)_T - T\left(\frac{dS}{dV}\right)_T \tag{4.26}$$

$$\left(\frac{d^2 A}{dV^2}\right)_T = \left(\frac{d^2 E}{dV^2}\right)_T - T\left(\frac{d^2 S}{dV^2}\right)_T \tag{4.27}$$

$$\left(\frac{d^2 A}{dV^2}\right)_T = -\left(\frac{dP}{dV}\right)_T \quad \text{from Eq. (4.25)}$$

Using Eqs. (4.25) and (4.27) into Eq. (4.23)

$$\overline{(V - V_0)^2} = -kT\left(\frac{dV}{dP}\right)_T$$

$$\frac{\overline{(V - V_0)^2}}{V_0} = \frac{1}{V_0} kT \left(-\frac{1}{V_0}\left(\frac{dV}{dP}\right)\right)_T \tag{4.28}$$

Quantity $\left(-\dfrac{1}{V_0}\left(\dfrac{dV}{dP}\right)\right)_T$ is isothermally compressibility which is independent of $V_0$, so relative mean square fluctuation of volume is inversely proportional to volume itself becoming small for large volume.

Let us check Eq. (4.28) by application to perfect gas in Boltzmann statistical mechanics using $PV = nRT$, we have

$$\left(-\frac{1}{V_0}\left(\frac{dV}{dP}\right)\right)_T = \frac{1}{P} \quad \text{At constant } T, PdV + VdP = 0$$

and

$$\frac{\overline{(V - V_0)^2}}{V_0} = \frac{kT}{PV_0} = \frac{kT}{N_0 kT} = \frac{1}{N_0} \tag{4.29}$$

where $N_0$ = mean number of molecules in $V_0$.

For the substance other than perfect gas the compressibility is ordinarily less than for a perfect gas so Eq. (4.23) predicts smaller relative fluctuations of density, a perfect incompressible solid would have low density fluctuation. On the other hand, in some cases, the compressibility can be greater than for a perfect gas, e.g. a real gas (not perfect) near the critical point.

## 4.12 MAXWELL–BOLTZMANN DISTRIBUTION LAW

Consider a system made up of a large number of similar molecules enclosed in a vessel of constant volume. Suppose, all that is known about the system is that its energy $E$ is constant. For this purpose, a microcanonical ensemble is considered. The points representing the system are distributed uniformly through a thin shell lying between two surfaces in the $\gamma$-space representing constant energies of $E$ and $E + dE$, respectively.

Suppose the system of $N$ similar but distinguishable molecules considered in the preceding section, has a total energy which lies within the range of $E$ to $E + dE$. It follows then from the aforementioned postulate that the probability of finding the system in the state in which $n_1$ molecules are in the first cell, $n_2$ in the second cell and so on, with $n_i$ in the $i$th cell, is proportional to $G$, the number of unit cells, i.e. the total volume, occupied by the system in the $\gamma$-space, thus

$$G = \frac{N!}{n_1! n_2! \cdots n_i!} \times \text{constant}$$

where $G$ is a required probability. In each group the $n_i$ elements may be distributed among $g_i$ elementary wave functions. Since symmetry considerations do not arise, there is no restriction as to the number of elements associated with each wavefunction and so to the total number of ways in which the distribution may be made is equal to $g_i^{n_i}$ in each group. To completely define number of eigenstates for the system as a whole, it is therefore given by,

$$G = N! \frac{g_1^{n_1}}{n_1!} \frac{g_2^{n_2}}{n_2!} \cdots \frac{g_i^{n_i}}{n_i!} \cdots$$

$$= N! \prod \frac{g_i^{n_i}}{n_i!}$$

Now according to postulate of equal a priori probability of distribution of particles, the probability $W$ of the system having a particular distribution of particles is

$$W \propto G$$
$$W = G \times \text{constant}$$

$$W = N! \prod \frac{g_i^{n_i}}{n_i!} \times \text{constant} \qquad (4.30)$$

But according to condition of maximum probability, we have,

$$\delta \ln W = 0 \qquad (4.31)$$

Taking log of Eq. (4.30), we get

$$\ln W = \ln N! + \sum_i [n_i \ln g_i - \ln n_i !] + \text{constant}$$

Applying Stirling's formula, i.e. $\log N! = N \log N - N$

$$\ln W = N \ln N - N + \sum_i [n_i \ln g_i - n_i \ln n_i + n_i] + \text{constant}$$

$$\ln W = N \ln N - N + \sum_i n_i \ln g_i - \sum_i n_i \ln n_i + \sum_i n_i + \text{constant}$$

$$\ln W = N \ln N - N + \sum_i n_i \ln g_i - \sum_i n_i \ln n_i + N + \text{constant}$$

$$\ln W = N \ln N + \sum_i n_i \ln g_i - \sum_i n_i \ln n_i + \text{constant}$$

$$\ln W = N \ln N + \sum_i [n_i \ln g_i - n_i \ln n_i] + \text{constant}$$

Differentiating with respect to $n_i$, we get

$$d \ln W = \sum_i \left[ \ln g_i \delta n_i - n_i \frac{1}{n_i} \delta n_i - \ln n_i \delta n_i \right]$$

$$d \ln W = \sum_i [\ln g_i \delta n_i - \delta n_i - \ln n_i \delta n_i]$$

$$d \ln W = \sum_i \ln g_i \delta n_i - \sum_i \delta n_i - \sum_i \ln n_i \delta n_i \quad (4.32)$$

Since, the total number of particle $N$ remains constant, we have

$$\sum_i n_i = N = \text{constant} \quad (4.33)$$

From Eqs. (4.32) and (4.33), we get

$$d \ln W = \sum_i \ln g_i \delta n_i - \sum_i \ln n_i \delta n_i$$

$$d \ln W = \sum_i (\ln g_i - \ln n_i) \delta n_i \quad (4.34)$$

Comparing Eqs. (4.31) and (4.34), we get

$$\sum_i (\ln g_i - \ln n_i) \delta n_i = 0$$

or

$$\sum_i (\ln n_i - \ln g_i) \delta n_i = 0$$

$$\sum_i (\ln n_i / g_i) \delta n_i = 0 \qquad (4.35)$$

Now we have $\sum n_i \varepsilon_i = E = \text{constant} \quad \left( \because \sum_i \varepsilon_i \delta n_i = 0 \right)$ (4.36)

Now applying the Lagrange method of undetermined constancy and multiplying Eq. (4.33) by constant $\alpha$ and Eq. (4.36) by constant $\beta$ and adding in Eq. (4.35), we get

$$\sum_i (\ln n_i/g_i + \alpha + \beta \varepsilon_i) \delta n_i = 0 \qquad (4.37)$$

It will be seen that Eq. (4.37) will be satisfied, in general, only if each term in the summation is zero, thus

$$\ln \frac{n_i}{g_i} + \alpha + \beta \varepsilon_i = 0 \qquad (4.38)$$

$$\ln \frac{n_i}{g_i} = -(\alpha + \beta \varepsilon_i)$$

$$\ln \frac{g_i}{n_i} = (\alpha + \beta \varepsilon_i)$$

$$\frac{g_i}{n_i} = e^{\alpha + \beta \varepsilon_i}$$

$$n_i = \frac{g_i}{e^{\alpha + \beta \varepsilon_i}}$$

or $\qquad n_i = g_i e^{-(\alpha + \beta \varepsilon_i)}$ (4.39)

Above equation is known as *Maxwell–Boltzmann distribution law* or *M.B. statistics*. This equation gives the most probable distribution of molecules among the various possible individual energy values, at statistical equilibrium, for a system of constant total energy. Attention may be called to the fact that in deriving the Maxwell–Boltzmann distribution law, no restriction was made as to the nature of the energy, i.e. translational, vibrational, rotational, etc. The equation may, thus, be regarded as applicable to the distribution of the total energy or of any form of energy, which has a constant value for the given state.

The attention must be given to the approximation and assumptions made in its derivation. In the first place, it has been assumed that, the molecules are distinguishable. Secondly, by the use of Stirling's approximation for $n_i!$ presupposes that all the $n_i$ values are very large. It must be remembered, of course, that in every case the identification of $\varepsilon_i$ with the actual energy of a molecule is the $i$th cell in the $\mu$-space presupposes the absence of forces acting between the molecules. The systems are thus assumed to consist of ideal gases, for it is only in these circumstances that the intermolecular forces are completely absent. However, under such conditions that the deviations from ideal behaviour are not large, the Maxwell–Boltzmann distribution law may be employed without incurring serious error.

## 4.13 EVALUATION OF CONSTANTS $\alpha$ AND $\beta$

From Eq. (4.39), we have

$$n_i = g_i e^{-(\alpha + \beta \varepsilon_i)} \tag{4.40}$$

Now if all the particles are present in ground state then $n_i = n_o$, $\varepsilon_i = 0$ and $g_i = 1$ and Eq. (4.40) gives

$$n_o = e^{-\alpha} \tag{4.41}$$

Putting the value of $e^{-\alpha}$ in Eq. (4.40), we get,

$$n_i = g_i \, n_o \, e^{-\beta \varepsilon_i} \tag{4.42}$$

Taking summation on both sides

$$\sum n_i = \sum g_i \, n_o \, e^{-\beta \varepsilon_i}$$

$$N = \sum g_i \, n_o \, e^{-\beta \varepsilon_i}$$

$$N = n_o \sum g_i \, e^{-\beta \varepsilon_i}$$

$$n_o = \frac{N}{\sum g_i \, e^{-\beta \varepsilon_i}}$$

Putting the value of $n_o$ in Eq. (4.42), we get

$$n_i = g_i \left( \frac{N}{\sum g_i \, e^{-\beta \varepsilon_i}} \right) e^{-\beta \varepsilon_i}$$

$$n_i = N \left( \frac{g_i \, e^{-\beta \varepsilon_i}}{\sum g_i \, e^{-\beta \varepsilon_i}} \right)$$

$$\frac{n_i}{N} = \left( \frac{g_i \, e^{-\beta \varepsilon_i}}{\sum g_i \, e^{-\beta \varepsilon_i}} \right) \tag{4.43}$$

Now, thermodynamically it can be shown that for most probable distribution of particles the entropy of system is given by

$$S = k \ln W_{\text{mp}} \tag{4.44}$$

(Here mp stands for most probable.)

But

$$W = N! \prod_i \frac{g_i^{n_i}}{n_i!}$$

$$S = k \ln \left( N! \prod_i \frac{g_i^{n_i}}{n_i!} \right) \tag{4.45}$$

Applying Stirling's approximation, we get

$$S = k\left(N\ln N - N + \sum_i n_i \ln g_i - \sum_i n_i \ln n_i + \sum_i n_i\right)$$

$$S = k\left(N\ln N - N + \sum_i n_i \ln g_i - \sum_i n_i \ln n_i + N\right)$$

$$S = k\left(N\ln N + \sum_i n_i \ln g_i - \sum_i n_i \ln n_i\right)$$

$$S = k\left[N\ln N + \sum_i n_i \ln \frac{g_i}{n_i}\right]$$

From Eq. (4.39) on rearrangement, we have

$$\ln \frac{g_i}{n_i} = \alpha + \beta \varepsilon_i$$

$$S = k\left[N\ln N + \sum n_i(\alpha + \beta\varepsilon_i)\right]$$

$$S = k\left[N\ln N + \sum n_i \alpha + \sum \beta \varepsilon_i n_i\right]$$

$$S = k[N\ln N + \alpha N + \beta E] \tag{4.46}$$

Differentiating equation with respect to $E$ at constant $V$, we get

$$\left(\frac{dS}{dE}\right)_V = k\beta \tag{4.47}$$

But

$$\left(\frac{dS}{dE}\right)_V = \frac{1}{T} \tag{4.48}$$

From Eq. (4.47) and (4.48), we get

$$k\beta = \frac{1}{T}$$

or

$$\beta = \frac{1}{kT}$$

Putting the value of $\beta$ in Eq. (4.43), we get

$$\frac{n_i}{N} = \frac{g_i\, e^{-\varepsilon_i/kT}}{\sum g_i\, e^{-\varepsilon_i/kT}} \tag{4.49}$$

Above equation is known as Maxwell–Boltzmann distribution law. This law gives the fraction of molecules which is in the most probable state of high temperature possessing the energy $E_i$.

**Note:** $g_i$ is the degeneracy or statistical weight of energy level $E_i$.

## 4.14 MAXWELL–BOLTZMANN DISTRIBUTION LAW OF VELOCITIES

This law can be derived from M.B. law by making following assumptions:

1. System under consideration is made up of non-interacting particles, so we can take each particle as monoatomic.
2. Translational energy is separable from other forms of energy.
3. Total energy of system = KE + PE (due to position $x, y, z$)

$$E_t = \text{KE} + \text{PE}$$
$$= 1/2\ mv^2 + \phi(x, y, z) \qquad (4.50)$$

Let a gas is introduced into a vessel. Molecules are moving with different velocities in different directions. Each molecule can be represented by coordinates: three position coordinates $x, y, z$ and three momenta coordinates $p_x, p_y, p_z$. Fraction of molecules lying in the range $x$ to $x + dx$, $y$ to $y + dy$ and $z$ to $z + dz$ is equal to number of particles in small volume element $dx$, $dy$ and $dz$, i.e. these are independent of position.

So
$$\text{PE} = \phi(x, y, z) = 0$$

Putting this value of PE in Eq. (4.50), we can write

$$E_t = 1/2\ mv^2 = 1/2\ mv^2\ m/m = 1/2\ m^2v^2/m = 1/2\ p^2/m \qquad (4.51)$$

According to M.B. law, we know

$$\frac{n_i}{N} = \frac{g_i\ e^{-\varepsilon_i/kT}}{\sum g_i\ e^{-\varepsilon_i/kT}} \qquad (4.52)$$

Degeneracy is unity, i.e. $g_i = 1$, then we get

$$\frac{n_i}{N} = \frac{e^{-\varepsilon_i/kT}}{\sum e^{-\varepsilon_i/kT}} \qquad (4.53)$$

Number of molecules in unit volume is

$$\frac{n_i}{N} = \frac{e^{-\varepsilon_i/kT}}{\sum e^{-\varepsilon_i/kT}}\ dx, dy, dz, dp_x, dp_y\ \text{and}\ dp_z$$

Now as $dx, dy$ and $dz = 1$, we get

$$\frac{n_i}{N} = \frac{e^{-\varepsilon_i/kT}}{\sum e^{-\varepsilon_i/kT}}\ dp_x, dp_y\ \text{and}\ dp_z \qquad (4.54)$$

Putting the value of $E$ from Eq. (4.51) into Eq. (4.54), we get

$$\frac{n_i}{N} = \frac{e^{-p^2/2mkT}}{\sum e^{-p^2/2mkT}}\ dp_x, dp_y\ \text{and}\ dp_z \qquad (4.55)$$

Now according to quantum mechanics

$$E_t = \frac{h^2}{8ml^2}\left[n_x^2 + n_y^2 + n_z^2\right]$$

where $l$ is length of cube, $n_x$ is the quantum number determining the possible value of the translational energy of the particle along x-axis, similarly $n_y$ and $n_z$ are values along y and z-axis.

Since $\frac{h^2}{8ml^2}$ is very small (of the order $10^{-30}$) for normal molecule.

It means that translational energy levels are very close to each other, therefore, summation of Eq. (4.55) may be replaced by integration

$$\frac{n_i}{N} = \frac{e^{-p^2/2mkT}dp_x dp_y dp_z}{\int_{-\infty}^{+\infty}\int_{-\infty}^{+\infty}\int_{-\infty}^{+\infty} e^{-p^2/2mkT}dp_x dp_y dp_z}$$

Putting $p^2 = p_x^2 + p_y^2 + p_z^2$, we get

$$\frac{n_i}{N} = \frac{e^{-p^2/2mkT}dp_x dp_y dp_z}{\int_{-\infty}^{+\infty}\int_{-\infty}^{+\infty}\int_{-\infty}^{+\infty} e^{-p_x^2 p_y^2 p_z^2/2mkT}dp_x dp_y dp_z}$$

$$\frac{n_i}{N} = \frac{e^{-p^2/2mkT}dp_x dp_y dp_z}{\int_{-\infty}^{+\infty} e^{-p_x^2/2mkT}dp_x \int_{-\infty}^{+\infty} e^{-p_y^2/2mkT}dp_y \int_{-\infty}^{+\infty} e^{-p_z^2/2mkT}dp_z}$$

Using standard integral $\int_{-\infty}^{+\infty} e^{-ax^2}dx = \left(\frac{\pi}{a}\right)^{1/2}$, we can write

$$\frac{n_i}{N} = \frac{e^{-p^2/2mkT}dp_x dp_y dp_z}{(2\pi mkT)^{1/2}(2\pi mkT)^{1/2}(2\pi mkT)^{1/2}}$$

$$\frac{n_i}{N} = (2\pi mkT)^{-3/2} e^{-p^2/2mkT}dp_x dp_y dp_z \qquad (4.56)$$

Above equation gives the number of molecules lying in volume element $dx$, $dy$ and $dz$ (Figure 4.4).

If we consider momentum of a molecule with mass $(m)$ is equal to $mv$. It can be shown that the number of molecules with velocities in the range $v$ to $v + dv$ is equal to number of molecules in small volume of a spherical shell between spheres of radii $(mv + mdv)$ and $(mv)$ as shown below (Figure 4.5).

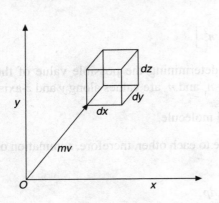
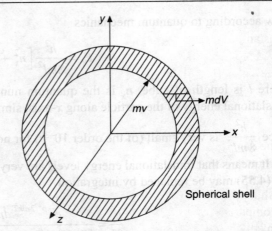

**FIGURE 4.4** Volume element $dx, dy, dz$.   **FIGURE 4.5** Volume of spherical shell at distance $mv$.

$$\text{Volume of spherical shell (shaded area)} = \frac{4}{3}\pi(mv+mdv)^3 - \frac{4}{3}\pi(mv)^3$$

$$\text{Volume of shell} = \frac{4}{3}\pi[m(v+dv)]^3 - \frac{4}{3}\pi(mv)^3$$

**Note:** In general volume of such a spherical shell $\simeq 4\pi r^2 a$ where $r$ is radius of inner sphere and it is $mv$ in this case

$$\text{Volume of shell} = 4\pi(mv)^2\, mdv$$

Now if we put, $dp_x, dp_y, dp_z = 4\pi m^2 v^2 mdv$ in Eq. (4.56), we get

$$\frac{n_i}{N} = (2\pi mkT)^{-3/2} e^{-p^2/2mkT} 4\pi m^2 v^2 mdv$$

$$\frac{n_i}{N} = 4\pi m^3 \left(\frac{1}{2\pi mkT}\right)^{3/2} e^{-p^2/2mkT} v^2 dv \tag{4.57}$$

Putting $p = mv$ in Eq. (4.57), we get

$$\frac{n_i}{N} = 4\pi \left(\frac{m}{2\pi kT}\right)^{3/2} e^{-m^2v^2/2mkT} v^2 dv$$

$$n_i = 4\pi N \left(\frac{m}{2\pi kT}\right)^{3/2} e^{-mv^2/2kT} v^2 dv \tag{4.58}$$

Above equation is called M.B. distribution law of velocities, i.e. this equation gives number of molecules with velocities in between $v$ and $v + dv$ in any direction.

## 4.15 BOSE–EINSTEIN STATISTICS

Suppose $N$ indistinguishable elements that constitute the system under consideration are divided into a series $n_1, n_2, \ldots, n_i$ in group, the energy is almost constant, viz. $E_1, E_2, \ldots, E_i$, respectively.

Then, since the total number of elements is constant and the total energy of the system is also constant, within narrow limits, it follows that

$$\sum_i n_i = N = \text{constant} \tag{4.59}$$

and

$$\sum_i \varepsilon_i n_i = E = \text{constant} \tag{4.60}$$

If the degeneracy, i.e. statistical weight of the $i$th level, in general, is equal to $g_i$, the total number of eigenstates for the group of $n_i$ elements is equal to the number of ways in which $n_i$ elements can be distributed among the $g_i$ wave functions. The required number of different ways is equivalent to that in which $n_i$ indistinguishable particles can be distributed in a box divided into $g_i$ compartments, without any restriction as to the number in each of the compartments. Imagine a box divided by $g_i - 1$ partitions into $g_i$ sections, the $n_i$ particles are then supposed to be distributed among these sections. The total number of permutations of the $n_i$ particles and the $g_i - 1$ permutation is $(n_i + g_i - 1)!$.

Since $n_i$ particles are indistinguishable, permutation among the particles themselves do not really produce a new arrangement. The total number just given should therefore be divided by $n_i!$. Further, the permutations among the $g_i - 1$ partitions do not alter the fact that there are still $g_i$ sections, hence, division of the total number of permutations by $(g_i - 1)!$ is also necessary. The number of ways

$$\frac{(n_i + g_i - 1)!}{n_i!(g_i - 1)!}$$

In each group the number of eigenstates is given by expression of the same type as that just desired for $i$th group, and hence the total number $G$ of eigenstate for the whole system, corresponding to the specified distribution of the $n$ elements is given by

$$G = \frac{(n_1 + g_1 - 1)!}{n_1!(g_1 - 1)!} \frac{(n_2 + g_2 - 1)!}{n_2!(g_2 - 1)!} \cdots \frac{(n_i + g_i - 1)!}{n_i!(g_i - 1)!} \cdots \tag{4.61}$$

$$= \prod_i \frac{(n_i + g_i - 1)!}{n_i!(g_i - 1)!} \cdots \tag{4.62}$$

Introducing now the postulate of the equal a priori probability of eigenstates, it follows that the probability $W$ of the system having the particular distribution specified, is proportional to the total number of eigenstates, hence for Bose–Einstein statistics.

$$W = \prod_i \frac{(n_i + g_i - 1)!}{n_i!(g_i - 1)!} \times \text{constant} \tag{4.63}$$

The condition of maximum probability of the system is found by setting $S \ln W$ equal to zero. Upon taking logarithms of Eq. (4.63), this becomes,

$$\ln W = \sum_i \{\ln(n_i + g_i - 1)! - \ln n_i! - \ln(g_i - 1)!\} + \text{constant} \tag{4.64}$$

and making use of the Stirling's formula,

$$\log N! = N \log N - N$$

$$\ln W = \sum_i \{(n_i + g_i) \ln(n_i + g_i) - n_i \ln n_i - g_i \ln g_i\} + \text{constant} \tag{4.65}$$

where $n_i + g_i - 1$ and $g_i - 1$ have been taken as equal to $n_i + g_i$ and $g_i$, respectively. Since the number $n_i$ is supposed to be very large, it may be treated as a continuous variable and hence differentiation of Eq. (4.65) with respect to $n_i$ gives for the most probable state of the system.

$$\delta \ln W = \sum_i [\ln n_i - \ln(n_i + g_i)] \delta n_i = 0$$

$$= \sum_i \left( \ln \frac{n_i}{n_i + g_i} \right) \delta n_i = 0 \tag{4.66}$$

Since $n_i$ and $g_i$ for each quantum group have been assumed to be large, the distribution of energy within a group may be regarded as virtually continuous, at least over a narrow range, it is thus possible to write

$$\delta n = \sum_i n_i = 0 \quad \text{and} \quad \sum_i \varepsilon_i n_i = \delta E = 0$$

Applying the Lagrange method of undetermined multiplier to these relationship in combinations with Eq. (4.66), the result is

$$= \sum_i \left( \ln \frac{n_i}{n_i + g_i} + \alpha + \beta \varepsilon_i \right) \delta n_i = 0 \tag{4.67}$$

Since the variations $\delta n_i$ are independent of one another

$$\ln \frac{n_i}{n_i + g_i} + \alpha + \beta \varepsilon_i = 0 \tag{4.68}$$

$$\ln \frac{g_i}{n_i} + 1 = \alpha + \beta \varepsilon_i$$

$$\frac{g_i}{n_i} + 1 = e^{\alpha + \beta \varepsilon_i} \tag{4.69}$$

$$n_i = \frac{g_i}{e^{\alpha + \beta \varepsilon_i} - 1} \tag{4.70}$$

Above equation is the mathematical representation of the Bose–Einstein statistics for the most probable distribution of elements among energy levels.

## 4.16 APPLICATION OF BOSE–EINSTEIN STATISTICS FOR A PHOTON GAS

Derivation of Planck's equation for the distribution of energy in black body radiation is possible with the help of Bose–Einstein equation. It appears that radiation in thermal equilibrium in a box with walls, which do not absorb any of the radiations, may be treated as a system of elements, viz. photons, obeying the Bose–Einstein statistics. The energy of photon, according to quantum theory, is equal to $h\nu$, the exchange of a photon between the two states of different frequency would thus result in an energy change. It is postulated, however, that the energy of the system remains constant, consequently there must be a change in the number of photons as the result of two states of different frequency. If the restrictions $\Sigma \delta n_i = 0$ is no longer applicable, the term involving the undetermined multiplier $\alpha$ in the derivation of the Bose–Einstein distribution law does not enter into the argument, the final result, instead of Eq. (4.70), is then

$$n_i = \frac{g_i}{e^{\varepsilon_i/kT} - 1} \tag{4.71}$$

$$\therefore \quad \beta = \frac{1}{kT}$$

Writing $h\nu$ for the energy $E_i$, this expression may be put in a form

$$dn = \frac{g(d\nu)}{e^{h\nu/kT} - 1} \tag{4.72}$$

This is distribution law, where $dn$ is the number of photons at equilibrium in the $g(d\nu)$ elemental states lying in the frequency range $\nu$ to $d\nu$.

Since radiation may be regarded as consisting of electromagnet.

The equation of propagation is given as,

$$\frac{\partial^2 \phi}{\partial x^2} + \frac{\partial^2 \phi}{\partial y^2} + \frac{\partial^2 \phi}{\partial z^2} = \frac{1}{c^2} \frac{\partial^2 \phi}{\partial t^2} \tag{4.73}$$

where $c$ is velocity of light.

The radiation in the box under consideration can be regarded as consisting of standing waves and hence a solution of equation will be

$$\phi = \psi(x, y, z) \, A \sin 2\pi \nu t + B \cos 2\pi \nu t$$

where $\psi(x, y, z)$ is a function of the coordinates $x, y, z$ only. Insertion of this solution into Eq. (4.73). We get

$$\frac{\partial^2 \psi}{\partial x^2} + \frac{\partial^2 \psi}{\partial y^2} + \frac{\partial^2 \psi}{\partial z^2} = -\frac{4\pi^2 \nu^2}{c^2} \psi$$

$$\frac{\partial^2 \psi}{\partial x^2} + \frac{\partial^2 \psi}{\partial y^2} + \frac{\partial^2 \psi}{\partial z^2} = -\frac{4\pi^2 \varepsilon^2}{h^2 c^2} \psi \tag{4.74}$$

Since $\varepsilon$ may be put equal to $h\nu$ by the quantum theory. The wave equation for photon thus becomes,

$$-\frac{h^2 c^2}{4\pi^2} \left( \frac{\partial^2 \psi}{\partial x^2} + \frac{\partial^2 \psi}{\partial y^2} + \frac{\partial^2 \psi}{\partial z^2} \right) = \varepsilon^2 \psi \tag{4.75}$$

Treating the photon as a particle, in a cubical box of side $l$, with components in three directions parallel to the axis $x, y, z$. The energy value in $x$ direction,

$$\varepsilon_x = \frac{hc}{2l} n_x^2 \tag{4.76}$$

For the energy of the component in the direction, similar equation for $\varepsilon_y$ and $\varepsilon_z$. It follows, therefore, that the total energy $\varepsilon$ will be

$$\varepsilon = \frac{hc}{2l} r^2 \tag{4.77}$$

where the resultant quantum number $r$ is defined by

$$r^2 = n_x^2 + n_y^2 + n_z^2 \tag{4.78}$$

Replacing $\varepsilon$ in Eq. (4.77) by $h\nu$ and $l$ by $V^{1/3}$, where $V$ is volume of the cubical box, it is found that

$$r^2 = 2V^{1/3}\, \nu/c$$

The number of eigenstates for systems with frequencies from zero to $\nu$ is seen to be

$$g = \frac{1}{8}\left(\frac{4}{3}\pi r^3\right)$$

$$g = \frac{4}{3}\pi V \frac{\nu^3}{c^3}$$

and hence by differentiation

$$g(d\nu) = 4\pi V \frac{\nu^2}{c^3} d\nu \tag{4.79}$$

For the number of eigenstates lying in the frequency range $\nu$ to $\nu + d\nu$. This result is not complete, however, since radiation of each frequency has two independent directions of polarization, it is necessary to multiply by two giving

$$g(d\nu) = 8\pi V \frac{\nu^2}{c^3} d\nu \tag{4.80}$$

Combining this expression with Eq. (4.72), which is appropriate form of the Bose–Einstein distribution law the result is

$$\frac{dn}{V} = 8\pi \frac{\nu^2}{c^3} \frac{1}{e^{h\nu/kT} - 1} d\nu \tag{4.81}$$

The left-hand side of the equation is the number of photon per unit volume, if this is multiplied by $h\nu$, the energy of a photon, the result is the energy per unit volume, i.e. the energy density $du$ in the given frequency range thus,

$$du = 8\pi \frac{h\nu^3}{c^3} \frac{1}{e^{h\nu/kT} - 1} d\nu \tag{4.82}$$

This is the Planck's equation for the distribution of the energy density over the wavelength (or frequency) of black body radiation at a given temperature.

## 4.17 THE FERMI–DIRAC STATISTICS

Fermi–Dirac statics (F.D.S.) postulates that the $n$ similar elements constituting a given system are distinguishable and that only antisymmetric solutions of the wave equations are permitted. It is assumed that the number of elements is constant and so also is the total energy. In the determination of the number of eigenstates for the system of $n_i$ elements in the group, where $g_i$ eigen-functions are available to each element, it must be remembered that because of the restriction of the Fermi–Dirac statistics to antisymmetric states, only one of the $n_i$ elements can be associated with each of $g_i$ elementary wavefunctions.

It follows, of course, that $g_i$ must be greater than or equal to $n_i$ since these must be at least one elementary wave function for every element in the group. The required number of eigenstates is equivalent to the number of combinations of $g_i$ particles taken $n_i$ at a time, this gives

$$\frac{g_i!}{n_i!(g_i - n_i)!}$$

For the number of different eigenstates in the $i$th group. The total number of eigenstates $G$ for the whole system is then,

$$G = \frac{g_1!}{n_1!(g_1 - n_1)!} \cdot \frac{g_2!}{n_2!(g_2 - n_2)!} \cdots \frac{g_i!}{n_i!(g_i - n_i)!} \cdots \quad (4.83)$$

$$G = \prod_i \frac{g_i!}{n_i!(g_i - n_i)!} \quad (4.84)$$

and the probability of the given state is

$$W = \prod_i \frac{g_i!}{n_i!(g_i - n_i)!} \times \text{constant} \quad (4.85)$$

Proceeding as in the previous case, it is seen that,

$$\ln W = \sum_i \{\ln g_i! - \ln n_i! - \ln (g_i - n_i)!\} + \text{constant}$$

and an introduction of Stirling's approximation, assuming $g_i - n_i$ is very large, as well as $n_i$ and $g_i$, gives

$$\ln W = \sum_i [(g_i - n_i) \ln (g_i - n_i) - n_i \ln n_i + g_i \ln g_i] + \text{constant} \quad (4.86)$$

The condition of most probable state is then

$$\delta \ln W = \sum_i [\ln n_i - \ln(g_i - n_i)]\delta n_i = 0$$

$$= \sum_i \left(\ln \frac{n_i}{g_i - n_i}\right)\delta n_i = 0 \quad (4.87)$$

Upon introducing the conditions of constant number of elements in the system and constant total energy and using the method of undetermined multiplier, it is found that

$$\sum_i \left( \ln \frac{n_i}{g_i - n_i} + \alpha + \beta \varepsilon_i \right) \delta n_i = 0$$

and since the $\delta n_i$ are arbitrary, it follows that

$$\ln \frac{n_i}{g_i - n_i} + \alpha + \beta \varepsilon_i = 0 \qquad (4.88)$$

$$\ln \left( \frac{g_i}{n_i} - 1 \right) = \alpha + \beta \varepsilon_i$$

and

$$\frac{g_i}{n_i} - 1 = e^{\alpha + \beta \varepsilon_i} \qquad (4.89)$$

$$n_i = \frac{g_i}{e^{\alpha + \beta \varepsilon_i} + 1} \qquad (4.90)$$

The most probable distribution, among the various energy levels of the elements of a system obeying the Fermi–Dirac statistics is given by Eq. (4.90).

### 4.17.1 Applications of Fermi–Dirac Statistics

These are

1. Slight gas degeneracy
2. Extreme gas degeneracy
3. Electron gas in metals
4. Thermoionic emission of electron from metals

*Slight gas degeneracy*

The distribution law derired from the Fermi–Dirac statistics is

$$n_i = \frac{g_i}{e^{\alpha + \beta \varepsilon_i} + 1} \qquad (4.91)$$

and this may be the point in the form

$$dn = \frac{g(d\varepsilon)}{Be^{\varepsilon/kT} + 1} \qquad (4.92)$$

$$g(d\varepsilon) = 2Q \sqrt{\frac{x}{\pi}} \, dx$$

where $Q = 2\pi V \dfrac{v^2}{c^2}$

$$g(dE) = 2Q \sqrt{\frac{x}{\pi}} \, dx$$

where $Q = 2\pi V \dfrac{v^2}{c^3}$

Using the value of $g(d\varepsilon)$ given by equation and introducing the additional degeneracy factor $Q_i$, it follows that for point masses with weak interaction

$$dn = \frac{2Q}{\pi^{1/2}} \frac{x^{1/2} dx}{Be^x + 1} \qquad (4.93)$$

where $x = \varepsilon/KT$ and $Q$ is an additional statistical weight factor to allow for other forms of degeneracy. Above equation, is same for Bose–Einstein statistics, except that $+1$ replaces $-1$ in the usual manner.

Integration of above equation gives

$$n = \frac{2Q}{\pi^{1/2}} \int_0^\infty \frac{x^{1/2} dx}{Be^x + 1} \qquad (4.94)$$

and, further

$$E = \frac{2kTQ}{\pi^{1/2}} \int_0^\infty \frac{x^{3/2} dx}{Be^x + 1} \qquad (4.95)$$

The evaluation of the integrals is again simplified if $B > 1$, but this is not a necessity for Fermi–Dirac statistics, as it was for the Bose–Einstein case. The value of $\alpha$ in the distribution Eq. (4.90) may be positive or negative without introducing any difficulties. If $\alpha$ is positive so that $B$ is greater than unity, the condition is referred to as slight gas degeneracy, since the treatment is then quite similar to that used for the Bose–Einstein case, it will be given first. If $B > 1$, it is possible to write

$$(Be^x + 1)^{-1} = \frac{e^{-x}}{B}\left(1 - \frac{e^{-x}}{B} + \frac{e^{-2x}}{B^2} - \cdots\right) \qquad (4.96)$$

which is seen to be similar to the expression for $(Be^x - 1)^{-1}$, except that the signs of alternate terms in the parentheses on the right-hand side are changed. Without going through the details of the argument, it is readily found that for 1 mol of a monoatomic Fermi–Dirac gas,

$$E = \frac{3}{2} RT \left(1 + \frac{1}{2^{5/2} B} - \frac{1}{3^{5/2} B^2} + \cdots\right) \qquad (4.97)$$

$$E = \frac{3}{2} RT \left(1 + \frac{1}{2^{5/2}}\left(\frac{N}{Q}\right) - \frac{1}{3^{5/2}}\left(\frac{N}{Q}\right)^2 + \cdots\right) \qquad (4.98)$$

$$P = \frac{RT}{V}\left(1 + \frac{1}{2^{3/2} B} - \frac{1}{3^{5/2} B^2} + \cdots\right) \qquad (4.99)$$

$$P = \frac{RT}{V}\left(1 + \frac{1}{2^{3/2}}\left(\frac{N}{Q}\right) - \frac{1}{3^{5/2}}\left(\frac{N}{Q}\right)^2 + \cdots\right) \qquad (4.100)$$

where $B$ is defined as

$$\frac{1}{B} = \left(\frac{N}{Q}\right) = \frac{Nh^2}{Q_i (2\pi mkT)^{3/2} V} \qquad (4.101)$$

These equations will give the total energy and pressure, respectively, under condition of slight gas degeneracy.

### Extreme gas degeneration

At low temperature, the value of $B$ may well become much less than unity, the condition of extreme gas degeneration will then arise. It $B$ is assumed to be extremely small or zero as it will be at the absolute zero, then the term $Be^x + 1$ in the denominators of the integrals in Eq. (4.94) and (4.95) will be equal to unity.

Equation (4.94) may then be written as,

$$n = \frac{2Q}{\pi^{1/2}} \int_0^{1/B} x^{1/2} dx \tag{4.102}$$

where the upper limit of integration has been changed from infinity to $1/B$, since $B$ has been taken as zero. Carrying out the integration, the result

$$n = \frac{4Q}{3\pi^{1/2}} \frac{1}{B^{3/2}}$$

$$\frac{1}{B} = \left(\frac{N}{Q}\right) = \left(\frac{3n\pi^{1/2}}{4Q}\right)^{2/3} \tag{4.103}$$

But Eqs. (4.94) and (4.95) in application of Bose–Einstein statistics

$$Q = Q_i \frac{(2\pi mkT)^{3/2}}{h^3} V$$

Putting the value of $Q$ into Eq. (4.103)

$$\frac{1}{B} = \frac{h^2}{2mkT} \left(\frac{3n}{4\pi V Q_i}\right)^{2/3} \tag{4.104}$$

where $V$ is the volume occupied by the $n$ particles.

Just as Eq. (4.94) reduces to Eq. (4.102) in the special case of extreme gas degeneration, i.e. when $B$ is virtually zero. So, Eq. (4.95) becomes,

$$E_0 = \frac{2kTQ}{\pi^{1/2}} \int_0^{1/B} x^{3/2} dx \tag{4.105}$$

$$E_0 = \frac{4kTQ}{5\pi^{1/2}} \frac{1}{B^{5/2}} \tag{4.106}$$

The symbol, $E_0$ being used because the energy is that for the absolute zero. Introducing the value for $1/B$ given by Eq. (4.104) into Eq. (4.106), the result is

$$E_0 = \frac{3nh^2}{10m} \left(\frac{3n}{4\pi V Q_i}\right)^{2/3} \tag{4.107}$$

It is observed that both classical and Bose–Einstein statistics lead to value of zero for the energy of a monoatomic gas at the absolute zero. A Fermi–Dirac gas, however, possesses appreciable energy, i.e. zero point energy at this temperature. The expression for the zero point energy of a highly degenerate Fermi–Dirac gas can be put in an alternative form that is of interest. According to equation

$$g(\Delta \varepsilon) = \frac{4\pi m V}{h^3}(2m\varepsilon)^{1/2} d\varepsilon$$

The number of eigenstates of a point mass with energy between $\varepsilon$ and $\varepsilon + d\varepsilon$ is given by

$$g(\Delta \varepsilon) = Q_i \frac{4\pi m V}{h^3}(2m\varepsilon)^{1/2} d\varepsilon \qquad (4.108)$$

where, $Q_i$ has been introduced to allow for other degeneracy factors. The total number $g^*$ of eigenstates with energy does not exceed a specified value $\varepsilon^*$, is then obtained by integration, thus

$$g^* = Q_i \frac{2\pi V}{h^3}(2m)^{3/2} \int_0^{\varepsilon^*} \varepsilon^{1/2} d\varepsilon$$

$$g^* = Q_i \frac{4\pi V}{3h^3}(2m\varepsilon^*)^{3/2} \qquad (4.109)$$

At the absolute zero all the particles, e.g. electrons will collect in the lowest possible energy states, but since the Fermi–Dirac statistics do not permit more than one element to each eigenstates, it follows that the number of elements is equal to the number of eigenstates. This result can be derived directly from Eq. (4.102), since $B$ is zero at the absolute zero, it is evident that $n_i$ is equal to $g_i$. It follows therefore that the value of $g^*$ as given by Eq. (4.109) may be identified with $n$, the number or elements, hence

$$n = Q_i \frac{4\pi V}{3h^3}(2m\varepsilon^*)^{3/2} \qquad (4.110)$$

Therefore,
$$\varepsilon^* = \frac{h^2}{2m}\left(\frac{3n}{4\pi V Q_i}\right)^{2/3} \qquad (4.111)$$

It is evident, therefore by comparison with Eq. (4.104) that

$$\frac{\varepsilon^*}{kT} = \frac{1}{B} \qquad (4.112)$$

and, from Eq. (4.106)

$$E_0 \propto \frac{3}{5} n\varepsilon^* \qquad (4.113)$$

The quantity $\varepsilon^*$ is sometimes called the *Fermi energy*, it represents the energy of the highest level filled at the absolute zero for the given system. The average energy per single particle, i.e. $E_0/n$ at the absolute zero is thus equal to three-fifth of the value for the particle of highest energy at this temperature, i.e. $3/5\ \varepsilon^*$.

The Fermi–Dirac statistics as already indicated apply to electrons and it is in fact in connection with the properties of the so called "electron gas" that their statistics have proved most useful. For electrons $Q_i$ is equal to 2 because of the two possible spin orientations hence Eq. (4.107) becomes

$$E_0 = \frac{3nh^2}{40m}\left(\frac{3n}{\pi V}\right)^{2/3} \quad (4.114)$$

This gives the energy of a system of $n$ electrons occupying a volume $V$ at the absolute zero.

It has been seen earlier that the pressure $P$ of any gas, irrespective of the type of statistics it obeys is related to the energy $E$ by the equation $P = 2E/3V$ and hence for the Fermi–Dirac gas using equation (4.113)

$$P_0 = \frac{2n}{5V}\varepsilon^* \quad (4.115)$$

for the pressure at the absolute zero. The behaviour of a gas obeying the Fermi–Dirac statistics is thus, quite different in this respect also from Bose–Einstein or classical gases. For the latter two, the pressure of the absolute zero of temperature should be zero, where as for a Fermi–Dirac gas there should be an appreciable zero point pressure.

At the temperature above the absolute zero at which there is still considerable gas degeneration, i.e. $B$ is still less than unity, the evaluation of the integrals presents some difficulties; the only results will be given here. The expression for the energy can be stated in the form of a series that converges rapidly at low temperature, viz.

$$E = \frac{3}{5}n\varepsilon^*\left\{1 + \frac{5\pi^2}{12}\left(\frac{kT}{\varepsilon^*}\right)^2 - \frac{\pi^4}{16}\left(\frac{kT}{\varepsilon^*}\right)^4 + \cdots\right\} \quad (4.116)$$

which obviously reduces to Eq. (4.113) when $T$ is zero. The pressure can be calculated from the energy in the usual manner. The heat capacity of an ideal Fermi–Dirac gas is important, this can be derived from Eq. (4.116) by making of the familiar thermodynamic relationship, viz.

$$C_V = \left(\frac{\partial E}{\partial T}\right)_V$$

$$C_V = \frac{1}{2}nk\pi^2\left(\frac{kT}{\varepsilon^*}\right)\left\{1 - \frac{3\pi^2}{10}\left(\frac{kT}{\varepsilon^*}\right)^2 + \cdots\right\} \quad (4.117)$$

If the temperature is not too high, the quantity in the brackets may be set equal to unity so that

$$C_V = \frac{1}{2}nk\pi^2\left(\frac{kT}{\varepsilon^*}\right) \quad (4.118)$$

The heat capacity of a highly degenerate Fermi–Dirac gas at low temperature is, thus, seen to be directly proportional to the absolute temperature.

### Electron gas in metals

The basis of this theory is that a metal consists of a system of fixed positive nuclei and a number of mobile electrons—generally referred to as electron gas. There is no priori method of estimating how many of the electrons in the metal may be regarded as free. For alkali metal, it is reasonable to suppose that there is one free electron for each atom and this same assumption leads to satisfactory results for other metals. In the treatment of the gas in metals, it is postulated that the free electrons remain in the potential box constituted by the metal, its potential energy will have this value, but in order to leave the box, that is, for the electrons to be emitted from the metal, it must acquire additional energy.

Theoretically, the Fermi–Dirac statistics must of course, apply to electrons but it is of interest to see what the conditions are under which the classical statistics would represent an adequate approximations. Information in these connections can be obtained by calculating the value of $B$; according to Eq. (4.104), $B$ is given by

$$B = \frac{2mkT}{h^2}\left(\frac{4\pi V Q_i}{3n}\right)^{2/3} \tag{4.119}$$

For electrons, the mass $m = 9.1 \times 10^{-28}$ g and $Q_i = 2$. Consider an average metal of atomic weight 100 and density 10 g/cc the volume $V$ of 1 g atom is 10 cc and the number of atom and hence of electrons $n$, assuming one free electron for each atom is $6.02 \times 10^{23}$. Under these conditions Eq. (4.119) gives

$$B = 1.5 \times 10^{-5} T$$

It is at once evident that if $B$, i.e. $e^\alpha$ is to be of the order of unity, so that the classical distribution might be applicable, the temperature would have to be extremely high. It is apparent therefore, that at all reasonable temperature and especially at low temperature, it is necessary to use the equations of the Fermi–Dirac statistics in connection with the study of the electron gas in metals. A simple illustration of this fact is provided by a consideration of the contribution of electrons to the heat capacity of a metal; at all reasonable temperature, this is given by Eq. (4.118) and remembering that $kT/\varepsilon^*$ is equal to $B$. It follows that the electron heat capacity per gram atom of metal is then

$$(C_V)_{\text{elec}} = 1/2\, R\, \pi^2 B \tag{4.120}$$

where $R$ is molar gas constant. Utilizing the average value of $B$ derived above, it is found that

$$(C_V)_{\text{elec}} = 1.5 \times 10^{-4}\, T \text{ cal per g atom} \tag{4.120}$$

At an ordinary temperature, viz. 300 K, the electron contribution to the atomic heat capacity of a metal will be about 0.05 cal. This is small in comparison with the total heat capacity of more than 6 cal and so there is little hope of observing the electronic heat capacity under these conditions. It will be shown by means of the Debye theory of the specific heat of solids that if the electronic contribution is ignored, the heat capacity of a solid element at very low temperatures should be proportional to $T^3$. Subtracting the heat capacity calculated by the Debye equation from the observed value should be the heat capacity due to the electron gas. The results are in general agreement with the Eq. (4.120) in particular, the electronic contribution to the heat capacity has been found to be proportional to the absolute temperature in agreement with expectation from Fermi–Dirac statistics.

The pressure of the electron gas in a metal at the absolute zero may be calculated by means of Eq. (4.115) for an average metal, i.e. atomic weight 100 and density 10 g/cc this pressure is found to be of the order of $10^5$ atm. At this normal temperature the pressure would be even higher. The reason why the free electrons of these high pressures remain within metal is because of the electrostatic attraction between the electrons and the fixed positive ions that constitute the crystal lattice.

### Thermoionic emission of electrons from metals

The translational energy $\varepsilon_x$ in the $x$ direction of a particle of mass $m$ moving in a cubical box of side $l$ is given as,

$$\varepsilon_x = \frac{h^2}{8ml^2} n_x^2 \qquad (4.121)$$

where $n_x$ is the appropriate quantum number. Further, it was seen earlier that $n_x$ also gives the number of eigenstates with energy lying between zero and $\varepsilon_x$ in the direction parallel to $x$-axis. Since the energy is entirely kinetic, it is possible to replace $\varepsilon_x$ by $p_x^2/2m$, where $p_x$ is the $x$ component of the momentum, making this substitution, Eq. (4.121) yields

$$p_x^2 = \frac{h^2}{4l^2} n_x^2$$

Therefore, 
$$n_x = \frac{2l}{h} p_x \qquad (4.122)$$

Since $n_x$ is directly proportional to $g(dp_x)$ because it gives eigenstates of the energy with respect to a particular quantum states

Therefore, 
$$n_x = g(dp_x) = \frac{2l}{h} dp_x \qquad (4.123)$$

The number of eigenstates in the range $dp_x$, $dp_y$ and $dp_z$, i.e. the element of momentum space having this extension would be

$$g(dp) = \frac{8l^3}{h^3} dp_x dp_y dp_z \qquad (4.124)$$

So far no account has been taken of the degeneracy of the energy states due to other causes, such as spin for electrons, which are now being considered. It is necessary to multiply by two, as seen above, hence Eq. (4.124) becomes,

$$g(dp) = \frac{16V}{h^3} dp_x dp_y dp_z \qquad (4.125)$$

where, $l^3$ has been replaced by $V$, volume of container. It has been mentioned that the terms $dp_x$, etc., refer to absolute values irrespective of sign, it is now required to take the sign of the momentum into account, since in the subsequent treatment it is necessary for the components of the momentum of the various electrons to have the same sign. For this purpose, it may be supposed that half the total number of electrons have positive $p_x$ while the other half have negative $p_x$, the same assumption may also be made for the components $p_y$ and $p_z$. The number of eigenstates will then be one-eight as many as that given by Eq. (4.125), thus for the present case it is possible to write

$$g(dp) = \frac{2V}{h^3} dp_x dp_y dp_z \qquad (4.126)$$

Introducing this result into the Fermi–Dirac distribution law Eq. (4.92) gives

$$dn = \frac{2V}{h^3} \frac{1}{Be^{\varepsilon/kT} + 1} dp_x dp_y dp_z$$

For the number of electrons in the volume $V$, each having a total kinetic energy $\varepsilon$ and momenta within the specified range. The corresponding number of unit volume is then

$$dn = \frac{2}{h^3} \frac{1}{Be^{\varepsilon/kT} + 1} dp_x dp_y dp_z \qquad (4.127)$$

The number of electrons $n(dp_x)$ with the $x$ component of momentum in the range $dp_x$ that are incident in unit time on a unit area, normal to the $x$-axis is obtained by integrating Eq. (4.127) over all possible values of $p_y$ and $p_z$. This satisfy the specified condition and the result must be multiplied by the speed of the electrons $p_x/m$ in the direction, thus

$$n(dp_x) = \frac{2p_x}{h^3 m} dp_x \int_{-\infty}^{+\infty} \int_{-\infty}^{+\infty} \frac{1}{Be^{\varepsilon/kT} + 1} dp_y dp_z \qquad (4.128)$$

The total kinetic energy $\varepsilon$ of the electrons can be represented by

$$E = \frac{1}{2m}(p_x^2 + p_y^2 + p_z^2)$$

and if quantity $p_p^2$ is defined by

$$p_p^2 = p_y^2 + p_z^2$$

it follows that $E = \dfrac{1}{2m}(p_x^2 + p_p^2)$ \hfill (4.129)

Making this substitution in Eq. (4.128) and at the same time changing the variables, the result obtained is

$$n(dp_x) = \frac{4\pi}{h^3 m} p_x dp_x \int_0^{+\infty} \frac{1}{Be^{(p_x^2 + p_p^2)/2mkT} + 1} p_p dp_p$$

and on evaluation of the integral, it is found that

$$n(dp_x) = \frac{4\pi kT}{h^3} \ln\left(\frac{e^{-p_x^2/2mkT}}{B} + 1\right) p_x dp_x \qquad (4.130)$$

The kinetic energy in the $x$ direction is given by

$$E = \frac{p_x^2}{2m}$$

Therefore, $md\varepsilon_x = p_x dp_x$.

The number $n(d\varepsilon_x)$ of electrons, with kinetic energy in the range $\varepsilon_x$ is the $\varepsilon_x + d\varepsilon_x$ striking the surface referred to above, is then obtained from Eq. (4.130) as

$$n(d\varepsilon_x) = \frac{4\pi mkT}{h^3} \ln\left(\frac{e^{-\varepsilon_x/kT}}{B} + 1\right) d\varepsilon_x \qquad (4.131)$$

According to Eq. (4.112) $\frac{\varepsilon^*}{kT} = \frac{1}{B}$, where $\varepsilon^*$ is the so-called Fermi energy for a highly degenerate gas at very low temperature. For the present problem, however, it is possible to write as a rough approximation

$$\frac{\varepsilon^*}{kT} \cong \frac{1}{B}$$

and introduction of this result into equation (4.131) gives

$$n(d\varepsilon_x) = \frac{4\pi mkT}{h^3} \ln\left(\frac{e^{-(\varepsilon_x - \varepsilon^*)/kT}}{B} + 1\right) d\varepsilon_x \qquad (4.132)$$

If the quantity $\varepsilon_x - \varepsilon^*$ is large in comparison with $kT$, all terms beyond the first may be neglected in the expansion of logarithm in the form of a power series, i.e. $\ln(x+1) \cong x$ if $x$ is small, hence

$$n(d\varepsilon_x) = \frac{4\pi mkT}{h^3} e^{-(\varepsilon_x - \varepsilon^*)/kT} d\varepsilon_x \qquad (4.133)$$

It will now be supposed that the electrons under consideration constitute the electron gas in a metal. Equation (4.133) then gives the number of these electrons coming from the inside of the metal that are incident on unit area of its surface in unit time with kinetic energy, normal to the surface, lying in the range $\varepsilon_x$ to $\varepsilon_x + d\varepsilon_x$. If $\eta$ is the work required to take a single electron from rest inside the metal and bring it outside the metal, also in a state of rest, the number of electrons $Z$ striking the surface of the metal, from inside, that succeed in escaping is given by

$$Z = \int_\eta^\infty n(d\varepsilon_x)$$

$$= \frac{4\pi mkT}{h^3} \int_\eta^\infty e^{-(\varepsilon_x - \varepsilon^*)/kT} d\varepsilon_x$$

$$= \frac{4\pi mk^2T^2}{h^3} e^{-(\eta - \varepsilon^*)/kT} \qquad (4.134)$$

The Fermi energy $\varepsilon^*$ is the energy of an electron in the highest occupied level in the metal, at the absolute zero, hence the quantity $\eta - \varepsilon^*$, which appears in the experimental term in Eq. (4.134) may be regarded as the average energy required to remove an electron from the interior of the metal, at constant temperature and volume and bring it to rest outside the metal. This energy quantity, which may be given the symbol $\chi$ and is generally expressed in electron volts is known as the *thermoionic work function*, making this substitution of $\chi$ form $\eta - \varepsilon^*$ in Eq. (4.134), the latter becomes

$$Z = \frac{4\pi m k^2 T^2}{h^3} e^{-\chi/kT} \qquad (4.135)$$

This expression gives the rate at which electrons leave unit area of a metal, as a function, of temperature. It is therefore, the equation for the phenomenon of thermoionic emission. In practice, the emission is measured as the current flowing from unit area of the heated metal to a collecting electrode, the magnitude of the current $I$ is obtained by multiplying the number of electrons leaving the unit area of the surface in unit time by the electronic charge $e$, thus

$$I = Ze = \frac{4\pi m k^2 e}{h^3} T^2 e^{-\chi/kT} \qquad (4.136)$$

$$= A T^2 e^{-\chi/kT} \qquad (4.137)$$

where $A$ is a universal constant. This result is identical in form with the empirical Richardson equation for the thermoionic emission.

If the magnitude of the constant $A$ is evaluated, it is found that it should theoretically be equal to 120 A cm$^{-2}$deg$^{-2}$. Experimental observation, however, has led in a number of reliably cases to a figure of approximately 60 A cm$^{-2}$ deg$^{-2}$ and several explanations have been proposed to account for the discrepancy. It will be noted, in the first place, that in the derivation of Eq. (4.136) the tactic assumption was made that every electron reaching the surface of the metal from the interior with energy in excess of the amount $\eta$ would actually succeed in escaping. It is possible, however, some of the electrons which theoretically have sufficient energy to permit their escape, are reflected at the surface, the result obtained in Eq. (4.136) should thus be multiplied by transmission coefficient equal to the fraction of electrons reaching the surface that are not reflected. According to the experimental results, this transmission coefficient should have value of about 0.5, in several cases although quantum mechanical calculations indicate that a figure much closer to unity is to be expected. Another possibility is that the thermoionic work function is not independent of temperature, as is assumed in the calculation of the value of $A$ from the experimental results. Such a temperature variation, if not allowed for, would influence the apparent magnitude of $A$.

## 4.18 COMPARISON OF THREE STATISTICS

For the purpose of comparison of the three forms of statistics, the essential equations obtained in each case will be repeated here, they are

1. *Bose–Einstein statistics:*

$$\frac{g_i}{n_i} + 1 = e^{\alpha + \beta \varepsilon_i}$$

or

$$n_i = \frac{g_i}{e^{\alpha + \beta \varepsilon_i} - 1}$$

2. *Fermi–Dirac statistics:*

$$\frac{g_i}{n_i} - 1 = e^{\alpha + \beta \varepsilon_i}$$

or
$$n_i = \frac{g_i}{e^{\alpha+\beta\varepsilon_i}+1}$$

3. *Maxwell–Boltzmann statistics:*

$$\frac{g_i}{n_i} = e^{\alpha+\beta\varepsilon_i}$$

or
$$n_i = \frac{g_i}{e^{\alpha+\beta\varepsilon_i}}$$

The condition under which Bose–Einstein and Fermi–Dirac statistics yield result virtually identical to those given by the Maxwell–Boltzmann statistics. This will occur if $g_i/n_i$ is very large in comparison with unity, so that

$$\frac{g_i}{n_i}+1 \approx \frac{g_i}{n_i}-1 \approx \frac{g_i}{n_i}$$

In these circumstances, all the three expressions for the distribution law will become equivalent to that for the Maxwell–Bolzmann statistics.

In general, provided the temperature is not too low or the pressure too high, the number of available eigenstates $g_i$ is large in comparison with the number of elements $n_i$ so that $g_i/n_i$ is then much greater than unity. It follows, therefore, that for almost all conditions under which normal gases exist the classical distribution law should be adequate to describe their actual behaviour, within the possible limits of experimental observation. There are a limited number of cases in which the classical distribution law is not applicable, but there are very few, they are chiefly three, viz. radiation, liquid helium II and the electron gas in metals apart from conditions of extremely low temperature or high pressure.

If laws of classical mechanics are applied according to which the individual atoms or molecules are supposed to have definite position and momenta, it is called classical statistical mechanics. If laws of quantum mechanics are applied, according to which the individual atoms or molecules are supposed to have only quantized statistical mechanics. The only classical mechanics is that of Maxwell and Boltzmann while there are two different quantum statistical mechanics, namely, Bose–Einstein and Fermi–Dirac statistics.

Another important difference between the classical statistics and quantum statistics lies in the fact that the former treats the different particles as distinguishable whereas the latter treats them as indistinguishable.

Further, in Maxwell–Boltzmann statistics and Bose–Einstein statistics, any number of particles may occupy the same energy level, whereas in Fermi–Dirac statistics only one particle is supposed to occupy a particular energy level. The particles obeying M.B. statistics are called *Maxwellons* or *Boltzmannos* whereas those following Bose–Einstein and Fermi–Dirac statistics are respectively called *Bosons* and *Fermions*.

Bose–Einstein statistics is applied generally to those systems, which have symmetric wavefunction i.e. wavefunctions whose sign does not change on interchange of the coordinates of any two particles in the system. Systems possessing such wavefunctions are photons and atoms or molecules whose nuclei contain total even number of protons and neutrons, e.g. $He^4$, $D_2$, $N_2$, etc.) On the other hand, Fermi–Dirac statistics is applied to systems having asymmetric wavefunctions i.e. wavefunctions whose sign changes when the co-ordinates of any two particles are changed. Systems possessing such wavefunctions are electrons, protons and atoms or molecules whose nuclei contain total odd number of protons and neutrons.

## EXERCISES

1. What is ensemble? Define various types of ensembles.
2. Define Sterling's formula of approximation.
3. Derive an equation for most probable distribution in a macroscopic system.
4. Define the term fluctuation. What do you understand by (i) mean square fluctuation, (ii) relative fluctuation and (iii) mean square relative fluctuation.
5. Derive relation for energy fluctuation in canonical ensemble.
6. Derive relation for fluctuation in density.
7. What is Maxwell–Boltzmann distribution law? Derive Maxwell–Boltzmann distribution equation.
8. Derive equation for Maxwell–Boltzmann distribution law of velocities.
9. What is Bose–Einstein statistics? Derive relation for Bose–Einstein statistics.
10. Explain application of Bose–Einstein statistics for a photon gas.
11. What is Fermi–Dirac statistics? Derive relation for Fermi–Dirac statistics.
12. Explain applications of Fermi–Dirac statistics in case of slight gas degeneracy.
13. Explain applications of Fermi–Dirac statistics in case of extreme gas degeneracy.
14. Explain applications of Fermi–Dirac statistics in case of electron gas in metals.
15. Compare three types of statistics i.e. Fermi–Dirac statistics, Bose–Einstein statistics and Maxwell–Boltzmann statistics.

## SUGGESTED READINGS

Atkins, P. and J.D. Paula, *Atkins Physical Chemistry*, Oxford University Press, 2004.

Baielein, R., *An Introduction to Statistical Mechanics: Atoms and Information Theory*, Freeman Publication, 1971.

Baracca, A., R. Livi, and S. Ruffio, *Statistical Mechanics: Foundation, Problems, Perspectives*, World Scientific, 2002.

Berry, S.R., S.A. Rice, and J. Ross, *Matter in Equilibrium: Statistical Mechanics and Thermodynamics*, 2nd ed., Oxford University Press, 2001.

Betts D.S. and R.E. Turner, *Introductory Statistical Mechanics*, Addison-Wesley, 1993.

Bowley, R. and M. Sanchez, *Introductory Statistical Mechanics*, 2nd ed., Oxford University Press, 2000.

Bowley, R. and M. Sanchez, *Introductory Statistical Mechanics*, Oxford University Press, 1996.

Chandler, D., *Introduction to Model Statistical Mechanics*, Oxford University Press, 1987.

Chandler, D., *Introduction to Modern Statistical Mechanics*, Oxford University Press, 1987.

Cowan, B., *Topics in Statistical Mechanics*, World Scientific, 2005.

Darlas, T.C., *Statistical Mechanics*, IOP Publishing, 1999.

Davidson, N., *Statistical Mechanics*, Dover Publications, 2003.

Gale, J.D. and J.M. Seddon, *Thermodynamics and Statistical Mechanics*, Wiley Interscience, New York, 2002.

Garvod, C., *Statistical Mechanics and Thermodynamics*, Oxford University Press, 1995.

Glazer, A.M. and J.S. Wark, *Statistical Mechanics: A Survival Guide*, Oxford University Press, 2002.

Greiner, W., L. Neise, and H. Stocker, *Thermodynamics and Statistical Mechanics*, Springer Verlag, 1995.

Haung, K., *Thermodynamics and Statistical Mechanics*, 2nd ed., Wiley, 1987.

Huang, K., *Statistical Mechanics*, John Wiley, 1987.

Kubo, R., *Statistical Mechanics*, North Holland, 1965.

Ma, S.K., *Statistical Mechanics*, World Scientific Publication, 1985.

McQuarrie, D.A., *Statistical Mechanics*, Harper and Row, 1976.

McQuarrie, D.A., *Statistical Mechanics*, Universal Science Books, 2000.

Metiu, H., *Physical Chemistry: Statistical Mechanics*, Taylor and Francis, 2005.

Morandi, G., E. Ercolessi, and F. Napoli, *Statistical Mechanics: An Intermediate Course*, 2nd ed., World Scientific, 2001.

Pathria, R.K., *Statistical Mechanics*, 2nd ed., Butterworth Heinemann, 1996.

Phiiles, G., *Elementary Lectures in Statistical Mechanics*, Springer-Verlag, 2000.

Stowe, K.S., *Introduction to Statistical Mechanics and Thermodynamics*, John Wiley, 1984.

Tolman, R.C., *Principles of Statistical Mechanics*, 2nd ed., Dover Publications, 1979.

Trevena, D.H., *Statistical Mechanics: An Introduction*, Prentice Hall, 1993.

Widom, B., *Statistical Mechanics: A Concise Introduction for Chemists*, Cambridge University Press, 2002.

# Part II
# Spectroscopy

# Quantum Mechanics

Classical mechanics based on Newton's laws of motion and Maxwell's electromagnetic wave theory was able to explain phenomenon related to large size objects (macroscopic particles). However, it fails when applied to small particles such as electrons, atoms, molecules, etc. (microscopic particles). For example, according to classical mechanics, it should be possible to determine simultaneously the position and velocity (or momentum) of a moving particle, but this is contradicted by Heisenberg's uncertainty principle. Similarly, classical mechanics assumes that the energy is emitted or absorbed continuously whereas Planck's quantum theory postulates that energy is emitted or absorbed not continuously, but discontinuously in the form of packets of energy, called quanta. Further, the concept of quantum numbers was introduced arbitrarily to explain the atomic spectra. A few phenomena for which the classical mechanics fails to give satisfactory explanation are: black body radiation, photoelectric effect, heat capacities of solids, and atomic and molecular spectra. To explain the above phenomena, quantum mechanics comes into existence. In view of the failure of classical mechanics to explain the phenomenon associated with the small particles, a new mechanics has been put forward to explain these phenomena. One of these is called the *matrix mechanics* put forward by Heisenberg in 1925. It is purely mathematical and does not assume any atomic model. The other is called the *wave mechanics* put forward by Schrödinger in 1926. It is based on de Broglie concept of dual character of matter and thus takes into account the particle as well as wave nature of material particles. However, it has been shown that both the mechanics are essentially equivalent so far as the basic physical concepts are concerned. Wave mechanics being comparatively simpler and more useful in application to chemistry, and will be discussed here. It is also called *particle mechanics* or *quantum mechanics* because it deals with the problems that arise when particles such as electrons, nuclei, atoms, molecules, etc. are subjected to a force.

The branch of science which takes into consideration of de Broglie concept of dual nature of matter and Planck's quantum theory, and is able to explain the phenomena related to small particles is known as *quantum mechanics*.

The heart of the quantum mechanics is an equation called Schrödinger wave equation. It can be derived directly or on the basis of certain postulates of quantum mechanics.

## 5.1 SCHRÖDINGER WAVE EQUATION (SWE)

In 1926, Erwin Schröndinger gave Schröndinger wave equation (SWE), basic fundamental equation of quantum mechanics. The wave equation is,

$$\frac{\partial^2 \psi}{\partial x^2} + \frac{\partial^2 \psi}{\partial y^2} + \frac{\partial^2 \psi}{\partial z^2} + \frac{8\pi^2 m (E-V)}{h^2} \psi = 0 \tag{5.1}$$

In terms of Laplacian operators

$$\nabla^2 \psi + \frac{8\pi^2 m (E-V)}{h^2} \psi = 0$$

$$\nabla^2 \psi = \frac{-8\pi^2 m (E-V)}{h^2} \psi \tag{5.2}$$

Multiplying both sides by $\frac{-h^2}{8\pi^2 m}$, we get

$$-\frac{h^2}{8\pi^2 m} \nabla^2 \psi = E\psi - V\psi$$

$$-\frac{h^2}{8\pi^2 m} \nabla^2 \psi + V\psi = E\psi$$

$$\left( -\frac{h^2}{8\pi^2 m} \nabla^2 + V \right) \psi = E\psi \tag{5.3}$$

This equation implies that the operation $\left( \frac{-h^2}{8\pi^2 m} \nabla^2 + V \right)$ carried on the function $\psi$ is equal to the total energy multiplied with the function $\psi$. The operator $\left( \frac{-h^2}{8\pi^2 m} \nabla^2 + V \right)$ is called *Hamiltonian operator* and is represented by $\hat{H}$. Thus,

$$\hat{H}\psi = E\psi \tag{5.4}$$

This is another short form of Schrödinger wave equation. Here $\psi$ is called the eigenfunction and $E$ is called eigenvalue.

# Quantum Mechanics

**Note:**

1. Eigenfunction is said to be normalized set, if $\int \psi_n^* \psi_m d\tau = 1$
2. Eigenfunction is said to be orthogonal set, if $\int \psi_n^* \psi_m d\tau = 0$
3. Eigenfunction is said to be orthonormal set, if $\int \psi_n^* \psi_m d\tau = \begin{pmatrix} 0, & n \neq m \\ 1, & n = m \end{pmatrix}$
4. Addition of operator follows the rule, $(\hat{A} + \hat{B})f = \hat{A}f + \hat{B}f$
5. Subtraction of operator follows the rule, $(\hat{A} - \hat{B})f = \hat{A}f - \hat{B}f$
6. Multiplication of operator follows the rule, $\hat{A} \cdot \hat{B}f = \hat{B}f = f'; \hat{A}f' = f''$
7. $\hat{A} \cdot \hat{B}f \neq \hat{B} \cdot \hat{A}f$. If $\hat{A} \cdot \hat{B}f = \hat{B} \cdot \hat{A}f$ than two operators are said to commute.

## 5.1.1 Postulates of Quantum Mechanics

SWE can be derived from two concepts. One is using the wave motion of electrons given jointly by de Broglie and Heisenberg. Another method is by making use of certain postulates of quantum mechanics. These postulates are:

1. To each observable quantity in classical mechanics like position, velocity, momentum, energy, etc., there corresponds a certain mathematical operators in quantum mechanics, the nature of which depends upon the classical expression for the observable quantity. For example (see Table 5.1):

**TABLE 5.1** Few examples of mathematical operators for various observable quantities

| Operation | Operator | Result of operation on $x^3$ |
|---|---|---|
| Taking the square | $()^2$ | $x^6$ |
| Taking the square root | Root | $x^{3/2}$ |
| Multiplication by a constant $K$ | $K$ | $Kx^3$ |
| Differentiation w.r.t. $x$ | $d/dx$ | $3x^2$ |
| Integration w.r.t. $x$ | $\int () dx$ | $x^4/4 + C$ |

(a) Operator corresponding to a position co-ordinate is multiplication by the value of that co-ordinate, i.e. operator for a position co-ordinate $x$ is the multiplier $x$.

(b) The operation representing the momentum $p$ in the direction of any co-ordinate $x$ is the differential operator.

$$\frac{h}{2\pi r} \frac{\partial}{\partial x} \quad \text{(Bohr permitted orbital)}$$

or

$$-\frac{ih}{2\pi} \frac{\partial}{\partial x} = p_x$$

where $i$ is iota and is given by $\sqrt{-1}$.

2. If $\psi$ is a well behaved function for a given state of a system and $\hat{A}$ is a suitable operator for the observable quantity or property, then the operation on $\psi$ by the operator $\hat{A}$ gives $\psi$ multiplied by a constant value (say $a$) of the observable property, i.e.

$$\hat{A}\psi = a\psi$$

The given state is called the *eigenstate of the system*, $\psi$ is called the eigenfunction and $a$ is called the eigenvalue.

3. If a number of measurements are made over the configuration space, then the average value of the quantity (represented by $\bar{a}$) is given by

$$\bar{a} = \frac{\oint \psi^* \hat{A}\psi \, d\tau}{\oint \psi^* \psi \, d\tau}$$

The basis of this postulate is as follows:

$$a = \frac{\hat{A}\psi}{\psi}$$

But this form of expression is not used in quantum mechanics because of the expression $A\psi/\psi$ will be a function of the coordinates of the system and as such, it varies from place to place and cannot be equated to a constant $\lambda$. If, however, both the numerator and the denominator on the right-hand side of above equation are multiplied with $\psi$ and integrated over the entire space accessible to the system, then the expression $\dfrac{\oint \psi^* \hat{A}\psi \, d\tau}{\oint \psi^* \psi \, d\tau}$ no longer becomes a function of the co-ordinates and can then appropriately be equated to an average value of the constant $\lambda$.

4. If the wavefunction $\psi$ of a system is given at a certain instant, not only properties of the system at that instant are described, but its behaviour at all subsequent instants is also determined. The mathematical expression of this fact is that $(\partial\psi/\partial t)$ at any instant must be determined by the value of the function itself at that instant.

The time development of a wavefunction is given by

$$\frac{ih}{2\pi} \frac{\partial \psi}{\partial t} = \hat{H}\psi \tag{5.5}$$

which is a Schröndinger's equation. By substituting $\hat{H}$.

$$\left(-\frac{h^2}{8\pi^2 m}\nabla^2 + V(r)\right)\psi = \frac{ih}{2\pi}\frac{\partial \psi}{\partial t} \tag{5.6}$$

The wavefunction $\psi$ in Eq. (5.6) is obviously a function of the space co-ordinates of a particle, i.e. $x, y, z$ and time $t$. Using a collective symbol $r$ for these space co-ordinates $x, y$ and $z$, one can write equation as

$$-\frac{h^2}{8\pi^2 m}\frac{\partial^2 \psi}{\partial r^2} + V(r)\psi = \frac{ih}{2\pi}\frac{\partial \psi}{\partial t} \qquad (5.7)$$

The wavefunction $\psi$ may be factorized into two functions, one depending only on $r$ and the other on $t$.

$$\psi(r,t) = \psi(r)\,\psi'(t)$$

$$\frac{\partial \psi_{(r,t)}}{\partial r} = \psi'(t)\frac{\partial \psi_{(r)}}{\partial r}$$

$$\frac{\partial^2 \psi_{(r,t)}}{\partial r^2} = \psi'(t)\frac{\partial^2 \psi_{(r)}}{\partial r^2} \qquad (5.8)$$

Similarly, $\dfrac{\partial \psi_{(r,t)}}{\partial t} = \psi(r)\dfrac{\partial \psi'_{(t)}}{\partial t}$ \qquad (5.9)

Using Eqs. (5.8) and (5.9) in Eq. (5.7), one obtains

$$-\frac{h^2}{8\pi^2 m}\psi'(t)\frac{\partial^2 \psi(r)}{\partial r^2} + V(r)\psi(r)\psi'(t) = \frac{ih}{2\pi}\psi(r)\frac{\partial \psi'(t)}{\partial t} \qquad (5.10)$$

Dividing Eq. (5.10) by $\psi(r)\,\psi'(t)$, we have

$$\left\{-\frac{h^2}{8\pi^2 m}\frac{1}{\psi(r)}\frac{\partial^2 \psi(r)}{\partial r^2} + V(r)\right\} = \frac{ih}{2\pi}\frac{1}{\psi'(t)}\frac{\partial \psi'(t)}{\partial t} \qquad (5.11)$$

In Eq. (5.11), terms in the parenthesis on the left-hand side are independent of $t$, while the terms on the right-hand side is independent of $r$. Thus, the variables $r$ and $t$ are separated and each side of Eq. (5.11) may be equated to a constant, say $E$, giving

$$-\frac{h^2}{8\pi^2 m}\frac{\partial^2 \psi(r)}{\partial r^2} + V(r)\psi(r) = E\cdot\psi(r) \qquad (5.12a)$$

$$\frac{ih}{2\pi}\frac{\partial \psi'(t)}{\partial t} = E\cdot\psi'(t) \qquad (5.12b)$$

Equation (5.12a) is a time independent Schröndinger wave equation and Eq. (5.12b) is a time dependent Schrödinger wave equation. Hence, fourth postulate is defined as for every time independent state of system, a function $\psi$ of the co-ordinates can be written which is single valued, continuous and finite throughout the whole configuration space. This function describes completely the state of the system.

## 5.1.2 Derivation of SWE Based on de Broglie Dual Nature and Heisenberg Uncertainty Principle

Consider a vibration of a stretched string travelling along the $x$-axis with a velocity $u$. If $\omega$ is the amplitude of the wave at any point whose co-ordinate is $x$ at any time $t$, then the equation for such a wave motion is

$$\frac{\partial^2 \omega}{\partial x^2} = \frac{1}{u^2} \frac{\partial^2 \omega}{\partial t^2} \tag{5.13}$$

$\omega$ is a function of $x$ and $t$. Hence, we may write

$$\omega = f(x) f'(t) \tag{5.14}$$

But for the stationary waves, as occur in stretched string, we know that

$$f'(t) = A \sin 2\pi \nu t \tag{5.15}$$

where $\nu$ is the frequency of vibration and $A$ is a constant, equal to the maximum amplitude of the wave.

Substituting the value of $f'(t)$ from Eq. (5.15) into Eq. (5.14)

$$\omega = f(x) A \sin 2\pi \nu t \tag{5.16}$$

Differentiating this equation twice w.r.t. $t$, we get

$$\frac{\partial^2 \omega}{\partial t^2} = -f(x) 4\pi^2 \nu^2 (A \sin 2\pi \nu t)$$

$$= -4\pi^2 \nu^2 f(x) f'(t) \tag{5.17}$$

Differentiating Eq. (5.14) twice w.r.t. $x$, we get

$$\frac{\partial^2 \omega}{\partial x^2} = \frac{\partial^2 f(x)}{\partial x^2} f'(t) \tag{5.18}$$

Substituting the values of $\partial^2 \omega/\partial t^2$ and $\partial^2 \omega/\partial x^2$ from Eqs. (5.17) and (5.18) into the Eq. (5.13), we get

$$\frac{\partial^2 f(x)}{\partial x^2} f'(t) = \frac{1}{u^2} - 4\pi^2 \nu^2 f(x) f'(t)$$

$$\frac{\partial^2 f(x)}{\partial x^2} = \frac{-4\pi^2 \nu^2}{u^2} f(x) \tag{5.19}$$

This equation is time independent and hence, represents the variation of the amplitude function $f(x)$ with $x$ only.

Further, velocity $u$ is related to the frequency $\nu$ and the wavelength $\lambda$ of the wave by the expression

$$u = \nu \lambda \tag{5.20}$$

Substituting this value in Eq. (5.19)

$$\frac{\partial^2 f(x)}{\partial x^2} = \frac{-4\pi^2}{\lambda^2} f(x) \quad (5.21)$$

This equation has been derived for the wave motion in one direction only. The wave equation for wave motion in three dimensions will be

$$\frac{\partial^2 \psi}{\partial x^2} + \frac{\partial^2 \psi}{\partial y^2} + \frac{\partial^2 \psi}{\partial z^2} = \frac{-4\pi^2}{\lambda^2} \psi \quad (5.22)$$

According to de Broglie equation $\lambda = \frac{h}{mv}$ Eq. (5.22) becomes

$$\frac{\partial^2 \psi}{\partial x^2} + \frac{\partial^2 \psi}{\partial y^2} + \frac{\partial^2 \psi}{\partial z^2} = \frac{-4\pi^2 m^2 v^2}{h^2} \psi \quad (5.23)$$

Further, we know that
Total energy $E$ of the particle = K.E. + P.E.

$$E = 1/2\, mv^2 + V$$
$$1/2\, mv^2 = E - V$$
$$mv^2 = 2(E - V) \quad (5.24)$$

Substituting the value of $mv^2$ from Eq. (5.24) in Eq. (5.23), we get

$$\frac{\partial^2 \psi}{\partial x^2} + \frac{\partial^2 \psi}{\partial y^2} + \frac{\partial^2 \psi}{\partial z^2} + \frac{8\pi^2 m(E-V)}{h^2} \psi = 0$$

This equation which represents the wave motion of the particle in three dimensions is called Schrödinger wave equation.

### 5.1.3 Derivation of SWE Based on Postulates of Quantum Mechanics

As we know
$$E = T + V \quad (5.25)$$
$$T = 1/2\, mv^2 = p^2/2m \quad (5.26)$$

where $p$ represents the total momentum of the particle.
Further, we know that

$$p^2 = p_x^2 + p_y^2 + p_z^2 \quad (5.27)$$

$$\therefore \quad T = \frac{p_x^2 + p_y^2 + p_z^2}{2m} \quad (5.28)$$

Substituting this value in Eq. (5.25), we get

$$E = \frac{p_x^2 + p_y^2 + p_z^2}{2m} + V \quad (5.29)$$

But according to the first postulate of quantum mechanics, the operators for the momenta $p_x$, $p_y$ and $p_z$ are,

$$p_x = \frac{h}{2\pi i} \frac{\partial}{\partial x}, \quad p_y = \frac{h}{2\pi i} \frac{\partial}{\partial y} \quad \text{and} \quad p_z = \frac{h}{2\pi i} \frac{\partial}{\partial z} \tag{5.30}$$

$V$ is a function of position co-ordinates and hence, the operator for $V$ is $V$ itself. The operator for the energy $E$ is the Hamiltonian operator $H$.

Hence Eq. (5.29) takes the form,

$$\hat{H} = \frac{1}{2m}\left[\left(\frac{h}{2\pi i}\frac{\partial}{\partial x}\right)^2 + \left(\frac{h}{2\pi i}\frac{\partial}{\partial y}\right)^2 + \left(\frac{h}{2\pi i}\frac{\partial}{\partial z}\right)^2\right] + V$$

$$\hat{H} = \frac{-h^2}{8\pi^2 m}\left(\frac{\partial^2}{\partial x^2} + \frac{\partial^2}{\partial y^2} + \frac{\partial^2}{\partial z^2}\right) + V$$

$$\hat{H} = \frac{-h^2}{8\pi^2 m}\nabla^2 + V \tag{5.31}$$

Further according to the second postulate, we must have

$$\hat{H}\psi = E\psi$$

or $$\hat{H}\psi - E\psi = 0 \tag{5.32}$$

Substituting the value of $\hat{H}$ from Eq. (5.31) into Eq. (5.32), we get

$$\left(\frac{-h^2}{8\pi^2 m}\nabla^2 + V\right)\psi - E\psi = 0$$

which can be rearranged and written in the form

$$\nabla^2\psi + \frac{8\pi^2 m}{h^2}(E - V)\psi = 0$$

which is the required Schrödinger wave equation in one of the common forms.

## 5.2 SOLUTION OF SCHRÖDINGER WAVE EQUATION TO PARTICLE IN A BOX

This is the simplest application of SWE to the translational motion of a particle (electron, atom or molecule) in space. The results obtained can explain as to why the energies are quantized, i.e. can have only discrete values unlike the classical mechanics according to which the energies associated with the motion of a particle can vary continuously, i.e. can have any value.

### 5.2.1 Particle in a One-dimensional Box

The motion of a particle in a one-dimensional box is like the flow of an electron in a wire. Let us assume that a single electron of mass $m$ is restricted to move in a region of space from $x = 0$ to

$x = a$ and, that its potential energy with in the box is constant and taken as equal to zero for the sake of convenience (Figure 5.1). In order that the particle remain with in the box, it is essential to assume that potential energy on or outside the walls is very high ($= \infty$) so that as soon as the particle reaches the walls, it is reflected back into the box.

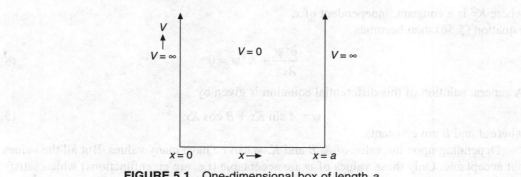

**FIGURE 5.1** One-dimensional box of length $a$.

Now the SWE for one-dimensional box is,

$$\frac{\partial^2 \psi}{\partial x^2} + \frac{8\pi^2 m(E-V)}{h^2} \psi = 0 \tag{5.33}$$

As outside the box, $V = \infty$, therefore for the outside the box, Eq. (5.33) becomes

$$\frac{\partial^2 \psi}{\partial x^2} + \frac{8\pi^2 m(E-\infty)}{h^2} \psi = 0 \tag{5.34}$$

Neglecting $E$ in comparison to $\infty$, Eq. (5.34) reduces to

$$\frac{\partial^2 \psi}{\partial x^2} - \frac{8\pi^2 m}{h^2} \infty \psi = 0$$

$$\frac{\partial^2 \psi}{\partial x^2} - \infty \psi = 0$$

or

$$\frac{\partial^2 \psi}{\partial x^2} = \infty \psi$$

$$\psi = \frac{1}{\infty} \frac{\partial^2 \psi}{\partial x^2} = 0 \tag{5.35}$$

This proves that outside the box, $\psi = 0$, which implies that the particles cannot go outside the box.

For the particle (electron) within the box, $V = 0$, therefore, the SWE Eq. (5.33) takes the form

$$\frac{\partial^2 \psi}{\partial x^2} + \frac{8\pi^2 m}{h^2} E\psi = 0 \tag{5.36}$$

As for the given state of the system, the energy $E$ is constant (postulate of quantum mechanics) therefore, we put

$$\frac{8\pi^2 m}{h^2} E = K^2 \tag{5.37}$$

where $K^2$ is a constant, independent of $x$.
Equation (5.36) then becomes

$$\frac{\partial^2 \psi}{\partial x^2} + K^2 \psi = 0 \tag{5.38}$$

A general solution of this differential equation is given by

$$\psi = A \sin Kx + B \cos Kx \tag{5.39}$$

where $A$ and $B$ are constants.

Depending upon the value of $A$, $B$ and $K$, $\psi$ can be have many values. But all the values are not acceptable. Only those values of $\psi$ are acceptable (i.e. are eigenfunctions) which satisfy the following condition, viz.

$$\psi = 0 \text{ at } x = 0 \text{ and } x = a$$

Putting $\psi = 0$, when $x = 0$, Eq. (5.39) becomes

$$0 = A \sin 0 + B \cos 0$$
$$0 = 0 + B \qquad \text{(Because } \sin \theta = 0 \text{ and } \cos \theta \text{ is 1)}$$

i.e. $\qquad B = 0$

Thus, when $x = 0$, Eq. (5.39) becomes (by putting $B = 0$)

$$\psi = A \sin Kx \tag{5.40}$$

Now putting $\psi = 0$, when $x = a$, Eq. (5.40), becomes

$$0 = A \sin Ka$$

i.e. $\qquad \sin Ka = 0 \tag{5.41}$

This equation holds good only when the values of $Ka$ are integral multiplies of $\pi$, i.e.

$$Ka = n\pi \tag{5.42}$$

However, value of $n = 0$ may be excluded because it makes $K = 0$. Hence, $\psi = 0$ for any value of $a$ between 0 and $a$ i.e. with in the box. This is not true because the particle is always assumed to be present within the box.

From Eq. (5.42),

$$K = n \pi/a \tag{5.43}$$

$$\psi = A \sin\left(\frac{n\pi}{a}\right) x \tag{5.44}$$

This gives the expression for the eigenfunction $\psi$. The expression for the eigenvalue of the energy may be obtained as follows:
From Eq. (5.37),

$$E = \frac{K^2 h^2}{8\pi^2 m} \tag{5.45}$$

Substituting the value of $K$ from Eq. (5.43), we get

$$E = \frac{(n\pi/a)^2 h^2}{8\pi^2 m}$$

i.e.
$$E = \frac{n^2 h^2}{8ma^2} \qquad (5.46)$$

Equations (5.44) and (5.46) are the solution of SWE for a particle in one-dimensional box. However, Eq. (5.44) contains the undetermined constants $A$. Its value can be obtained by the process of normalization of the wavefunction as follows:

$$\int_0^a \psi\psi^* \, d\tau = 1$$

Substituting the value of $\psi (=\psi^*)$ from Eq. (5.44), we get

$$A^2 \int_0^a \sin^2 \frac{n\pi}{a} x \, dx = 1$$

Because
$$\int_0^a \sin^2\left(\frac{n\pi}{a} x\right) = \frac{a}{2}$$

Therefore,
$$A^2 \frac{a}{2} = 1$$

or
$$A = \sqrt{\frac{2}{a}}$$

Hence, the normalized wavefunction (which will also be a solution to the SWE) is

$$\psi = \sqrt{\frac{2}{a}} \sin\left(\frac{n\pi}{a} x\right) \qquad (5.47)$$

## 5.3 CHARACTERISTICS OF THE WAVEFUNCTION OR QUANTIZATION OF ENERGY

A few energy levels and the corresponding wavefunctions are shown graphically in Figure 5.2. Besides the points on the wall of the box, there are points inside the box where the wavefunction is zero. These points are called *nodes*. It is evident from the Figure 5.2 that as the quantum number $n$ increases, the number of nodes on the wave increases. For example, state whose wavefunction is $\psi_n$ has $(n-1)$ nodes inside the box. This type of behaviour is general for all systems. Increasing the number of nodes decreases the wavelength, which corresponds to increasing the kinetic energy.

Since, $n$ can have only integral values equal to 1, 2, 3, etc. therefore, it follows that the energy $E$ associated with the motion of a particle in a box can have only discrete values, i.e. energy is quantized. The integer $n$ is called the quantum number of the particle. It is important to note that as the quantum number $n$ increases, the separation between them increases. It may also be noted that separation of energy levels also depends upon the box length $a$. As $a$ increases, i.e. the space available to a particle increases, energy quanta become smaller and energy levels move close together. If the box length becomes very large, quantization disappears and there is a smooth transition from quantum behaviour to classical behaviour.

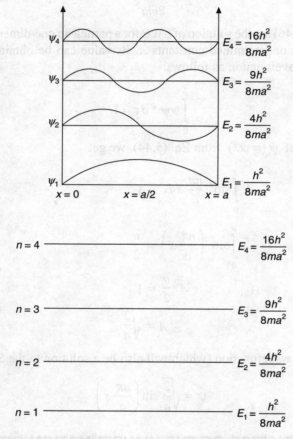

**FIGURE 5.2** Energy levels and their corresponding wavefunctions for first four quantum numbers.

**Problem 5.1** An electron is confined in a one-dimensional box of length 1 Å. Calculate its ground state energy in electron volts. Is quantization of energy level observable?

*Solution:* For ground state, $n = 1$

$$E = \frac{n^2 h^2}{8ma^2} \Rightarrow \frac{1^2 \times (6.626 \times 10^{-34}\, \text{Js})^2}{8 \times (9.1 \times 10^{-31}\, \text{kg}) \times (1.6 \times 10^{-19}\, \text{J/eV})} = 37.66$$

To see if energy levels of electron are quantized, let us calculate the energy of the first excited state $n = 2$.

$$E_2 = 150.4 \text{ eV}$$
$$\Delta E = E_2 - E_1 \Rightarrow 150.4 - 37.6 \Rightarrow 112.8 \text{ eV}$$

This means that if electron falls from first excited state to the ground state, it will emit an energy of 112.8 eV. Thus, it is possible to observe quantization in the energy levels of microscopic system like the electron.

**Problem 5.2** A ball of mass 1 g, confined to a one-dimensional box of length 0.1 m moves with a velocity of 0.01 ms$^{-1}$. Calculate the quantum number $n$. Is it possible to observe the quantization of energy levels of the ball.

*Solution:* K.E. of the ball equal to total energy since, the potential energy within the box is zero. In the ground state $n = 1$.

Therefore, $E_1 = \dfrac{1}{2} m v_x^2 = \dfrac{1}{2} (0.001 \text{ kg}) (0.01 \text{ m s}^{-1})^2 \Rightarrow 5 \times 10^{-8} \text{ J}$

$$E = \dfrac{n^2 h^2}{8 m a^2}$$

Hence, $n = \left[ \dfrac{8 m a^2 n^2 E_n}{h^2} \right]^{1/2} = (8 m E_n)^{1/2} \dfrac{a}{h}$

$= \left[ 8(0.001 \text{ kg})(5 \times 10^{-8} \text{ J}) \right]^{1/2} \left[ \dfrac{0.1 \text{m}}{6.626 \times 10^{-34} \text{ Js}} \right] = 3 \times 10^{27}$

To see if the energy levels of the ball are quantized, let us calculate the difference in the successive energy levels, say $n$ and $n + 1$.

$$\Delta E = E_{n+1} - E_n \Rightarrow [(n+1)^2 - n^2] h^2/8ma^2 \Rightarrow (2n+1) E_0$$
$$n = 3 \times 10^{27}; n + 1 = 3 \times 10^{27} + 1 \approx 3 \times 10^{27}$$
$$\Delta E = (2n+1) E_0 \Rightarrow [2 \times (3 \times 10^{27}) + 1] (6.62 \times 10^{-34})^2/8 \times 0.001 \text{ kg} \times (0.1)^2$$
$$\Delta E = 3.3 \times 10^{-35} \text{ J}$$

Since $E_1 = 5 \times 10^{-8}$ J and $\Delta E = E_2 - E_1$ ∴ $E_2 = \Delta E + E_1$
$E_2 = 5 \times 10^{-8}$ J $+ 3.3 \times 10^{-35}$ J
$E_2 \approx 5 \times 10^{-8}$ J

Since $E_1$ and $E_2$ are practically same, it is not possible to observe quantization in the energy levels of macroscopic systems like the ball.

**Problem 5.3** What will happen if the walls of one-dimensional box are suddenly removed.

*Solution:* Electron will become free to move without any restriction on the value of the potential energy. If the potential energy is taken to be zero, then the solution to SWE $\partial^2 y / dx^2 + 2m/h^2$ $(E - 0), y = 0$, and is given by equation $\psi = A \sin Kx$. Since, $\psi = 0$ at $x = 0$ and $x = a$, we have $A$ $\sin Ka = 0$ or $\sin Ka = 0$ or $Ka = n\pi$ so that, $K = n\pi/a$. However, arbitrary constants $A$ and $K$ can have any values, which may be assigned to them. Thus, the energy eigenvalues given by $E = K^2h^2/8\pi^2m$ can not quantized. Hence, we conclude that a freely moving particle (electron) without any restriction has continuous energy spectrum.

**Problem 5.4** What is zero point energy of a particle in a one-dimensional box of a infinite height. Is the occurrence of zero point energy in accordance with the Heisenberg uncertainty principle.

*Solution:* Since $E_n = n^2 h^2/8ma^2$, the energy corresponding to $n = 1$, viz., $E_1 = h^2/8ma^2$ is called the *zero point energy* (ZPE) of the particle. Since, ZPE is finite (and not equal to zero), it means that the particle inside the box is not at rest even at 0 K. This being so, the position of the particle cannot be precisely known. Thus, occurrence of ZPE implies uncertainty in the position, $\Delta x$ and also uncertainty in the $x$ component of the linear momentum $\Delta p_x$. This means that occurrence of ZPE is in accordance with the Heisenberg uncertainty principle.

## 5.4 THE HARMONIC OSCILLATOR

### 5.4.1 Linear Harmonic Oscillator

For a particle undergoing simple harmonic oscillations in one dimension the restoring force is proportional to the displacement $x$ from the equilibrium position (Hooke's law) (Figure 5.3),

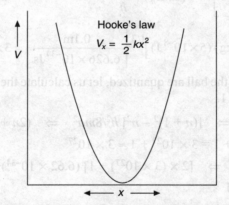

**FIGURE 5.3** Variation of potential energy ($V$) with displacement ($x$) in case of linear harmonic oscillator.

this condition may be expressed mathematically in the form of the equation,

$$\frac{md^2x}{dt^2} = -kx \tag{5.48}$$

$$-\frac{md^2x}{xdt^2} = k \tag{5.49}$$

where $m$ is the mass of the oscillating particle and $k$ is a constant. Eq. (5.49) can be solved in the form of equation given below

$$x = A \sin(Bt + C) \tag{5.50}$$

where $A$, $B$ and $C$ are constants. Since, the motion is periodic in character, the condition repeating themselves exactly after each oscillation, it follows that

$$B = 2\pi v \tag{5.51}$$

where $v$ is the frequency of oscillation of the particle under consideration. Inserting value of $B$ in Eq. (5.50), we get

$$x = A \sin(2\pi vt + C) \tag{5.52}$$

On taking second derivative of Eq. (5.52), we get

$$\frac{d^2x}{dt^2} = -4v^2\pi^2 A \sin(2\pi vt + C)$$

By substituting value of $x$ and $\frac{d^2x}{dt^2}$ in Eq. (5.49), we get

$$k = -\frac{m}{A\sin(2\pi vt + C)} - 4v^2\pi^2 A \sin(2\pi vt + C)$$
$$k = 4\pi^2 v^2 m \tag{5.53}$$

The potential energy $V$ of the oscillator is equal to $\frac{1}{2}kx^2$ and hence

$$V = 2\pi^2 v^2 m x^2 \tag{5.54}$$

This expression, derived by the methods of classical mechanics, gives the potential function required for the solution of the wave equation in terms of the frequency, $v$ of a classical harmonic oscillator.

For a particle oscillating in one direction only, e.g. parallel to the $x$-axis, the co-ordinates $y$ and $z$ remain constant, hence in this case, the wave equation reduces to the one-dimensional form.

$$\frac{d^2\psi}{dx^2} + \frac{8\pi^2 m}{h^2}(E - V)\psi = 0 \tag{5.55}$$

Inserting the value of the potential energy given by Eq. (5.54), this equation becomes

$$\frac{d^2\psi}{dx^2} + \frac{8\pi^2 m}{h^2}(E - 2\pi^2 v^2 m x^2)\psi = 0 \tag{5.56}$$

To simplify the from of this differential equation, let

$$a = \frac{8\pi^2 mE}{h^2} \quad \text{and} \quad b = \left(\frac{4\pi^2 vm}{h}\right)^2 \tag{5.57}$$

Then Eq. (5.56) can be written as

$$\frac{d^2\psi}{dx^2} + (a - bx^2)\psi = 0 \tag{5.58}$$

Introducing in place of $x$ a new dimensionless variable $q$ defined by

$$q = x\sqrt{b} \qquad (5.59)$$

$$q^2 = x^2 b \qquad (5.60)$$

$$q^2 \propto x^2$$

By dividing Eq. (5.58) with $b$, we get

$$\frac{d^2\psi}{dx^2} + \left(\frac{a}{b} - x^2\right)\psi = 0 \qquad (5.61)$$

And hence, on replacing $x^2 = q^2$, we get

$$\frac{d^2\psi}{dq^2} + \left(\frac{a}{b} - q^2\right)\psi = 0 \qquad (5.62)$$

The solutions $\psi(x)$ of this equation must of course, satisfy the condition of a well behaved function that is to say, the eigenfunctions must be continuous, single valued and finite for all values of $x$ from $-\infty$ to $+\infty$.

### 5.4.2 Harmonic Oscillator: Asymptotic Solution

Asymptotic solution of Eq. (5.62) can be obtained by assuming $q \gg \sqrt{a/b}$ under these conditions, the Eq. (5.62) becomes

$$\frac{d^2\psi}{dq^2} - q^2\psi = 0 \qquad (5.63)$$

Solution of which are

$$\psi = e^{\pm q^2/2} \qquad (5.64)$$

This may be verified by differentiation with respect to $q$, thus

$$\frac{d^2\psi}{dq^2} = q^2 e^{\pm q^2/2} \pm e^{\pm q^2/2} \qquad (5.65)$$

$$\frac{d^2\psi}{dq^2} \Rightarrow (q^2 \pm 1) e^{\pm q^2/2} \qquad (5.66)$$

Since $q^2$ is very large, $q^2 \pm 1$ is not appreciably different from $q^2$ and so Eq. (5.66) becomes

$$\frac{d^2\psi}{dq^2} = q^2 e^{\pm q^2/2} = q^2 \psi \qquad (5.67)$$

$$\therefore \qquad \psi = e^{\pm q^2/2} \qquad \text{proved.}$$

It is obvious, therefore, that of the two possible solutions given by Eq. (5.64), viz.,

$$\psi = e^{q^2/2} \quad \text{and} \quad \psi = e^{-q^2/2} \tag{5.68}$$

The former is not an acceptable wavefunction for it, $\psi$ increases rapidly with increasing $q$ and hence with increasing numerical value of $x$, the alternative solution, i.e. $\psi = e^{-q^2/2}$, however satisfies the necessary condition and so it may be regarded as satisfactory asymptotic solution of the wave equation.

### Recursion Formula

Solution of Eq. (5.62) will contain the term $e^{-q^2/2}$ as a factor, a possible form of solution will be

$$\psi = e^{-q^2/2} f(q) \tag{5.69}$$

where $f(q)$ is a function of $q$, and hence of $x$, the nature of which is to be determined. On differentiating Eq. (5.69), it is seen that

$$\frac{d^2\psi}{dq^2} = e^{-q^2/2}\left\{\frac{d^2f}{dq^2} - 2q\frac{df}{dq} + (q^2-1)f\right\} \tag{5.70}$$

In which $f$ is written for $f(q)$ substitution for $d^2\psi/dq^2$ and $\psi$ as given by Eqs. (5.69) and (5.70) into Eq. (5.62) yields the result

$$e^{-q^2/2}\left(\frac{d^2f}{dq^2} - 2q\frac{df}{dq} + (q^2-1)f\right) + e^{-q^2/2}\left(\frac{a}{b} - q^2\right)f = 0 \tag{5.71}$$

$$\therefore \quad e^{-q^2/2}\left[\frac{d^2f}{dq^2} - 2q\frac{df}{dq} + \left(\frac{a}{b}-1\right)\right]f = 0 \tag{5.72}$$

Since $e^{-q^2/2}$ is not zero, except for $q = \pm\infty$, it follows that the expression in the brackets must be zero, that is

$$\frac{d^2f}{dq^2} - 2q\frac{df}{dq} + \left(\frac{a}{b}-1\right)f = 0 \tag{5.73}$$

The assumption is now made that $f(q)$ may be expressed in the form of a power series in $q$; thus

$$f(q) = \alpha_0 + \alpha_1 q + \alpha_2 q^2 + \alpha_3 q^3 + \alpha_4 q^4 + \cdots \tag{5.74}$$

Then

$$\frac{df}{dq} = \alpha_1 + 2\alpha_2 q + 3\alpha_3 q^2 + 4\alpha_4 q^3 + 5\alpha_5 q^4 + \cdots \tag{5.75}$$

$$\frac{d^2f}{dq^2} = 2\alpha_2 + 6\alpha_3 q + 12\alpha_4 q^2 + 20\alpha_5 q^3 + 30\alpha_6 q^4 + \cdots \tag{5.76}$$

Putting these values in Eq. (5.73), then it gives

$$2\alpha_2 + 6\alpha_3 q + 12\alpha_4 q^2 + 20\alpha_5 q^3 + \cdots - 2\alpha_1 q - 4\alpha_2 q^2 - 6\alpha_3 q^3 + \cdots$$
$$+ \left(\frac{a}{b}-1\right)\alpha_0 + \left(\frac{a}{b}-1\right)\alpha_1 q + \left(\frac{a}{b}-1\right)\alpha_2 q^2 + \left(\frac{a}{b}-1\right)\alpha_3 q^3 + \cdots \quad (5.77)$$

In order for this series to vanish for all values of $q$ as it necessary if $f(q)$ is to be solution of Eq. (5.73) the coefficient of individual powers of $q$ must vanish separately. Hence the coefficient of $q^0$,

$$2\alpha_2 + \left(\frac{a}{b}-1\right)\alpha_0 = 0 \quad (5.78)$$

For the coefficient of $q^1$

$$6\alpha_3 - 2\alpha_1 + \left(\frac{a}{b}-1\right)\alpha_1 = 0 \quad (5.79)$$

For the coefficient of $q^2$

$$12\alpha_4 - 4\alpha_2 + \left(\frac{a}{b}-1\right)\alpha_2 = 0 \quad (5.80)$$

In general, for the coefficient of $q^n$ to vanish,

$$(n+1)(n+2)\alpha_{n+2} - 2n\alpha_n + \left(\frac{a}{b}-1\right)\alpha_n = 0 \quad (5.81)$$

and consequently

$$\frac{\alpha_{n+2}}{\alpha_n} = \frac{2n - \frac{a}{b} + 1}{(n+1)(n+2)} \quad (5.82)$$

where $n$ is an integer or zero. This recursion formula, permits the calculation of the coefficient $\alpha_{n+2}$ of the term involving $q^{n+2}$ in the power series for $f(q)$, i.e. Eq. (5.74) provided the coefficient $a_n$ of $q^n$ is known. Thus, from the coefficient of $q^0$, it is possible to derive that of $q^2$ and hence of $q^4$, $q^6$, etc. Similarly, if the coefficient of $q^1$ is known, those of $q^3$, $q^5$, etc. can be know.

## 5.4.3 Energy Levels of Harmonic Oscillator

If no restriction is placed on the values of the fraction $a/b$ which is related to the energy $E$ of the oscillator by the Eq. (5.57), the function $\psi$ as defined by $e^{q^2/2} f(q)$ may prove to be unacceptable, for it may increase rapidly with increasing $q$ and consequently with increasing $x$. This may be seen in the following manner, consider the series for $e^{q^2}$, viz.

$$e^{q^2} = 1 + q^2 + \frac{q^4}{2!} + \cdots \frac{q^n}{\left(\frac{1}{2}n\right)!} + \cdots \quad (5.83)$$

The coefficient of $q^n$, which may be represented by $\beta_n$ is $1 \big/ \left(\frac{1}{2}n\right)!$ while that of $q^{n+2}$ represented by $\beta_{n+2}$ is $1 \big/ \left(\frac{1}{2}n+1\right)!$. The recursion formula for the exponential series for $e^{q^2}$ is thus given by

$$\frac{\beta_{n+2}}{\beta_n} = \frac{\left(\frac{1}{2}n\right)!}{\left(\frac{1}{2}n+1\right)!} = \frac{1}{\frac{1}{2}n+1} \tag{5.84}$$

In the limit, when $n$ is very large, so that unity is negligible in comparison with $1/2n$, this becomes

$$\lim_{n\to\infty} \frac{\beta_{n+2}}{\beta_n} = \frac{2}{n} \tag{5.85}$$

Examination of Eq. (5.82) shows that when $n$ is very large, the recursion formula for the power series representing $f(q)$ also becomes

$$\lim_{n\to\infty} \frac{\alpha_{n+2}}{\alpha_n} = \frac{2}{n} \tag{5.86}$$

So that for large values of the exponent $n$, the series for $f(q)$ will behave like that for $e^{q^2}$. If this is the case, then according to Eq. (5.69) the eigenfunction $\psi$ for large values of $n$ will become equivalent to $e^{-q^2/2}\, e^{q^2}$ and hence to $e^{q^2/2}$. It has already seen that a function of this kind cannot be an acceptable wavefunction, since this quantity increase as $q$ increases. The series governed by the recursion formula of Eq. (5.76) cannot therefore form the part of satisfactory eigenfunction unless some restriction is introduced which makes the series break off after a finite number of terms. In other words, the expression for $f(q)$ as given by Eq. (5.74), should be restricted in such a manner as to make it polynomial rather than power series. The eigenfunction $\psi$ can then be set equal to the product of this polynomial and the factor $e^{-q^2/2}$. Since, the value of latter decreases with increasing $q$ and hence with increasing $x$, the eigenfunction as a whole will behave in the same manner. Such a function will evidently satisfy the conditions for a suitable wavefunction.

By setting the numerator in the recursion formula (5.82) equal to zero, that is if

$$\frac{a}{b} - 1 - 2n = 0 \tag{5.87}$$

$$\frac{a}{b} = 2n + 1 \tag{5.88}$$

The series for $f(q)$ will break off after the term $q^n$, since the coefficient of $q^{n+2}$, and of all higher terms, will be zero. This is the condition therefore, that makes the series $f(q)$ a polynomial so that $e^{-q^2/2} f(q)$ will be satisfactory wavefunction. Substituting the values of $a$ and $b$ given by Eq. (5.57) into Eq. (5.88) it is seen that the restriction that makes the wavefunction suitable is

$$\frac{2E}{h\nu} = 2n + 1 \tag{5.89}$$

$$E_n = \left(n + \frac{1}{2}\right)h\nu \tag{5.90}$$

This result implies that the Schrödinger equation for a linear harmonic oscillator can have physically acceptable solutions only for certain discrete values of the energy. These values represented by $E_n$ given by Eq. (5.90) for $n = 0, 1, 2,\ldots$ are the appropriate eigenvalues.

### 5.4.4 Eigenfunction of Harmonic Oscillator

The corresponding eigenfunction $\psi_n$ can be derived from Eq. (5.56) as a function of $x$.
The Eq. (5.56) is

$$\frac{d^2\psi}{dx^2} + \frac{8\pi^2 m}{h^2}(E - 2\pi^2\nu^2 m x^2)\psi = 0$$

Value of eigenfunction comes out to be

$$\psi_n = N_n e^{-q^2/2} H_n(q) \tag{5.91}$$

where $N_n$ is normalizing factor, and $H_n(q)$ represents a Hermite polynomial of degree $n$, defined by

$$H_n(q) = (-1)^n e^{q^2} \frac{d^n e^{-q^2}}{dq^n} \tag{5.92}$$

The first few Hermite polynomials, for $n = 0, 1, 2$ and 3 are as follows

$n = 0$    $H_0(q) = 1$    $H_0 = (-1)^0 e^{q^2} \dfrac{d^0 e^{-q^2}}{dq^0} \Rightarrow e^{q^2} e^{-q^2} = 1$

$n = 1$    $H_1(q) = 2q$    $H_1 = (-1)^1 e^{q^2} \dfrac{d(e^{-q^2})}{dq^1} \Rightarrow -1 e^{q^2} e^{-q^2} \cdot -2q = 2q$

$n = 2$    $H_2(q) = 4q^2 - 2$    $H_2 = (-1)^2 e^{q^2} \dfrac{d^2(e^{-q^2})}{dq^2} \Rightarrow 1 e^{q^2} e^{-q^2}(4q^2 - 2)$

$n = 3$    $H_3(q) = 8q^3 - 12q$                                                                                                        (5.93)

## 5.5 ZERO POINT ENERGY OR RESIDUAL ENERGY

The lowest possible energy $E_0$ of an oscillator occurs when $n$ is zero; thus

$$E_0 = \left(0 + \frac{1}{2}\right)h\nu \qquad E_0 = \frac{1}{2}h\nu \tag{5.94}$$

This is sometimes written as $E_0 = \dfrac{1}{2} hc\nu$, where $\nu$ is now the frequency in wavenumbers. It is seen from Eq. (5.94) that even in the vicinity of absolute zero of temperature when the vibrational energy of a molecular oscillator has its lowest possible value, that value would still not be less than that zero point energy of $\frac{1}{2}h\nu$ per oscillator whose frequency is $\nu$ (s$^{-1}$). The existence of this zero point energy is, in fact, an aspect of uncertainty principle. It might be imagined that at the absolute zero, at least internal motion of molecule would cease entirely, so that the position of constituent atoms could be identified exactly. This is not the case however, since the molecule still has vibrational energy equal to $\frac{1}{2}h\nu$ for each oscillator, the atoms are in a state of vibration even at the absolute zero and so their precise position cannot be defined.

It is of interest here that the older quantum theory led to the equation $E_n = nh\nu$ for the energy levels of a linear oscillator. It was evident from a study of the vibrational spectra of molecules, however, that an expression of this kind did not define correctly the energy of a harmonic oscillator of two atoms vibrating relative to each other. By utilizing Eq. (5.90) derived from quantum mechanics, the conclusions reached are in satisfactory agreement with the results of spectroscopic studies.

**Problem 5.5** What is restriction on $\alpha/\beta$ if $n = 1$ wavefunction of a one-dimensional SHO has to satisfy the wave equation $\frac{d^2\psi}{dq^2} + \left(\frac{\alpha}{\beta} - q^2\right)\psi_1 = 0$. Given that $\psi_1 = Nq\, e^{-q^2/2}$, where $N$ is a constant.

*Solution:*

$$\psi_1 = Nq\, e^{-q^2/2}$$

$$\frac{d\psi_1}{dq} = Nq\, e^{-q^2/2}(-q) + Ne^{-q^2/2}$$

$$\Rightarrow -Nq^2\, e^{-q^2/2} + Ne^{-q^2/2}$$

$$\frac{d^2\psi_1}{dq^2} \Rightarrow -Nq^2(-q)e^{-q^2/2} - 2Nqe^{-q^2/2} - Nqe^{-q^2/2}$$

$$\Rightarrow Nq^3 e^{-q^2/2} - 3Nqe^{-q^2/2}$$

Now,

$$\frac{d^2\psi_1}{dq^2} + \left(\frac{\alpha}{\beta} - q^2\right)\psi = 0$$

Putting value of $\psi$ and $\frac{d^2\psi_1}{dq^2}$

$$Nq^3 e^{-q^2/2} - 3Nqe^{-q^2/2} + \frac{\alpha}{\beta} Nqe^{-q^2/2} - Nq^3 e^{-q^2/2} = 0$$

$$-3Nqe^{-q^2/2} + \frac{\alpha}{\beta} Nqe^{-q^2/2} = 0$$

Hence $\frac{\alpha}{\beta} = 3$ which is desired restriction.

**Problem 5.6** Verify the Heisenberg's uncertainty relation with respect to the motion of a particle in one-dimensional box of length $L$.

*Solution:* Since the particle can be anywhere from $x = 0$ to $x = L$, the uncertainty in position $\Delta x = L$.

The momentum can have any value from $\dfrac{nh}{2L}$ to $\dfrac{-nh}{2L}$, therefore,

$$\Delta p_x = \frac{nh}{2L} - \left(\frac{-nh}{2L}\right) \approx \frac{nh}{L} \qquad \left[\frac{P^2}{2M} = \frac{n^2 h^2}{8ma^2} = p^2 = \frac{h^2}{L^2}\right]$$

For the state $n = 1$, $\Delta x \cdot \Delta p_x \approx L \cdot \dfrac{h}{L} \approx h$

**Problem 5.7** For a particle of mass ($M$) executing harmonic oscillation. The potential energy is given as $V = \dfrac{1}{2} K x^2$ and oscillation frequency as $v = \dfrac{1}{2\pi}\sqrt{\dfrac{K}{M}}$.

Using these classical relation for $V$ and $v$, and definitions $\alpha = \dfrac{8\pi^2 M E}{h^2}$ and $\beta = \dfrac{4\pi M v}{h}$. Write the Schrödinger equation in the simple from and (ii) find asymptotic form of the solution.

*Solution:*

(i) The Hamiltonian operator is

$$\hat{H} = \frac{-h^2}{8\pi^2 M}\frac{d^2}{dx^2} + \frac{1}{2} K x^2 \quad \text{has been already solved.}$$

$$\hat{H} = \frac{-h^2}{8\pi^2 M}\frac{d^2}{dx^2} + 2\pi^2 M v^2 x^2 \quad \text{using } v = \frac{1}{2\pi}\sqrt{\frac{K}{M}}$$

The Schrödinger equation is

$$\left[-\frac{h^2}{8\pi^2 M}\frac{d^2}{dx^2} + 2\pi^2 M v^2 x^2\right]\psi(x) = E\psi(x)$$

Multiplying by $\dfrac{-8\pi^2 M}{h^2}$ and using relations $\alpha = \dfrac{8\pi^2 M E}{h^2}$ and $\beta = \dfrac{4\pi^2 M v}{h}$. The Schrödinger equation assumes the form

$$\left[\frac{d^2}{dx^2}\left(\frac{-16\pi^4 M^2 v^2}{h^2}\right)x^2\right]\psi(x) = \frac{-8\pi^2 M E}{h^2}\psi(x)$$

or

$$\left[\frac{d^2}{dx^2} + \frac{8\pi^2 M E}{h^2} - \left(\frac{16\pi^4 M^2 v^2}{h^2}\right)x^2\right]\psi(x) = 0$$

or
$$\left[\frac{d^2}{dx^2} + \alpha - \beta^2 x^2\right]\psi(x) = 0$$

A change of variable $y = \sqrt{\beta}\, x$ so that $dy/dx = \sqrt{\beta}$ further simplifies the equation. By the chain rule of successive diffentiation.

$$\frac{d\psi}{dx} = \frac{d\psi}{dy}\frac{dy}{dx} = \left(\frac{d\psi}{dy}\right)\sqrt{\beta}$$

and
$$\frac{d^2\psi}{dx^2} = \frac{d}{dx}\left(\frac{d\psi}{dx}\right) = \frac{d}{dx}\left(\sqrt{\beta}\,\frac{d\psi}{dy}\right)$$

$$\Rightarrow \sqrt{\beta}\,\frac{d}{dy}\left(\frac{d\psi}{dy}\right)\frac{dy}{dx} = \beta\,\frac{d^2\psi}{dy^2}$$

Thus,
$$\left[\frac{d^2}{dy^2} + \left(\frac{\alpha}{\beta} - y^2\right)\right]\psi(y) = 0$$

(ii) *Asymptotic Solution:* The solutions $\psi(y)$ have to be continuous and finite. The asymptotic form of the solution may be obtained by examining the equation above for $y \to \pm\infty$. In that case $\alpha/\beta$ will be negligible before $y^2$ so that the equation above becomes

$$\frac{d^2\psi}{dy^2} = +y^2\psi(y)$$

For which general solution is

$$\psi(y) = Ae^{y^2/2} + Be^{-y^2/2}$$

Since $\psi(y)$ has to remain finite for all values of $y$, $A$ must be zero; thus

$$\psi(y) = Be^{-y^2/2} \quad \Rightarrow \quad Be^{-\beta x^2/2}$$

**Problem 5.8** Derive the Hermite polynomials for the first few values of the vibrational quantum number.

*Solution:* To derive the Hermite polynomials $H(y)$ we have to determine the coefficients of the series:

$$H(y) = a_0 + a_1 y + a_2 y^2 + a_3 y^3 + \cdots + a_k y^k$$

With the help of recursion formula

$$\frac{a_{k+2}}{a_k} = \frac{2k+1-\dfrac{\alpha}{\beta}}{(k+2)(k+1)} \tag{1}$$

And with the condition that for some value of $K (= n$, say) the numerator in Eq. (1) must vanish. Thus, setting $a_1 = 0$ we have the coefficients

$$a_0, \quad a_1 = \frac{1 - \dfrac{\alpha}{\beta}}{2} a_0, \quad a_4 = \frac{5 - \dfrac{\alpha}{\beta}}{12} a_2, \quad a_6 = \frac{9 - \dfrac{\alpha}{\beta}}{30} a_4 \quad \text{and so on.}$$

And setting $a_0 = 0$, we have

$$a_1, \quad a_3 = \frac{3 - \dfrac{\alpha}{\beta}}{6} a_1, \quad a_5 = \frac{7 - \dfrac{\alpha}{\beta}}{20} a_3 \quad \text{and so on.}$$

Now the permissible values of $\dfrac{\alpha}{\beta}$ are $2n + 1$ i.e. 1, 3, 5, 7, ... Corresponding to $n = 0, 1, 2, 3,...$

For $n = 0$, $\quad \dfrac{\alpha}{\beta} = 1, \quad E = \dfrac{1}{2} h\nu, \quad a_2 = a_4 = \cdots = 0 \quad$ and $\quad H_0(y) = a_0$

For $n = 1$, $\quad \dfrac{\alpha}{\beta} = 3, \quad E = \dfrac{3}{2} h\nu, \quad a_3 = a_5 = \cdots = 0 \quad$ and $\quad H_1(y) = a_1 y$

For $n = 2$, $\quad \dfrac{\alpha}{\beta} = 5, \quad E = \dfrac{5}{2} h\nu, \quad a_2 = -2a_0, a_4 = a_6 = \cdots = 0 \quad$ and $\quad H_2(y) = a_0(1 - 2y^2)$

The polynomials are indexed by the vibrational quantum number $n$.

**Problem 5.9** A diatomic molecule can be treated as a simple quantum mechanical oscillator. How is the SWE modified?

Show that (i) $\psi = e^{-\beta x^2}$ is a solution. (ii) $E = \dfrac{1}{4\pi} \sqrt{\dfrac{K}{\mu}}$, $K$ being the force constant and $\mu$ the reduced mass of oscillator.

**Solution:** The Schrödinger equation is

$$\frac{-h^2}{8\pi^2} \left( \frac{1}{M_1} + \frac{1}{M_2} \right) \frac{d^2\psi}{dx^2} + \frac{1}{2} Kr^2 \psi = E\psi$$

$$\frac{-h^2}{8\pi^2 \mu} \frac{d^2\psi}{dx^2} + \frac{1}{2} Kr^2 \psi = E\psi \tag{1}$$

where $\dfrac{1}{\mu} = \dfrac{1}{M_1} + \dfrac{1}{M_2}$, $M_1$ and $M_2$ being the masses of two atoms and $r = |x_1 - x_2|$, $x_1$ and $x_2$ being their displacements.

(i) using $\psi = e^{-\beta x^2}$ in Eq. (1)

$$\frac{-h^2}{8\pi^2 \mu} (-2\beta\psi + 4\beta^2 x^2 \psi) + \frac{1}{2} Kr^2 \psi = E\psi \tag{2}$$

Rearranging the Eq. (2)

$$(-2\beta + 4\beta^2 x^2)\psi + \frac{8\pi^2 \mu}{h^2}\left(E - \frac{1}{2}Kr^2\right)\psi = 0$$

or

$$\left[\left(4\beta^2 - \frac{4\pi^2 \mu K}{h^2}\right)x^2 + \left(\frac{8\pi^2 \mu E}{h^2} - 2\beta\right)\right]\psi = 0 \tag{3}$$

since $\psi \neq 0$, the Eq. (3) is satisfied, if both

$$4\beta^2 - \frac{4\pi^2 \mu K}{h^2} = 0 \quad \text{and} \quad \frac{8\pi^2 \mu E}{h^2} - 2\beta = 0$$

i.e. if $\quad \beta = \dfrac{\pi\sqrt{\mu K}}{h} \quad$ and $\quad E = \dfrac{\beta h^2}{4\pi^2 \mu} = \dfrac{\pi\sqrt{\mu K}}{h} \times \dfrac{h^2}{4\pi^2 \mu} = \dfrac{h}{4\pi}\sqrt{\dfrac{K}{\mu}}$

**Problem 5.10** The complete wavefunction of simple harmonic oscillator is given as $\psi_n(x) = N_n H_n(q)e^{-y^2/2}$, where $N_n$ is the normalization factor, $H_n(y)$. The Hermite polynomial in $y$ and $y = \sqrt{\beta}\,x$; $n$ may be 0, 1, 2, 3,...

Given $N_n = \left(\dfrac{1}{2^n n!}\right)^{1/2}\left(\dfrac{\beta}{\pi}\right)^{1/4}\quad$ and $\quad\left[H_n(q) = (-1)^n \dfrac{e^{q^2} d^n e^{-q^2}}{dq^n}\right]$

Calculate the first four wavefunctions; sketch them and also sketch $\psi^2$. Comment on the sketches. Shows that the wavefunctions are alternately even and odd.

*Solution:* In all functions $e^{-q^2/2} = e^{-\beta x^2/2}$ is common factor.

For $n = 0$, $\quad N_0 = \left(\dfrac{\beta}{\pi}\right)^{1/4}$; $\quad H_0(y) = 1 \quad$ or $\quad H_0(x) = 1$

$$\psi_0(x) = \left(\dfrac{\beta}{\pi}\right)^{1/4} e^{-\beta x^2/2}$$

$n = 1$, $\quad N_1 = \dfrac{1}{\sqrt{2}}\left(\dfrac{\beta}{\pi}\right)^{1/4}$; $\quad H_1(y) = 2y \quad$ or $\quad H_1(x) = 2\sqrt{\beta}\,x$

$\therefore \quad\quad\quad\quad\quad \psi_1(x) = \dfrac{1}{\sqrt{2}}\left(\dfrac{\beta}{\pi}\right)^{1/4} 2\sqrt{\beta}\,x\,e^{-\beta x^2/2}$

$n = 2$, $\quad N_2 = \dfrac{1}{\sqrt{8}}\left(\dfrac{\beta}{\pi}\right)^{1/4}$; $\quad H_2(y) = 4y^2 - 2 \quad$ or $\quad H_2(x) = 4\beta x^2 - 2$

$$\psi_2(x) = \dfrac{1}{\sqrt{8}}\left(\dfrac{\beta}{\pi}\right)^{1/4}(4\beta x^2 - 2)e^{-\beta x^2/2}$$

$$n = 3, \quad N_3 = \frac{1}{\sqrt{48}} \left( \frac{\beta}{\pi} \right)^{1/4}; \quad H_3(y) = 8y^3 - 12y$$

$$\therefore \quad \psi_3(x) = \frac{1}{\sqrt{48}} \left( \frac{\beta}{\pi} \right)^{1/4} \sqrt{\beta} \, (8\beta x^3 - 12x) \, e^{-\beta x^2/2}$$

A function is even (+) if $\psi(-x) = \psi(x)$ and odd (−) if $\psi(-x) = -\psi(x)$. The factor $e^{\beta x^2/2}$ is always even (+) being a function of $x^2$. Hence, the over all character is determined by the even or odd character of the Hermite polynomial $H_n(x)$ which in turn, depends upon the integer $n$; for $n$ even (0, 2, 4, …) it is even and for $n$ odd (1, 3, 5, …) it is odd. Thus, the even and odd $\psi_n(x)$ alternates (Figure 5.4).

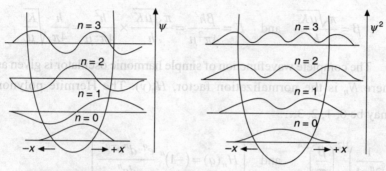

**FIGURE 5.4** A plot of $\psi$ and $\psi^2$ of linear harmonic oscillator.

*Comments*:

1. $\psi_n(x)$ is alternately even (or odd); $\psi_n^2(x)$ is always even.
2. Nodes exist in the excited states indicating points of zero probability.
3. Finite probability exists for the oscillator even beyond the classical turning points (e.g. $P$ and $Q$ in $\psi_0$ or $\psi_0^2$ ). This may be attributed to the existence of finial potential $1/2 Kx^2$ at the turning points of the oscillation.

## 5.6 THE RIGID ROTATOR

Consider two spherical particles of masses $m_1$ and $m_2$ situated at the fixed distances $r_1$ and $r_2$, respectively, from the centre of gravity of the system. The distance between the mass centres of the particles then has the constant value $r_0$ which is equal to the sum of $r_1$ and $r_2$. It is supposed that this system rotates about an axis passing through the centre of gravity, and normal to plane containing two particles. Such a system constitutes a rigid rotator, the term rigid being employed because of the fixed distance between the particles (Figure 5.5).

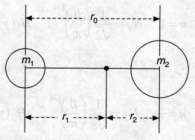

**FIGURE 5.5** A rigid diatomic rotator.

### 5.6.1 Energy and Moment of Inertia of Rotator

The kinetic energy $T$ of a particle, equal to $1/2\, mv^2$, may also be expressed as

$$T = \frac{1}{2}mv^2 = \frac{1}{2}m(\dot{x}^2 + \dot{y}^2 + \dot{z}^2) \tag{5.95}$$

where $\dot{x}$, $\dot{y}$ and $\dot{z}$ represent the components of the velocity $v$ parallel to three axes at right angles. In order to express this result in terms of spherical coordinates, the usual transformations are made, namely (Figure 5.6).

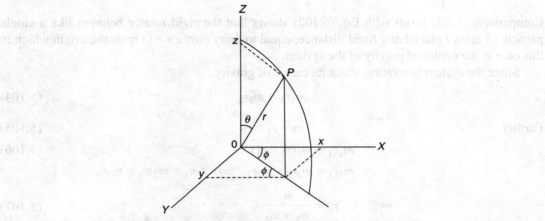

**FIGURE 5.6** Three spherical co-ordinates in case of rigid rotator.

$$x = r \sin\theta \cos\phi \tag{5.96}$$
$$y = r \sin\theta \sin\phi \tag{5.97}$$
$$z = r \cos\theta \tag{5.98}$$

And it is found that

$$T = \frac{1}{2}m(\dot{r}^2 + r^2\dot{\theta}^2 + r^2\dot{\phi}^2 \sin^2\theta) \tag{5.99}$$

If the distance $r$ of the particle from the origin is fixed the derivatives $\dot{r}$ and $\dot{r}^2$ is zero; hence the equation for the kinetic energy becomes

$$T = \frac{1}{2}mr^2(\dot{\theta}^2 + \dot{\phi}^2 \sin^2\theta) \tag{5.100}$$

For the two particles, forming a rigid rotator the total kinetic energy is given by the sum of two expressions similar to equation above

$$T = \frac{1}{2}m_1 r_1^2(\dot{\theta}^2 + \dot{\phi}^2 \sin^2\theta) + \frac{1}{2}m_2 r_2^2(\dot{\theta}^2 + \dot{\phi}^2 \sin^2\theta) \tag{5.101}$$

It should be noted that the rigid rotator has no potential energy and so equation above gives total energy $E$ as well as the kinetic energy. Writing this result in a slightly different form, it follows that

$$E = \frac{1}{2}(m_1 r_1^2 + m_2 r_2^2)(\dot{\theta}^2 + \dot{\phi}^2 \sin^2 \theta) \qquad (5.102)$$

The expression in the first parenthesis is that for the moment of inertia ($I$) of the system of two particles; hence above equation

$$E = \frac{1}{2} I(\dot{\theta}^2 + \dot{\phi}^2 \sin^2 \theta) \qquad (5.103)$$

Comparison of this result with Eq. (5.100) shows that the rigid rotator behaves like a single particle of mass $I$ placed at a fixed distance, equal to unity (since $r = 1$) from the origin which in this case is the centre of gravity of the system.

Since the system is rotating about its centre of gravity.

$$m_1 r_1 = m_2 r_2 \qquad (5.104)$$

Further
$$r_1 + r_2 = r_0 \qquad (5.105)$$

$\therefore \qquad m_1 r_1 = m_2(r_0 - r_1) \qquad (5.106)$

$$m_1 r_1 = m_2 r_0 - m_2 r_1 \quad \Rightarrow \quad (m_1 + m_2) r_1 = m_2 r_0$$

$\therefore \qquad r_1 = \dfrac{m_2}{m_1 + m_2} r_0 \qquad (5.107)$

It is also since after making some changes in above equation

$$m_1 r_1^2 + m_2 r_2^2 = m_1 r_1 r_0 \quad \text{or} \quad \left[ m_1 r_1^2 + m_2 r_2^2 = m_1 \left( \frac{m_2}{m_1 + m_2} \right) r_0^2 = \mu r_0^2 = I \right] \qquad (5.108)$$

By putting the value of $r_1$, we get

$$m_1 r_1^2 + m_2 r_2^2 = \frac{m_1 m_2}{m_1 + m_2} r_0^2 \qquad (5.109)$$

$$= \mu r_0^2 \qquad (5.110)$$

where $\mu$ is known as *reduced mass of the system*. The left-hand side of Eq. (5.110) is equal to the moment of inertia $I$, and hence the latter may be represented as

$$I = \mu r_0^2$$

## 5.6.2 Wave Equation for Rotator

SWE in three dimensions is

$$\frac{d^2 \psi}{dx^2} + \frac{d^2 \psi}{dy^2} + \frac{d^2 \psi}{dz^2} + \frac{8\pi^2 m}{h^2}(E - V)\psi = 0$$

and if this is converted into a spherical coordinates (Figure 5.6), then the result is

$$\frac{1}{r^2}\frac{d}{dr}\left(r^2\frac{d\psi}{dr}\right)+\frac{1}{r^2\sin\theta}\frac{d}{d\theta}\left(\sin\theta\frac{d\psi}{d\theta}\right)+\frac{1}{r^2\sin^2\theta}\frac{d^2\psi}{d\phi^2}+\frac{8\pi^2 m}{h^2}(E-V)\psi=0 \quad (5.111)$$

For a rigid rotator, as seen above, the mass $m$ may be replaced by the moment of inertia $I$, and $r$ by unity. Further, the potential energy $V$ is zero, and so wave equation reduces to

$$\frac{1}{\sin\theta}\frac{d}{d\theta}\left(\sin\theta\frac{d\psi}{d\theta}\right)+\frac{1}{\sin^2\theta}\frac{d^2\psi}{d\phi^2}+\frac{8\pi^2 IE}{h^2}\psi=0 \quad (5.112)$$

This is differential equation with two independent variable $\theta$ and $\phi$ representing the rotational motion of the system and the precessional motion, respectively.

Equation (5.112) contains two angular variables $\theta$ and $\phi$. It can be solved by the method of separation of variables, i.e. we look for a solution of the form

$$\psi(\theta,\phi) \Rightarrow Y(\theta)Z(\phi) \quad (5.113)$$

Substituting the value of $\psi$ in Eq. (5.112) and multiplying by $\sin^2\theta$ and dividing by $Y(\theta)Z(\phi)$, we get

$$\frac{\sin\theta}{Y}\frac{d}{d\theta}\left(\sin\theta\frac{dY}{d\theta}\right)+\frac{8\pi^2 IE}{h^2}\sin^2\theta=-\frac{1}{Z}\frac{d^2Z}{d\phi^2} \quad (5.114)$$

We can set both sides of Eq. (5.114) equal to a constant, say $m^2$, therefore obtaining two differential equations each in one variable. These equations are

$$-\frac{1}{Z}\frac{d^2Z}{d\phi^2}=m^2$$

By multiplying both sides by $Z$, we get

$$-\frac{d^2Z}{d\phi^2}=m^2 Z \quad (5.115)$$

and by dividing with $\sin^2\theta$ and multiply by $Y$ to the left hand side of Eq. (5.114), we get

$$\frac{1}{\sin\theta}\frac{d}{d\theta}\left(\sin\theta\frac{dY}{d\theta}\right)+\left(\beta-\frac{m^2}{\sin^2\theta}\right)Y=0 \quad (5.116)$$

where $\beta = 8\pi^2 IE/h^2$ \quad (5.116a)

Equation (5.115) has the solution

$$Z(\phi)=N\exp(\pm im\phi), \quad \text{where } i=\sqrt{-1} \quad (5.117)$$

where, $N$ is normalization constant.

This wavefunction is acceptable provided $m$ is an integer. This condition arises because $Z$ must be single valued.

Thus, 
$$Z(\phi)=Z(\phi+2\pi) \quad (5.118)$$

It follows therefore that $\exp(2\pi mi)=1$ \quad (5.119)

Since $e^x = \cos x + i\sin x$ \quad (5.120)

$\therefore \quad \cos 2\pi m + i\sin 2\pi m = 1$ \quad (5.121)

This can be true only if $m = 0, \pm 1, \pm 2, \pm 3, \ldots$ etc.

Let us now normalize the wavefunction $Z(\phi)$ to determine the normalization constant $N$.

$$\int_0^{2\pi} Z^* Z \, d\phi = 1 \qquad (0 \leq \phi \leq 2\pi) \tag{5.122}$$

or

$$N^2 \int_0^{2\pi} e^{im\phi} \times e^{-im\phi} \, d\phi = 1 \tag{5.123}$$

or

$$N^2 \int_0^{2\pi} d\phi = 1 \text{ i.e. } N^2 (2\pi) = 1, \text{ so that}$$

$$N = \frac{1}{\sqrt{2\pi}} = (2\pi)^{-1/2}$$

Hence normalized wavefunction becomes

$$Z_{\pm m}(\phi) = 2\pi^{-1/2} \exp(\pm im\phi); \quad m = 0, 1, 2, 3 \tag{5.124}$$

$$Z = \frac{1}{\sqrt{2\pi}} e^{im\phi} \tag{5.125}$$

### 5.6.3 Legendre's Equation for Rigid Rotator

Taking Eq. (5.116)

$$\frac{1}{\sin\theta} \frac{d}{d\theta}\left(\sin\theta \frac{dY}{d\theta}\right) + \left(\beta - \frac{m^2}{\sin^2\theta}\right) Y = 0$$

A new variable $x$ is now defined by

$$x = \cos\theta \text{ so that}$$
$$1 - x = 1 - \cos\theta$$
$$1^2 - x^2 = \sin^2\theta$$
$$\sin\theta = \sqrt{1 - x^2} \tag{5.126}$$

Consequently

$$\frac{dY}{d\theta} = \frac{dx}{d\theta} \cdot \frac{dY}{dx} = -\sin\theta \frac{dY}{dx} \tag{5.127}$$

And hence, in general

$$\frac{d}{d\theta} = -\sin\theta \frac{d}{dx} \tag{5.128}$$

Further,

$$\sin\theta \frac{dY}{d\theta} = -\sin^2\theta \frac{dY}{dx} \tag{5.129}$$

$$= -(1 - x^2) \frac{dY}{dx} \tag{5.130}$$

Making use of Eq. (5.128) and (5.130) to change the variable from $\theta$ to $x$, Eq. (5.116) becomes

$$\frac{d}{dx}\left\{(1-x^2)\frac{dY}{dx}\right\} + \left(\beta - \frac{m^2}{1-x^2}\right)Y = 0 \tag{5.131}$$

This equation is known as *Legendre's equation*. In order to solve above equation conveniently by the polynomial method in a similar manner as the case with harmonic oscillator, the function $Y$ should be replaced by another function of $x$, namely $G(x)$, defined by

$$Y = (1-x^2)^{\frac{1}{2}m} G \tag{5.132}$$

$$\frac{dY}{dx} = -mx(1-x^2)^{\frac{1}{2}m-1} G + (1-x^2)^{\frac{1}{2}m}\frac{dG}{dx} \tag{5.133}$$

$$\therefore \quad (1-x^2)\frac{dY}{dx} = -mx(1-x^2)^{\frac{1}{2}m} G + (1-x^2)^{\frac{1}{2}m+1}\frac{dG}{dx} \tag{5.134}$$

Hence,

$$\frac{d}{dx}\left\{(1-x^2)\frac{dy}{dx}\right\} = \left\{-m(1-x^2)^{\frac{1}{2}m} + m^2x^2(1-x^2)^{\frac{1}{2}m-1}\right\}G - \left\{2x(m+1)(1-x^2)^{\frac{1}{2}m}\right\}$$

$$\cdot G' + (1-x^2)^{\frac{1}{2}m-1} G'' \tag{5.135}$$

Here $\quad G' = \dfrac{dG}{dx} \quad$ and $\quad G'' = \dfrac{d^2G}{dx^2} \tag{5.136}$

Substituting this value in Eq. (5.131) and dividing through by $(1-x^2)^{\frac{1}{2}m}$, which is non-zero except at the limits noted, i.e. when $x$ is $+1$ or $-1$, resultant equation after few changes will be

$$(1-x^2)G'' - 2(m+1)xG' + \{\beta - m(m+1)\}G = 0 \tag{5.137}$$

$$(1-x^2)G'' - 2axG' + bG \tag{5.138}$$

where $a = m+1 \quad$ and $\quad b = \beta - m(m+1) \tag{5.139}$

As before, the assumption is now made that $G(x)$ can be expressed as a power series; thus

$$G = \alpha_0 + \alpha_1 x + \alpha_2 x^2 + \alpha_3 x^3 + \cdots \tag{5.140}$$

$$G' = \alpha_1 + 2\alpha_2 x + 3\alpha_3 x^2 + 4\alpha_4 x^3 + \cdots \tag{5.141}$$

$$G'' = 2\alpha_2 + 6\alpha_3 x + 12\alpha_4 x^2 + 20\alpha_5 x^3 + \cdots \tag{5.142}$$

Introducing these expressions into Eq, (5.138) and arranging the terms vertically according to increasing powers of $x$ the result is

$$2\alpha_2 + 6\alpha_3 x + 12\alpha_4 x^2 + 20\alpha_5 x^3 + \cdots - 2\alpha_2 x^2 - 6\alpha_3 x^3 - \cdots - 2a\alpha_1 x$$
$$-4a\alpha_2 x^2 - 6a\alpha_3 x^3 - \cdots + b\alpha_0 + b\alpha_1 x + b\alpha_2 x^2 + b\alpha_3 x^3 + \cdots = 0 \tag{5.143}$$

In order that this series may be zero for all possible values of $x$, the coefficients of individual powers of $x$ must vanish separately; therefore

$$2\alpha_2 + b\alpha_0 = 0$$
$$6\alpha_3 + (b - 2a)\alpha_1 = 0$$
$$12\alpha_4 + (b - 4a - 2)\alpha_2 = 0$$
$$20\alpha_5 + (b - 6a - 6)\alpha_3 = 0$$

where $n$ is zero or an integer. Inserting the values for $a$ and $b$, from Eq. (5.139), *recursion formula* is obtained for the coefficients in the power series for $G(x)$.

$$(n+1)(n+2)\alpha_{n+2} + [b - 2na - n(n+1)]\alpha_n = 0 \qquad (5.144)$$

$$\frac{\alpha_{n+2}}{\alpha_n} = \frac{(n+m)(n+m+1) - \beta}{(n+1)(n+2)} \qquad (5.145)$$

Now considering Eq. (5.116), if $\beta = l(l+1)$, where $l$ is the rotational quantum number, then this equation becomes a standard mathematical equation whose solution are known to be associated Legendre's polynomials $P_l^{|m|} \cos\theta$ where $l$ is either zero or a positive integer and $l > |m|$. The normalized solutions are

$$Y(\theta) = Y_{e,\pm m}(\phi) = \left[\frac{2l+1}{2} \times \frac{(l-|m|)!}{(l+|m|)!}\right]^{1/2} P_l^{|m|} \cos\theta \qquad (5.146)$$

### 5.6.4 Energy Levels of Rigid Rotator

In order that $G(x)$ may form the part of an acceptable wavefunction, $\psi$. It is necessary that it should be a polynomial, breaking off after a finite number of terms. The condition for this can be found, as before, by equating the numerator of the recursion formula to zero, i.e.

$$(n+m)(n+m+1) - \beta = 0$$
$$\therefore \quad \beta = (n+m)(n+m+1) \qquad (5.147)$$

It has been already seen that $m$ must be zero or an integer, and since the same condition applies to $n$, the sum $n + m$, which may be replaced by $l$, is also zero or integral. It follows, therefore, that the condition for a satisfactory wavefunction may be written as

$$\beta = l(l+1)$$

where $l$ equals to $(m + n)$ may be 0, 1, 2, 3, etc. Introducing the value of $\beta$ from Eq. (5.116a)

$$\frac{8\pi^2 IE}{h^2} = l(l+1)$$

or

$$E = l(l+1)\frac{h^2}{8\pi^2 I} \qquad (5.148)$$

which gives the permitted values (eigenvalues) for the energy of a rigid rotator with free axis.

**Problem 5.11** Calculate the angular momentum of a rigid diatomic rotating molecule (say HCl) in the second rotational energy level. Compare it with the angular momentum of an electron in the 2p atomic orbital.

*Solution:* For a rigid diatomic rotating molecule

$$|\vec{L}| = [J(J+1)]^{1/2}\, \hbar; \quad J = 0, 1, 2, 3, \ldots$$

where $J$ is rotational quantum number. For second rotational state $J = 1$

$$|\vec{L}| = \sqrt{2}\, \hbar = \sqrt{2} \times 1.055 \times 10^{-34}\, \text{Js} = 1.49 \times 10^{-34}\, \text{Js}$$

Again for an electron in an atomic orbital

$$|\vec{L}| = [l(l+1)]^{1/2}\, \hbar; \quad l = 0, 1, 2, 3, \ldots$$

For 2p electron, $l = 1$

$$|\vec{L}| = \sqrt{2}\, \hbar \quad \Rightarrow \quad 1.49 \times 10^{-34}\, \text{Js}$$

We see that the two values are the same.

**Problem 5.12** For a particle rotating along the circumference of a circle of radius $r$, setup Schrödinger equation in the appropriate form and obtain expression for the energy eigenvalues and normalized eigenfunctions.

*Solution:* The single degree of freedom for this problem can be chosen to be either the length ($S$) of the arc or the angle $\phi$.

For the particle of mass $m$ moving along the arc, the linear momentums $P\left(= \dfrac{mdS}{dt}\right)$ has the operator $\dfrac{h}{2\pi i}\dfrac{d}{dS}$. Since, there is no potential energy, the energy of the particle is wholly kinetic and the Hamiltonian operator $\hat{H} = \dfrac{P^2}{2m} = \dfrac{-h^2}{8\pi^2 m}\dfrac{d^2}{dS^2}$

Thus, the Schrödinger equation $\hat{H}\psi = E\psi$ takes the form

$$\frac{-h^2}{8\pi^2 m}\frac{d^2\psi}{dS^2} = E\psi$$

Since $S = r\phi, \quad \dfrac{d\phi}{dS} = \dfrac{1}{r}$

$$\frac{d\psi}{dS} = \frac{d\psi}{d\varphi}\frac{d\varphi}{dS} = \frac{1}{r}\frac{d\psi}{d\varphi} = \frac{u}{r} \qquad \left(\text{Putting } u = \frac{d\psi}{d\varphi}\right)$$

$$\frac{d^2\psi}{dS^2} = \frac{1}{r}\left(\frac{du}{dS}\right) = \frac{1}{r}\left(\frac{du}{d\phi}\cdot\frac{d\phi}{ds}\right) = \frac{1}{r^2}\frac{du}{d\varphi}$$

$$= \frac{1}{r^2}\frac{d^2\psi}{d\varphi^2}$$

The Schrödinger equation in the angular variable $\varphi$ is

$$\frac{-h^2}{8\pi^2 mr^2} \frac{d^2\psi}{d\varphi^2} = E\psi$$

$$\frac{d^2\psi}{d\varphi^2} = \frac{-8\pi^2 mr^2}{h^2} E\psi = -m^2\psi \quad \left(\text{Putting } m^2 = \frac{8p^2 mr^2 E}{h^2}\right)$$

This gives $\quad \psi = Ae^{\pm im\varphi}$

Since $\varphi$ and $\varphi + 2\pi$ are physically the same point, the single valuedness of $\psi$ requires that

$$\psi(\varphi) = \psi(\varphi + 2\pi)$$

or $\quad Ae^{im\varphi} = Ae^{im(\varphi+2\pi)}$

$\Rightarrow \quad Ae^{im\varphi} \cdot e^{im2\pi}$

For this $e^{im2\pi}$ has to be 1, or $\cos(m2\pi) + i\sin(m2\pi) = 1$
This is possible only if $m = 0, \pm 1, \pm 2, \pm 3, \ldots$
Solving for $E$,

$$E = \frac{m^2 h^2}{8\pi^2 mr^2} = \frac{m^2 h^2}{8\pi^2 I}, (I = mr^2)$$

$m$ is a quantum number, called *rotational quantum number* whose values determines the rotational energy.

Applying the normalization conditions

$$\int_0^{2\pi} \psi^*\psi\, d\varphi = \int_0^{2\pi} A^* e^{-im\phi} Ae^{-im\phi} d\varphi = 1$$

or $\quad |A|^2 \int_0^{2\pi} d\varphi = |A|^2 \cdot 2\pi = 1$

or $\quad A = \frac{1}{\sqrt{2\pi}}$

$\therefore \quad \psi = \frac{1}{\sqrt{2\pi}} e^{im\varphi}$

To get the real functions we take the linear combinations of the degenerate $e^{\pm im\varphi}$ functions

$$\psi_+ = \frac{1}{\sqrt{2\pi}} \left[\frac{1}{\sqrt{2}} (e^{im\varphi} + e^{-im\varphi})\right] = \frac{1}{\sqrt{\pi}} \cos(m\varphi)$$

$$\psi_- = \frac{1}{\sqrt{2\pi}} \left[\frac{1}{i\sqrt{2}} (e^{im\varphi} - e^{-im\varphi})\right] = \frac{1}{\sqrt{\pi}} \sin(m\varphi)$$

The factor $\frac{1}{\sqrt{2}}$ and $\frac{1}{i\sqrt{2}}$ in $\psi_+$ and $\psi_-$ have been introduced to normalize the functions.

**Problem 5.13** Use the function $\psi = \dfrac{1}{\sqrt{2\pi}} e^{im\varphi}$ to determine the energy eigenvalue and the expectation value for a particle in a ring of radius $r$.

Solution:
$$\hat{H}\psi = E\psi$$

$$\dfrac{-h^2}{8\pi^2 mr^2} \dfrac{d^2}{d\varphi^2}\left(\dfrac{1}{\sqrt{2\pi}} e^{im\varphi}\right) \dfrac{-h^2}{8\pi^2 mr^2} \dfrac{1}{\sqrt{2\pi}} i^2 M^2 e^{im\varphi} = \dfrac{m^2 h^2}{8\pi^2 mr^2}\left(\dfrac{1}{\sqrt{2\pi}} e^{im\varphi}\right)$$

$$\bar{E} = \int \psi^* \hat{H}\psi d\tau \quad \Rightarrow \quad \dfrac{1}{2\pi}\int_0^{2\pi} (e^{-im\varphi}) \dfrac{-h^2}{8\pi^2 mr^2} \dfrac{d^2}{d\varphi^2}(e^{im\varphi}) d\varphi$$

$$\Rightarrow \quad \dfrac{1}{2\pi}\times\left(\dfrac{-h^2}{8\pi^2 mr^2}\right)\times iM^2 \int_0^{2\pi} d\varphi \quad \Rightarrow \quad \dfrac{M^2 h^2}{2\pi \times 8\pi^2 mr^2}\times (2\pi) = \dfrac{M^2 h^2}{8\pi^2 mr^2}$$

**Problem 5.14** Two masses $m_1$ and $m_2$ connected by a rod to turn a rigid rotator, are restricted to rotate in a plane. Find the expression for its eigenvalue and eigenfunctions.

Solution: Let the centre of gravity of the rotator be at the origin and let the distances of $m_1$ and $m_2$ from the origin be $r_1$ and $r_2$, respectively,

then
$$m_1 r_1 = m_2 r_2 \quad \text{and} \quad r_1 + r_2 = r \text{ (say)}$$

This gives
$$r_1 = \dfrac{m_2 r}{m_1 + m_2} \quad \text{and} \quad r_2 = \dfrac{m_1 r}{m_1 + m_2}$$

Kinetic energy of rotation $= \dfrac{1}{2} m V_1^2 + \dfrac{1}{2} m_2 V_2^2$, where $V_1$ and $V_2$ are linear velocities of $m_1$ and $m_2$. For a small angle, $d\varphi$ of rotation, $dS = r d\varphi$, where $dS$ is small arc.

The linear velocity are, therefore, $V_1 = \dfrac{dS}{dt} = \dfrac{r_1 d\varphi}{dt} = r_1 \omega$

And
$$V_2 = \dfrac{r_2 d\varphi}{dt} = r_2 \omega$$

Hence,
$$KE = \dfrac{1}{2} m_1 r_1^2 \omega^2 + \dfrac{1}{2} m_2 r_2^2 \omega^2$$

Substituting for $r_1$ and $r_2$, $KE = \dfrac{1}{2}\omega^2 \left(\dfrac{m_1 m_2}{m_1 + m_2}\right) r^2$

$$\Rightarrow \qquad \dfrac{1}{2}\mu r^2 \omega^2 = \dfrac{1}{2} I \omega^2$$

This result permits the rotator to be treated as a single particle of mass $\mu$ rotating in a circle of radius $r$. The Schrödinger equation will be

$$\frac{-h^2}{8\pi^2 \mu r^2} \frac{d^2\psi}{d\varphi^2} = E\psi$$

For expressions of $E$ and $\psi$, see problem 5.12.

**Problem 5.15** An electron rotates in a circular orbits of radius 0.5 Å. Calculate the first three energy levels and point out the degeneracy.

Solution: $\quad E_m = \dfrac{m^2 h^2}{8\pi^2 mr^2};\quad m = 0,\quad E_0 = 0$

$m = \pm 1,\quad E_{+1} = E_{-1} = \dfrac{h^2}{8\pi^2 mr^2}$

$\Rightarrow \quad \dfrac{(6.624 \times 10^{-34}\, \text{Js})^2}{8\pi^2 \times 9.1 \times 10^{-31} \times (0.5 \times 10^{-10}\, \text{m})^2}$

$\Rightarrow \quad 2.21 \times 10^{-17}\, \text{J}$

$m = \pm 2,\quad E_{+2} = E_{-2} = \dfrac{4h^2}{8\pi^2 mr^2}$

$\Rightarrow \quad \dfrac{4 \times (6.62 \times 10^{-34})^2}{8\pi^2 \times 9.1 \times 10^{-31} \times (0.5 \times 10^{-10})^2}$

$\Rightarrow \quad 4 \times 2.21 \times 10^{-17}\, \text{J} = 8.84 \times 10^{-17}\, \text{J}$

The states with $m = \pm 1$ and $\pm 2$ are each doubly degenerate.

**Problem 5.16** A cyclic polyene, in the free electron approximation is a ring in which the $\pi$ electron move independent of each other (i.e. $V = 0$). Use this model to calculate (i) the total ground state energy of benzene, and (ii) the wavelength of the longest electronic transitons, given each C–C bond length is 1.42 Å and mass of an electron is $9 \times 10^{-31}$ kg.

Solution: The circumference of the ring, $2\pi r = nd$, where $r$ = radius, $d$ = C–C bond length and $n$ = number of C–C bonds. For benzene

$$r = \frac{6 \times 1.42 \times 10^{-10}\, \text{m}}{2\pi} = \frac{8.521 \times 0^{-10}\, \text{m}}{2\pi}$$

$$E_M = \frac{M^2 \times (6.62 \times 10^{-34}\, \text{Js})^2}{8\pi^2 \times 9.1 \times 10^{-31}\, \text{kg} \times \left(\dfrac{8.52 \times 10^{-10}\, \text{m}}{2\pi}\right)^2}$$

$E_M = M^2 \times 3.31 \times 10^{-19}\, \text{J}$

(i) State $M = 0$ has $E_0 = 0$ and the states $M = \pm 1$ have
$$E_{+1} = E_{-1} = 3.31 \times 10^{-19} \text{ J} \quad \text{(Doubly degenerate)}$$
According to Paulis principle, not more than 2 electrons remain in any state. Thus, the total ground state energy $E$ of benzene having $6\pi$ electrons is
$$E = (2 \times 0 + 2 \times 3.31 \times 10^{-19} + 2 \times 3.31 \times 10^{-19} \text{ J})$$
$$E = 13.24 \times 10^{-19} \text{ J}$$

(ii) Longest wavelength electronic transiton corresponds to
$$M = \pm 1 \quad \text{to} \quad M = \pm 2$$
The resulting excited state has the energy
$$E = 2 \times 0 + (2 \times 3.31 \times 10^{-19} + 1 \times 3.31 \times 10^{-19}) + (1 \times 4 \times 3.31 \times 10^{-19})$$
$$E = 23.17 \times 10^{-19} \text{ J}$$
$$E^1 - E = \frac{hc}{\lambda} \quad \text{or} \quad \lambda = \frac{hc}{E^1 - E} = \frac{6.62 \times 10^{-34} \text{ Js} \times 3 \times 10^{10} \text{ cm s}^{-1}}{(23.17 - 13.24) \times 10^{-19} \text{ J}}$$
$$\lambda = 200 \times 10^{-7} \text{ cm} \quad \Rightarrow \quad 200 \text{ nm}$$

## 5.7 HYDROGEN ATOM OR HYDROGEN LIKE ATOM: A CENTRAL FORCE FIELD PROBLEM

A *H*-like atom (Figure 5.7) may be regarded as a system of two charge particles, positive charge (nucleus) and negative charge (electron) between which columbic forces are operating. For a two or more particle system Schrödinger wave equation is most conveniently written in form of equation,

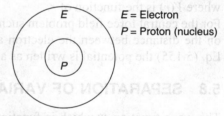

**FIGURE 5.7** H-atom.

$E$ = Electron
$P$ = Proton (nucleus)

$$H\psi = E\psi \quad (5.149)$$

Hemiltonian operator $H$ is defined as:
$$H = -\frac{h^2}{8\pi^2} \sum \frac{1}{m_i} \nabla_i^2 + V \quad (5.150)$$

For *H*-like atom consisting of two particles, the Hamiltonian is given by
$$H = -\frac{h^2}{8\pi^2}\left(\frac{1}{m_1}\nabla_1^2 + \frac{1}{m_2}\nabla_2^2\right) + V \quad (5.151)$$

If $\psi_T$ and $E_T$ are used to represent the total eigenfunction and total energy, respectively for the whole system then combination of Eqs. (5.149) and (5.151) gives two wave equations for the *H*-like system as

$$\frac{1}{m_1}\nabla_1^2 \psi_T + \frac{1}{m_2}\nabla_2^2 \psi_T + \frac{8\pi^2}{h^2}(E_T - V)\psi_T = 0 \quad (5.152)$$

The total eigen function $\psi_T$ includes the contribution due to translational motion of the system as a whole. This factor may be removed in the following manner, if it is supposed that $\psi_T$ is product of two eigenfunction namely $\psi_T$ and $\psi$ where $\psi_T$ depends upon the position of centre of mass of the system and $\psi$ which is a function only of the distance between the particles, then it is possible to show that

$$\psi_T \Rightarrow \psi_t \text{ and } \psi$$

$\psi$ is the function of distance between proton and electron.

$$\nabla^2 \psi + \frac{8\pi^2 \mu}{h^2}(E - V)\psi = 0 \tag{5.153}$$

i.e.

$$\frac{\partial^2 \psi}{\partial x^2} + \frac{\partial^2 \psi}{\partial y^2} + \frac{\partial^2 \psi}{\partial z^2} + \frac{8\pi^2 \mu}{h^2}(E - V)\psi = 0 \tag{5.154}$$

where $x$, $y$, and $z$ are the co-ordinates of the centre of the mass of the system. $E$ is total energy, $V$ is potential energy exclusive of the translational energy and $\mu$ is reduced mass.

The potential energy $V$ is a function of position of two particles relative to one another and is independent of the translational motion of the system as a whole. The wave Eq. (5.154) is thus identical with that for a single particle of mass $m$ in a field of potential $V$ transforming to spherical co-ordinates, it will become

$$\frac{1}{r^2}\frac{\partial}{\partial r}\left(r^2 \frac{d\psi}{dr}\right) + \frac{1}{r^2 \sin\theta}\frac{\partial}{\partial \theta}\left(\sin\theta \frac{\partial \psi}{\partial \theta}\right) + \frac{1}{r^2}\frac{1}{\sin^2\theta}\frac{\partial^2 \psi}{\partial \phi^2} + \frac{8\pi^2 \mu}{h^2}(E - V(r))\psi = 0 \tag{5.155}$$

where $V(r)$ is the function of $r$.

For the central force field problem such that as under consideration, the potential $V$ depends only on the distance between the electron and the proton and not on the angle $\theta$ and $\phi$. Hence, in Eq. (5.155) the potential is written as a function of $r$ only.

## 5.8 SEPARATION OF VARIABLES

The wave function $\psi$ which is function of $r$, $\theta$ and $\phi$ can be expressed as the product of three functions, i.e. $R(r)$, $Y(\theta)$ and $Z(\phi)$, each of which is a function of the variables i.e.

$$\psi(r, \theta, \phi) = R(r)\, Y(\theta)\, Z(\phi) \tag{5.156}$$

Making this substitution in Eq. (5.159) and multiplying throughout by $\dfrac{r^2 \sin^2 \theta}{RYZ}$ the result is:

$$\frac{\sin^2 \theta}{R}\frac{d}{dr}\left(r^2 \frac{dR}{dr}\right) + \frac{1}{Z}\frac{d^2 Z}{d\phi^2} + \frac{\sin\theta}{Y}\frac{d}{d\theta}\left(\sin\theta \frac{dY}{d\theta}\right) + r^2 \sin^2\theta \frac{8\pi^2 \mu}{h^2}((E - V)r) = 0 \tag{5.157}$$

In this expression, the 2nd term is a function of variable $\phi$ and hence it can be shown by an argument similar to that with the rigid rotator problem that this term must be a constant representing the constant by $-m^2$. It is possible to write

$$\frac{1}{Z}\frac{d^2 Z}{d\phi^2} = -m^2 \tag{5.158}$$

and
$$\frac{d^2 Z}{d\phi^2} + m^2 Z = 0 \qquad (5.159)$$

By putting value of $-m^2$ in place of second term in Eq. (5.157) gives,

$$\frac{\sin^2\theta}{R}\frac{d}{dr}\left(r^2\frac{dR}{dr}\right) - m^2 + \frac{\sin\theta}{Y}\frac{d}{d\theta}\left(\sin\theta\frac{dY}{d\theta}\right) + r^2\sin^2\theta\frac{8\pi^2\mu}{h^2}(E - V) = 0 \qquad (5.160)$$

And if this is divided throughout by $\sin^2\theta$, final result after rearrangement is

$$\underbrace{\frac{1}{R}\frac{d}{dr}\left(r^2\frac{dR}{dr}\right) + \frac{r^2 8\pi^2\mu}{h^2}(E - V(r))}_{r\text{ dependent only}} = \underbrace{\frac{m^2}{\sin^2\theta} - \frac{1}{Y\sin\theta}\frac{d}{d\theta}\left(\sin\theta\frac{dY}{d\theta}\right)}_{\theta\text{ dependent only}} \qquad (5.161)$$

LHS on this equation is seen to be dependent on the variable $r$ only while RHS is dependent on variable $\theta$ only. It can be shown therefore by similar argument that each side is a constant. Setting each side equal to $\beta$ and multiplying LHS of equation by $\dfrac{R}{r^2}$

$$\frac{1}{r^2}\frac{d}{dr}\left(r^2\frac{dR}{dr}\right) - \frac{\beta R}{r^2} + \frac{8\pi^2 m}{h^2}(E - V(r))R = 0 \qquad (5.162)$$

$$\frac{1}{\sin\theta}\frac{d}{d\theta}\left(\sin\theta\frac{dY}{d\theta}\right) + \left(\beta - \frac{m^2}{\sin^2\theta}\right)Y = 0 \qquad (5.163)$$

$\theta$ and $\phi$ are called spherical co-ordinates or eigenfunctions.

## 5.9 SPHERICAL EIGENFUNCTION

It will now be apparent that wave equation has been separated into three equations namely (5.159, 5.162 and 5.163) each of which involve one variable i.e. $\phi$, $r$, and $\theta$, respectively. It remains therefore to solve each of these equations individually. Equations (5.159) and (5.113) are identical with the equation for rigid rotator. Hence it follows that, as for the latter, solution given by normalized eigen function

$$Z(\phi) = \frac{1}{\sqrt{2\pi}} e^{im\phi} \qquad (5.164)$$

where $m = 0, \pm 1, \pm 2 \ldots$

The +ve and –ve value representing different possible solutions and solution of Eq. (5.163)

$$Y(\theta) = \sqrt{\frac{2l+1}{2}\frac{(l-m)!}{(l+m)!}}\, p_l^m \cos\theta \qquad (5.165)$$

The functions $Y(\theta)$ and $Z(\phi)$ are referred to as the spherical ion functions, since they give spherically distinct function from the radical function, i.e. distribution of electron in three-dimensional space. Value of $\beta$ in Eq. (5.163) is similar to that derived for rigid rotator and hence for the present case

$$\beta = l(l+1) \qquad (5.166)$$

If the eigenfunction are to be finite integral value of $m$ cannot exceed $l$.

## 5.10 SOLUTION OF RADICAL WAVE FUNCTION

Upon inserting the value of $\beta = l(l+1)$ into Eq. (5.162) it becomes,

$$\frac{1}{r^2}\frac{d}{dr}\left(r^2\frac{dR}{dr}\right) + \left[-\frac{l(l+1)}{r^2} + \frac{8\pi^2 m}{h^2}(E - V(r))\right]R = 0 \qquad (5.167)$$

$$\frac{d^2R}{dr^2} + \frac{2}{r}\frac{dR}{dr} + \left[-\frac{l(l+1)}{r^2} + \frac{8\pi^2 m}{h^2}(E - V(r))\right]R = 0 \qquad (5.168)$$

$$\frac{d^2R}{dr^2} + \frac{2}{r}\frac{dR}{dr} + \left[-\frac{l(l+1)}{r^2} + \frac{8\pi^2 mE}{h^2} - \frac{8\pi^2 m}{h^2}V(r)\right]R = 0 \qquad (5.169)$$

The system consists of an electronic charge $e$ moving in the central force field of a nucleus of effective charge $Ze$ at a distance $r$. Value of $V(r)$ is thus given by

$$V(r) = \frac{-ze^2}{r} \qquad (5.170)$$

Now putting the value of $V(r)$ in Eq. (5.169), we get

$$\frac{d^2R}{dr^2} + \frac{2}{r}\frac{dR}{dr} + \left[-\frac{8\pi^2 mE}{h^2} + \frac{8\pi^2 mZe^2}{h^2 r} - \frac{l(l+1)}{r^2}\right]R = 0 \qquad (5.171)$$

With $E$ (–ve), it is convenient to introduce a new parameter defined by

$$n^2 = \frac{-2\pi^2 mZ^2 e^4}{h^2 E} \qquad (5.172)$$

As well as a new variable $x$ in place of $r$ defined as

$$x = \frac{2Z}{na_0}r \qquad (5.173)$$

where $a_0$ is $\dfrac{h^2}{4\pi^2 \mu e_0}$.

Therefore,

$$x = \frac{2Zr 4\pi^2 me_0}{nh^2}$$

$$= \frac{8\pi^2 mZ\varepsilon_o r}{nh^2} \qquad (5.174)$$

It will be observed that the quantity $a_0$ is identical with so called *Bohr orbit* of the H-atom, i.e. smallest orbit of the normal H-atom as given by Bohr's theory. Now introducing value of $n$ and $x$ into Eq. (5.171) and replacing of $R(r)$ by a new function $X(x)$ leads to

$$\frac{d^2X}{dx^2} + \frac{2}{x}\frac{dX}{dx} + \left[-\frac{1}{4} + \frac{n}{x} - \frac{l(l+1)}{x^2}\right]X = 0 \qquad (5.175)$$

Solution of this equation may be obtained by taking $x$ very large so that corresponding asymptotic solution of this above equation is

$$\frac{d^2X}{dx^2} - \frac{1}{4}X = 0 \qquad (5.176)$$

For which solutions are

$$X(x) = e^{x/2} \quad \text{and} \quad X(x) = e^{-x/2} \qquad (5.177)$$

Since $x$ may vary from 0 to infinity, the $X(x)$ increases as $x$ increases, this will lead to an unacceptable wavefunction. On the other hand, the second solution will be satisfactory. A possible solution of Eq. (5.175) is thus given by

$$X(x) = e^{-x/2} F(x) \qquad (5.178)$$

where $F(x)$ is another function of a variable $x$. From various considerations, it appears that $F(x)$ effects may be split into factors i.e. $x^l$ and $G^x$, where, $l$ has same significance as in Eq. (5.175). Hence, it is possible to write

$$R(r) = X(x)$$

$$X(x) = e^{-x/2} x^l G(x) \qquad (5.179)$$

From this relationship $\dfrac{d^2X}{dx^2}$ and $\dfrac{dX}{dx}$ can be evaluated in terms $G(x)$. If the results are substituted in Eq. (5.175), then

$$\frac{xd^2G}{dx^2} + [(2l+1)+1-x]\frac{dG}{dx} + (n-l-1)G = 0 \qquad (5.180)$$

Comparison of this equation with equation of Legendre's polynomial shows that $G(x)$ may be identified with associated Legendre's polynomials.

$$G(x) = L_k^P(x) \text{ with } P = 2l+1 \quad \text{or} \quad k = n+l \qquad (5.181)$$

This polynomial or more correctively, the polynomial multiplied by a constant factor will thus be a solution of Eq. (5.180). So that it is possible to write

$$G(x) = C L_{n+l}^{2l+1}(x) \qquad (5.182)$$

where $C$ is a constant, which may be made equal to normalization factor. Therefore, the complete expression $R(r)$ is thus seen from Eq. (5.179) to be given as

$$R(r) = C e^{-x/2} x^l L_{n+l}^{2l+1}(x) \qquad (5.183)$$

This is clearly an acceptable function for the associated Legendre's polynomials, as already seen, has finite number of turns only.

## 5.11 ATOMIC ENERGY LEVEL AND QUANTUM NUMBER

In the associate Legendre's polynomial $L_k^P(x)$, which is an acceptable solution of Eq. (5.180). Both $p$ and $k$ must be zero or an integer. Utilizing the equivalent value of $p$ and $k$, i.e. $(2l + 1)$ and $(n + l)$, respectively, it follows that $(2l + 1)$ and $(n + l)$ must be zero or an integer. It has been previously established that $l$ must be zero or integer and since $p$ is less than or equal to $k$ $(p \le k)$ and $n = 1, 2, 3,\dots$ . It follows, therefore, from the definition of $n$ given by Eq. (5.172) that the eigenvalue $E_n$ of the energy for the case under consideration is,

$$E_n = \frac{-2\pi^2 \mu Z e^4}{n^2 h^2} \tag{5.184}$$

where $n$ is integer or zero. This result is same as that obtained by Bohr energy level of H-like atom. It is seen therefore that the acceptable solution of wave equation leads to permitted energy value which are same as those derived from Bohr theory. The integer nature of $n$, which can now be identified with the principal quantum number of the atom, is clearly a necessary condition for acceptable solution rather than a postulate as in Bohr treatment. Since $k \ge p$, it follows using value of $k$ and $p$

$$(n + l) \ge 2l + 1$$
$$\therefore \quad l \le n - 1 \tag{5.185}$$

$$S\ 0 = 1 - 1 \quad \text{if } n = 1$$
$$P\ 1 = 2 - 1 \quad \text{if } n = 2$$
$$D\ 2 = 3 - 1 \quad \text{if } n = 3$$

Although principal quantum number $n$ may theoretically be zero value, has no physical significance. For it $E_n$ is numerically equal to $\infty$. The possible value of $n$ are thus 1, 2, 3, 4,... and hence it follows that $l$, which is equivalent to Azimuthal quantum number can be 0, 1, 2, 3,... $(n-1)$. It has been seen in connection with the problem of rigid rotator that the possible value of $m$ (magnetic quantum number) 0, ±1, ±2, ±3,..., ± $l$ and for the same condition applied to the present case, provided there is no perturbing fields, these $(2l + 1)$ values correspond to same energy. But in magnetic field, the levels are separated and $(2l + 1)$ different orientations are possible.

Wave mechanics thus provide a satisfactory basis for the occurrence of three quantum numbers $(n, l$ and $m)$.

## 5.12 NORMALIZED H-LIKE EIGENFUNCTION

The complete solution $\psi(\theta, \phi, r)$ of the wave equation for H-like atom is obtained by multiplying the spherical eigen function $Y(\theta)$ and $Z(\phi)$ by the radial eigen function $R(r)$. Value of $C$ in Eq. (5.183) for $R(r)$ can be found by applying normalizing condition for the physically significant interval 0 to $\infty$

$$\int_0^\infty R(r) R^*(r) r^2 dr = 1 \tag{5.186}$$

The factor $r^2$ being necessary to convert the length $dr$ into an elemental volume. Introduction of values of $R(r)$ and $R^*(r)$, which are identical in this case, leads to the result.

$$C^2 = \left(\frac{x}{r}\right)^3 \frac{(n-l-1)!}{[2n(n+l)!]^3}$$

$$C = \sqrt{\left(\frac{x}{r}\right)^3 \frac{(n-l-1)!}{[2n(n+l)!]^3}}$$

Inserting this value in Eq. (5.183) and making use of the definition $x$ given by Eq. (5.173), it follows that normalized radial part eigenfunction for a H-like atom is given by

$$R_{nl}(r) = \sqrt{\left(\frac{2Z}{na_0}\right)^3 \frac{(n-l-1)!}{[2n(n+l)!]^3}} \, e^{-Zr/na_0} \left(\frac{2Zr}{na_0}\right)^l L_{n+l}^{2l+1}\left(\frac{2Zr}{na_0}\right) \quad (5.187)$$

The subscript $nl$ has been added to indicate that this function involve the quantum numbers $n$ and $l$. Similarly, $Y_{lm}(\theta)$ depends upon $l$ and $m$ while $Z_m(\phi)$ contains only $m$. These two functions are repeated here for the sake of completeness.

$$Y_{lm}(\theta) = \sqrt{\frac{2l+1}{2} \frac{(l-m)!}{(l+m)!}} \, P_l^m \cos\theta \quad (5.188)$$

$$Z_m(\phi) = \frac{1}{\sqrt{2\pi}} e^{im\phi} \quad (5.189)$$

Although these function appears to be very complicated. They actually reduce to very simple form, specially for low values of quantum number $n$, $m$ and $l$. As these are the basis of immediate interest. The expression for $R(r)$, $Y(\theta)$ and $Z(\phi)$ for $n = 1$ and 2, for $l = 0$ and 1, and for $m = 0$ and $\pm 1$ are given in Tables 5.2 and 5.3.

**TABLE 5.2** Normalized radial function for H-like atoms

| $n$ | $l$ | $R_{nl}(r)$ |
|---|---|---|
| 1 | 0 | $2\left(\dfrac{Z}{a_0}\right)^{3/2} e^{-Zr/a_0}$ |
| 2 | 0 | $\dfrac{1}{2\sqrt{2}}\left(\dfrac{Z}{a_0}\right)^{3/2}\left(2 - \dfrac{Zr}{a_0}\right) e^{-Zr/a_0}$ |
| 2 | 1 | $\dfrac{1}{2\sqrt{2}}\left(\dfrac{Z}{a_0}\right)^{3/2}\left(\dfrac{Zr}{a_0}\right) e^{-Zr/2a_0}$ |

**TABLE 5.3** Normalized spherical functions for hydrogen like atom

| $l$ | $m$ | $Y_{lm}(\theta)$ | $Z_m(\phi)$ |
|---|---|---|---|
| 0 | 0 | $\dfrac{1}{\sqrt{2}}$ | $\dfrac{1}{\sqrt{2\pi}}$ |
| 1 | 0 | $\sqrt{\dfrac{3}{2}}\cos\theta$ | $\dfrac{1}{\sqrt{2}}$ |
| 1 | $\pm 1$ | $\sqrt{\dfrac{3}{2}}\sin\theta$ | $\dfrac{1}{\sqrt{2\pi}}e^{\pm i\phi}$ or $\dfrac{1}{\sqrt{\pi}}\cos\phi$ or $\dfrac{1}{\sqrt{\pi}}\sin\phi$ |

In Table 5.3, both the complex and real values of the eigenfunctions $Z_m(\phi)$ are even for $m = \pm 1$.

The real form of this complete eigen function exclusive of spin for H-like atom with electron in either K shell, i.e. $n = 1$ shell or i.e. $n = 2$ are given in Table 5.4. In accordance with the limitation on the values of $l$ and $m$, the only possibility for $n$ is $l$, is

$$n = 1, \quad l = 0 \quad \text{and} \quad m = 0$$

The electron can then be described by symbol 1s.

When $n = 2$, there are four possible electronic states, electron is 2s. $m = 1, 0, -1$ which corresponds to three possible $2p_x$, $2p_y$ and $2p_z$ states.

**TABLE 5.4** Complete eigen function for H-like atom

| $n$ | $l$ | $m$ | States | Eigen function (complete) |
|---|---|---|---|---|
| 1 | 0 | 0 | 1s | $\psi_{100} = \dfrac{1}{\sqrt{\pi}}\left(\dfrac{Z}{a_0}\right)^{3/2} e^{-Zr/a_0}$ |
| 2 | 0 | 0 | 2s | $\psi_{200} = \dfrac{1}{4\sqrt{2\pi}}\left(\dfrac{Z}{a_0}\right)^{3/2}\left(2 - \dfrac{Z_r}{a_0}\right) e^{-Zr/2a_0}$ |
| 2 | 1 | 0 | $2p_z$ | $\psi_{210} = \dfrac{1}{4\sqrt{2\pi}}\left(\dfrac{Z}{a_0}\right)^{5/2} (e^{-Zr/2a_0}) r\cos\theta$ |
| 2 | 1 | 1 | $2p_x, 2p_y$ | $\psi_{211} = \dfrac{1}{4\sqrt{2\pi}}\left(\dfrac{Z}{a_0}\right)^{5/2} (e^{-Zr/2a_0}) r\sin\theta\cos\phi$ |
|  |  |  |  | $\psi_{211} = \dfrac{1}{4\sqrt{2\pi}}\left(\dfrac{Z}{a_0}\right)^{5/2} (e^{-Zr/2a_0}) r\sin\theta\sin\phi$ |

## 5.13 EIGENFUNCTION AND PROBABILITY DISTRIBUTION

Value of square of radial function is a measure of the probability of finding of an electron in an element of unit length at a distance $r$ from the nucleus. According to classical theory, which requires that electron to be confined to a definite orbit, this probability should be zero at a distance except that representing the position of the orbit. The radial function of quantum mechanics such as those given in Table 5.2, however leads to quite different conclusions. There is generally a finite probability of finding the electron at all distances from the nucleus from 0 to ∞ as is to be expected. This probability falls of rapidly to very small values when $r$ is large. So, instead of studying $R(r)^2$ a more useful property known as distribution function $4\pi r^2 R(r)^2$ which is a measure of probability that electron will be found in a spherical shell at a distance $r$ from the nucleus. Some examples of this distribution function are given in Figure 5.8.

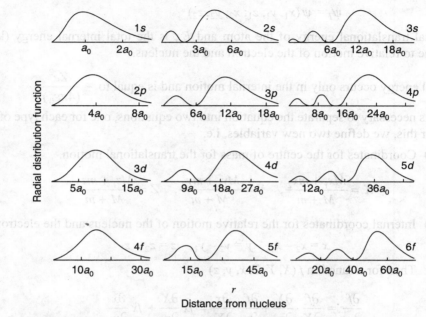

**FIGURE 5.8** Radial distribution function in case of hydrogen atom.

## Conclusions

The curve for $1s$, $2p$, $3d$, and $4f$, etc. electrons shows one maxima only and it is of interest to note that in these cases, the maxima occurs at distances from nucleus equal to the corresponding Bohr orbit. As a general rule, $s$ and $p$ eigen function, i.e. $l = 0$ and 1 for same value of $n$, for example, $\psi_{200}$, $\psi_{210}$ and $\psi_{211}$ show approximately same dependence on $r$ and the difference between these eigen functions lies essentially in their dependence on angle $\theta$ and $\phi$, i.e. to say the radial parts of the eigen function are approximately the same, but spherical part differ appreciably.

**Problem 5.17** A hydrogen-like atom is assumed to be a system of a positive nucleus and a negative electron interacting through coulomb's law.

(i) Write down complete SWE,
(ii) Separate the equation into an equation for the translational motion of the atom as a whole and an equation for the motion of the electron relative to the nucleus.

*Solution:* (i) There are two types of motions involved in this system, the motion of the atom as a whole and the relative motion of the nucleus and the electron. If $x_1$, $y_1$ and $z_1$ be the coordinates of the nucleus and $x_2$, $y_2$ and $z_2$ be those of the electrons, then the SWE can be written in a straight-forward manner as:

$$-\frac{h^2}{8\pi^2}\left(\frac{1}{M}\nabla_1^2 + \frac{1}{m}\nabla_2^2\right) - \frac{Ze^2}{(4\pi\varepsilon_0)r}\psi_T = E_T\psi_T \quad (1)$$

where $M$ is mass of nucleus, $m$ is mass of electron, $Z$ is charge on the nucleus and $e$ is charge on the electron.

$$r = [(x_2 - x_1)^2 + (y_2 - y_1)^2 + (z_2 - z_1)^2]^{1/2}$$
$$\psi_T = \psi(x_1, y_1, z_1; x_2, y_2, z_2)$$

$E_T$ is the total translational energy of the atom and $E_{Tr}$ is the total internal energy (kinetic + potential) due to relative motion of the electron and the nucleus.

Potential energy occurs only in the internal motion and is equal to $-\dfrac{Ze^2}{(4\pi\varepsilon_0)r}$.

(ii) It is necessary to separate the equation into two equations, one for each type of motion. For this, we define two new variables, i.e.

(a) Coordinates for the centre of mass for the translational motion.

$$X = \frac{Mx_1 + mx_2}{M + m}, \quad Y = \frac{My_1 + my_2}{M + m}, \quad Z = \frac{Mz_1 + mz_2}{M + m}$$

(b) Internal coordinates for the relative motion of the nucleus and the electron.

$$x = x_2 - x_1, \quad y = y_2 - y_1, \quad z = z_2 - z_1$$

Then for a function $f(X, Y, Z, x, y, z)$

$$\frac{\partial f}{\partial x_1} = \frac{\partial f}{\partial X}\cdot\frac{\partial X}{\partial x_1} + \frac{\partial f}{\partial x}\cdot\frac{\partial x}{\partial X_1} = fx\frac{\partial X}{\partial x_1} + fx\frac{\partial x}{\partial x_1}$$

and

$$\frac{\partial^2 f}{\partial x_1^2} = \left[\left(\frac{\partial fx}{\partial x}\cdot\frac{\partial X}{\partial x_1} + \frac{\partial fx}{\partial x}\cdot\frac{\partial x}{\partial x_1}\right)\frac{\partial X}{\partial x_1} + fx\frac{\partial^2 X}{\partial x_1}\right]$$
$$+ \left(\frac{\partial fx}{\partial X}\cdot\frac{\partial X}{\partial x_1} + \frac{\partial fx}{\partial x}\cdot\frac{\partial x}{\partial X_1}\right)\frac{\partial X}{\partial x_1} + fx\frac{\partial^2 x}{\partial x_1^2}$$

Now $\dfrac{\partial fX}{\partial X} = \dfrac{\partial^2 f}{\partial X^2}, \quad \dfrac{\partial X}{\partial x_1} = \dfrac{M}{M + m}, \quad \dfrac{\partial x}{\partial x_1} = -1$

$\dfrac{\partial fx}{\partial x} - \dfrac{\partial}{\partial x}\left(\dfrac{\partial f}{\partial X}\right), \quad \dfrac{\partial^2 X}{\partial x_1^2} = 0, \quad \dfrac{\partial fx}{\partial X} = \dfrac{\partial}{\partial x}\left(\dfrac{\partial fx}{\partial x}\right)$

$\dfrac{\partial fx}{\partial x} = \dfrac{\partial^2 f}{\partial x^2}, \quad \dfrac{\partial^2 x}{\partial x_1^2} = 0$

$$\frac{\partial^2 x}{\partial x_1^2} = \left[\left(\frac{M}{M+m}\right)\frac{\partial^2}{\partial x^2} - \frac{\partial^2}{\partial x \partial X}\right]\frac{M}{M+m} + \left[\left(\frac{M}{M+m}\right)\frac{\partial^2}{\partial X \partial x} - \frac{\partial^2}{\partial x^2}\right]$$

$$= \left[\left(\frac{M}{M+m}\right)^2 \frac{\partial^2}{\partial x^2} + \frac{2M}{M+m}\left(\frac{\partial^2}{\partial x \partial X}\right) + \frac{\partial^2}{\partial x^2}\right]$$

By a similar procedure it is found that

$$\frac{\partial^2}{\partial x^2} = \left(\frac{M}{M+m}\right)^2 \frac{\partial^2}{\partial X^2} + \frac{2M}{M+m}\left(\frac{\partial^2}{\partial X \partial x}\right) + \frac{\partial^2}{\partial x^2}$$

The procedure is repeated to obtain the expression for

$$\frac{\partial^2}{\partial y_1^2}, \quad \frac{\partial^2}{\partial y_2^2}, \quad \frac{\partial^2}{\partial z_1^2}, \quad \frac{\partial^2}{\partial z_2^2}$$

By substitution these values into SWE (1), yields

$$\left[\frac{-h^2}{8\pi^2}\left\{\frac{1}{M}\left(\frac{M}{M+m}\right)^2 \nabla_e^2 + \frac{1}{m}\left(\frac{m}{M+m}\right)^2 \nabla_e^2 + \frac{1}{M}\nabla_r^2 + \frac{1}{M}\nabla_r^2\right\}\right.$$
$$\left. \frac{-Ze^2}{(4\pi\varepsilon_0)r}\right]\psi_T = E_T\psi_T$$

$$\left[\frac{-h^2}{8\pi^2}\left\{\frac{1}{m+M}\nabla_e^2 + \frac{1}{M}\nabla_r^2\right\}\frac{-Ze^2}{(4\pi\varepsilon_0)r}\right]\psi_T = E_T\psi_T \qquad (2)$$

where $\nabla_e^2 = \frac{\partial_2}{\partial X^2} + \frac{\partial_2}{\partial Y^2} + \frac{\partial_2}{\partial Z^2}$, $\Delta_r^2 = \frac{\partial_2}{\partial x^2} + \frac{\partial_2}{\partial y^2} + \frac{\partial_2}{\partial z^2}$

where $M \gg m$, $m + M = M$ and $\mu = \frac{Mm}{M+m} \approx m$ so that above equation can be written

$$\left[-\frac{h^2}{8\pi^2}\left\{\frac{1}{M}\nabla_e^2 + \frac{1}{m}\nabla_r^2\right\}\frac{-Ze^2}{(4\pi\varepsilon_0)r}\right]\psi_T = E_T\psi_T \qquad (3)$$

Now, let $\psi_T = \Phi(c) \cdot \psi(r)$ where $c$ and $r$ denoting the coordinates collectively of the centre of mass and of the relative motion. By substituting this product for $\psi_T$ in Eq. (3) and carrying out the differentiation and dividing by $\psi_T$, we get

$$\frac{-h^2}{8\pi^2 M}\frac{1}{\phi(c)}\nabla_c^2 \phi(c) - \frac{h^2}{8\pi^2 M}\frac{1}{\psi(r)}\nabla_r^2 - \frac{Ze^2}{(4\pi\varepsilon_0)r} = E_T$$

$$\frac{-h^2}{8\pi^2 M}\frac{1}{\phi(c)}\nabla_c^2 \phi(c) = \frac{h^2}{8\pi^2 M}\frac{1}{\psi(r)}\nabla_r^2 + \frac{Ze^2}{(4\pi\varepsilon_0)r} + E_T$$

The coordinates are thus separated, and for LHS to be equal to RHS for the all values of the coordinates, both sides must be equal to the some constant, (say) $W$. On rearranging,

$$\therefore \quad -\frac{h^2}{8\pi^2 M} \nabla_c^2 \, \phi(c) = W \phi(c) \tag{4}$$

And
$$-\frac{h^2}{8\pi^2 m} \nabla_r^2 \, \psi_{(r)} - \frac{Ze^2}{(4\pi\varepsilon_0)r} \psi_{(r)} = (E_T - W)\, \psi_{(r)} = E_r \, \psi_{(r)} \tag{5}$$

Equation (4) represents translational motion of the electron and constant $W$ is the translation energy of the centre of the mass, while Eq. (5) is the electronic wave equation and $E_T - W$ is equal to $E_r$, is the internal energy due to motion of the electron relative to the nucleus.

**Problem 5.18** The unnormalized angular function, which is a solution of SWE of hydrogen atom is $\phi_{(\phi)} = N \exp(im\phi)$. Normalize it.

*Solution:* The wave function is normalized if

$$\langle \phi_{(\phi)} / \phi_{(\phi)} \rangle = \int_0^{2\pi} \phi^* \phi \, d\phi = 1$$

Substituting the value of $\phi_{(\phi)}$, we have

$$N^2 \int_0^{2\pi} e^{im\phi} e^{-im\phi} d\phi = N^2 \int_0^{2\pi} d\phi = N^2 2\pi = 1$$

$$N = \left(\frac{1}{2\pi}\right)^{1/2}$$

Hence, the normalized angular function is

$$\phi_{(\phi)} = \frac{1}{\sqrt{2\pi}} \exp(im\phi)$$

**Problem 5.19** Obtain the real functions from the angular functions defined by

$$\phi_{\pm m}(\phi) = \sqrt{\frac{1}{2\pi}} \exp(\pm im\phi) \quad \text{for } m = 1.$$

*Solution:*
$$\phi_{+1}(\phi) = \frac{1}{\sqrt{2\pi}} e^{i\phi}; \quad \phi_{-1}(\phi) = \frac{1}{\sqrt{2\pi}} e^{-i\phi}$$

Taking the linear combinations of these functions, we have

$$\phi_+ = \frac{1}{\sqrt{2\pi}} (e^{i\phi} + e^{-i\phi}) = \frac{1}{\sqrt{2\pi}} (2 \cos \phi) \quad \text{(Euler's formula } e^{\pm i\phi} = \cos \phi \pm i \sin \phi)$$

$$\phi_- = \frac{1}{\sqrt{2\pi}}(e^{i\phi} - e^{-i\phi}) = \frac{1}{\sqrt{2\pi}}(2\sin\phi)$$

It $\phi_+$ and $\phi_-$ are further normalized, the normalization constant is found to be $\frac{1}{\sqrt{2}}$. Hence, we have

$$\phi_+ = \frac{1}{\sqrt{\pi}}\cos\phi \quad \text{and} \quad \phi_- = \frac{1}{\sqrt{\pi}}\sin\phi$$

**Problem 5.20** The unnormalized 1s wavefunction for hydrogen atom is given by $\psi_{1s} = \psi_{100} = N\exp(-r/a_0)$, where, $a_0$ is the Bohr radius. Determine the normalized wavefunction.

*Solution:* Here, we have to determine the normalization constant $N$. For the wavefunction to be normalized, the following condition holds

$$\int \psi_{1s}\psi_{1s}\, d\tau = 1$$

In polar coordinate $(r, \theta, \phi)$ the volume element $d\tau = r^2\, dr\, \sin\theta\, d\theta\, d\phi$
Using this expression for $d\tau$ and integrating between the appropriate limits, we have

$$\int \psi_{1s}\psi_{1s}\, d\tau = N^2 \int e^{-2r/a_0}\, d\tau \;\Rightarrow\; N^2 \int_0^\infty r^2 e^{-2r/a_0}\, dr \int_0^\pi \sin\theta\, d\theta \int_0^{2\pi} d\phi = 1$$

Each of these integrals can be evaluated separately as follows

$$\int_0^{2\pi} d\phi = 2\pi \quad \text{and} \quad \int_0^\pi \sin\theta\, d\theta = [-\cos\theta]_0^\pi = -(-1)-(-1) = 2$$

Using the result $\int_0^\infty x^n e^{-ax}\, dx = \dfrac{n!}{a^{n+1}}$; $n$ is a positive integer; $a > 0$, we obtain

$$\int_0^\infty r^2 e^{-2r/a_0}\, dr = \frac{2!}{\left(\dfrac{2}{a_0}\right)^3} = \frac{a_0^3}{4}$$

Hence, $\quad N^2\left(\dfrac{a_0^3}{4}\right)(2)(2\pi) = 1 \quad \therefore\; N = \left(\dfrac{1}{\pi a_0^3}\right)^{1/2} = \dfrac{1}{\sqrt{\pi}\, a_0^{3/2}}$

Hence, the normalized ground state wavefunction is

$$\psi_{1s} = \frac{1}{\sqrt{\pi}\, a_0^{3/2}} \exp(-r/a_0)$$

**Problem 5.21** Using the normalized wavefunction $\psi_{1s} = \left(\dfrac{1}{\sqrt{\pi}\, a_0^3}\right)^{1/2} \exp(-r/a_0)$ for the ground state of hydrogen atom. Calculate the probability for the electron to be confined in a sphere of radius $r = a_0$, the Bohr radius.

*Solution:* We know that from Born's interpretation of the wave function, the probability is given by

$$P(r)\, dr = \psi^* \psi\, d\tau$$

$\therefore$ Integration will leads to

$$\int P(r)\, dr = \int \psi^* \psi\, d\tau \quad \text{where } d\tau = r^2\, dr\, \sin\theta\, d\theta\, d\phi$$

$$P = \frac{1}{\pi a_0^3} \int_0^{a_0} r^2 e^{-2r}\, dr \int_0^{\pi} \sin\theta\, d\theta \int_0^{2\pi} d\phi = \frac{4}{a_0^3} \int_0^{a_0} r^2 e^{-2r}\, dr$$

carrying out integration by parts, we obtain

$$P = \frac{4}{a_0^3} \left[ \left(\frac{a_0}{2} r^2 e^{\frac{-2r}{a_0}}\right)_0^{a_0} + \left(-\frac{a_0^2}{2} r e^{\frac{-2r}{a_0}}\right)_0^{a_0} + \left(-\frac{a_0^3}{4} r e^{\frac{-2r}{a_0}}\right)_0^{a_0} \right]$$

$$P = -5e^{-2} + 1 \implies 0.323$$

**Problem 5.22** Calculate the most probable distance $r_{mp}$ of the electron from the nucleus in the ground state of hydrogen atom, given that normalized ground state wave function is

$$\psi_{1s} = \frac{1}{\sqrt{\pi}\, a_0^{3/2}} \exp(-r/a_0)$$

*Solution:* Quantum mechanically the probability of finding the electron in a spherical shell of the thickness $dr$ at a distance $r$ from the nucleus is given by

$$P(r)\, dr = 4\pi |\psi_{1s}|^2 r^2\, dr \implies 4\pi r^2 \left(\frac{1}{\pi a_0^3}\right) e^{-2r/a_0}\, dr$$

For the probability to be a maximum, $\dfrac{dP(r)}{dr} = 0$. Thus,

$$\frac{dP(r)}{dr} = \frac{4}{a_0^3} \left(\frac{-2r^2}{a_0} + 2r\right) e^{-2r/a_0} = 0$$

Since, the Bohr radius $a_0 \neq 0$ and the exponential is also $\neq 0$, hence

$$\frac{-2r^2}{a_0} + 2r = 0 \quad \therefore \quad r = r_{mP} = a_0$$

**Problem 5.23** Calculate the average distance of the electron from the nucleus in the ground state of hydrogen atom, given that the normalized ground state wavefunction is,

$$\psi_{1s} = \frac{1}{\sqrt{\pi}\, a_0^{3/2}} \exp(-r/a_0).$$

*Solution:* In quantum mechanics, the average (or expectation) value of dynamical variable $r$ is given by $\langle r \rangle = \int \psi_n r \psi_n d\tau$. In the present case, the average distance of the electron from the nucleus is given by $\langle r \rangle = \int \psi_{1s} r \psi_{1s}\, d\tau$. Since $d\tau = r^2 dr \sin\theta\, d\theta\, d\phi$.

Hence on integrating between appropriate limits, we get

$$\langle r \rangle = \frac{1}{\pi a_0^3} \int_0^\infty r^3 e^{-2r/a_0}\, dr \int_0^\pi \sin\theta\, d\theta \int_0^{2\pi} d\phi$$

$$= \frac{1}{\pi a_0^3} \cdot \frac{3!}{\left(\dfrac{2}{a_0}\right)^4} (2)(2\pi) = \frac{3}{2} a_0$$

**Note:** According to Bohr's theory of hydrogen atom, distance of electron from the nucleus in the ground state is exactly equal to $a_0$, the Bohr radius. But according to Heisenberg uncertainty principle, the electron cannot be exactly located in its orbit. Hence, quantum mechanically, while the most probable distance is $a_0$, the average distance is $3/2\, a_0$, i.e. the average distance is 50% greater than Bohr theory result. The inescapable conclusion we draw from this is that we really do not know where the electron is located in its orbit at a given moment.

**Problem 5.24** Show that the average value of $1/r$ for an electron in the 1s orbital of hydrogen atom is $1/a_0$ where $a_0$ is the Bohr radius, given that $\psi_{1s} = \dfrac{1}{\sqrt{\pi a_0^{3/2}}} \exp(-r/a_0)$.

*Solution:*

$$\left\langle \frac{1}{r} \right\rangle = \int \psi_{1s} \left(\frac{1}{r}\right) \psi_{1s} d\tau \quad \text{since} \quad d\tau = r^2 d\sin\theta\, d\theta d\phi,$$

Hence integrating between the appropriate limits,

$$\left\langle \frac{1}{r} \right\rangle = \frac{1}{\pi a_0^{3/2}} \int_0^\infty r e^{-2r/a_0}\, dr \int_0^\pi \sin\theta\, d\theta \int_0^{2\pi} d\phi$$

$$\left\langle \frac{1}{r} \right\rangle \Rightarrow \frac{1}{\pi a_0^3}\left[\frac{1}{(2/a_0)^2}(2)(2\pi)\right] = \frac{1}{a_0}$$

**Problem 5.25** The energy of atoms and molecules, as obtained by solving the SWE is generally expressed in units of hartrees (ha), named after Hartree, who has done fundamental work in quantum mechanics. Varify that 1 hartree = 27.2 eV.

**Solution:**

$$1 \text{ hartree (ha)} = \frac{e^2}{a_0}, \quad \text{where} \quad a_0 = \text{Bohr radius} = \frac{\hbar^2}{me^2}$$

$$\therefore \quad 1 \text{ hartree} = \frac{me^4}{\hbar^2} \text{ (in cgs units)}$$

$$= \frac{me^4}{(4\pi\varepsilon_0)^2 \hbar^2} \text{ (in SI units)}$$

Let us solve this problem in both the units
(i) In cgs units

$$1 \text{ hartree} = \frac{9.1 \times 10^{-23} \text{ g} \times (4.8 \times 10^{-10} \text{ esu})^4}{(1.05 \times 10^{-27} \text{ erg})^2 (1.6 \times 10^{-12} \text{ erg/eV})}$$

$$= 27.2 \text{ eV}$$

(ii) In SI units

$$1 \text{ hartree} = \frac{9.1 \times 10^{-31} \text{ kg} \times (1.6 \times 10^{-19} \text{ e})^4}{(1.11 \times 10^{-10} \text{ C}^2 \text{N}^{-1} \text{m}^{-2})^2 (1.05 \times 10^{-34} \text{ Js})^2 (1.6 \times 10^{-19} \text{ J eV}^{-1})}$$

$$\Rightarrow \quad 27.2 \text{ eV}$$

Also, 1 hartree = 2 Rydberg

**Problem 5.26** Write the Hamiltonian for the following quantum systems: (a) Helium atom, (b) Lithium atom (c) $H_2^+$ (d) $H_2$ and (e) $H_2^-$. Which of the terms is hardest to handle in solving SWE in each case.

**Solution:**

(a) *Helium atom:* It contains one nucleus and two electrons, numbered 1 and 2. Here, $r_1$ is the distance of electron 1 from the nucleus, $r_2$ is the distance of electron 2 from the nucleus and $r_{12}$ is the inter electron distance. The charge on the electron is $e$ while that on the nucleus is $Ze$ (here $Z = 2$).

The Hamiltion for the system is given by

$$\hat{H} = \frac{-\hbar^2}{2m}\nabla_1^2 - \frac{\hbar^2}{2m}\nabla_2^2 - \frac{2e^2}{4\pi\varepsilon_0 r_1} - \frac{2e^2}{4\pi\varepsilon_0 r_2} + \frac{e^2}{4\pi\varepsilon_0 r_{12}} \quad (1)$$

$$\Rightarrow \quad \frac{-\hbar^2}{2m}(\nabla_1^2 + \nabla_2^2) - 2e^2\left[\frac{1}{4\pi\varepsilon_0 r_1} + \frac{1}{4\pi\varepsilon_0 r_2}\right] + \frac{e^2}{4\pi\varepsilon_0 r_{12}}$$

In Eq. (1), the first two terms are the kinetic energy (Laplacian operator) for electrons 1 and 2, respectively, third and fourth terms give the attraction of two electrons to the nucleus and the last term gives the repulsion between two electrons. It is called the inter electron repulsion energy term.

(b) *Lithium atom:* It contains one nucleus and three electrons 1, 2, 3 (Here $Z = 3$). Hence

$$\hat{H} = \frac{-\hbar^2}{2m}(\nabla_1^2 + \nabla_2^2 + \nabla_3^2) + \frac{1}{4\pi\varepsilon_0}\left[-\frac{3e^2}{r_1} - \frac{3e^2}{r_2} - \frac{3e^2}{r_3} + \frac{e^2}{r_{12}} + \frac{e^2}{r_{23}} + \frac{e^2}{r_{31}}\right]$$

(c) $H_2^+$: It contains two nuclei A and B and one electron $Z = 1$
The Hamiltonian is given by

$$\hat{H} = \frac{-\hbar^2}{2m}\nabla^2 + \frac{1}{4\pi\varepsilon_0}\left[-\frac{e^2}{r_a} - \frac{e^2}{r_b} + \frac{e^2}{R_{ab}}\right]$$

There is no inter-electron repulsion term here. The last term is inter-nuclear repulsion energy term.

(d) $H_2$: It contains two nuclei $A$ and $B$ and two electron 1 and 2.

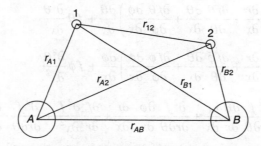

**FIGURE 5.9** Structure of hydrogen molecule.

$Z = 1$, the Hamiltonian is given by

$$\hat{H} = \frac{-\hbar^2}{2m}(\nabla_1^2 + \nabla_2^2) + \frac{1}{4\pi\varepsilon_0}\left[-\frac{e^2}{r_{A1}} - \frac{e^2}{r_{a1}} - \frac{e^2}{r_{B1}} - \frac{e^2}{r_{B2}} + \frac{e^2}{r_{12}} + \frac{e^2}{R_{AB}}\right]$$

Here the term $\frac{e^2}{r_{12}}$ is the inter-electron repulsion term and the last term is the inter-nuclear repulsion.

(e) $H_2^-$: It contains three electrons 1, 2, 3 and two nuclei $A$ and $B$. Hence

$$\hat{H} = \frac{-\hbar^2}{2m}(\nabla_1^2 + \nabla_2^2 + \nabla_3^2) + \frac{1}{4\pi\varepsilon_0}\left[-\frac{e^2}{r_{A1}} - \frac{e^2}{r_{A2}} - \frac{e^2}{r_{A3}} - \frac{e^2}{r_{B1}} - \frac{e^2}{r_{B2}} + \frac{e^2}{r_{12}} + \frac{e^2}{r_{23}} + \frac{e^2}{R_{AB}}\right]$$

In all the above cases, the inter-electron repulsion term is the most difficult to handle in solving the Schrödinger wave equation. For molecule, the inter-nuclear term is considered a constant, since the nuclei are infinitely heavier than the electrons.

**Problem 5.27** The electronic Schrödinger equation for hydrogen-like atoms is

$$\frac{-h^2}{8\pi^2 M}\nabla^2\psi - \frac{Ze^2}{(4\pi\varepsilon_0)r} = E\psi,$$

where subscript $r$ has been dropped for convenience. Transform the Laplacian operator $\nabla^2$ into spherical coordinates and write the SWE accordingly.

*Solution:* The Cartesian coordinates $(x, y, z)$ are related to spherical coordinates as

$$x = f(r, \theta, \phi), \quad y = f(r, \theta, \phi) \quad \text{and} \quad z = f(r, \theta)$$

$$\frac{\partial f}{\partial x} = \frac{\partial f}{\partial r}\frac{\partial r}{\partial x} + \frac{\partial f}{\partial \theta}\frac{\partial \theta}{\partial x} + \frac{\partial f}{\partial \phi}\frac{\partial \phi}{\partial x} \Rightarrow fr\frac{\partial r}{\partial x} + f\theta\frac{\partial \theta}{\partial x} + f\phi\frac{\partial \phi}{\partial x}$$

And $\quad \dfrac{\partial^2 f}{\partial x^2} = \left(\dfrac{\partial fr}{\partial r}\cdot\dfrac{\partial r}{\partial x} + \dfrac{\partial fr}{\partial \theta}\dfrac{\partial \theta}{\partial x} + \dfrac{\partial fr}{\partial \phi}\dfrac{\partial \phi}{\partial x}\right)\dfrac{\partial r}{\partial x} + fr\dfrac{\partial^2 r}{\partial x^2}$

$$+ \left(\frac{\partial f\theta}{\partial r}\frac{\partial r}{\partial x} + \frac{\partial f\theta}{\partial \theta}\frac{\partial \theta}{\partial x} + \frac{\partial f\theta}{\partial \phi}\frac{\partial \phi}{\partial x}\right)\frac{\partial \theta}{\partial x} + f\theta\frac{\partial^2 \theta}{\partial x^2}$$

$$+ \left(\frac{\partial f\phi}{\partial r}\frac{\partial r}{\partial x} + \frac{\partial f\phi}{\partial \theta}\frac{\partial \theta}{\partial x} + \frac{\partial f\phi}{\partial \phi}\frac{\partial \phi}{\partial x}\right)\frac{\partial \phi}{\partial x} + f\phi\frac{\partial^2 \theta}{\partial x^2}$$

$$\Rightarrow \frac{\partial^2 f}{\partial r^2}\left(\frac{\partial r}{\partial x}\right)^2 + \frac{\partial^2 f}{\partial r\partial\theta}\frac{\partial \theta}{\partial x}\frac{\partial r}{\partial x} + \frac{\partial^2 f}{\partial r\partial\theta}\frac{\partial \phi}{\partial x}\cdot\frac{\partial r}{\partial x} + \frac{\partial f}{\partial r}\frac{\partial^2 f}{\partial x^2} + \frac{\partial^2 f}{\partial r\partial\theta}\cdot\frac{\partial r}{\partial x}\cdot\frac{\partial \theta}{\partial x} + \frac{\partial^2 f}{\partial \theta^2}\left(\frac{\partial \theta}{\partial x}\right)^2$$

$$+ \frac{\partial^2 f}{\partial\phi\partial\theta}\frac{\partial \phi}{\partial x}\frac{\partial \theta}{\partial x} + \frac{\partial f}{\partial \theta}\frac{\partial^2 \theta}{\partial x^2} + \frac{\partial^2 f}{\partial r\partial\phi}\frac{\partial r}{\partial x}\cdot\frac{\partial \phi}{\partial x} + \frac{\partial^2 f}{\partial\phi\partial\theta}\frac{\partial \theta}{\partial x}\frac{\partial \phi}{\partial x} + \frac{\partial^2 f}{\partial \phi^2}\left(\frac{\partial \phi}{\partial x}\right)^2 + \frac{\partial f}{\partial \phi}\frac{\partial^2 \phi}{\partial x^2}$$

Similar expressions are obtained for $\dfrac{\partial^2 f}{\partial y^2}$ and $\dfrac{\partial^2 f}{\partial z^2}$.

Therefore, $\quad \nabla^2 f = \dfrac{\partial^2 f}{\partial r^2}\sum_K\left(\dfrac{\partial r}{\partial K}\right)^2 + \dfrac{\partial^2 f}{\partial \theta^2}\sum_K\left(\dfrac{\partial \theta}{\partial K}\right)^2 + \dfrac{\partial^2 f}{\partial \phi^2}\sum_K\left(\dfrac{\partial \phi}{\partial K}\right)^2$

$$+ 2\left[\frac{\partial^2 f}{\partial\theta\partial r}\sum_K\frac{\partial \theta}{\partial K}\frac{\partial r}{\partial K} + \frac{\partial^2 f}{\partial\phi\partial r}\sum_K\frac{\partial \phi}{\partial K}\frac{\partial r}{\partial K} + \frac{\partial^2 f}{\partial\phi\partial\theta}\sum_K\frac{\partial \phi}{\partial K}\cdot\frac{\partial \theta}{\partial K}\right]$$

$$+ \frac{\partial f}{\partial r}\sum_K\frac{\partial^2 r}{\partial K^2} + \frac{\partial f}{\partial \phi}\sum_K\frac{\partial^2 \theta}{\partial K^2} + \frac{\partial f}{\partial \phi}\sum_K\frac{\partial^2 \phi}{\partial K^2} \tag{1}$$

where $K = x, y, z$.

The necessary derivatives are found with the aid of the following transformations

$$x = r\sin\theta\cos\phi, \quad y = r\sin\theta\sin\phi, \quad z = r\cos\theta$$

$$x^2 + y^2 + z^2 = r^2, \quad \theta = \frac{1}{\cos}\left(\frac{z}{r}\right), \quad \phi = \frac{1}{\tan}\left(\frac{y}{x}\right)$$

$$\frac{\partial r}{\partial K} \quad \frac{\partial^2 r}{\partial K^2} \quad \frac{\partial \theta}{\partial K} \quad \frac{\partial^2 \theta}{\partial K^2} \quad \frac{\partial \phi}{\partial K} \quad \frac{\partial^2 \phi}{\partial K^2}$$

$$K = x \quad \frac{x}{r} \quad \frac{y^2+z^2}{r^3} \quad \frac{1}{r^2}\frac{xz}{\sqrt{x^2+y^2}} \quad \frac{z[y^2-2x^4-y^2-y^2(x^2-z^2)]}{r^4(x^2+y^2)^{3/2}} \quad \frac{-y}{x^2+y^2} \quad \frac{-2xy}{(x^2+y^2)^2}$$

$$K = y \quad \frac{y}{r} \quad \frac{x+z^2}{r^3} \quad \frac{1}{r^2}\frac{yz}{\sqrt{x^2+y^2}} \quad \frac{z[x^4-2y^4-x^2(y^2-z^2)]}{r^4(x^2+y^2)^{3/2}} \quad \frac{-x}{x^2+y^2} \quad \frac{-2xy}{(x^2+y^2)^2}$$

$$K = z \quad \frac{z}{r} \quad \frac{x^2+y^2}{r^3} \quad \frac{-\sqrt{x^2+y^2}}{r^2} \quad \frac{2z\sqrt{x^2+y^2}}{r^4} \quad 0 \quad 0 \tag{2}$$

Substituting Eq. (2) for the derivatives in Eq. (1), gives

$$\nabla^2 f = \frac{\partial^2 f}{\partial r^2} + \frac{z}{r}\frac{\partial f}{\partial r} + \frac{1}{r^2}\frac{\partial^2 f}{\partial \theta^2} + \frac{1}{r^2}\cot\theta\frac{\partial f}{\partial r} + \frac{1}{r^2\sin^2\theta}\frac{\partial^2 f}{\partial \phi^2}$$

Noting that $\dfrac{\partial}{\partial r}(r^2 df) = \dfrac{r^2\partial^2 f}{\partial r^2} + 2r\dfrac{\partial f}{\partial r}$

And $\dfrac{\partial}{\partial \theta}\left(\sin\theta\dfrac{\partial f}{\partial \theta}\right) = \sin\theta\dfrac{\partial^2 f}{\partial \theta^2} + \cos\theta\dfrac{\partial f}{\partial \theta}$

We have $\nabla^2 = \dfrac{1}{r^2}\dfrac{\partial}{\partial r}\left(r^2\dfrac{\partial}{\partial r}\right) + \dfrac{1}{r^2\sin\theta}\dfrac{\partial}{\partial \theta}\left(\sin\theta\dfrac{\partial}{\partial \theta}\right) + \dfrac{1}{r^2\sin^2\theta}\dfrac{\partial^2}{\partial \phi^2}$ \hfill (3)

The Schrödinger wave equation in spherical coordinates is therefore

$$-\frac{h^2}{8\pi^2 m}\left[\frac{1}{r^2}\frac{\partial}{\partial r}\left(r^2\frac{\partial}{\partial r}\right) + \frac{1}{r^2\sin\theta}\frac{\partial}{\partial \theta}\left(\sin\theta\frac{\partial}{\partial \theta}\right) + \frac{1}{r^2\sin^2\theta}\frac{\partial^2}{\partial \phi^2}\right]\psi \; \frac{-Ze^2}{(4\pi\varepsilon_0)r}\psi = E\psi$$

**Problem 5.28** The Schrödinger equation for hydrogen-like atoms in spherical coordinates is

$$-\frac{h^2}{8\pi^2 m}\left[\frac{1}{r^2}\frac{\partial}{\partial r}\left(r^2\frac{\partial}{\partial r}\right) + \frac{1}{r^2\sin\theta}\frac{\partial}{\partial \theta}\left(\sin\theta\frac{\partial}{\partial \theta}\right) + \frac{1}{r^2\sin^2\theta}\frac{\partial^2}{\partial \phi^2}\right]\psi - \frac{Ze^2}{(4\pi\varepsilon_0)r}\psi = E\psi$$

Assuming $\psi = R(r)\,P(\theta)\,F(\phi)$ attempt to separate the equation into three independent equations with only one variable in each.

*Solution:*
$$\psi = R(r)\,P(\theta)\,F(\phi) \tag{1}$$

(a) Substitute Eq. (1) in SWE to get

$$\frac{-h^2}{8\pi^2 M}\left[PF\frac{1}{r^2}\frac{\partial}{\partial r}\left(r^2\frac{\partial R}{\partial r}\right) + RF\frac{1}{r^2\sin\theta}\frac{\partial}{\partial \theta}\left(\sin\theta\frac{\partial P}{\partial \theta}\right) + RP\frac{1}{r^2\sin^2\theta}\frac{\partial^2 F}{\partial \phi^2}\right]$$
$$-\frac{Ze^2}{(4\pi\varepsilon_0)r}RPF = ERPF \tag{2}$$

(b) Multiply Eq. (2) throughout by $\dfrac{-8\pi^2 mr^2 \sin^2\theta}{h^2 RPR}$ and rearrange to get

$$\frac{\sin^2\theta}{R}\frac{\partial}{\partial r}\left(r^2\frac{\partial R}{\partial r}\right) + \frac{\sin\theta}{P}\frac{\partial}{\partial\theta}\left(\sin\theta\frac{\partial P}{\partial\theta}\right)$$

$$+ \frac{8\pi^2 mr^2 \sin^2\theta}{h^2}\left(E + \frac{Ze^2}{(4\pi\varepsilon_0)r}\right) + \frac{1}{F}\frac{\partial^2 F}{\partial\phi^2} = 0 \qquad (3)$$

(c) When $\phi$ changes, the first three terms do not change so that the equation can be written as

$$\frac{1}{F}\frac{d^2 F}{d\phi^2} = -\text{constant (say } M^2) \qquad (4)$$

(d) Substituting Eq. (4) into Eq. (3) and dividing by $\sin^2\theta$, we have

$$\frac{1}{R}\frac{d}{dr}\left(r^2\frac{dR}{dr}\right) + \frac{1}{P\sin\theta}\frac{d}{d\theta}\left(\sin\theta\frac{dP}{d\theta}\right) + \frac{8\pi^2 mr^2}{h^2}\left(E + \frac{Ze^2}{(4\pi\varepsilon_0)r}\right) - \frac{M^2}{\sin^2\theta} = 0 \qquad (5)$$

This separates $r$ and $\theta$ parts; the first and third terms in Eq. (5) depend on $r$ only while second and fourth terms depend on $\theta$ only. Hence, by the previous argument, the sum of the first and the third terms may be set equal to a constant, say $\beta$, so that the sum of the second and fourth terms equals $-\beta$. Thus, we have the following three separate equations.

1. Radial ($r$) equation

$$\frac{d}{dr}\left(r^2\frac{dR}{dr}\right) + \frac{8\pi^2 mr^2}{h^2}\left(E + \frac{Ze^2}{(4\pi\varepsilon_0)r}\right)R = \beta R \qquad (6)$$

2. $\theta$ equation

$$\frac{1}{\sin\theta}\frac{d}{d\theta}\left(\sin\theta\frac{dP}{d\theta}\right) - \frac{M^2 P}{\sin^2\theta} = \beta P \qquad (7)$$

3. $\phi$ equation

$$\frac{1}{F}\frac{d^2 F}{d\phi^2} = -M^2 \qquad (8)$$

The numerical parameters $M$ and $\beta$ can be determined after the boundary conditions are specified. Energy ($E$) appears in the radial Eq. (6) and is expressed in units of rhydberg (1 rhydberg = ionization potential of H atom) or Hartree (1 hartree = 2 rhydbergs). Radial distance $r$ is expressed in units of Bohr radius $a_0 = (0.529 \times 10^{-8}\,\text{cm})$. The $\theta$ and $\phi$ are the Zenith and Azimuthal angle variables.

## 5.14 APPROXIMATION METHODS

The Schrödinger wave equation can be solved exactly for any atom or molecule and is more complicated than the hydrogen atom. For this, approximation methods can be used to solve the

Schrödinger equation to almost any desired accuracy. Here, we will discuss the two most widely used of these methods: the *variation method* and *perturbation theory*.

To deal with the time independent SWE for systems that contains interacting particles, we need approximation methods. Two most common approximation methods are variation method and perturbation method.

## 5.14.1 Perturbation Method: An Overview

In order to understand perturbation method, let us consider a system with a time independent Hamiltonian operator, $\hat{H}$ for which we are facing difficulty to solve the Schrödinger wave equation,

$$\hat{H}\psi_n = E_n\psi_n \tag{5.190}$$

$\hat{H}\psi_n = E_n\psi_n$ is the simplest form of SWE for the systems of fixed stationary states. Let us also assume that Hamiltonian operator of the energy, $\hat{H}$ is only slightly different from the Hamiltonian operator for the ground state of a system, i.e. $\hat{H}^0$ whose Schrödinger wave equation is given as:

$$\hat{H}^0\psi_n^{(0)} = E_n^{(0)}\psi_n^{(0)} \tag{5.191}$$

Solution of this equation can be obtained with the help of one-dimensional anharmonic oscillator, according to which Hamiltonian operator can be defined as:

$$\hat{H} = -\frac{\hbar^2}{2m}\frac{d^2}{dx^2} + \frac{1}{2}kx^2 + cx^3 + dx^4 \tag{5.192}$$

The above Hamiltonian Eq. (5.192) resembles very closely to the Hamiltonian of the ground state of one-dimensional harmonic oscillator which is given as:

$$\hat{H}^0 = -\frac{\hbar^2}{2m}\frac{d^2}{dx^2} + \frac{1}{2}kx^2 \tag{5.193}$$

The eigenfunction and eigenvalues of the anharmonic oscillator are to be closely related to those of the harmonic oscillator, if the constants $c$ and $d$ in Eq. (5.192) are considered to be very small. Such a system with Hamiltonian $\hat{H}^0$ can be termed as the *unperturbed system*. The system with Hamiltonian $\hat{H}$ is termed as *perturbed system*. The difference between the two Hamiltonians, i.e. unperturbed, $\hat{H}^0$ and perturbed, $\hat{H}$ depends upon the magnitude of perturbation;

$$\hat{H}' = \hat{H} - \hat{H}^0 \tag{5.194}$$

It is very important to note here that prime does not refer to differentiation as we generally do in mathematics. On rearranging Eq. (5.194), we get

$$\hat{H} = \hat{H}^0 + \hat{H}' \tag{5.195}$$

For the anharmonic oscillator with Hamiltonian Eq. (5.192), the perturbation on the related harmonic oscillator is $\hat{H}^1 = cx^3 + dx^4$. In $\hat{H}^0\psi_n^{(0)} = E_n^{(0)}\psi_n^{(0)}$ Eq. (5.191), $E_n^{(0)}$ and $\psi_n^{(0)}$ are called *unperturbed energy* and *unperturbed wave function* of a particulate energy state say $n$. For $\hat{H}^0$ equal to the harmonic oscillator Hamiltonian [Eq. (5.193)], $E_n^{(0)}$ is $(n + 1/2)h\nu$, where $n$ is non-negative integer. It is very important to note here that superscript (0) does not mean to the ground state. Perturbation theory can be applied to any energy state. Here the superscript (0) denotes the unperturbed system which can be any either ground or first or second excited.

Our task is to relate the unknown eigenvalues and eigenfunctions of the perturbed system to the known eigenvalues and eigenfunctions of the unperturbed system. For this, we shall imagine that the perturbation is applied gradually, giving a continuous change from the unperturbed to the perturbed system. Mathematically, this corresponds to introducing a parameter $\lambda$ into the Hamiltonian, so that $\hat{H} = \hat{H}^0 + \lambda \hat{H}'$ when $\lambda$ is zero, we have the unperturbed system. As $\lambda$ increases, the perturbation grows larger, and at $\lambda = 1$ the perturbation is fully turned on. We have introduced $\lambda$ as a convenience in relating the perturbed and unperturbed eigenfunctions and ultimately we shall set $\lambda = 1$, thereby eliminating it.

### 5.14.2 Perturbation Theory to Solve Eigenfunction and Eigenvalue

Let us assume that we want to solve the Schrödinger wave equation for the problem of the linear harmonic motion of a system whose Hamiltonian operator $H$ for the perturbed system is only slightly different from the Hamiltonian operator $H_0$ for unpurturbed system. For the Hamiltonian operator $H_0$, we have a set of associated eigenvalues $E_1^0, E_2^0, \ldots E_n^0$ and the corresponding eigenfunctions $\psi_1^0, \psi_2^0, \ldots \psi_n^0$ which satisfy the Schrödinger wave equation as given below,

$$H_0 \psi_n^0 = E_n^0 \psi_n^0 \tag{5.196}$$

As we have assumed that $H$ (for perturbed system) is only slightly different from $H_0$ (for unpurturbed system), we can write

$$H = H_0 + \lambda H^{(1)} \tag{5.197}$$

where, $\lambda$ is defined as magnitude of small amount of perturbation given to the system, and the term $\lambda H^{(1)}$ which is called a small perturbation applied to the system. The final Schrödinger equation after applying small perturbation, which we actually want to solve is therefore given as:

$$(H_0 + \lambda H^{(1)})\psi_n = E_n \psi_n \tag{5.198}$$

Let us assume that if magnitude of applied perturbutation is zero, i.e. $\lambda$ is placed equal to zero and then Eq. (5.198) becomes equal to Eq. (5.196). For this case, value of eigenfunction $\psi_n$ and eigenvalue $E_n$ approach to that of $\psi_n^0$ and $E_n^0$ as magnitude of perturbation approaches zero, i.e., $\lambda \to 0$. Since, values of $\psi_n$ and $E_n$ are functions of magnitude of perturbation, i.e. $\lambda$, hence we can expand them in the form of power series as given in Eqs. (5.199) and (5.200).

$$\psi_n = \psi_n^0 + \lambda \psi_n^1 + \lambda^2 \psi_n^2 + \cdots \tag{5.199}$$

$$E_n = E_n^0 + \lambda E_n^1 + \lambda^2 E_n^2 + \cdots \tag{5.200}$$

On putting those values of $\psi_n$ and $E_n$ in Eq. (5.198), we get

$$H_0\psi_n^0 + \lambda H^{(1)}\psi_n^0 + H_0\psi_n^1 + \lambda^2 H^{(1)}\psi_n^1 + H_0\psi_n^2 + \ldots$$
$$= E_n^0\psi_n^0 + \lambda(E_n^1\psi_n^0 + E_n^0\psi_n^1) + \lambda^2(E_n^2\psi_n^0 + E_n^1\psi_n^1 + E_n^0\psi_n^2) + \ldots \quad (5.201)$$

The coefficients of the various powers of $\lambda$ on the two sides of the equation must be equal, in order to satisfy for all values of perturbations. Hence, on equating the coefficients of the various powers of $\lambda$, it gives the series of equations given as follows:

$$H_0\psi_n^0 = E_n^0\psi_n^0 \quad (5.202)$$

$$(H_0 - E_n^0)\psi_n^1 = E_n^1\psi_n^0 - H^{(1)}\psi_n^0 \quad (5.203)$$

$$(H_0 - E_n^0)\psi_n^2 = E_n^2\psi_n^0 + E_n^1\psi_n^1 - H^{(1)}\psi_n^1 \quad (5.204)$$

Equation (5.203) is already solved for unpurturbed system. If we solve Eq. (5.203), we can find the values of $\psi_n^1$ and $E_n^1$. The solution of Eq. (5.204) will give values of $\psi_n^2$ and $E_n^2$ and so on.

If we want to solve Eq. (5.203), let us assume that the expansion of the function $\psi_n^1$ in terms of normalized and orthogonal set of function as $\psi_1^0, \psi_2^0, \ldots \psi_n^0$ is

$$\psi_n^1 = A_1\psi_1^0 + A_2\psi_2^0 + \cdots + A_m\psi_m^0 \quad (5.205)$$

where, $A_m$ are the small amount of increments imposed on wave function $\psi_m^0$ which is to be determined. The function $H^{(1)}\psi_n^0$ can similarly also be expanded into the series,

$$H^{(1)}\psi_n^0 = H_{1n}^{(1)}\psi_1^0 + H_{2n}^{(1)}\psi_2^0 + \cdots + H_{mn}^{(1)}\psi_m^0 \quad (5.206)$$

By substituting the values of these expansion series into Eq. (5.203), gives

$$(H_0 - E_n^0)(A_1\psi_1^0 + A_2\psi_2^0 + \cdots) = E_n^1\psi_n^0 - H_{1n}^{(1)}\psi_1^0 + H_{2n}^{(1)}\psi_2^0 + \cdots \quad (5.207)$$

By the use of Eq. (5.202) into Eq. (5.207), we get

$$(E_1^0 - E_n^0)A_1\psi_1^0 + (E_2^0 - E_n^0)A_2\psi_2^0 + \cdots = E_n^1\psi_n^0 - H_{1n}^{(1)}\psi_1^0 + H_{2n}^{(1)}\psi_2^0 + \cdots \quad (5.208)$$

On the both sides of the equation, the coefficient of each $\psi_n^0$ must be equal. For example, the coefficient of $\psi_n^0$ is zero on the left side but on the right side it is $(E_n^1 - H_{nm}^{(1)})$, so that

$$(E_n^1 - H_{nm}^{(1)}) = 0 \quad (5.209)$$

Hence, the first order of perturbation energy is found to be

$$E_n^1 = H_{nm}^{(1)} = \int \psi_n^{0*} H^{(1)} \psi_n^0 d\tau \quad (5.210)$$

On comparing the coefficient of $\psi_n^0$ ($m = n$), we obtain $A_m$ as:

$$(E_m^0 - E_n^0)A_m = -H_{mn}^{(1)} \quad \text{or} \quad A_m = \frac{H_{nm}^{(1)}}{(E_n^0 - E_m^0)} \tag{5.211}$$

Above Eq. (5.211) gives us the values of all the increments except $A_n$. The coefficient $A_n$ may be determined by the requirement that $\psi_n$ must be normalized. Wave function $\psi_n$ may expressed as:

$$\psi_n = \psi_n^0 + \lambda \sum_m{}' A_m \psi_m^0 + \lambda A_n \psi_n^0 + \lambda^2 (...)$$

Here the symbol $\sum_m{}'$ signifies that one have to take sum of the all values of $m$ except $n$. Hence,

$$\int \psi_n^* \psi_n d\tau = \int \psi_n^{0*} \psi_n^0 d\tau + \lambda \sum_m{}' A_m \int \psi_m^{0*} \psi_n^0 d\tau + \lambda \sum_m{}' A_m \int \psi_n^{0*} \psi_m^0 d\tau + 2\lambda A_n \int \psi_n^{0*} \psi_m^0 d\tau + \lambda^2 (.....) \tag{5.212}$$

As we know, the eigenfunctions $\psi_m^0$ are both normalized and orthogonal, and hence Eq. (5.212) reduces to

$$\int \psi_n^* \psi_n d\tau = 1 + 2\lambda A_n + \lambda^2 (...) \tag{5.213}$$

In order that the wave function, $\psi_n$ is to be normalized then the right side of Eq. (5.213) must be taken as unity for all values of $\lambda$ so that we must put all $A_n$ equal to zero. The result of the first order of increment in $\lambda$ are therefore given as:

$$E_n = E_n^0 + \lambda H_{nn}^{(1)} + \lambda^2 (...) \tag{5.214}$$

$$\psi_n = \psi_n^0 + \lambda \sum_m{}' \frac{H_{nm}^{(1)}}{(E_n^0 - E_m^0)} \psi_m^0 + \lambda^2 (...) \tag{5.215}$$

Similarly, we can find the values of $\psi_n^{(2)}$ and $E_n^{(2)}$ by using above process. For this we have to assume that

$$\psi_n^{(2)} = B_1 \psi_1^0 + B_2 \psi_2^0 + \cdots + B_m \psi_m^0 + \cdots \tag{5.216}$$

In Eq. (5.216), the coefficients $B_m$ are to be determined,

$$(H_0 - E_n^0)\psi_n^{(2)} = (E_1^0 - E_n^0)B_1 \psi_1^0 + (E_2 - E_n^0)B_2 \psi_2^0 + \cdots \tag{5.217}$$

By the use of Eqs. (5.205), (5.206), (5.210) and (5.211), we can find the values of eigenfunction and eigenvalue as:

$$E_n^1 \psi_n^{(1)} = \sum_m{}' \frac{H_{nn}^{(1)} H_{mm}^{(1)}}{(E_n^0 - E_m^0)} \psi_m^0 \tag{5.218}$$

$$H^1\psi_n^{(1)} = \sum_m{}' \frac{H_{mn}^{(1)}}{(E_n^0 - E_m^0)} H^1 \psi_m^0 = \sum_m{}' \frac{H_{mn}^{(1)}}{(E_n^0 - E_m^0)} \sum_K H_{Km}^{(1)} \psi_K^0$$

$$= \sum_K \sum_m{}' \frac{H_{Km}^{(1)} H_{mn}^{(1)}}{(E_n^0 - E_m^0)} \psi_K^0 \tag{5.219}$$

By the use of above equations, Eq. (5.204) becomes,

$$\sum_K (E_K^0 - E_n^0) B_K \psi_K^0 = E_n^2 \psi_n^0 + \sum_m{}' \frac{H_{nm}^{(1)} H_{mn}^{(1)}}{(E_n^0 - E_m^0)} \psi_m^0 - \sum_K \sum_m{}' \frac{H_{Km}^{(1)} H_{mn}^{(1)}}{(E_n^0 - E_m^0)} \psi_K^0 \tag{5.220}$$

On equating the coefficient of $\psi_n^0$ on both sides of Eq. (5.220), we obtain that

$$0 = E_n^{(2)} - \sum_m{}' \frac{H_{nm}^{(1)} H_{mm}^{(1)}}{(E_n^0 - E_m^0)} \tag{5.221}$$

$$E_n^{(2)} = \sum_m{}' \frac{H_{nm}^{(1)} H_{mm}^{(1)}}{(E_n^0 - E_m^0)} \tag{5.222}$$

On equating coefficients of $\psi_K^0$ where $K \neq n$ gives

$$(E_K^0 - E_n^0) B_K = \frac{H_{nm}^{(1)} H_{Kn}^{(1)}}{(E_n^0 - E_K^0)} - \sum_m{}' \frac{H_{Km}^{(1)} H_{mn}^{(1)}}{(E_n^0 - E_m^0)} \tag{5.223}$$

On rearranging Eq. (5.223), we get

$$B_K = \sum_m{}' \frac{H_{Km}^{(1)} H_{mn}^{(1)}}{(E_n^0 - E_m^0)(E_n^0 - E_K^0)} - \frac{H_{nn}^{(1)} H_{Kn}^{(1)}}{(E_n^0 - E_K^0)^2} \tag{5.224}$$

To fulfill the condition of normalization of $\psi_n$, it is essential that $B_K$ must vanish. The results are corrected to the second order in increment of $\lambda$, for the different energy levels and corresponding wave functions whose values are therefore given as:

$$E_n = E_n^0 + \lambda H_{nn}^{(1)} + \lambda^2 \sum_m{}' \left[ \frac{H_{nm}^{(1)} H_{mn}^{(1)}}{E_n^0 - E_m^0} \right] \psi_K^0 + \lambda^3 (...) + \cdots \tag{5.225}$$

$$\psi_n = \psi_n^0 + \lambda \sum_m{}' \frac{H_{mn}^{(1)}}{(E_n^0 - E_m^0)} \psi_m^0 + \lambda^2 \sum_K{}' \left[ \sum_m{}' \frac{H_{Km}^{(1)} H_{mn}^{(1)}}{(E_n^0 - E_m^0)(E_n^0 - E_K^0)} - \frac{H_{nn}^{(1)} H_{Kn}^{(1)}}{(E_n^0 - E_K^0)^2} \right] \psi_K^0 + \lambda^3 (...) + \cdots \tag{5.226}$$

In majority of the applications of Schrödinger wave equation, it is found to be convenient to absorb the parameter $\lambda$ into the function $H^{(1)}$ or by assuming $\lambda$ equal to unity in Eqs. (5.225) and (5.226).

### 5.14.3 Variation Method

Variation method is the another completely different method of finding approximate solution of the Schrödinger wave equation which is based upon the following theorem: Let us assume that $H$ operator for energy operates on any function, $\psi$ which is function of class $Q$ such that $\int \psi^* \psi \, d\tau = 1$ and if the lowest possible eigenvalue of the operator $H$ is $E_0$, then the theorem is given as:

$$\int \psi^* H \psi \, d\tau \geq E_0 \qquad (5.227)$$

In order to prove, above very simple theorem [Eq. (5.227)], let us consider the integral

$$\int \psi^* (H - E_0) \psi \, d\tau = \int \psi^* H \psi \, d\tau - E_0 \int \psi^* \psi \, d\tau \qquad (5.228)$$

$$\Rightarrow \int \psi^* H \psi \, d\tau - E_0$$

On expanding the eigenfunction $\psi$ in a series of the eigenfunctions $\psi_1, \psi_2 \ldots \psi_i$ of Hamiltonian operator for energy $H$, Then Eq. (5.228) will modify as:

$$\int \psi^* (H - E_0) \psi \, d\tau = \int \left( \sum_i C_i^* \psi_i^* \right) (H - E_0) \left( \sum_i C_i \psi_i \right) d\tau \qquad (5.229)$$

As we know, the eigenfunction $\psi_i$ are eigenfunction of Hamiltonian operator $H$, i.e. $H\psi_i = E_i \psi_i$ and therefore Eq. (5.229) will become

$$\int \psi^* (H - E_0) \psi \, d\tau = \int \left( \sum_i C_i^* \psi_i^* \right) \left( \sum_i (E_i - E_0) C_i \psi_i \right) d\tau$$

$$\sum_i C_i^* C_i (E_i - E_0) \qquad (5.230)$$

As we know, the product of two normalization constants, i.e. is a positive number and also by definition

$$E_i \geq E_0$$

Hence,

$$\int \psi^* (H - E_0) \psi \, d\tau \geq 0 \qquad (5.231)$$

$$\int \psi^* H \psi \, d\tau \geq E_0 \qquad (5.232)$$

The sign of equality in above two equations, i.e. Eqs. (5.231) and (5.232), can be applicable only when $\psi = \psi_0$ where, $\psi_0$ is the eigenfunction with the eigenvalue $E_0$. The method of applying this variation theorem, i.e. Eq. (5.232) is very simple in principle. For the application of variation theorem, a trial eigenfunction, $\psi(\lambda_1, \lambda_2, \ldots)$ normalized to unity is taken. The value of integral in the left hand side of Eq. (5.232), i.e. $J = \int \psi^* H \psi \, d\tau$ is then calculated. The result thus obtained will of course also be a function of parameters $\lambda_1, \lambda_2,\ldots$ The value of integral $J$ is then set equal to minimum with respect to parameters, $\lambda_1, \lambda_2,\ldots$ The result thus obtained is an approximation to the corresponding eigenfunction. On expanding the variation theorem for a large number of parameters in a function of a well chosen form, a very close approximation to the correct eigenvalue and eigenfunction can thus be obtained.

As an example of the variation method, let us consider the harmonic oscillator for which the Hamiltonian operator, $H$ is defined as:

$$H = -\frac{h^2}{8\pi^2 m}\frac{d^2}{dx^2} + \frac{k}{2}x^2 \qquad (5.233)$$

In order to fulfill the condition of normalization of eigenfunction, If we put $\psi = Ce^{-\lambda x^2}$, the condition that $\psi$ to be normalized is fulfilled, where normalization constant $c = \sqrt[4]{\dfrac{2\lambda}{\pi}}$.

Therefore, $\quad H\psi = -\dfrac{ch^2}{8\pi^2 m}(4\lambda^2 x^2 - 2\lambda)e^{-\lambda x^2} + \dfrac{cK}{2}x^2 e^{-\lambda x^2}$

and the integral, $\quad J = \displaystyle\int_{-\infty}^{+\infty}\left[-\dfrac{ch^2}{8\pi^2 m}(4\lambda^2 x^2 - 2\lambda)e^{-2\lambda x^2} + \dfrac{c^2 K}{2}x^2 e^{-2\lambda x^2}\right]dx \qquad (5.234)$

On solving these integrals, we obtain

$$J = \frac{h^2 \lambda}{8\pi^2 m} + \frac{K}{8\lambda} \qquad (5.235)$$

The condition that the integral $J$ will be a minimum is to be fulfilled, hence

$$\frac{dJ}{d\lambda} = \frac{h^2}{8\pi^2 m} - \frac{K}{8\lambda^2} = 0 \quad \text{or} \quad \lambda = \frac{\pi}{h}\sqrt{mK}$$

Hence, lowest possible eigenvalue for the ground state of energy is

$$E_0 \leq \frac{h}{4\pi}\sqrt{\frac{K}{m}} = \frac{1}{2}h\nu \qquad (5.236)$$

The value of corresponding eigenfunction is $\psi = \sqrt[4]{\dfrac{2}{h}\sqrt{mK}}\, e^{-\frac{\pi}{h}\sqrt{mk}x^2} \qquad (5.237)$

The variation method can also be applied for the calculation of energy levels and eigenfunctions for excited states. Once we have solved an approximation to the ground state by the above variation method, we can select a second trial eigenfunction which is orthogonal to one that we obtained for the ground state. Similar procedure can be applied to find solution of an approximation to the first excited state and so on.

## 5.14.4 Approximation Method for the Ground State of the Helium Atom

In order to solve these approximate methods for multi electron atoms like helium atom, $Li^+$, $Be^{2+}$ etc. we shall try to calculate the energy of the ground state of the helium atom by using both methods, i.e. by the application of the first order perturbation theory and by the variation method. First of all we will try to solve approximation method for the ground state of helium atom. The Hamiltonian operator, $H$ for the helium atom (or two-electron atoms such as $Li^+$, $Be^{2+}$ etc.) is (if we do not consider the terms arising from the motion of the nucleus).

$$H = -\frac{h^2}{8\pi^2 m}\left(\nabla_1^2 + \nabla_2^2\right) - \frac{Ze^2}{r_1} - \frac{Ze^2}{r_2} + \frac{e^2}{r_{12}} \qquad (5.238)$$

where $\nabla_1$ and $\nabla_2$ are the well-known Laplacian operators for electrons 1 and 2. The distances of these electrons from the nucleus is given by $r_1$ and $r_2$. The distance between the two electrons is given by $r_{12}$ and $Z$ is the effective nuclear charge. It is here recommended to use atomic units for the purpose of simplifying the calculation of integrals involved in problems of this type, i.e. two-electron atoms. This is done by expressing distances in terms of the Bohr radius, $a_0 = \frac{h^2}{4\pi m e^2}$.

The distances in terms of Bohr radius modifies as $r_1 = a_0 R_1$, $r_2 = a_0 R_2$, $r_{12} = a_0 R_{12}$

$\frac{\partial^2}{\partial x_1^2} = \frac{1}{a_0^2}\frac{\partial^2}{\partial X_1^2}$. The Hamiltonian operator thus becomes,

$$H = -\frac{h^2}{8\pi^2 m}\frac{1}{a_0^2}\left(\nabla_1^2 + \nabla_2^2\right) - \frac{Ze^2}{a_0 R_1} - \frac{Ze^2}{a_0 R_2} + \frac{e^2}{a_0 R_{12}} \qquad (5.239)$$

or in units of $\frac{e^2}{a_0}$

$$H = -\frac{1}{2}\left(\nabla_1^2 + \nabla_2^2\right) - \frac{Z}{R_1} - \frac{Z}{R_2} + \frac{1}{R_{12}} \qquad (5.240)$$

On applying the methods of the perturbation theory to helium atom, we obtain

$$H = H_0 + H^{(1)}$$

where

$$H_0 = -\frac{1}{2}\left(\nabla_1^2 + \nabla_2^2\right) - \frac{Z}{R_1} - \frac{Z}{R_2} \qquad (5.241)$$

$$H^{(1)} = \frac{1}{R_{12}} \qquad (5.242)$$

The eigenfunction for the ground state, i.e. unperturbed system or zeroth eigenfunctions can be separated as:

$$H_0 \psi^0 = E^0 \psi^0 \qquad (5.243)$$

If eigenfunction and eigenvalue for unperturbed system are set equal to $\psi^0 = \psi^0_{(1)} \psi^0_{(2)}$, $E^0 = E^0_{(1)} + E^0_{(2)}$, then Eq. (5.243) is immediately separable into two equations,

$$\frac{1}{2}\nabla_1^2 \psi^0_{(1)} + \left(E^0_{(1)} + \frac{Z}{R_1}\right)\psi^0_{(1)} = 0 \qquad (5.244a)$$

$$\frac{1}{2}\nabla_2^2 \psi^0_{(2)} + \left(E^0_{(2)} + \frac{Z}{R_2}\right)\psi^0_{(2)} = 0 \qquad (5.244b)$$

The above two equations, i.e., Eqs. (5.244a) and (5.244b) are similar to equation for a hydrogen like atom with nuclear charge $Z$. For the ground state of the helium atom, we therefore have

$$\psi^0_{(1)} = \frac{1}{\sqrt{\pi}} Z^{3/2} e^{-ZR_1}, \quad \psi^0_{(2)} = \frac{1}{\sqrt{\pi}} Z^{3/2} e^{-ZR_2}$$

$$\psi^0 = \psi^0_{(1)} \psi^0_{(2)} = \frac{Z^2}{\pi} e^{-Z(R_1+R_2)}$$

$$E^0 = E^0_{(1)} + E^0_{(2)} = 2Z^2 E_{1S}(H) \qquad (5.245)$$

Here $E_{1S}(H)$ is defined as the energy of the ground state of the hydrogen atom and in ordinary units its value is $-\frac{1}{2}\frac{e^2}{a_0}$. The first order correction to the energy according to Eq. (5.240) in perturbation theorem is given by

$$E^{(1)} = \iint \psi^{0*} H^{(1)} \psi^0 d\tau_1 d\tau_2$$

$$\Rightarrow \quad \frac{e^2}{a_0} \frac{Z^6}{\pi^2} \iint \frac{e^{-2ZR_1} e^{-2ZR_2}}{R_{12}} d\tau_1 d\tau_2 \qquad (5.246)$$

where $d\tau_1 = R_1^2 \sin\theta_1 dR_1 d\theta_1 d\phi_1$ and $d\tau_2 = R_2^2 \sin\theta_2 dR_2 d\theta_2 d\phi_2$

The value of energy in Eq. (5.246) cannot be evaluated because of the presence of the term $1/R_{12}$ in the integral. In a very similar manner as in harmonic oscillator, the quantity $1/R_2$ can be expanded in terms of the associated Legendre's polynomial as

$$\frac{1}{R_{12}} = \sum_l \sum_m \frac{(l-|m|)!}{(l+|m|)!} \frac{R_<^l}{R_>^{l+1}} P_L^{|m|}(\cos\theta_1) P_L^{|m|}(\cos\theta_2) e^{im(\phi_1-\phi_2)} \qquad (5.247)$$

Here $R_<$ signifies that it is smaller and $R_>$ signifies that it is larger of the qualities $R_1$ and $R_2$, respectively. Here the wave functions do not involve the angles unambiguously. Only the constant functions, i.e. $P_0^0(\cos\theta_1)$ and $P_0^0(\cos\theta_2)$ are involved in the wave functions. As we know, the associated Legendre's polynomial are orthogonal to each other, all the terms in the summation will vanish except for those $l = 0$, $m = 0$. For these terms $P_0^0(\cos\theta_1) = 1$ so that Eq. (5.246) becomes

$$E^{(1)} = \frac{e^2}{a_0}\frac{Z^6}{\pi^2}\iint \frac{e^{-2ZR_1}e^{-2ZR_2}}{R_>}d\tau_1 d\tau_2 \tag{5.248}$$

The solution of integration over the angles in above equation gives a factor $4\pi^2$, so we have left only the integral over $R_1$ and $R_2$ which may be solved in a simple manner to give

$$E^{(1)} = \frac{5}{8}Z\frac{e^2}{a_0} \tag{5.249}$$

The energy for the lowest ground state of helium atom is therefore given as

$$E = E^0 + E^{(1)} = \left(2Z^2 - \frac{5}{4}Z\right)\left(-\frac{1}{2}\frac{e^2}{a_0}\right)$$

$$= \left(2Z^2 - \frac{5}{4}Z\right)E_{1S}(H) \tag{5.250}$$

The value of first ionization energy to remove an electron form the valence shell of helium that is the minimum energy required to remove an electron from the helium atom is given as

$$\left(2Z^2 - \frac{5}{4}Z\right)E_{1S}(H) = \frac{3}{2}E_{1S}(H) = \frac{3}{2}(13.60) = 20.40 \text{ eV}$$

But the actual value is 24.58 eV (observed value) so that our calculated value is in error by 4.18 eV or about 16 percent. The results of first order perturbation theory are seems to be plausible if we compare the total binding energy. The values of calculated and observed total binding energy are: calculated $\frac{11}{2}(13.6) = 74.8$ eV, observed 78.98 eV. An error of 4.18 eV, i.e. about 5 percent. Although the magnitude of actual error is the same in both cases but the percentage error is of course decreased in case of total binding energy.

## 5.15 VARIATION METHOD APPLIED TO HELIUM ATOM

Let us apply variation method to solve SWE for helium atom. In order to solve variation method for helium atom, we may choose a reasonable trial eigenfunction for use in the variation method. Let us consider that one of the electrons of the helium atom is in an excited state and the other in the ground state. The first electron is subjected to the full attractive force of the nucleus because it is in ground state and considered to be very close to nucleus, so that $\psi_{(1)}^0$ should be essentially the same as in Eq. (5.245). As the (1s) electron screens the nucleus more or less completely and hence the electron in the excited state, moves essentially in the field of nucleus of charge $e$ and so $\psi_{(2)}^0$ for an excited state should approximate more closely to a hydrogen wave function with $Z = 1$. Above said generalization suggests that a good trial eigenfunction would be such that

$$\psi = \frac{Z'^3}{\pi} e^{-Z'(R_1+R_2)} \tag{5.251}$$

where $Z'$ is effective nuclear charge between the electron 1 and 2, the best value of Eq. (5.251) can be determined by solving $\psi$ for energy. The eigenfunction written above in Eq. (5.251) is the solution of the equation:

$$H_0^1 \psi = E_0^1 \psi \tag{5.252}$$

In units of $\dfrac{e^2}{a_0}$

$$H_0^1 = -\frac{1}{2}(\nabla_1^2 + \nabla_2^2) - \frac{Z'}{R_1} - \frac{Z'}{R_2}$$

$$E_0^1 = 2Z'^2 E_{1s}(H)$$

We can solve for the integral given as

$$E = \iint \psi^* H \psi \, d\tau_1 d\tau_2 \tag{5.253}$$

Here $H$ is defined as

$$H = -\frac{1}{2}(\nabla_1^2 + \nabla_2^2) - \frac{Z}{R_1} - \frac{Z}{R_2} + \frac{1}{R_{12}}$$

We can define $H\psi$ as

$$H\psi = \left[-\frac{1}{2}(\nabla_1^2 + \nabla_2^2) - \frac{Z'}{R_1} - \frac{Z'}{R_2}\right]\psi - (Z-Z')\left(\frac{1}{R_1} + \frac{1}{R_2}\right)\psi + \frac{1}{R_{12}}\psi \tag{5.254}$$

$$\Rightarrow \qquad E_0'\psi - (Z-Z')\left(\frac{1}{R_1} + \frac{1}{R_2}\right)\psi + \frac{1}{R_{12}}\psi$$

The integral in Eq. (5.253) is limited to

$$E = E_0'\psi - (Z-Z')\left[\iint \frac{\psi^2}{R_1} d\tau_1 d\tau_2 + \iint \frac{\psi^2}{R_2} d\tau_1 d\tau_2\right] + \iint \frac{\psi^2}{R_{12}} d\tau_1 d\tau_2 \tag{5.255}$$

The values of the two integrals in the brackets of the above equation are equal, as they differ only in the exchange of the subscripts 1 and 2. The value of the first of the integral is

$$16\pi^2 \frac{Z'^6}{\pi^2} \int_0^\infty e^{-2Z'R_1} R_1^2 dR_1 \int_0^\infty e^{-2Z'R_2} R_2^2 dR_2 = Z' \tag{5.256}$$

The last integral is identical with Eq. (5.246) if $Z$ is replaced by $Z'$. The result for the energy is thus given as

$$E = E_0' - 2Z'(Z-Z')\frac{e^2}{a_0} + \frac{5}{8}Z'\frac{e^2}{a_0}$$

$$= \left[2Z'^2 + 4Z'(Z-Z') - \frac{5}{4}Z'\right]E_{1S}(H)$$

$$= \left[-2Z'^2 + 4ZZ' - \frac{5}{4}Z'\right]E_{1S}(H) \qquad (5.257)$$

As per variation theorem, one can obtain the best approximation to the true energy by giving $Z'$ the value which will make the energy a minimum. Hence,

$$\frac{\partial E}{\partial Z'} = \left(-4Z' + 4Z - \frac{5}{4}\right)E_{1S}(H) = 0$$

$$Z' = Z - \frac{5}{16} \qquad (5.258)$$

Substituting this result in to Eq. (5.257), i.e. equation for the energy of helium atom, we obtain

$$E = 2Z'^2 E_{1S}(H) = 2\left(Z - \frac{5}{16}\right)^2 E_{1S}(H) \qquad (5.259)$$

The first ionization potential of helium atom is thus given as

$$\left[2\left(\frac{27}{16}\right)^2 - 4\right]E_{1S}(H) = 1.695(13.6) = 23.05 \text{ eV}$$

Here the difference between calculated and observed value of first ionization potential in case of helium atom is thus reduced to 1.53 eV or about 6 percent. The total binding energy is calculated to be 77.45 eV as compared to the experimental value 78.98 eV or an error of about 2 percent which is much less as compared to perturbation method applied to helium atom.

By suitably modifying the variation method, i.e. introducing more parameters into the trial eigenfunction $\psi$, we can approach more and more closely to the experimental result. In fact very good results are obtained, if the variable $R_{12}$ is unequivocally introduced into the trial eigenfunction. For example, Hylleras has used the trial eigenfunction as

$$\psi = A\left\{e^{-Z'(R_1+R_2)}(1+CR_{12})\right\} \qquad (5.260)$$

where, $A$ is the normalization factor and $Z'$ and $C$ are adjustable constants. Value for the first ionization potential of the helium atom thus obtained by variation method is in error by only 0.34 eV. In order to reduce the error, more general functions were used that is,

$$\psi = A\{e^{-Z'(R_1+R_2)}(\text{Polynomial } R_1 R_2 R_3)\} \tag{5.261}$$

By using above more general function, the error is further reduced which is in close agreement with the experimental value. By using the trial eigenfunction involving a polynomial of fourteen terms give a value for the first ionization energy which agreed with the experimental value within 0.002 eV.

## 5.16 ANGULAR MOMENTUM IN QUANTUM MECHANICS

The angular momentum may be possessed by: (1) a rotating molecule, (2) an electron orbiting around an atom, (3) the spinning electrons, and (4) certain spinning nuclei. For a particle of mass $m$ revolving around a point at a distance $r$, the angular momentum $L$ is given by

$$\mathbf{L} = m\mathbf{u}r = mr^2\omega = I\omega \tag{5.262}$$

Here $\mathbf{u}$ is the linear velocity, $\omega$ is the angular velocity and $I$ is the moment of inertia. The particle has kinetic energy $E_k$ given by

$$E_k = \frac{1}{2} mu^2 = \frac{1}{2} mr^2\omega^2 = \frac{1}{2} I\omega^2 = \frac{L^2}{2I} \tag{5.263}$$

In three dimensions, the angular momentum is represented by a vector, $\vec{\mathbf{L}}$. Consider a mass $m$ rotating about a fixed point $P$ with a linear velocity $\mathbf{u}$. Then, the angular momentum is given by

$$\vec{\mathbf{L}} = \vec{\mathbf{r}} \times \overrightarrow{m\mathbf{u}} = \vec{\mathbf{r}} \times \vec{\mathbf{p}} \tag{5.264}$$

where $\vec{\mathbf{r}}$ is the vector from the fixed point $P$ to the mass point and $\vec{\mathbf{p}}$ is the linear momentum vector. The vector $\vec{\mathbf{L}}$ is perpendicular to the plane defined by $\vec{\mathbf{r}}$ and $\vec{\mathbf{p}}$. From classical mechanics, it can be shown that the components of the classical angular momentum vector are given by

$$L_x = yp_z - zp_y \tag{5.265}$$

$$L_y = zp_x - xp_z \tag{5.266}$$

$$L_z = xp_y - yp_x \tag{5.267}$$

The square of the angular momentum is given by the scalar product of $\vec{\mathbf{L}}$ with itself. Thus,

$$L^2 = \vec{\mathbf{L}}\vec{\mathbf{L}} = L_x^2 + L_y^2 + L_z^2 \tag{5.268}$$

The square of the angular momentum is a scalar. If no torque is acting on the particle, its angular momentum remains constant in classical mechanics. Classical mechanics permits all possible values of $L$. In quantum mechanics, the angular momentum is represented by an operator. The quantum mechanical operators for the components of angular momentum are obtained by replacing the quantities in Eqs. (5.265), (5.266) and (5.267) with their corresponding quantum mechanical operators. Thus,

$$\hat{P}_x = -i\hbar \frac{\partial}{\partial x} \text{ and so on. Hence,}$$

$$\hat{L}_x = -i\hbar\left(y\frac{\partial}{\partial z} - z\frac{\partial}{\partial y}\right) \tag{5.269}$$

$$\hat{L}_y = -i\hbar\left(z\frac{\partial}{\partial x} - x\frac{\partial}{\partial z}\right) \tag{5.270}$$

$$\hat{L}_z = -i\hbar\left(x\frac{\partial}{\partial y} - y\frac{\partial}{\partial x}\right) \tag{5.271}$$

The operator for the square of the momentum is given by

$$\hat{L}^2 = \left|\vec{L}\right|^2 = \vec{L}\vec{L} = \hat{L}_x^2 + \hat{L}_y^2 + \hat{L}_z^2 \tag{5.272}$$

One may ask now if we can measure the angular momentum $L$ and its component $L_x$, $L_y$ and $L_z$ simultaneously. It is found from quantum mechanics that, according to Heisenberg's uncertainty principle, only square of the angular momentum, $L^2$ and $L_x$ or $L^2$ and $L_y$ or $L^2$ and $L_z$ can be simultaneously measured. This fact is expressed by saying that $L^2$ commutes with of its components, i.e.,

$$[\hat{L}^2, \hat{L}_x] = [\hat{L}^2, \hat{L}_y] = [\hat{L}^2, \hat{L}_z] = 0 \tag{5.273}$$

On the other hand, $\hat{L}_x$ and $\hat{L}_y$, $\hat{L}_y$ and $\hat{L}_z$, and $\hat{L}_z$ and $\hat{L}_x$ cannot be simultaneously measured. In quantum mechanical language, these operators do not commute with each other, i.e.

$$[\hat{L}_x, \hat{L}_y] \neq 0; [\hat{L}_y, \hat{L}_z] \neq 0; [\hat{L}_z, \hat{L}_x] \neq 0 \tag{5.274}$$

Again, in the time independent Schröndinger equation $\hat{H}\psi = E\psi$, we know that $\psi$ is the eigenfunction and $E$ is the eigenvalue of the Hamiltonian operator, $\hat{H}$. Naturally, we want to know the eigenfunction and eigenvalues of the angular momentum operators. There is another important quantum mechanical result according to which if two operators commute with each other, they have same eigenfunctions. Thus, according to Eq. (5.273), $L^2$ and $\hat{L}_z$ (or $\hat{L}^2$ and $\hat{L}_x$, etc.) have the same eigenfunctions.

Keeping in view the above considerations, it can be easily shown that the eigenvalues of $\hat{L}^2$ and $\hat{L}_z$ are

$$|\vec{L}^2| = [l(l+1)]\hbar^2; \quad l = 0, 1, 2, 3 \tag{5.275}$$

$$|\vec{L}_z| = m_l \hbar; \quad m_l = -l, -l+1, \ldots, l-1, l \tag{5.276}$$

where $m_l$ is the component of $l$ along a suitable axis; in this case, $z$-axis. From Eq. (5.275),

$$|\vec{L}| = [l(l+1)]^{1/2}\hbar \tag{5.277}$$

## 5.17 ANGULAR MOMENTUM OPERATOR

The cartesian components of the angular momentum operator are given by

$$\hat{L}_x = -i\hbar \left[ y \frac{\partial}{\partial z} - z \frac{\partial}{\partial y} \right] \quad (5.278a)$$

$$\hat{L}_y = -i\hbar \left[ z \frac{\partial}{\partial x} - x \frac{\partial}{\partial z} \right] \quad (5.278b)$$

$$\hat{L}_z = -i\hbar \left[ x \frac{\partial}{\partial y} - y \frac{\partial}{\partial x} \right] \quad (5.278c)$$

In order to determine whether $L_x$, $L_y$ and $L_z$ can be precisely specified simultaneously, we must determine whether each of the pairs of the corresponding operators commute. Let us evaluate the commutator $[\hat{L}_x, \hat{L}_y]$

$$[\hat{L}_x, \hat{L}_y] = \hat{L}_x \hat{L}_y - \hat{L}_y \hat{L}_x$$

$$= -i\hbar \left[ y \frac{\partial}{\partial z} - z \frac{\partial}{\partial y} \right] (-i\hbar) \left[ z \frac{\partial}{\partial x} - x \frac{\partial}{\partial z} \right]$$

$$= -\hbar^2 \left[ y \frac{\partial}{\partial z} z \frac{\partial}{\partial x} - y \frac{\partial}{\partial z} x \frac{\partial}{\partial z} - z \frac{\partial}{\partial y} z \frac{\partial}{\partial x} + z \frac{\partial}{\partial y} x \frac{\partial}{\partial z} \right]$$

$$= -\hbar^2 \left[ y \frac{\partial}{\partial x} + yz \frac{\partial^2}{\partial z \partial x} - yx \frac{\partial^2}{\partial z^2} - z^2 \frac{\partial^2}{\partial y \partial x} + zx \frac{\partial^2}{\partial y \partial z} \right]$$

$\left( \text{because, terms like } \frac{\partial z}{\partial x} = 0 \text{ and } \frac{\partial z}{\partial z} = 1 \right)$

Again, $\quad [\hat{L}_x, \hat{L}_y] = \hat{L}_x \hat{L}_y - \hat{L}_y \hat{L}_x$

$$= -\hbar^2 \left[ z \frac{\partial}{\partial x} - x \frac{\partial}{\partial z} \right] \left[ y \frac{\partial}{\partial z} - z \frac{\partial}{\partial y} \right]$$

$$= -\hbar^2 \left[ z \frac{\partial}{\partial x} y \frac{\partial}{\partial z} - z \frac{\partial}{\partial x} z \frac{\partial}{\partial y} - x \frac{\partial}{\partial z} y \frac{\partial}{\partial z} + x \frac{\partial}{\partial z} z \frac{\partial}{\partial y} \right]$$

$$= -\hbar^2 \left[ zy \frac{\partial^2}{\partial x \partial z} - z^2 \frac{\partial^2}{\partial x \partial y} - xy \frac{\partial^2}{\partial z^2} + x \frac{\partial}{\partial y} + xz \frac{\partial^2}{\partial z \partial y} \right]$$

Hence, $\quad \hat{L}_x \hat{L}_y - \hat{L}_y \hat{L}_x = -\hbar^2 \left[ y \frac{\partial}{\partial x} - x \frac{\partial}{\partial y} \right] = i\hbar \left[ i\hbar \left( y \frac{\partial}{\partial x} - x \frac{\partial}{\partial y} \right) \right]$

$$i\hbar \left[ -i\hbar \left( x \frac{\partial}{\partial y} - y \frac{\partial}{\partial x} \right) \right] = i\hbar \hat{L}_z \quad (5.279)$$

Thus,
$$[\hat{L}_x, \hat{L}_y] = i\hbar \hat{L}_z \quad (5.280a)$$

Similarly, we can show that
$$[\hat{L}_y, \hat{L}_z] = i\hbar \hat{L}_x; \quad \text{and} \quad [\hat{L}_z, \hat{L}_x] = i\hbar \hat{L}_y \quad (5.280b)$$

None of the relation in Eqs. (5.279) and (5.280a and 5.280b) is zero, which means that the three components $L_x$, $L_y$, $L_z$ of the angular momentum cannot be specified simultaneously. If we specify one component, the other two become uncertain. By convention, the component $L_z$ is chosen to have a specific value so that $L_x$ and $L_y$ cannot have specific values. However, the z-axis itself is chosen arbitrarily. But when the atom or molecule is placed in a magnetic or electric field, the direction of the field is chosen as the z-axis.

Next we are interested in knowing whether $L$ and $L_z$ (its z-component) can be specified simultaneously or not. To find it out, we recall that $L$ is given by
$$L = (L_x^2 + L_y^2 + L_z^2)^{1/2}$$

This means that the operator $\hat{L}$ has to be defined as the square root of another operator. Ignoring the meaning of the square root of an operator, let us find out whether the operators $\hat{L}^2$ and $\hat{L}_z$ commute.

$$\begin{aligned}[] [\hat{L}^2, \hat{L}_z] &= [\hat{L}_x^2 + \hat{L}_y^2 + \hat{L}_z^2, \hat{L}_z] \\ &= \hat{L}_x^2 \hat{L}_z + \hat{L}_y^2 \hat{L}_z + \hat{L}_z^2 \hat{L}_z - \hat{L}_z \hat{L}_x^2 - \hat{L}_z \hat{L}_y^2 - \hat{L}_z \hat{L}_z^2 \\ &= (\hat{L}_x^2 \hat{L}_z - \hat{L}_z \hat{L}_x^2) + (\hat{L}_y^2 \hat{L}_z - \hat{L}_z \hat{L}_y^2) + (\hat{L}_z^2 \hat{L}_z - \hat{L}_z \hat{L}_z^2) \\ &= (\hat{L}_x^2, \hat{L}_z) + (\hat{L}_y^2, \hat{L}_z) + (\hat{L}_z^2, \hat{L}_z) \end{aligned} \quad (5.281)$$

The first term in Eq. (5.281), viz.

$$\begin{aligned} (\hat{L}_x^2, \hat{L}_z) &= \hat{L}_x \hat{L}_x \hat{L}_z - \hat{L}_z \hat{L}_x \hat{L}_x \\ &= \hat{L}_x \hat{L}_x \hat{L}_z - \hat{L}_x \hat{L}_z \hat{L}_x + \hat{L}_x \hat{L}_z \hat{L}_x - \hat{L}_z \hat{L}_x \hat{L}_x \\ &= \hat{L}_x (\hat{L}_x \hat{L}_z - \hat{L}_z \hat{L}_x) + (\hat{L}_x \hat{L}_z - \hat{L}_z \hat{L}_x) \hat{L}_x \\ &= \hat{L}_x (\hat{L}_x, \hat{L}_z) + (\hat{L}_x, \hat{L}_z) \hat{L}_x \\ &= i\hbar \hat{L}_x \hat{L}_y - i\hbar \hat{L}_y \hat{L}_x \qquad (\because [\hat{L}_x, \hat{L}_z] = -[\hat{L}_z, \hat{L}_x]) \\ &= -i\hbar (\hat{L}_x \hat{L}_y + \hat{L}_y \hat{L}_x) \end{aligned} \quad (5.282)$$

Similarly, it can be shown that

$$(\hat{L}_y^2, \hat{L}_z) = i\hbar (\hat{L}_x \hat{L}_y + \hat{L}_z \hat{L}_x) \quad (5.283)$$

$$(\hat{L}_z^2, \hat{L}_z) = 0 \quad (5.284)$$

Adding Eqs. (5.282), (5.283) and (5.284), we have

$$[\hat{L}^2, \hat{L}_z] = 0 \tag{5.285}$$

which means that the square of the angular momentum and its z-component can be specified simultaneously. We can similarly show that $\hat{L}^2$ commutes with $\hat{L}_x$ and $\hat{L}_y$. Commutation relations mean that the angular momentum $L$ of a relating particle can be represented by a vector of precise length **L** with one and only one of its components (conversely $L_z$) precisely defined. Since the other two components ($L_x$ and $L_y$) are uncertain, the vector may lie anywhere on the surface of a cone whose axis coincides with the z-axis. In the classical sense, it corresponds to vector **L** precessing about z-axis.

The relations (5.280a), (5.280b) and (5.285) constitute the quantum mechanical definition of angular momentum. In fact, any vector **J** may be called angular momentum if it satisfies the following relations:

$$[\hat{J}_x, \hat{J}_y] = i\hbar \hat{J}_z; [\hat{J}_y, \hat{J}_z]$$

$$= i\hbar \hat{J}_x; [\hat{J}_z, \hat{J}_x] = i\hbar \hat{J}_y \tag{5.286}$$

$$[\hat{J}^2, \hat{L}_k] = 0, \tag{5.287}$$

where $k = x, y, z$

and 

$$\hat{J}^2 = (\hat{J}_x^2 + \hat{J}_y^2 + \hat{J}_z^2)^{1/2} \tag{5.288}$$

A rotating system is thus characterized by two operators $\hat{J}^2$ and $\hat{J}_z$ and their eigenvalues. Before we proceeds to find the eigenvalues and eigenfunctions, we shall introduce two new operators $\hat{J}_+$ and $\hat{J}_-$, called the *ladder operators* (or step-up and step-down operators or raising and lowering operator's) and examine some of their commutation properties.

## 5.18 THE LADDER OPERATOR

These operators are defined as

$$\hat{J}_+ = \hat{J}_x + i\hat{J}_y$$

$$\hat{J}_- = \hat{J}_x - i\hat{J}_y \tag{5.289}$$

They are so called because, as we shall see later, they have the effect of raising and lowering eigenvalues, respectively. These operators commute with $\hat{J}^2$, but not with $\hat{J}_z$; they also do not commute between themselves as described below:

$$[\hat{J}^2, \hat{J}_+] = [\hat{J}^2, (\hat{J}_x + i\hat{J}_y)] = \hat{J}^2(\hat{J}_x + i\hat{J}_y) - (\hat{J}_x + i\hat{J}_y)\hat{J}^2$$

$$= \hat{J}^2\hat{J}_x + i\hat{J}^2\hat{J}_y - \hat{J}_x\hat{J}^2 - i\hat{J}_y\hat{J}^2 = (\hat{J}^2\hat{J}_x - \hat{J}_x\hat{J}^2) + i(\hat{J}^2\hat{J}_y - \hat{J}_y\hat{J}^2)$$

$$= [\hat{J}^2, \hat{J}_x] + i[\hat{J}^2, \hat{J}_y] = 0 + 0 = 0 \tag{5.290}$$

Similarly, $[\hat{J}^2, \hat{J}_-] = 0$ \hfill (5.291)

Again, 
$$[\hat{J}_+, \hat{J}_-] = [(\hat{J}_x + i\hat{J}_y), \hat{J}_z] = (\hat{J}_x + i\hat{J}_y)\hat{J}_z - \hat{J}_z(\hat{J}_x + i\hat{J}_y)$$
$$= \hat{J}_x\hat{J}_z + i\hat{J}_y\hat{J}_z - \hat{J}_z\hat{J}_x - i\hat{J}_z\hat{J}_y$$
$$= (\hat{J}_x\hat{J}_z - \hat{J}_z\hat{J}_x) + i(\hat{J}_y\hat{J}_z - \hat{J}_z\hat{J}_y)$$
$$= [\hat{J}_x, \hat{J}_z] + i[\hat{J}_y, \hat{J}_z] = -i\hbar\hat{J}_y - \hbar\hat{J}_x$$
$$= -\hbar(\hat{J}_x + i\hat{J}_y) = -\hbar\hat{J}_+ \tag{5.292}$$

Similarly, we can show that
$$[\hat{J}_-, \hat{J}_z] = -\hbar\hat{J}_- \tag{5.293}$$
$$[\hat{J}_+, \hat{J}_-] = -2\hbar\hat{J}_z \tag{5.294}$$

## 5.19 EIGENVALUES $\hat{J}^2$ AND $\hat{J}_z$

The commutation relations provide an elegant means of arriving at the eigenvalues of the angular momentum operators $\hat{J}^2$ and $\hat{J}_z$. Since $\hat{J}^2$ and $\hat{J}_z$ commute, they have common set of eigenfunctions. Denoting these eigenfunctions as $\psi_{j,m}$ (or Dirac notation, as $|j, m\rangle$), the two eigenvalue equations are

$$\hat{J}^2|j, m\rangle = k_j|j, m\rangle \tag{5.295}$$
$$\hat{J}_z|j, m\rangle = k_m|j, m\rangle \tag{5.296}$$

where, $j$ and $m$ are the quantum numbers, which specify the eigenvalues $k_j$ and $k_m$, respectively. Operating on both sides of Eq. (5.296) by $\hat{J}_z$, we obtain

$$\hat{J}_z^2|j, m\rangle = \hat{J}_z(k_m|j, m\rangle) = k_m\hat{J}_z|j, m\rangle = k_m^2|j, m\rangle \tag{5.297}$$

Subtracting Eq. (5.297) from Eq. (5.295), we get

$$(\hat{J}^2 - \hat{J}_z^2)|j, m\rangle = (k_j - k_m^2)|j, m\rangle \tag{5.298}$$

or
$$(\hat{J}_x^2 + \hat{J}_y^2)|j, m\rangle = (k_j - k_m^2)|j, m\rangle \tag{5.299}$$

(because $\hat{J}^2 = \hat{J}_x^2 + \hat{J}_y^2 + \hat{J}_z^2$).

Now, $(\hat{J}_x^2 + \hat{J}_y^2)$ is the operator for the physical quantity $(J_x^2 + J_y^2)$ which can never be negative; hence, the eigenvalue $(k_j - k_m^2)$ in Eq. (5.299) must always be positive, that is

$$k_j \geq k_m^2 \text{ for any value of } k_m \tag{5.300}$$

Let us now consider the effect of the ladder operators and to know why they are called by this name. We shall first consider the effect of $\hat{J}_+$ on the eigenvalue of $\hat{J}^2$. Since, $\hat{J}^2$ commutes with $\hat{J}_+$ Eq. (5.290), we have

$$\hat{J}^2 \hat{J}_+ | j, m \rangle = \hat{J}_+ \hat{J}^2 | j, m \rangle = \hat{J}_+ (k_j | j, m \rangle) = k_j (\hat{J}_+ | j, m \rangle) \quad (5.301)$$

This means that $\hat{J}_+ | j, m \rangle$ is also an eigenfunction of $\hat{J}^2$ with the same eigenvalue $(k_j)$, that is, there is no effect of $J_+$ on the eigenvalue of $\hat{J}^2$.

Let us consider the effect on $k_m$. By adding and subtracting $\hat{J}_+ \hat{J}_z$, we obtain

$$\hat{J}_+ \hat{J}_z | j, m \rangle = (\hat{J}_+ \hat{J}_z + \hat{J}_z \hat{J}_+ - \hat{J}_+ \hat{J}_z) | j, m \rangle = \{\hat{J}_+ \hat{J}_z + [\hat{J}_z, \hat{J}_+]\} | j, m \rangle$$

$$= \{\hat{J}_+ \hat{J}_z | j, m \rangle | [\hat{J}_z \hat{J}_+] | j, m \rangle \}$$

$$= \hat{J}_+ k_m | j, m \rangle + \hbar \hat{J}_+ | j, m \rangle = \hat{J}_+ (k_m + \hbar) | j, m \rangle$$

$$= (k_m + \hbar) \hat{J}_+ | j, m \rangle$$

(by Eq. (5.296) and (5.292), and from the fact that $[\hat{J}_z, \hat{J}_+] = -[\hat{J}_+, \hat{J}_z]$)

This means that $\hat{J}_+ | j, m \rangle$ is also an eigenfunction of $\hat{J}_z$, but its eigenvalue is raised from $k_m$ to $(k_m + \hbar)$, i.e. by one atomic unit. The clue to this derivation is that $\hat{J}_+$ does not commute with $\hat{J}_z$.

Similarly, it can be shown that $\hat{J}_-$ lowers the eigenvalue of $\hat{J}_z$ from $k_m$ to $(k_m - \hbar)$, i.e. by one atomic unit, but keeps that of $\hat{J}^2$ unaltered, that is,

$$\hat{J}^2 (\hat{J}_- | j, m \rangle) = k_j (\hat{J}_- | j, m \rangle)$$

$$\hat{J}_z (\hat{J}_- | j, m \rangle) = (k_m - \hbar)(\hat{J}_- | j, m \rangle)$$

It is now easy to show that

$$\hat{J}_+ (\hat{J}_\pm | j, m \rangle), \hat{J}_\pm [\hat{J}_\pm | j, m \rangle], \ldots$$

are all eigenfunctions of $\hat{J}_z$ with altered eigenvalues and also of $\hat{J}^2$ with unaltered eigenvalues. Successive applications of the operators $\hat{J}_\pm$ will, therefore, generate a series of eigenfunction of $\hat{J}_z$ whose eigenvalues constitute the series

$$k'_m, k'_m + \hbar, k'_m + 2\hbar, \ldots, k''_m - 2\hbar, k''_m - \hbar, k''_m$$

or $\qquad k'_m, k'_m + 1, k'_m + 2, \ldots, k''_m - 2, k''_m - 1, k''_m \qquad$ (in atomic units)

where for a given value of $k_j$, $k'_m$ is the lowest and $k''_m$ is the highest eigenvalue. The lowest and the highest values of $k_m$ are governed by the condition in Eq. (5.300), viz., $k_m^2 \leq k_j$. It also follows that the maximum and minimum values are related as

$$k''_m = k'_m + n\hbar$$

or
$$k_m'' = k_m' + n \quad \text{(in a.u.)}$$
where $n$ is a positive integer.

## 5.20 THE PAULI EXCLUSION PRINCIPLE

Lithium with $Z = 3$, has three electrons. The first two occupy a 1s orbital drawn even more closely than in He around the more highly charged nucleus. The third electron, however, does not join the first two in the 1s orbital because that configuration is forbidden by the Pauli exclusion principle.

*No more than two electrons may occupy any given orbital and, if two do occupy one orbital, then their spin must be paired.*

Electrons with paired spins, which we denote as $\uparrow\downarrow$, have zero net spin angular momentum because the spin of one electron is cancelled by the spin of the other. Specifically, one electron has $m_s = -\frac{1}{2}$ and they are oriented on their respective cones so that the resultant spin is zero. The exclusion principle is the key to the structure of complex atoms, to chemical periodicity, and to molecular structure. It was proposed by Wolfgang Pauli in 1924 when he was trying to account for the absence of some lines in the spectrum of helium. Later he was able to derive a very general form of the principle from theoretical considerations.

The Pauli exclusion principle in fact applies to any pair of identical fermions (particles with half integral spin). Thus, it applies to protons, neutrons and $^{13}C$ nuclei (all of which have spin 1/2) and to $^{35}Cl$ nuclei (which have spin 3/2). It does not apply to identical bosons (particles with integral spin), which include photons (spin 1), $^{12}C$ nuclei (spin 0). Any number of identical bosons may occupy the same orbital. The Pauli exclusion principle is a special case of a general statement called the Pauli principle:

*When the labels of any two identical fermions are exchanged, the total wavefunction changes sign. When the labels of any two identical bosons are exchanged, the total wavefunction retain the same sign.*

By total wavefunction is meant the entire wavefunction, including the spin of the particles. Consider the wave function for two electrons $\psi(1, 2)$. The Pauli principle implies that it is a fact of nature (which has its root in the theory of relativity) that the wavefunction must change sign if we interchange the labels 1 and 2 wherever they occur in the function:

$$\psi(2, 1) = -\psi(1, 2)$$

Suppose the two electrons in an atom occupy an orbital $\psi$, then in the orbital approximation the overall wavefunction is $\psi(1)\,\psi(2)$. To apply the Pauli principle, we must deal with the total wavefunction, the wavefunction including spin. There are several possibilities for two spins: both $\alpha$, denoted $\alpha(1)\,\alpha(2)$, both $\beta$, denoted $\beta(1)\,\beta(2)$, and one $\alpha$ and the other $\beta$, denoted either $\alpha(1)\beta(2)$ or $\alpha(2)\beta(1)$. Because we cannot tell which electron is $\alpha$ and which is $\beta$, in the last case it is appropriate to express the spin states as the (normalized) linear combinations,

$$\sigma_+(1, 2) = \left(\sqrt{\frac{1}{2}}\right)[\alpha(1)\beta(2) + \beta(1)\alpha(2)]$$

$$\sigma_-(1,2) = \left(\sqrt{\frac{1}{2}}\right)[\alpha(1)\beta(2) - \beta(1)\alpha(2)]$$

Because these allow one spin to be $\alpha$ and other $\beta$ with equal probability. The total wavefunction of the system is therefore the product of the orbital part and one of the four spin states:

$$\psi(1)\psi(2)\,\alpha(1)\alpha(2) \quad \psi(1)\psi(2)\,\beta(1)\beta(2) \quad \psi(1)\psi(2)\sigma_+(1,2) \quad \psi(1)\psi(2)\sigma_-(1,2)$$

The Pauli principle says that for a wavefunction to be acceptable (for electrons), it must change sign when the electrons are exchanged. In each case, exchanging the labels 1 and 2 converts the factor $\psi(1)\psi(2)$ into $\psi(2)\psi(1)$, which is the same, because the order of multiplying the functions does not change the value of the product. The same is true of $\alpha(1)\alpha(2)$ and $\beta(1)\beta(2)$. Therefore, the first two overall products are not allowed, because they do not change sign. The combination $\sigma_+(1,2)$ changes to

$$\sigma_+(2,1) = \left(\sqrt{\frac{1}{2}}\right)[\alpha(2)\beta(1) + \beta(2)\alpha(1)]$$

Because it is simply the original function written in a different order. The third overall product is therefore also disallowed. Finally, consider $\sigma_-(1,2)$:

$$\sigma_-(2,1) = \left(\sqrt{\frac{1}{2}}\right)[\alpha(2)\beta(1) - \beta(2)\alpha(1)] = -\left(\sqrt{\frac{1}{2}}\right)[\alpha(1)\beta(2) - \beta(1)\alpha(2)] = -\sigma_-(1,2)$$

This combination does not change sign (it is antisymmetric). The product $\psi(1)\psi(2)\,\sigma_-(1,2)$ also changes sign under particle exchange, and therefore it is acceptable.

Now, we see that only one of the four possible states is allowed by the Pauli principle, and the one that survives has paired $\alpha$ and $\beta$ spins. This is the content of the Pauli exclusion principle. The exclusion principle is irrevelent when the orbital occupied by the electrons are different, and both electrons may then have (but need not have) the same spin state. Nevertheless, even then the overall wavefunction must still be antisymmetric overall, and must still satisfy the Pauli principle itself.

## 5.21 ELECTRONIC STRUCTURE OF ATOMS: RUSSELL–SAUNDER COUPLING (R–S TERM)

Consider the ground state electronic configuration of a carbon atom, $1s^2 2s^2 2p^2$. The two electrons present in three 2p orbitals in carbon atom may be present in any of the three 2p orbitals $(2p_x, 2p_y, 2p_z)$ and have any spins, i.e. clockwise or anticlockwise in accordance with the Pauli exclusion principle. Hence, energies of these different possible states in case of carbon atoms may varies. This is the reason a more detailed designation of the electronic states of atoms. For this, R-S term was introduced. In order to assign R-S term, first of all, determine total orbital angular momentum **L** and total spin angular momentum **S** and then adding **L** and **S** together vectorically to obtain the total angular momentum, **J**. The total angular momentum **J** is the sum of

**L** and **S**. Such a designation of the electronic states of atoms is called *Russell-Saunders coupling*, and is presented as an atomic term symbol, which has the form

$$^{2S+1}L_J$$

Here subscript $J$ designate total angular momentum, and superscript $2s + 1$ designate the degeneracy. $L$ may necessarily have values such as 0, 1, 2, 3,.... Here, orbital angular momentum, $L = 0$ means $S$, 1 means $P$, 2 means $D$ and 3 means $F$ orbitals and so on. One can make the correspondence as

$$L = \begin{matrix} 0 & 1 & 2 & 3 & 4 & 5 & ... \\ S & P & D & F & G & H & ... \end{matrix}$$

The total spin quantum number $S$ can have values such as $0, \frac{1}{2}, 1, \frac{3}{2}, ...$ and correspondingly left superscript on a term symbol, i.e. $2S + 1$ (spin multiplicity) will have values such as 1,2,3,.... Hence in this stage once, ignoring for the subscript $J$, then term symbol will be written as

$$^3S \quad ^2D \quad ^1P$$

Total orbital angular momentum and total spin angular momentum may be represented by the vector sums over all the electrons in an atoms as

$$L = \sum_i l_i \quad \text{and} \quad S = \sum_i s_i$$

For example, for the electron configuration $ns^2$ (two electron in a $ns$ orbital), there is only one possible set of values of $m_{l1}, m_{s1}, m_{l2},$ and $m_{s2}$ (Table 5.5):

**TABLE 5.5** Possible set of values of $m_{l1}, m_{s1}, m_{l2},$ and $m_{s2}$

| $m_{l1}$, | $m_{s1}$, | $m_{l2}$, | $m_{s2}$ | $M_L$ | $M_S$ |
|---|---|---|---|---|---|
| 0 | $+\frac{1}{2}$ | 0 | $-\frac{1}{2}$ | 0 | 0 |

It is concluded that the only values of $M_L$ is $M_L = 0$ which means that $L = 0$. Similarly, it is also concluded that the only value of $M_S$ is $M_S = 0$ means that $S = 0$. The value of total angular momentum **J** is thus given by

$$\mathbf{J} = \mathbf{L} + \mathbf{S}$$

Hence, for any system having two electrons in $ns^2$ electron configuration, $L = 0$, $S = 0$ and $J = 0$. The value of total orbital angular momentum $L = 0$ which can be written as $S$ in the term symbol, and so we find that the term symbol corresponding to $ns^2$ electron configuration is $^1S_0$ which may be read as singlet $S$ zero. As we know the two electrons present in $ns^2$ orbitals must have opposite spins, hence the total spin angular momentum will be zero. Also, both electrons occupy an orbital that has no angular momentum, thus the total angular momentum must be zero.

Similarly, system having a $np^6$ and $nd^{10}$ electronic configuration will also have a $^1S_0$ term symbol. An electron configuration that has a term symbol other than $^1S_0$ is $ns^1n's^1$, where $n \neq n'$.

For example, helium atom having excited state electronic configuration $1s^1 2s^1$. In order to find the possible values of $m_{l1}$, $m_{s1}$, $m_{l2}$, and $m_{s2}$, one can formulate a table as: because $m_{l1}$ and $m_{l2}$ can both have a maximum value of 0, the maximum value of $M_L = 0$ and 0 is its only possible value. Similarly, because $m_{s1}$ and $m_{s2}$ can both have values of $\pm(1/2)$, can be $-1$, 0, or 1. Table can be set in a manner that columns headed by the possible values of $M_S$ and its rows headed by the possible values of $M_L$, and we then fill in the sets of values of $m_{l1}$, $m_{s1}$, $m_{l2}$ and $m_{s2}$ as shown in Table 5.6.

**TABLE 5.6** Possible values of $M_s$

| $M_L$ | $M_S$ | | |
|---|---|---|---|
| | 1 | 0 | $-1$ |
| 0 | $0^+, 0^+$ | $0^+, 0^-$; $0^-, 0^+$ | $0^-, 0^-$ |

Here the symbol $0^+$ in the table signifies that $m_l = 0$ and $m_s = +\dfrac{1}{2}$ and $0^-$ means that $m_l = 0$ and $m_s = -\dfrac{1}{2}$. The possible sets of values of $m_{l1}$, $m_{s1}$, $m_{l2}$, and $m_{s2}$ that are consistent with each value of $M_L$ and $M_S$ are called *microstates*. There are two possible spins $\left(\pm\dfrac{1}{2}\right)$ for the electron in the $ns$ orbitals and two possible spins for the electron in the $n's$ orbital and four micro states in the table.

It is very important to mention here that we include both $0^+$, $0^-$ and $0^-$ and $0^+$ because the electrons are in non-equivalent orbitals, i.e. they are present in different $s$ orbitals (e.g., 1s and 2s). Here note that all the values of $M_L$ in the above table are zero, so they all must correspond to $L = 0$. In addition, the largest value of $M_S$ is 1. Consequently, $S$ must equal 1 and the values $M_S = 1$, 0, and $-1$ correspond to, $L = 0$, $S = 1$, correspond to a $^3S$ state. The middle column in above table consists of two microstates, but it makes no difference which one we choose. These two pairs of $L = 0$, $S = 1$ and $L = 0$, $S = 0$ along with their possible values of $M_J$ may be briefed as:

$L = 0, S = 1$  $\qquad\qquad\qquad$  $L = 0, S = 0$

$M_L = 0, M_S = 1, 0, -1$  $\qquad\qquad$  $M_L = 0, M_S = 0$

$M_J = M_L + M_S = 1, 0, -1$  $\qquad\quad$  $M_J = M_L + M_S = 0$

The values of $M_J$ here imply that $J = 1$ for the $L = 0$, $S = 1$ and that $J = 0$ for the $L = 0$, $S = 0$. The two term symbols corresponding to the electron configuration $ns^1\, n's^1$ are

$$^3S_1 \quad \text{and} \quad ^1S_0$$

The above two-term symbol corresponds to two different electronic states with different energies. One can observe that triplet state ($^3S_1$) has lower energy than the singlet state ($^1S_0$).

## EXERCISES

1. Set up and solve the Schrödinger wave equation for a particle in an infinite one-dimensional box, with potential energy zero inside the box. Normalize the wavefunctions.
2. Show that for a particle in a one-dimensional box of width $a$,

    (i) $\langle x \rangle = a/2$,    (ii) $\langle p_x \rangle = 0$,    (iii) $\langle x^2 \rangle = a^2 \left[ \dfrac{1}{3} - \dfrac{1}{2\pi^2 n^2} \right]$.

3. Discuss the solution of Schrödinger wave equation for a particle in a three-dimensional box with edges of length $a$ assuming that the potential is zero within the box and infinite outside the box. What is meant by degeneracy of energy levels?
4. Set up Schrödinger wave equation for a simple harmonic oscillator, and solve it for the energy eigenvalues.
5. Derive the expression for the energy of a rigid rotator using the Schrödinger wave equation.
6. Write the Schrödinger wave equation for hydrogen atom in terms of polar co-ordinates. Separate the resultant equation in three equations using the technique of separation of variables. How do the quantum numbers $n$, $l$ and $m$ emerge from the solution of the wave equation?
7. Show that the following radial wavefunction of hydrogen atom is normalized:

$$R_{1,0}(r) = \left( \dfrac{2}{a_0} \right)^{3/2} \exp\left( -\dfrac{r}{a_0} \right).$$

8. Using first order time independent perturbation theory, solve the Schrödinger wave equation for the ground state energy of helium atom.
9. Using the variation method solve the Schrödinger wave equation for the ground state energy of helium atom.
10. Calculate the degeneracy of the energy level with energy equal to

    (i) $11(h^2/8ma^2)$    (ii) $12(h^2/8ma^2)$ for a particle in a cubical box.

    [**Ans.** (i) 3 and (ii) 1]

11. Calculate the wavelength of the first three lines of Pashen series for hydrogen atom.
12. Calculate the de Broglie wavelength of a body of mass 1 mg moving with a velocity of $10 \text{ m s}^{-1}$.

## SUGGESTED READINGS

Atkins, P. and de Paula J., *Atkins Physical Chemistry*, Oxford University Press, New York, 2004.

Atkins, P.W. and R.S. Friedman, *Molecular Quantum Mechanics*, 3rd ed., Oxford University Press, 1997.

Chandra, A.K., *Quantum Chemistry*, Tata McGraw-Hill, New Delhi, 2004.

Engel, T. and P. Reid, *Quantum Chemistry and Spectroscopy*, Pearson, 2006.

House, J.E., *Fundamentals of Quantum Mechanics*, Academic Press, 2003.

Levine, I.N., *Quantum Chemistry*, 4th ed., Prentice Hall College Division, 1991.

McQuarrie, D., *Quantum Chemistry*, Universal Science Books, 1983.

McWeeny, R. and B.T. Sutcliffe, *Methods of Molecular Quantum Mechanics*, Academic Press, 1969.

Pilar, F.L., *Elementary Quantum Chemistry*, Dover Publications, 2001.

Prasad, R.K., *Quantum Chemistry*, 3rd ed., New Age International, 2006.

Puri, B.R., L.R. Sharma, and M.S. Pathania, *Principles of Physical Chemistry*, Vishal Publishing, Jalandhar, 2004.

Schtz, G.C. and M.A. Ratner, *Quantum Mechanics in Chemistry*, Dover Publications, 2002.

Silbey, R.J. and R.A. Alberty, *Physical Chemistry*, 3rd ed., John Wiley, 2003.

Simons, J. and J. Nichols, *Quantum Mechanics in Chemistry*, Oxford university Press, 1976.

Szabo, A. and N.S. Ostlund, *Modern Quantum Chemistry: Introduction to Advanced Electronic Structure Theory*, Dover Publication, 1996.

Szabo, A. and S.N. Ostund, *Modern Quantum Chemistry*, McGraw-Hill, New York, 1989.

# 6

# Introduction to Molecular Spectroscopy

Spectroscopy deals with the study of how chemical species (i.e. atoms, molecules, solutions) react or respond to light. Some studies depend on how much light an atom absorbs. The electromagnetic radiation absorbed, emitted or scattered by the molecule is analyzed. Typically, a beam of radiation from a source such as a laser is passed through a sample, and the radiation exiting the sample is measured. Some, like Raman, depend on a molecule's vibrations in relation to the scattering of light.

*The microwave spectroscopy is defined as the study of the interaction of matter with electromagnetic radiation of the spectrum.*

## 6.1 INTRODUCTION

When any matter is exposed to electromagnetic radiation, the atoms and molecules present in the substance respond to the energy absorbed by them. The interaction of electromagnetic radiation with matter can be detected by detectors. This can be done by observing either the decrease or phase shift of electromagnetic radiation as it comes out of the matter. The absorption of electromagnetic radiation may also cause emission of an optical photon by the atoms or molecules present in the substance or the deflection/vibration of atoms. There is relative population difference between the ground state and the excited states which can be detected by the detector.

In case of microwave spectroscopy, interaction of electromagnetic radiation with the atoms/ molecules present in the substance takes place in the microwave region. All molecules having permanent dipole moment when exposed to microwave radiation starts rotating and rotation gives rotational spectra and molecules are said to be microwave active. The molecules which do not have permanent dipole moment are microwave inactive. The energy splitting in the microwave region of the spectrum takes place due to the magnetic dipole and electric quadrupole interactions between the nuclei and electrons in atoms and molecules. Thus, microwave spectroscopy helps in the study of precision determinations of spins and moments of nuclei.

The molecules start rotating by absorbing microwave radiation within the microwave range, and thus microwave spectroscopy provides a great deal of information about the moments of inertia, mechanisms of spin-rotation coupling, electron-spin resonance, paramagnetic resonance, and in the study of other physical properties of rotating molecules like bond length, bond strength, centre of gravity and force constant.

## 6.2 ELECTROMAGNETIC SPECTRUM

The electromagnetic spectrum covers a wide range of wavelength and photon energies. Light used to *see* an object, must have a wavelength about the same size or smaller than the object. The advanced light source (ALS) generates light in the far ultraviolet and soft X-ray regions, which span the wavelengths suited to studying molecules and atoms (Figure 6.1).

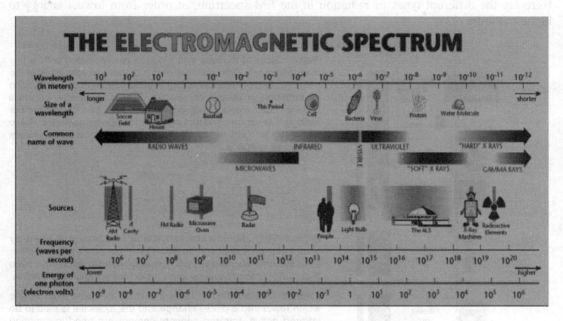

**FIGURE 6.1** Spectrum of electromagnetic radiation.

The electromagnetic (EM) spectrum is just a name that scientists give a bunch of types of radiation when they want to talk about them as a group. Radiation is energy that travels and spreads out as it grows—visible light that comes from a lamp in your house and radio waves that come from a radio station are two types of electromagnetic radiation. Other examples of EM radiation are microwaves, infrared and ultraviolet light, X-rays and gamma-rays . Hotter, more energetic objects and events create higher energy radiation than cool objects. Only extremely hot objects or particles moving at very high velocities can create high-energy radiation like X-rays and gamma-rays (Figure 6.2).

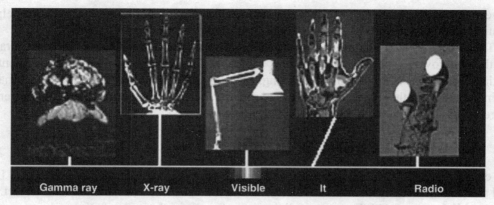

**FIGURE 6.2** Different uses of electromagnetic spectrum.

Here are the different types of radiation in the EM spectrum, in order from lowest energy to highest (Figure 6.3):

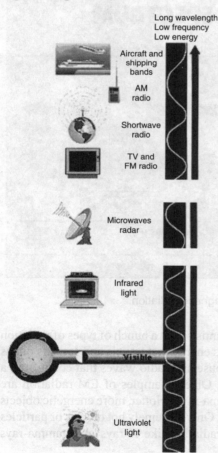

*Radio:* Spin of nucleus and electron is associated with a tiny magnetic dipole. The reversal of this dipole can interact with the magnetic field of radiowaves region of the spectrum thus producing an emission or absorption spectra. Similar types of energy are used in radio stations that emit radiowaves into the air for our radio sets to capture and turn into your favorite tunes/music. Radio waves are also emitted by others like heavy objects sun and stars and gases in space.

*Microwaves:* Microwave radiations in space are used by astronouts to know about the structure of our sun, stars and other solar systems. Microwaves are nowadays used in microwave oven in cooking our food because water molecules are microwave active.

*Infrared:* For a molecule to be IR active, its atoms must vibrate which result into a dipole change and the molecule is said to be infrared active. Infrared sensors camera are used in shooting animals in night because skin emits infrared light, which can be detected. In space, IR light maps are used.

*Visible:* The movement of electronic charge in the molecule is possible due to excitation of a valence electron which results in change in electric dipole and thus gives rise to visible spectrum by its interaction with the undulatory electric field of radiation.

*Ultraviolet:* Sun is the ultimate source of ultraviolet (UV) radiation. UV radiation causes skin cancer. UV radiation can also be generated artificially in laboratory.

**FIGURE 6.3** Contd.

# Introduction to Molecular Spectroscopy

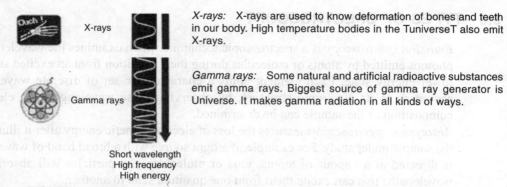

**FIGURE 6.3** Different types of electromagnetic radiation with their source.

Out of complete spectrum of electromagnetic radiation, all radiations are unable to reach the surface of the earth. The radiation that reach the surface of earth are visible spectrum, radio frequencies, and few ultraviolet radiations. If all radiations of the electromagnetic spectrum reach the surface of the earth then no life is possible on the earth. Ozone layer is beneficial for us as it acts as a screen to selectively filter useful radiations for us and does not allow harmful radiations to reach the surface of the earth.

Astronomers can get enough—above of the earth's atmosphere or observe some infrared wavelengths on mountain tops or by flying their telescopes in an aircraft. Experiments can also be taken up to altitudes as high as 35 km by balloons which can operate for months. Rocket flights can take instruments all the way above the earth's atmosphere for just a few minutes before they fall back to earth, but a great many important first results in as astronomy and astrophysics came from just those few minutes of observations. For long-term observations, however, it is best to have your detector on an orbiting satellite (Figure 6.4).

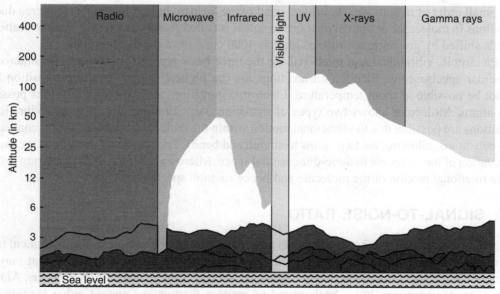

**FIGURE 6.4** Electromagnetic spectrum with altitude from sea level.

## 6.3 TYPES OF SPECTROSCOPY

1. *Emission spectroscopy* is a spectroscopic technique which examines the wavelengths of photons emitted by atoms or molecules during their transition from an excited state to a lower energy state. Each element emits a characteristic set of discrete wavelengths according to its electronic structure, by observing these wavelengths the elemental composition of the sample can be determined.
2. *Absorption spectroscopy* measures the loss of electromagnetic energy after it illuminates the sample under study. For example, if a light source with a broad band of wavelengths is directed at a vapour of atoms, ions or molecules, the particles will absorb those wavelengths that can excite them from one quantum state to another.

If energy is transferred to the radiation field or nonradiative decay if no radiation is emitted. Redirection of light due to its interaction with matter is called *scattering*, and may or may not occur with transfer of energy, i.e. the scattered radiation has a slightly different or the same wavelength.

## 6.4 SCATTERING

What happen when electromagnetic radiations are bombarded on a substance? Some of the radiations pass through matter, some are absorbed within the substance, some of them are reflected back in its original direction (Rayleigh scattering) and a small fraction is scattered in other directions. Light that is scattered at the same wavelength as the incoming light in backward direction is called *Rayleigh scattering*. Radiations that are scattered at a different wavelength than incident radiation are called *strokes* and anti-strokes radiation. Strokes, anti-strokes and Rayleigh radiation are known as scattering of light. Transparent solids like glass and some polymers scatter light due to vibrations (phonons) is called *Brillouin scattering*. Brillouin scattering is typically shifted by a very small order of magnitude, i.e. 0.1 to 1 cm$^{-1}$ of the incident light. Light that is scattered due to vibrations in molecules or optical phonons in solids is called *Raman scattering*. Raman scattered light is shifted by greater magnitude as high as 4000 cm$^{-1}$ from the incident light.

Electronic, vibrational and rotational are the three basic types of transitions encountered in molecular spectroscopy. Electronic transitions are the highest energy requiring transition that cannot be possible at room temperature. Electronic transitions are the only transitions possible with atoms. Molecules shows two types of transition, i.e. vibrations and rotations. Vibrational transitions are possible due to vibrational motion within the molecules (just like spring connected between atoms, allowing the two atoms to stretch and bend). The rotational transitions are due to the rotation of the molecule in three-dimensional space. Microwave region of the spectrum results in the rotational motion of the molecule and hence rotation spectroscopy.

## 6.5 SIGNAL-TO-NOISE RATIO

A signal produced due to other components present along with the sample or the instrument used in the measurement is known as *noise*. Noise is unwanted (meaningless) signal that may be generated in the spectrum either due to impurities, solvent, source, detector or amplifier. Almost every recorded spectrum has a background of random fluctuations caused either by spurious electronic signals produced by the source or the detector, or generated in amplifying equipment. These fluctuations are usually referred as *noise*. The requirement of good spectrum is that the intensity of real spectrum peak should be some three or four times that of the noise fluctuations (a signal-to-noise ratio of three or four order of magnitude). This requirement places a lower limit on the intensity of observable signals. The ability of an instrument to distinguish between signals and noise is usually expressed in terms of signal-to-noise ratio (S/N) which in case is given by

$$\frac{S}{N} = \frac{\text{Average signal amplitude}}{\text{Average noise amplitude}} \qquad (6.1)$$

A higher value of $S/N$ implies greater noise reduction and hence a more precise measurement. It may be noted that the $S/N$ ratio cannot be increased by simply amplification because increase in the magnitude of signal is accompanied by a corresponding increase in the magnitude of noise.

## 6.6 RESOLVING POWER

The resolving power of a spectrometer is its ability to distinguish adjacent absorption bands or two very close spectral lines as separate entities.

Molecular absorption spectra may have a large number of signals at different wavelengths. No molecular absorption takes place at a single frequency only, but always over a spread of wavelengths, usually with very narrow but sometimes with quite large spectra. It is for this reason that we draw spectra with broadened line shapes.

Resolving power of the instrument depends upon the width of the slit. The width of the slit should not be larger than the separation between the lines. However, it must be remembered that a narrower slit allows less total energy from the beam to reach the detector and consequently the intrinsic signal strength will be less. There comes a point when decreasing the slit width results in such a weak signals that they become indistinguishable from the background noise. Thus, spectroscopy is a continual battle to find the minimum slit width consistent with acceptable signal-to-noise values.

## 6.7 WIDTH OF THE SPECTRAL LINES

Most of the absorption and emissions spectra are more or less broad peaks instead of sharp lines. One reason for this is that mechanical slits in spectrometers are not infinitely narrow and thus allowing a range of frequencies rather than a single frequency to pass through it and to fall on the detector, hence blurring the diffraction pattern. By making improvements in spectrometer design one can improve the resolving power of the instrument, however, there is a minimum width inherent in any atomic or molecular transition, i.e. natural line width, beyond which no instrument, however superior, will show a sharpening. This width arises essentially because the energy levels of atomic and molecular systems are not precisely determined, but have certain fuzziness or imprecision. In addition to this, there are several factors which contribute to broadening of spectral lines are like collision broadening in liquid and gaseous phase, Doppler broadening due to Doppler shift responsible for natural line width, Heisenberg uncertainty principle broadening which depends upon lifetime of an excited electronic state compared to lower energy state, etc.

### 6.7.1 Doppler Broadening

This type of broadening occurs due to Doppler effect, according to which in a gaseous sample, frequency of the radiation emitted or absorbed changes when the molecule is moving towards or away from the observer or the observing instrument. Doppler effect is applicable to gaseous samples because according to kinetic theory of gases, the molecules of a gas are continually moving in a zig-zag manner. If the molecules emitting the frequency $v$ (source) is moving away from the observer with a velocity $v$, the observer detects radiation of frequency

$$v' = \frac{v}{1 + \frac{v}{c}} \tag{6.2}$$

where $c$ is the velocity of light.

If the molecule is approaching the observer, then it appears to emit radiation of frequency

$$v' = \frac{v}{1 - \frac{v}{c}} \tag{6.3}$$

The difference $v - v' = \delta v$ is called *Doppler broadening* or *Doppler shift*.

It may also be noted that Doppler broadening increases with increase in temperature because the molecular speed increase with increase in temperature. Hence, to obtain spectra of maximum sharpness, it is best to work at low temperatures.

### 6.7.2 Lifetime Broadening

This type of broadening occurs in all types of samples, i.e. gaseous, liquids, solids or solutions. It arises because in the solution of time dependent Schrödinger wave equation, it is impossible to specify the energies of the levels exactly (as they are changing with time). If on the average, a system survives in a state for time $\tau$ (called the lifetime of the state) then its energy levels are blurred to an extent of $\delta E$,

$$\delta E = \frac{h}{2\pi\tau} = \frac{\hbar}{\tau} \tag{6.4}$$

The quantity $\delta E$ (the corresponding quantity $\delta \bar{v}$) is called lifetime broadening. Eq. (6.4) is similar to Heisenberg uncertainty principle and quantity $\delta E$ or $\delta \bar{v}$ is called *uncertainty broadening*.

From Eq. (6.4), we observe that $\delta E = \frac{1}{\tau}$. Thus, shorter the lifetime of the states involved in the transition, greater is the broadening of the spectral line. As no excited state has an infinite lifetime, therefore, all states are subjected to some lifetime broadening.

The main reason for the finite lifetime of the excited states is the collisions between the molecules or with the walls of the container. If the mean time between collisions (called *collisional lifetime*) is $\tau_{coll}$ then width of the resulting line (called *collisional line width*) is

$$\delta E_{coll} = \frac{h}{2\pi\tau_{coll}} = \frac{\hbar}{\tau_{coll}} \tag{6.5}$$

In a gaseous sample, the collisional lifetime can be increased and hence broadening minimized by working at low pressure.

## 6.8 INTENSITY OF SPECTRAL LINES

The intensity of spectral lines are mainly due to the concentration or path length of the sample, but other factors also contribute to the intensity of spectral lines—are transition probability (likelihood of a system in one state changing to another state) and population of the states (the number of atoms or molecules initially in the state from which the transition occurs). However, it is often possible to decide whether a particular transition is forbidden or allowed. This process is essentially the deduction of selection rules, which allows us to decide between which levels transitions will give rise to spectral lines. If we have two levels from which transitions to a third are equally probable, then obviously the most intense spectral line will arise from the level which initially has the greater population. There is a simple statistical rule (Boltzmann distribution rule) governing the population of a set of energy levels.

The strength or intensity of a spectral line depends upon:

1. the population of states (population density of state),
2. the strength of the incident radiation, and
3. the probability of that transition

Selection rule governed the probability of transition. Any transitions that are allowed according to selection rule are strong and intense spectral transition. Any transition that are not allowed according to selection rule are weak and low intensity spectral transition or either absent.

## 6.9 DEGREES OF FREEDOM OF MOTION

Consider a molecule made up of $N$ atoms. Since almost all the mass of the atom is concentrated in its nucleus and the nucleus is very small in size (the radius being of the order of $10^{-15}$ m whereas molecule is of the order $10^{-10}$ m), the atoms may be considered as mass points. To represent each mass point, we require three coordinates. Hence to represent the instantaneous position of $N$ mass points in space, we require $3N$ co-rdinates.

*The number of coordinates required to specify the position of all the mass points, i.e. atoms in a molecule is called the number of degrees of freedom.*

Thus, a molecule made up of $N$ atoms has $3N$ degrees of freedom. According to kinetic theory of matter, when thermal energy is absorbed by a molecule, it is stored within the molecule in the form of:

1. translational motion of the molecule and
2. internal movement of the atoms within the molecules, i.e. rotational motion and vibrational motion.

The translational motion of the molecule means the motion of the centre of mass of the molecules as a whole. The centre of mass of the molecule can be represented by three coordinates. Hence, there are three translational degrees of freedom. The remaining $(3N-3)$ coordinates represent the internal degrees of freedom.

However, as already mentioned the internal degrees of freedom may be subdivided into:
1. rotational degrees of freedom, and
2. vibrational degrees of freedom.

For rotational motion, there are two degrees of freedom for a linear molecule and three for a nonlinear molecule.

Remaining degrees of freedom for linear molecules are equal to $(3N-5)$ and for the nonlinear molecules, they are equal to $(3N-6)$. These describe the motion of the nuclei with respect to one another and thus represent the number of vibrational degrees of freedom. Hence,

1. Vibrational degrees of freedom of linear molecule containing $N$ atoms = $(3N-5)$.
2. Vibrational degrees of freedom of non-linear molecule containing $N$ atoms = $(3N-6)$.

## 6.10 ROTATIONAL SPECTROSCOPY

*Rotational spectroscopy* or microwave spectroscopy is defined as the study of absorption of electromagnetic radiation in microwave region of the spectrum by the molecules. Rotational spectroscopy is applicable to the gaseous system in which atoms/molecules absorb in the gas phase and here the rotational motion of the molecule is quantized. In solids or liquids, the rotational motion is usually quenched due to molecular collisions.

### 6.10.1 Rotational Active Molecules

For a molecule to be rotationally active, it must have a permanent dipole moment. A molecule is said to be have permanent dipole movement if it have a difference between the centre of charge and the centre of mass, or it have unequivalent separation from centre of gravity of two unlike charges. A molecule possessing permanent dipole moment can interact with the electromagnetic radiation in the microwave region and can absorb or emit a photon and hence microwave active. Permanent dipole moment of the molecules possessing two unlike charges enables the electric field of the microwave radiation to exert a torque on the molecule causing it to rotate. Thus, polar molecules possessing permanent dipole moment like $H_2O$, $NO$, $N_2O$ etc. are microwave active and give rotational spectrum. Homonuclear diatomic molecules like $H_2$, $N_2$, $O_2$ etc. and symmetrical linear molecules like $CO_2$, $C_6H_6$ do not give rotational spectra.

An easiest and simplest rotational spectrum is shown by carbon monoxide molecule. The next simplest rotation spectrum is shown by hydrogen cyanide (linear triatomic molecule) and next is by NCH (non-linear triatomic molecule). The order of simplest rotational spectrum is CO < HCN < NCH.

The pattern of energy levels in the rotational spectrum of a molecule is determined by its symmetry. For convenience, molecules are divided into three classes: (1) Linear molecules, (2) Symmetric tops, and (3) Asymmetric tops.

1. *Linear molecules:* For a linear molecule, the separation of lines in the rotational spectrum can be related directly to the moment of inertia of the molecule, and for a molecule of known atomic masses, can be used to determine the bond lengths (structure) directly. For diatomic molecules, this process is trivial, and can be made from a single measurement of the rotational spectrum. For linear molecules with more atoms, rather more work is

required, and it is necessary to measure molecules in which more than one isotope of each atom have been substituted (effectively this gives rise to a set of simultaneous equations which can be solved for the bond lengths).
2. *Symmetric tops:* A symmetric top is a molecule in which two moments of inertia are the same, and the third is different. There are two types of symmetric tops molecules, i.e. *oblate symmetric tops* (saucer shaped) and *prolate symmetric tops* (rugby football). Oblate and prolate symmetry tops molecules can be distinguished by different types of rotational spectra.
3. *Asymmetric tops:* In asymmetric tops type of molecules, all three moments of inertia along three molecular axes ($I_A$, $I_B$ and $I_C$) are different. Sometimes asymmetric tops molecules have spectra that are similar to that of a linear molecule or a symmetric top molecule. But in this case there must be some structural resemblance between the two.

## 6.10.2 Energy Level of a Rigid Rotator

If a diatomic rigid rotator is considered to be joined along its line of centre by a bond equal in length to the distance $r_0$ between the two nuclei (Figure 6.5). Then the allowed rotational energy of the molecule around the axis passing through the centre of gravity and perpendicular to the line joining the nuclei are given by

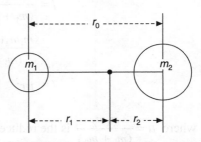

**FIGURE 6.5** Diatomic rigid rotator.

$$E_r = \frac{h^2}{8\pi^2 I} J(J+1) \quad (6.6)$$

where $J$ is rotational quantum number that can have values 0, 1, 2, 3, etc. and $I$ is the moment of inertia of the molecule about the axis of rotation, i.e.

$$I = \left(\frac{m_1 m_2}{m_1 + m_2}\right) r_0^2 = \mu r_0^2 \quad (6.7)$$

where $m_1$ and $m_2$ are the atomic masses of the two atoms of the diatomic molecule and

$$\left(\frac{m_1 m_2}{m_1 + m_2}\right) = \mu \quad (6.8)$$

is called *reduced mass*.

Centre of gravity of a diatomic molecule is the point which satisfies the following condition:

$$m_1 r_1 = m_2 r_2 \quad (6.9)$$

The moment of inertia of the diatomic molecule is given by

$$I = m_1 r_1^2 + m_2 r_2^2 \quad (6.10)$$
$$= m_2 r_2 r_1 + m_1 r_1 r_2$$

$$= r_1 r_2 (m_1 + m_2) \qquad (6.11)$$

But
$$r_1 + r_2 = r_0 \qquad (6.12)$$

Therefore, from Eq. (6.9)
$$m_1 r_1 = m_2 (r_0 - r_1)$$

or
$$r_1 = \frac{m_2 r_0}{m_1 + m_2} \qquad (6.13)$$

Similarly,
$$r_2 = \frac{m_1 r_0}{m_1 + m_2} \qquad (6.14)$$

Substituting these values in Eq. (6.10), we get

$$I = \frac{m_1 m_2^2}{(m_1 + m_2)^2} r_0^2 + \frac{m_2 m_1^2}{(m_1 + m_2)^2} r_0^2$$

$$= \frac{m_1 m_2 (m_1 + m_2)}{(m_1 + m_2)^2} r_0^2 = \frac{m_1 m_2}{(m_1 + m_2)} r_0^2$$

$$I = \mu r_0^2 \qquad (6.15)$$

where $\mu = \dfrac{m_1 m_2}{(m_1 + m_2)}$ is the reduced mass of diatomic molecule.

Now, by definition, the angular momentum of rotating molecule is given by
$$L = I\omega \qquad (6.16)$$
where $\omega$ is the angular velocity.

But we know that angular momentum is quantized whose values are given by
$$L = \sqrt{J(J+1)} \, \frac{h}{2\pi} \qquad (6.17)$$

where $J = 0, 1, 2, 3, \ldots$, called the *rotational quantum numbers*.
Further, the energy of a rotating molecule is given by
$$E = I\omega^2$$

Therefore, the quantized value of the rotational energy will be given by
$$E_r = \frac{1}{2} I\omega^2 = \frac{(I\omega)^2}{2I} = \frac{L^2}{2I} \qquad (6.18)$$

Substituting the value of $L$ from Eq. (6.17), we get,
$$E_r = J(J+1) \frac{h^2}{4\pi^2} \times \frac{1}{2I}$$

or
$$E_r = \frac{h^2}{8\pi^2 I} J(J+1) \qquad (6.19)$$

Putting $J = 0, 1, 2, 3$, etc. in Eq. (6.6), the pattern of the rotational energy level obtained will be as shown in Figure 6.6. Evidently spacing between the energy levels increases as $J$ increases (Figure 6.6).

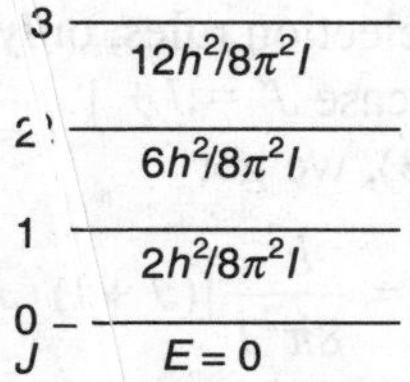

```
3 ―――――――――
       12h²/8π²I
2 ―――――――――
       6h²/8π²I
1 ―――――――――
       2h²/8π²I
0 ― ― ― ― ― ―
J       E = 0
```

**FIGURE 6.6** Various rotational levels and their corresponding energies.

### 6.10.3 Rotational Selection Rules

According to selection rules, the transition takes place only between those rotational energy levels for which
$$\Delta J = \pm 1$$
That is change in the rotational quantum number is unity.

The transition $\Delta J = +1$ corresponds to absorption and $\Delta J = -1$ correspond to emission. In other words, a transition can take place from a particular rotational level to just the next higher or just the lower rotational level. All these transitions can take place with equal probability whereas all other transitions are forbidden.

### 6.10.4 Rotational Spectra of the Diatomic Molecules

The allowed rotational energies of diatomic molecules is given by Eq. (6.19) as below
$$E_r = \frac{h^2}{8\pi^2 I} J(J+1)$$
In terms of frequency, we can write,
$$v = \frac{h}{8\pi^2 I} J(J+1) \qquad (6.20)$$
Further, in terms of wavenumber, Eq. (6.20) can be written as
$$\bar{v} = \frac{h}{8\pi^2 Ic} J(J+1) \qquad (6.21)$$
$$= BJ(J+1) \qquad (6.22)$$
where $B = \dfrac{h}{8\pi^2 Ic}$ is called *rotational constant*.

Putting $J = 0, 1, 2, 3$, etc. in Eq. (6.22), the wavenumbers of the different rotational levels will be $2B, 6B, 12B, 20B, 30B, \ldots$ and so on.

When a transition takes place from a lower rotational level with rotational quantum number $J$ to a higher rotational level with rotational quantum number $J'$, the energy absorbed will be given by
$$\Delta E_r = E_r' - E_r \quad \text{or} \quad (E_{J'} - E_J)$$
$$= \frac{h^2}{8\pi^2 I} J'(J'+1) - \frac{h^2}{8\pi^2 I} J(J+1)$$

$$= \frac{h^2}{8\pi^2 I}[J'(J'+1) - J(J+1)] \tag{6.23}$$

However, according to the rotational selection rules, only those rotational transitions are allowed for which $\Delta J = \pm 1$. Hence, in present case $J' = J + 1$.

Substituting this value in Eq. (6.23), we get

$$\Delta E_r = \frac{h^2}{8\pi^2 I}[(J+1)(J+2) - J(J+1)]$$

$$= \frac{h^2}{8\pi^2 I} 2(J+1) \tag{6.24}$$

To express in terms of wavenumbers, we put

$$\Delta E_r = h\nu = h\frac{c}{\lambda} = hc\bar{\nu}$$

Hence, $$hc\bar{\nu} = \frac{h^2}{8\pi^2 I} 2(J+1)$$

$$\bar{\nu} = \frac{h}{8\pi^2 Ic} 2(J+1)$$

$$\bar{\nu} = 2B(J+1) \tag{6.25}$$

$\bar{\nu}$ in above equation represents the wavenumbers of the spectral lines which will be obtained as a result of the transitions between the rotational levels. Putting $J = 0, 1, 2, 3$, etc. or in general from $J$ to $J + 1$ involving absorption of energy, the wavenumbers of the lines obtained will be $2B, 4B, 6B, 8B,\ldots$ and so on (Figure 6.7).

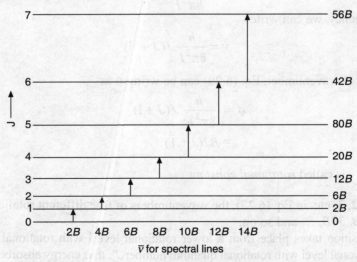

**FIGURE 6.7** Rotational spectrum of rigid diatomic rotator.

We find that the most important feature of the rotational spectrum is that every two successive lines have a constant difference of wavenumber equal to 2B. This is called *frequency separation* or *wavenumber separation*. Thus, various lines in the rotational spectra will be equally spaced.

Thus separation between any two successive lines will have the difference in wavenumber given by

$$\Delta \bar{v} = 2B = \frac{h}{4\pi^2 Ic}$$

Another important feature of the spectra is that intensities of different transitions are not equal. The intensities increase with increasing $J$, passes through maximum and then decreases as $J$ further increases (Figure 6.8).

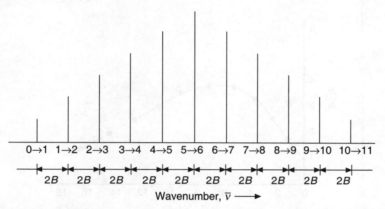

**FIGURE 6.8** Relative intensity of rotational lines in rotational spectra of diatomic rigid rotator.

### 6.10.5 Relative Intensity of Rotational Spectral Lines

It is expected that the relative intensities of the spectral lines will depend upon the relative population of the energy levels. Greater is the population of an energy level, greater is the number of molecules that can be promoted to the next higher level and hence greater is the intensity of absorption. But the population of a rotational energy level with quantum number $J$ (relative to the ground level for which $J = 0$) is given according to Boltzmann distribution law by the equation

$$\frac{N_J}{N_0} = e^{-E_J/kT} \qquad (6.26)$$

However, rotational energy levels are degenerate and the degeneracy of any two rotational levels with quantum number $J$ is given by

$$g_J = 2J + 1 \qquad (6.27)$$

Thus, the relative population of an energy level and hence intensity of the rotational lines will be proportional to the product of both the above facts, i.e.

$$\text{Intensity} \propto \frac{N_J}{N_0} = (2J+1)e^{-E_J/kT}$$

Putting $E_J = h\nu = h\dfrac{c}{\lambda} = hc\bar{\nu} = hcBJ(J+1)$, we get

$$\dfrac{N_J}{N_0} = (2J+1)e^{-hcBJ(J+1)/kT}$$

As $J$ increases, the first factor, viz. $(2J + 1)$ increases whereas the exponential factor, viz. $e^{-hcBJ(J+1)/kT}$ decreases slowly in the beginning and then rapidly as the values of $J$ become higher and higher. Hence, plot of $\dfrac{N_J}{N_0}$ versus $J$ for a rigid diatomic molecule is obtained as shown in Figure 6.9.

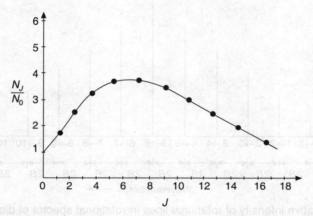

**FIGURE 6.9** Relative population of rotational energy level.

The graph clearly shows that relative population and hence the intensity of transitions first increases, reaches a maximum value and then decreases.

### 6.10.6 Hyperfine Interactions

The magnetic and electrostatic interactions in the molecule are responsible for small changes in the spectra, i.e. additional splittings and shifts of rotational lines takes place. Extent of splitting and shift of rotational line in particular depends upon strength of magnetic and electrostatic interactions and it differs in different molecules. In general, the strength of these hyperfine interactions depend upon:

(i) Electron–electron spin interaction.
(ii) Magnetic/magnetic–dipole interaction.
(iii) Electron spin–molecular rotation.
(iv) Electron spin–nuclear spin interaction.
(v) Electron charge–nuclear electric quadrupole interaction.
(vi) Nuclear spin–nuclear spin interaction.

Above said hyperfine interactions are responsible for characteristic energy levels that are detected in NMR and ESR spectroscopic techniques.

## 6.11 ROTATIONAL-VIBRATIONAL SPECTROSCOPY

Vibrational transitions are high energy transitions and are due to vibrational motion of the molecule. Vibrational motion is also accompanied by rotational motion. Vibrational-rotational spectra are observed in infrared and Raman spectroscopy.

The energy of vibrational transitions ($10^{-20}$ J/mol) is ten times higher than rotational transitions ($10^{-21}$ J/mol). Highly-resolved vibrational spectra will show fine rotational transitions structure. However, molecular vibrational and rotational motion do have some effect on one other, and hence, the rotational and vibrational contributions to the energy of the molecule can be studied independently:

$$E_{\text{vib,rot}} = E_{\text{vib}} + E_{\text{rot}} = \left(v + \frac{1}{2}\right)h\nu_0 + hcBJ(J+1) \quad (6.28)$$

where $v$ is the vibrational quantum number, $J$ is the rotational quantum number, $\nu_0$ is the frequency of the vibration, and $B$ is the rotational constant.

### 6.11.1 Evaluating Vibrational Spectra

Figure 6.10 shows the infrared spectrum of DCl molecule. A number of lines are observed in the vibrational spectra is due to single vibrational transition from ground state ($v = 0$) to first excited vibrational state ($v = 1$). This single vibrational transition corresponds to a number of rotational transitions and hence multiple peaks are observed. The peaks on the left are known as the *R-branch* ($\Delta J = +1$). The peaks on the right are called the *P-branch* ($\Delta J = -1$).

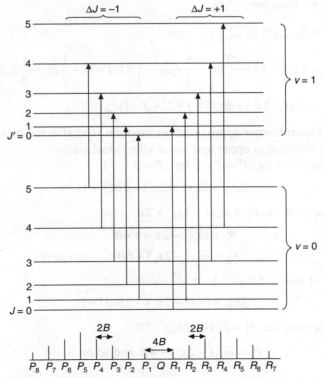

**FIGURE 6.10** Infrared spectrum of a linear diatomic molecule (DCl).

In the centre of vibrational spectra, there is $Q$-branch (gap between the $R$- and $P$-branch). A peak would have appear here in the centre for a vibrational transition in which the rotational energy did not change ($\Delta J = 0$). But according to the selection rule of rotational spectroscopy, transitions correspond to $\Delta J = 0$ are not allowed and hence there is no $Q$-branch. This selection rule explains why the $P$- and $R$-branches are observed ($\Delta J = \pm 1$), but not the $Q$-branch ($\Delta J = 0$). Transitions correspond to branches for which, ($\Delta J = \pm 2, \pm 3, \ldots$) are also allowed and can be observed in vibrational-rotational spectroscopy.

The separation between each of the vibrational lines, i.e. $P$- and $R$-branch should be $2B$ where $B$ is the rotational constant for a given molecule. Practically, it is seen that the spacing between the $R$-branch peaks decreases and between $P$-branch peaks increases as the frequency increases. This is possible due to the reason that the bonds between the atoms in a molecule not being entirely rigid and they behave elastically.

Selection rule for combined vibrational and rotation transitions are the same as those for individual vibrational and rotational transitions, i.e.

$$\Delta v = \pm 1, \pm 2, \pm 3, \ldots \quad \Delta J = \pm 1 \tag{6.29}$$

If $\Delta v = 0$, then there is pure rotational transitions. If $\Delta J = 0$, it represents absence of rotational transitions or hence no rotational motion but this rarely happens.

Let us assume $J$ represents rotational quantum number of vibrational ground state ($v = 0$) and $J'$ represents rotational quantum number of first vibrational excited state ($v = 1$). Let us take an example of transition from ground state ($v = 0$) to the first excited vibrational level, i.e. $v = 1$, then different transitions we have are

$$\Delta E = E_{J', v=1} - E_{J, v=0}$$

$$= BJ'(J'+1) + \frac{3}{2}\bar{\omega}_e - \frac{5}{4}x_e\bar{\omega}_e - \left\{ BJ(J+1) + \frac{1}{2}\bar{\omega}_e - \frac{1}{4}x_e\bar{\omega}_e \right\}$$

$$= \bar{\omega}_e(1 - 2x_e) + B(J' - J)(J' + J + 1) \tag{6.30}$$

According to Born-Oppenheimer approximation, rotation is unaffected by vibrational changes, i.e. the value of $B$ is identical in upper and lower vibrational states.

For $R$-branch, $\Delta J = +1$ i.e. $J' = J + 1$ or $J' - J = +1$

$$\Delta E_R = \bar{\omega}_e(1 - 2x_e) + 2B(J+1) \tag{6.31}$$

Hence, $R_0$ line will appear, i.e. $R_0 = \bar{\omega}_e(1 - 2x_e) + 2B$

Similarly, $$R_1 = \bar{\omega}_e(1 - 2x_e) + 4B$$
$$R_2 = \bar{\omega}_e(1 - 2x_e) + 6B \quad \text{and so on} \ldots$$

For $P$-branch, $\Delta J = -1$ that is $J = J' + 1$ or $J' - J = -1$

$$\Delta E_P = \bar{\omega}_e(1 - 2x_e) - 2B(J'+1) \tag{6.32}$$

Hence, $P_1$ line will appear, i.e. $P_1 = \bar{\omega}_e(1 - 2x_e) - 2B$

$$P_2 = \bar{\omega}_e(1 - 2x_e) - 4B \text{ and so on}\ldots.$$

## 6.12 WHEN DOES A MOLECULE ABSORB INFRARED LIGHT?

For both IR absorption and Raman scattering, we need to discuss what in molecules produces electric fields. The important point is that molecules have dipoles. Permanent dipole moments arise in a molecule because the electrons are not evenly distributed among all atoms. Consider the HCl and CO molecules. In HCl, H is relatively positive and Cl is relatively negative. In CO, C is relatively positive and O is relatively negative. HCl and CO molecules will act as a dipole.

When two particles of charges $+q$ and $-q$ are separated by a distance $r$, the permanent electric dipole moment $\mu$ is given by

$$\mu = q^2 r$$

To see what the magnitude of molecular dipoles is, we consider a proton and electron separated by one Angstrom. The dipole moment would be $\mu = 1.602 \times 10^{-19}$ coulomb × angstrom = $1.602 \times 10^{-29}$ coulomb-m. In cgs units, this corresponds to $4.809 \times 10^{-18}$ esu-cm. Molecular dipoles are in this range.

Hetero-diatomic molecules such as HCl, NO and CO are examples of dipoles. All heterodiatomic molecules have permanent dipole moments because one atom is necessarily more electronegative than the other. Homo-nuclear diatomic molecules, such as $N_2$ and $O_2$ do not have permanent dipole moments since both nuclei attract the electrons equally.

What happens when a heteronuclear diatomic molecule vibrates at a particular frequency?

The molecular dipole moment also oscillates about its equilibrium value as the two atoms move back and forth. In the simplest case, we consider the vibration of the molecule as a harmonic oscillator. To a first approximation we consider the atoms to be oscillating like a spring. The harmonic potential describes this:

$$V(x) = \frac{kx^2}{2} \tag{6.33}$$

The oscillator case for a diatomic molecule is shown in Figure 6.11.

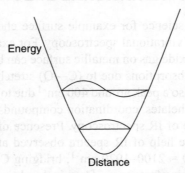

**FIGURE 6.11** Oscillator case for a diatomic molecule.

This oscillating dipole can absorb energy from the oscillating electric field of light only in the field that oscillates at the same frequency. If the frequencies of light and the vibrations were not the same then this transition is non-resonant with the light—the molecule does not absorb. The transition dipole energy is given by the frequency of light. The lowest vibrational level is called the *zero point energy* level. At temperatures approaching absolute zero the molecule is at the lowest energy level. There is still motion at this temperature—this motion is related to Heisenberg uncertainty principle. Note that it is the difference in the energies of the zero point vibration and the vibration of the first excited state that gives the transition dipole energy.

A molecule may have no net dipole moment, but still absorb IR light since groups within larger molecules also have dipole moments. For example, the molecules ethylene and hexaflorobenzene have no net dipole moment. But each C–H bond and C–F bond does have a dipole moment, and both of these molecules have IR absorption.

## 6.13 OVERTONE, COMBINATION FREQUENCIES AND FERMI RESONANCE

There is the possibility of first, second, third etc. overtone at a frequency near $2v_1, 3v_1,...,2v_2, 3v_2,...,2v_3, 3v_3,...$etc. where $v_1$ is the fundamental vibrational frequency. The intensity of $2v_2,...,3v_2,...,2v_3, 3v_3,...$ etc. falls rapidly. In addition to the overtone, selection rule permits combination and difference bands. The combination bands arise simply from the addition of two or more fundamental frequencies or overtones. Such combinational bands as $v_1 + v_2, 2v_1 + v_2, v_1 + v_2 + v_3$, etc. become allowed, although their intensities are normally small.

Similarly, there are difference bands such as $v_1 - v_2, 2v_1 - v_2, v_1 + v_2 - v_3$, also have small intensity but are often found in complex spectra. The intensity of overtone or combination bands may sometimes be considered enhanced by a resonance phenomenon. They are sometimes termed as accident degeneracy. Accident degeneracy is found most often between a fundamental and some overtones or combinational bands. Sometimes two close molecular vibrations frequency resonate and exchange energy, this phenomenon is known as *Fermi resonance* when a fundamental resonates with an overstone.

## 6.14 APPLICATIONS OF VIBRATIONAL SPECTROSCOPY

1. In the study of surface science for example surface chemistry of an adsorbate can be studied with the help of vibrational spectroscopy. For example, adsorption of molecular poisonous carbon monoxide gas on metallic surface can be investigated from IR spectra. IR spectra give strong absorptions due to (C—O) stretching frequency at a wavelength of 2143 cm$^{-1}$. There is also a peak around 400 cm$^{-1}$ due to metal–carbon stretching mode.
2. The structure of metal chelates, coordination compounds and cage compounds can also be studied with the help of IR spectroscopy. Presence of CO in these metal complexes can be detected with the help of IR spectra observed at, molecular CO (gas phase) = 2143 cm$^{-1}$, terminal CO = 2100–1920 cm$^{-1}$, bridging CO = 1920–1800 cm$^{-1}$. Further, the decrease in the stretching frequency of terminal and molecular CO gas can be explained in terms of the *Dewar-Chatt* or *Blyholder model* for the bonding of CO to metals.
3. The atmospheric composition of giant planets like Jupiter, Saturn, Uranus, Neptune etc. can be investigated with the help of IR spectroscopy.
4. The value force constant can be determined with the help of IR spectroscopy which in turn determines the strength of the bond. Higher is the strength of the bond, higher is the value of mean vibrational frequency.

5. The value of mean vibrational frequency ($\bar{\omega}$) can be estimated with the help of force constant ($k$) and reduced mass ($\mu$) as given below:

$$\bar{\omega} = \frac{1}{2\pi c}\sqrt{\frac{k}{\mu}} \qquad (6.34)$$

6. In the finger print region, i.e. 100–1500 cm$^{-1}$, there are skeletal vibrations in the region 1400–700 cm$^{-1}$ which may arise from linear or branched chain structures. The region from 4000–1500 cm$^{-1}$ is known as *characteristics functional group region*. Table 6.1 gives characteristics functional group stretching frequency of some $=$ molecular groups.

**TABLE 6.1** Characteristics functional group stretching frequency

| Functional group | IR absorption frequency (cm$^{-1}$) |
|---|---|
| Alcohol (—OH) | 3600 |
| Amine (—NH$_2$) | 3400 |
| Acetylene ($\equiv$CH) | 3300 |
| Benzene ($=$CH) | 3060 |
| Venyl ($=$CH$_2$) | 3030 |
| Methyl (—CH$_3$) | 2850 |
| Thioalcohol (—SH) | 2600 |
| Cyanide (—C$\equiv$N) | 2250 |
| Acetylene (C$\equiv$C) | 2220 |
| Aldehyde and Ketones (C$=$O) | 1600–1750 |
| Venyl (C$=$C) | 1650 |
| (C—C), (C—N) and (C—O) | 1200–1000 |
| C$=$S | 1100 |
| C—F | 1050 |
| C—Cl | 725 |
| C—Br | 650 |
| C—I | 550 |

# EXERCISES

1. Which of the following molecules would show: (a) rotational spectrum, (b) vibrational spectrum: Br$_2$, HBr, CS$_2$?

2. A certain transition involves an energy change of $2.6 \times 10^{-24}$ J molecule$^{-1}$. If there are 1000 molecules in the ground state, what is the approximate equilibrium population of the excited state at temperature of (a) 28 K, (b) 295 K, and (c) 2900 K?

3. The rotational spectrum of BrF shows a series of equidistant lines 0.7143 cm$^{-1}$ apart. Calculate the rotational constant $B$, and hence the moment of inertia and bond length of the molecule.

4. How many normal modes of vibration are possible for the following molecules: HBr, O$_2$, OCS (linear), SO$_2$ (bent), BCl$_3$, acytelene, CH$_4$ CH$_3$I, C$_6$H$_6$?

5. Explain why the C = O stretching vibration of an aldehyde gives rise to a strong absorption in the infrared, yet the absorption due to the C = C vibration in an alkene is normally very weak.
6. State the selection rule for rotational spectra, vibrational spectra and Raman scattering.
7. Write down applications of vibrational and Raman spectroscopy.
8. Calculate the energies of the first four rotational levels of HI free to rotate in three dimensions using bond length 160 pm.
9. For a harmonic oscillator of effective mass $2.88 \times 10^{-25}$ kg, the difference in adjacent energy is 3.17 J. Calculate the force constant of the oscillator.
10. Which of the vibration of an $AB_3$ molecule are infrared or Raman active when it is: (a) trigonal planar and (b) trigonal pyramidal?

## SUGGESTED READINGS

Atkins, P.W. and R.S. Friedman, *Molecular Quantum Mechanics*, Oxford University Press, 2005.

Atkins, P.W., *Quanta: A Handbook of Concepts*, 2nd ed., Oxford University Press, 1991.

Bernath, P.F., *Spectra of Atoms and Molecules*, Oxford University Press, 1995.

Bonin, K. and W. Happer, *Atomic Spectroscopy*, In Encyclopedia of Applied Physics (Ed., G.L. Trigg) 2, 245, VCH, New York, 1991.

Condon, U.K. and H. Odabasi, *Atomic Structure*, Cambridge University Press, 1980.

Haigh, C.W., "The Theory of Atomic Spectroscopy: JJ coupling, intermediate coupling and configuration interaction," *J. Chem. Educ.*, **72,** 206, 1995.

Johnson Jr, C.S. and L.G. Pedersen, *Problems and Solutions in Quantum Chemistry and Physics*, Dover Publications, New York, 1986.

# 7

# X-Ray Crystallography

X-ray crystallography is a powerful and unambiguous method for the structural determination and characterization of crystalline substances—from simple salts and small molecules to complex proteins. There is no technique that can match the speed and reliability of single crystal X-ray diffraction for determining the structure of molecules in the solid state. X-ray provides fast, reliable, secure access to small molecule crystal structure determination.

W.H. Bragg and his son Sir W.L. Bragg gave a relation in 1913 to explain why the cleavage faces of crystals reflect X-ray beams at certain angles ($\theta$) which is popularly known as Bragg's law as given below

$$n\lambda = 2d \sin\theta \tag{7.1}$$

Here, $d$ is the distance between atomic layers in a crystal, and $\lambda$ is the wavelength of the incident X-ray beam and $n$ is an integer which represents plane which is deflecting the X-ray.

Braggs were awarded the Nobel Prize in Physics in 1915 for their work in crystal structures determination starting with NaCl, ZnS and C (diamond).

Now the question arises, why XRD of crystal are important to us? The XRD of crystals are important to us because of following four reasons:

1. It measures the average spacings between layers or rows of atoms.
2. It determines the orientation of a single crystal or grain.
3. It finds the crystal structure of an unknown material.
4. It measures the size, shape and internal stress of small crystalline region.

## 7.1 SPACE LATTICE AND UNIT CELL

The regular arrangement of points, i.e. atoms, molecules or ions, constituting the crystal in three-dimensional space within the crystal is called the *space lattice* or *crystal lattice*.

If a big size crystal is broken more and more, ultimately a stage is reached when we get the smallest possible crystal. The smallest crystal obtained will be a small cube. This small cubic crystal has all the elements of symmetry as possessed by the big cubic crystal (Figure 7.1). Moreover, it is evident that the complete lattice has been obtained by the repetition of this smallest unit in different directions (Figure 7.2).

**FIGURE 7.1** A cubic crystal.

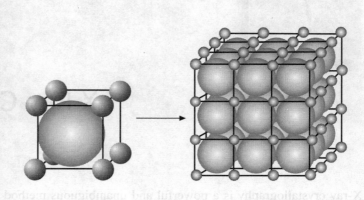

**FIGURE 7.2** A unit cell and space lattice.

The smallest portion of the complete space lattice which when repeated over and over again in different directions produces the complete space lattice is called *unit cell*.

A crystal consists of a periodic arrangement of the unit cell into a lattice. The unit cell can contain a single atom or atoms in a fixed arrangement. Crystals consist of planes of atoms that are spaced a distance $d$ apart, but can be resolved into many atomic planes, each with a different $d$ spacing. $a$, $b$ and $c$ (length) and $\alpha$, $\beta$ and $\gamma$ (angles) between $a$, $b$ and $c$ are lattice constants or parameters which can be determined by XRD.

Seven crystal systems are:

| *Crystal class* | *Axis system* |
|---|---|
| Cubic | $a = b = c$, $\alpha = \beta = \gamma = 90°$ |
| Tetragonal | $a = b \neq c$, $\alpha = \beta = \gamma = 90°$ |
| Hexagonal | $a = b \neq c$, $\alpha = \beta = 90°$, $\gamma = 120°$ |
| Rhombohedral | $a = b = c$, $\alpha = \beta = \gamma \neq 90°$ |
| Orthorhombic | $a \neq b \neq c$, $\alpha = \beta = \gamma = 90°$ |
| Monoclinic | $a \neq b \neq c$, $\alpha = \gamma = 90°$, $\beta \neq 90°$ |
| Triclinic | $a \neq b \neq c$, $\alpha \neq \beta \neq \gamma \neq 90°$ |

## 7.2 HAUY'S LAW OF RATIONALITY OF INDICES

To explain the relative direction, orientation of faces and the planes present in a crystal, a set of three-coordinate axes are required which depend upon the symmetry of the crystal. These are lines coinciding the edges or parallel to edges between principal axes such that the faces of the crystal will either intercept these axes at certain distances from the origin or parallel to some of the axes, the intercepts in this case are infinity.

In cubic crystals, the three axes $x$, $y$, and $z$ are mutually at right angle to each other and these meet at the point $O$ called the origin. These three lines (axes) are called *crystallographic axes*. A suitable plane which cut these axes is selected and this is called a *unit plane* or *standard plane* (Figure 7.3).

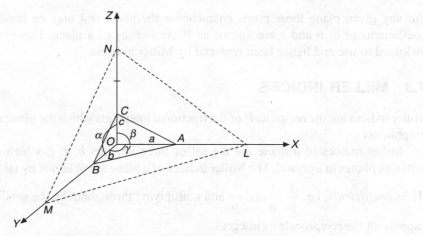

**FIGURE 7.3** The three crystallographic axes.

Consider three axes $X$, $Y$, and $Z$, which represent crystallographic axes. Let a unit plane $ABC$ cut these axes at $A$, $B$ and $C$ at distances $OA$, $OB$ and $OC$, respectively from origin. Let $OA = a$, $OB = b$, and $OC = c$.

The ratio of the intercepts is $a : b : c$ and this ratio is called *axial ratio*. It is the ratio of the intercepts made by the plane of the crystal on the three crystallographic axes.

For a plane $LMN$ (say) the intercepts are $2a$, $2b$, $3c$, these are simple whole number of the unit plane. For any plane $UVW$ (say) let the intercepts are $u$, $v$ and $w$. Thus

$$u = m\,a$$
$$v = n\,b$$
$$w = p\,c$$

$m$, $n$ and $p$ are called *Weiss coefficients*. Hauy's law of rationality of indices or intercepts was proposed by Hauy. It states that it is possible to choose along three-coordinate axes unit distances $(a, b, c)$ not necessarily of the same length such that the ratio of the three intercepts of any plane in the crystal is given by;

$$m\,a : n\,b : p\,c$$

where $m$, $n$, and $p$ are either integral whole numbers including infinity or fraction of whole numbers.

For any crystal a set of three-coordinate axes can be chosen in such a way that all the faces of the crystal will either intercept these axes at definite distance from the origin or be parallel to some of the axes in which case the intercepts are at infinity.

In case of crystal of topaz mineral, i.e. $Al_2(F, OH)_2 SiO_4$ for which four different parameters have the following values:

(i) $m = 1, n = 1, p = 1$  
(ii) $m = 1, n = 1, p = \infty$  
(iii) $m = 1, n = 1, p = 2/3$  
(iv) $m = 2, n = 1, p = \infty$

The ratio of intercepts are:

(i) $a : b : c$  
(ii) $a : b : \infty c$  
(iii) $a : b : : 2/3\, c$  
(iv) $2a : b : \infty c$

For any given plane these ratios characterize the plane and may be used to represent it. The coefficients of $a$, $b$ and $c$ are known as *Weiss indices* of a plane. However, Weiss indices are awkward to use and hence been replaced by Miller indices.

## 7.3 MILLER INDICES

Miller indices are the reciprocals of the fractional intercepts which the plane makes with crystallographic axes.

Miller indices of a plane are the set of three integers $h$, $k$, $l$, which are used to describe different planes in a crystal. The Miller indices of a plane are obtained by taking reciprocals of the *Weiss coefficients*, i.e. $\dfrac{1}{m}, \dfrac{1}{n}$ and $\dfrac{1}{p}$ and multiplying throughout by the smallest number, this will express all the reciprocals as integers.

Therefore, Miller indices are the reciprocal of the coefficient of the intercepts expressed as integers. The ratio

$$\frac{a}{ma} : \frac{b}{nb} : \frac{c}{pc}$$

or

$$\frac{1}{m} : \frac{1}{n} : \frac{1}{p}$$

are Miller indices and are represented by $h$, $k$, $l$.

These are rbeciprocal intercepts of that face on the crystallographic axes.

For a plane in which Weis notation is $1 : 1 : \infty$ becomes in Miller notation $\dfrac{1}{1}, \dfrac{1}{1}, \dfrac{1}{\infty}$ or $(1, 1, 0)$.

Similarly, the four planes mentioned above in case of topaz crystal have Miller indices $(1, 1, 1)$, $(1, 1, 0)$, $(2, 2, 3)$, $(1, 2, 0)$ of the crystal face notation.

If the unit plane makes intercepts on the negative side say $-a$, $-b$ and $\infty c$, the Miller indices of the planes would be $(-1, -1, 0)$ or $(\bar{1}, \bar{1}, 0)$. The bar indicates the intersection of the plane on the negative side of the crystallographic axes.

The law of rational indices may be stated as:

*For any crystalline substance, there is a set of axes in terms of which all natural occurring faces have reciprocal intercepts proportional to the integers (h, k, l).*

In case of octahedron, Miller notations of some of the faces will have Miller indices (1, 1, 1), $(\bar{1}, 1, 1), (1, \bar{1}, 1), (1, 1, \bar{1}), (\bar{1}, \bar{1}, 1), (\bar{1}, 1, \bar{1}), (\bar{1}, \bar{1}, \bar{1})$. The crystallographic axes being at right angles with origin at the centre of the crystal.

The distance between the parallel faces is represented by $d_{h,k,l}$. For na cubic lattice of length a, the interplanar spacing are given by

$$d_{h,k,l} = \sqrt{\frac{a}{h^2 + k^2 + l^2}}$$

where h, k, and l are Miller indices and a is the length of the sides of the cube.

## 7.4 DERIVING BRAGG'S LAW

English physicists W.H. Bragg and W.L. Bragg studied crystal structure in detail. They exposed different crystals to X-ray at different angles of the incident. They observed that the atomic planes of a crystal cause an incident beam of X-rays to interfere with one another as they leave the crystal. The phenomenon is called *X-ray diffraction*.

In order to derive Bragg's law, they exposed X-ray radiation over a crystal. Here two beam of X-ray are taken for an example (Figure 7.4). It is important to note here that the second beam of X-ray has to travel some extra distance as compared to first beam of X-ray. The second beam continues to travel the next layer of the crystal where it is scattered by atom B. Thus, the second beam of X-ray must travel the extra distance $AB + BC$ if the two beams are to continue travelling adjacent and parallel. Braggs suggested that this extra distance travelled by second beam must be an integral (n) multiple of the wavelength ($\lambda$):

$$n\lambda = AB + BC \quad (7.2)$$

The second beam of X-ray must travel the extra distance ($AB + BC$) to continue travelling parallel and adjacent to the first (top) beam of X-ray.

Here d is the hypotenuse of the right angle triangle ABz. By using trigonometric relation, we can show that

$$AB = d \sin\theta \quad (7.3)$$

It is clear from Figure 7.4 that $AB = BC$, therefore Eq. (7.2) becomes,

$$n\lambda = 2AB \quad (7.4)$$

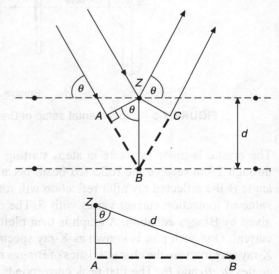

**FIGURE 7.4** Deriving Bragg's law using the reflection geometry and applying trigonometry.

By inserting the value of AB from Eq. (7.3) in to Eq. (7.4) we get,

$$n\lambda = 2d\sin\theta \qquad (7.5)$$

Equation (7.5) is popularly known as *Bragg's Law*.

## 7.5 EXPERIMENTAL SETUP OF BRAGGS METHOD OF STRUCTURE DETERMINATION

X-rays are generated in the X-ray tube $A$. These are made monochromatic by passing through the absorbing screens $B$. Then the beam of X-ray of definite wavelength are made to pass through the slit $L$ which collimates it into a fine beam and falls upon the face of the crystal $C$ mounted on the rotating table. The table can be made to rotate about a vertical axis and the rotation can be read on a circular graduated scale $S$ independent of turn table $T$. The rays after reflection from the crystal enter the detector D, which is either ionization chamber or Giegler counter. The ionization chamber is generally filled with easily ionisable gases like $SO_3$, $CH_3$, $Br_2$, etc. X-Ray ionizes the gas and ionization current is measured by using quadrant electrometer $E$ (Figure 7.5). The strength of the current will be proportional to the intensity of the entering X-rays.

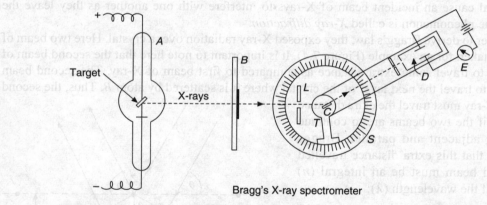

**FIGURE 7.5** Experimental setup of Bragg's method for structure determination.

The crystal is made to rotate in steps starting from $\theta = 0$. The ionization chamber is rotated through $2\theta$ to receive the reflected beam because when the incident ray is rotated through an angle $\theta$, the reflected ray after reflection will rotate double the angle, i.e. $2\theta$. It is found that the value of ionization current varies with $\theta$. The maximum ionization corresponds to the value $\theta$ given by Braggs equation. A graph is then plotted between the glancing angle $\theta$ and ionization current. One such plot is known as X-ray spectrum of the crystals (Figure 7.6). The peak in the X-ray spectrum are the characteristics of Braggs reflection corresponding to different order Braggs angles $\theta_1$, $\theta_2$ and $\theta_3$. The first peak corresponds to strong reflection. From the observed value of glancing angle $\theta$ and with known values of $d$ and $n$, we can calculate the values of $\lambda$ with the help of Bragg's equation.

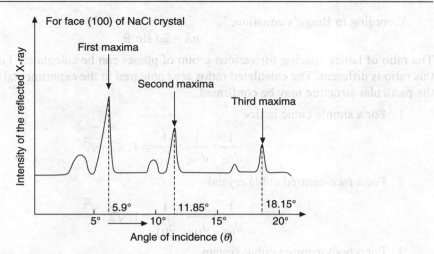

**FIGURE 7.6** X-ray plot of angle of incidence versus ionization current.

## Calculation of wavelength

In case of cubic lattice $d$ is related to the cell edge (a) by the following relations.

1. For simple cubic lattice
$$d = a\frac{\sqrt{2}}{2}$$

2. For face centred cubic lattice
$$d = \frac{a}{2}$$

3. For body centred cubic
$$d = a\frac{\sqrt{3}}{2}$$

The value of $a$, i.e. length of the cubic cell is given by

$$a = \left[\frac{\text{No. of atoms in unit cell} \times \text{Mol. Wt.}}{\text{Avogadro's No.} \times \text{Density}}\right]^{1/3}$$

According to the Bragg's equation

$$n\lambda = 2d \sin\theta$$

Knowing the value of $\theta$ and $n$, i.e. order of reflection, $\lambda$ can be calculated.

## 7.6 DETERMINATION OF CRYSTAL STRUCTURE BY BRAGG'S METHOD

To determine the crystal structure, X-rays are allowed to fall on crystal surface and the crystal is rotated. X-rays are reflected from various lattice planes. The reflections are recorded by Bragg's X-ray spectrometer. The glancing angle for each intense reflection is also recorded.

According to Bragg's equation,
$$n\lambda = 2d \sin \theta$$

The ratio of lattice spacing for various group of planes can be calculated. For different crystals, this ratio is different. The calculated ratios are compared in the experimental ratios and from this the particular structure may be confirmed.

1. For a simple cubic lattice

$$\frac{1}{d_{100}} : \frac{1}{d_{110}} : \frac{1}{d_{111}} = 1 : \sqrt{2} : \sqrt{3}$$

2. For a face-centred cubic crystal

$$\frac{1}{d_{100}} : \frac{1}{d_{110}} : \frac{1}{d_{111}} = 1 : \sqrt{2} : \frac{\sqrt{3}}{2}$$

3. For a body-centred cubic system

$$\frac{1}{d_{100}} : \frac{1}{d_{110}} : \frac{1}{d_{111}} = 1 : \frac{\sqrt{2}}{2} : \sqrt{3}$$

Let from first order intensity, reflection for planes (1, 0, 0), (1, 1, 0) and (1, 1, 1) the glancing angles be $\theta_1$, $\theta_2$, and $\theta_3$ respectively.

For the same value of $\lambda$, we have

$$2d_{100} \sin \theta_1 = 2d_{110} \sin \theta_2 = 2d_{111} \sin \theta_3 = 1 \times \lambda$$

$$\frac{1}{d_{100}} : \frac{1}{d_{110}} : \frac{1}{d_{111}} :: \sin \theta_1 : \sin \theta_2 : \sin \theta_3$$

In case of NaCl crystal

$$\theta_1 = 5.9°$$
$$\theta_2 = 8.4°$$
$$\theta_3 = 5.2°$$

$$\frac{1}{d_{100}} : \frac{1}{d_{110}} : \frac{1}{d_{111}} = \sin 5.9° : \sin 8.4° : \sin 5.2°$$

$$\frac{1}{d_{100}} : \frac{1}{d_{110}} : \frac{1}{d_{111}} = 0.1027 : 0.1461 : 0.0906$$

$$= 1 : \frac{0.1461}{0.1027} : \frac{0.0906}{0.1027}$$

$$= 1 : 1.422 : 0.8821$$

$$\frac{1}{d_{100}} : \frac{1}{d_{110}} : \frac{1}{d_{111}} :: 1 : \sqrt{2} : \frac{\sqrt{3}}{2}$$

This shows that crystal of NaCl has face-centred cubic lattice.

## 7.7 SEPARATION BETWEEN LATTICE PLANES

The distance between successive lattice planes in cubic, tetragonal and other rhombic crystal systems in which the edges are mutually at right angle to each other can be calculated.

Two-dimensional crystal lattice is shown in Figure 7.7. The origin of coordinates is represented by $O$. Consider a plane $LM$ which makes intercepts $2a$ and $3b$ on $X$ and $Y$ axes, respectively. The entire lattice may be considered as consisting of lattice planes which are parallel to $LM$ such that all the lattice points are covered by these planes. Draw $ON$ perpendicular from $O$ on $LM$.

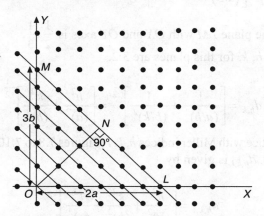

**FIGURE 7.7** Two-dimensional crystal lattice.

The angles $LON$ and $LOM$ are similar, therefore

$$\frac{ON}{OL} = \frac{OM}{LM}$$

Therefore,

$$ON = \frac{OM}{LM} \times OL$$

$$ON = \frac{3b \times 2a}{\sqrt{(2a)^2 + (3b)^2}}$$

$$ON = \frac{1}{\sqrt{\left(\frac{1}{3b}\right)^2 + \left(\frac{1}{2a}\right)^2}}$$

$$ON = \left[\frac{1}{(2a)^2} + \frac{1}{(3b)^2}\right]^{-\frac{1}{2}}$$

It is clear from Figure 7.7, that there are five more planes between $O$ and $LM$. In all, there are six planes between $O$ and $LM$.

In other words, between $O$ and $LM$ there are six planes and the plane $AB$ is to 6th plane from $O$.

The separation between successive lattice planes is given by

$$d = \frac{ON}{6} = \frac{1}{6}\left[\frac{1}{(2a)^2} + \frac{1}{(3b)^2}\right]^{-\frac{1}{2}} = \left[\frac{36}{(2a)^2} + \frac{36}{(3b)^2}\right]^{-\frac{1}{2}}$$

$$d = \left[\frac{1}{(a/3)^2} + \frac{1}{(b/2)^2}\right]^{-\frac{1}{2}}$$

The separation ratio of the plane $LM$ with $OX$ and $OY$ axes is $\frac{1}{2} : \frac{1}{3}$.

Thus, Miller indices $h, k$, for this planes are 3:2.

$$d_{h,k} = \left[\frac{1}{(a/h)^2} + \frac{1}{(b/k)^2}\right]^{-\frac{1}{2}} = \left[\frac{h^2}{(a)^2} + \frac{k^2}{(b)^2}\right]^{-\frac{1}{2}}$$

For three-dimensional lattice with Miller indices $h, k, l$ (Figures 7.8 to 7.10), the separation between the successive planes, i.e. $d_{h,k,l}$ is given by

$$d_{h,k,l} = \left[\frac{h^2}{(a)^2} + \frac{k^2}{(b)^2} + \frac{l^2}{(c)^2}\right]^{-\frac{1}{2}}$$

In case of cubic crystal $a = b = c$

$$d_{h,k,l} = \frac{1}{\sqrt{h^2 + k^2 + l^2}}$$

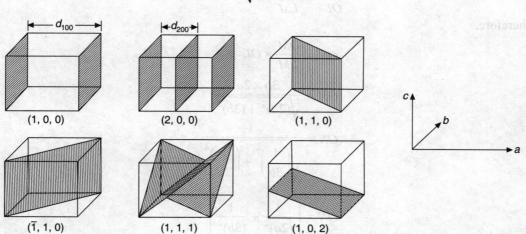

**FIGURE 7.8** Showing various planes and spacing in crystals.

X-Ray Crystallography    **257**

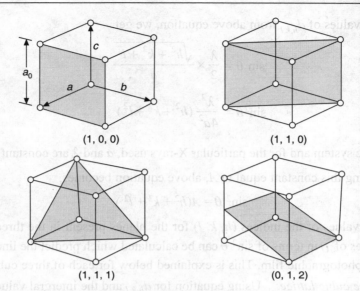

**FIGURE 7.9** Several atomic planes and their $d$-spacings in a simple cubic crystal.

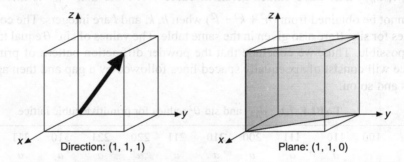

**FIGURE 7.10** Indexing of planes and directions in a cubic crystal.

## 7.8 X-RAY DIFFRACTION PATTERNS OF CUBIC LATTICES

Cubic system is the simplest of all the systems because all lengths of the unit cell are equal and each angle is equal to 90°. For any set of planes with Miller indices ($h$, $k$, $l$), the interplanar distances can be calculated using the expressions

$$d_{h,k,l} = \frac{a}{\sqrt{h^2 + k^2 + l^2}}$$

Also, according to Bragg's equation, for the glancing reflection, i.e. for the first order diffraction,

$$2 d_{h,k,l} \sin \theta = \lambda$$

$$\sin \theta = \frac{\lambda}{2 d_{h,k,l}}$$

Substituting the values of $d_{h,k,l}$ from above equation, we get

$$\sin\theta = \frac{\lambda}{2} \times \frac{\sqrt{h^2 + k^2 + l^2}}{a}$$

$$\sin^2\theta = \frac{\lambda^2}{4a^2}(h^2 + k^2 + l^2)$$

For a given cubic system and for the particular X-rays used, $a$ and $\lambda$ are constant and hence $\frac{\lambda^2}{4a^2}$ is constant. Putting this constant equal to $A$, above equation becomes,

$$\sin^2\theta = A(h^2 + k^2 + l^2)$$

Substituting the values of the indices ($h$, $k$, $l$) for the planes present in the three types of cubic lattices, the values of $\theta$ in terms of $\sin^2\theta$ can be calculated which predict the lines that would be obtained on the photographic film. This is explained below for each of three cubic lattices.

1. *Primitive cubic lattice:* Using equation for $d_{h,k,l}$ and the intergral value for the indices $h$, $k$, $l$, $d_{h,k,l}$ may have values given in Table 7.1. Note that $\frac{a}{\sqrt{7}}$ is missing because 7 cannot be obtained from $(h^2 + k^2 + l^2)$ when $h$, $k$, and $l$ are integers. The corresponding values for $\sin^2\theta$ are also given in the same table. The values of $\sin^2\theta$ equal to $7A$ are also not possible. Thus, we conclude that the powder diffraction pattern of primitive cubic lattice will consist of six equally spaced lines followed by a gap and then again a set of lines and so on.

**TABLE 7.1** $d_{h,k,l}$ and $\sin\theta^2$ values for primitive cubic lattice

| hkl | 100 | 110 | 111 | 200 | 210 | 211 | 220 | 221 | 310 | 311 | 222 | 320 |
|---|---|---|---|---|---|---|---|---|---|---|---|---|
| $d_{h,k,l}$ | $a$ | $\frac{a}{\sqrt{2}}$ | $\frac{a}{\sqrt{3}}$ | $\frac{a}{\sqrt{4}}$ | $\frac{a}{\sqrt{5}}$ | $\frac{a}{\sqrt{6}}$ | $\frac{a}{\sqrt{8}}$ | $\frac{a}{\sqrt{9}}$ | $\frac{a}{\sqrt{10}}$ | $\frac{a}{\sqrt{11}}$ | $\frac{a}{\sqrt{12}}$ | $\frac{a}{\sqrt{13}}$ |
| $\sin\theta^2$ | $A$ | $2A$ | $3A$ | $4A$ | $5A$ | $6A$ | $8A$ | $9A$ | $10A$ | $11A$ | $12A$ | $13A$ |

2. *Face-centred cubic lattice:* For this type of lattice, it is found that reflections in phase take place only from planes for which $hkl$ values are either all odd or all even. Carrying out the calculations as above, the values of $d_{h,k,l}$ and $\sin^2\theta$ are given in Table 7.2. Thus lines will be observed at the angles shown in Table 7.2.

**TABLE 7.2** $d_{h,k,l}$ and $\sin\theta^2$ values for face-centred cubic lattice

| hkl | 100 | 110 | 111 | 200 | 210 | 211 | 220 | 221 | 310 | 311 | 222 | 320 |
|---|---|---|---|---|---|---|---|---|---|---|---|---|
| $d_{h,k,l}$ | | | $\frac{a}{\sqrt{3}}$ | $\frac{a}{\sqrt{4}}$ | | | $\frac{a}{\sqrt{8}}$ | | | $\frac{a}{\sqrt{11}}$ | $\frac{a}{\sqrt{12}}$ | |
| $\sin\theta^2$ | | | $3A$ | $4A$ | | | $8A$ | | | $11A$ | $12A$ | |

3. *Body-centred cubic crystal:* For this type of lattice it is observed that reflections are not observed from the planes for which $(h^2 + k^2 + l^2)$ is odd. Carrying out the calculations as

before, the values of $d_{h,k,l}$ and $\sin^2\theta$ are given in Table 7.3. Thus, lines will be observed at the angles shown in Table 7.3. It may be noted that all of them will be equally spaced.

**TABLE 7.3** $d_{h,k,l}$ and $\sin\theta^2$ values for body-centred cubic lattice

| hkl | 100 | 110 | 111 | 200 | 210 | 211 | 220 | 221 | 310 | 311 | 222 | 320 |
|---|---|---|---|---|---|---|---|---|---|---|---|---|
| $d_{h,k,l}$ | | $\dfrac{a}{\sqrt{2}}$ | | $\dfrac{a}{\sqrt{4}}$ | | $\dfrac{a}{\sqrt{6}}$ | $\dfrac{a}{\sqrt{8}}$ | | $\dfrac{a}{\sqrt{10}}$ | | $\dfrac{a}{\sqrt{12}}$ | |
| $\sin\theta^2$ | | $2A$ | | $4A$ | | $6A$ | $8A$ | | $10A$ | | $12A$ | |

The diffraction pattern of the three types of cubic lattices, as explained above may be represented together as shown in Figure 7.11.

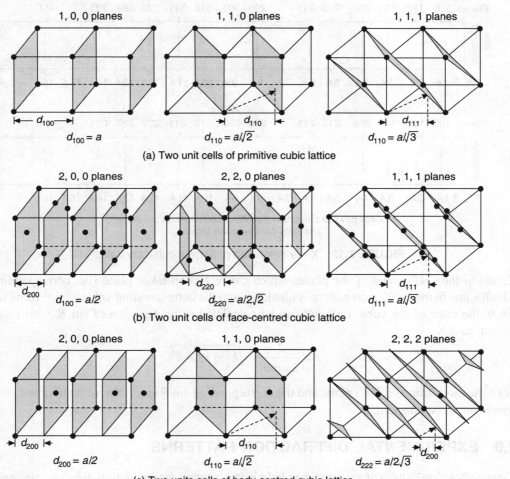

(a) Two unit cells of primitive cubic lattice

(b) Two unit cells of face-centred cubic lattice

(c) Two units cells of body-centred cubic lattice

Planes passing through lattice points of (a) simple (b) face-centred and (c) body-centred cubic lattices and their interplanar distances

**FIGURE 7.11** Different planes in a cubic crystal passing through lattice points.

Thus, the types of cubic lattice can be determined by observing the X-ray diffraction pattern produced from the sample. The missing lines extend a great help in ascertaining the type of lattice (Figure 7.12).

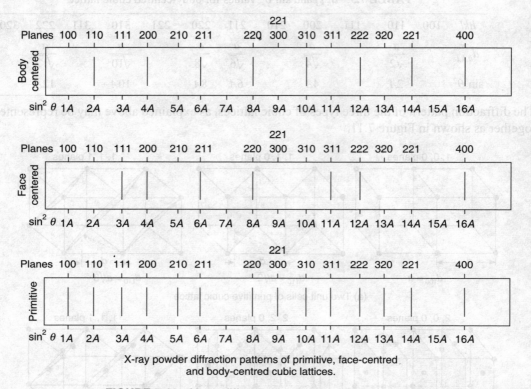

X-ray powder diffraction patterns of primitive, face-centred and body-centred cubic lattices.

**FIGURE 7.12** X-ray diffraction pattern of cubic crystals.

Knowing the $(hkl)$ value of the planes which the reflection takes place (i.e. corresponding to which a line in the diffraction pattern is obtained) and the corresponding value of $\sin^2 \theta$ and hence $\sin \theta$, the edge of the cubic unit cell can be calculated using equation of $\sin \theta$, which can be written as

$$a = \frac{\lambda}{2 \sin \theta} \times \sqrt{h^2 + k^2 + l^2}$$

For reflection from and $(hkl)$ plane and the corresponding $\sin \theta$ value, the value of $a$ will always comes out to be constant.

## 7.9 EXPERIMENTAL DIFFRACTION PATTERNS

X-ray diffraction pattern of any specimen help in its structure determination, for example, particle size determination and cubic structure determination. Exact particle size is determined by Scherrer equation which is given as

$$d = \frac{0.9\lambda}{\beta \cdot \cos\theta}$$

where $d$ is the mean diameter of the nanoparticles, $\lambda$ is the wavelength of X-ray radiation source, $\beta$ is the angular full width at half maximum of the X-ray diffraction peak at the diffraction angle $\theta$.

Figure 7.13 show X-ray diffraction patterns of cubic SiC. Geometry or morphology of the crystal is determined by indexing the different diffraction pattern or assigning Miller indices to different peaks.

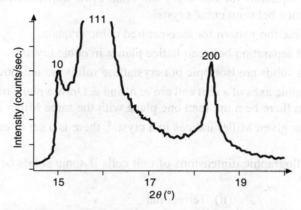

**FIGURE 7.13** X-ray diffraction patterns of cubic SiC.

## 7.10 FIRST DISCOVERY OF X-RAY DIFFRACTION

Friedrich and Knipping first observed X-ray diffraction in 1912 after they got a hint from their research advisor, Max von Laue, at the University of Munich. Von Laue's description of X-ray interference further simplified Bragg's Law. In their experiments, Braggs used crystals in the reflection geometry to analyze the intensity and wavelengths of X-rays generated by different materials. Later instrument used for characterizing X-ray spectra was named as Bragg spectrometer.

Laue knew that X-rays had wavelengths in the order of 1.0 Å. At that time, Paul Ewald's described optical theories which describes approximate distance between atoms in a crystal by the same length. Laue stated that X-rays would diffract, in a similar manner as that of diffraction of light from small periodic scratches drawn on a solid surface. In 1918, Ewald gave a theory, in a similar manner to his earlier optical theory, quantitatively explaining the fundamental physical interactions associated with XRD.

## 7.11 DETECTION OF DIAMONDS WITH THE HELP OF X-RAY DIFFRACTION STUDY

For the detection of presence of diamond in a crystal, we have to take X-ray diffraction pattern of the crystal. On exposing the crystal with X-rays of wavelength ($\lambda$) 1.54Å, if one finds peaks on

X-ray diffraction pattern at $\theta$ values that correspond to each of the $d$-spacings that characterize diamond, i.e. 1.075Å, 1.261Å, and 2.06Å. Presence of peaks at 1.075Å, 1.261Å, and 2.06 Å confirm the presence of diamond in the cubic crystal.

## EXERCISES

1. Define Hauy's law of rationality of indices in detail.
2. Write a short note on Miller indices. How they are related to structure of cubic crystals?
3. Derive Bragg's equation for cubic crystal. Also draw experimental setup to determine interplanar distance between cubic crystals.
4. Draw X-ray diffraction pattern for face-centred cubic crystal.
5. Discuss in detail separation between lattice planes in cubic crystal.
6. Why amorphous solids are isotropic but crystalline solids are anisotropic?
7. The crystallographic axes of a unit cell are $a$, $b$, and $c$. Draw a plane of which Miller indices are (2, 2, 1). Can there be more than one plane with the same Miller indices?
8. Show that for the given Miller indices in a crystal, there is a set of equally spaced parallel planes.
9. What are crystallographic dimensions of unit cells if some solids belong to the following crystal systems?
    (i) Triclinic        (ii) Tetragonal
10. What types of $hkl$ planes give X-ray reflection in phase for face-centred cubic lattice?
11. Both NaCl and KCl have similar structures, yet their X-ray diffraction patterns are remarkably different. Why?
12. How is a unit plane in a crystal is decided? Define Hauy's law of rational indices.
13. How are Miller indices of a plane or a face determined? What are the Miller indices of the faces of a simple cube?
14. Show by calculation what types of diffraction patterns will be obtained for three types of cubic lattices? What type of lattices NaCl, KCl and CsCl have?
15. Justify why the lines in the X-ray diffraction pattern of a body-centred cubic lattice are equally spaced.

## SUGGESTED READINGS

Azaroff, L.V., *Introduction to Solids*, Tata McGraw-Hill, New Delhi, 1960.

Berg, R.J. and G.J. Dienes, *Physical Chemistry of Solids*, Academic Press, 1992.

Chakrabarty, D.K., *Solid State Chemistry*, New Age International, New Delhi, 1996.

Cheetham, A.K. and P. Day, *Solid State Chemistry Compound*, Oxford University Press, Oxford, 1992.

Glusker, J.P. and K.N. Trueblood, *Crystal Structure Analysis*, Oxford University Press, 1985.

Harrison, W.A., *Electronic Structure and Properties of Solids: The Physics of the Chemical Bond*, W.H. Freeman, 1980.

Hoffmann, R., *Solids and Surfaces: A Chemist View of Bonding in Extended Structures*, VCH Publishers, 1988.

Kachhava, C.M., *Solid State Physics*, Tata McGraw-Hill, New Delhi, 1990.

Kroger, F.A., *The Chemistry of Imperfect Crystals*, North Holland, 1973.

Nelson, D.L. and T.F. George, *Chemistry of High Temperature Superconductor* II, Oxford University Press, Oxford, 1999.

Rao, C.N.R. and J. Gopalakrishnan, *New Directions in Solid State Chemistry*, 2nd ed., Cambridge University Press, 1997.

Smart, L. and E. Moore, *Solid State Chemistry: An Introduction*, Chapman % Hall, 1995.

Stout, G.H. and L.H. Jensen, *X-Ray Structure Determination*, Macmillan Press, London, 1968.

Titley, R.J.D., *Crystals and Crystal Structures*, Wiley, 2006.

West, A.R., *Solid State Chemistry and its Applications*, John Wiley, 1987.

West, A.R., *Basic Solid State Chemistry*, 2nd ed., John Wiley, 1999.

Ziman, J., *Theory of Solids*, Cambridge University Press, 1964.

# 8
# Solid State Chemistry

## 8.1 INTRODUCTION

Upon heating a solid material, many types of reactions can takes place. So, the classification of solid-state reaction is an important and interesting concept. If the main purpose of the reaction is the preparation of a new homogeneous solid phase, then, it may not be necessary that the reaction is solid–solid type only. By the reaction of a solid with liquid or any gas many useful solids can be prepared. So, the growth of crystals may be from a liquid or vapour phase.

So, we may classify the solid state reactions, on the bases of physical state of the reactants and the products involved in the reactions as shown below:

| | | | |
|---|---|---|---|
| Type I | A(s) | → | B(s) + C(g) |
| Type II | A(s) + B(s) | → | C(s) |
| Type III | A(s) + B(s) | → | C(s) + D(g) |
| Type IV | A(s) + B(g) | → | C(s) |
| Type V | A(s) + B(g) | → | C(s) + D(g) |
| Type VI | A(s) | → | B(g) + C(g) |

Most of these reactions take place at high temperatures.

The most widely used method for the preparation of a solid is the direct reaction of mixture of solid starting materials, i.e. the materials in solid phase. Usually, solids do not react together at room temperature over normal time scales and it is necessary to heat them to much higher temperatures often 1000–1500°C. This shows that both thermodynamic and kinetic considerations are important in solid state reactions. From the above discussion, we can say that the solid state reactions take place with great difficulty and only at high temperatures it is because of,

1. differences in the structure between reactants and products, and
2. the large amount of reorganization that involved in forming the product.

For the solid state reactions, essential requirement is, bond must be broken and reformed and atoms must move towards each other for recombination to form desired product. The reacting ions are normally trapped on their appropriate lattice sites in the crystal and it is difficult to go into empty sites. Thus, only at very high temperatures such ions have sufficient thermal energy to enable them to jump out of their normal lattice sites and diffuse through the crystals.

## 8.2  WAGNER REACTION MECHANISM OF SOLIDS

To explain the mechanism of solid state reaction, let us take an example of the reaction of MgO and $Al_2O_3$ to form $MgAl_2O_4$ spinel. Generally, $MgAl_2O_4$ has a structure, which shows some similarities and differences to the structures of both MgO and $Al_2O_3$. Both MgO and $MgAl_2O_4$ spinal have a cubic closed packed oxides ions. On the other hand, in $Al_2O_3$ the oxide ions are in distorted hexagonal close packing. On other side, $Al^{3+}$ occupy octahedral sites in both $Al_2O_3$ and $MgAl_2O_4$ whereas the $Mg^{2+}$ ions are octahedral in MgO, but tetrahedral in $MgAl_2O_4$ spinal. Structural consideration is very important for solid state reactions.

The mechanism of reaction between MgO and $Al_2O_3$ involving the counter diffusion of $Mg^{2+}$ and $Al^{3+}$ ions through the product layer followed by further reaction at two reactant–product interfaces. This mechanism is known as *Wagner mechanism*. The detailed mechanism is shown in Figure 8.1. Initially, two crystals MgO and $Al_2O_3$ are in intimate contact across one shared face.

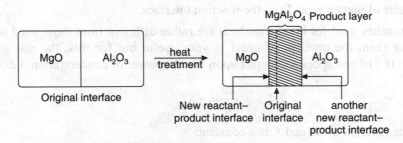

**FIGURE 8.1**  Wagner reaction mechanism.

But, after an appropriate heat treatment, the crystals have partially reacted to form a layer of $MgAl_2O_4$ at the interface. So, the first stage of the reaction is the formation of $MgAl_2O_4$ nucleic and this process is known as *nucleation*. In order to further reaction to occur and the $MgAl_2O_4$ layer to grow thicker, counter diffusion of $Mg^{2+}$ and $Al^{3+}$ ions must pass or cross right through the existing $MgAl_2O_4$ product layer to the new reaction interfaces. At this stage, there are two reaction interfaces: one that between MgO and $MgAl_2O_4$ and another between $MgAl_2O_4$ and $Al_2O_3$ layers.

In order to maintain charge balance, for every three $Mg^{2+}$ ions, which diffuse to the right-hand interface, two $Al^{3+}$ ions must diffuse to the left-hand interface. The reactions that occur at these two interfaces may be written as:

1. Interface $MgO/MgAl_2O_4$

$$2Al^{3+} + 3Mg^{2+} + 4MgO \longrightarrow MgAl_2O_4$$

2. Interface $MgAl_2O_4/Al_2O_3$

$$3Mg^{2+} + 2Al^{3+} + 4Al_2O_3 \longrightarrow 3MgAl_2O_4$$

*Overall reaction:*

$$4MgO + 4Al_2O_3 \longrightarrow 4MgAl_2O_4$$

So, it can be seen that reactions (2) gives three times as much spinal product as reaction (1); and hence right-hand interface should grow or move at three times faster rate than that of left-hand interface. This mechanism has been experimentally tested and found to be satisfactory.

## 8.3 KINETICS OF SOLID STATE REACTIONS

Many factors are usually involved in a particular solid-state reaction. These are mainly:

1. the area of contact between the reacting solids and hence their surface areas,
2. the rate of nucleation of the product phase, and
3. the rates of diffusion of ions through the various phases and specially, through the product phase.

So the analysis of kinetic data may be difficult. In kinetics studies of the reaction rates, it is desired to know, which is the slow and rate determining step in a reaction, so there are at least three possibilities for this:

1. transport of matter to the reaction interface,
2. reaction at interface, and
3. transfer of matter away from the reaction interface.

But, the approaches used for kinetic analysis are rather different from those used for gas phase reactions. For them the concept of order is very useful but for this, the rate of change of concentration ($C$) of one species depends upon the $n$th power of concentration, i.e.

$$\frac{dc}{dt} = -KC^n$$

where, $n$ is the reaction order and $K$ is a constant.

From the value of $n$, important insight into the reaction mechanism may be obtained such as number of molecules involved in a particular reaction step.

For most of the solid state reactions, it is usually incorrect and misleading to think in terms of reaction order, since the reactions do not involve molecules. It is sometimes, found that, $n$ is not a simple integer but may be fractional.

## 8.4 THERMAL DECOMPOSITION REACTION

Reactions of a solid, which gives one or more solid or gaseous products are known as *thermal decomposition reaction*. Reaction of types I and VI belongs to thermal decomposition reaction as explained above. Let us first consider the reaction of type I, i.e.

$$A(s) \longrightarrow B(s) + C(g)$$

In this reaction, decomposition of solid A, give a new solid B and a gaseous product C. Upon heating A, a stage is achieved when the nuclei of B will be formed and embedded in A. If, once the nuclei is formed and after this the rapid growth or nucleation of product layer occur, then it will be an autocatalytic in nature. If the activation energy of the reaction is less than that of formation of nuclei, the reaction will proceed only at few nuclei and in this case, many dehydration reactions take place. On the other hand, if the activation energy of nucleation and growth are comparable, then there will be practically no induction period. These cases are shown, by the following plots, between the fraction of A decomposed ($\alpha$) versus time (Figure 8.2).

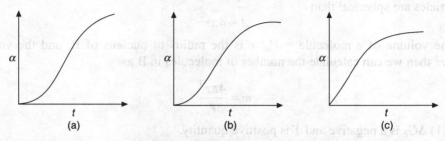

**FIGURE 8.2** Plots between fraction of A decomposed versus time.

But, in an another case, in which initially a number of nuclei are formed and after which growth proceeds rapidly all over.

Figure (8.3) shows two stages: Such that $OX$ represents the formation of few nuclei slowly and after there formation, the rapid growth of the nucleation is seen along the $XY$ portion of the curve.

So, we can see that $\alpha$ versus $t$ plots can be explained in terms of nucleation and growth of the newly formed solid phase. It should be realized that for reaction type I, nucleation can occur only at the surface, so that the gaseous product can escape.

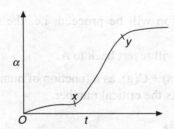

**FIGURE 8.3** After a slow nucleation there is faster decomposion ($XY$).

## 8.5 FREE ENERGY OF NUCLEATION

In type I reaction, on heating A, nuclei of B starts forming and embedded in A. The structure of B is generally different from the structure of A, so, there will arise some strain energy. The size of B plays an important role in its formation. As the growth of crystal B increases, the stability increases. So, there exists a critical size of B at which the thermodynamic stability of B will overcome the strain energy created by its formation. Below this size, nuclei of B will revert back to A. If the nuclei of B are bigger than the critical size, transformation of A to B will continue.

Free energy of the reaction for this reaction is given by:

$$\Delta G = m \, \Delta G_b + aY \tag{8.1}$$

where
- $m$ = number of molecules in the nucleus of B,
- $\Delta G_b$ = bulk free energy of the reaction,
- $a$ = shape factor,
- $Y$ = strain energy per unit area of the interface between A and B.

If the particles are spherical than

$$A = 4\pi r^2$$

and, if the volume of a molecule = $V_m$, $r$ is the radius of nucleus of B, and the volume of B = $4/3 \pi r^3$ then we can calculate the number of molecules in B as

$$m = \frac{4\pi r^3}{3V_m}$$

In Eq. (8.1) $\Delta G_b$ is a negative and $Y$ is positive quantity.

So, the growth of B makes opposite contribution to the free energy $\Delta G$ of the reaction.

At the critical mass, i.e. the critical value of $m$ ($m^*$), free energy change, $\Delta G$ will be zero, i.e. at this, the condition the formation of nuclei of B must be comparable with their inversion back to A.

The value of $m^*$ is given by Jacobs and Thomkins equation as given below:

$$m^* = -\left(\frac{2Y}{3\Delta G_b}\right)^3 36\pi V_m^2 \qquad (8.2)$$

when $m > m^*$,

$\dfrac{\partial \Delta G}{\partial m}$ is negative and the reaction will be proceed, i.e. the nucleus of B will grow and if $m < m^*$, $\dfrac{\partial \Delta G}{\partial m}$ is positive and B will revert back to A.

$\Delta G$ of the reaction A(s) → B(s) + C(g), as a function of number of molecules per nuclei of B as shown in Figure 8.4. Here $m^*$ is the critical number.

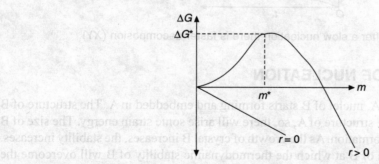

**FIGURE 8.4** $\Delta G$ of the reaction A(s) → B(s) + C(g), as a function of number of molecules/nuclei.

## 8.6 LAWS GOVERNING NUCLEATION

Nuclei of the product (B), will be formed in the reactant solid matrix (A) at different points. If the fluctuations of the atoms are high due to the motion atoms, the molecules possess high energy and if this energy is sufficient to overcome the energy of activation of nucleation then more nuclei of the products will be formed. So, we can say that due to fluctuations of the atoms, nucleation proceeds. So, generally nucleation occurs at some lattice defects mostly at point defects like vacancies, interstisitial or dislocation.

If the nucleation requires decomposition of a single molecule, it can be shown that

$$\frac{dN}{dT} = k_1 N_0 \exp(-K_1 t) \tag{8.3}$$

where
 $N$ = number of nuclei present at the time $t$,
 $N_0$ = number of potential nucleation sites, and
 $K_1$ = rate constant

If the energy of activation of nucleation is large, $K_1$ is small and equation becomes

$$\frac{dn}{dt} = k_1 N_0 \tag{8.4}$$

Under these conditions, the number of nuclei increases linearly with time.

But when $K_1$ is very small, i.e. $N = N_0$
At this situation, when nucleation is instantaneous and if the nucleation involves stepwise decomposition of the two or three molecules, the rate follows a power law such as,

$$\frac{dn}{dt} = A t^{\beta-1} \tag{8.5}$$

where $A$ is a constant and $\beta$ is equal to two or three.

## 8.7 GROWTH OF NUCLEI

If the nucleation follows power law and growth is linear, it can be shown that,

$$\alpha = C\, t^n \tag{8.6}$$

where, $C$ and $n$ are constants.

But for reaction of type I, the pressure $P$ developed in a closed system is proportional to $\alpha$, hence

$$P \propto \alpha$$

i.e.
$$P = Y \alpha \tag{8.7}$$

where $Y$ is proportionality constant
Putting the value of Eq. (8.6) in Eq. (8.7), we get

$$P = Y C\, t^n$$

i.e.
$$P = D\, t^n \tag{8.8}$$

where, $D$ is a new constant.

This power law equation holds good for many reactions of type I, such as, decomposition of barium and calcium azides, aged mercury fulminate and aged silver oxalate. It has been found that as the temperature of decomposition is varied as the value of $n$ changes. The growth of nuclei is also affected by the size of the crystals. This can be explained by assuming that the growth is much slower, if the crystals are small. In order to explain the observed high values of $n$ (between 11.2 and 22.8) in case of the decomposition of mercury fulminate, Garner and Hailes assumed that the nuclei are linear branching chains. This assumption led to the expression.

$$\alpha = C\, e^{kt} \tag{8.9}$$

where $k$ is branching constant or coefficient.

The pressure developed during the decomposition inside a closed system is

$$P = D\, e^{kt} \tag{8.10}$$

If the branched chains interfere with each other, the fraction decomposed can be expressed as

$$\log \frac{\alpha}{1-\alpha} = kt + C \tag{8.11}$$

which gives,

$$\log \frac{P}{P_f - P} = kt + C \tag{8.12}$$

where, $P_f$ is final pressure of system after complete decomposition of the solid and $C$ is constant.

Several other equation were developed assuming different modes of nucleation and growth. Towards the end of the reaction, the difference of the molecular volumes of the reactant and the product solids may being about to collapse of the interference between them. This will give blocks of the reactant in which there may not be any product nuclei. If each molecule in such blocks has some probability of decomposition, the rate of the reaction will be proportional to the amount of the reactant left. Hence,

$$\frac{d\alpha}{dt} = K(1-\alpha) \tag{8.13}$$

which gives

$$\log \frac{1}{1-\alpha} = Kt \tag{8.14}$$

## 8.8 ELECTRONIC PROPERTIES AND BAND THEORY OF METALS, INSULATORS AND SEMICONDUCTORS

At a first glance, the main difference between metals, semiconductors and insulators is the magnitudes of their conductivity, i.e. the ability to conduct electricity. Their ability is measured by a physical term, *conductivity*. So, the difference between them is shown as:

1. Metal conduct electricity very easily, i.e. their magnitude of conductivity is $(\sigma) \approx 10^4$ to $10^6$ ohm$^{-1}$cm$^{-1}$.
2. Insulator conduct electricity very poorly or mostly not at all, i.e. their magnitude of conductivity is $(\sigma) \leq 10^{-15}$ ohm$^{-1}$cm$^{-1}$ and,
3. The semiconductor that conduct electricity mildly, i.e. their magnitude of conductivity lie in between $(\sigma) \approx 10^{-5}$ to $10^3$ ohm$^{-1}$cm$^{-1}$.

On the other hand, we can distinguish between them, on the basis of their fundamental difference between the mechanism of conduction, which is explained as:

Conductivity of most semiconductors and insulators increases rapidly with increasing temperature, whereas the conductivity of metal shows a gradual or slight decrease with rise in temperature, which is explained by the conductivity relationship, i.e.

$$\sigma = n e \mu$$

where $n$ is the number of charge carriers, $e$ is the electronic charge and $\mu$ is the mobility of charge carriers.

As we know that the value of $e$ is constant and temperature independent and the mobility term $\mu$ is similar in most materials in that usually decreases slightly with increasing temperature due to collisions between the moving electrons and photons, i.e. lattice vibrations. So, the main source of difference between the metal, insulator and semiconductor lies in the value of $n$ and its temperature dependence, which is given below:

1. For *metals*, $n$ is large and essentially unchanged with change in temperature, so the only variable is $\mu$. Since $\mu$ decrease slightly with the increase in temperature, so $\sigma$ also decrease.
2. For *semiconductors* and *insulators*, $n$ usually increases exponentially with the increase in temperature. So, the effect of this dramatic increase in the value of $n$ is more than the effect of small decrease in $\mu$. Hence $\sigma$ increases rapidly with the increases in temperature.
3. Insulators are extreme examples of semiconductor in which, $n$ is very small at normal temperatures. Thus, some insulator becomes semiconductor at high temperatures, where $n$ becomes appreciably high.

## 8.9 ELECTRONIC STRUCTURE OF SOLIDS—BAND THEORY

The bonding in metals can also be explained on the basis of molecular orbital theory; also called *Band theory*. This was developed by Helix Block in 1928. We have learnt that in the combination of atomic orbital belonging to two atoms, two molecular orbitals are formed. Out of these two, one is bonding molecular orbital while the other is antibonding molecular orbital. But in case, atomic orbitals belonging to three atoms are involved, then three molecular orbitals are formed. In addition to bonding and antibonding molecular orbitals, there is another molecular orbital called nonbonding molecular orbital. The nonbonding molecular orbital lies between bonding and antibonding molecular orbitals and has exactly the same energy as the participating atomic orbitals. Let us illustrate by taking an example of lithium ($1s^2, 2s^1$) atom (Figures 8.5 and 8.6).

But the nonbonding molecular orbital has no contribution towards the energy state of the molecule. Now let us extend this theory and consider the combination of four atomic orbitals belonging to four atoms of lithium. As expected, four molecular orbitals will emerge out of which two will be bonding and the other two will be antibonding.

From the above picture, it is quite evident that the lower bonding molecular orbital is slightly more bonding than other. The same is also true for the antibonding molecular orbitals also.

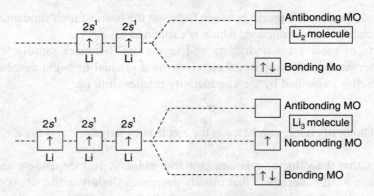

**FIGURE 8.5** Band theory in case of lithium atom.

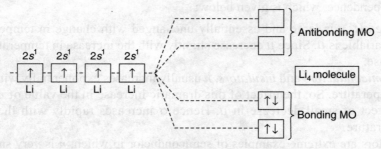

**FIGURE 8.6** Four atomic orbitals of lithium forming four molecular orbitals.

In other words, the different bonding and antibonding molecular orbitals are not at the same energy state. We are aware of the fact that a large number of metal atoms are closely packed in space. Thus, in case $n$ atomic orbitals of the atoms take part in the bond formation, the $n$ molecular orbitals will result. They will be so close in energy that they begin to blur and will not present a distinct view. A continuous energy bands involving a number of such molecular orbitals will be outcome as given below (Figure 8.7).

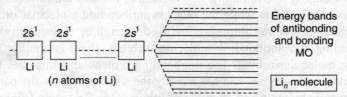

**FIGURE 8.7** Energy band of $n$ lithium atoms.

The molecular orbitals can no longer be treated as discrete energy levels, but as energy bands. Each energy level in the band can accommodate two electrons with opposite spins. Now each atom of lithium has only one valence electron per atom and two electrons can be accommodated in a molecular orbital. Thus, it is quite obvious that for lithium, only half of the molecular orbitals in the 2s valence band are occupied, while the other half are vacant or empty (Figure 8.8). So, the band which holds the electrons is termed as *valence band* and the other band without any electrons, but responsible for the conductivity, is called *conduction band*.

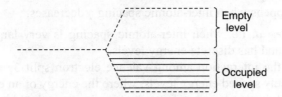

**FIGURE 8.8** Valence band and conduction band.

## 8.10 ENERGY BANDS IN SOLIDS

Electrons of each isolated atom have discrete energy levels. When two similar atoms are brought closer then there is an interaction between the valence energy levels. The atom in a solid are so close to each other that the energy levels produced after splitting, due to interaction between the various atoms, will be so close to each other, that they appear as continuous. These closely spaced energy levels form an energy band. If the energy band contains the number of electrons equal to the number of electrons permitted by Pauli's exclusion principle, then the energy band is said to be *completely filled*. In a completely filled band, there is no free electron for the conduction of current.

## 8.11 FORMATION OF BANDS

Consider a silicon crystal having $N$ atoms. The electronic configuration of silicon ($Z = 14$) is $1s^2$, $2s^2$, $2p^6$, $3s^2$, $3p^2$. The energy levels of electrons in the silicon atom are shown in Figure 8.9.

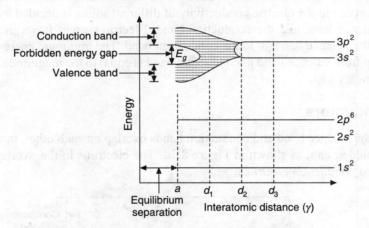

**FIGURE 8.9** The energy levels of electrons in the silicon atom.

The levels are $1s$, $2s$, $2p$ and $3s$ are completely filled and the level $3p$ contains only two electrons, whereas it can accommodate six electrons.

In a silicon crystal, there are $14N$ electrons. For each silicon atom, there are two states in energy level $1s$. So, there are $2N$ states in energy level $1s$, $2N$ states in energy level $2s$, $6N$ energy states in energy level $2p$, $2N$ states in energy level $3s$ and $6N$ states in $3p$ energy levels. In $3p$ energy levels only $2N$ states are filled and $4N$ states are empty.

Let us see what happens when inter-atomic spacing $\gamma$ decreases.

1. When $\gamma = d_3 \gg a$, i.e. when inter-atomic spacing is very large, each atom behaves independently and has discrete energy levels.
2. When $\gamma = d_2$, the interaction among valence electron split $3s$ and $3p$ level into large number of closely spaced energy levels, where the energy of an electron may be slightly less or more than the energy of an electron in an isolated atom. Thus, two bands corresponding to $3s$ and $3p$ states are formed. As the inter-ionic spacing $\gamma$ is further decreased, the energy bands corresponding to $3s$ and $3p$ states spread more and hence energy gap between these bands decreases.
3. When $\gamma = d_1$, i.e. when inter-atomic spacing is further reduced, the $3s$ and $3p$ bands overlap and the energy gap between them disappears. In this case, all $8N$ levels ($2N$ corresponding to $3s$ energy level and $6N$ corresponding to $3p$ energy level) are now continuously distributed. Out of these $8N$ levels, $4N$ levels are filled and $4N$ is empty.
4. When $\gamma = a$, i.e. at equilibrium separation, the filled and unfilled energy levels are separated by an energy gap called *forbidden energy gap*. The energy of forbidden energy gap is denoted by $E_g$. The lower energy band is called *valence band* and the upper unfilled energy band is called *conduction band*.

## 8.12 ENERGY BANDS IN CONDUCTORS, INSULATORS AND SEMICONDUCTORS

Conductors, insulators and semiconductors can be distinguished, on the basis of band theory of solids. The difference in the electric conductivity of different solids is decided by the energy gap between the valence band and the conduction band. Electric current flows in the solid, when electrons from the valence band go to the conduction band. The minimum energy required by an electron to jump from valence band to conduction band is equal to the magnitude of the energy of the forbidden energy gap.

### 8.12.1 Conductors

In conductors, the valence band and conduction bands overlap on each other. In the other words, there is no forbidden gap, as shown in Figure 8.10. The electrons in the overlapping region of energy are called *conduction electrons*.

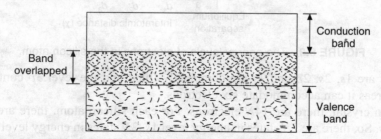

**FIGURE 8.10** Overlap of valence band and conduction band in metals.

As there are a large number of conduction electrons, so the metal is a good conductor. In metals, the higher occupied band, i.e. valence band is only partly full as shown in Figure 8.11.

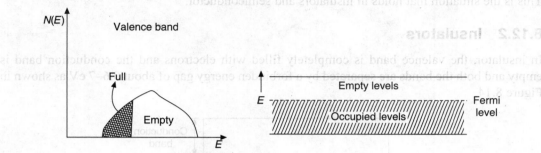

**FIGURE 8.11** Valence band is only partly full.

In this, some energy levels below the Fermi level, ($E_f$), i.e. in occupied level are vacant, whereas some above the $E_f$ are occupied, i.e. there is overlapping of valence band and conduction band is occurred or some electrons can jump from occupied or valence band to conduction band. So due to this, electrons in singly occupied states close to $E_f$ are able to move and are responsible for the high conductivity of metals. Energy band of a metal, e.g. Na is given in Figure 8.12.

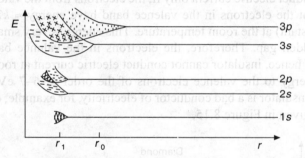

**FIGURE 8.12** Energy band of sodium metal.

In this, both $3s$ and $3p$ bands contain electrons. Overlapping of bands is responsible for the metallic properties of the alkaline earth metals.

Band structure of beryllium is shown in Figure 8.13.

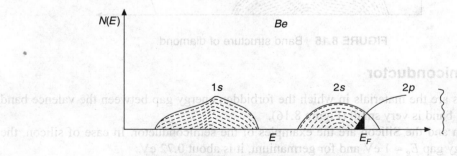

**FIGURE 8.13** Band structure of beryllium.

It has overlapping of 2s, 2p bands, both of which are only partly full. If the 2s and 2p did not overlap then the 2s band would be full, the 2p band empty and beryllium would not be metallic. This is the situation that holds in insulators and semiconductor.

## 8.12.2 Insulators

In insulator, the valence band is completely filled with electrons and the conduction band is empty and both the bands are separated by a forbidden energy gap of about 0.6–7 eV as shown in Figure 8.14.

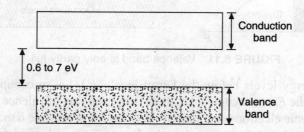

**FIGURE 8.14** Forbidden energy gap in case of insulator.

The insulator can conduct electric current only if, the electrons from the valence band move to the conduction band. But the electrons in the valence band have energy $\approx kT = 0.025$ eV (where $k$ = Boltzmann's constant) at the room temperature. This energy is very small as compared to the energy of the forbidden gap. Therefore, the electrons in the valence band cannot go to the conduction band and hence, insulator cannot conduct electric current at room temerature. It may be noted that the energy to the valence electrons of the order of 6–7 eV cannot be provided conveniently, hence insulator is a bad conductor of electricity, for example, diamond. The Energy band of diamond is given in Figure 8.15.

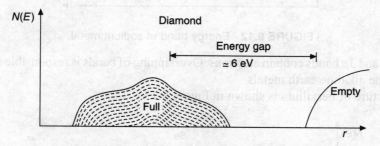

**FIGURE 8.15** Band structure of diamond.

## 8.12.3 Semiconductor

Semiconductors are the materials in which the forbidden energy gap between the valence band and conduction band is very small (Figure 8.16).

Germanium and the Silicon are the examples of the semiconductor. In case of silicon, the forbidden energy gap $E_g \approx 1$ eV and for germanium, it is about 0.72 eV.

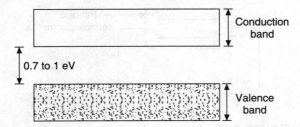

**FIGURE 8.16**  A small energy gap in case of semiconductor.

At 0 K, the electrons in the valence band do not have sufficient energy to jump to the conduction band and hence semiconductor behaves as insulator at 0 K. Even at room temperature, some of the valence electrons have sufficient thermal energy to jump to the conduction band. Hence, semiconductor may conduct at room temperature.

Semiconductor may be divided into two groups:

1. Intrinsic semiconductor and
2. Extrinsic Semiconductors.

## 8.13  INTRINSIC SEMICONDUCTOR

The semiconductor in which the current carrier, holes and electrons are created due to thermal excitation across the energy gap is called an *intrinsic semiconductor*.

In these semiconductors, the valence band and conduction band are separated by forbidden gap $E_g \approx 1$ eV. The valence band is completely filled with electrons while conduction band is empty. At the room temperature, the electrons at the top of the valence band gain thermal energy and jump over the energy gap to reach the conduction band. The electron reaching the conduction band leave equal number of holes in the valence band as shown in Figure 8.17.

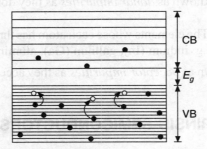

**FIGURE 8.17**  Number of holes in the valence band.

For this, the number of electrons $n$ in the conduction band is governed entirely by:

1. The magnitude of the band gap and
2. Temperature.

Pure silicon is intrinsic semiconductor, when the semiconductor is connected to a battery, a potential difference $V$ is developed across the ends of the semiconductor. The electrons move towards the positive terminal and holes move towards the negative terminal of the battery as shown in Figure 8.18.

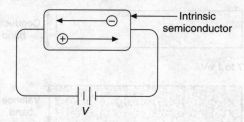

**FIGURE 8.18** A picture of intrinsic semiconductor.

## 8.14 EXTRINSIC SEMICONDUCTORS

Intrinsic semiconductors have thermally generated current carriers (holes and electrons) and hence the conductivity of an intrinsic semiconductor is low.

To increase the conductivity of an intrinsic semiconductor, some suitable impurities can be added in it. Small amount of impurity increases the conductivity of the semiconductor.

*Doping:* The process of adding the impurities in the intrinsic semiconductor is called *doping*.

*Doped or extrinsic semiconductor:* A semiconductor obtained after adding impurity in the intrinsic semiconductor is called *extrinsic* or *doped semiconductor*.

## 8.15 TYPES OF IMPURITIES

There are two types of impurities that can be added in the intrinsic semiconductor to increase its conductivity.

1. *Pantavalent Impurity:* The elements whose, each atom has five valence electrons are called *pentavalent impurities*, e.g. arsenic (As), antimony (Sb), phosphorous (P), etc. These impurities are known as *donar impurities* as they donate extra free electrons to the semiconductor.
2. *Trivalent Impurity:* The elements whose each atom has three valence electrons are called *trivalent impurities*, e.g. indium (In), gallium (Ga), aluminum (Al), boron (B), etc.

These impurities are known as *acceptor impurities* as they accept electrons from the covalent bonds of the semiconductor.

## 8.16 TYPES OF EXTRINSIC SEMICONDUCTORS

Depending upon the type of the impurity added in the intrinsic semiconductor, the extrinsic semiconductor is classified into two categories.

### 8.16.1 p-Type or p-Type Semiconductor

When trivalent impurity is added to pure germanium or silicon crystal, we get extrinsic semiconductor, known as *p-type semiconductor*. Trivalent impurity atom, say indium, has three valence electrons. When an atom of indium is added to silicon crystal, the atom replaces one of the silicon atom and settles in a lattice site. This indium atom forms three covalent bonds with the neighbouring silicon atom. The fourth bond remains incomplete. This covalent bond has a deficiency of one electron. This deficiency of an electron is called *hole* and behaves like a positively charged particle. The indium atom at the lattice site attracts an electron from the neighbouring silicon atom to fill the hole. Now a new hole is created in the atom from which the electron has been attracted to fill the hole as shown in Figure 8.19. In this way, a large number of holes are formed by adding more and more trivalent (indium) atoms in the silicon crystals.

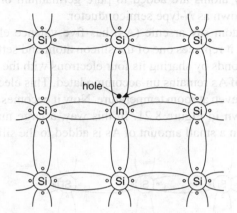

**FIGURE 8.19** A *p*-type semiconductor with a hole.

The crystals of this type has an excess of holes or positive charge carriers and hence known as p-type semiconductors or positive-type semiconductor. Majority charge carriers in p-type semiconductors are holes and the minority carriers are electrons which are thermally generated. Since, each trivalent impurity atom accepts one electron from the neighbouring silicon atom, so it is known as acceptor impurity.

### Energy band diagram of p-type semiconductor

In p-type semiconductor, an electron from one band jumps to fill the hole in the neighbouring incomplete band. As a result of this, another hole is created at the position from where the electron had jumped.

The energy needed by the electron to jump to the hole is provided by the thermal agitation of the crystal. The unfilled energy state created is located just above the valence band of semiconductor. This energy state or level is called *acceptor level* and is shown by dotted line in Figure 8.20.

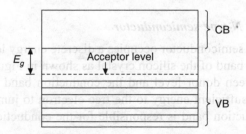

**FIGURE 8.20** Acceptor level in case of p-type semiconductor.

The electrons from the valence band can move easily into the unfilled energy level or acceptor level at room temperature. Consequently, holes are created in the valence band. These holes are mobile so act as current carriers and hence contribute to the electric current.

## 8.16.2 n-Type or n-Type Semiconductor

When pentavalent impurity atoms are added to pure germanium or silicon crystal, we get an extrinsic semiconductor known as *N*-type semiconductor.

Pantavalent impurity atom say arsenic (As) has five valence electrons. When As atom is added to the silicon crystal, it replaces one of the silicon atom and settles in a lattice site. This As atom forms four covalent bonds by sharing its four electrons with the neighbouring silicon atom. The fifth valence electron of As remains un-accommodated. This electron is loosely bound to its atom and is detached easily even at room temperature. Now it becomes a free electron and wanders through the crystal as shown in Figure 8.21. In this way, a large number of free electrons are available in the crystal when a small amount of As is added to the silicon crystal.

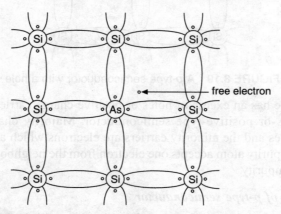

**FIGURE 8.21** An n-type semiconductor with excess electrons.

The crystal of this type has excess number of electrons or negatively charged carriers and hence is known as n-*type semiconductor* or *negative-type semiconductor*. Majority charge carriers in n-type semiconductors are electrons and minority charge carriers are holes, which are thermally generated. Since each pentavalent impurity atom donates one electron to the crystal, so, it is known as *donor impurity*.

### *Energy band diagram in N-type semiconductor*

The 5th electron in n-type semiconductor occupies a discrete energy level known as *donor level* just below the conduction band of the silicon crystal as shown in Figure 8.22.

The energy gap between donor level and the conduction band is very small. Even room temperature provides the sufficient energy to the free electron to jump to the conduction band. This electron in the conduction band is responsible for the conduction of current in the n-type semiconductor.

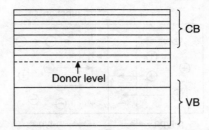

**FIGURE 8.22** Donar level in *n*-type semiconductor.

## 8.17 p–n JUNCTION

When a p-type semiconductor is joined to n-type semiconductor through a suitable means, the contact surface of both the semiconductors is known as p–n *junction*.

### 8.17.1 Formation of p–n Junction

A small sphere of indium (trivalent impurity) is pressed on a thin wafer of *n*-type germanium or silicon slab. The system is heated, so that the indium fuses to the surface of germanium and produces p-type germanium just below the surface of contact as shown in Figure 8.23A. This p-type along with n-type Germanium wafer (a thin layer) form a p–n junction as shown in Figure 8.23B. In case of Figure 8.23B both the upper and lower portion of the system have metallic contacts.

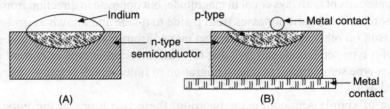

**FIGURE 8.23** (A) A *p–n* Junction (B) both the upper and lower portion of p–n junction have metallic contact.

### 8.17.2 Formation of p–n Junction Diode

There is a high concentration of holes in the *p*-type semiconductor and high concentration of electrons in n-type semiconductor. When both are in contact, electrons move towards p-region and holes towards n-region after diffusing through the junction. The electrons that move to the p-region re-combine with holes. As a result of this recombination, holes disappear and an excess negative charge appears on this side of junctions. When holes diffuse across the junction, an excess positive charge will appear in the n-side of the junction as shown in Figure 8.24. These positive and negative charges on both the sides of the junction are immobile.

Thus, a potential differences develops across the junction, which opposes the further diffusion of electrons and holes. This potential difference is called potential barriers ($V_b$). The magnitude of this potential barrier is about a 0.7 V for silicon and 0.38 V for the germanium crystal at room temperature. The narrow region at the p–n junction, which has positive and negative immobile charges is called *depletion layer* whose thickness is about $10^{-3}$ mm or $10^{-6}$ m.

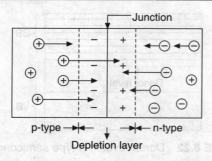

**FIGURE 8.24** A p–n junction diode.

Symbolic representation of *p–n* junction diode is shown as; ──▷|──

## 8.18 EXPLANATION AND ENERGY BAND DIAGRAM OF p–n JUNCTION

To explain the p–n junction we take the help of Figure 8.24. When two types of semiconductor comes in contact with each other, then the positive charged particles, i.e. holes from the p-type are moved towards n-type and the current carriers of n-type, i.e. electrons tried to move from n-type to p-type side, i.e. flow of current carriers occurs in between these two types. Firstly, when electron moves from right side to the left side then electronic current is passed from left to the right side because electronic current is always equal in magnitude, but opposite in direction from conventional current. Here conventional current passes from p-side to n-side and known as *generation current* denoted as $I_g$. But, on other hand, holes are also moved from p-side to n-side and combine with the electrons of n-type semiconductor. So, due to which an another type of current is also passed from p-type to n-type semiconductor due to migration of holes which is known as *recombination current* denoted as $I_r$.

So, in case of simple semiconductor junction, these two types of currents are same i.e. $I_g = I_r$ and no net current flow through the semiconductor. But when $I_g > I_r$, there is some amount of current is passed through the semiconductor which is very small in amount, i.e. in mA. So, due to migration of holes and electrons there is a net accumulation of positive and negative on the junction which form a barrier for further movement of holes and electrons. This barrier is known as *potential barrier* ($V_b$) and thin layer is formed which is called *Depletion layer*.

During the formation of junction between p-type and n-type, there is the development of space charge. The fermi levels of the two materials equalize and the energy level diagram is shown in Figure 8.25.

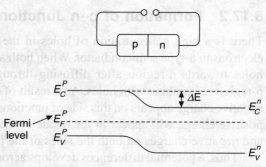

**FIGURE 8.25** Energy level diagram of *p–n* junction.

*Types*: There are mainly two-types of p–n junctions, which are:

1. Forward bias, and
2. Reversed bias

## 8.18.1 Forward Bias

When a battery of EMF greater than the barrier potential is connected to a p–n junction diode in such a way that the positive terminal of the battery is connected to p-region and the negative terminal of battery is connected to n-region of the junction diode, than the p–n junction diode is said to be forward biased. Forward biasing of a p–n junction diode is shown in Figure 8.26.

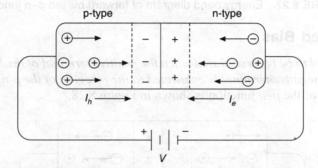

**FIGURE 8.26** Forwarding biasing of a *p–n* junction diode.

When a battery is connected to the p–n junction diode, a potential difference ($V$) develops across the diode. This potential difference reduces the barrier potential and hence the junction resistance becomes very low. The holes (majority carrier) in the p-region and electrons (majority carriers) in the n-region are repelled towards the junction. The majority carrier in the both the region acquire sufficient energy to cross over the barrier potential across the junction, thus the flow of electron to the left side and the holes to the right side of the junction begins. The movement of holes and electrons constitute the hole current ($I_h$) and the electron current ($I_e$), respectively.

In the region of p–n junction, electrons and holes recombine. For every electron–hole recombination near the junction, a covalent bond of p-type semiconductor connected to the +ve terminal of the battery breaks. The electrons liberated enters the positive terminal of the battery and holes moves to the right side of the junction. Similarly, more electrons reach from the negative terminal of the battery and enters the *n*-region to compensate the electrons lost by the combination with the holes at the junction. These electrons diffuse through the junction and enter the *p*-region. Here, they again combine with the holes. Thus, a large current flows through the junction.

### *Energy band diagram for forward biased p–n junction*

In the forward bias, the barrier for the flow of electrons from *n*-side is reduced and hence, $I_r$ will increase with $V_{ex}$. The net current in the forward bias will increase with voltage. The energy band diagram of forward biased p–n junction assumes the shape and is given in Figure 8.27.

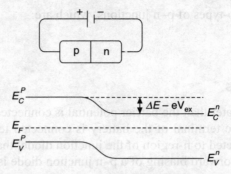

**FIGURE 8.27** Energy band diagram of forward biased p–n junction.

## 8.18.2 Reversed Bias

A *p–n junction is said to be reverse biased when the positive terminal of the battery is connected to the n-region and negative terminal is connected to the p-region of the p-n junction diode.*

Reverse biasing of the *p–n* junction is shown in Figure 8.28.

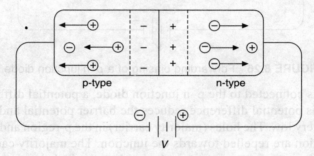

**FIGURE 8.28** Reverse biasing of p–n junction.

When a battery is connected in reverse mode to the p–n junction diode, a potential difference of $V$ volts is applied across the diode.

This potential difference is added in the barrier potential which in turn increases the barrier potential. Hence, the junction resistance increases. The holes and electrons in the p-region and n-region, respectively, are attracted by the negative and positive terminals of the battery. Thus, both holes and electrons are displaced away from the junction. As a result of this, holes in the p-region and electrons in the n-region cannot cross through the junction. Therefore, the flow of current in diode is almost stopped.

There is small saturation current due to minority carriers in p-region and n-region. This current is not affected by the applied voltage, but increases with the increases in temperature.

If the reverse bias is increased to a high value, the covalent bond-near the junction break down and a large number of electron–hole pairs are liberated. Thus, the reverse current increases abruptly to a very high value. This phenomenon is called *break down* and the value of reverse voltage is called *Avalanche voltage* or *Zenar voltage*.

In the reverse biasing, the thickness of depletion layer and the junction resistance increases.

*Energy band diagram of reverse biased p–n junction:* During the reverse biasing of the p–n junction diode, the energy barriers increases by an amount of $eV_{ex}$ and, hence, a very small constant current $I_r$ can flow, $I_g$ is not affected. The net current $I_g - I_r$ will be very small and constant. The energy band of the reverse biased p–n junction diode acquires the shape and form as shown in Figure 8.29.

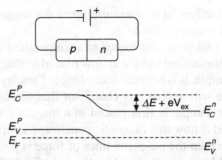

**FIGURE 8.29** Energy band of the reverse biased p–n. junction diode.

## 8.19 SUPERCONDUCTIVITY

Some metals and alloys exhibit almost zero resistivity and infinite conductivity when cooled to sufficiently low temperatures. These superconducting specimens are referred to as *superconductors* and the phenomenon is known as *superconductivity*. The sufficient low temperature at which a conductor become superconductor is known as *critical temperature* or *transition temperature* of that specimen.

It was discovered by the Dutch Physicist, Heike Kamerlingh Onnes in Leiden in 1911. He studied the variation of the electrical resistivity of mercury with temperature. According to him, the value of resistance of mercury used was 172.7 ohms in the liquid condition at 0°C or 273 K. At 4.3 K this had sunk to 0.084 ohms, i.e. 0.0021 times the resistance which the solid Hg would have at 0°C. As the temperature sank further to 1.5 K, this value remained the upper limit of resistance. This observation marked the phenomenon of superconductivity.

It was expected that at such low temperature, the incidence of scattering being low and the resistance of the sample would be very small.

### 8.19.1 Some Basic Facts About Superconductivity

Following are some basic facts which were observed and are very necessary to explain the fact of superconductivity:

*Infinite conductivity*

When a specimen is cooled to a temperature below its critical temperature the resistivity of the sample becomes zero and the sample behaves like superconductor. Direct measurement shows that the resistance must be less than, at least a factor of $10^{17}$. So, the conductivity $\sigma$ becomes infinity. X-ray study reveals that there is no change in the crystal structure at this temperature. This is not a magnetic transition as can be inferred from neutron scattering experiment. The state originates from electronic transition is a completely new thermodynamic state. Relationship between current and electric field is given by microscopic form of Ohms law:

$$I = \sigma E$$

### Meissner effect

In 1931, Meissner noted that effect of temperature upon the magnetic field or the relationship between the temperature and magnetic field.

According to this law, if a metal is first cooled to a temperature below its critical temperature making it superconductor. If the cooled sample is now placed in magnetic field, the magnetic flux lines will not penetrate the sample. It is because according to Faraday's law of magnetic induction, current induced in the sample will oppose any change of flux through the specimen.

On the other hand, if the sample is first placed in a magnetic field, then the magnetic flux lines penetrate the sample and if now this magnetic penetrated sample is cooled to a temperature below its critical temperature then the magnetic lines of force is pushed out from the sample.

This effect is known as *Meissner effect* and is shown in Figure 8.30.

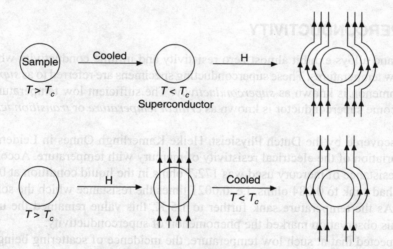

**FIGURE 8.30** Meissner effect.

So, we can say that an applying the magnetic field upon a superconductor, it is shown that the superconductor is a diamagnetic in nature or when a metal is cooled at its critical temperature, it will become a superconductor and a diamagnet. This effect is known as *Meissner effect*. This diamagnetic character is much higher than that of the normal diamagnetic substance, i.e. a superconductor is a perfect diamagnet.

### Critical field

It is observed that the Meissner effect is only observed at low applied magnetic fields. But, when this applied magnetic field exceeds a particular limit the superconducting state gets destroyed and the superconductor becomes a normal conductor. This limiting value of the external applied magnetic field is called *critical field* and is denoted by $B_c$ and this value depends upon the low temperature of the superconductor. If the strength of the applied field is increased and exceeds the critical value of the magnetic field, superconductivity is breaks down and the magnetic flux lines penetrate the specimen placed in that magnetic field.

As we have discussed that the value of critical field for a sample depends upon the temperature of the superconductor sample. It is observed that the lower is the temperature of sample, higher will be the value of magnetic field required to destroy the superconductivity. This effect has been extensively studied experimentally for different elements.

If the $H_c(T)$ is the strength of applied magnetic field to destroy the superconductivity in a specimen at temperature $T$.

Then the difference in the free energy per unit volume between the superconductivity state and a normal state at this temperature is given by the energy density of magnetic field as given:

$$A_n(T) - A_s(T) = \frac{H_c^2(T)}{8\pi}$$

Energy difference is called the *condensation energy*. Dependence of the condensation energy on the temperature is given by the following empirical relationship (Figure 8.31):

$$H_c(T) = H_c(O)\left[1 - \frac{T^2}{T_c^2}\right]$$

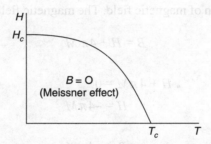

**FIGURE 8.31** Variation of critical field with the temperature.

## 8.19.2 Type I and Type II Superconductors

Zero resistivity and the Meissner effect have been found as the essential properties of a superconductor. We have also seen that the strong magnetic field, when applied on the superconductor, destroys the superconductivity of that superconductor. These properties of the superconductor lead to the characterization of the superconductors into two categories as:

1. Type I superconductor and
2. Type II superconductor

*Type I superconductor*

In such type of superconductors, the magnetic behaviour is of the nature as shown in Figure 8.32. This trend shows that type I superconductors are completely diamagnetic up to $B_a = H_c$ and becomes nonmagnetic just thereafter as $B_a$ is increased. Besides this, the change from magnetic to nonmagnetic state is abrupt. In other words, a superconductor which exhibit a complete

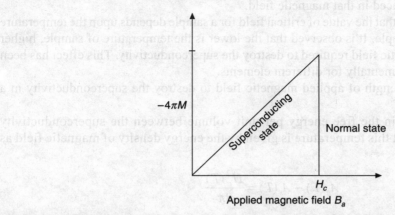

**FIGURE 8.32** Figure showing variation of magnetization property (*M*) of sample with applied magnetic field.

Meissner effect or perfect diamagnetism for a bulk superconductor is known as *type I superconductor*.

This behaviour of the superconductor depends upon the variation of magnetization property (*M*) of the sample as a function of magnetic field. The magnetic field of induction (*B*) is related to the magnetization through

$$B = H + 4\pi M$$

But for superconductor $B = 0$

So,
$$H + 4\pi M = 0$$
$$H = -4\pi M$$

But when $B_a \leq H_c$, then

$$B_a = -4\pi M$$

And for $B_a > H_c$

$$M = 0$$

### Type II superconductor

In such a type of superconductor the change over from diamagnetic state to nonmagnetic is not abrupt or sharp, but it occurs gradually as $B_a$ passes, across $H_c$. This behaviour of type II superconductor is shown in Figure 8.33. Here, we see that the magnetic flux starts penetrating the material at $B_a = H_{c1} < H_c$ and finally Meissner effect vanishes at $B_a = H_{c2}$, i.e. thereafter the material becomes a normal conductor. One may also note in this figure that the loss of area by the curve between $H_{c1}$ and $H_c$ gets restored between $H_c$ between $H_{c2}$. Another point to note is that in the range $H_{c1}$–$H_{c2}$ the state of superconductor is called *vortex state*, and that the value of $H_{c2}$ is much higher than that of $H_c$, it could be even 100 times higher. Such a property of superconductor alloys has assumed a lot of importance in the superconductor technology. Because of this variety of alloys of Nb, Al and Ge have been made available.

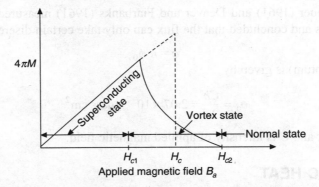

**FIGURE 8.33** Type II semiconductor showing vortex state.

### 8.19.3 Isotope Effect

Another distinct property of superconductor is the isotope effect. It has been observed that the critical temperature $T_c$ of a particular element varies from isotope to isotope. That is $T_c$ depends upon the mass of the isotope of the element. For instance, $T_c$ of Hg varies from 4.185 K to 4.146 K as its atomic mass ($M$) varies from 199.5 to 203.4 a.m.u. $T_c$ even changes when different isotopes are mixed. This phenomenon is termed as *isotope effect*.

The experimental values of $T_c$ usually follows:

$$M^\alpha T_c = \text{constant} \tag{8.15}$$

where parameter $\alpha$ is less than unity. It is about 1/2. So, greater the atomic mass $M$, smaller the critical temperature $T_c$. So putting the value of $\alpha$, Eq. (8.15) becomes

$$M^{1/2} T_c = \text{constant}$$
$$T_c = \text{constant} \times M^{-1/2}$$
$$T_c \propto M^{-1/2}$$

i.e.
$$T_c \propto \frac{1}{\sqrt{M}}$$

So, we can define the isotope effect as:

*The critical temperature $T_c$ of a superconductor is inversely proportional to the square root of their atomic mass.*

## 8.20 PERSISTENT CURRENT AND FLUX QUANTIZATION

As we have discussed earlier that in Meissner effect, when a superconductor is put in a magnetic field, surface current is generated which keeps the flux out of the sample. But, when sample is in toroid shape and subjected to the magnetic field in normal state, flux lines are inside the sample and if this sample is now cooled to a temperature below $T_c$ the flux lines are expelled. But in this situation, some flux lines pass through the hole in the toroid and remains into the hole and go to infinity if the magnetic field is switched off. So, the magnetic flux inside the toroid sample form the closed loops. This flux (trapped) remains constant and known as *persistent current*.

Doll and Nabouer (1961) and Deaver and Fiarbanks (1961) measured the trap flux using different techniques and concluded that the flux can only take certain discrete value.

$$\phi = n\,\phi_0$$

where $\phi_0$ (flux quantum) is given by

$$\phi_0 = \frac{Ch}{2e} = 2.07 \times 10^{-7}\ \text{guass cm}^2$$

and $n$ is any integer and $C$ is intensity of applied magnetic field.

## 8.21 SPECIFIC HEAT

At low temperature, the variation of specific heat of a metal is given by the relation:

$$AT + BT^3$$

where linear form ($AT$) is due to the contribution to the specific heat by the conduction electrons and the cubic term ($BT^3$) is due to lattice vibrations.

As the temperature of a normal metal is lowered, a discontinuity in the specific heat is observed at $T = T_c$ with the specific heat in the super conducting state being higher than the expected value for a normal metal. As the temperature is lowered further, the specific heat decreases following an exponential law,

$$C_V = \exp\left(\frac{-\Delta}{k_B T}\right)$$

where $\Delta$ represent separation between electronic energy level of superconductor.

Rather than the linear behaviour expected of a metal. The specific heat, therefore falls below the expected normal state value as shown in Figure 8.34.

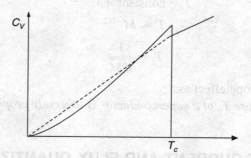

**FIGURE 8.34** Exponential dependence of specific heat on temperature.

The exponential dependence of the specific heat with temperature indicates that the energy spectrum of a superconductor has excited levels, which are separated from the ground state by an energy $2\Delta$. Later we will see that there are many direct evidences of the existence of such a gap in the electronic spectrum. It may be remarked that, while the specific heat shows a finite discontinuity at $T = T_c$, there is no associated latent heat. This is indicative of a second order phase transition at $T = T_c$.

## 8.22 TWO TYPES OF SUPERCONDUCTIVITY

Superconductivity or superconduction is of two types:
1. Low temperature superconductivity (LTSC), and
2. High temperature superconductivity (HTSC)

### 8.22.1 Low Temperature Superconductivity (LTSC)

Low Temperature Superconductivity (LTSC) was first discovered in 1911 by the Dutch Physicist, H. Kamerlingh Onnes, who was awarded the Nobel Prize in 1913 for his contributions to low temperature physics, which led to the production of liquid helium. He found that down to 4.15 K the resistance of Hg decreased with decrease in temperature as in case of most metals. However, at critical temperature $T_c$ = 4.15 K, the resistance fall sharply close to zero. Thus, at or below the critical temperature, Hg becomes a superconductor. At very low temperatures, many metals, alloys and certain compounds become superconductors. The critical temperature for superconductivity lies between 0.1 K and 10 K. Since, the superconductor has almost zero resistance, it can carry an electric current without losing energy and the current can flow for ever. Superconductor also exhibit Meissner effect which states that a superconductor does not allow the magnetic field to pass through it. In the other words, it behaves like a perfectly diamagnetic substance. The Meissner effect gives rise to levitation: levitation occurs when the objects float in air. This can be achieved by the mutual repulsion between a permanent magnet and a superconductor.

### 8.22.2 High Temperature Superconductivity (HTSC)

Because of the very low temperature, at which most materials become superconductors, i.e. when $T_c$ is very low, LTSC has not found widespread use. The higher value of $T_c$ known until 1986 was about 23 K. In that year, G. Bendroz and A. Muller discovered a cuprate (a mixed oxide Ba–La–Cu–O system) which had a $T_c$ of about 35 K. Bendroz and Muller were awarded in the year 1987 Physics Nobel prize, for the discovery of HTSE. Another high $T_c$ superconductor $YBa_2Cu_3O_{7-x}$ ($x \leq 0.1$) was discovered in 1987 by Chinese American physicists, Wu. Chu and their co-workers, the $T_c$ being 98 K. This temperature is significant because it allows liquid nitrogen (boiling point 77 K) to be used as a coolant rather than the more expensive liquid helium. This superconductor called yttrium–barium–cuprate, is the 1–2–3 system because of the ratio of the metal present. The non-stoichiometry appears to be necessary for HTSC. It appears that copper is necessary for superconductivity. This cuprate has the perovskite structure. This comprises of three cubic perovskite units stacked one on the top of the other, giving an elongated (tetragonal) unit cell. The upper and lower cubes have $Ba^{2+}$ ion at the body centered position and the smaller $Cu^{2+}$ ion at each corner. The middle cube is smaller and has a $Y^{3+}$ ion at the body centre. As pervoskite structure has the formula $ABO_3$ and the stoichiometry of this compound would be $YBa_2Cu_3O_9$. Since, the formula actually found in $YBa_2Cu_3O_{7-x}$. There is evidently a massive oxygen deficiency: about 25% of oxygen sites in the crystal are vacant.

The LTSC of metals and alloys has been explained by the BCS theory purposed by the American physicists J. Barden, L. Cooper and J. Schreifer, who were awarded the 1972 physics Nobel Prize, but there is no satisfactory theory of HTSC. Nevertheless, the following features may be born in mind:

1. Many warm superconductors contain copper, which is known to exist in three oxidation states +1, +2 and +3. Also, $Cu^{2+}$ forms many tetragonally distorted complexes. Both these facts may be important.
2. The high $T_c$ superconductors are related to the perovskite structure.
3. The oxygen deficiency is important. Neutron diffraction studies reveal that the vacancies left by the missing oxygen atoms are well ordered. Since, Cu is normally octahedrally surrounded by six oxygen atoms, when an oxygen vacancy occurs, the two Cu atoms may interact directly with each other. Interaction such as $Cu^{2+} - Cu^{3+}$ or $Cu^{+} - Cu^{2+}$ may occur by the transfer of an electron between the two Cu atoms.

The preparation of warm superconductor is still much of an art involving grinding, heating, annealing or slow cooling, etc. and each research laboratory has its own recipe. The 1–2–3 superconductor can be prepared by the pH–adjusted precipitation and high temperature decomposition of the carbonates:

$$2Y^{3+} + 3HCO_3^{-} \xrightarrow{-H^+} Y_2(CO_3)_3$$

$$Ba^{2+} + HCO_3^{-} \xrightarrow{-H^+} BaCO_3$$

$$Cu^{2+} + HCO_3^{-} \xrightarrow{-H^+} CuCO_3$$

$$Y_2(CO_3)_3 + 4BaCO_3 + 6CuCO_3 \xrightarrow{\sim 950°C} YBa_2Cu_3O_{7-x} + 13CO_2 \uparrow$$

## 8.23 OCCURRENCE OF SUPERCONDUCTIVITY

It occurs in very large elements and compounds. Scientist tried to get material with higher value of critical field and critical temperature.

Some of them are:

### 8.23.1 Convential Superconductor

Alloy also exhibit the property of superconductivity. Among elements Nb has highest critical temperature $T_c$ of about 9.4 K at atmospheric pressure. So, the compounds and alloys of Nb have been extensively investigated. $Nb_3$–Sn has critical temperature $T_c$ = 18.5 K and $H_c$ (0), i.e. zero temperature critical field = 28 $T$, i.e. $28 \times 10^4$ guass. The mechanical property of $Nb_3$–Sn possesses serious problem in its fabrication, since it is brittle. So far, only two materials $V_3Ga$ and $Nb_3$–Al have been manufactured in sufficient quantities for use as coil. The $V_3$–Ga has a slightly lower critical field ($H_c$) and lower critical temperature ($T_c$) than $Nb_3$–Sn while the $Nb_3$–Al is a shade better on both counts. So, among the convential superconductors, $Nb_3$–Ge has the highest critical temperature $T_c$ = 23.3 K and a critical field $H_c$ = 38 T.

It was realized that the critical temperature is pushed upwards when Nb is alloyed with smaller and smaller atoms to form a class of compounds, A–15. However, as we progress along the series by alloying Nb with Sn, Al, Ga (31), Ge (32) and finally Si (14) the alloys become more and more difficult to prepare. $Nb_3Si$ has not been isolated in pure form and the samples which have obtained critical temperature around 18 K. A class of Mo compounds called *cheveral compounds* have high critical field. The highest measured critical field of 54 T at 4.2 K occurs in a cheveral compound $PbGb_{0.2}Mo_6S_{0.8}$.

## 8.23.2 Organic Superconductor

Mostly organic and organometallic compounds are insulators, through a number of them exhibit high electrical conductivity and other properties typical of a metal. First synthesized superconductor was in 1973, when tetrathiofulvalene-tetracyanoquinodimethane (TTF–TCNQ), a material with unusual conductivity was synthesized. It is an excellent conductor at 85 K with a conductivity of 5000 $\Omega^{-1}$ cm$^{-1}$. It exhibits in a class of selenium-based organic compound called *Bechgaard Salt* which have critical temperature $T_c$ around 1.5 K. There are charge transfer salt of type (TMTSF)$_2$X where TMTSF stands for *Tetramethyltetra selenium fulvaline* and $X$ is an important inorganic anion such as PF$_6$, TaF$_6$, ReO$_6$, etc. The structure of TMTSF is given in Figure 8.35.

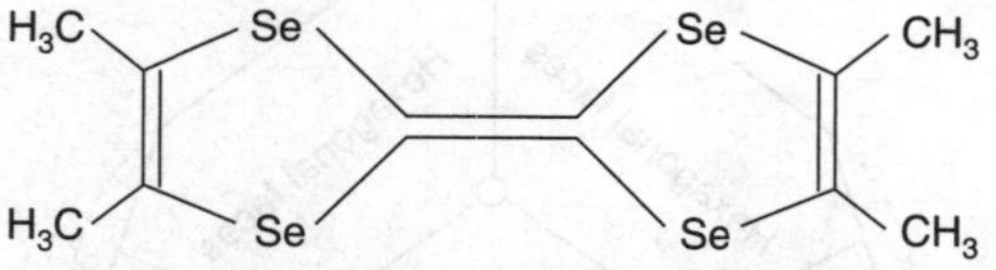

Tetramethyltetraseleniumfulvaline (TMTSF)

**FIGURE 8.35** Structure of tetramethyltetraseleniumfulvaline (TMTSF).

Another known class of organic metal belongs to (BEDT-TTA)$_2$X has been also discovered. Where, cation (BEDT-TTF) represents *bis-ethylenedithiolotetrafulvaline* which is given in Figure 8.36.

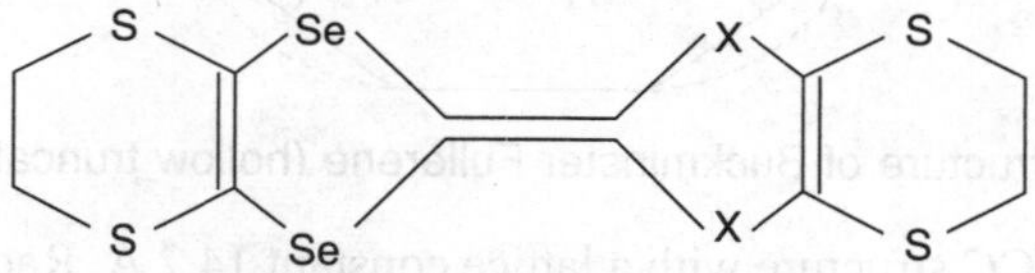

Bis-ethylenedithiolotetrafulvalence (BEDT–TTF)

**FIGURE 8.36** Structure of bisethylenedithiolotetrafulvalene.

Several members of this family show superconducting behaviour. This family has electron donor species. In the structural β-phase, the anions resides within a cavity of ethylene group hydrogen atoms formed by conjugated sheet layers of BEDT-TTF. Increase in conductivity results by increasing the length of anion used. The longest tri-atomic anion containing terminal halide atom $I_3^-$ gives a $T_c$ of about 8 K. New species of longer anion length such as (NCS–M–SCN)$^-$ and (NC–M–CN)$^-$, where M is a metal are being investigated for higher critical temperatures. Attempts to prepare longer anion lengths using Ag and Au are being made which may push $T_c$ to about 40 K.

There are several another groups of organic superconductors. DMET is formed by joining halved TMTSF and BEDT-TTF molecules, which is shown in Figure 8.37.

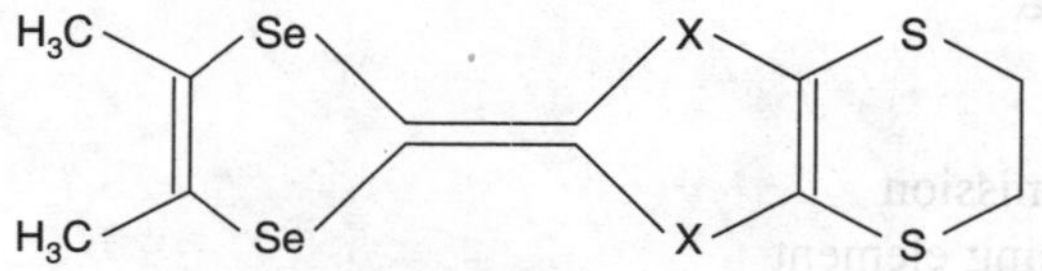

(DMET)

**FIGURE 8.37** Structure of DMET superconductor.

### 8.23.3 Fullerenes

One of the most unlikely candidates for superconductivity is carbon. Till recently, carbon known to exist in two basic structural forms, i.e. graphite and diamond. But in 1985, H. W. Kroto discovered an entirely new form of carbon cluster containing 60 C atoms. This carbon cluster is extremely stable and can survive hostile environment of the interstellar space. It has been suggested that in $C_{60}$, the C atoms are at 60 vertices of a hollow truncated icosahedron having 20 hexagonal faces and 12 pentagonal faces. The structure resembles like soccer ball. These have been named Buckminster Fullerene after the famous architect of R. Buckminster Fuller (Figure 8.38).

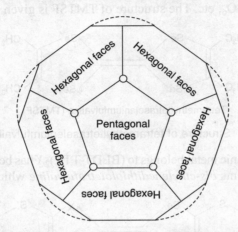

**FIGURE 8.38** Structure of Buckminster Fullerene (hollow truncated icosahedron).

$C_{60}$ molecule pack in an FCC structure with a lattice constant 14.2 Å. Radius of molecule is 3.54 Å So, there is large enough space for dopents of small radii. Doping with K gives the maximum conductivity of 450 $\Omega^{-1}cm^{-1}$ starting with $x = 0$ in $K_xC_{60}$, the conductivity rises as $x$ increases up to a value of 3. For higher value of $x$ conductivity decreases.

Both bulk and thin samples of $K_3C_{60}$ shows onset of superconductivity below 18 K. Both samples of $Rb_3C_{60}$ show superconductivity with a critical temperature of about 30 K. A high value of $T_c$ ($\approx 45$ K) has been reported for $Rb_{2.7}Tl_{2.2}C_{60}$.

## 8.24 APPLICATIONS OF SUPERCONDUCTOR

Main areas towards the use of superconductivity are:

1. High field magnets
2. Flux trapping and shielding
3. Magnetic bearing
4. Energy storage
5. DC transformer
6. DC power transmission
7. Electrical switching element
8. Superconducting fuse and Breaks.

## 8.25 ORGANIC SOLID STATE CHEMISTRY

In organic chemistry, chemical reactivity and the products of chemical reactions are governed by the molecular structure of the compounds involved. In case of solid state, an additional factor comes into play. This is the crystal structure of the starting materials and the structural arrangement of constituent particles in the crystal. As in most crystals, the molecules are packed in a highly ordered and regular manner, so there is a very limited number of orientation for it and they are unable to move. So, the adjacent molecules are able to react together only if certain criteria are met, having suitable reactive centres should be present on adjacent molecules in right orientation and sufficiently close together.

$$n\left( {>}C = C{<} \right) \xrightarrow{\text{Polymerization}} \left( \begin{array}{c} | \\ C \\ | \end{array} = \begin{array}{c} | \\ C \\ | \end{array} \right)_n$$

Polymerization reactions may occur by adding together the olefinic groups. In case of solid for polymerization, the double bonds of the olefinic groups should appear parallel and not farther than ~4Å apart. Addition of two unit will give cyclobutane linking.

The reaction in solid state do not usually occurs spontaneously, but need a catalyst such as UV light. The product obtained in this case is quite different from that of the solution form, as a symmetrically pure compounds will be formed or obtained. Much work had been done in this field by Schmidt and co-workers in 1960.

Stereochemical control of organic reactions is also important. There are additional steric factors that are imposed by particular arrangement of molecules in the crystal structure. Two types of effects are present which are given below:

1. *Intramolecular Effect:* In which internal rearrangements, cyclizations, eliminations reaction, etc., occur within a single molecule.
2. *Intermolecular Effects:* In which adjacent molecules react together.

### 8.25.1 Intramolecular Reactions

Mostly organic molecules are flexible and can change their shape in liquid and gaseous state. As cylohexane can exist in boat form as well as in chair form. The molecules can readily flip from one form to the other by without breaking of bonds, are called conformational effects. But in case of solid state such flexibility is not possible, instead only one particular shape or conformation is observed, e.g. dimethyl meso-$\beta$, $\beta'$-dibromoadipate (DMMDBA) (Figure 8.39).

Figure 8.39 Molecule DMMDBA adopts the conformation shown in the crystalline state by the reaction with gaseous $NH_3$, two molecules of HBr are eliminated from solid (1) to give solid dimethyl transmuconate (2). Both reactant and product are centro symmetric. If the reaction is carried out in liquid phase, molecule (1) gives a variety of products as (3), (4) and (5).

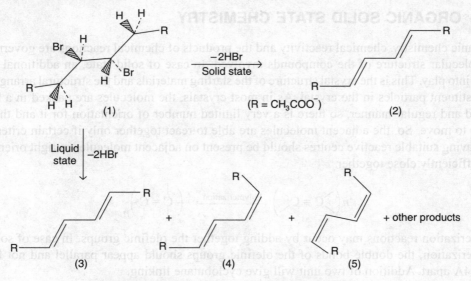

**FIGURE 8.39** Solid state reactions of DMMDBA.

### 8.25.2 Intermolecular Reactions

In liquid state, molecules are having random motion in various directions. So, one particular molecule can react with other of same or different kinds in many different directions. In solid state, molecules are in fixed positions. It places a severe and specific limitation for the way in which such fixed molecules can react together. It give rise to stereochemically pure products. It brings the molecule more closer and reactions occur sometimes which otherwise would not occur at all.

## 8.26 ELECTRICALLY CONDUCTING ORGANIC SOLIDS

It is a difficult process to prepare electrically conducting polymers of organic solids. Such materials would combine or related to the mechanical properties of polymers such as flexibility and ease of fabrication as thin films. Because of problems of atmospheric instability and degradation, none have been commercially exploited as yet.

Some main materials are:

### 8.26.1 Conjugated Systems

*Doped polyacetylene*

Mostly organic solids are electrical insulator. Electrons cannot move freely from one molecule to another or within the molecule in a crystal. But, in case of conjugated systems that contain a skeleton of alternate double and single bond, case is opposite, e.g. polyethylene are insulator because it does not contain double bond and single bond alternatively, although the polymer precursor or its monomer, i.e. ethylene contains a double bond. In the other words, polyethylene itself is saturated, and contains only single bond.

$$n \left( \begin{array}{c} H \\ H \end{array} C = C \begin{array}{c} H \\ H \end{array} \right) \longrightarrow$$

Ethylene → Polyethylene

But, in case of polyacetylene, there is alternative double and single bond showing the properties of electrical conductance. As acetylene consists of triple bond and polyacetylene consists of alternate double and single bond. Polyacetylene has a modest electrical conductivity having the value of $\sigma = 10^{-9}$ $\Omega^{-1}$ cm$^{-1}$ for *cis* form and $10^{-5}$ $\Omega^{-1}$ cm$^{-1}$ for *trans* form, which is compared to that of semiconductors such as silicon.

$n(H-C \equiv C-H)$ → (trans) / (cis)

Polyacetylene

The value of $\sigma$ is low because $\pi$ electrons are not completely delocalized in polyacetylene. It has a band gap of 1.9 eV. But its conductivity can be enhanced by adding impurity in it, as called *doping*. Mac Diarmid, Heeger and co-workers discovered that on doping polyacetylene with suitable inorganic compounds, its conductivity increases.

Dopent may be of two types:

1. $Br_2$, $SbF_5$, $WF_6$, $H_2SO_4$ all of which act as electron acceptor.
2. Any alkali metal which may act as electron donor.

By these dopents, conductivity increases as high as $10^3$ $\Omega^{-1}$ cm$^{-1}$ in transpolyacetylene. This is comparable to the metals and have been termed as *synthetic metals*.

*Preparation:* It is prepared by catalytic polymerization of acetylene in the absence of oxygen. Catalyst used is: Zeiglor–Natta which is a mixture of $Al(CH_2CH_3)_3$ and $Ti(OC_4H_9)_4$ or $TiCl_4$. In one method, acetylene is bubbled through a solution of the catalyst and a solid polyacetylene precipitates are formed.

In another method, acetylene gas is introduced into a glass tube whose inner surface is coated with catalyst then a layer of polyacetylene is formed at the surface of catalyst. The amount formed from acetylene to *cis* and transpolyacetylene depends also upon the temperature. *Trans* form is more stable and forms at higher temperature ~100°C. The *cis* form is obtained at low temperature approx. –80°C. At room temperature a mixture is formed. As the *trans* form have more conductance it is desired to prepare. It can be prepared directly at 100°C or by heating the *cis* form or *cis*–*trans* mixture form.

Doping can be achieved by exposing polyacetylene to gaseous or liquid dopent, e.g. $Br_2$ acts as electron acceptor, bromine-doped polyacetylene may be regarded as $(CH)_n^{s+}Br^-$ electron transfer takes place during its formation. Mechanism of electron transfer is still not known. The (PA) film have a complex morphology in which polyacetylene chain of polyacetylene folds up to form plates and the plates overlap to form fibre.

*Applications:* It has various applications: by analogy with convential semiconductor, p–n junctions diode may be fabricated. In this phenomenon, two films of PA are brought into contact. One is doped with an electron donor and is n-type, the other is doped with an electron acceptor and is p-type. Such devices are easy to make with large surface area. They have potential application in solar energy conversion. However, one problem that remained to be solved is the sensitivity of PA to oxygen. Perhaps modified or substituted PA will be made in future which retain high conductivity and not liable to atmospheric attack.

With electronic conductivity, ionic conductivity is also possible and that certain doped polyacetylene may be used as reversible electrons in new type of batteries. The doping is carried out electrochemically.

The possibilities of using polymeric electrode in batteries, especially in solid state batteries, is very attractive for at least two reasons:

1. The polymers are very light, and
2. Polymers have a flexible structures and this should reduce the problems of contact resistances at the electrode–solid electrolyte interface.

## *Poly-para-phenylene*

It consists a long chain of benzene rings and is also a semiconductor, it has been less studied than PA but has been doped successfully to give much increased electronic conductivity, e.g. on dopping with $FeCl_3$, a product of appropriate composition $[C_6H_4(FeCl_3)_{0.16}]x$ has been obtained with the conductance value $0.3 \, \Omega^{-1} cm^{-1}$ at room temperature.

## *Polypyrrole*

Pyrrole is heterocyclic molecule with a five membered ring, $C_4H_5N$. It can be polymerized to give a long chain structure effectively has alternate double and single bonds, giving a delocalized $\pi$-electron system. Polypyrole itself has a low conductivity, but by adding dopant (perchlorate) to give p-type conductivity as high as $10^2 \, \Omega^{-1} \, cm^{-1}$. It is stable in air and able to withstand temperature up to 250°C.

## 8.27 ORGANIC CHARGE TRANSFER COMPLEXES: NEW SUPERCONDUCTORS

In this, there are two components, simultaneously organic systems, in which one component is a π-electron donor and the other is an electron acceptor. In this, some behave as synthetic metals and a few are superconducting at very low temperature. The superconductivity at a higher temperature, e.g. at 50 or 100 K attracts to research on related new materials.

Examples of *strong π-electron acceptors* are:
Tetra-cyano-quinodimethane (TCNQ), chloronil

*π-electron donors:*
Paraphenylenediamine (PD) or
Tetramethyl paraphenylene diamine (TMPD)
Tetrathiofulvalene(TTF)

R = H, CH$_3$
(TMPD)          (TTF)

In crystalline complexes of π-donors and acceptors, the molecules form stacks in which donors and acceptors alternate, e.g. TMPD and chloronil form mixed stacks which can be represented as

(TMPD)$^+$ (chloronil)$^-$ (TMPD)$^+$ (chloranil)$^-$ ...

In this case, both organic aromatic compound are taken, however, aromatic complexes are also possible in which only one component as an aromatic system is taken, e.g. TCNQ with alkali metals (TCNQ)$_3$ Cs$_2$ and TTF with halogens (TTF)$_2$Br.

Although many of them are semiconductors, some such as TTF–TCNQ and NMP–TCNQ (N-methyl phenazine) have very high conductivity ~ $10^3 \Omega^{-1} cm^{-1}$. The complex (TTF)$_2$Br is superconducting at low temperature below 4.2 K and high pressure 25 K bar.

## 8.28 DEFECTS IN SOLIDS: PERFECT AND IMPERFECT CRYSTALS

A perfect crystal may be defined as one in which all the atoms are at rest on their correct lattice positions in the crystal structure.

In the other words, an ideal crystal is one which has same unit cell containing the same lattice points, across the whole of the crystal. Such a perfect crystal can be obtained, hypothetically, only at absolute zero (0 K or –273°C). So, as the temperature increases, the chance that a lattice site may be unoccupied by an ion increases. This constitutes a defect. Since the number of defects depends upon the temperature, so they are, sometimes, called *thermodynamic defects*. So we can say that, at all the real temperatures, crystals are imperfect. In some crystals, the number of defects present may be very small, i.e. less than 1%, as in high purity diamond or quartz crystals. In other crystals, very high defect concentration, i.e. larger than 1% may be present.

### Why crystals are imperfects?

Crystals are imperfect because the presence of the defects, up to a certain concentration, leads to a reduction of free energy which can be explained as:

Consider, $\Delta G$ is the free energy of a perfect crystal in which defect will be occurred. $\Delta H$ is the amount of energy required to create a single defect in a perfect crystal.

For a single defect, there are so many probabilities of vacant cationic sites, e.g. if crystal is one mole then approximately $10^{23}$ vacant cationic sites are available. Or we can say that for a defect, there are so many choice of positions to occupy, i.e. approximately $10^{23}$.

The entropy gained by having this choice of positions is called *configurational entropy* and is given by the Boltzmann formula such as,

$$S = K \ln W$$

where $W$ is the probability. So by increase in $W$, entropy is increased.

Consequently, the free energy $\Delta G = \Delta H - T\Delta S$ decreases because $\Delta H$ and $\Delta S$ are independent of temperature. So, $-T\Delta S$ term is larger with increasing temperature. So, the free energy is lowered with the defect and finally to lower its energy, crystals show imperfecticity.

## 8.29 TYPES OF DEFECTS

Defects can be broadly divided into two groups:
1. *Stoichiometric Defects:* Two types: (1) Schottky defect and (2) Frenkel defect.
2. *Nonstoichiometric defects:* Two types: (1) metal excess defect and (2) metal deficiency defects.

On the other hand, according to size and shape of defects, crystal defects are classified into three catagories:
1. *Point defect:* Includes both, i.e. stoichiometric defect and non-stoichiometric defect.
2. *Line defect:* two types: (1) edge dislocation and (2) screw dislocation.
3. Plane defect.

Now we will be discussed these defects in detail under two groups explained above.

### 8.29.1 Stoichiometric Defects

In stoichiometric defect, the crystal composition remains unchanged on introducing the defect. In stoichiometric compounds, irregularity in the arrangement of ions in a lattice can occur due to vacancy at a cation and an anion site or by the migration of an ion to some other interstitial site. Two types of such defects are common which are Schottky and Frenkel defects. Both are one-dimensional or point defects, so we will study these defects under point defect discussion.

### 8.29.2 Nonstoichiometric Defects

In nonstoichiometric defects, the crystal composition is changed on introducing the defects. Non-stoichiometric compounds are those where the ratio of the number of cations to the number of anions does not correspond to a simple whole number as suggested by formula. Nonstoichiometric compounds does not obey the law of constant composition. This makes the structure irregular in many ways. Nonstoichiometric defects are of two types depending upon whether positive charge in excess or negative charge in excess. These are known as *metal excess defects* and *metal deficiency defects*, respectively, which will be discussed latter.

Now according to the shape and size of defects or according to dimensions, crystal defects can be classified as:

### 8.29.3 Point Defects

Deviations occur due to missing atoms, displaced atoms or extra atoms, the imperfections is termed as *point defects*. These are due to imperfect packing during the original crystallization or they may arise from thermal vibrations of atoms at elevated temperatures due to excitation of atoms. Common point defects are:

1. Schottky defect and
2. Frenkel defect

Some less common point defects are:

3. Metal excess defect and
4. Metal deficiency defect.

## 8.30 SCHOTTKY DEFECT

This defect arises if some lattice points are not occupied in a crystal. These unoccupied points are called *lattice vacancies* or *holes*. Schottky defect is a stoichiometric defect. In this defect, there is always a pair of vacant sites, in which one for cationic vacancy and other for anionic vacancy (Figure 8.40). To compensate these vacancies, there should be two extra atoms at the surface of crystal for each Schottky defect.

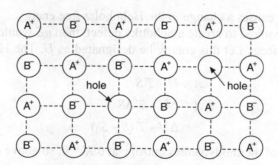

**FIGURE 8.40** Schottky defect.

In other words, defect which arises when equal number of cations and anions are missing from the crystal lattice is known as *Schottky defects*.

Schottky defect is the principal point defect in the alkali halides. This sort of defect tends to be formed in compounds with high co-ordination numbers and where the positive and negative ions are of similar size, e.g. NaCl, KBr, AgBr, KCl and CsCl furnish good examples of ionic solids in which Schottky defects appear.

Consequences of Schottky defects are:
1. It lower the density of crystal by missing the species.
2. Crystal as a whole remains neutral because number of missing positive and negative ions are same.
3. Crystal can conduct electricity to some extend by this defect. As the electric field is applied, nearby ion move from its lattice site to occupy the vacancy and itself generate a new vacancy and procedure goes on and on.

### 8.30.1 Number of Schottky Defects

Consider an ionic crystal containing $N$ ions in which $n$ Schottky defects are produced by the removal of $n$ cations and $n$ anions from the interior of the crystal. The different ways in which each kind of ion can be removed is given by

$$\frac{N(N-1)(N-2)(N-3)\ldots(N-n+1)}{n!} = \frac{N!}{(N-n)!\,n!} \tag{8.16}$$

The different ways in which $n$ Schottky defects can be produced is obtained by squaring the Eq. (8.16) because the number of cation and anion vacancies are equal. Creation of defect in a crystal means creation of disorder. Since the entropy is a measure of the disorder of the system. So with the creation of defects there is increase in the entropy of crystal. But the entropy is related to thermodynamic probability $W$ by the Boltzmann equation.

$$S = k \ln W \tag{8.17}$$

where $k$ is Boltzmann constant. In present case,

$$W = \left[\frac{N!}{(N-n)!\,n!}\right]^2 \tag{8.18}$$

The increase in entropy causes a change in the Helmholtz free energy.

If $E$ is the energy required to create a Schottky defect, then $nE$ would be the energy required to create $n$ Schottky defects. Let this energy be designated as $U$. The Helmholtz free energy is given by

$$A = U - TS$$

$$\Delta A = \Delta U - T \Delta S$$

$$= \Delta U - T(S - S_0)$$

According to third law of thermodynamics entropy at 0 K = 0, so above equation will modify as

$$= \Delta U - TS$$

$$\Delta A = E - TS \tag{8.19}$$

where $\Delta U$ has been replaced by $E$.

Thus, from Eqs. (8.17), (8.18) and (8.19), we get

$$\Delta A = E - T(k \ln W)$$

$$\Delta A = E - kT \ln\left[\frac{N!}{(N-n)!n!}\right]^2 \tag{8.20}$$

Using the Stirlings approximation, i.e.

$$\ln x! = x \ln x - x$$

We find that

$$\ln\left[\frac{N!}{(N-n)!n!}\right]^2 = 2[\ln N! - \ln(N-n)! - \ln n!]$$

$$= 2\left[N \ln N - N - \{(N-n)\ln(N-n) - (N-n)\} - n \ln n - n\right]$$

$$= 2\left[N \ln N - N - (N-n)\ln(N-n) + N + n - n \ln n - n\right]$$

$$= 2\left[N \ln N(N-n)\ln(N-n) - n \ln n\right]$$

Hence, Eq. (8.20), becomes

$$\Delta A = E - 2kT\left[N \ln N(N-n)\ln(N-n) - n \ln n\right] \tag{8.21}$$

At equilibrium, at a given temperature,

$$\left[\frac{\partial(\Delta A)}{\partial n}\right]_T = 0$$

Also since $N$ is constant,

$$\left[\frac{\partial N}{\partial n}\right]_T = 0$$

Hence, from Eq. (8.21)

$$\left[\frac{\partial(\Delta A)}{\partial n}\right]_T = E - 2kT \ln\frac{(N-n)!}{(n)!} = 0$$

$$E = 2kT \ln\frac{(N-n)!}{(n)!}$$

$$\frac{(N-n)!}{(n)!} = \exp\frac{E}{2kT} \tag{8.22}$$

Since, the number of Schottky defects, $n$ is much smaller than the number of ions $N$ in a crystal, i.e.

$$n \lll N$$

Hence, $N - n = N$
So, Eq. (8.22) may be written as

$$\frac{N}{n} = \exp\frac{E}{2kT}$$

$$\frac{n}{N} = -\exp\frac{E}{2kT}$$

$$n = N \exp\left(-\frac{E}{2kT}\right) \tag{8.23}$$

Hence, Eq. (8.23) gives the number of Schottky defects in a crystal.

## 8.31 FRENKEL DEFECT

This defect arise when an ion occupies an interstitial position between the lattice points. This is also a stoichiometric defect, in which an atom is displaced off its lattice site into an interstitial site that is normally empty.

In other words, we can say that, "those defects which are raised when ions are missing from their crystal lattice sites and go into the interstitial spaces is called *Frenkel defect*." In this defect a hole may exist because an ion occupies an interstitial position rather than its correct lattice site.

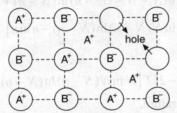

**FIGURE 8.41** A crystal showing Frankel defect.

Frenkel defect is commonly found in ZnS and AgBr. In AgBr, some of $Ag^+$ ions are generally missing from their regular positions and occupies position in between the other ions in the lattice. Similarly in ZnS, $Zn^{2+}$ occupies interstitial space in between the $S^{2-}$ ion in the lattice.

Frenkel defects is generally favoured in following types of compounds:

1. Which have low co-ordination number. In compounds with low co-ordination number, the attractive forces are lesser which are easy to overcome so that the cations can easily move into the interstitial space.
2. Which have highly polarizing cations and an easily polarisable anion.
3. Which have large difference in the size of cations and anions.

*Consequences of Frenkel defects*

1. It does not change the density of crystal because there is no any net loss of ions.
2. Crystal as a whole remains neutral.
3. Crystal can also conduct electricity to some extent by this effect.

*AgBr is ionic solid which shows both schottky and frankel defect.

*Number of Frenkel defects*

Consider an ionic crystal having $N$ ions and $N_i$ interstitial spaces or positions in its structure. The number of ways in which $n$ Frenkel defects can be formed is given by

$$W = \left[\frac{N!}{(N-n)!n!}\right] \times \left[\frac{N_i!}{(N_i-n)!n!}\right] \tag{8.24}$$

Let the energy required to displace an ion from its proper position to an interstitial position be $E$. Then the energy required to produce $n$ Frenkel defects would be $nE$. Let this energy be designated as $E$.

Proceeding as before, by using Boltzmann entropy equation, viz. $S = k \ln W$, the Helmholtz free energy equation viz. $\Delta A = \Delta E - T\Delta S$ and the Stirling approximation for evaluating the factorial terms, we arrive at the conclusion that

$$n = (NN_i)^{1/2} \exp\left(-\frac{E}{2kT}\right) \quad (8.25)$$

This Eq. (8.25) gives the number of Frenkel defects in a crystal.

## 8.32 THERMODYNAMICS OF POINT DEFECTS

Both Schottky and Frenkel defects are intrinsic defects, i.e. they are present in the pure material and a certain number of these defects must be present from thermodynamic considerations. These defects arise because the crystals are usually prepared at high temperatures. So the more defects are present at the higher temperatures. On cooling the crystals at room temperature a small number of the defects may be eliminated, but in general the cooling rate is extremely slow so the defects are preserved in cooling and are then present in excess at the equilibrium concentration.

Two approaches are used to study point defects at equilibrium.

1. Statistical thermodynamic and
2. Law of mass action

In statistical thermodynamics, the complete partition function for a rational model of defected crystal is constructed. The free energy is expressed in terms of partition function and then the free energy is minimized in order to obtain the equilibrium condition. This method can also be applied to nonstoichiometric defects.

Alternatively, the law of mass action can be applied to Schottky and Frenkel equilibria and the concentration of defects are expressed as an exponential function of temperature. Because the law of mass action is simple and easy for application to stoichiometric crystals, this is used for Schottky and Frenkel defects.

### 8.32.1 Thermodynamics of Schottky Defect

The Schottky equilibria in a common NaCl crystal can be represented as:

$$Na^+ + Cl^- + V_{Na}^s + V_{Cl}^s \rightleftharpoons V_{Na} + V_{Cl} + Na^{+,s} + Cl^{-,s}$$

where  $Na^+$ and $Cl^-$ represent normally occupied cation and anion sites,

$V_{Na}^s$ and $V_{Cl}^s$ represent vacant cation and anion surface sites, respectively,

$V_{Na}$ and $V_{Cl}$ represent cation and anion vacancies, and

$Na^{+,s}$ and $Cl^{-,s}$ represent occupied cation and anion surface site

Thus, the equilibrium constant for formation of Schottky defects is given by:

$$K = \frac{[V_{Na}][V_{Cl}][Na^{+,s}][Cl^{-,s}]}{[Na^+][Cl^-][V_{Na}^s][V_{Cl}^s]} \quad (8.26)$$

Since the number of surface sites is always constant in a crystal of constant total surface area. Hence the number of $Na^+$ and $Cl^-$ ions that occupy surface sites is always constant. In the formation of Schottky defect, Na$^+$ and Cl$^-$ ions move out of the crystal to occupy surface sites, but at the same time an equal number of fresh surface sites are created.

Therefore,
$$[Na^{+,s}] = [V_{Na}^s] \text{ and } [Cl^{-,s}] = [V_{Cl}^s]$$

So from this, Eq, (8.26) reduces to

$$K = \frac{[V_{Na}][V_{Cl}]}{[Na^+][Cl^-]} \tag{8.27}$$

Let $N$ be the total number of sites of each kind, and $N_V$ be the number of vacancies of each kind. The number of occupied sites of each kind is $N - N_V$. So, by substituting these into Eq. (8.27) we get

$$K = \frac{(N_V)^2}{(N - N_V)^2} \tag{8.28}$$

For small concentration of defects:

$$N - N_V \approx N$$

So from Eq. (8.28)

$$K = \frac{(N_V)^2}{N^2}$$

$$N_V = N\sqrt{K} \tag{8.29}$$

But the equilibrium constant $K$ can be expressed as an exponent function of temperature.

$$K \propto \exp\left(\frac{-\Delta G}{RT}\right)$$

$$\propto \exp\left(-\frac{(\Delta H - T\Delta S)}{RT}\right)$$

$$\propto \exp\left(\frac{-\Delta H}{RT}\right) \exp\left(\frac{\Delta S}{R}\right)$$

$$K = \text{constant} \times \exp\left(\frac{-\Delta H}{RT}\right) \tag{8.30}$$

Now from Eqs. (8.29) and (8.30), we get

$$N_v = N\left[\text{constant} \times \exp\left(\frac{-\Delta H}{RT}\right)\right]^{1/2}$$

$$N_v = N \times \text{consant} \times \exp\left(\frac{-\Delta H}{2RT}\right) \tag{8.31}$$

A factor of 2 is appeared in the exponent part of the expression for the number of defects. This is because there are two defective sites per defect, i.e two vacancies per Schottky defect.

## 8.32.2 Thermodynamics of Frenkel Defects

Frenkel equilibria in a crystal of AgBr can be represented as:

$$Ag^+ + V_i \rightleftharpoons Ag_i^+ + V_{Ag}$$

where $Ag^+$ represents occupied lattice position by $Ag^+$ ion.

$V_i$ represents the interstitial vacant site

$Ag_i^+$ represents the occupied interstitial site and

$V_{Ag}$ represents the vacant lattice site of $Ag^+$ ion.

The equilibrium constant for this is given by:

$$K = \frac{[Ag_i^+][V_{Ag}]}{[Ag^+][V_i]} \qquad (8.32)$$

Let $N$ be the number of lattice sites that would be occupied in a perfect crystal and $N_i$ be the number of occupied interstitial sites, i.e

$$[V_{Ag}] = [Ag_i^+] = N_i$$

and

$$[Ag^+] = N - N_i$$

and for most of the regular crystal structures

$$[V_i] = \alpha N$$

That is the number of available interstitial sites is simply related to the number of occupied lattice sites. For AgCl, $\alpha = 2$ because there are two tetrahedral interstitial sites for every octahedral site that is occupied by $Ag^+$.

Substituting these considerations in Eq. (8.32), we get

$$K = \frac{(N_i)^2}{(N - N_i)(\alpha N)} \simeq \frac{N_i^2}{\alpha N^2} \qquad (8.33)$$

Using Arrhenius expression for temperature dependence of the number of Frenkel defects, we get

$$[V_{Ag}] = [Ag_i^+] = [N_i] = N\sqrt{\alpha} \exp\left(\frac{-\Delta G}{2RT}\right)$$

$$[V_{Ag}] = \text{constant} \times N \exp\left(\frac{-\Delta H}{2RT}\right) \qquad (8.34)$$

Similarly, a factor 2 is also appeared in the exponent part of the expression for the number of defects. This is because there are also two defective sites for Frenkel defect, one for vacancy and other for one interstitial site per Frenkel defect.

## 8.33 METAL EXCESS DEFECTS

Those defects in which positive charge is in excess are called *metal excess defects*. These defects may arise in two ways:

1. When a negative ion may be missing from its lattice site leaving a hole, which is occupied by an extra electron to maintain the electrical balance as shown in Figure 8.42.

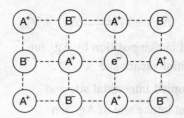

**FIGURE 8.42** Metal excess defect due to missing of anion.

There is evidently an excess of positive metal ion although the crystal as a whole is neutral. This defect is somewhat similar to Schottky defect, but differs in having only one hole and not a pair as in case of Schottky defect. This type of defect is not very common.

For example, when NaCl is treated with Na vapour, a yellow nonstoichiometric form of NaCl is obtained in which there is excess of $Na^+$ ions. In a case, the extra positive charge is balanced by the presence of free electron in the lattice.

2. The metal excess defect may also appear in another way.

When an extra positive ion may occupy an interstitial position in lattice and to maintain electricity neutrality, an electron is present in the interstitial space (Figure 8.43).

Although this type of defect is some what similar to Frenkel defect, yet it differs

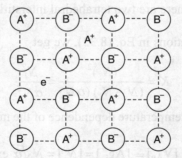

**FIGURE 8.43** Metal excess defect due to extra cation in interstitial site.

from that in having no holes and in having interstitial electrons. This defect is much more common than the first type of metal excess defect. This type of defect is shown by such crystals, which are likely to develop Frenkel defect. An interesting example is ZnO crystal.

## Consequences of metal excess defects

1. As the crystals associated with metal excess defects of the first or second kind contain free electrons, they can conduct electricity to some extent because the number of defects and the number of electrons are very small. Such crystals are generally termed as *semiconductors*.

2. The crystals showing metal defects are generally coloured. This is due to the presence of free electrons. When these electrons are excited to higher energy levels by absorption of certain wavelengths from the visible white light, these compounds appear coloured. For example, ZnO is white when cold, but appears yellow when hot.

## 8.34 METAL DEFICIENCY DEFECT

Those defects in which negative charge is in excess are called *metal deficiency*. In this defect, there is a deficiency of positive metal ions. This type of defect requires variable valency of the metal and might therefore be expected with the transition metals. These defects may arise in two ways.

1. In first way, a positive ion may be missing from its lattice site and the extra negative charge is balanced by an adjacent metal ion having two charges instead of one as shown in Figure 8.44.

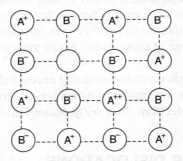

**FIGURE 8.44** Metal deficiency due to missing of cation from lattice site.

It means that the metal may be in position to show variable valency. Thus, this defect is generally shown by the compounds of transition metals.

Examples are FeO, FeS, NiO, etc.

2. In the second way, an extra negative ion may find an interstitial position and the charges are balanced by means of an extra charge on the adjacent metal ion as shown in Figure 8.45.

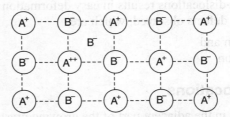

**FIGURE 8.45** Metal deficiency due to extra anion occupying interstitial site.

*Consequences of metal deficiency defects*

Crystals with metal deficiency defects are semiconductors. This property arises from the movement of an electron from one ion to another. The substances permitting this type of movement are known as p-*type semiconductors*.

*Colour centres*

Crystals of pure alkali metal halides such as NaCl, KCl, etc. are white. However, it is observed that if NaCl crystal is heated in the presence of Na vapours. It becomes yellow. Similarly, if KCl is heated in the presence of K vapour, it acquires magenta colour. These colours are produced due to some imperfection introduced in the crystal as explained below for the case of NaCl.

When NaCl is heated in an atmosphere of Na vapour, it becomes yellow. During heating the excess of Na atoms deposit on the surface of the NaCl crystal. $Cl^-$ ions then diffuse to the surface where they combine with the Na atoms, which become ionized by losing electrons. These electrons diffuse back into the crystal and occupy the vacant sites created by the $Cl^-$ ions. This electron is shared by all the $Na^+$ ions present around it and is thus considered as a delocalized electron. When light falls on it, it absorbs some energy for excitation from ground state to excited state. This give rise to colour. Such points are called *F-centers*, the name comes from the German word for colour, *Fabre*. Actually a range of energies can excite this electron so that an absorption band called *F-band* is observed.

It is possible to produce halide crystals containing excess halogen ions. Such excess halogen ions are accompanied by positive ion vacancies which serve to trap holes quite similar to the way the anion vacancies trapped the electrons. In order to distinguish the colour centres thus produced, they are called *V-centres*.

## 8.35 LINE DEFECTS OR DISLOCATIONS

If metals were perfectly crystalline, the stress needed to deform them would have been much higher ($10^2$–$10^4$ times) than actually observed. This have been explained by suggesting that the actual metallic crystal are having certain imperfections or defects which help them to undergo elastic deformation under small loads. These imperfections were postulated as *line defects* and are termed as *dislocations*.

In other words, If the periodicity of the atomic array is interrupted along certain direction in a crystal, such interruption occurs along the rows of a crystal structure and are called *line defects*. Under shear stress these dislocations results in easy deformation of a crystal.

There are two main line defects observed, which are,

1. Edge dislocation and
2. Screw dislocation

### 8.35.1 Edge Dislocations

If there is an irregularity in the adjacent part of the growing crystals resulting in the introduction of an extra row of the atoms. So due to this, the edge of the plane terminates within the crystal instead of passing all the way. Under the impact of shear, the dislocation moves across the crystal in such a way that the top half of the crystal is displaced one lattice distance with respect to the lower half.

At the edge dislocations, the atoms are pushed together above the edge and pulled a part below the edge. So the impurity is present below the edge (Figure 8.46). The atoms with larger diameters than the present atoms tends to concentrate below the edge and the impurity with smaller diameter tend to concentrate away from the edge.

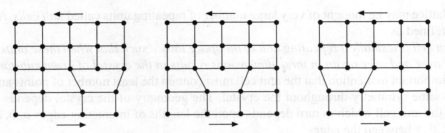

**FIGURE 8.46** Edge dislocation in crystal.

## 8.35.2 Screw Dislocations

To understand screw dislocations, imagine a cut in a crystal, with the particles to the left of the cut pushed up through the distance of one unit cell as shown in Figure 8.47. The unit cell now form a continuous spiral around the end of cut, the screw axis (dislocation line). In other words, the atomic planes around the screw axis form a spiral or helical ramp. Such a displacement of atoms is termed as *screw dislocation*. The existence of screw dislocation would help in the further growth of the crystal because atoms can easily settle down up the step. However, when the lower terrace of the crystal is completely covered, further growth ceases. Just like the edge dislocation, it can also move under the influence of appropriate shear force and thus help in the deformation.

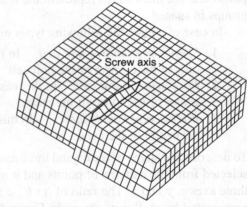

**FIGURE 8.47** Crystal showing screw dislocation.

## 8.36 PLANE DEFECTS

When whole layers in a crystal structure can be defective, defects are known as *plane defects*.

## 8.37 DIFFRACTION METHODS; SPACE LATTICE

A crystalline solid is made up of regular repeating three-dimensional pattern of constituent units in space. Every constituents unit in a crystal is represented by means of points. When these units are repeated systematically in space it results in some pattern of unit. This pattern is called *space lattice*. We may define space lattice as

Space lattice may be defined as the patterns of points in space such that the environment about each constituent unit is the same in crystal.

The space lattice is thought of extending in all directions throughout the entire crystal. The positions of the constituent particles in space are called *lattice points*.

## 8.38 UNIT CELL

A space lattice may be thought of very large number of repeating units called *unit cells*. A unit cell may be defined as

*A unit cell is a smallest repeating unit of the space lattice such that when these units cells are repeated over and over again in three dimensions results in the crystal of given substance.*

It is important to mention that the unit cell must contain the least number of points and should have the same symmetry throughout the crystal. The geometry of the crystal depends upon the shape of the unit cell which in turn depends upon the lengths of interacting edges $a$, $b$, $c$ and the angles $\alpha$, $\beta$, $\gamma$ between the edges.

It is not possible for a single unit cell to exist freely because the boundries of the unit cell are also the boundries of the neighbouring cell. When the crystal has the same lattice points throughout the three-dimensional network of the crystal: it is called an *ideal crystal*. It is the points and not lines which represent the space lattice. The lines only represent positions of point groups in space.

In case of a cube the following types of unit cells are possible:

1. *Simple or primitive unit cell:*  In this arrangement the constituent particles are present only at the corners of the unit cell.
2. *Face centred unit cell:*  In this case, the points are present at the corners as well as the centre of each face of the cube.
3. *Body-centred cubic lattice:*  In this case, the points are present at the corners as well as the centre of the cube.

To describe the shape of the crystal three axes are present, called *crystallographic axis*. A plane is selected from the patterns of points and it makes intercepts $a$, $b$, $c$ which are the co-ordinates of three axes $x$, $y$ and $z$. The ratio of $a : b : c$ is the axial ratio and the angle between the axes are represented by $\alpha$, $\beta$, $\gamma$ as shown in Figure 8.48. Generally the value of $b$ is taken as unity. The elements of crystal are the values of $a$, $b$, $c$ and $\alpha$, $\beta$, $\gamma$ which define the shaping of the crystal.

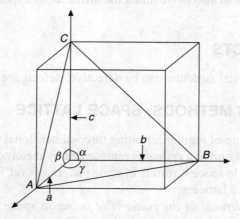

**FIGURE 8.48**  Elements of the crystals.

## 8.39 X-RAY CRYSTALLOGRAPHY

A crystal consists of a large number of building up units. These unit cells are uniformly distributed in three-dimensional space of the crystal. The inter-particle distance between the constituents are of the order of $10^{-8}$ cm. Lane and his co-workers suggest that the crystals can diffract X-ray in the same manner that a diffraction grating could diffract light rays because the inter-particle distance in a crystal was of the same order as that of the wavelength of the X-ray, i.e. $10^{-8}$ cm. The crystal could act as suitable natural grating for diffraction of X-rays.

Lane suggested that on passing a beam of X-rays in a crystal behind it a photographic plate is placed. An image was obtained showing series of the spots arranged in a geometric manner about the centre of beam. This provides a method to determine the internal structure of the crystal. For this purpose the following methods are used:

1. Lane photographic plate method
2. Bragg's X-ray spectrometric method
3. Powder method

Few methods will be discussed here.

### 8.39.1 Bragg's X-rays Spectrometric Method

According to Lane when a beam of X-ray is passed through a crystal and the impression is taken on a photographic plate, a pattern called *Lane's photograph* is obtained. By using this photograph, Bragg investigated the structure of crystals of NaCl, KCl, etc. Bragg deviced a spectrometer for measurement of the intensity of an X-ray beam. The diffraction pattern thus produced can be analyzed for inter planner distance by using Bragg's equation or Bragg's law.

*Bragg's law*

When a monochromatic X-ray falls on the atoms in a crystal, each atom acts as a source of scattering radiation of the same wavelength. The crystal source acts as a parallel reflecting plane. X-ray would be reflected according to the laws of reflection and the plane would be reflect X-ray at all angles. The intensity of reflected beam at certain angle will be maximum, when the two reflected waves from two different planes have path differences equal to an integral multiple of the wavelength of X-ray.

Therefore the reflected beam to be of maximum intensity,

$$2d \sin \theta = n\lambda$$

where $n = 1, 2, 3, \ldots$, etc.
The above equation is called Bragg's equation and it represents Bragg's law.
When $n = 1$

$$\lambda = 2d \sin \theta$$

The reflected beam of X-ray gives the spectrum of the first order. It is clear from Bragg's equation that if $\lambda$ of X-ray's which produces maximum intensity at the glacing angle $\theta$ is known, the inter-planner distance of a crystal, i.e. $d$ can be determined. Likewise if $d$ is known $\lambda$ can be calculated.

## Construction and working of X-ray spectrometer

A beam of X-ray of definite wavelength are made to pass through the slit S which collimates it into a fine beam and falls upon the face of the crystal C mounted on the rotating table. The table can be rotated about a vertical axis and the rotation can be read on a circular graduated scale as shown is Figure 8.49.

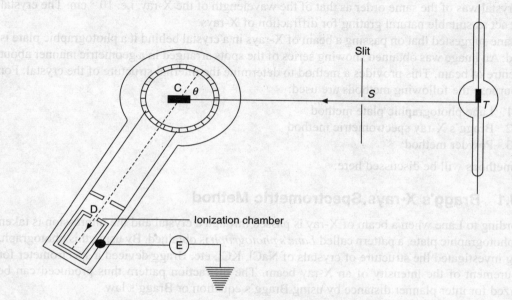

**FIGURE 8.49** Bragg's arrangement to measure interplaner distance d.

The rays after reflection from the crystal enter the detector D, which is either ionization chamber or Geiger counter.

Length of the unit cell (a) can be found from the relation,

$$a = \left(\frac{\text{No. of atoms in the unit cell} \times \text{mol. wt.}}{\text{Avogadro's no.} \times \text{density}}\right)^{1/3}$$

In case of cubic crystal like that of NaCl which has an FCC lattice, it has four atoms in the unit cell and has a density 2.18 g/cm$^{-1}$

Mol. Wt. of NaCl = 23 + 35.5 = 58.5
Avogadro's no. = 6.023 × 10$^{23}$
Length of the cubic cell, i.e.

$$a = \left[\frac{58.5 \times 4}{6.023 \times 10^{23} \times 2.18}\right]^{1/3}$$

$$a = [178.26 \times 10^{-24}]^{1/3}$$

$$a = 5.63 \times 10^{-8} \text{ cm}$$

For FCC lattice

$$d = \frac{a}{2} = \frac{5.36}{2} \times 10^{-8} \text{ cm} = 2.815 \times 10^{-8} \text{ cm}$$

$$d = 2.815 \text{ Å}$$

According to Bragg's equation

$$2d \sin \theta = n\lambda$$

Knowing the value of $\theta$ and $n$, i.e. order of reflection, $\lambda$ can be calculated.

### 8.39.2 Determination of Crystal Structure by Bragg's Method

To determine the crystal structure, X-rays are allowed to fall on the crystal surface and the crystal is rotated. X-rays are reflected from various lattice planes. The reflections are recorded by Bragg's X-ray spectrometer. The glancing angle for each intense reflection is also recorded.
According to Bragg's equation

$$2d \sin \theta = n\lambda$$

The ratio of lattice spacing for various group of planes can be calculated. For different crystals, this ratio is different. The calculated ratios are compared in the experimental ratios and from this the particular structure may be confirmed.

1. For a simple cubic lattice

$$\frac{1}{d_{100}} : \frac{1}{d_{110}} : \frac{1}{d_{111}} = 1 : \sqrt{2} : \sqrt{3}$$

2. For a FCC crystal lattice

$$\frac{1}{d_{100}} : \frac{1}{d_{110}} : \frac{1}{d_{111}} = 1 : \sqrt{2} : \frac{\sqrt{3}}{2}$$

3. For a BCC crystal lattice

$$\frac{1}{d_{100}} : \frac{1}{d_{110}} : \frac{1}{d_{111}} = 1 : \frac{\sqrt{2}}{2} : \sqrt{3}$$

Let from the first order intense reflections for planes (1, 0, 0), (1, 1, 0) and (1, 1, 1), the glancing angles be $\theta_1$, $\theta_2$ and $\theta_3$, respectively. For the same value $\lambda$, we have

$$2d_{100} \sin \theta_1 = 2d_{110} \sin \theta_2 = 2d_{111} \sin \theta_3 = 1 \times \lambda$$

$$\frac{1}{d_{100}} : \frac{1}{d_{110}} : \frac{1}{d_{111}} = \sin \theta_1 : \sin \theta_2 : \sin \theta_3$$

In case of NaCl crystal
$\theta_1 = 5.9°$
$\theta_2 = 8.4°$
$\theta_3 = 5.2°$

$$\frac{1}{d_{100}} : \frac{1}{d_{110}} : \frac{1}{d_{111}} = \sin 5.9° : \sin 8.4° : \sin 5.2°$$

$$= 0.1027 : 0.1461 : 0.0906$$

$$= 1 : \frac{0.1461}{0.1027} : \frac{0.0906}{0.1027}$$

$$= 1 : 1.422 : 0.8821$$

$$\frac{1}{d_{100}} : \frac{1}{d_{110}} : \frac{1}{d_{111}} = 1 : \sqrt{2} : \frac{\sqrt{3}}{2}$$

This shows that NaCl crystal has an FCC lattice structure. In other words it has face-centred cubic lattice.

### 8.39.3 Powdered Crystal Method

In earlier methods, a single crystal is required whose size is much larger than microscopic dimensions. However, in powder method, the crystal sample need not be taken in larger quantity, but as little as 1 mg of the material is sufficient for the study. The powder method was devised independently by Debye and Scherrer in Germany and Hull in America about the same time (1916) (Figure 8.50).

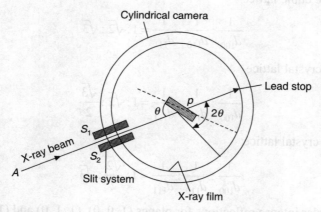

**FIGURE 8.50** The experimental arrangement of powder method.

*Arrangement*

The main features are outlined below:

1. $A$ is the source of X-rays which can be made approximately monochromatic by a filter $F$ not shown in the Figure 8.50.
2. Allow the X-ray beam to fall on the powdered specimen $p$ through the slits $S_1$ and $S_2$. The function of these slits is to get a narrow pencil of X-rays.
3. Fine powder $p$ stuck on a hair by means of gum, is suspended vertically in the axis of a cylindrical camera. This enables sharp lines to obtained on the photographic film which

is bent in the form of a circular arc surrounding the powder crystal.
4. The X-ray after falling on the powder pass out of the camera through a cut in the film so as to minimize the fogging produced by the scattering of the direct beam.
5. On a photographic plate the observed pattern consists of traces as shown in Figure 8.51.

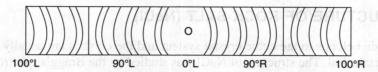

**FIGURE 8.51** Observed pattern of circular rings on photographic plate.

## Theory and calculations

When a monochromatic beam of X-rays is allowed to fall on the powder of a crystal then
1. There will be some particles out of the random orientation of small crystals in the fine powder, which lie with a given set of lattice planes (marking the correct angle with the incident beam) for reflection to occur.
2. While another fractions of the grains will have another set of planes in the correct position for the reflection to occur and so on.
3. Also, reflections are possible in the different orders for each set.

All the like orientations of the grains due to reflection for each set of planes and for each order will constitute a diffraction cone whose interactions with a photographic plate gives rise to a trace.

The crystal structure can be obtained from the arrangement of the traces and their relative intensities. We will now proceed to do this.

If the angle of incidence is $\theta$, then the angle of reflection will be $2\theta$ as shown in arrangement Figure 8.50.

If the film radius is $r$, the circumference $2\pi r$ corresponds to a scattering angle of $360°$. Then we can write.

$$\frac{1}{2\pi r} = \frac{2\theta}{360°} \Rightarrow \theta = 360° \times \frac{1}{4\pi r}$$

From the above equation, the value of $\theta$ can be calculated and substituting this value in the Bragg's equation

$$n\lambda = 2d \sin \theta$$

Then the value of $d$ can be calculated.

It is important to remark here that in order to index reflections, one must know the crystal system to which the specimen belongs. This is usually done by microscopic examination.

## Applications

1. This method is most useful for cubic crystals.
2. It is also used for determining the complex structure of metals and alloys. However, these structures could not be revealed by the earlier studies.

3. This method also helps to distinguish between the allotropic modifications of the same substance.
4. The structure of rubber was fully revealed by this method. Rubber has been seen to crystallize on stretching and de-crystallize on slackening.

## 8.40 STRUCTURE OF ROCK SALT (NaCl)

Sodium chloride belongs to the cubic crystal system and the crystals are usually cubic although they may be octahedral. The structure of NaCl was studied by the Bragg's spectrometer. For this purpose, the intensity of the ionization current was measured for different angles of the rotation of the crystal. Then a curve was plotted between the current intensity and glacing angles. The graph for (1, 0, 0), (1, 1, 0) and (1, 1, 1) is shown in Figure 8.52.

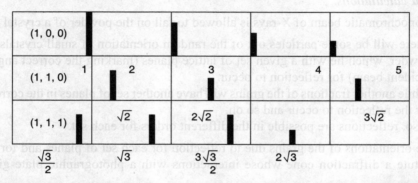

**FIGURE 8.52** Graph for NaCl crystal between current intensity and glacing angle.

From the graph it is evident that,
  (i) The intensities of current fall regularly for the faces (1, 0, 0) and (1, 1, 0), and
  (ii) There is difference for the (1,1,1) face. In this case, the first order spectrum is abnormally weak, second is abnormally strong, third is abnormally weak and fourth strong.

It was proved that the (1, 0, 0) face of NaCl crystal yielded reflection maxima at glancing angles of 5.9°, 11.85° and 18.15°. If we say that these angles are representing first, second and third order reflections, respectively, then their sine must be in the ratio of 1 : 2 : 3, i.e.

$$\sin 5.9° : \sin 11.85° : \sin 18.15° : : 0.1028 : 0 : 2057 : 0 : 3113$$
$$\sin 5.9° : \sin 11.85° : \sin 18.15° : : 1 : 2 : 3$$

So, these are in good agreement with our expectations. Let us now consider the Bragg's equation:

$$2d \sin \theta = n\lambda$$

$$d = \left(\frac{n\lambda}{2}\right) \sin \theta \qquad (8.35)$$

$$\frac{1}{d} = \left(\frac{2}{n\lambda}\right) \sin \theta$$

$$\frac{1}{d} \propto \sin \theta \qquad (8.36)$$

For planes (1, 0, 0), (1, 1, 0) and (1, 1, 1) for NaCl, the first order reflections are observed at glancing angles 5.9°, 8.4°, and 5.2°.

According to Eq. (8.36)

$$\frac{1}{d_{100}} : \frac{1}{d_{110}} : \frac{1}{d_{111}} = \sin 5.9° : \sin 8.4° : \sin 5.2°$$

$$= 0.1028 : 0.1461 : 0.0906$$

$$= 1 : \frac{0.1461}{0.1028} : \frac{0.0906}{0.1028}$$

$$= 1 : 1.4212 : 0.8813$$

$$\frac{1}{d_{100}} : \frac{1}{d_{110}} : \frac{1}{d_{111}} = 1 : \sqrt{2} : \frac{\sqrt{3}}{2}$$

The above ratio is same as is obtained for unit cell of a face-centred cubic lattice. Thus, NaCl has a face-centred cubic structure.

Now the problem is to decide about the structural units of the crystals. These units may be molecules, atoms or ions. This problem is solved by considering the plane (1, 1, 1). In plane (1, 1, 1), the first order spectrum is abnormally weak, second is strong, third is abnormally weak and fourth strong. The alternations of these intensities are due to alternate layers or planes of $Cl^-$ atoms separated by the layer of sodium atoms. Also, it is evident from the alternations in the intensities that the real unit of matter in crystal lattice is the atoms and not the molecules and not the ions (Figure 8.53).

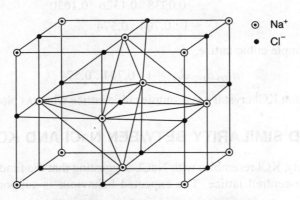

**FIGURE 8.53** Bragg structure of NaCl.

From the structure, it follows that:

1. The rock salt crystals have an FCC lattices with $Na^+$ and $Cl^-$ atoms arranged alternatively in all directions parallel to the three rectangular axes.
2. Each unit cell of NaCl consists of 14 $Na^+$ atoms and 13 $Cl^-$ atoms.
3. Each $Cl^-$ atom is surrounded by 6 $Na^+$ atoms and each $Na^+$ atom is surrounded by six $Cl^-$ atoms, i.e. the co-ordination number of $Na^+$ and $Cl^-$ is 6.

## 8.41 STRUCTURE OF SYLVINE (KCl)

The study of structure of sylvine was done by Bragg's X-ray spectrometer. The intensities of ionization current were determined for angles. By plotting the curve between the current for (1, 0, 0), (1, 1, 0) and (1, 1, 1) faces as shown in Figure (8.54).

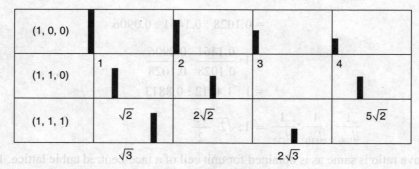

**FIGURE 8.54** Braggs Structure of KCl.

The first order spectrum from the (1, 0, 0), (1, 1, 0) and (1, 1, 1) planes of KCl was observed at the glancing angles 5.38°, 7.61° and 9.38°, respectively. Therefore,

$$d_{100}\ d_{110}\ d_{111} = \frac{1}{\sin 5.38°} : \frac{1}{\sin 7.61°} : \frac{1}{\sin 9.38°}$$

$$= \frac{1}{0.0938} : \frac{1}{0.1326} : \frac{1}{0.1630}$$

$$= 1 : 0.704 : 0.574$$

But, we know for a simple cubic lattice,

$$d_{100}\ d_{110}\ d_{111} = 1 : 0.704 : 0.574$$

Therefore, it follows that KCl crystal has a simple cubic lattice whereas NaCl has an FCC lattice.

## 8.42 EXPECTED SIMILARITY BETWEEN NaCl AND KCl

In almost every property, KCl resembles with NaCl, suggesting that the fundamental lattice should be the same, i.e. face-centred lattice. The expected behaviour is supported by the following observations:

The first order reflection from (1, 0, 0) planes of the rock salt and KCl occurred at 5.9° and 5.3°, respectively. As we know that the value of $d$ is inversely proportional to $\sin\theta$, it means

$$\frac{d_{100} \text{KCl}}{d_{100} \text{NaCl}} = \frac{\sin 5.9°}{\sin 5.3°} = 1.11$$

Since, the volume of a small cubic lattice is $(d_{111})^3$, it follows that if the KCl and NaCl have the same crystal lattice structure, the quantity $(1.11)^3 = 1.37$ should give the ratio of molecular volumes of the two salts. But the experimental value is 1.38, which suggests that the two must have the same space lattice, i.e face-centred cubic lattice. But, the Bragg's study indicated that KCl should have a simple cubic lattice. Thus, there exist some anomaly.

How to solve the above anomaly?

The above anomaly can be explained on the basis that,

*the X-ray scattering factor for an atom is equal to the number of extra planetary electron, viz. atomic number at small glancing angles*

The atomic number of K and Cl are 19 and 17, respectively. As these numbers are not very much different, the X-ray are unable to detect any difference between the two kinds of atoms. If we assume that all the atoms are identical, it means that the FCC arrangement of NaCl becomes a simple cubic arrangement for KCl.

In (1, 1, 1) face of KCl, the atomic numbers of K and Cl are so identical that the reflected intensities are nearly same and the odd orders are completely cut off.

Due to the above mentioned facts, KCl appears to be a simple cubic lattice. On the other hand, in the case of NaCl, there is difference between the atomic number of Na (11) and Cl (17). It means that their scattering factors are different and hence, give rise to an FCC lattice. From the above discussion we can safely claim that:

*Both the rock salt (NaCl) and Sylvine (KCl) have the same structure.*

Although the structures of NaCl and KCl are same, there is an important difference which may be observed by applying the expression of Bragg to both the crystals. Therefore,

$$\frac{d_{100} \text{NaCl}}{d_{100} \text{KCl}} = \frac{\sin 5.3°}{\sin 5.9°} = \frac{1}{1.11}$$

From the above expression, it is evident that the fundamental units are more widely spaced in the crystal of KCl than in the crystal of NaCl.

## 8.43 CLASSIFICATION OF SOLIDS

When a substance is at such a low temperature that the thermal agitation of its molecules is not strong. Intermolecular forces tends to hold the molecules together in more or less fixed positions. So these materials which acquire a fixed shape and size (volume), are called in solid state. So, these solid materials may be divided into two distinct classes (Table 8.1).

1. Crystalline solid
2. Amorphous solid

**TABLE 8.1** Two distinct classes of solids

| *Crystalline solids* | *Amorphous solids* |
|---|---|
| (i) The constituents particles are arranged in a regular fashion containing short range as well as long range order. | (i) The constituent particles are not arranged in regular fashion. There may be at the most some short range order only. |
| (ii) They have sharp melting points. | (ii) They melt over a range of temperatures. |
| (iii) They are anisotropic, i.e. properties like electrical conductivity, thermal expansion, etc. have different values in different directions. | (iii) They are isotropic, i.e properties like electrical conductivity, thermal expansion, etc. are same in different directions (just as in case of gases and liquids). |
| (iv) They undergo a clean cleavage. | (iv) They undergo an irregular cut. |

## 8.44 LATTICE ENERGY

The lattice energy of an ionic crystal is defined as the amount of energy released when cations and anions in their gaseous state are brought together from infinite distance to form a crystal i.e.

$$M^+(g) + X^-(g) \longrightarrow MX(s) + U$$

$U$ is the lattice energy

In other words, *the energy due to special arrangement of ions in a crystal* or the *energy required to separate the ions of a crystal is known as lattice energy of crystal.*

It is clear from above discussion that lattice energy is due to attraction of oppositely charged ions. So we can say that, in the formation of an ionic crystal energy is released, which is only due to attraction by it, is not satisfactory because as we know that ions are not any point charges, but they also consists of electron charge cloud. So, due to which there is also some repulsion between these ions. So the lattice energy is not only result of attractive forces between ions, but some contribution is also of the repulsive factors.

First of all we discuss about the attraction as given below,

*Attraction between ions OR Evaluation of Madelung Constant:*

As we know that, the coloumbic attraction between spherical cation and anion is

$$U = \frac{Z_i Z_j (e^2/4\pi\varepsilon_0)}{r_{ij}} \tag{8.37}$$

where, $i$ and $j$ are only for identifying ions which have charges $Z_i$ and $Z_j$, respectively, and $e$ is the electronic charge separated by a distance $r_{ij}$.

Consider, for a cation, which is interacting with all the other ions of the crystal, coloumbic interactions may be given as

$$U_+ = Z_+ \left(\frac{e^2}{4\pi\varepsilon_0}\right) \sum_i \frac{Z_i}{r_i^+} \tag{8.38}$$

where $r_i^+$ is the distance from the reference positive ion to $i$th ion. It is written by dimensionless ratio $\dfrac{r_i^+}{r}$ where $r$ is the shortest cation–anion distance in the crystal.

$$\frac{r_i^+}{r} = R_i^+ \tag{8.39}$$

Now multiplying and dividing Eq. (8.38) by $r$, we get

$$U_+ = \frac{Z_+}{r} \left(\frac{e^2}{4\pi\varepsilon_0}\right) \sum_i \frac{Z_i}{r_i^+/r}$$

From Eqs. (8.38) and (8.39), we get

$$U_+ = \frac{Z_+}{r}\left(\frac{e^2}{4\pi\varepsilon_0}\right)\sum_i \frac{Z_i}{R_i^+} \qquad (8.40)$$

By multiplying and dividing Eq. (8.40) by $Z_-$, we get

$$U_+ = \frac{Z_+ Z_-}{r}\left(\frac{e^2}{4\pi\varepsilon_0}\right)\sum_i \frac{Z_i/Z_-}{R_i^+} \qquad (8.41)$$

Similarly for an anion equation becomes

$$U_- = \frac{Z_+ Z_-}{r}\left(\frac{e^2}{4\pi\varepsilon_0}\right)\sum_i \frac{Z_i/Z_+}{R_i^-} \qquad (8.42)$$

Suppose in crystal $N$ cations and $N$ anions are present so the energy of crystal becomes

$$U = \frac{1}{2}N(U_+ + U_-) \qquad (8.43)$$

where term 1/2 prevents each ion–ion interaction from being counted twice. So from Eqs. (8.41), (8.42) and (8.43), we get

$$U = \frac{1}{2}N\left[\frac{Z_+ Z_-\left(\frac{e^2}{4\pi\varepsilon_0}\right)}{r}\sum_i \frac{Z_i/Z_+}{R_i^+} + \frac{Z_+ Z_-\left(\frac{e^2}{4\pi\varepsilon_0}\right)}{r}\sum_i \frac{Z_i/Z_-}{R_i^-}\right]$$

$$U_{(att.)} = \frac{NZ_+ Z_-\left(\frac{e^2}{4\pi\varepsilon_0}\right)}{r}\left[\frac{1}{2}\left(\sum_i \frac{Z_i/Z_-}{R_i^+} + \sum_i \frac{Z_i/Z_+}{R_i^-}\right)\right] \qquad (8.44)$$

So, in Eq. (8.44), the term present in square bracket depends upon the geometric arrangement of the ions in the crystal and the charges of these ions. This term is always constant and is known as *Madelung constant* denoted by $M$ and can be calculated and tabulated as in Table 8.2.

**TABLE 8.2** Madelung constant for various salts

| Rock salt | 1.7476 |
|---|---|
| Cesium chloride | 1.7627 |
| Zinc blende | 1.6380 |
| Flourite | 2.5194 |

*Repulsion between ions or calculation of repulsive potential exponent:*

In an ionic crystal there is some repulsion energy also to balance the attraction coloumbic energy. As considered the ions contain electron charge clouds so there is some repulsion as the ions come close to each other. This repulsion increases as the ions come very close to each other. So, according to Born, the repulsive energy is given by

$$U_{\text{rep}}(r) = \frac{B}{r^n} \tag{8.45}$$

where $B$ is a constant.

Experimentally Born exponent ($n$) can be determined from the compressibility data because the $n$ measures the resistance, which the ions exhibit when forced to approach each other very closely.

Thus, as we know that the total energy possessed by the crystal is due to the attraction and repulsion both. So, to calculate the repulsive constant and repulsive exponent, we discussed about energy with consideration that the crystal lattice consisting of Avogadro's number of ions.

So the total energy is given by

$$U_{\text{total}}(r) = U_{\text{att.}}(r) + U_{\text{rep.}}(r) \tag{8.46}$$

So by putting the value of $U_{\text{att}}(r)$ from Eq. (8.44) and $U_{\text{rep}}(r)$ from Eq. (8.45) in Eq. (8.46), we get

$$U_{\text{total}} = \frac{MN_A Z_+ Z_- e^2}{4\pi\varepsilon_0 r} + \frac{N_A B}{r^n} \tag{8.47}$$

To evaluate the repulsive constant, at the equilibrium lattice configuration, the attraction and repulsion are equal, so the rate of change of energy with respect to distance between ions is zero, we have

$$\frac{dU}{dr} = 0 \tag{8.48}$$

So, by differentiating Eq. (8.47) with respect to $r$ and compare with Eq. (8.48), we get

$$\Rightarrow \quad -\frac{MN_A Z_+ Z_- e^2}{4\pi\varepsilon_0 r} - \frac{nN_A B}{r^{n+1}} = 0 \tag{8.49}$$

So at equilibrium distance $r_0$, Eq. (8.49) becomes

$$\Rightarrow \quad -\frac{MN_A Z_+ Z_- e^2}{4\pi\varepsilon_0 r_0^2} - \frac{nN_A B}{r_0^{n+1}} = 0$$

$$\Rightarrow \quad -\frac{MN_A Z_+ Z_- e^2}{4\pi\varepsilon_0 r_0^2} = \frac{nN_A B}{r_0^{n+1}}$$

$$B = -\frac{MN_A Z_+ Z_- e^2}{4\pi\varepsilon_0 r_0^2} \times \frac{r_0^{n+1}}{nN_A}$$

$$B = -\frac{MZ_+ Z_- e^2 r_0^{n-1}}{4\pi\varepsilon_0 n} \tag{8.50}$$

So from Eq. (8.50), we have been calculating the repulsive constant $B$. But for getting the repulsive exponent, we must discuss about theoretical lattice energy.

## 8.45 THEORETICAL TREATMENT OF LATTICE ENERGY OR BORN–LANDE'S EQUATION

The theoretical treatment of ionic crystal lattice energy was given by M. Born and A. Lande. This treatment has been discussed below:

Consider the potential energy of an ion pair, $M^+X^-$, in a crystal separated by a distance $r$. The coloumbic electrostatic attraction is given by

$$U_{att}(r) = \frac{Z_+ Z_- e^2}{4\pi\varepsilon_0 r} \qquad (8.51)$$

In Eq. (8.51) $Z_+$ is a positive and $Z_-$ is a negative quantity and the product of a positive and negative quantity is always negative. So the sign of the above equation is always negative. This energy is more negative (or released) when the inter-ionic distance is very large and the energy is increased as the inter-ionic distance decreases. Note that $Z_+e$ is the charge of cation and $Z_-e$ is the charge of an anion. This situation is explained in Figure 8.55.

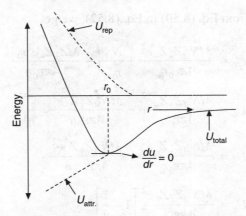

**FIGURE 8.55** Dependence of energy on interionic distance.

In a crystal lattice, there are more interactions between the ions than the simple one in an isolated ion pair. Thus, in NaCl lattice, each $Na^+$ ion experience attraction to the six nearest $Cl^-$ ions, repulsion by twelve nearest $Na^+$ ions, attraction to the next eight $Cl^-$ ions and repulsions by the next six $Na^+$ ions and so on.

The summation of all these geometrical interactions is known as *Medelung constant* ($M$). So, as we have discussed above that the energy of attraction in an ion pair in a crystal is given by Eqs. (8.44) and (8.51) as

$$U_{att}(r) = \frac{MZ_+ Z_- e^2}{4\pi\varepsilon_0 r}$$

The value of $M$ depends only on the geometry of the lattice and independent of ionic radius and charge. Thus, the value of Medelung constant in NaCl lattice is given by

$$M = 6 - \frac{12}{\sqrt{2}} + \frac{8}{\sqrt{3}} - \frac{6}{\sqrt{4}} + \cdots$$

A stable lattice can result only if there is also repulsive energy to balance the attractive columbic energy. The attractive energy becomes infinite at infinitesimally small distances as shown in Figure 8.55. However, the repulsion energy is also increased as the ions approach to each other very rapidly. As we have discussed the repulsive energy from Eq. (8.45) is

$$U_{(\text{rep})}(r) = \frac{B}{r^n}$$

Thus, total lattice energy is from Eq. (8.47)

$$U_{\text{total}} = \frac{MN_A Z_+ Z_- e^2}{4\pi\varepsilon_0 r} + \frac{N_A B}{r^n}$$

At equilibrium distance $r_0$ equation becomes

$$U_{0(\text{total})} = \frac{MN_A Z_+ Z_- e^2}{4\pi\varepsilon_0 r_0} + \frac{N_A B}{r_0^n} \tag{8.52}$$

By putting the value of $B$ from Eq. (8.50) in Eq. (8.52), we get

$$U_0 = \frac{MN_A Z_+ Z_- e^2}{4\pi\varepsilon_0 r_0} + \frac{N_A}{r_0^n}\left(-\frac{MZ_+ Z_- e^2 r_0^{n-1}}{4\pi\varepsilon_0 n}\right)$$

$$U_0 = \frac{MN_A Z_+ Z_- e^2}{4\pi\varepsilon_0 r_0} - \frac{MN_A Z_+ Z_- e^2}{4\pi\varepsilon_0 n} \cdot \frac{r_0^{n-1}}{r_0^n}$$

$$U_0 = \frac{MN_A Z_+ Z_- e^2}{4\pi\varepsilon_0 r_0} - \frac{MN_A Z_+ Z_- e^2}{4\pi\varepsilon_0 r_0 n}$$

$$U_0 = \frac{MN_A Z_+ Z_- e^2}{4\pi\varepsilon_0 r_0}\left[1 - \frac{1}{n}\right] \tag{8.53}$$

So given Eq. (8.53) is the Born–Lande's equation which gives the theoretical treatment of lattice energy.

### Calculation of repulsive potential exponent (n)

So from given Born–Lande's equation we can determine the value of $n$ as calculated as

$$U_0 = \frac{MN_A Z_+ Z_- e^2}{4\pi\varepsilon_0 r_0}\left[1 - \frac{1}{n}\right]$$

$$1 - \frac{1}{n} = \frac{U_0 \, 4\pi\varepsilon_0 r_0}{MN_A Z_+ Z_- e^2}$$

$$-\frac{1}{n} = \frac{U_0 \, 4\pi\varepsilon_0 r_0}{MN_A Z_+ Z_- e^2} - 1$$

$$\frac{1}{n} = 1 - \frac{U_0 4\pi\varepsilon_0 r_0}{MN_A Z_+ Z_- e^2}$$

$$n = 1 - \frac{MN_A Z_+ Z_- e^2}{U_0 4\pi\varepsilon_0 r_0} \tag{8.54}$$

So we can calculate the repulsive potential exponent. The value of $n$ depends upon the type of ion involved. Large ion having relatively higher value of $n$ because as $r_0$ increases, the value of $n$ become large.

## 8.46 LATTICE HEAT CAPACITY

Heat capacity of a substance is the energy required to raise the 1°C temperature of the substance. If substance refers to the solid material then the heat capacity termed as *lattice heat capacity*. So we can say that, if we want to raise the temperature of a lattice crystal through 1°C then the energy required is known as *lattice heat capacity*. If the substance or lattice crystal is one mole quantity then the term *molar heat capacity* is used.

Mathematically, for one mole of a system, the differentiation of lattice energy with respect to temperature at constant volume yields the molar heat capacity i.e.

$$C_V = \left(\frac{\partial U}{\partial T}\right)_V$$

In 1819, Dulong and Petit found that, the molar heat capacity at constant volume of most of the solid elements to room temperature was given by

$$C_V \approx 6 \text{ cal K}^{-1} \text{ mol}^{-1}$$

In order to account for this value, two theories of heat capacities were developed first by Albert Einstein in 1907 and second by Peter Debye in 1912.

## 8.47 EINSTEIN THEORY OF HEAT CAPACITY

Einstein theory have following postulates:

1. The atoms in a crystal lattice undergo small vibrations about their equilibrium configuration. In fact, an ideal crystal can be considered as a system of N atoms.
2. Each atom vibrates independently of the others and has three independent vibrational degrees of freedom. Thus, the crystal may be behave as a system of $3N$ independent and distinguishable hormonic oscillators.
3. There are no electronic, translational or rotational modes of motion in monoatomic crystal.

So, by using second and third postulates, the molar vibrational partition function of a crystal can be written as

$$Q_{\text{vib}} = (q_{\text{vib}})^{3N}$$
$$\ln Q_{\text{vib}} = 3N \ln q_{\text{vib}} \tag{8.55}$$

But

$$q_{\text{vib}} = \frac{e^{-h\nu/2kT}}{1 - e^{-h\nu/kT}} \tag{8.56}$$

where $\nu$ is the vibrational frequency of the oscillator.

and if $\dfrac{h\nu}{k} = \theta_{\text{vib}}$, i.e. characteristic Einstein temperature for vibration then Eq. (8.56) becomes

$$q_{\text{vib}} = \frac{e^{-\theta_{\text{vib}}/2T}}{1 - e^{-\theta_{\text{vib}}/T}} \tag{8.57}$$

By putting the value of Eq. (8.57) in Eq. (8.55), we get

$$\ln Q_{\text{vib}} = 3N \ln \frac{e^{-\theta_{\text{vib}}/2T}}{1 - e^{-\theta_{\text{vib}}/T}}$$

$$\ln Q_{\text{vib}} = 3N \ln \left[ e^{-\theta_{\text{vib}}/2T} - (1 - e^{-\theta_{\text{vib}}/T}) \right]$$

$$\ln Q_{\text{vib}} = 3N \ln e^{-\theta_{\text{vib}}/2T} - \ln(1 - e^{-\theta_{\text{vib}}/T}) 3N$$

$$\ln Q_{\text{vib}} = -\frac{3}{2} \frac{N\theta_{\text{vib}}}{T} - 3N \ln(1 - e^{-\theta_{\text{vib}}/T}) \tag{8.58}$$

Now the internal energy of an ideal Einstein crystal is given by

$$U = kT^2 \left[ \frac{\partial \ln Q_{\text{vib}}}{\partial T} \right]_{V,N} \tag{8.59}$$

By putting the value of Eq. (8.58), Eq. (8.59) becomes

$$U = kT^2 \left( \frac{\partial}{\partial T} \left[ -\frac{3}{2} \frac{N\theta_{\text{vib}}}{T} - 3N \ln(1 - e^{-\theta_{\text{vib}}/T}) \right] \right)$$

$$U = \frac{3}{2} Nh\nu + \frac{3Nk\theta_{\text{vib}}}{e^{\theta_{\text{vib}}/T} - 1} \tag{8.60}$$

But zero point energy is given by

$$U_0 = \frac{3}{2} Nh\nu = \frac{3}{2} R\theta_{\text{vib}} \tag{8.61}$$

So,

$$U - U_0 = \frac{3}{2} RT \frac{\theta_{\text{vib}}/T}{e^{\theta_{\text{vib}}/T} - 1} \tag{8.62}$$

But the molar heat capacity is given by

$$C_V = \left( \frac{\partial U}{\partial T} \right)_V$$

$$C_V = 3R \left( \frac{\theta_{\text{vib}}}{T} \right)^2 \frac{e^{\theta_{\text{vib}}/T}}{(e^{\theta_{\text{vib}}/T} - 1)^2} \tag{8.63}$$

As we considered that experimentally it is found that as very low temperature, i.e. at the limit on which $T \to 0$, $C_V$ approaches zero and at high temperature, i.e. at the limit on which $T \to \infty$, $C_V$ approaches the Dulong–Petit value of $3R$, i.e., $\approx 6$ cal K$^{-1}$ mol$^{-1}$.

Einstein Eq. (8.63) predicts these value successfully as $T \to 0$

$$e^{\theta_{vib}/T} - 1 \approx e^{\theta_{vib}/T}$$

Then $\lim\limits_{T \to \infty} C_V = 3R \left(\dfrac{\theta_{vib}}{T}\right)^2 e^{\theta_{vib}/T} = 0$ and again at $T \to \infty$

$$e^{\theta_{vib}/T} \approx 1 + \left(\dfrac{\theta_{vib}}{T}\right) \text{ Then}$$

$$\lim_{T \to \infty} C_V = 3R \left(\dfrac{\theta_{vib}}{T}\right)^2 \dfrac{e^{\theta_{vib}/T}}{1 + e^{\theta_{vib}/T} - 1}$$

$$C_V = 3R \left(\dfrac{\theta_{vib}}{T}\right)^2 \dfrac{e^{\theta_{vib}/T}}{(\theta_{vib}/T)^2}$$

$$C_V = 3R\, e^{\theta_{vib}/T}$$

$$C_V \approx 3R e^0$$

$$C_V \approx 3R \approx 6 \text{ cal K}^{-1} \text{ mol}^{-1}$$

Einstein's theory, however, is not successful in predicting the $C_v$ values in the lower and intermediate temperature ranges: the value predicted by it are lower than those actually observed.

## 8.48 DEBYE THEORY OF HEAT CAPACITIES OR DEBYE T-CUBED LAW

Peter Debye avoided the second postulate of the Einstein theory that the vibrations in a crystal lattice are independent. He recognized that the interionic forces in the crystal are very strong and hence atoms may not be treated as being independent. Debye assumed that the crystal behaves like a huge molecule, where in the motion of any one atom affects the motion of the neighbouring atoms.

According to Debye, a monoatomic crystal containing $N$ atoms must be considered as a system of $3N$ coupled harmonic oscillators. The frequency of each oscillator is not equal, but there is frequency distribution and the lowest frequency is taken as zero. If the temperature is not too high, the amplitudes of vibration are relatively small. Further, the collective modes of vibrations are the possible sound waves which can propagate through solids. The energy of sound wave inside a solid medium can be considered to be quantized in the form of phonons. This is similar to the energy in the form of photons. Phonons are not considered as true particles.

Assuming a continuous distribution of frequencies, the frequency distribution function is defined as

$$dN = f(\nu)\, d(\nu) \tag{8.64}$$

where $dN$ is the number of normal modes of vibrations in the frequency range from $v$ to $v+dv$
So it is thus required that

$$\int_0^{V_D} f(v)\,d(v) = 3N \tag{8.65}$$

where $V_D$ or $V_{max}$ is maximum possible oscillation frequency.

The partition function for the system of the Debye oscillator is given by

$$Q = q(v_1)^{f(v_1)} q(v_2)^{f(v_2)} \ldots q(v_D)^{f(v_D)}$$

$$\ln Q = \sum_{i=1}^{V_D} f(v_i) \ln q(v_i)$$

As a result of continuity, the summation may be replaced by integration, so that

$$\ln Q = \int_0^{V_D} f(v_i) \ln q(v)\,dv \tag{8.66}$$

where $q(v)$ is the partition function of the oscillator which is defined as

$$q(v) = q_{vib} = \frac{e^{-hv/2kT}}{1 - e^{-hv/kT}} \tag{8.67}$$

Debye used the Rayleigh–Jeans relation for the distribution function, viz.

$$f(v)\,d(v) = Cv^2\,dv \tag{8.68}$$

where $c$ is a constant and determined from the restriction that

$$\int_0^{V_D} f(v)\,d(v) = C\int_0^{V_D} v^2\,dv$$

$$\int_0^{V_D} f(v)\,d(v) = \frac{1}{3} CV_D^3 \tag{8.69}$$

By comparing Eqs. (8.65) and Eq. (8.69), we get

$$\frac{1}{3} CV_D^3 = 3N$$

$$C = \frac{9N}{V_D^3} \tag{8.70}$$

Now putting the value of $C$ in Eq. (8.68), we get

$$f(v)\,d(v) = \frac{9N}{V_D^3} V^2\,dv \tag{8.71}$$

By taking the log of Eq. (8.67), we get

$$\ln q(v) = -\frac{hv}{2kT} - \ln(1 - e^{-hv/kT}) \tag{8.72}$$

Now putting the value of Eqs. (8.71) and (8.72) in Eq. (8.66),

$$\ln Q = \int_0^{V_D} \left(\frac{9N}{V_D^3}\right) V^2 dv \left[-\frac{hv}{2kT} - \ln(1 - e^{-hv/kT})\right]$$

$$\ln Q = -\frac{9N}{V_D^3} \int_0^{V_D} \left[\frac{hv}{2kT} + \ln(1 - e^{-hv/kT})\right] V^2 dv \tag{8.73}$$

But

$$U = kT^2 \left(\frac{\partial \ln Q}{\partial T}\right)_{V,N}$$

$$U = \frac{9NkT}{V_D^3} \int_0^{V_D} \left[\frac{hv/kT}{e^{hv/kT} - 1} + \frac{hv}{2kT}\right] V^2 dv$$

$$U = \frac{9NkT}{V_D^3} \int_0^{V_D} \frac{hv}{2kT} V^2 dv + \frac{9NkT}{V_D^3} \int_0^{V_D} \frac{hv/kT}{e^{hv/kT} - 1} V^2 dv$$

$$U = \frac{9NkT}{V_D^3} \frac{h}{2kT} \int_0^{V_D} V^3 dv + \frac{9NkT}{V_D^3} \int_0^{V_D} \frac{hv/kT}{e^{hv/kT} - 1} V^2 dv$$

$$U = \frac{9Nh}{2V_D^3}\left[\frac{V_D^4}{4}\right] + \frac{9NkT}{V_D^3} \int_0^{V_D} \frac{hv/kT}{e^{hv/kT} - 1} V^2 dv$$

$$U = \frac{9NhV_D}{8} + \frac{9NkT}{V_D^3} \int_0^{V_D} \frac{hv/kT}{e^{hv/kT} - 1} V^2 dv \tag{8.74}$$

After adding $Nk = R$ and by putting $\frac{hv}{k} = \theta$ and $\frac{hV_D}{k} = \theta_D$, where $\theta_D$ is called the *characteristic Debye temperature of the crystal*. We obtain after multiplying and dividing first term by $k$

$$U = \frac{9R\theta_D}{8} + 9RT\left(\frac{T}{\theta_D}\right)^3 \int_0^{\theta_D/T} \left[\frac{(\theta/T)^3}{e^\theta - 1}\right] d\left(\frac{\theta}{T}\right) \tag{8.75}$$

Integral in Eq. (8.75) can be evaluated numerically. It is convenient to define the Debye function as

$$D_{(x)} = \frac{3}{x^3} \int_0^x \frac{Z^3}{e^\theta - 1} dZ \tag{8.76}$$

where, $x = \dfrac{\theta_D}{T}$ and $Z = \dfrac{\theta}{T}$

So, from Eq. (8.70), Eq. (8.75) becomes

$$U = \dfrac{9R\theta_D}{8} + 9RTD_{(\theta_D/T)} \qquad (8.77)$$

By differentiating $D_{(\theta_D/T)}$ and then carrying out differentiation by parts, we obtain

$$C_V = 3RD\left(\dfrac{\theta_D}{T}\right) + 3RT\dfrac{\partial}{\partial T}\left[D\left(\dfrac{\theta_D}{T}\right)\right]$$

$$C_V = 3R\left[4\left(\dfrac{\theta_D}{T}\right) - \dfrac{3(\theta_D/T)}{e^{\theta_D/T} - 1}\right] \qquad (8.78)$$

Now at higher temperature,

$$e^{\theta_D/T} - 1 \approx 1 + (\theta_D/T) - 1 \approx \left(\dfrac{\theta_D}{T}\right)$$

and $\lim\limits_{T \to \infty} D(\theta_D/T) \simeq 1$

Hence, $\lim\limits_{T \to \infty} C_V \simeq 3R$

Again at low temperature,

$$\lim\limits_{T \to \infty} D(\theta_D/T) \simeq \dfrac{\pi^4}{5}\left(\dfrac{T}{\theta_D}\right)^3$$

So that,

$$\lim\limits_{T \to \infty} C_V \cong \dfrac{12}{5}\pi^4 R\left(\dfrac{T}{\theta_D}\right)^3$$

$$C_V = 234\, R\left(\dfrac{T}{\theta_D}\right)^3 \qquad (8.79)$$

Since all the quantities on the RHS of Eq. (8.79) except $T$ are constant for a particular crystal then we can write this equation as

$$C_V = \alpha\, T^3 \qquad (8.80)$$

where $\alpha = \dfrac{234R}{\theta_D^3}$ and Eq. (8.80) is a famous *T-cubed law*.

## EXERCISES

1. Derive the Bragg's equation for X-ray crystallography.
2. What is lattice energy? Derive Born–Lande equation for the lattice energy of an ionic solid.

3. Describe electron sea model of metallic structure. How does this model explain the common properties of metals?
4. What are Miller Indices? How are they determined?
5. What are extrinsic semiconductors? Explain how the conductance of silicon (or germanium) can be increased by doping into it arsenic and indium.
6. What are $n$-type and $p$-type semiconductors. Explain how their combinations find application in the fabrication of transistors?
7. What is superconductivity? How would you explain superconductivity of metals.
8. What types of defects are commonly produced in crystals. Discuss briefly the following types of defects: (i) Schottky defect, (ii) Frenkel defect, (iii) Metal excess defect, (iv) Metal deficiency defects, (v) Line defects.
9. What is the cause of Schottky defects? Derive an expression for the number of Schottky defects in a crystal.
10. When do Frenkel defect arise? Derive an expression for the number of Frenkel defects in a crystal.
11. What do you mean by colour centres? How do they appear?
12. What is meant by dislocations? What are common types of dislocations?
13. NaCl crystallizes in a face centred cubic lattice. Calculate the number of unit cells in 1.0 g of the crystal. What is the number along each edge of the crystal?
[**Ans.** $2.57 \times 10^{21}$ unit cells, $1.37 \times 10^7$]
14. Calculate the angles at which first, second and third order reflections are obtained from planes 500 pm apart, using X-rays of wave length 100 pm.
[**Ans.** 5.74°, 11.54°, 17.46°]
15. Ag crystallizes in a cubic lattice. The density is $10.7 \times 10^3$ kg/m$^3$. If the edge length of the unit cell is 406 pm, determine the type of the lattice. [**Ans.** FCC]

## SUGGESTED READINGS

Azaroff, L.V., *Introduction to Solids*, Tata McGraw-Hill, New Delhi, 1960.

Berg, R.J. and G.J. Dienes, *Physical Chemistry of Solids*, Academic Press, 1992.

Chakrabarty, D.K., *Solid State Chemistry*, New Age International, New Delhi, 1996.

Cheetham, A.K. and P. Day, *Solid State Chemistry Compound*, Oxford University Press, Oxford, 1992.

Glusker, J.P. and K.N. Trueblood, *Crystal Structure Analysis*, Oxford University Press, Oxford, 1985.

Hannay, N.B., *Solid State Chemistry*, Prentice Hall, New Delhi, 1967.

Harrison, W.A., *Electronic Structure and Properties of Solids: The Physics of the Chemical Bond*, W.H. Freeman, 1980.

Hoffmann, R., *Solids and Surfaces: A Chemist View of Bonding in Extended Structures*, VCH Publishers, 1988.

Kachhava, C.M., *Solid State Physics*, Tata McGraw-Hill, New Delhi, 1990.

Keer, H.V., *Principles of Solid State*, Wiley Eastern, 1971.

Kroger, F.A., *The Chemistry of Imperfect Crystals*, North Holland, 1973.

Nelson, D.L. and T.F. George, *Chemistry of High Temperature Superconductor II*, Oxford University Press, Oxford, 1999.

Rao C.N.R. and J. Gopalakrishnan, *New Directions in Solid State Chemistry*, 2nd ed., Cambridge University Press, 1997.

Smart, L. and E. Moore, *Solid State Chemistry: An Introduction*, Chapman & Hall, 1995.

Stout, G.H. and L.H. Jensen, *X-Ray Structure Determination*, Macmillan Press, London, 1997.

West, A.R., *Basic Solid State Chemistry*, 2nd ed., John Wiley, 1999.

West, A.R., *Solid State Chemistry and its Applications*, John Wiley, 1987.

West, A.R., *Solid State Chemistry and its Application*, Plenum Press, New York.

Whitesell, J.K., *Organized Molecular Assemblies in the Solid State*, Wiley, New York, 1999.

Ziman, J., *Theory of Solids*, Cambridge University Press, 1964.

# Part III
# Instrumental Techniques

# 9

# Reference Electrodes and Potentiometric Methods

## 9.1 INTRODUCTION

In many electro-analytical methods, it is desirable that the half cell potential of one electrode be known, must be constant and completely insensitive to the composition of the solution under study. An electrode that fits this description is called a *reference electrode*. Employed in connection with the reference electrode will be an indicator electrode whose response is dependent upon the analyte concentration.

Various types of reference electrodes are:

## 9.2 CALOMEL ELECTRODE

Calomel half cell may be represented as follows:

$$\| Hg_2Cl_2(Sat.), KCl(XF)/Hg$$

where $X$ represents the formal concentration of KCl in the solution. The electrode reaction is given by the equation

$$Hg_2Cl_2(s) + 2e \rightleftharpoons 2Hg + 2Cl^-$$

The potential of this cell will vary with the chloride ion concentration $X$, and this quantity must be specified in describing the electrode (Table 9.1).

**TABLE 9.1** Specification of various electrodes available in the market

| Name | Conc. of $Hg_2Cl_2$ | Conc. of KCl | Reduction potential at 25°C |
|---|---|---|---|
| Saturated | Saturated | Saturated | + 0.242 V |
| Normal | Saturated | 1.0 F | + 0.280 V |
| Decinormal | Saturated | 0.1 F | + 0.334 V |

Note that each solution is saturated with mercury(I) chloride and that the cell differ only with respect to the KCl concentration. Note also that the potential of the normal calomel electrode is greater than the standard potential for half cell reaction because chloride ion activity in a 1 F solution of KCl is significantly smaller than 1.

Saturated calomel electrode (SCE) is most commonly used by the analytical chemists because of the ease with which it can be prepared. Compared with the other two, its temperature coefficient is somewhat larger; this is a disadvantage only where substantial changes in temperature occur during the measurement process.

A simple easily constructed saturated calomel electrode is shown in Figure 9.1. The salt bridge, simply a tube filled with saturated potassium chloride, supported over gelatine provides electric contact with the solution surrounding the indicator electrode. A fritted disk or a wad of cotton at one end of the salt bridge is often employed to prevent dropping of cell liquid and contamination of the solution by foreign ions.

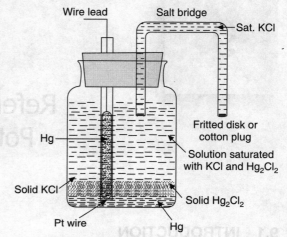

FIGURE 9.1  Saturated calomel electrode.

Several convenient calomel electrodes are available commercially. Typical of these is the one shown in Figure 9.2. It consists of a tube 5–10 cm in length and 0.5–1.0 cm in diameter. Mercury–mercury(I) chloride paste is concentrated in inner tube which is connected to the saturated KCl solution in the outer tube through a small opening. Contact with the second half cell is made by means of a fritted disk or porous fibre sealed in the end of the outer tubing. An electrode such as this has a relatively high resistance (2000–3000 $\Omega$) and a limited current carrying capacity. Convenient commercial electrodes are generally of two types: (1) fibre type and (2) sleeve type, depending upon construction of opening of outer tube. In fibre type, outer tube opening is of fine capillary plugged with asbestos fibre. In sleeve type, outer tube opening is of ground glass sleeve supported with washer on lower side (Figure 9.3).

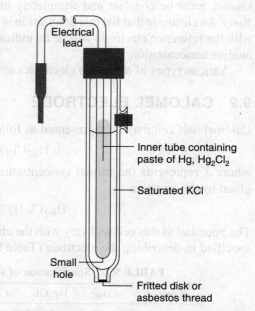

FIGURE 9.2  Saturated calomel electrode.

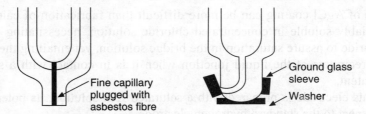

**FIGURE 9.3** Fibre type and sleeve type convenient commercial calomel electrodes.

Note: A reference electrode should satisfy following essential conditions:
1. The potential of the internal half cell must not be significantly altered if a small current passes through it.
2. The resistance of the entire electrode must not be too great.
3. The electrode should be easily assembled.
4. The components should be stable in contact with the atmosphere and at the operating temperature.

Various components of reference electrodes are:
1. The actual reference internal half cell, usually AgCl/Ag or calomel.
2. Salt bridge electrolyte.
3. A small channel in the tip of the electrode through which the salt bridge electrolyte flows very slowly and electrical contact is made with the other components of the electrochemical cell.

The standard potential of the reference electrodes includes the liquid junction potential of the electrochemical cell.

Calomel electrode comprises of a nonattackable element, such as platinum, in contact with Hg, mercury(I) chloride (calomel) and a neutral solution of KCl of known concentration and saturated with calomel.

A saturated calomel electrode exhibits a perceptible fluctuation following temperature changes, due in part to the time required for solubility equilibrium to be established. Those designed for measurements at elevated temperature have a large reservoir for KCl crystals. Calomel electrode becomes unstable at temperatures above 80°C and should be replaced with Ag/AgCl electrodes. In measurement in which any chloride ion contamination must be avoided, the mercury(I) sulphate and potassium sulphate electrode may be used.

## 9.3 SILVER/SILVER CHLORIDE ELECTRODE

The silver/sliver chloride electrode consists of metallic silver (wire, rod or gauze) coated with a layer of AgCl and immersed in a chloride solution of known concentration, i.e. also saturated with AgCl. The cell formed is,

$$|| AgCl\ (Sat.),\ KCl(XF)/Ag$$

The half cell reaction is $AgCl(s) + e^- \rightleftharpoons Ag + Cl^-$

It is a small compact electrode and can be used in any orientation. Electrode potentials are known up to 275°C.

Preparation of AgCl coating can be more difficult than fabrication of calomel electrode. AgCl is appreciably soluble in concentrated chloride solution, necessitating the addition of solid silver chloride to assure saturation in the bridge solution, yet entailing the risk that silver chloride may precipitate at the liquid junction when it is in contact with a solution of low chloride ion content.

Normally this electrode is prepared with a saturated KCl solution, its potential at 25°C is 0.197 V with respect to the standard hydrogen electrode.

A simple and easy constructed Ag/AgCl electrode is shown in Figure 9.4. The electrode is contained in a pyrex tube fitted with a 10 mm fritted glass disk. A layer of agar gel saturated with KCl is formed on the top of disk to prevent loss of solution from the half cell. The plug can be prepared by suspending about 5 g of pure Agar in 1000 ml of boiling water and then adding about 35 g of KCl. A portion of this suspension, while still warm is poured into tube; upon cooling, it solidifies to a gel with low electrical resistance. A layer of solid KCl is placed on the gel and the tube is filled with saturated solution of the salt. A drop or two drops of 1-F silver nitrate is then added and heavy gauze (1–2 mm diameter) silver wire is inserted in the solution.

In an aqueous titration, studies of this electrode occupied a pre-eminent position for many years, although the calomel electrode can and has been employed in virtually all types of solvent systems. Reproducibility of results vary from ±10 to ±20 mV in the more aqueous solvent mixtures, to ±50 mV in the nearly anhydrous media. Special salt bridges are often necessary.

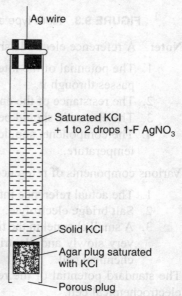

**FIGURE 9.4**  A Silver–silver chloride electrode.

*Tl/TlCl electrodes:* Construction, working and principle are same as that of calomel electrode with a few exceptions.

## 9.4 INDICATOR ELECTRODES

The measurement of the potential of an appropriate indicator electrode has been used for many years as a method of detecting the equivalence point in a variety of titrations. Interest is focussed on changes in the EMF of the electrochemical cell as a titrant of precisely known concentration is added to a solution of the test element.

EMF is directly reported as an activity of ions present in indicator electrodes. Requirements of reference electrodes is only necessary that the activity of one of the member of a pair of electrodes be substantial greater or faster than that of the other. Further, information about the sample and its reaction may be obtained by the complete recording of a potentiometric titration curve.

The chief advantage of potentio-metric titrations are applicability of turbid, fluorescent, opaque or coloured solutions or when suitable visual indicator electrodes are unavailable or inapplicable. Potentio-metric titration generally offer an increase in accuracy and precision at the cost of increased time and difficulty. Accuracy is increased because measured potential are used to detect rapid changes in activity that occur at equivalence point of the titration. Furthermore, it is the change in EMF versus titrant volume rather than absolute value of the EMF which is of interest. Thus, the influence of liquid junction potentials and activity coefficients have little or no effects.

### 9.4.1 Classification of Indicator Electrodes

1. *Electrodes of first order or (first kind):* Electrodes of the first kind are reversible with respect to the ions of the metal phase. The electrode is a piece of metal in contact with a solution of its ions, for example, silver dipping into a silver nitrate solution. One interface is involved. For the half cell

$$Ag^+ + e^- \rightleftharpoons Ag$$

The Nernst expression is

$$E = 0.799 - 0.0591 \log \frac{1}{[Ag^+]}$$

The metal must be thermodynamically stable with respect to air oxidation, especially at low ion activities. In neutral solutions, suitable electrodes are restricted to $Hg_2^{2+}/Hg$ and $Ag^+/Ag$. If oxygen is removed from the solution by de-aeration, other electrodes become feasible: $Cu^{2+}/Cu$, $Bi^{3+}/Bi$, $Pb^{2+}/Pb$, $Cd^{2+}/Cd$, $Sn^{2+}$, $Tl^+/Tl$ and $Zn^{2+}/Zn$.

A simple amalgam electrodes are also electrodes of first kind. For zinc, the reaction at the electrode is,

$$Zn^{2+} + Hg + 2e \rightleftharpoons Zn(Hg)$$

2. *Electrodes of second order or (second kind):* Electrode of second order involves two interfaces, such as metal coated with a layer of one of its sparingly soluble salts. The underlying electrode must be reversible. Consider a silver wire coated with a thin deposit of silver chloride. At the Ag/AgCl solution interface the electrochemical equilibrium is

$$AgCl(s) + e \rightleftharpoons Ag + Cl^-$$

In addition, there is a chemical equilibrium

$$AgCl(s) \rightleftharpoons Ag^+ + Cl^-; \quad K_{sp} = 1.8 \times 10^{-10}$$

Combining these two equations, we arrive at the Nernst expression

$$E = 0.799 - 0.0591 \log K_{sp} - 0.0591 \log [Cl^-]$$

This simplifies to

$$E = 0.222 - 0.0591 \log [Cl^-]$$

Electrodes of second order can be used for the direct determination of the activity of either the metal ion or the anion in the coating and also as an indicator electrode to follow titrations. Limitations on these electrodes are severe. They can be used only over a range of anion activities such that the solution remain saturated with respect to the metal coating. Interferences from other anions can occur if they too form an insoluble salt with the cation of the underlying electrodes.

3. *Electrodes of the third kind:* Reilley and co-workers showed how to use an electrode of known reversibility to measure activities of ions for which no electrode of the first kind exists. They used a small mercury electrode (or gold amalgam wire) in contact with a solution containing metal ions to be titrated with a chelon Y such as EDTA. A small added quantity of mercury(II) chelonate, $HgY^{2-}$ saturated the solution and established the half cell.

$$Hg/HgY^{2-}; MY^{(n-2)+}; M^{n+}$$

where, M is gold metal, and electrode potential is given by

$$E = E^0 + \frac{0.0591}{2} \log \frac{[M^{n+}][HgY^{2-}]}{[MY^{(n-2)+}]}$$

Because a fixed amount of $HgY^{2-}$ is present, the potential is dependent upon the ratio $\frac{[M^{n+}]}{[MY^{(n-2)+}]}$.

## 9.4.2 Metallic Redox Indicator Electrode

The redox electrode usually gold, platinum or carbon immersed in a solution containing both the oxidized and reduced states of a homogeneous and reversible oxidation–reduction system, develops a potential proportional to the ratio of two oxidation states. The only role of the redox electrode is to provide or accept electrons. An example is a platinum in contact with a solution of iron(III) and iron(II) ions. For the half reaction,

$$Fe^{3+} + e \rightleftharpoons Fe^{2+}; \quad E^0 = 0.771 \text{ V}$$

The Nernst expression is

$$E = 0.771 + \frac{0.0591}{1} \log \frac{[Fe^{2+}]}{[Fe^{3+}]}$$

Platinum electrodes are unsuitable for work with solutions containing powerful reducing agents, such as chromium(II), titanium(III) and vanadium(II) ions, because platinum catalyses the reduction of hydrogen ion by these reductants at the platinum surface. Consequently, the interfacial electrode potential will not reflect the changes in the composition of the solution. In these cases, a small pool of mercury can be used as an electrode because of high overpotential associated with the deposition of hydrogen gas on a mercury surface.

## 9.5 MEMBRANE ELECTRODE

Most convenient method for determining pH has involved the measurement of the potential that develops across a thin glass membrane separating the two solutions with different hydrogen ion concentrations. The phenomenon first reported by Cremer, has been extensively studied by many investigators; as a result the sensitivity and selectivity of glass membranes to pH is reasonably well understood. Furthermore, membrane electrodes have now been developed for the direct potentiometric determination of other ions, such as $K^+$, $Na^+$, $Li^+$, $F^-$ and $Ca^{2+}$.

It is convenient to divide membrane electrodes into four categories, based on membrane composition. These include (1) glass electrode, (2) liquid membrane electrode, (3) solid state or precipitate electrode and (4) gas sensing membrane. We shall discuss the properties and behaviour of a glass electrode in detail because of both its present and historical importance, greater availability, easy to operate and high stability.

## 9.6 THE GLASS ELECTRODE

Figure 9.5 shows a modern cell for the measurement of pH. It consists of commercially available calomel and glass electrode immersed in the solution, whose pH is to be measured. The calomel reference electrode is similar to the one described earlier.

The glass electrode is manufactured by sealing a thin, pH sensitive glass tip to the end of a heavy walled glass tubing. The resulting bulb is filled with a solution of hydrochloric acid (0.1 F) that is saturated with silver chloride. A silver wire is immersed in the solution and is connected via an external lead to one terminal of a potential measuring device; the calomel electrode is connected to the other terminal. The cell contains two reference electrodes each of whose potential is independent of pH; one is external calomel electrode and the other

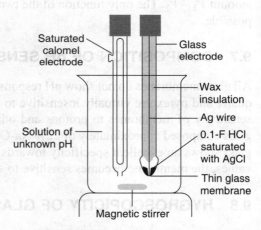

**FIGURE 9.5** The cell showing calomel and glass electrode.

is the internal silver–silver chloride reference electrode, which is a part of the glass electrode but is not the pH-sensitive component. In fact, it is the thin membrane at the tip of the electrode that responds to pH changes. Representation of a glass calomel cell for measurement of pH is given below.

$$\text{SCE} \parallel [H_3O]^+ = a_1 \left| \begin{array}{c} \text{Glass} \\ \text{membrane} \end{array} \right| [H_3O]^+ = a_2 [Cl^-] = 1.0 \text{ F, AgCl(Sat.)} \mid \text{Ag}$$

Here glass electrode consists of a silver–silver chloride reference electrode immersed in a solution of a fixed hydronium ion concentration and a glass membrane that separates these contents from the analyte solution.

Potential of a cell depends upon the hydrogen ion activities $a_1$ and $a_2$ at 25°C as follows:

$$E = Q + 0.0591 \log \frac{[a_1]}{[a_2]}$$

where $Q$ is a constant whose value is given by

$$Q = E_{Ag, AgCl} - E_{SCE} + j$$

where $E_{Ag,AgCl}$ and $E_{SCE}$ are the potential of two reference electrodes and $j$ is the asymmetry potential; the source and properties of $j$ will be considered later.

In practice, the hydrogen ion activity of the internal solution $a_2$ is fixed by its method of preparation and is constant. Therefore,

$$E = L + 0.0591 \log a_1$$
$$E = L - 0.0591 \text{ pH}$$

where $L$ is a new constant containing the logarithmic function of $a_2$.

It is important to note that the potential $E_{SCE}$, $E_{Ag,AgCl}$ and $j$ remain constant during a pH measurement. Thus, the source of the pH-dependent variation in $E$ must lie in the glass membrane. That is, when $a_1$ and $a_2$ differs, the two surfaces of membrane must differ in potential by some amount $V_1 - V_2$. The only function of the two electrodes is to make observation of this difference possible.

## 9.7 COMPOSITION OF pH SENSITIVE GLASS MEMBRANES

All glass membranes do not show pH response as shown in representation of calomel cell. Indeed, quartz and pyrex are virtually insensitive to pH variations. Many glass electrodes were tried for sensitivity of membranes to protons and other cations, and it has been found that corning 015 glass composed approximately of 22% $Na_2O$, 6% $CaO$ and 72% $SiO_2$ has been widely used. This glass shows an excellent specificity towards hydrogen ions up to a pH of about 9. At higher pH values, the membrane becomes sensitive to sodium and other alkali ions as well.

## 9.8 HYGROSCOPICITY OF GLASS MEMBRANES

It has been shown that the surface of a glass membrane must be hydrated in order to function as a pH electrode. Even corning 015 glass shows little pH response after dehydration by storage over a desiccant; however its sensitivity is restored after standing for a few hours in acidulated water. The hydration involves absorption of approximately 50 mg of water per cubic centimetre of glass.

## 9.9 RESISTANCE OF GLASS MEMBRANES

A typical commercial glass electrode which ranges in thickness between 0.03 and 0.1 mm, has an electrical resistance of 50–500 MΩ. Current conduction through the dry glass region is ionic and involves movement of the alkali ions from one site to another. Within the two gel layers the current is carried by both alkali and hydrogen ions. At each gel–solution interface, current passage involves the transfer of protons. The direction of transfer will be from gel to solution at the one,

$$\underset{\text{Solid}}{H^+Gl^-} \rightleftharpoons \underset{\text{Solid}}{Gl^-} + \underset{\text{Soln.}}{H^+} \tag{9.1}$$

and from the solution to gel at the other.

$$\underset{\text{Solid}}{Gl^-} + \underset{\text{Soln.}}{H^+} \rightleftharpoons \underset{\text{Solid}}{H^+Gl^-} \tag{9.2}$$

| External solution surface site occupied by $H^+$ | Hydrated gel ~$10^{-4}$ mm sites occupied by $H^+$ and $Na^+$ | Dry glass layer 0.1 mm. All sites occupied by $Na^+$ | Hydrated gel ~$10^{-4}$ mm sites occupied by $H^+$ and $Na^+$ | Internal solution surface sites occupied by $H^+$. $[H^+] = a_2$ |
|---|---|---|---|---|

## 9.10 THEORY OF GLASS–ELECTRODE POTENTIAL

The potential across a glass membrane consists of a boundary potential and a diffusion potential. Under ideal condition only the boundary potential is affected by pH.

The boundary potential for the membrane of a glass electrode contains two components, each associated with one of the gel–solution interfaces. If $V_1$ is the component arising at the interface between the external solution and the gel and if $V_2$ is the corresponding potential at the inner surface (i.e. internal solution and gel) then the boundary potential $E_b$ for the membrane is given by

$$E_b = V_1 - V_2 \quad (9.3)$$

The potential $V_1$ is determined by the hydrogen ion activities both in the external solution and upon the surface of the gel; it can be considered, a measure of the driving force of the reaction shown by Eq. (9.2). In the same way, the potential $V_2$ is related to the hydrogen ion activities in the internal solution and on the corresponding gel surface.

It can be demonstrated from thermodynamic considerations that $V_1$ and $V_2$ are related to hydrogen ion activities at each interface as follows.

$$V_1 = j_1 + \frac{RT}{F} \ln \frac{a_1}{a_1'} \quad (9.4)$$

$$V_2 = j_2 + \frac{RT}{F} \ln \frac{a_2}{a_2'} \quad (9.5)$$

where $a_1$ and $a_2$ are activities of the hydrogen ion in the solutions on either side of the membrane; and $a_1'$ and $a_2'$ are the hydrogen ion activities in each of the gel layers contacting the two solutions. If the two gel surfaces have the same number of sites from which protons can leave, then the two constants $j_1$ and $j_2$ will be identical, so also will be the two activities $a_2'$ and $a_2'$ in the gel layers, provided that, all original sodium ions on the surface have been replaced by protons. If these equalities are justified, substitution of Eqs. (9.4) and (9.5) into Eq. (9.3) yields,

$$E_b = V_1 - V_2 = \frac{RT}{F} \ln \frac{a_1}{a_2} \quad (9.6)$$

Thus, provided the two gel surfaces are identical, the boundary potential $E_b$ depends only upon the activities of the hydrogen ions in the solution on either side of the membrane. If one of these activities $a_2$ is kept constants, then Eq. (9.6) simplifies to

$$E_b = V_1 - V_2 = \text{Constant} + \frac{RT}{F} \ln a_1 \quad (9.7)$$

and, the potential becomes a measure of the hydrogen ion activities in the external solution.

In addition to the boundary potential, so-called diffusion potential, also develops in each of the two gel layers. Its source lies in the difference between the mobilities of hydrogen ions to the alkali metal ions in the membrane. The two diffusion potentials are equal and opposite in sign insofar as the two solution–gel interfaces are the same. Under these conditions, then the net diffusion potential is zero, and the EMF across the membrane depends only on the boundary potential as given by Eq. (9.6).

## 9.11 ASYMMETRIC POTENTIAL

If identical solutions and identical reference electrodes are placed on either side of the membrane, $V_1 - V_2$ should be zero. However, it is found that a small potential, called the *asymmetric potential*, often does develop when this experiment is performed. Moreover, the asymmetry potential associated with a given glass electrode changes slowly with time.

The cause for the asymmetry potential may includes factors such as difference in strains established within the two surfaces during manufacture of the membrane, mechanical and chemical attack or other contamination of the outer surface during use or differential orientation of ions or molecules across the membranes. The effect of the asymmetric potential on a pH measurement is eliminated by frequent calibration of the electrode against a standard buffer of known pH.

## 9.12 GLASS ELECTRODES FOR THE DETERMINATION OF OTHER IONS

A number of investigators have demonstrated that the presence of $Al_2O_3$ or $B_2O_3$ in glass causes the desired effect. Eisenman and co-workers carried out a systematic study of glasses containing $Na_2O$, $Al_2O_3$ and $SiO_2$ in various proportions, they have demonstrated clearly that, it is practical to prepare membranes for the selective measurement of several cations in the presence of others. Glass electrodes for $K^+$ ions and $Na^+$ ions are now available commercially.

Studies have led to the development of glass membranes that are suitable for the direct potentio metric measurements of the concentrations of $Na^+$, $K^+$, $NH_4^+$, $Rb^+$, $Cs^+$, $Li^+$ and $Ag^+$. Some of these glasses are reasonably selective, as shown in Table 9.2.

**TABLE 9.2** Properties of certain cation sensitive glasses

| Principal cation to be measured | Glass composition | Remarks |
|---|---|---|
| $Li^+$ | 15% $Li_2O$<br>25% $Al_2O_3$<br>60% $SiO_2$ | Best for $Li^+$ in presence of $H^+$ and $Na^+$ ions. |
| $Na^+$ | 11% $Na_2O$<br>18% $Al_2O_3$<br>71% $SiO_2$ | Nernst type of response to ~$10^{-5}$ m $Na^+$ ions. |
| $Na^+$ | 10.4 $Li_2O$<br>22.6 $Al_2O_3$<br>67 $SiO_2$ | Highly $Na^+$ selective, but very time dependent. |

*(Contd.)*

**TABLE 9.2** Properties of certain cation sensitive glasses (*Contd.*)

| Principal cation to be measured | Glass composition | Remarks |
|---|---|---|
| $K^+$ | 27 $Na_2O$<br>5 $Al_2O_3$<br>68 $SiO_2$ | Nernst type of response to < $10^{-4}$ m potassium ion. |
| $Ag^+$ | 28.8 $Na_2O$<br>19.1 $Al_2O_3$<br>52.1% $SiO_2$ | Highly sensitive and selective to $Ag^+$ but poor stability. |
| $Ag^+$ | 11 $Na_2O$<br>18 $Al_2O_3$<br>71 $SiO_2$ | Less selective for $Ag^+$ but more reliable. |

## 9.13 LIQUID MEMBRANE ELECTRODES

Liquid membrane electrode (Figure 9.6) measures the potential that is established across the interface between the solution to be analyzed and an immiscible liquid that selectively bonds with the ion being determined. Liquid membrane electrodes permit the direct potentio-metric determination of the activities of several polyvalent cations and certain anions as well.

A liquid membrane electrode differs from a glass electrode, only in the respect that the solution of known and fixed activity is separated from the analyte by a thin layer of an immiscible organic liquid instead of a thin glass membrane.

A porous, hydrophobic plastic disk serves to hold the organic layer between the two aqueous solutions. By wick action, the pores of the disk or membrane are maintained full of the organic liquid from the reservoir in the outer of the two concentric tubes. The inner tubes contains an aqueous standard solution of $MCl_2$, where $M^{2+}$ is the cation whose activity is to be determined. This solution is also saturated with AgCl to form a Ag–AgCl reference electrode with silver lead wire. Organic liquid when in contact with an aqueous solution of a divalent cation, an exchange equilibrium is established that can be represented as,

$$\underset{\text{Organic phase}}{RH_{2x}} + \underset{\text{aq phase}}{xM^{2+}} \rightleftharpoons \underset{\text{Organic phase}}{RM_x} + \underset{\text{aq phase}}{2XH^+}$$

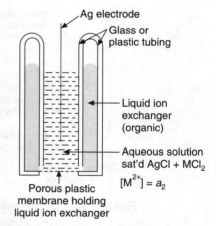

**FIGURE 9.6** Liquid membrane electrode.

By repeated treatment, the exchange liquid can be converted essentially completely to the cationic form $RM_x$, it is this form that is employed in electrodes for the determination of $M^{2+}$.

In order to determine $p$M of a solution, the electrode is immersed in a solution of the analyte which also contains a reference electrode, usually a saturated calomel electrode. The potential between the external and internal reference electrodes is proportional to the $p$M of the analyte.

Figure 9.7 shows construction details of a commercial liquid-membrane electrode that is selective for calcium ions. The ion exchanger is an aliphatic diester of phosphoric acid dissolved in a polar solvent. The chain length of the aliphatic group in the former range from 8 to 16 carbon atoms. The diester contain a single dissociable proton, thus two molecules are required to bond a divalent cation, here calcium. It is this affinity for calcium ions that imparts the selective properties of the electrode. The internal aqueous solution in contact with the exchanger contains a fixed concentration of calcium chloride, a silver–silver chloride reference electrode is immersed in this solution. When used for a calcium ion determination, the porous disk containing the ion exchange liquid separates the solution to be analyzed from the reference calcium chloride solution. The equilibrium established at each interface can be represented as

$$[(RO)_2 POO]_2 Ca \rightleftharpoons 2(RO)_2 POO^- + Ca^{2+}_{aq.}$$
$$\text{Organic} \quad\quad\quad \text{Organic}$$

The potential of the electrode is $E = K + \dfrac{0.0591}{2} \log a_1$

Here, $a_1$ is the activity of $Ca^{2+}$ ions.

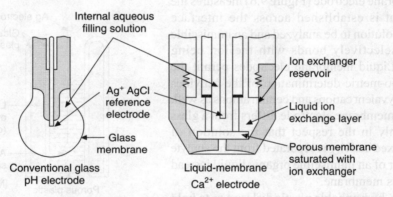

**FIGURE 9.7** Liquid-membrane electrode that is selective for calcium ions.

The performance of the calcium electrode just described is reported to be independent of pH in the range between 5.5 and 11. At lower pH levels, hydrogen ions undoubtedly exchange with calcium ions on the exchanger to a significant extent, the electrode then becomes pH as well as $p$Ca dependent. Its sensitivity to calcium ion exceeds that for magnesium ion by a factor of 50 and that for sodium or potassium by a factor of 1000. It can be employed to measure $Ca^{2+}$ activities as low as $10^{-5}$ m.

The calcium ion membrane electrode has proved to be a valuable tool for physiological studies, because this ion plays important roles in nerve conduction, bone formation, muscle contraction, cardiac conduction and contraction, and renal tabular function. Another specific ion electrode of great significance in physiological studies is that for potassium, because of nerve signals appears to involve movement of this ion across nerve membranes. Study of the process requires an electrode which can detect small concentration of potassium ion in the presence of much larger concentrations of sodium.

## 9.14 SOLID STATE OR PRECIPITATE ELECTRODES

As we have seen the selectivity of glass membrane in glass electrode is due to the presence of anionic sites on its surface that shows particular affinity towards certain positively charged ions. By analogy, a membrane having similar cationic sites might be expected to respond selectivity towards anions. To exploit this possibilities, attempts have been made to prepare membranes of salts containing the anion of interest and a cation that selectivity precipitate that anion from aqueous solution, for example, barium sulphate has been proposed for sulphate ion and silver halide for various halide ions.

The problem encountered in this approach has been in finding methods for fabricating membrane from the desired salt with adequate physical strength, conductivity and resistance to abrasion and corrosion.

A solid state electrodes selective for fluoride ion has been described. The membrane consists of a single crystal of lanthanum fluoride that has been doped with a rare earth to increase its electrical conductivity. The membrane, supported between a reference solution and the solution to be measured, shows the theoretical response to changes in fluoride ion activity to as low as $10^{-6}$ m. The electrode is supposed to be selective for fluoride ions over other common anions by several order of magnitude. Only hydroxide ion appears to offer serious interference.

Membranes prepared from cast pellets of silver halides have been successfully used in electrodes for the selective determination of chloride, bromide and iodide ions.

## 9.15 GAS-SENSING ELECTRODES

Gas-sensing electrode consists of a reference electrode, a membrane and a solution housed in a cylindrical plastic tube. A thin replaceable, gas-permeable membrane that serves to separate the internal electrolyte solution from the analyte solution is attached to one end of the tube (Figure 9.8).

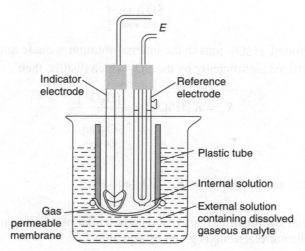

**FIGURE 9.8** Schematic diagram of a gas-sensing electrode.

The membrane is thin, microporous film manufactured from a hydrophobic plastic. Water and electrolytes are prevented from entering and passing through the pores of the film owing to its water repellent properties. Thus, pores contain only air or other gases that the membrane is exposed to and not liquid water. When a solution containing a gaseous analyte such as $SO_2$ is placed in contact with the membrane, distillation of the $SO_2$ into the pores occurs as shown by the reaction.

$$\underset{\text{Ext. soln.}}{SO_2(aq.)} \rightleftharpoons \underset{\text{Memb. pores}}{SO_2(gas)}$$

Because, the pores are numerous, a state of equilibrium is rapidly approached. The $SO_2$ in the pores, however, is also in contact with the internal solution and a second reaction can readily take place i.e.

$$\underset{\text{Memb. pores}}{SO_2(gas)} \rightleftharpoons \underset{\text{Int. soln.}}{SO_2(aq.)}$$

As a consequence of two reactions, the film of internal solution adjacent to the membrane rapidly (in a few seconds) equilibrates with the external solution. Further, another equilibrium is established that causes the pH of the internal surface film to change, namely,

$$SO_2(aq.) + 2H_2O \rightleftharpoons HSO_3^- + H_3O^+$$

A glass electrode immersed in this film is then employed to detect the pH change. The overall reaction for the process just described is obtained by adding the three chemical equations to give

$$\underset{\text{Ext. soln.}}{SO_2(aq.)} + 2H_2O \rightleftharpoons \underset{\text{Int. soln.}}{HSO_3^-} + H_3O^+$$

The equilibrium constant for the reaction is given by

$$K = \frac{[H_3O^+][HSO_3^-]}{\left[\underset{\text{Ext. soln.}}{SO_2(aq.)}\right]}$$

If now the concentration of $HSO_3^-$ ions in the internal solution is made relatively high so that its concentration is not altered significantly by the $SO_2$ which distills, then

$$K_g = K[HSO_3^-] = \frac{[H_3O^+]}{\left[\underset{\text{Ext. soln.}}{SO_2(aq.)}\right]}$$

which may be rewritten as $a_1 = [H_3O^+] = K_g \left[\underset{\text{Ext. soln.}}{SO_2(aq.)}\right]$

where $a_1$ is the internal hydrogen ion activity. The potential of a cell consisting of a reference electrode and any indicator electrode is given by equation

$$-\log a_1 = \frac{E_{obs} - K'}{0.0591/n}$$

where $a_1$ is the activity of the ion to which the indicator electrode is sensitive and $K'$ is a constant dependent upon the characteristics of both the indicator and reference electrodes. In this cell the indicator electrode is a pH-sensitive glass electrode for which $n = 1$. Thus, combining last two equations gives

$$-\log a_1 = -\log \left( K_g \left[ SO_2(aq.) \right]_{\text{Ext. soln.}} \right) = \frac{E_{obs} - K'}{0.0591}$$

$$-\log [SO_2(aq.)]_{\text{Ext.}} = \frac{E_{obs} - K'}{0.0591} + \log K_g$$

Finally combining the two constants $K'$ and $K_g$ gives

$$-\log \left[ SO_2(aq.) \right]_{\text{Ext. soln.}} = pSO_2 = \frac{E_{obs} - K''}{0.0591}$$

where $K'' = K' - 0.0591 \log K_g$.

The constants $K''$ can be evaluated by measuring $E_{obs}$ for a solution having a known $SO_2$ concentration. Thus, the potential of the cell consisting of the internal reference and indicator electrode is determined by $SO_2$ concentration of external solution.

The possibility exists for increasing the selectivity of the gas-sensing electrode by employing an internal electrode that is sensitive to species other than $SO_2$, for example, a nitrate-sensing electrode could be used to provide a cell that would be sensitive to nitrogen dioxide. Here, the equilibrium would be

$$\underset{\text{Ext. soln.}}{2NO_2(aq.)} + H_2O \rightleftharpoons \underset{\text{Int. soln.}}{NO_3^- + NO_2^-} + 2H^+$$

This electrode should permit the determination of $NO_2$ in the presence of gases such as $SO_2$, $CO_2$ and $NH_3$ which also alter the pH of the internal solution.

Gas-sensing electrodes are commercially available for $SO_2$, $NO_2$ and $NH_3$. Others will undoubtedly be developed in the near future.

## 9.16 LIQUID JUNCTION POTENTIAL

The junction potential arises from an unequal distribution of cations and anions across the boundary between the two electrolyte solutions, is known as *liquid junction potential*. This difference results from the difference in the rates with which various charged species migrate under the effects of diffusion. Consider, for example, the events occurring at the interface between 1 F and 0.01 F hydrochloric acid solutions. We may symbolize this interface as

$$HCl\ (1\ F) \mid HCl\ (0.01\ F)$$

Both hydrogen and chloride ions tend to diffuse across this boundary from the more concentrated to the more dilute solution, the driving force for this migration being proportional to the concentration difference. The rate at which ions move under the influence of a fixed force, varies considerably (i.e., their mobilities are different) in the present example, hydrogen ions are several times more mobile than chloride ions. As a consequence, there is a tendency for the hydrogen ions to outstrip chloride ions as diffusion takes place; a separation of charge is the net result.

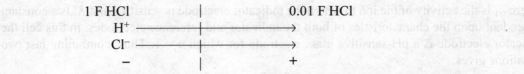

The more dilute side of the boundary becomes positively charged, owing to the more rapid migration of hydrogen ions; the concentrated side therefore acquires a negative charge from the slower moving chloride ions. The charge that develops tend to counteract the difference in mobilities between the two ions, as a consequence, an equilibrium condition develops. The potential difference resulting from this charge separation may amount to several hundredths of a volt or occasionally more.

In a simple system such as this, where the behaviour of only two ions need to be considered, the magnitude of the junction potential can be calculated from a knowledge of the mobilities of the ions involved. However, it is seldom that a cell of analytical importance has a sufficiently simple composition to permit such a computation.

## 9.17 OHMIC POTENTIAL: IR DROP

Current passage through a galvanic or an electrolyte cell requires a driving force or a potential to overcome the resistance of the ion movements towards the anode or the cathode. Just as metallic conduction, this force follows Ohm's law and is equal to the product of the current in amperes and the resistance of the cell in ohms. The force is generally referred to as the *ohmic potential* or the *IR drop*.

In general, the net effect of *IR* drop is to increase the potential required to operate an electrolytic cell and to decrease the measured potential of a galvanic cell. Therefore, the *IR* drop is always subtracted from the theoretical cell potential. That is

$$E_{cell} = E_{cathode} - E_{anode} - IR_{drop}$$

**Problem 9.1** Calculate the potential when 0.1 A is drawn from the galvanic cell

$$Cd|Cd^{2+}(0.1\text{ m})\|Cu^{2+}(1.0\text{ m})|Cu$$

Assume a cell resistance of 4.0 $\Omega$, $E^0_{Cu^{2+}/Cu} = 0.337$ V and $E^0_{Cd^{2+}/Cd} = 0.403$ V

*Solution:* Since both cation concentrations are 1.0 m, the respective half cell potentials are equal to the standard potentials, thus

$$E = E^0_{Cu} - E^0_{Cd} = 0.337 - (-0.403) = 0.740 \text{ V}$$
$$E_{Cell} = 0.740 - IR \Rightarrow 0.740 - 0.1 \times 4.0 \Rightarrow 0.340 \text{ V}$$

**Problem 9.2** Calculate the potential required to cause a current of 0.1 A to pass in the reverse direction in the foregoing cell.

*Solution:*
$$E = E^0_{Cd} - E^0_{Cu} = 0.403 - 0.337 = -0.740 \text{ V}$$
$$E_{Cell} = -0.740 - IR \Rightarrow -0.740 - 0.1 \times 4.0 \Rightarrow -1.140 \text{ V}$$

Here, an external potential greater than 1.140 V would be needed to cause $Cd^{2+}$ to deposit and to dissolve at a rate corresponding to 0.1 A.

*Polarized electrode:* An electrode is said to be polarized if its potential deviates from the reversible or equilibrium value. An electrode is said to be depolarized by a compound if that compound lowers the amount of polarization.

## 9.18 POLARIZATION

When a voltaic cell is set up and a current is taken from it, the EMF of the cell readily decreases. It so happens because the copper electrode becomes covered with the bubbles of the gas which make a gas electrode with EMF opposite to that of the cell. If the bubbles of the gas are removed mechanically or chemically, the EMF of the cell remains constant. This phenomenon of back EMF, brought about by the product of electrolysis is termed as *polarization,* of course, exactly the same phenomenon takes place within electrolysis, in which the reactions are just the reverse of those occurring in a cell. Thus, when a current of electricity is allowed to pass between two platinum electrodes immersed in a solution of dilute $H_2SO_4$, electrolysis takes place and $H_2$ and $O_2$ gases are evolved at the cathode and anode, respectively.

$$H_2SO_4 \rightleftharpoons 2H^+ + SO_4^{2-}$$

$$SO_4^{2-} + H_2O \rightleftharpoons H_2SO_4 + O^{2-}$$

$$2H^+ + 2e \rightleftharpoons H_2 \text{ (Cathode)}$$

$$2O^{2-} + 4e \rightleftharpoons O_2 \text{ (Anode)}$$

Outside the solution, the current flows from cathode to anode, while inside the solution the direction of current flows is the reverse. Now remove the battery and join the two electrodes through a galvanometer. It will be observed that a small current is flowing between the two electrodes and the direction of which in this case is opposite to that during the process of electrolysis, i.e. the current flows from anode to cathode outside the solution. Since these new electrodes have an EMF opposite to that of the cell, it is called the *back EMF*. Thus, unless applied EMF is greater than the polarization EMF, electrolysis almost stops.

### 9.18.1 Causes of Polarization

1. The liberated gases offer resistance to the normal flow of current through the cell and hence a high voltage is necessary to keep the normal flow.
2. The product of electrolysis may convert the inner platinum electrodes into active electrodes which can exercise a back EMF.
3. The deposited metals may also form a cell functioning in the opposite direction similar to the gas electrodes.
4. A simple type of polarization due essentially to the slowness of the diffusion of ions in the solution is concentration polarization.

### 9.18.2 Polarization Effects

The linear relationship between the potential and the instantaneous current being drawn from or forced through a cell is frequently observed experimentally when $I$ is small, at high currents, however, marked departure from linear behaviour occur. Under these circumstances, the cell is said to be polarized. Thus, a polarized electrolytic cell requires application of potentials larger than theoretical for a given current flow, similarly, a polarized galvanic cell develops potentials that are smaller than predicted. Under some conditions, the polarization of a cell may be so extreme that the current becomes essentially independent of the voltage, under these circumstances polarization is said to be complete (Figure 9.9).

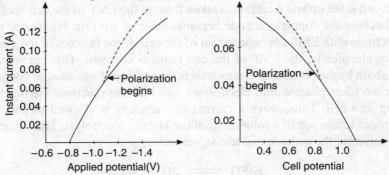

**FIGURE 9.9**  Current voltage curve for the cell $Cu + Cd^{2+} \rightarrow Cu^{2+} + Cd$ and $Cu^{2+} + Cd \rightarrow Cu + Cd^{2+}$, respectively.

Polarization is an electrode phenomenon, either or both electrodes in a cell can be affected. Included among the factors influencing the extent of polarization are the size, shape and composition of the electrodes, the composition of the electrolyte solution, the temperature and rate of stirring, the magnitude of the current, and the physical states of the species involved in the cell reaction. Some of these factors are sufficiently understood to permit quantitative statements concerning their effects upon cell processes. Others, however, can be accounted for an empirical basis only.

For purpose of discussion, polarization phenomenon are conveniently classified into the two categories of concentration polarization and kinetic polarization (overvoltage).

## 9.19 CONCENTRATION POLARIZATION

When the reaction at an electrode is rapid and reversible, the concentration of the reacting species in the layer of solution immediately adjacent to the electrode is always that which would be predicted from the Nernst equation. Thus, the $Cd^{2+}$ ion concentration in the immediate vicinity of a cadmium electrode will be always given by

$$E = E^0_{Cd} - \frac{0.0591}{2} \log \frac{1}{[Cd^{2+}]}$$

irrespective of the concentration of this cation in the bulk of the solution. Since, the reduction of cadmium ion is rapid and reversible, the concentration of this ion in the film of liquid surrounding the electrode is determined at any instant by the potential of the cadmium electrode at that instant. If the potential changes, there is an essentially instantaneous alteration of concentration in the film as required by the Nernst equation. This change will involve either deposition of cadmium or dissolution of the electrode.

## 9.20 DECOMPOSITION POTENTIAL

When two smooth platinum electrodes are placed in dilute solution of sulphuric acid and low voltage is applied, practically no current flows through the circuit. But as the applied voltage is gradually increased by means of an external battery so as to electrolyze the solution, a galvanometer in the circuit will show that the current increases in the manner represented by the curve shown in Figure 9.10.

It appears that an appreciable voltage must be applied before current can flow freely through the cell. The EMF applied at the point $D$ at which electrolysis restarts and proceed continuously is called *decomposition voltage* of the particular solution, with the given electrode material. It may also be defined as the minimum voltage that is required to bring about electrolysis of an electrolyte without any hindrance. It is at the point $D$, that is general, the steady evolution of hydrogen and oxygen gas bubbles is observed.

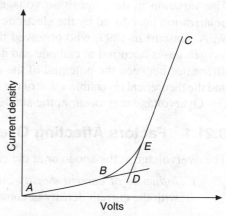

**FIGURE 9.10** Variation of current density with potential.

Decomposition potential vary with the conditions, but is the characteristic of definite electrolysis. Thus, decomposition potential of $CuSO_4$ is 1.6, of $Pb(NO_3)_2$ 1.8 and of $Cd(NO_3)_2$ 2.3 volts. It should be noted that most strong acids and bases have decomposition potential of 1.7 volts.

### 9.20.1 Importance and Significance of Decomposition Potential

Zinc and iron will have high decomposition potential than that of copper. This fact is of extensive importance in electro-refining and electrometallurgy of metals. The decomposition voltage of $CuSO_4$ and $ZnSO_4$ are 1.5 and 2.55 volts, respectively. When an EMF of less than 2.5 volts is applied across the copper electrodes, copper will be deposited on the cathode, while zinc remains in the solution.

The decomposition potentials are of great importance in the controlled deposition of metals or other electrolyte products when a potential difference is applied to the electrodes of an electrolytic cell containing several different electrolytes. The one having the lowest decomposition potential is electrolyzed first, and as the potential is increased, other salts are electrolyzed.

It is seen that the products of electrolysis can be controlled not only by altering the applied voltage, but also by changing the concentration. It follows, therefore, that when different kinds of ions are competing at an electrode for the loss or gain of electrons, two factors are involved: (i) decomposition potential and (ii) concentration.

The former is a measure of the energy requirements and the latter is a measure of collision with the electrode surface.

## 9.21 OVERVOLTAGE (KINETIC POLARIZATION)

The variation in decomposition voltage with the electrode material results from difference in polarization introduced by the electrode material. A detailed study of this problem was made by W. A. Gaspari in 1899, who observed the potential at which visible evolution of hydrogen and oxygen gases occurred at cathode and anode, respectively, of a number of different metals. The difference between the potential of the electrode when the gas evolution was actually observed and the theoretical reversible value of the same solution was called the *overvoltage or overpotential*.

Overvoltage may occur at the anode as well as at the cathode.

### 9.21.1 Factors Affecting Overvoltage

The overvoltage at the anode or at the cathode is influenced by the following factors:

1. *Influence of current density on overvoltage:* The variation of hydrogen overvoltage with the current density at constant temperature can be represented by the equation

$$\eta = A + B \log I$$

   where $\eta$ is overvoltage, $I$ = current density and $A$ and $B$ are constants. Above equation is equation of straight line, a plot of $\eta$ against log $I$ should be a straight line. Slope $B$, which corresponds very closely to $2 \times \dfrac{2.303 RT}{F}$.

2. *Influence of pH on overvoltage:* pH of solution does not influence overvoltage to a significant extent untill and unless very strong acidic or alkaline solution are present.

3. *Influence of temperature on overvoltage:* Increase of temperature will decrease the overvoltage. Generally change is 2 milli volts per degree approximately.

4. *Influence of pressure on overvoltage:* At higher pressures, the overvoltage is slightly affected while at the lower pressure, it increases rapidly on copper, nickel and mercury cathodes.

5. *Influence of surface on overvoltage:* Increasing the effective area of the surface causes a decrease in overvoltage. On smooth, shiny and polished surfaces the overvoltage is invariably greater than rough, pitted or etched surfaces.

6. *Influence of impurity on overvoltage:* Impurities affects overvoltage over cathodic areas.

### 9.21.2 Importance of Overvoltage

It explains why Pb is used in lead accumulator? It is due to high overvoltage of lead during charging that the metal is deposited on the cathode instead of the hydrogen being evolved. This view is further confirmed by the fact that if the lead is covered with a layer of a metal with a very low overvoltage, i.e. platinum, electrolysis is carried out and no lead is deposited, but hydrogen is evolved.

The phenomenon of overvoltage is utilized in several electrolytic processes. If electrodes with high overvoltage are used in electrolytic reduction it counts to the same thing as increasing the reducing power of the electrode.

It is only due to the existence of hydrogen overvoltage, that makes it possible to deposit electrolytically metals, which have a more negative potential than hydrogen (such as zinc, Cd, tin, etc.) from an acid solution.

It has been observed that overvoltage also plays an important role in the industrial production of $Cl_2$ and NaOH by the electrolysis of NaCl solution.

Overvoltage occurs also at the electrodes upon which gases other than hydrogen are evolved.

## 9.22 APPLICATION OF POTENTIOMETRIC TITRATIONS

1. *In study of acid–base or neutralization process:* Change in $H^+$ ion concentration (i.e. pH change) is studied by hydrogen electrode. Vinegar and mixture of acid can be titrated without indicator (Figures 9.11, 9.12 and 9.13).

$$E(H^+, H_2) = \frac{2.303\ RT}{F} \log [H^+]$$

$$E(H^+, H_2) = -0.0591\ \text{pH at } 25°C$$

$$E_{cell} = E^0_{cell} - (-0.0591)\ \text{pH}$$

$$= 0.242 + 0.0591\ \text{pH}$$

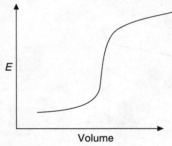

**FIGURE 9.11** Variation of EMF with volume of base added from the burette in acid–base neutralization reaction.

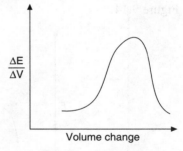

**FIGURE 9.12** Single derivative curve of potential against change in volume.

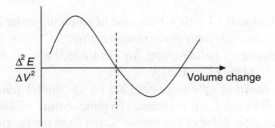

**FIGURE 9.13** Double derivative acid–base neutralization curve to detect end point.

2. *Oxidation-reduction titrations:* Indicator electrode now consists of an inert metal, such as Pt wire. The metal acts as an oxidation–reduction indicator electrode, EMF is determined by the activity ratio of the substance being oxidized or reduced.

The EMF of the cell $Pt|Fe^{2+}, Fe^{3+}|KCl\,(aq.)\cdot Hg_2Cl_2\,(s), Hg$

$$E = E^0 - \frac{RT}{F}\log\frac{[\text{Oxid. state}]}{[\text{Reduc. state}]} \Rightarrow E^0 - \frac{RT}{F}\log\frac{a_{Fe^{3+}}}{a_{Fe^{2+}}}$$

*Advantage:* Coloured solution and several substances can be simultaneously determined in one titration. No special indicator electrode required.

3. *Precipitation titration:* An electrode reversible to one of the ions is used as an indicator electrode.

$AgNO_3$ vs. NaCl: A silver electrode is used.

$Ag\,|\,AgNO_3\,(aq)\,KNO_3\,\|\,KCl, Hg_2Cl_2\,(s), Hg$

$Ag^+ + NO_3^- + K^+ + Cl^- = AgCl + KNO_3$

Concentration of $Ag^+$ ions goes on decreasing and hence potential of indicator electrode goes on increasing continuously. At the end point, concentration of $Ag^+$ ions becomes very small due to slight solubility of AgCl, potential changes abruptly. Precipitation titration is used for titration of halides, cyanides and other anions forming sparingly soluble salts of silver. Analysis of solutions composed of chlorides, bromides and iodides is shown in Figure 9.14.

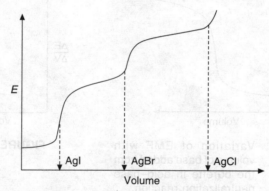

**FIGURE 9.14** Precipitation titration silver halides potentiometrically.

Order of precipitation is Cl < Br < I because of solubility order is Cl > Br > I.

4. *Complexometric titration (only potentiometrically):* $CN^-$ vs. $Ag^+$ to form $[Ag(CN)_2]$ even in the presence of halide using Ag electrode. $Fe^{3+}$ vs. EDTA (chelating agent) $Hg|HgY^{2-}, MY^{2-}, M^{2+}$. $Y^{2-}$ is EDTA.

5. *Potentiometric titration of phosphoric acid with NaOH (only potentiometrically):* $H_3PO_4$ is polybasic acid, i.e. it is tribasic. To remove third proton from the acid, a pinch of $CaCl_2$ is added after 2/3rd of the base is added from the burette. The removal of third proton is possible with $CaCl_2$ due to following reaction.

$$Na_2HPO_4 + CaCl_2 \rightarrow CaPO_4 + HCl$$

Presence of HCl reduces the pH of the solution from $X$ to $Y$ (Figure 9.15).

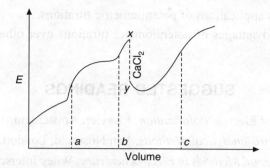

**FIGURE 9.15** Potentiometric titration of phosphoric acid with NaOH.

6. *Special Titrations:*

   (a) *Titration of amino acids:*  2–3 milli equivalent of amino acid is dissolved in 50 ml of 0.1 m perchloric acid. The excess of acid is back titrated with 0.1 m $CH_3COONa$.

   (b) *Titration of amines:*  Organic amines in 50 ml of glacial acetic acid (0.1 N) can be titrated potentiometrically with 0.1 M NaOH.

   (c) *Titration of Zn:*  Take 50 mg of Zn in beaker having pH adjusted to 1–3. Add few drops of $K_4Fe(CN)_6$ to this solution and titrate against standard potassium ferrocyanide $(K_3Fe(CN)_6)$.

   (d) *Titration of Mn:*  Sample containing Mn is dissolved in acid. Add 250 ml of saturated solution of sodium pyrophosphate. Adjust the pH between 5–7 with $H_2SO_4$ or NaOH. Now titrate it against standard 0.025 m $KMnO_4$ solution. The $Mn^{2+}$ is oxidized to $Mn^{3+}$ and permagnate is reduced to Mn(III).

## EXERCISES

1. What are reference electrodes? Describe briefly various types of reference electrodes.
2. Describe the construction, working and principle of a calomel electrode. Also describes various types of calomel electrodes available in the market.
3. Describe the construction, working and principle of a silver–silver chloride electrode. How it is different from calomel electrode?
4. What is an indicator electrode? Describes various types of indicator electrodes.
5. What is a membrane electrode? Explain it with some suitable examples.
6. Explain the construction, principle and working of a glass electrode. What are various applications of glass electrode?
7. What are various characteristics of a glass electrode?
8. Explain construction, principle and working of a gas-sensing electrode.
9. What is liquid junction potential?
10. What is polarization? Explain polarization effects and various types of polarization.
11. What are various factors affecting the polarization?
12. Write down the importance and significance of decomposition potential.

13. Write down various applications of potentiometric titrations.
14. What are various advantages of potentiometer titrations over other types of instrumental titrations?

## SUGGESTED READINGS

Bottcher, C.J.F., *Theory of Electric Polarization*, Elsevier, Amsterdam, 1952.

Charlot, G., *Modern Electro-analytical Methods*, Van Nostrand, London, 1958.

Delahay, P., *New Instrumental Methods in Electrochemistry*, Wiley Interscience, New York, 1954.

Donbrow, M., *Instrumental Methods in Analytical Chemistry their Application and Practice*, Vol. 1, Pitman, London, 1966.

Lingane, J.J., *Electro-Analytical Chemistry*, 2nd ed., Interscience, New York, 1958.

Sharma, B.K., *Instrumental Methods of Chemical Analysis*, Goel Publishing House, Meerut.

Skoog, D.A. and D.M. West, *Fundamentals of Analytical Chemistry*, 3rd ed., Rinehart & Winston, New York.

Willard, H.H., L.L. Merrit, J.A. Dean, and F.A. Settle, *Instrumental Methods of Analysis*, 6th ed., CBS Publishers & Distributors, Delhi.

# 10

# Polarography and Voltammetry

## 10.1 THEORETICAL CONSIDERATION OF CLASSICAL POLAROGRAPHY

Voltammetry comprises a group of electro-analytical procedures that are based upon the potential–current behaviour of a small, easily polarized electrode in the solution being analyzed.

Historically, voltametry developed from the discovery of polarography in the early 1920s. Later, Heyrovsky and co-workers adopted the principles of polarography for the detection of end points in volumetric analysis, such methods have come to be known as *amperometric titration's*.

Most polarographic analysis is performed in aqueous solution, but other solvent systems may be substituted if necessary for quantitative analysis, the optimum concentration range lies between $10^{-2}$ and $10^{-4}$ m. An analysis can be easily performed on 1–2 ml of solution, with a little effort, a volume as small as one drop is sufficient. The polarographic method is thus particularly useful for the determination of quantities in the milligram to microgram range.

Polarographic data are obtained by measuring current as a function of the potential applied to a special type of electrolytic cell (Figure 10.1). A plot of data gives current–voltage curves, called *polarograms* (Figure 10.2), which provides both qualitative and quantitative information about the composition of the solution in which the electrodes are immersed.

## 10.2 POLAROGRAPHIC CELL

A polarographic cell shown in Figure (10.1), consists of a small, easily polarized microelectrode, a large nonpolarizable reference electrode and the solution to be analyzed.

The microelectrode at which the analytical reaction occurs, is an inert metal surface with an area of a few square millimetres. The dropping mercury electrode is the most common type. Here mercury is forced by gravity through a very fine capillary to provide a continuous flow of identical droplets with a maximum diameter range between 0.5 and 1 mm. The lifetime of a drop is typically 2–6 s. Other microelectrodes, consisting of small diameter wires or disks of platinum or other metal, can be used instead.

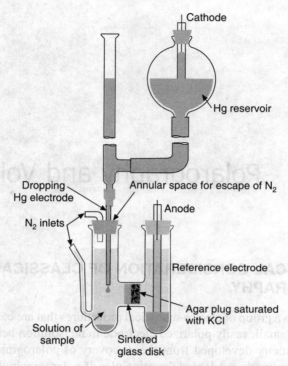

**FIGURE 10.1** A dropping mercury electrode and the reference electrode (polarographic cell).

The reference electrode should be massive relative to the microelectrode so that its behaviour remains essentially constant with the passage of small currents, i.e. it should remain unpolarized during the analysis. A saturated calomel electrode and salt bridge is frequently employed, another common reference electrode consists of simply of a large pool of mercury.

## 10.3 POLAROGRAM

Polarogram is a plot of current as a function of the potential applied to a polarographic cell. The microelectrode is ordinarily connected to the negative terminal of the power supply, by convention, the applied potential is given a negative sign under these circumstances. By convention also, the current is designated as positive when the flow of electrons is from the power supply into the microelectrode that is when that electrode behaves as a cathode.

The lower one is for a solution that is 0.1 F in Potassium chloride, the upper one is for a solution that is additionally $1 \times 10^{-3}$ F in cadmium chloride. A step-shaped current-voltage curve, called a *polarographic wave*, is produced as a result of the reaction.

$$Cd^{2+} + 2e^- + Hg \rightleftharpoons Cd(Hg)$$

The sharp increase in current at about −2.0 V in both plots is associated with reduction of $K^+$ ions to give a potassium amalgam. A polarographic wave suitable for analysis is obtained only in the presence of a large excess of a supporting electrolyte. KCl serves this function in the present example. Examination of polarogram for the supporting electrolyte alone reveals that a small current, called the residual current passes through the cell even in the absence of cadmium ions. The voltage at which the polarogram for the electrode reactive species departs from the residual current curve is called the *decomposition potential*.

A characteristic feature of the polarographic wave is the region in which the current after increasing sharply, becomes essentially independent of the applied voltage, under these circumstances it is called a *limiting current*. We shall see that the limiting current is the result of a restriction in the rate at which the participant in the electrode process can be brought to the surface of the microelectrode, with proper control over experimental conditions. This rate is determined exclusively for all points on the wave by the velocity at which the reactant diffuses. A diffusion controlled limiting current is given a special name, the *diffusion current* and is assigned the symbol $i_d$. Ordinarily, the diffusion current is directly proportional to the concentration of the reactive constituents and is thus of prime importance from the stand point of analysis. As shown in Figure 10.2 the diffusion current is the difference between the limiting and the residual current.

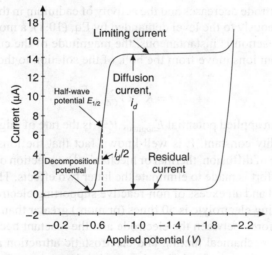

**FIGURE 10.2** Polarogram for cadmium ion.

One other important quantity, that half wave potential, is the potential at which the current is equal to one half the diffusion current. The half wave potential is usually given by the symbol $E_{1/2}$. It may permit qualitative identification of the reactant.

## 10.4 INTERPRETATION OF POLAROGRAPHIC WAVES

The reduction of $Cd^{2+}$ ions to yield Cd amalgam at a dropping mercury electrode (DME) as shown in Figure 10.1, is said to be reversible.

The half cell reaction will be

$$Cd^{2+} + Hg + 2e \rightleftharpoons Cd(Hg)$$

Here, reversible implies that the electron transfer process is sufficiently rapid that the activities of reactants and products in the liquid film at the interface, between the solution and the mercury electrode, are determined by the electrode potential alone. Thus, for the reversible reduction of cadmium ions, it may be assumed that at any instant the activities of reactant at this interface are given by

$$E_{\text{applied}} = E_A^0 - \frac{0.0591}{2} \log \frac{[\text{Cd}]^0}{[\text{Cd}^{2+}]^0} - E_{\text{ref}} \qquad (10.1)$$

Here, $[\text{Cd}]^0$ is the activity of metallic cadmium dissolved in the surface film of the mercury and $[\text{Cd}^{2+}]^0$ is the activity of the ion in the aqueous solvent. Note that the superscript zero has been employed for the activities term to emphasize that this relationship applies to surface films of the two media only. The activity of cadmium ion in the bulk of the solution and of elemental cadmium in the interior of mercury drop will be quite different from the surface activity. The term $E_{\text{applied}}$ is the potential applied to the dropping electrode and $E_A^0$ is the standard potential for the half reaction in which a saturated cadmium amalgam is the product.

Consider now what will happen when $E_{\text{applied}}$ is sufficiently negative to cause appreciable reduction of cadmium ion. Because the reaction is reversible, the activity of cadmium ion in the film surrounding the electrode decreases and the activity of cadmium in the outer layer of mercury drop increases instantaneously to the level demanded by Eq. (10.1), a momentary current results. Because the reduction reaction is instantaneous, the magnitude of the current depends upon the rate at which the cadmium ions move from the bulk of the solution to the surface where reaction occurs. That is

$$i = K' \times V_{\text{Cd}}^{2+}$$

where $i$ is the current at an applied potential $E_{\text{applied}}$, $V_{\text{Cd}}^{2+}$ is the rate of migration of cadmium ions, and $K'$ is a proportionality constant. It is well-known fact that the ions or molecules in a cell migrate as a consequence of diffusion, thermal or mechanical convection or electrostatic attraction. In polarography every effort is made to eliminate the latter two effects. Thus, vibration or stirring of the solution is avoided and an excess of non-reactive supporting electrolyte is employed. If the concentration of supporting electrolyte is 50 times (or more) greater than that of one reactant, the attractive (or repulsive) force between the electrode and the reactant becomes negligible.

When the forces of mechanical mixing and electrostatic attraction are eliminated, the only force responsible for the transport of cadmium ions to the electrode surface is diffusion. Since the rate of diffusion is directly proportional to the concentration (strictly active) difference between two parts of the solution. We may write

$$V_{\text{Cd}}^{2+} = K''([\text{Cd}^{2+}]) - [\text{Cd}^{2+}]_0$$

where $[\text{Cd}^{2+}]$ is the concentration in the bulk of the solution from which ions are diffusing and $[\text{Cd}^{2+}]_0$ is the concentration in the aqueous film surrounding the working electrode. The diffusion is the only process bringing cadmium ions to the surface, it follows that

$$i = K' \, V_{\text{Cd}^{+2}} = K'K''([\text{Cd}^{2+}] - [\text{Cd}^{2+}]_0)$$

$$\Rightarrow \qquad = K([\text{Cd}^{2+}] - [\text{Cd}^{2+}]_0)$$

Note that $[Cd^{2+}]_0$ becomes smaller as $E_{applied}$ is made more negative. Thus, rate of diffusion as well as the current increases with increase in applied potential. When the applied potential becomes sufficiently negative, however, the concentration of cadmium ion in the surface film approaches zero with respect to the concentration in the bulk of the solution. Under these circumstances the rate of diffusion, and thus the current becomes constant. That is, when

$$[Cd^{2+}]_0 \ll [Cd^{2+}]$$

the expression for current becomes

$$i_d = K[Cd^{2+}]$$

where $i_d$ is the potential independent diffusion current. Note that the magnitude of the diffusion current is directly proportional to the concentration of the reactant in the bulk of the solution. Quantitative polarography is based upon the fact that when the current in a cell is limited by the rate at which a reactant can be brought to the surface of an electrode, a state of complete polarization is said to exist.

Half wave potential is an equation relating the applied potential $E_{applied}$, and current $i$ can readily be derived. For the reduction of cadmium ion to the cadmium amalgam, the equation takes the form,

$$E_{applied} = E_{1/2} - \frac{0.0591}{n} \log \frac{i}{i_d - i} \tag{10.2}$$

where

$$E_{1/2} = E_A^0 - \frac{0.0591}{n} \log \frac{f_{Cd} K_{Cd}}{f_{Cd^{2+}} K_{Cd^{2+}}} - E_{ref} \tag{10.3}$$

where $f_{Cd}$ and $f_{Cd^{2+}}$ are activity coefficients for the metal in the amalgam and the ion in the solution; $K_{Cd}$ and $K_{Cd^{2+}}$ are proportionality constants related to the rates at which the two species diffuse in the respective media.

## 10.5 EFFECT OF COMPLEX FORMATION ON POLAROGRAPHIC WAVES

The data in Table 10.1 indicate that the half waves potential for the reduction of a metal complex is generally more negative than that for reduction of the corresponding simple metal ion (Table 10.1).

**TABLE 10.1** Effects of complexing agents on polarographic half wave potentials at the dropping mercury electrode

| Ion | Non-complexing media | 1-F KCN | 1-F KCl | 1-F $NH_3$, 1-F $NH_4Cl$ |
|---|---|---|---|---|
| $Cd^{2+}$ | −0.59 | −1.18 | −0.64 | −0.81 |
| $Zn^{2+}$ | −1.00 | Non-reducing | −1.00 | −1.35 |
| $Pb^{2+}$ | −0.40 | −0.72 | −0.44 | −0.67 |
| $Ni^{2+}$ | — | −1.36 | −1.20 | −1.10 |
| $Co^{2+}$ | — | −1.45 | −1.20 | −1.29 |
| $Cu^{2+}$ | +0.02 | Non-reducing | +0.04 | −1.24 |

Lingane has shown that the shift in half wave potential as a function of concentration of complexing agent can be employed to determine the formula and the formation constant for complexes, provided that the cation involved reacts reversibly at the dropping electrode. Thus, for the reactions.

$$M^{n+} + Hg + ne^- \Leftrightarrow M(Hg)$$

And
$$M^{n+} + XA^- \Leftrightarrow MA^{(n-x)+}$$

He derived the relationship

$$(E_{1/2})_c - E_{1/2} = -\frac{0.0591}{n} \log K_f - \frac{0.0591X}{n} \log F_A \qquad (10.4)$$

where $(E_{1/2})_c$ is the half wave potential when the formal concentration of $A$ is $F_A$ while $E_{1/2}$ is the half wave potential in the absence of complexing agents, and $K_f$ is the formation constant of the complex. Equation (10.4) can be employed for determination of the formula of the complex. Thus, a plot of the half wave potential against $\log F_A$ for several formal concentrations of the complexing agent gives a straight line, the slope of which is $\frac{0.0591X}{n}$. If $n$ is known, the combining ratio of ligand to metal ion is thus obtained. Equation (10.4) can then be employed to calculate $K_f$.

## 10.6 DROPPING MERCURY ELECTRODE (DME) OR CURRENT VARIATIONS DURING THE LIFETIME OF A DROP

The current passing through a cell containing a dropping electrode undergoes periodic fluctuations in frequency corresponding to the drop rate. As a drop breaks, current falls to zero, it then increases rapidly as the electrode area grows because of the greater surface to which diffusion can occur. For convenience, well-damped galvanometer is generally employed for current measurement. As shown in Figure 10.3, the oscillations under these circumstances are limited to a reasonable magnitude and the average current is readily determined provided the drop rate is reproducible. Note that the effect of irregular drops in the centre of the limiting current region, probably caused by vibration of apparatus.

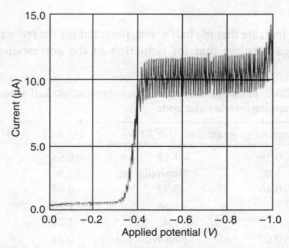

**FIGURE 10.3** Current variation during the lifetime of a drop.

### 10.6.1 Advantages of DME

1. Large overvoltage for the formation of hydrogen from hydrogen ions. As a consequence, the reduction of many substances from acidic solutions can be studied without interference.
2. Because a new metal surface is continuously generated, the behaviour of the electrode is independent of its past history. Thus, reproducible current voltage curves are obtained regardless of how the electrode has been used previously.
3. The third useful feature of a dropping mercury electrode is that reproducible average currents are immediately achieved at any given applied potential.

### 10.6.2 Disadvantage of DME

Most serious is the ease with which mercury is oxidized, this property severely restricts the use of mercury as an anode. At applied potentials much above +0.4 V (versus saturated calomel electrode) formation of mercury(I) occurs, the resulting current masks the polarographic waves of other oxidizable species in the solution. Thus, DME can be employed only for the analysis of reducible or very easily oxidizable substances.

## 10.7 POLAROGRAPHIC DIFFUSION CURRENTS

In 1934, D. Ilkovic determine the magnitude of the diffusion currents at 25°C

$$i_d = 607 \, n \, D^{1/2} \, m^{2/3} \, t^{1/6} \, C \tag{10.5}$$

where $n$ is the number of Faradays per mole of reactant, $D$ is the diffusion coefficients for the reactive species expressed in units of square centimetres per second, $m$ is the rate of mercury flow in milligrams per second, $t$ is the drop time in seconds and $C$ is the concentration of reactant in millimoles per litre. The quantity 607 represents the combination of several constants.

## 10.8 CAPILLARY CHARACTERISTICS AND ITS EFFECT ON DIFFUSION CURRENTS

The product $m^{2/3} \, t^{1/6}$ in the Ilkovic equation called the *capillary constant*, describes the influence of DME characteristics upon the diffusion current, since both $m$ and $t$ are readily evaluated experimentally, comparison of diffusion currents from different capillaries is thus possible. Two factors, other than the geometry of the capillary itself, play a part in determining the magnitude of the capillary constant. The head that forces the mercury through the capillary, influences both $m$ and $t$ such that the diffusion current is directly proportional to the square root of the mercury column height. The drop time $t$ of a given electrode is also affected by the applied potential since the interfacial tension between the mercury and the solution varies with the charge on the drop. Generally, it passes through a maximum at about –0.4 V (versus saturated calomel electrode) and then falls off rapidly, at –0.2 V, it may be only half of its maximum value. Fortunately, the diffusion current varies only as the one-sixth power of the drop time so that, over small potential ranges, the decrease in current due to this variation is negligibly small.

## 10.9  MIXED ANODIC–CATHODIC WAVES

Anode waves as well as cathodic waves are encountered in polarography. The former are less common because of the relatively small range of anodic potential that can be covered with the DME before oxidation of the electrode itself commences. An example of anodic wave is illustrated in curve $A$ of Figure 10.4, where the electrode reaction involves the oxidation of iron(II) to iron(III) in the presence of citrate ions. A diffusion current is obtained at zero volt (versus the saturated calomel electrode) which is due to the half reaction.

$$Fe^{2+} \Leftrightarrow Fe^{3+} + e$$

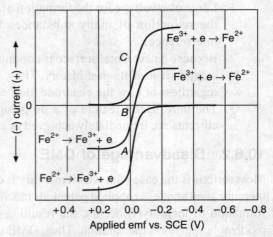

**FIGURE 10.4**  Mixed anode–cathode waves.

As the potential is made more negative, a decrease in the anodic current occurs at about –0.2 V and the current becomes zero because of the oxidation of iron(II) ion has ceased.

Figure 10.4 shows polarographic behaviour of iron(II) and iron(III) in a citrate medium. Curve $A$ is the anodic wave for a solution in which $[Fe^{2+}] = 1 \times 10^{-3} m$. Curve $B$ is the anodic–cathodic wave for a solution in which $[Fe^{2+}] = [Fe^{3+}] = 0.5 \times 10^{-3} m$. Curve $C$ is the cathodic wave for a solution in which $[Fe^{3+}] = 1 \times 10^{-3} m$.

Curve $C$ represents the polarogram for a solution of iron(III) in the same medium. Here, a cathodic wave results from the reduction of the iron(III) to the divalent state. The half wave potential is identical with that for the anodic waves indicating that the oxidation and reduction of the two iron species are perfectly reversible at the dropping electrode.

Curve $B$ is the polarogram of an equiformal mixture of iron(II) and iron(III). The portion of the curve below the zero current line corresponds to the oxidation of the iron(II). This ceases at an applied potential equal to the half wave potential. The upper portion of the curve is due to the reduction of iron(III).

## 10.10  OXYGEN WAVE

Dissolved oxygen is readily reduced at the dropping mercury electrode. An aqueous solution saturated with air exhibits two different waves attributable to this element. The first results from the reduction of oxygen to peroxide.

$$O_2(g) + 2H^+ + 2e \Leftrightarrow H_2O_2$$

The lower curve is for 0.1 F KCl alone (Figure 10.5). The second corresponds to the further reduction of hydrogen peroxide

$$H_2O_2 + 2H^+ + 2e^- \Leftrightarrow 2H_2O$$

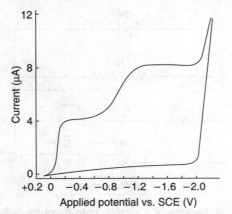

**FIGURE 10.5** Applied potential vs. SCE, (V) – polarogram for the reduction of oxygen in an air sat. 0.1 F KCl solution.

As would be expected from stoichiometric considerations, the two waves are of equal height.

While the polarographic waves are convenient for the determination of the concentration of dissolved oxygen, the presence of this element often interferes with the accurate determination of other species. Thus, oxygen removal is ordinarily the first step in a polarographic analysis. Aeration of the solution for several minutes with an inert gas accomplishes this end, a stream of the same gas usually nitrogen, is passed over the surface during the analysis to prevent reabsorption.

## 10.11 APPLICATION OF POLAROGRAPHY

### 10.11.1 Determination of Diffusion Currents

The diffusion current can be evaluated from the difference between the two i.e. limiting current and residual current at some potential in the limiting current region. Because the residual current usually increases nearly with applied voltage. It is often possible to make the correction by extrapolation as shown in Figure 10.6. Figure 10.7 shows simple circuit diagram of a simple instrument for polarographic measurement. Where, $R_1$ is variable resistance, $R_2$ is constant resistance, 1 and 2 are two keys.

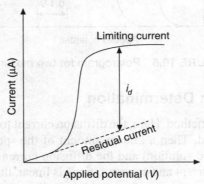

**FIGURE 10.6** Polarogram showing diffusion current.

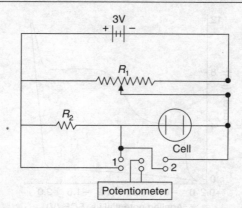

**FIGURE 10.7** Simple circuit diagram of a simple instrument for polarographic measurements.

## 10.11.2 Analysis of Mixtures

The quantitative determination of each species in a multicomponent mixture from a single polarogram is theoretically feasible provided the half wave potentials of the various species are sufficiently different where a major constituent is more easily reduced than a minor one. However, the accuracy with which the later can be determined may be poor because its diffusion current can occupy only a small fraction of the current scale. This problem does not arise when the minor constituent is the more easily reduced component.

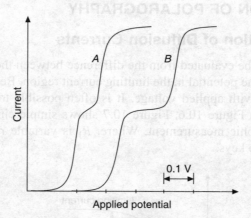

**FIGURE 10.8** Polarogram for two mixtures.

## 10.11.3 Concentration Determination

It is done by standard addition method. Here, the diffusion current for an accurately known volume of sample solution is measured. Then a known amount of the species of interest is introduced (at a known volume of standard solution) and the diffusion current is again evaluated. Provided the relationship between the current and concentrations is linear, the increase in wave height will permit calculation of the concentration in the original solution.

## 10.11.4 Inorganic Polarographic Analysis

Most metallic cations are reduced at DME to form metal amalgam or an ion of lower oxidation state. Even the alkali metal and alkaline earth metal are reducible, provided the supporting electrolyte does not react at the high potentials required, here the tetralkyl ammonium halides are useful.

The successful polarographic analysis of cations frequently depends upon the supporting electrolyte that is used. For example, with KCl as a supporting electrolyte, the waves of iron(III) and copper(II) interfere with one another, in a fluoride medium, however, the half wave potential of the former is shifted by about –0.5 V while that for the latter is altered by a few hundredth of a volt. The presence of fluoride thus results in the appearance of separate waves for the two ions.

The polarographic method is also applicable to the analysis of such inorganic anions as bromate, iodate, dichromate, vandate, selenite and nitrite. In general, polarograms for these substances are affected by the pH of the solution because the hydrogen ion is a participant in the reduction process. As a consequence, strong buffering of the solutions of some fixed pH is necessary to obtain reproducible results.

Certain inorganic anions that form complexes or precipitate with the ions of mercury are responsible for anodic waves and can be studied with the help of polarogram.

## 10.11.5 Organic Polarographic Analysis

Several common functional groups are oxidized or reduced at the dropping electrode. Compounds containing these groups are thus subject to polarographic analysis.

In general, the reactions of organic compounds at a microelectrode are slower and more complex than those for inorganic cations. Thus, theoretical interpretation of data is more difficult or even impossible. Despite these handicaps, organic polarography has proved fruitful for the determination of structure, the qualitative identification of compounds and the quantitative analysis of the mixtures.

Organic electrode processes ordinarily involve hydrogen ions, the most common reaction being represented as:

$$R + nH^+ + ne^- = RH_n$$

where R and $RH_n$ are the oxidized and reduced forms of the organic molecule. Half wave potentials for organic compounds are therefore markedly pH dependent. Thus in organic polarography good buffering is vital for the generation of reproducible half wave potentials and diffusion currents.

Instead of using pure water, other solvents are used for organic polarography. Aqueous mixture containing various amounts of such miscible solvents as glycols, dioxane, alcohols, cellosolve or glacial acetic acid have been employed. Anhydrous media, such as acetic acid, formamide and ethylene glycol have also been investigated. Supporting electrolyte are often lithium salts or tetra-alkyl ammonium salts.

Organic compounds containing any of the following functional group can be expected to produce one or more polarographic waves.

1. Carbonyl group and quinines produce polarographic waves, in general, aldehyde are reduced at lower potentials than ketones.
2. Certain carboxylic acids are reduced polarographically, although simple aliphatic and aromatic monocarboxylic acids are not. Dicarboxylic acids such as fumaric acid, maleic

or pthalic acid, in which carboxylic groups are conjugated with one another give characteristic polarograms.
3. Most peroxides and epoxides yield polarograms.
4. Nitro, nitroso, amine oxide and azo groups are generally reduced at the dropping electrode.
5. Most organic halogen groups produce a polarographic wave which results from replacement of the halogen group with an atom of hydrogen.
6. The carbon–carbon double bond is reduced when it is conjugated with another double bond, an aromatic ring or an unsaturated group.
7. Hydroquinones and mercaptans produce anodic waves.

## 10.12 AMPEROMETRIC TITRATIONS

The current passing through a polarographic cell at some fixed potential is recorded as a function of reagent volume. Plots of data on either side of the equivalence point are straight lines with differing slopes. The end point is established by extrapolation of their intersection. The amperometric method is inherently more accurate than the polarographic method and less dependent upon the characteristic of the capillary and the supporting electrolyte, although it must be kept constant during the titration. Finally, the substance being determined need not be reactive at the electrode, a reactive reagent or product is equally satisfactory.

## 10.13 AMPEROMETRIC TITRATION CURVES

Amperometric titration curves typically take one of the forms shown in Figure 10.9. Curve (a) represents a titration in which the analyte reacts at the electrode while the reagent does not. The titration of lead with sulphate or oxalate ions may be cited as an example. A linear decrease in current is observed as lead ions are removed from the solution by precipitation. The curvature near the equivalence points reflects the incompleteness of the precipitation reaction in this region. The end point is obtained by extrapolation of the linear portions as shown in Figure 10.9.

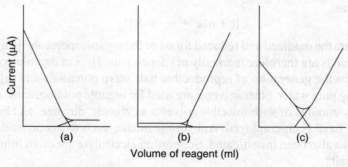

**FIGURE 10.9** Amperometric titration curves.

Curve (b) is typical of a titration in which the reagent reacts at the microelectrode and the analyte does not. An example would be the titration of magnesium with 8-hydroquinoline.

Curve (c) corresponds to the titration of lead ion with chromate solution at an applied potential greater than −1.0 V. Both lead and chromate ions give diffusion currents and a minimum in the curve signals the end point.

## 10.14 APPARATUS AND TECHNIQUES OF AMPEROMETRIC TITRATION

Figure 10.10 shows a typical cell for an amperometric titration, a calomel half cell is usually employed as the non-polarizable electrode. The indicator electrode may be dropping mercury electrode or a micro-electrode wire electrode as shown in Figure (10.10).

The cell should have a capacity of 75–100 ml. For linear plots, it is necessary to correct for volume changes due to the added titrant. The measured currents can be multiplied by the quantity $\dfrac{(V+U)}{V}$, where $V$ is the original volume of analyte and $U$ is the volume of the reagent. Thus, all measured currents are corrected back to the original volume. An alternative, which is often satisfactory, is to use a reagent that is 20 (or more) times as concentrated as the solution being titrated. Under these circumstances, $U$ is so small with respect to $V$ that the correction is negligible. This approach requires the use of microburrete so that a total reagent volume of 1–2 ml can be measured with a suitable accuracy. The microburrete should be so arranged that its tip can be touched to the surface of the solution after each addition of reagent to permit removal of the fraction of a drop that tends to remain attached.

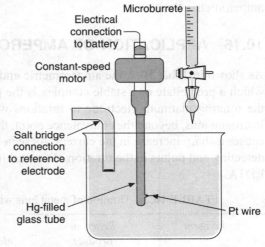

**FIGURE 10.10** A typical cell for an amperometric titration.

## 10.15 MICROELECTRODES; THE ROTATING PLATINUM ELECTRODE

Many amperometric titrations can be carried out conveniently with a DME. For reactions involving oxidizing agent that attack mercury (bromine, silver ion, iron(II), among others) a rotating platinum electrode is preferable. This microelectrode consists of a short length of a platinum wire sealed into the side of a glass tube. Mercury within the glass tube provides an electrical contact between the wire and the lead to the polarograph. The tube is held in the hollow chuck of a synchronous motor and rotated at a constant speed in excess of 600 rpm. Commercial models of the rotating platinum electrode are available.

Polarographic waves that are similar to those observed with the dropping mercury electrode can be obtained with the rotating platinum electrode. Here, however, the reactive species is brought to the electrode surface not only by the diffusion but also by mechanical mixing. As a consequence, the limiting current are as much as 20 times larger than those obtained with the microelectrode that is supplied by the diffusion only. With a rotating electrode, steady currents are instantaneously obtained. This behaviour is in distinct contrast to the behaviour of a solid microelectrode in the absence of stirring.

Several limitations restrict the widespread application of the rotating platinum electrode to polarography. The low hydrogen overvoltage prevents its use as a cathode in acidic solutions. In addition, the high current obtained cause the electrode to be practically sensitive to the traces of oxygen in the solution. These two factors have largely limits its use to anodic reactions. Limiting currents from a rotating electrode are often influenced by the previous history of the electrode and are seldom as reproducible as the diffusion currents obtained with a dropping electrode. These limitations, however, do not seriously restrict the use of the rotating electrode for amperometric titrations.

## 10.16 APPLICATION OF AMPEROMETRIC TITRATIONS

As shown in Table 10.2, the amperometric end points have been largely confined to titrations in which a precipitate or a stable complex is the product. A notable exception is the application of the rotating platinum electrode to titrations with bromate ion in the presence of bromide and hydrogen ions, beyond the equivalence point, the bromine concentration, which increases rapidly, causes a sharp increase in the current. As seen in table, the amperometric end points is useful for detecting end points in the titration of several metal ions with organic precipitating agent or with EDTA.

**TABLE 10.2** Titration of metal ions with organic precipitating agent or with EDTA

| Reagent | Reaction product | Type electrode | Substance to be determined |
|---|---|---|---|
| $K_2CrO_4$ | Precipitation | DME | $Pb^{2+}$, $Ba^{2+}$ |
| $Pb(NO_3)_2$ | Precipitation | DME | $SO_4^{2-}$, $MoO_4^{2-}$, $F^-$, $Cl^-$ |
| 8-Hydroxyquinoline | Precipitation | DME | $Mg^{2+}$, $Zn^{2+}$, $Cu^{2+}$, $Al^{3+}$, $Bi^{3+}$, $Fe^{3+}$ |
| Cupferron | Precipitation | DME | $Cu^{2+}$, $Fe^{3+}$ |
| Dimethylglyoxime | Precipitation | DME | $Ni^{2+}$ |
| $\alpha$-Nitroso-$\beta$-napthol | Precipitation | DME | $Co^{2+}$, $Cu^{2+}$, $Pd^{2+}$ |
| $K_4Fe(CN)_6$ | Precipitation | DME | $Zn^{2+}$ |
| $AgNO_3$ | Precipitation | Rotating Pt. electrode | $Cl^-$, $Br^-$, $I^-$, $CN^-$, RSH |
| EDTA | Complex | DME | $Bi^{3+}$, $Cd^{2+}$, $Cu^{2+}$, $Ca^{2+}$, etc. |
| $KBrO_3$, KBr | Substitution, addition or oxidation | Rotating Pt electrode | Certain phenols, aromatic amines, olefins, $N_2H_2$, As(III), Sb(III) |

A convenient modification of the amperometric method involves the use of two identical, stationary micro electrodes in a well stirred solution of the sample. A small potential (say 0.1 to 0.2 volt) is applied between these electrodes and current that flows is followed as a function of the volume of added reagent. The end point is marked by a sudden current rise from zero, a decrease in the current to zero, or a minimum (at zero), in a V-shaped curve. Although the use of two polarized electrodes for end point detection was proposed before 1900, but it comes into real existence around 1950. The name *dead-stop technique* was used to describe the technique.

Twin-polarized platinum microelectrodes are conveniently used for end point detection for oxidation–reduction titrations.

Twin–silver microelectrodes have been employed for end points detection in precipitation titration involving silver ion with a standard solution of the chloride ion. The current proportional to the metal ion concentration would result from reaction

$$[\text{Cathode } Ag^+ + e^- \Leftrightarrow Ag$$
$$\text{Anode } Ag \Leftrightarrow Ag^+ + e^-]$$

With the effective removal of $Ag^+$ ion by the analyte reaction, cathodic polarization would occur, and the current would approach zero at the end point.

## 10.17 MODIFIED VOLTAMETRIC METHODS

All modified techniques are applicable to electrodes other than the standard DME. In any electrochemical system, an equilibrated diffusion layer is established between the electrode and the bulk of the solution. Once the reaction potential have been reached, the concentration of electroactive species of interest changes from the bulk concentration value at position far removed from the electrode surface to essentially zero at the electrode surface. When the classical slow scan rates are employed, the slope of the concentration gradient within the diffusion layer is determined primarily by the rate of depletion of the electroactive species at the electrode surface. It varies from essentially zero at potential significantly more positive than the reduction potential, to a value governed by the concentration and diffusion coefficient of the electroactive species at a potential well past the reduction potential. However, when the polarographic method uses rapid changes in potential either because of the scan rate is fast or because a pulse modulation of some type is employed, the slope of the concentration gradient at any particular potential will be greater than in the slow scan instance. The bulk concentration will be closer to the electrode surface, the number of electroactive species arriving at the surface per unit time will be greater, and larger current signals will result. A few modified voltametric methods are pulse polarography, current sample polarography, stripping analysis, etc.

### 10.17.1 Pulse Polarography

Pulse polarography takes advantage of the fact that, following a sudden change in applied potential, the capacitative current surge decays much more rapidly than does the Faradaic current. In this technique, a small amplitude voltage pulse, in addition to the linearly increasing DC ramp (about 1 mVs$^{-1}$) normally used for DC polarography, is applied to the polarographic cell. As each mercury drop forms it is allowed to grow for a period of time, perhaps 1.9 s at the DC ramp potential, after which a sudden voltage pulse of perhaps 50 ms duration is applied. The pulse is synchronized with the maximum growth of the mercury drop, if a dropping mercury electrode is used. The current is measured 40 ms after the application of the pulse to allow time for the charging current to decay to a very low value. The capacitative current actually delays exponentially at a rate governed by the magnitude of the capacitance and series resistance of the system. During this time interval the Faradaic current also delays somewhat but, does not reach the diffusion controlled level because the concentration gradient at the instant of current measurement is considerably larger. Each succeeding drop is polarized with a somewhat larger pulse. This method gives a current voltage curve similar to that obtained in DC polarography except for the cancellation of the capacitive components. The measured signal is the Faradaic current that flows at the pulse potential minus any Faradaic current flowing due to the fixed DC potential. The limiting current is given by the Cottrell Equation.

$$I_{\lim} = n\,FC\,A\sqrt{\frac{D}{\pi t}}$$

In comparison with classical DC polarography, the sensitivity is about 6.5 times better. A much larger gain in real sensitivity is achieved through the virtual elimination of the charging current.

Derivative pulse polarography employs two current sampling intervals of equal time periods. The first sample period occurs just before the pulse application and the second sample period occurs at the end of the pulse. The current samples are stored on memory capacitors and the difference is displayed on a recorder in a display and hold manner. Difference in the two stored samples occurs only in the region of the half wave potential where the current is changing rapidly with the potential. The voltamogram recorded in this way is a peak-shaped incremental derivative of a conventional DC polarogram. Because the applied pulse is held for an appreciable length of time, the system response is not strongly dependent on the electrode kinetics, at least not to the extent expected of AC polarography. This implies that pulse polarographic techniques may be readily used for organic and other electro chemically irreversible systems. For pulse of small amplitudes, $\Delta E$, the peak current is given by

$$\Delta i_{\max} = n^2 F^2/4RT \sqrt{\frac{D}{\pi t}}\,AC\,(\Delta E)$$

whereas for pulse of large amplitude, it is given by

$$\Delta i_{\max} = nFAC\sqrt{\frac{D}{\pi t}}\left(\frac{\sigma - 1}{\sigma + 1}\right)$$

where $\sigma = \exp[(\Delta E)\,nF/2RT]$. In the limit, for $\Delta E \gg RT/nF$, $(\sigma - 1)/(\sigma + 1)$ approaches unity and $i_{\max}$ becomes the limiting current as given by the Cottrel equation. The use of a large pulse amplitude allows one to obtain large signals from extremely dilute solutions, with some distortions in the curve shape, because the pulse amplitude is an appreciable portion of the total polarographic wave.

Figure 10.11 Curve $A$ shows a linearly increasing scan voltage upon which a 35 mV pulse is superimposed during the last 50 ms of the 2 s drop time. Curve $B$, the overall current flowing through the cell as a result of applied voltage. Curve $C$, the capacitive component of the cell current. Curve $D$, the Faradaic component of the cell current measured above the DC background. The Curve $E$, the net current signal measured during the last 10 ms of the life of the drop after the capacitive current has decayed to near zero.

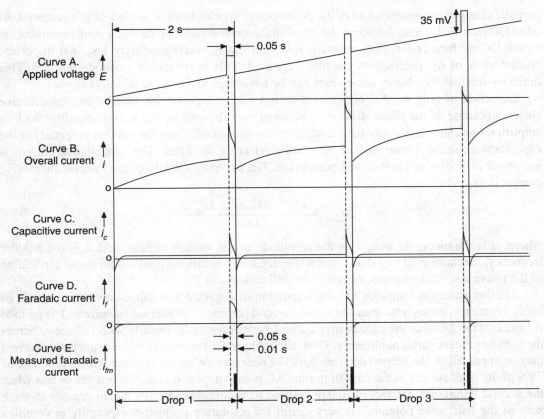

**FIGURE 10.11** Current–voltage diagram illustrating the operation of a pulse polarograph.

In principle, the derivative mode is less sensitive than the normal mode, but the resolution is better, about $50/n$ mV. It is capable of detecting $10^{-8}$ m, a 100-fold increase over classical polarography and a 10-fold increase over derivative DC polarography.

### 10.17.2 Current Sampled Polarography or AC Polarography

In current sampled polarographic technique, a potential period in time such as a sine wave, and of relatively small amplitude (1–35 mV) and low frequency (10–60 Hz), is superimposed upon the slow linear voltage sweep of DC polarography. This sinusoidal variation in potential is similar to that employed in cyclic voltammerty with the exception that the potential excursion are of much smaller magnitude, typically 10 mV peak to peak. The direct current component of the total current is blocked out and only the rectified and damped alternating component is displayed as a function of DC potential. By looking only at the alternating portion of the current that flows and detecting its amplitude, one is in effect looking at the difference in current that flows between the minimum and maximum applied potentials during the modulation period. The current is sampled just before the mercury drop is dislodged. A peak output signal with its maximum amplitude at the half wave potential rather than a step-shaped wave, is produced, because the reversible wave attains its maximum slope at the half wave potential, therefore, a given AC signal causes the periodic changes in concentration of the electroactive species to be maximum at this potential. At other DC potentials along the wave, the sinusoidal potential variation causes less of a perturbation of the DC surface concentration and the AC current is correspondingly less. An important characteristic of AC polarography is that it responds only to reversible electrode reaction. This limits its applicability, but in some cases can be advantageous in avoiding interferences.

Detection of only the AC component allows one to separate the Faradic and capacitative currents because of the phase difference between them by employing a phase sensitive lock-in amplifier, one can select either the Faradaic (phase shifted 45° from the applied potential) or the capacitative current (phase shifted 90°) while rejecting the other. The capacitative current is important in studies of kinetics and adsorption. The maximum height of the Faradic alternating current is given by

$$\Delta i_{max} = \frac{n2F^2\ AVC\omega^{1/2}\ D_{ox}^{1/2}}{4RT}$$

where, $A$ is the electrode area, $V$ is the amplitude of the voltage signals, and $\omega$ is the angular frequency. [Because even a moderate cell resistance serves to mix the phases, successful application of the phase discriminator requires very low cell resistances.]

Another successful approach to the separation of Faradaic and capacitative components of the cell current is through the measurement of second harmonics of alternating current. The method is based on the fact that the capacitative current varies essentially linearly with voltage, whereas the Faradic process varies nonlinearly. Thus, although the first derivative of the capacitative current may be appreciable, the second derivative will be near zero. Whereas the base line current may be 75% of the peak current in the case of normal AC polarography. It is only about 5% or less when the second harmonics are recorded. The second harmonic, because, its signal crosses through zero of the half wave potential, is very useful for resolution problems, especially in complex mixtures.

### 10.17.3 Stripping Analysis

The technique of stripping analysis, also called *linear-potential sweep stripping chronoamperometry*, involves two steps: a concentration or pre-electrolysis step in which the desired component is deposited cathodically or anodically, followed by a reverse electrolysis in which the component is determined. In the anodic variant of stripping analysis, the metal concerned is reduced at a controlled potential for a definite time under fixed conditions of geometry and stirring. The working electrode may be a hanging mercury drop or a mercury film on a wax-imprignated graphite electrode or in some cases an inert solid electrode such as platinum or carbon. The final anodic dissolution or stripping process involves a linear anodic scan in which the metal is oxidized. The resulting stripping voltammogram shows peaks, the heights of which are generally proportional to the concentration of the corresponding electro-active metal ions and the potentials of which have the same qualitative interpretation as their half wave potentials in polarography. Standards are carried through identical pre-electrolysis and stripping steps. A typical anodic scan is shown in Figure 10.12 for 0.100 ml sample of whole blood prepared for analysis by digestion with perchloric acid and brought to 5.0 ml volume with sodium acetate. As little as $10^{-8}$ m cadmium (corresponding to about $10^{-6}$ wt%) has been determined with a precision of $\pm 3.0\%$ using 15 min of pre-electrolysis and linear scan rate of 21 mVs$^{-1}$. Because the standard addition method is usually used for evaluating the unknown concentration, the reproducibility will be the same as the precision. By extending the pre-electrolysis time to 60 min, the sensitivity is extended to $10^{-9}$ m but deviations will be about 10–20% at this concentration level.

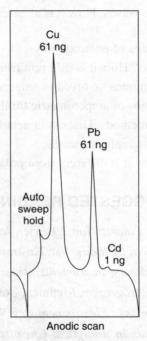

**FIGURE 10.12** Plating time 30 min, plating pot –1.0 V, sweep rate 60 mV/s, chart speed 12.7 cm/mm, blood sample 0.2 ml, current range 200 µA, stripping time ~16 s.

Cathodic stripping analysis consists of forming an insoluble layer at the electrode surface during an anodic pre-electrolysis and stripping it off by reverse electrolysis. Unlike anodic stripping analysis, the deposition potential depends on the concentration to be determined and shifts about $60/n$ mV to more positive values for each order of magnitude decrease in the anion concentration. It is not possible to carry the deposition to completion because of the limitation imposed by finite solubility of the layer, usually salts of mercury(I) precipitates. Such a method has been used for determination of the halides, tungustate and molybdate, lower concentration limits are about $5 \times 10^{-6}$m.

# EXERCISES

1. Define polarography. How it is different from amperometric titrations.
2. Write down the advantage of polarographic estimation over amperometric titrations.
3. Describe construction, working and principle of polarographic cell.

4. What is a polarogram? What types of information does it provide?
5. Discuss the effect of complex formation on polarographic wave.
6. Define capillary characteristics of dropping mercury electrode.
7. Describe construction and working of droping mercury electrode.
8. Write down advantage and disadvantage of dropping mercury electrode.
9. Define mixed anodic cathodic waves. How it is affected by the presence of impurity in the analyte solution.
10. Write down various applications of polarography.
11. What is amperometric titration? How it is different from polarography?
12. Describes apparatus and techniques involved in amperometric titrations.
13. Write down various applications of amperometric titrations.
14. Define modified voltametric methods. Discuss in detail pulse polarography.
15. Define current sampled polarography in detail.
16. What is striping analysis? How it is different from polarography?

## SUGGESTED READINGS

Bottcher, C.J.F., *Theory of Electric Polarization*, Elsevier, Amsterdam, 1952.

Charlot, G., *Modern Electroanalytical Methods*, Van Nostrand, London, 1958.

Collins, J.R., *Electrochemical Measuring Instruments*, Rider, New York, 1962.

Crow, D.R. and J.V. Westwood, *Polarography*, Methuen, London, 1968.

Delahay, P., *New Instrumental Methods in Electrochemistry*, Wiley Interscience, New York, 1954.

Donbrow, M., *Instrumental Methods in Analytical Chemistry their Application and Practice*, Vol. 1, Pitman, London, 1966.

Heyrovsky, J., *Principles of Polarography*, Academic Press, 1966.

Kambara, T., *Modern Aspects of Polarography*, Plenum Press, New York, 1966.

Kolthoff, I.M. and J.J. Lingane, *Polarography*, Vols. 1 and 2, 2nd ed., Interscience, New York, 1952.

Lingane, J.J., *Electroanalytical Chemistry*, 2nd ed., Interscience, New York, 1958.

Meites, L., *Polarographic Technique*, Interscience, New York, 1955.

Meites, L., *Polarographic Technique*, 2nd ed., Interscience, New York, 1965.

Milner, G.W.C., *Principles and Applications of Polarography and other Electroanalytical Process*, Longmans, Green, New York, 1951.

Muller, O.H., *Polarographic Methods of Analysis*, 2nd ed., Chem Education Publication, Easton, 1951.

Schmidt, H. and M. Stackelberg, *Modern Polarographic Methods*, Academic Press, New York, 1963.

Scholander, A., *Introduction to Practical Polarography*, Radiometer, Lopenhangeer, 1950.

Sharma, B.K., *Instrumental Methods of Chemical analysis*, Goel Publishing House, Meerut.

Skoog, D.A. and D.M. West, *Fundamentals of Analytical Chemistry*, 3rd ed., Rinehart & Winston, New York.

Stock, J.T., *Amperometric Titrations*, Vol. 20, Interscience, New York, 1965.

Willard, H.H., L.L. Merrit, J.A. Dean, and F.A. Settle, *Instrumental Methods of Analysis*, 6th ed., CBS Publishers, Delhi.

Zuman, P. and I.M. Kolthoff, *Progress in Polarography*, Vols. 1 and 2, Wiley, New York, 1962.

# 11

# Conductometric Methods

## 11.1 INTRODUCTION

Conductance measurements were among the first to be used for determining solubility product, dissociation constant and other properties of electrolytic solution.

Conductance is an additive property of a solution depending upon all the ions present in it. Solution conductometric measurement therefore, are nonspecific. This nonspecificity restricts the quantitative analytical use of this technique to situations, where only a single electrolyte is present or where the total ionic species needs to be ascertained. Changes in the slope of conductance versus titrant volume occur because ionic mobility vary and also because of the formation of insoluble or non-ionized materials.

## 11.2 ELECTROLYTIC CONDUCTIVITY

Electrolytic conductivity is a measure of the ability of a solution to carry an electric current. Solutions of electrolytes conduct an electric current by the migration of ions under the influence of an electric field like a metallic conductor, they obey Ohm's law. Exceptions to this law occur only under abnormal conditions, for example, very high voltage or high frequency currents. The current $I$ flowing between the electrodes immersed in the electrolyte will vary inversely with the resistance of the electrolytic solution $R$. The reciprocal of resistance $1/R$ is called the conductance and is expressed in reciprocal ohms or mhos.

The standard unit of conductance is specific conductance $(K)$ which is defined as the reciprocal of the resistance in ohms of a 1 cm cube of liquid at a specified temperature. The unit of specific conductance is the reciprocal $\Omega$ cm. The observed conductance of a solution depends inversely on the distance between the electrodes and directly upon their area $A$,

$$\frac{1}{R} = K \frac{A}{d} \tag{11.1}$$

The electrical conductance depends upon the number of ions per unit volume of the solution and upon the velocities with which these ions move under the influence of the applied electromotive force.

As a solution of an electrolyte is diluted, the specific conductance in Eq. (11.1) will decrease. Fewer ions to carry the electric current are present in each cubic centimetre of solution. However, in order to express the ability of individual ions to conduct, a function, called the *equivalent conductance* is employed. It may be derived from Eq. (11.1), where $A$ is equal to the area of two large parallel electrodes set 1 cm apart and holding between them a solution containing one equivalent of solute. If $C_s$ the concentration of solution in gram equivalents per litre, then the volume of solution in cubic centimetres per equivalent is equal to $\frac{1000}{C_s}$ so that Eq. (11.1) becomes

$$\Lambda_{eq} = \frac{K}{C_s} \times 1000 \tag{11.2}$$

At infinite dilution the ions theoretically are independent of each other and each ion contributes its part to the total conductance, thus

$$\Lambda_\infty = \sum (\lambda_+) + \sum (\lambda_-) \tag{11.3}$$

where $\lambda_+$ and $\lambda_-$ are the ionic conductance of cations and anions, respectively at infinite dilution. Electrolytic conductance as well as equivalent conductance and molar conductance increase with dilution whereas specific conductivity of an electrolyte solution decreases with dilution. Specific conductance of an electrolyte falls with dilution because the number of current carrying particles, i.e. ions present per centimetre cube of the solution become less and less on dilution (Figure 11.1).

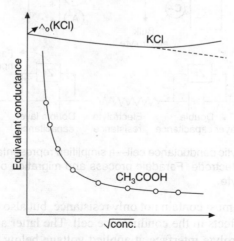

**FIGURE 11.1** Variation of equivalent conductance with square root of concentration.

However, increase of $\wedge_{eq}$ or $\wedge_M$ conductivity on dilution is due to the fact that these are the product of specific conductivity and the volume $V$ of the solution containing one gram equivalent or one gram molecule of the electrolyte. As the decreasing value of specific conductivity is more than compensated by the increasing value of $V$ so the value of $\wedge_{eq}$ and $\wedge_M$ increases with dilution.

### 11.2.1 Measurement of Electrolytic Conductance

Electrolytic conductance measurements usually involve determination of the resistance of a segment in solution between two parallel electrodes by means of Ohm's law. These electrodes are made up of platinum metal that has been coated with a deposit of platinum black to increase the surface area and reduce the polarization resistance. Some of the more important phenomena associated with the application of a voltage between the electrodes immersed in a liquid electrolyte is indicated in Figure 11.2 for an idealized system. To eliminate the effects of processes associated with the electrodes, measurements are made with an alternating current at 50, 100, 1000 Hz. Some variations of wheat-stone bridge are generally employed. The evaluation involves a comparative procedure. A conductance cell is calibrated by determining its cell constant, using a solution of known conductivity.

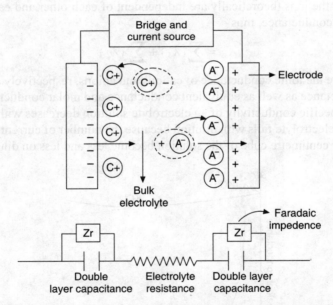

**FIGURE 11.2** Electrolytic conductance cell—a simplified representation of the double layer at the electrode, Faradaic process and migration of ions through the bulk electrolyte.

Generally the bridge circuit must contain not only resistance, but also capacitance (or inductance) to balance the capacitive effects in the conductance cell. The latter arises from electrical double layer at the electrode–electrolyte interface at applied voltage below the decomposition voltage, and from the frequency dependent resistance (impedance) associated with the Faradic processes at voltage above the decomposition voltage. For resistance of less than $10^4 \Omega$, the model of a conductance cell as the electrolyte resistance in series with the double layer capacitance is a reasonable physical approximation. The magnitude of impedance at 1000 Hz is of the order of

$$\frac{1}{2\pi fc} = 1.6 \text{ to } 16 \,\Omega$$

for capacitance value of 10–100 µF/cm² of electrode surface. Introduction of a variable capacitance (or inductance) into the bridge circuit permits compensation of the phase shift between current and voltage caused by the capacitance in the electrolytic cell.

Parallel resistance capacitance (*RC*) balancing values are needed and small capacitances can be obtained with higher accuracy and less frequency dependence than larger ones. For example, the parallel capacitance requires to compensate a series capacitance of 100 µF with a resistance of 1000 Ω is only 300 pF.

### 11.2.2 Applications of Conductivity Measurements

There are many applications of conductivity measurements, some of them are given below.

1. *Solubility of sparingly soluble salt:* Salt like AgCl, BaSO$_4$ PbSO$_4$, etc. are sparingly soluble and their solubility cannot be determined by any chemical method. By conductometric methods, its solubility is determined by the relation:

$$S = \frac{K \times 1000 \times E}{\lambda\infty} \text{ g/l}$$

where, $E$ is an equivalent weight of a substance
At 25°C

$$S = \frac{K \times 1000 \times 143.5}{138.27} \text{ g/l for AgCl}$$

2. *Ionic product of water:* Pure water ionizes to a very slight extent and so we have the equilibrium

$$H_2O \rightleftharpoons H^+ + OH^-$$

$$K = \frac{[H^+][OH^-]}{H_2O}, \text{ where } K \text{ is an ionization constant of } H_2O.$$

Since water is supposed to be slightly ionized, the concentration of unionized water may be considered as practically constant.

$$K[H_2O] = [H^+][OH^-]$$
$$K_w = [H^+][OH^-]$$

The product of ionic concentration of H$^+$ and OH$^-$ expressed in gram moles per litre is constant at constant temperature and is known as *ionic product of water*. According to Kohlrausch law

$$[H^+] \text{ or } [OH^-] = \frac{5.54 \times 10^{-8} \times 1000}{\wedge_{eq} \times 548.3} = 1.01 \times 10^{-7}$$

$$\therefore \quad K_w = (1.01 \times 10^{-7})(1.01 \times 10^{-7})$$
$$K_w = 1.02 \times 10^{-14} \text{ at } 25°C$$

3. *Basicity of an acid:* According to Ostwald, the basicity of an acid is given by

$$B = \frac{\lambda_{1024} - \lambda_{32}}{10.8}$$

where $\lambda_{1024}$ and $\lambda_{32}$ denote the equivalent conductivities of sodium salts of an acid of dilution of $\lambda_{1024}$ litres and $\lambda_{32}$ litres per gram equivalent.

To prepare N/32 solution of sodium salt of an acid, a solution of N/16 NaOH is first prepared and 100 ml of this solution is then taken in a measuring flask and neutralized by adding a concentrated solution of the acid using suitable litmus paper. At the end point, the volume is made exactly 200 ml. This is N/32 solution of sodium salt of the acid. N/1024 solution can be prepared by dilution of N/32 solution. Hence, measuring the equivalent conductivity of N/32 and N/1024 solution of the salt, basicity $B$ can be calculated from the above relation.

4. *Conductometric titrations:* The determination of the end points of a titration by mean of conductivity measurement is known as *conductometric titrations*. Conductometric titrations are based on the principle that the equivalent conductivity depends upon the number and mobility of the ions. The method has advantage that, it can be used with coloured solution and will work where no indicator is found to be satisfactory. In order to get accurate results, i.e., accurate end-point readings, it is necessary to keep the temperature of the solution constant and to have one of the constituents fairly concentrated to avoid diluting of the solution. In general, mobility increase by about 2% for each 1°C increase in temperature. In order to keep the volume change small, generally the titrant should be about 10 times as concentrated as the solution being titrated. Correction can be accomplished by multiplying the observed conductance by the factor $(V + v)/V$, where $V$ = initial volume and $V + v$ is the final volume.

Thus,
$$C_{\text{corrected}} = \frac{(v + V)}{v} \times C_{\text{observed}}$$

However, this may not be necessary, provided that the total volume of reagent added does not exceed 1–2% of the solution titrated.

(a) *Conductometric titration of strong acid against strong base:* Consider the titration of a dilute solution of HCl with NaOH.

$$HCl + NaOH \rightarrow NaCl + H_2O$$

As NaOH is added, the concentration of $H^+$ ions is decreased and although $H^+$ ions are replaced by $Na^+$ ions, the mobility of $Na^+$ is much less than $H^+$ ions, so that conductivity of the solution decreases rapidly. The solution at neutralization that is at the end point contains only $Na^+$ and $Cl^-$ ions and will have minimum conductance. Now if a little NaOH is added after neutralization, the conductivity again increases owing to the presence of $OH^-$ ions, since the latter has the 2nd greatest mobility (Figure 11.3).

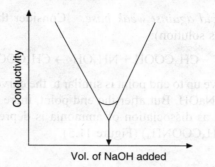

**FIGURE 11.3** Conductometric titration of strong acid against strong base.

In the actual titration of this type, the lines are likely to be slightly curved due to:
   (i) Variation in temperature, at least, due to the heat of neutralization.
   (ii) Inter-ionic effect.
   (iii) Increase in volume of solution due to added reagent.
   (iv) Due to presence of foreign ions present.
(b) *Strong acid with a weak base:*   Consider the titration of HCl with $NH_4OH$.

$$HCl + NH_4OH \rightarrow NH_4Cl + H_2O$$

After the end point is passed, further addition of $NH_4OH$ will cause no change in the conductance as $NH_4OH$, a weakly ionized electrolyte has a very small conductivity compared with that of the acid or the salt.

(c) *Weak acid against a strong base:*   Consider the titration of acetic acid against NaOH

$$CH_3COOH + NaOH \rightarrow CH_3COONa + H_2O$$

when small amount of NaOH is added to $CH_3COOH$, the conductivity decreases, but since the concentration of $H^+$ ions in acetic acid is small, the conductance of solution soon increases due to the formation of sodium ions and acetate ions. After the complete neutralization of acid, further addition of alkali introduces excess of $OH^-$ ions. The conductance of the solution therefore begins to increase more sharply (Figure 11.4).

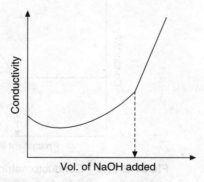

**FIGURE 11.4** Conductometric titration of weak acid against strong base.

(d) *Weak acid against weak base:* Consider the titration of acetic acid with $NH_3$ (aqueous solution):

$$CH_3COOH + NH_4OH \rightarrow CH_3COONH_4 + H_2O$$

The curve up to end point is similar to the curve obtained when acetic acid is titrated against NaOH. But after the end point, there is little effect on conductivity of the solution as dissociation of ammonia is depressed by the presence of ammonium salts ($CH_3COONH_4$) (Figure 11.5).

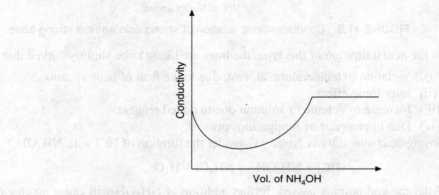

**FIGURE 11.5** Conductometric titration of weak acid against weak base.

(e) *Conductometric precipitation titrations:* Consider the reaction between $AgNO_3$ and $KCl$:

$$AgNO_3 + KCl \rightarrow AgCl \downarrow + KNO_3$$

Here one salt (KCl) is replaced by an equivalent amount of another salt ($KNO_3$), so the conductance remains almost constant in the early stages of the titration. After the end point passed, however, excess of the added salt cause a sharp increase in the conductance (Figure 11.6).

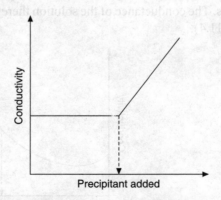

**FIGURE 11.6** Conductometric precipitation titration.

When both the products of the reaction are sparingly soluble, for example, in the titration of $MgSO_4$ with $Ba(OH)_2$ the curve obtained takes the form shown in Figure 11.7.

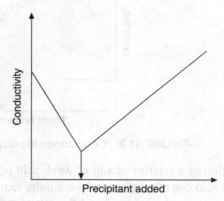

**FIGURE 11.7** Conductometric precipitation titration of $MgSO_4$ with $Ba(OH)_2$.

$$MgSO_4 + Ba(OH)_2 \rightarrow Mg(OH)_2 + BaSO_4$$

In this case, conductivity decreases in the beginning, but increases after the end point because of free barium hydroxide.

Generally, precipitation titrations are not as accurate as acid/base titrations when carried out conductometrically. This is due to:

(i) The slow separation of the precipitate with consequent super-saturation of the solution.

(ii) The removal of titrated solute by the adsorption on the precipitates. The best results can, however, be obtained by working with dilute solution in the presence of a relatively large amount of alcohol which causes a decrease in the solubility of the precipitate and there is also less adsorption.

(f) *Conductometric displacements titrations:* Salts of strong acid and weak base can be conductometrically titrated against a strong base and salts of strong bases and weak acid can be titrated against a strong acid. Consider the titration of sodium acetate with HCl.

$$CH_3COONa + HCl \rightarrow CH_3COOH + NaCl$$

In this titration only slight increase in conductance is obtained at the end point. This is because of fact that the $Cl^-$ ions, have a somewhat higher conductance than do the acetate ions.

In a similar manner, it is also possible to titrate the salt of a weak base ($NH_4Cl$) and strong acid against a strong base (NaOH).

$$NH_4Cl + NaOH \rightarrow NH_4OH + NaCl$$

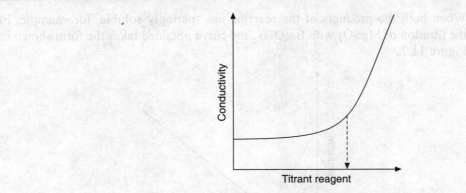

**FIGURE 11.8** Conductometric displacements titrations.

The titration of a mixture of salt of weak acid ($CH_3COONa$) and weak base ($NH_3$) by strong acid can also be carried out conductometrically. Similarly, it is possible to titrate, a mixture of weak acid and the salt of a weak base against a strong base.

(g) *Conductometric redox titrations:* Most oxidation–reduction titrations involve a decrease in the hydrogen ion concentration. For example:

$$6Fe^{2+} + Cr_2O_7^{2-} + 14H^+ \rightarrow 6Fe^{3+} + 2Cr^{3+} + 7H_2O$$

Due to the high mobility of $H^+$ ions this would indicate a sharp decrease in conductance during the initial part as shown in Figure 11.9.

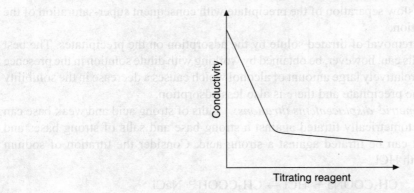

**FIGURE 11.9** Conductometric redox titrations.

(h) *Complexometric titrations:* These are similar to precipitation titrations in that total conductance change is usually small. A good example is the titration of KCl with $Hg(ClO_4)_2$, two breaks in the curve are obtained, one due to the formation of $HgCl_4^{2-}$ and other principal one at the end. Hall et al. have studied EDTA titrations conductometrically and were able to determine accurately a large number of cations in dilute acetate buffer solutions, which had formed stable EDTA complexes in the pH range 5–6 (Figure 11.10).

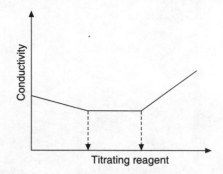

**FIGURE 11.10** Conductometric complexometric titrations.

(i) *Non-aqueous acid–base titrations:* Samples already in non-aqueous solvent can be conductometrically titrated using reagent such as alcoholic NaOH, perchloric acid, etc. as titrant. Van Merun and Dahen have titrated a large number of weak organic acids as methanol, dimethyl fomamide and pyridine with good results using tetramethyl ammonium hydroxide and potassium methoxide in methanol–benzene and pyridine–benzene.

## EXERCISES

1. Define various types of conductance. Explain the variation of equivalent and molar conductance with concentration.
2. Write the procedure involved in electrolytic conductance measurement.
3. Write down various applications of conductometric estimation.
4. Write down various advantages and disadvantages of conductometric measurement over potentiometric estimation.
5. How conductometric estimation is different from volumetric estimation.

## SUGGESTED READINGS

Charlot, G., *Modern Electroanalytical Methods*, Van Nostrand, London, 1958.

Delahay, P., *New Instrumental Methods in Electrochemistry*, Wiley Interscience, New York, 1954.

Donbrow, M., *Instrumental Methods in Analytical Chemistry Their Applications and Practice*, Vol. 1, Pitman, London, 1966.

Donbrow, M., *Instrumental Methods in Analytical Chemistry Their Principles and Practice*, Vol. 1, Pitman, London, 1966.

Lingane, J.J., *Electroanalytical Chemistry*, 2nd ed., Interscience, New York, 1958.

Pungor, E., *Oscillometry and Conductometry*, Pergamon Press, New York, 1965.

Skoog, D.A. and D. M. West, *Fundamentals of Analytical Chemistry*, 3rd ed., Rinehart & Winston, New York.

Sharma, B.K., *Instrumental Methods of Chemical Analysis*, Goel Publishing House, Meerut.

Willard, H.H., L.L. Merrit, J.A. Dean, and F.A. Settle, *Instrumental Methods of Analysis*, 6th ed., CBS Publishers, Delhi.

# Coulometric Analysis

## 12.1 INTRODUCTION

Coulometer is a device for measuring quantity of electricity determining the amount of chemical change brought about by the current. Various types of coulometer, such as oxygen hydrogen coulometer, silver coulometer, iodine coulometer, etc. are in use and, in fact, each coulometer is supposed to proceed at 100% current efficiency and put in series with the reaction cell. The current efficiency is defined as $\dfrac{100\, N_r}{N_t}$, where $N_r$ is the number of coulombs used to promote the desired reaction and $N_t$ is the total number of coulombs passed. For example, in the electrolysis of a solution containing a copper(II) salt, if all the electrons flowing from the cathode combine with $Cu^{2+}$ ion to the plate (copper metal) the current efficiently is 100%.

Coulometric method of analysis is based on the exact measurement of quantity of electricity that passes through a solution during the occurrence of an electrochemical reaction. The principle of coulometric analysis is based on Faradays's laws which may be expressed in the form that the extent of chemical reaction at an electrode is directly proportional to the quantity of electricity passing though the electrode. The substance of interest may be oxidized or reduced at one of the electrodes. This analysis is called *primary coulometric analysis* and in this analysis, the substance to be estimated reacts at an electrode which is maintained at a constant potential with respect to the solution and the current decreases as this substance is removed from the solution. The substance of interest may also react quantitatively in solution with a single product of electrolysis and the method is called *secondary coulometric method*. In this analysis, one of the product of electrolysis react with the substance to be estimated and the process is often carried out under constant current conditions and an end-point indicator is required. In either case, the fundamental requirement of coulometric analysis is that only a single overall reaction takes place for which the electrolytic reaction used for the determination proceeds with 100% current efficiency.

In an experiment, weight of the element $W$ deposited will be,

$$W = \frac{AQ}{n \times 96,500}$$

where $A$ is the atomic weight, $Q$ is the coulombs of electricity and $n$ is the valency of the element.

## 12.2 HYDROGEN–OXYGEN COULOMETER

It consists of a glass tube having a length of about 40 cm and an external diameter of about 20 mm, 100 ml of a mixture of hydrogen and oxygen which corresponds to about 500 coulombs can be collected above the sheets of bright platinum having an area of about 1.7 cm² welded to short length of stout platinum wires. A calibrated tube is connected to the electrolysis tube by a length of pressure rubber tube and is movable through a vertical distance approximately equal to its length. It is, therefore, possible to adjust the pressure of the collected gases to atmospheric pressure before measuring the volumes of the gases. The electrolyte solution in the coulometer is generally a solution of $K_2SO_4$ (Figure 12.1).

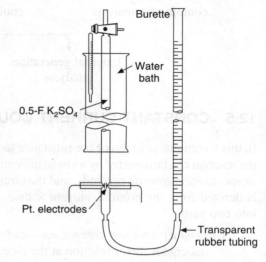

**FIGURE 12.1** A hydrogen–oxygen coulometer.

## 12.3 SILVER COULOMETER

It consists of a platinum or a silver vessel which acts as a cathode and contains a solution of pure silver nitrate as an electrolyte. The silver nitrate is purified by repeated crystallization from acidified solutions, followed by fusion. The solution of silver nitrate used for actual measurement should contain between 10 and 20 g of the salt in 100 ml. A rod of pure silver enclosed in a porous pot acts as the anode. The current density at the anode should not exceed 0.2 A cm⁻². After electrolysis, the electrolyte is taken out and the platinum vessel is washed, dried and weighed. The increase in the weight gives the amount of silver deposited. As we know that 96,500 C of electricity deposits 107.88 g of silver, the quantity of electricity that passes can be calculated easily.

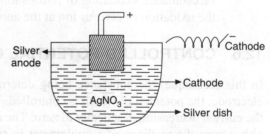

**FIGURE 12.2** A silver coulometer.

## 12.4 IODINE COULOMETER

This coulometer consists in reducing iodine at a platinum electrode acting as cathode or generating iodine at an anode and titrating it against standard arsenous solution.

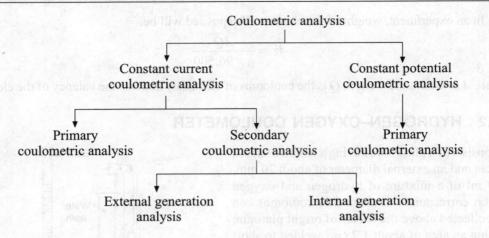

## 12.5 CONSTANT CURRENT COULOMETRIC ANALYSIS

In this technique, solution of the substance to be determined is electrolyzed. The completeness of the reaction can be detected by a visual indicator in the solution or by amperometric, potentiometric or spectro-photometric methods, and the circuit is then opened. The amount of electricity passed is derived from the product, *current × time*. Constant current technique can be further divided into two parts.

1. *Primary constant current coulometry:* In this technique, the element to be estimated undergoes direct reaction at the electrode with 100% efficiency.
2. *Secondary constant current coulometry:* In this technique, one of the titrants is quantitatively produced at an electrode, which then stochiometrically reacts with ions to be estimated. Oxidation of ferrous ion into ferric ion by cerric ion which is produced by the oxidation of cerrous ion at the anode, provides a good example.

## 12.6 CONTROLLED POTENTIAL COULOMETRIC ANALYSIS

In this technique, the substance being determined reacts with 100% efficiency at a working electrode, the potential of which is controlled. The completion of the reaction can be obtained by the current decreasing practically to zero. The amount of the substance reacted can be completed either from the readings of a coulometer in series with the cell or by means of a current time integrating device.

## 12.7 CHARACTERISTICS OF COULOMETRIC ANALYSIS

1. As the concentration level is decreased, coulometric methods become more accurate and precise than classical methods, because of the fact that electrical currents can be controlled and measured with great precision than can volumes of standard solution.
2. In addition to many neutralization, redox, precipitates and complexation titration, many others titrations which are performed by coulometric methods are titrations in molten salt

media, titration of highly hazardous materials and titrations which utilize unstable or difficulty prepared titrants, such as bromine, chlorine, silver(II), uranium(IV) or (V), copper(I), tin(II), titanium(II) and chromium(II), etc.

3. Controlled current technique has widely been used than the controlled potential technique, because the former is faster, requires simpler instrumentation and less expensive. It can be applied to many reactions with excellent accuracy and precision.
4. It is not essential that species must necessarily participate directly in the electron transfer process at the electrode. The substance being determined may involve wholly or in a part in a reaction, i.e. secondary to the electrode reaction. For example, at the outset of the oxidation of Fe(II) at platinum anode, all current transfer results from the reaction, $Fe^{2+} \Leftrightarrow Fe^{3+} + e^-$. As the concentration of iron(II) decreases, however, concentration polarization may cause the anode potential to rise until decomposition of water occurs as a competing process. That is

$$2H_2O \Leftrightarrow O_2(g) + 4H^+ + 4e^-$$

To avoid the consequent error, an excess of cerium(III) can be introduced at the start of electrolysis. This ion is oxidized at a lower anode potential than water

$$Ce^{3+} \Leftrightarrow Ce^{4+} + e^-$$

The cerium(IV) produced diffuses rapidly from the electrode surface, where it can then oxidize an equivalent amount of Fe(II).

$$Ce^{4+} + Fe^{2+} \Leftrightarrow Ce^{3+} + Fe^{3+}$$

The net effect is an electrochemical oxidation of iron(II) with 100% current efficiency even through only a fraction of the iron(II) ions are directly oxidized at the electrode surface.

## 12.8 CONSTANT CURRENT COULOMETRY OR COULOMETRIC TITRATIONS

This technique was first introduced by Somegyi and Twebelledy in 1938. Here, the titrant substance is generated, either internally or externally by the passage of a constant current through an electrolyte with 100% current efficiency.

Coulometric titration has the advantage over a conventional burette titration in that a standard solution is not required. The method is particularly useful for titrations of micro and semimicro amounts.

An excess of KI is added to the thiosulphate solution taken in an electrolytic cell and a constant current is allowed to pass through the cell. Iodine liberated at the anode immediately reacts with thiosulphate. The titration can be performed by adding starch whereby a blue colour is obtained. The amount of thiosulphate can then be calculated by the amount of electricity used in generating iodine, i.e. $i \times t$. A coulometric titration, in common with the more conventional titration, requires some means of detecting the point of chemical equivalence. Most of the end points applicable to volumetric analysis are equally satisfactory here, colour change of indicators, potentiometric, amperometric and conductometric measurements have all been successfully applied.

The analogy between a volumetric and a coulometric titration extends well beyond the common requirement of an observable end point. In both, amount of unknown is determined through evaluation of its combining capacity–in one case for a standard solution and in the other for a quantity of electricity. Similar demands are made of the reactions, i.e. they must be rapid, essentially complete and are free of side reactions.

## 12.9 CELLS FOR COULOMETRIC TITRATIONS

A typical coulometric titration cell is shown in Figure 12.3. It consists of a generator electrode at which the reagent is formed, and a second electrode to complete the circuit. The generator electrode which should have a relatively large surface area, is often a rectangular strip or a wire coil of platinum, a guaze electrode such as that shown in Figure 12.3 can also be employed. The product formed at the second electrode frequently represents potential source of interference. For example, anodic generation of oxidizing agents is frequently employed by the evolution of hydrogen from the cathode unless this gas is allowed to escape from the solution, reaction with oxidizing agent becomes a likelihood. To eliminate this type of difficulty, the second electrode is isolated by a sintered glass disk or some other porous medium.

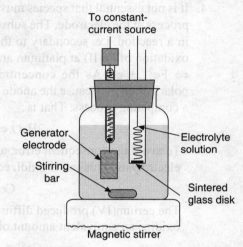

**FIGURE 12.3** A coulometric titration cell.

## 12.10 APPARATUS AND METHOD

A simple apparatus for coulometric titrations is shown in Figure 12.4. The power supply consists of two or more high capacity 45V B-batteries, the current from which passes through a calibrated standard resistance, $R_1$. A potentiometer is connected across $R_1$ to permit accurate measurement of the potential drop, from which current is calculated by Ohm's law. The resistance of $R_1$ should be chosen so that $IR_1$ is about 1 V, with this arrangement a precise determination of current can be obtained with even a relatively simple potentiometer. The variable resistance $R_2$ has a maximum value of about 20,000 Ω.

When the circuit is completed by throwing the switch to the number 2 position, the current $I$ passing through the cell is;

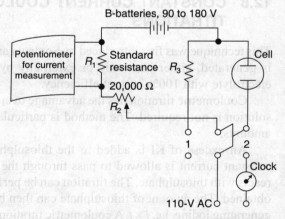

**FIGURE 12.4** Apparatus used in coulometric titration.

$$I = \frac{E_B + E_{cell}}{R_1 + R_2 + R_B + R_{cell}}$$

where $E_B$ is the potential of the batteries and $E_{cell}$ comprises the cathode and anode potential of the titration cell plus any overvoltage or junction potential associated with its operation. The resistance of batteries and cell are symbolized by $R_B$ and $R_{cell}$ respectively.

The potential of dry cells remain reasonably constant for short periods of time, provided the current drawn is not too large, it is safe to assume, therefore, that $E_B$ as well as $R_B$ will remain unchanged during any given titration. Variations in $I$, then arise only from changes in $E_{cell}$ and $R_{cell}$. Ordinarily, however $R_{cell}$ will be of the order 10–20 $\Omega$ while $R_2$ is perhaps 10,000 $\Omega$. Thus, even if $R_{cell}$ were to change by as much as 10 $\Omega$, which is highly unlikely, the effect on the current would be less than 1 ppt (part per thousand).

Changes in $E_{cell}$ during a titration usually have a greater effect on the current, for the cell potential may be altered by as much as 0.5 V during the electrolysis. This change will cause a variation of 0.5–0.6% in $I$ if $E_B$ is 90 V. The same change would cause a variation of only about 0.3% if $E_B$ was 180 V. Experience has shown these to be fairly realistic figures for this sample power source, provided current is drawn more or less continuously from the battery. To achieve this condition, the switching arrangement as shown in figure causes the imposition a resistance $R_3$ in the circuit whenever current is not passing through the cell, the magnitude of $R_3$ is selected to be comparable with $R_{cell}$.

## 12.11 EXTERNAL GENERATION OF REAGENTS

Sometimes, the internal generation of titrant interferes with the titration, it is therefore necessary to generate the titrant externally. For this purpose an assembly of the type shown in Figure 12.5, is generally used. For example, the coulometric titration of acids involves the formation of base at the cathode $2H_2O + 2\ e^- \Leftrightarrow H_2(g) + OH^-$. During electrolysis, an electrolyte solution such as sodium sulphate is fed through the tubing at a rate of about 0.2 ml/s. The hydrogen ions formed at the anode are washed down one arm of the T-tube (along with an equivalent number of sulphate ions) while the hydroxyl ions produced at the cathode are transported through the other. Both electrode reactions shown proceed with 100% current efficiency, thus solution emerging from the left arm of the apparatus can be used for the titration of acids and that from the right arm for the titration of bases. A current of about 250 mA is satisfactory for the titration of various acids or bases in the range between 0.2 and 2 meq (milliequivalent). End points can be detected with a glass calomel electrode system. This apparatus has also been used for the electrolytic generation of iodine from iodine solution.

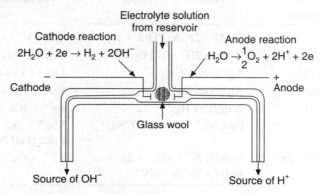

**FIGURE 12.5** Coulometric titration cell for the external generation of the titrant.

## 12.12 APPLICATION OF COULOMETRIC TITRATIONS

1. *Neutralization titrations:* Both weak as well a strong acids can be titrated with a high degree of accuracy using electro-generated hydroxide ions. Both potentiometric and indicator end points can be used for these titrations. Coulometric titration of strong and weak bases can be performed with $H^+$ ion generation at a platinum anode.

$$H_2O \Leftrightarrow \frac{1}{2}O_2 + 2H^+ + 2e^-$$

The generation is possible either internally or externally. In the internal generation, the cathode must be isolated from the solution in order to prevent interference from the hydroxide ions produced at the electrode.

2. *Precipitation and complex–formation titrations:* A large number of coulometric precipitation titrations are based upon anodically generated silver ions. Similar application using mercury(I) ions produced at a mercury anode have also been reported. The end point can be detected either by adsorption indicators or the potentiometric method.

The coulometric method has also been applied to the titration of several cation with EDTA ($HY^{3-}$) ion generated at a mercury cathode. In this application, an excess of mercury(II)–EDTA complex is mixed with an ammonical solution of the sample. The EDTA anion is then released by electrochemical reduction of mercury(II),

$$HgNH_3Y^{2-} + NH_4^+ + 2e^- \rightarrow Hg(l) + 2NH_3 + HY^{3-}$$

The liberated $HY^{3-}$ then reacts with the cation being determined. For example, with $Ca^{2+}$ the reactions is

$$Ca^{2+} + HY^{3-} + NH_3 \rightarrow CaY^{2-} + NH_4^+$$

Because the mercury chelate is more stable than the corresponding complexes with calcium, zinc, lead or copper, complexation of these ions cannot occur until the electrode process free the complexing agents (Table 12.1).

**TABLE 12.1** Typical application of coulometric titration involving neutralization, precipitates and complex formation reactions

| Species determined | Generator electrode reaction | Secondary analytical reaction |
|---|---|---|
| Acids | $2H_2O + 2e^- \Leftrightarrow 2OH^- + H_2$ | $OH^- + H^+ \Leftrightarrow H_2O$ |
| Bases | $H_2O \Leftrightarrow 2H^+ + 1/2\,O_2 + 2e^-$ | $H^+ + OH^- \Leftrightarrow H_2O$ |
| $Cl^-$, $Br^-$, $I^-$ | $Ag(s)\ Ag^+ + e^-$ | $Ag^+ + Cl^- \Leftrightarrow AgCl(s)$ |
| Mercaptans | $Ag(s) \Leftrightarrow Ag^+ + e^-$ | $Ag^+ + RSH \Leftrightarrow AgSR(s) + H^+$ |
| $Cl^-$, $Br^-$, $I^-$ | $2Hg(l) \Leftrightarrow Hg_2^{2+} + 2e^-$ | $Hg^{2+} + 2Cl^- \Leftrightarrow Hg_2Cl_2(s)$ |
| $Zn^{2+}$ | $Fe(CN)_6^{3-} + e^- \Leftrightarrow Fe(CN)_6^{4-}$ | $3Zn^{2+} + 2K^+ + 2Fe(CN)_6^{4-} \Leftrightarrow K_2Zn_3[Fe(CN)_6]_2(s)$ |
| $Ca^{2+}$, $Cu^{2+}$, $Zn^{2+}$ and $Pb^{2+}$ | $HgNH_3Y^{2-} + NH_4^+ + 2e^- \Leftrightarrow Hg(l) + 2NH_3 + HY^{3-}$ | $HY^{3-} + Ca^{2+} \Leftrightarrow CaY^{2-} + H^+$ |

3. *Oxidation–reduction titrations:* Table 12.2 shows oxidizing and reducing agents that can be generated by making use of coulometric method and the analysis to which they have been applied. Electro-generated bromine has been found to be very much useful among the oxidizing agent. The titration of arsenic(II), antimony(III) and $H_2S$ can be carried out by electrolyte generation of bromine or iodine.

$$2 Br^- \rightarrow Br_2 + 2 e^- \text{ (anode)}$$

$$2H_2O + 2 e^- \rightarrow 2OH^- + H_2 \text{ (cathode)}$$

$$AsO_3^{3-} + Br_2 + 2OH^- \rightarrow AsO_4^{3-} + 2Br^- + H_2O$$

The coulometric titration of an oxidizing agent (e.g. $K_2Cr_2O_7$, $KMnO_4$, etc.) can be carried out by having present a considerable excess of the ferric salts in the cathodic compartment containing oxidizing agent. The reactions are:

$$H_2O \rightarrow 2H^+ + 1/2\ O_2 + 2 e^- \text{ (anode)}$$

$$2Fe^{3+} + 2 e^- \rightarrow 2Fe^{2+} \text{ (cathode)}$$

**TABLE 12.2** Oxidizing and reducing agent generated by coulometric method

| Reagent | Generator electrode reaction | Substance determined |
|---|---|---|
| $Br_2$ | $2Br^- \Leftrightarrow Br_2 + 2e^-$ | As(III), Sb(III), U(IV), Tl(I), $I^-$, $SCN^-$, $NH_3$, $N_2H_4$, $NH_2OH$, Phenol, aniline, mustard gas, 8-hydroxyquinolive. |
| $Cl_2$ | $2Cl^- \Leftrightarrow Cl_2 + 2e^-$ | As(III), $I^-$ |
| $I_2$ | $2I^- \Leftrightarrow I_2 + 2e^-$ | As(III), Sb(III), $S_2O_3^{2-}$, $H_2S$ |
| $Ce^{4+}$ | $Ce^{3+} \Leftrightarrow Ce^{4+} + e^-$ | Fe(II), $T_1$(III), V(IV), As(III), I, Fe(CN)$_6^{4-}$ |
| $Mn^{3+}$ | $Mn^{2+} \Leftrightarrow Mn^{3+} + e^-$ | $H_2C_2O_4$, Fe(II), As(III) |
| $Ag^{2+}$ | $Ag^+ \Leftrightarrow Ag^{2+} + e^-$ | Ce(III), V(IV), $H_2C_2O_4$, As(III) |
| $Fe^{2+}$ | $Fe^{3+} + e^- \Leftrightarrow Fe^{2+}$ | Cr(VI), Mn(VII), V(V), Ce(IV) |
| $Ti^{3+}$ | $TiO^{2+} + 2H^+ + e^- \Leftrightarrow Ti^{3+} + H_2O$ | Fe(III), V(V), Ce(IV), U(VI) |
| $CuCl_3^{2-}$ | $Cu^{2+} + 3Cl^- + e^- \Leftrightarrow CuCl_3^{2-}$ | V(V), Cr(VI), $IO_3^-$ |
| $U^{4+}$ | $UO_2^{2+} + 4H^+ + 2 e^- \Leftrightarrow U^{4+} + 2H_2O$ | Cr(VI), Ce(IV) |

## 12.13 CONTROLLED POTENTIAL COULOMETRIC ANALYSIS

The technique is similar to electro-gravimetric methods using potential control. They differ only in that a quantity of electricity is measured rather than a weight of deposit. In contrast to Coulometric titration, a single reaction of the working electrode is required, although the species being determined nead not react directly.

A controlled potential coulometric analysis requires a potentiostat and a chemical coulometer, placed in series with the working electrode.

The chemical coulometer, such as oxygen hydrogen coulometer can be used in the circuit to measure the quantity of electricity. A schematic diagram of the essential components of a potentiostat is shown in Figure 12.6.

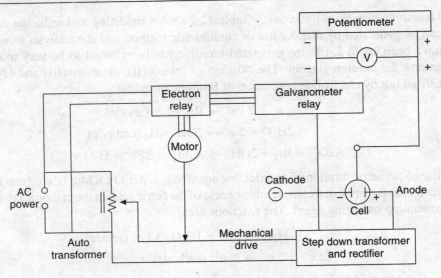

**FIGURE 12.6** Schematic diagram of a potentiostat.

Passage of current through the chemical coulometer liberates hydrogen at the cathode and oxygen at the anode. Both gases are collected and their total volume is measured by determining the volume of liquid displaced. The gas temperature can be ascertained by providing the water jacket and thermometer.

The total quantity of electricity can also be measured graphically by careful measurement of the current passing through the cell at known time intervals. The area under the curve that relates those two variables yields the desired quantity.

Various mechanical and electronic integrators have also been developed that can be used for the evaluation of current time integrals in a coulometric analysis.

Oxygen–hydrogen coulometer gives results with an accuracy of 0.1%, but it is not accurate in electrolysis less than 10 coulombs. Lingane used mercury cathode, because the central potential at a mercury cathode can be pre-assessed from polarographic data, but mercury is not good as an anode. He used a silver helix as an anode.

The supporting electrolyte is added to the cell and nitrogen is allowed to pass through for about 5 min. The potential is controlled and the electrolysis is allowed to proceed until the current reaches to a negligibly small constant value. The titration of chloride, bromide and iodide have successfully been carried out by making use of mercury pool cathode and silver helix anode.

In coulometric procedures, one attempts to attain as closely as possible a net current efficiency 100%. The so-called background current, however, prevents the attainment of exactly 100% current efficiency even when the electrolyzed solution is completely free from interfering reactants.

The background current which itself is a type of interference, is a collective term that refers to any current that flows, but does not accomplish the desired reaction when a pure solution is electrolyzed.

The simplest way to minimize background current errors is to utilize a relatively large quantity of electro-active species. As a result, contribution of background current to the total $Q$ becomes insignificant.

According to Meites and Moros, the background current may be divided into five component currents:

1. Charging current (current required to charge electrical double layer).
2. Impurity Faradaic current (current arises from electrolysis of impurities present in medium or electrode).
3. Continuous Faradaic current (current due to the electrolysis of some component of medium itself).
4. Kinetic background current (current arises when product of electrolysis is slowly converted to some other form).
5. Induced background current (current arises when the electrode reaction induces another reaction to take place).

## 12.14 SELECTION OF EXPERIMENTAL CONDITION IN POTENTIOSTATIC COULOMETRY

In potentiostatic coulometric, the experimental conditions are so chosen as to obtain a good compromise between the desired accuracy, selectivity and speed of analysis. These are: (1) cell geometry, (2) stirring efficiency, (3) electrode potential, (4) electrode material, (5) solvent, (6) nature and concentration of supporting electrolyte, (7) complexing agent or surface active agents, and (8) Temperature.

It should be noted that all the above types of currents also pertain to controlled current technique, but to different extent. The charging current is larger in controlled current coulometry than in potentiostatic coulometry, because of relatively larger variation in generator electrode potential during a controlled current electrolysis. The continuous Faradaic current, the kinetic and induced background currents, on the other hand, make a smaller contribution when the electrolysis is carried out at constant current. These are time dependent quantities and the total electrolysis time is generally smaller in a controlled current electrolysis than in potentiostatic electrolysis. The Faradaic impurity current is expected to be of same magnitude in both the techniques and can be minimized by electrolysis of the medium and reagents to the end point before adding the sample test portion.

In conclusion, constant current coulometric is subjected to less error from background current contribution than constant potential coulometric. Thus, constant current Coulometric can be employed for the determination of small quantities with more convenience and greater accuracy than can potentiostatic coulometry.

## EXERCISES

1. Define coulometric analysis. What is the role of coulometric analysis in today's life?
2. Define hydrogen–oxygen coulomter.
3. Define silver coulometer. How it is different from hydrogen–oxygen coulometer and iodine coulometer.
4. Classify various types of coulometric analysis.
5. Define constant current coulometry.

6. Define constant potential coulometry.
7. Write down characteristics of coulometric analysis.
8. Define construction and working of cell for coulometric analysis.
9. Explain apparatus and method involved in coulometric analysis.
10. Write down various applications of coulometric analysis.
11. Explain apparatus and method involved in controlled potential (potentiostatic) coulometric analysis.

## SUGGESTED READINGS

Albresch, K. and I. Classen, *Coulometric Analysis*, Chapman & Hall, London, 1961.

Charlot, G., *Modern Electroanalytical Methods*, Van Nostrand, London, 1958.

Delahay, P., *New Instrumental Methods in Electrochemistry*, Wiley Interscience, New York, 1954.

Donbrow, M., *Instrumental Methods in Analytical Chemistry Their Application and Practice*, Vol. 1, Pitman, London, 1966.

Donbrow, M., *Instrumental Methods in Analytical Chemistry Their Principles and Practice* Vol. 1, Pitman, London, 1966.

Lingane, J.J., *Electroanalytical Chemistry*, 2nd ed., Interscience, New York, 1958.

Milner, G.W.C. and G. Phillips, *Coulometry in Analytical Chemistry*, Pergamon Press, New York, 1967.

Sharma, B.K., *Instrumental Methods of Chemical Analysis*, Goel Publishing House, Meerut.

Skoog, D.A. and D.M. West, *Fundamentals of Analytical Chemistry*, 3rd ed., Rinehart & Winston, New York.

Willard, H.H., L.L. Merrit, J.A. Dean, and F.A. Settle, *Instrumental Methods of Analysis*, 6th ed., CBS Publishers and Distributers, Delhi.

# Part IV
# Electrochemistry

# 13
# Electrified Interface

## 13.1 INTRODUCTION

In the bulk of the electrolyte there is perfect isotropy and homogeneity and there are number of preferentially directed electric fields or dipoles.

The forces operating on particles near the phase boundary are therefore, anisotropic. The arrangement of particles in the interface region is a compromise between the structure demanded by both the phases, i.e. metal phase and bulk phase.

The term *electrical double layer* or just a *double layer*, is used to describe the arrangement of charges and oriented dipoles constituting the interface region at the boundary of an electrolyte. Even, if the material consists not of free charges, but of permanent dipoles or of molecules, in which, dipoles can be induced, a potential difference across the boundary can arise from a net orientation of the dipoles constituting it.

The redistribution is the structural basis of the potential difference across the interface (Figure 13.1).

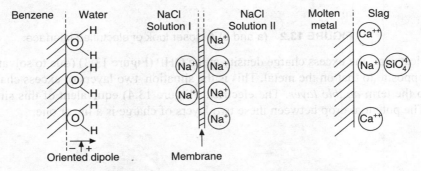

**FIGURE 13.1** Some examples of electrified interfaces.

*A closer look into an electrified interface:* The metal is made up of a lattice of positive ions (kernels) and free mobile electrons. The charged particles in solution unevenly distributed owing to the presence of a metal-solution phase boundary, may give rise to the excess charge density $q_m$ in the metal, with which they interact or an excess charge density $q_m$ generated by an external source (battery), may effect the charge distribution on the solution side of the interface (Figure 13.2).

The first row is largely occupied by water dipoles. This is the hydration sheath of the electrode. The second row is largely reserved for solvated ions. The locus of centres of these solvated ions is called *outer Helmholtz plane* referred as OHP. On the top of the first row of water molecules, a sort of secondary hydration sheath, feebly bound to the electrode may be present.

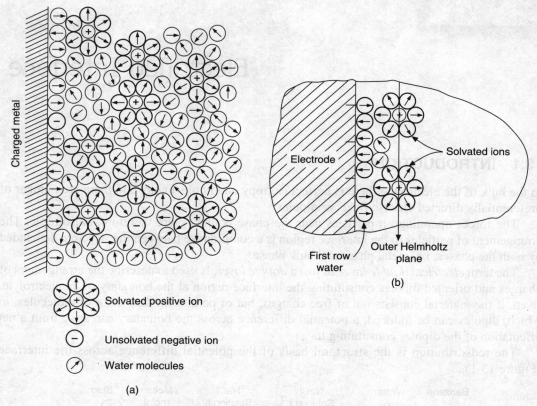

**FIGURE 13.2** (a and b) A closer look of electrified interface.

In the simplest case, the excess charge density at the OHP (Figure 13.3) (due to solvated ions) is equal and opposite to that on the metal. This is the situation–two layers of excess charge, which gave rise to the term *double layer*. The electrical (Figure 13.4) equivalent of this situation is a capacitor. The potential drop between these two layers of charge is a linear one.

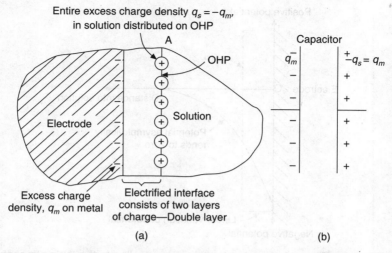

**FIGURE 13.3** Electrified interface (a) is just like parallel plate capacitor (b).

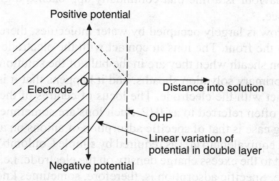

**FIGURE 13.4** Linear variation of potential in double layer.

The solvation sheaths of these ions and the first row of water molecules on the electrode are not shown in the Figure 13.3.

A case which is not so simple, is the one, in which the excess charge density on the OHP is not equivalent to that on the metal, but less. Some of the solvated ions leave their second row seats and random walk about in the solution. In this case, the excess charge density in the solution deceases with the distance from the electrode (Figure 13.5).

The potential falls off into the solution at first sharply and then asymptotically tends to zero (Figure 13.6).

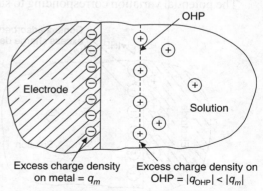

**FIGURE 13.5** A case of $q_m > q_s$.

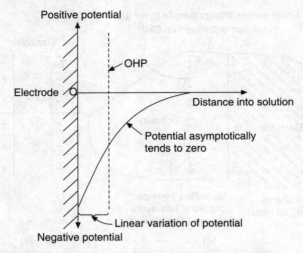

**FIGURE 13.6** The variation of potential corresponding to the ionic atmosphere.

Asymptotically behaviour is a line that continually approaches a given curve but does not meet it at a finite distance.

Although the first row is largely occupied by water molecules, there are some ionic species which find their way to the front. The ions in contact with the electrode are those which do not have a primary hydration sheath when they are in the bulk of the solution. Most anions and large cations do not possess primary solvation sheaths, and it happens that it is anions and big cations which do achieve contact with the electrode. The locus of centres of these ions is known as the *inner Helmholtz plane*, often referred to as IHP. Such ions are sometimes said to be specifically adsorbed. An interesting case is that of specific adsorption of anions on a metal charged opposite in sign. Here $|q_{CA}|$ is not equivalent to $|q_m|$ as required by simple Coulomb's law forces, but greater than or super-equivalent to the excess charge density on the electrode, i.e., $|q_{CA}| > |q_m|$ specifically adsorbs on the electrode. Specific adsorption is, therefore, sometimes known as super-equivalent adsorption.

The potential variation corresponding to super-equivalent adsorption is shown in Figure 13.7.

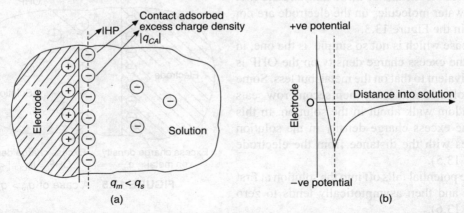

**FIGURE 13.7** Super equivalent adsorption (a) potential variation corresponding to superequivalent adsorption (b).

The potential falls linearly up to a plane through the locus of centres of specifically adsorbed ions, the IHP and then changes course and turning up goes asymptotically to the value in the bulk of the solution. The double layer formed at a boundary between phases containing charged entities has two fundamental aspects, the electrical aspects and structural aspects.

The electrical aspects concern the magnitude of the excess charge densities on each phase. It also concerns the variation of potential with distance from the interface. The structural aspect is a matter of knowing how the particles of the two phases (ions, electrodes, dipoles, neutral molecules) are arranged in the interface region, so as to electrify the interface.

The electrical and structural aspects of the double layer are intimately related. The charge or potential difference is the characteristics of particular structure and vice versa.

The formation of an electrified interface has been described in the following steps.

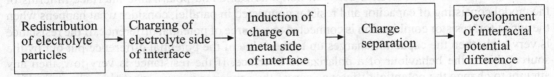

In systems in which one of the phases, e.g. a metal electrode can be connected to an external source of charge, the formation of an electrified interface can be conceived in the following way.

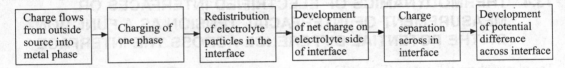

## 13.2 IMPORTANCE OF ELECTRIFIED INTERFACE

1. Electric field can be used to guide the colloidal particles to deposit upon metals and colour them. The hues formed in this way may be more permanent than paint.
2. The friction between two solids, which in the presence of liquid films may depend on the double layers at their interfaces. Thus, efficiency of a wetted rock drill depends on the double layer structure at the metal drill–aqueous solution interface.
3. In the process of metal deposition, the ions from the solution must be electrically energized to cross the interface region and deposit on the metal. This electrical energy must be picked up from the fields at the interface, which itself depends upon the double layer structure.
4. Corrosion rate depends partly on the structure of the double layer, i.e. on the electric field across the interface, which in turn governs the rate of metal dissolution.
5. In biology too, the mechanism by which nerves carry messages from brain to muscles is based on the potential difference across the membrane, which separates a nerve cell from the environment.

## 13.3 POLARIZABLE AND NONPOLARIZABLE INTERFACE

To polarize an interface means to alter the potential difference across it, to be polarizable means to be susceptible to changes in potential difference.

An example of a nonpolarizable interface is the well known calomel electrode and classic example of a polarizable interface is the interface between pure mercury and an aqueous solution (e.g. KCl).

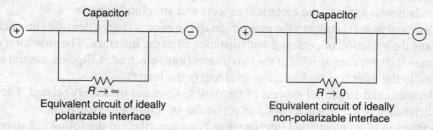

Equivalent circuit of ideally polarizable interface

Equivalent circuit of ideally non-polarizable interface

To understand the difference between nonpolarizable and polarizable interface in terms of the model consisting of capacitor and resistor connected in parallel, consider what happens when the capacitor–resistor combination is connected to a source of potential difference. If the resistance is very high, then the capacitor charges up to the value of the potential difference put out by the source, this is the behaviour of a polarizable interface. If the resistance is very low, then any attempt to change the potential difference across the capacitor is compensated by charge leaking through the low resistance path. This is the behaviour of a nonpolarizable interface.

## 13.4 THERMODYNAMICS OF ELECTRIFIED INTERFACES OR MEASUREMENT OF INTERFACIAL TENSION AS A FUNCTION OF THE POTENTIAL DIFFERENCE ACROSS THE INTERFACE

The electrochemical system (Figure 13.8) has four essential parts. These are:

1. a mercury–solution polarizable interface,
2. a nonpolarizable interface,
3. An external source of variable potential difference $V$ and
4. an arrangement to measure the surface tension of the mercury in contact with the solution.

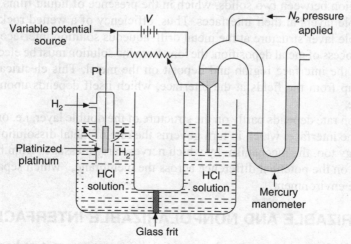

**FIGURE 13.8** Schematic apparatus for the measurement of surface tension ($\gamma$) of mercury as a function of cell potential $V$.

The system permits electro-capillary measurements, i.e. measurement of the surface tension of the mercury in contact with the solution as a function of the electrical potential difference across the interface. The measurement of surface tension is achieved by using a fine capillary and adjusting the height of a mercury column so that the mercury in the capillary is stationary.

When the mercury column in the capillary is stationary, and therefore in mechanical equilibrium. Its weight is exactly balanced by the total force of its surface tension. The weight of the mercury column acts downward and is equal to the density times the volume of the column $\pi r^2 h$, times the gravitational constant $g$. This weight compensates the surface tension force which is equal to the perimeter of the contact $2\pi r$ times the upward component $\gamma \cos \theta$ of the surface tension (Figure 13.9). Since $2\pi r \gamma \cos \theta = \pi r^2 h \rho g$. It follows that $\gamma = \dfrac{rh\rho g}{2 \cos \theta}$ for the mercury glass interface $\theta \approx 0$. Hence, $\gamma = \dfrac{rh\rho g}{2}$.

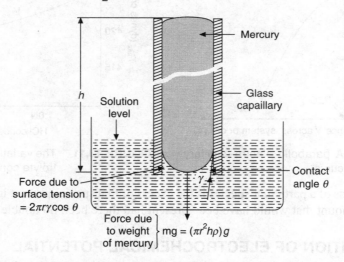

**FIGURE 13.9** An arrangement to measure surface tension of mercury in contact with solution.

A lot of useful information can be gained from the form of electro-capillary curves, i.e. plots of interfacial tension ($\gamma$) versus changes in interfacial potential difference $V$. The interfacial tension depends on the forces arising from the particles present in the interface region. If the arrangement of these particles, i.e. composition of the interface is altered by varying the forces at the interface, would change and thus cause a change in the interfacial tension. One would expect therefore that the surface tension $\gamma$ of the metal solution interface should vary with the potential difference $V$ supplied by the external source. The experimental $\gamma$ versus $V$ curves obtained by electro-capillary measurements demonstrate this variation of surface tension $\gamma$ with the potential difference $V$ across the cell. What is informative, however, is the nature of the variation. A typical $\gamma$ versus $V$ electro- capillary curve is almost a parabola (Figure 13.10). The potential at which the surface tension is maximum, is known as that of the *electro-capillary maximum* (ECM). The measurements also show that surface tension varies with the composition of the electrotype (Figure 13.11). This is easily seen by comparing electro-capillary curves obtained in solutions of different electrolyte concentration. As the solution is diluted, the maximum of surface tension rises because with dilution charge density decreases and hence surface tension increases. The surface tension was found to be related to the surface excess of species in the interface. The surface excess in turn represents in some way the structure of the interface. It follows therefore that electro-capillary curves must contain many interesting messages about the double layer at the electrode–electrolyte interface. But to understand such messages, one must learn to decode the electro-capillary data. It is necessary to derive quantitative relations among surface tension, excess charge on the metal, cell potential, surface excess and solution composition (Figure 13.11).

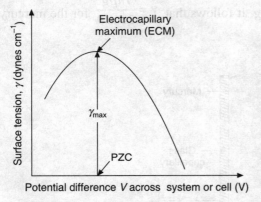

**FIGURE 13.10** A parabolic electro-capillary curve.

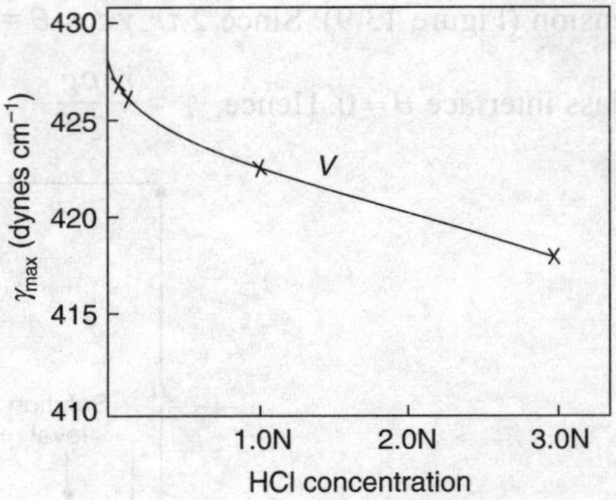

**FIGURE 13.11** The variation of $\gamma_{max}$ with electrolyte concentration.

Surface excess of a particular species is the excess of that species present in the surface phase relative to the amount that would have been present had there been no double layer.

## 13.5 DEFINITION OF ELECTROCHEMICAL POTENTIAL

The work done in moving unit test charge from bulk to surface is accompanied by the following potential involvements. (1) The outer potential $\psi$ arising from the work done to transport a unit test charge to a point just outside a charged but dipole layer free phase. (2) The surface potential $\chi$ arising from the work done to carry the unit charge across the dipole layer at the surface of an uncharged phase. (3) Inner potential $\phi$ arising from the work done to carry the test charge from infinity up to and across the dipole layer at a charged phase. In defining the $\chi$ and $\phi$ potentials the test charges were prohibited from interacting with the bulk of the phases.

In addition to electrical interaction, test charge also feels chemical interaction to take it in the bulk of solution. These are, for instance, ion–solvent interactions, ion–ion interactions and the solvent–solvent interactions. Therefore, chemical work, i.e. work done against all these interactions with the particles of the material phase. The chemical potential ($\mu_i$) of a particular species $i$ is the work done to bring a mole of $i$ particles from infinity into the bulk of an uncharged dipole layer free material phase. The total work is the sum of the chemical work and the electrical work. Thus,

$$\bar{\mu} = \mu + ZF\phi \quad \Rightarrow \quad \mu + ZF(\psi + \chi)$$

Here $\bar{\mu}$ is the total work of electrochemical potential, $\mu$ is chemical work when no charge is present, i.e. bulk phase, $ZF\phi$ is the electrical work, i.e. when charge is present at the interface region.

If the particles are uncharged then $Z_i = 0$. Therefore,

$$\bar{\mu}_i = \mu_i$$

The electrochemical potential $\bar{\mu}_i$ includes all types of work involved in bringing particles of charge $Z_i e_0$ into material phases. The $\bar{\mu}$ includes the chemical work $\mu$, the charge contribution $\psi$ and the dipole contribution $\chi$. Just as the gradient of the chemical potential $(\partial \mu/\partial x)$ for the $x$ component of this gradient acts as driving force on pure diffusion and the gradient of the electric potential $(d\phi/dx)$ acts as driving forces in pure conduction, the gradient of the electrochemical potential $(\partial \bar{\mu}/\partial x)$ can be considered the total driving force for the transport of a charged species. The total transport process consisting of both diffusion and conduction.

## 13.5.1 Can the Chemical and Electrical Work be Determined Separately

The separation into chemical and electrical terms is possible with the gradients but not with quantities, i.e. $\mu$ and $\phi$ themselves. The reason is simple. The electrochemical potential $\bar{\mu}_j$ was only conceptually separated into a chemical term $\mu_j$ and an electrical term $Z_j F\phi$. The conceptual separation was based on thought experiments. In practice, no experimental arrangement can be devised to correspond to the thought experiment. Thus, for example, one cannot switch off the charges and dipole layer at the surface of a solution as one can switch off the externally applied field in a transport experiment. Only the combined effect of $\mu_j$ and $Z_j F\phi$ can be determined.

## 13.5.2 Criterion of Thermodynamic Equilibrium between Two Phases

For a system to be at equilibrium, it is essential that there is no drift of any species, hence there should be zero gradients for the electrochemical potentials of all the species. It follows, therefore, that for an interface to be at equilibrium, the gradients of electrochemical potential of the various species must be zero across the phase boundary, i.e.

$$\frac{d\bar{\mu}_j}{dx} = 0$$

It follows that the value of the electrochemical potential of a species $j$ must be the same on both sides of the interface, i.e., $(\bar{\mu}_j)_m = (\bar{\mu}_j)_s$. In other words, the change in electrochemical potential in transporting the species from one phase to the other must be zero, i.e. $\Delta \bar{\mu}_j = 0$. Now the electrochemical potential $\bar{\mu}_j$ of the species $j$ in a particular phase is the change in free energy of the system resulting from the introduction of a mole of $j$ particles into the phase, while keeping the other conditions constants, i.e.

$$\bar{\mu}_j = \left(\frac{\partial G}{\partial n_j}\right)_{T,P,n_i=n_j}$$

Hence, the equality of electrochemical potentials on either side of the phase boundary implies that the change in free energy of the system resulting from the transfer of particles from one phase to the other should be the same on that due to the transfer in the other direction. But, this is only another way of stating that when a thermodynamic system is at equilibrium its free energy is a minimum, i.e. $d\overline{G} = 0$.

### 13.5.3 Nonpolarizable Interface and Thermodynamic Equilibrium

Since the difference in electrochemical potential of a species $i$ between two phases is the work done to carry a mole of this species from one phase to the other, it must be same as the work in the opposite direction. This implies a free flow of species across the interface. But an interface which maintains an open border is none than a nonpolarizable interface. Thermodynamic equilibrium exists at a nonpolarizable interface. Hence, one can immediately apply the criteria of thermodynamic equilibrium to a nonpolarizable interface.

$$^s\Delta^m \overline{\mu}_j = {}^s\Delta^m(\mu_j + Z_j F\phi)$$

$$^s\Delta^m \overline{\mu}_j = {}^s\Delta^m \mu_j + Z_j F {}^s\Delta^m \phi = 0$$

$$Z_j F {}^s\Delta^m \phi = -{}^s\Delta^m \mu_j$$

$$^s\Delta^m \phi = -\frac{1}{Z_j F} {}^s\Delta^m \mu_j$$

$$d({}^s\Delta^m \phi) = -\frac{1}{Z_j F} d\mu_j$$

This is the equation for a nonpolarizable interface.

## 13.6 SOME THERMODYNAMICS THOUGHTS ON ELECTRIFIED INTERFACE

Combined form of first and second law is

$$dU = TdS - W \qquad (13.1)$$

where $dU$ is the internal energy, $TdS$ is the heat supplied, $W$ is the work done by the system reversibly.

To introduce a mole of the species $i$, chemical work done on the system is $\mu_i$. Hence, to alter the number of moles of $i$ in the system by $dn_i$, the work done by the system is $-\mu_i dn_i$. Hence,

$$dU = TdS - W - \Sigma \mu_i dn_i \qquad (13.2)$$

Here $\Sigma \mu_i dn_i$ is the work done by the system in expelling $dn_i$ moles of species $i$, $\mu_i$ being the work of transfer per mole.

In an electrode–electrolyte interface $m_1/s$, what are the various possible types of work? First is work of expansion $PdV$, and second is the work of increasing the area of the interface $\gamma dA$, where $\gamma$ is the interfacial tension, and finally if metallic phase is connected to an external source of electricity then electrical work of transporting the charge $dq'_m$ is ${}^{m_1}\Delta^s \phi dq'_m$.

Introducing these works in place of $W$ in above equation,

$$dU = \overset{o}{T} d\overset{*}{S} - \overset{o}{P} d\overset{*}{V} - \overset{o}{\gamma} d\overset{*}{A} - {}^{m_1}\Delta^s \overset{*}{\phi} dq'_m - \sum \overset{o}{\mu}_i d\overset{*}{n}_i \qquad (13.3)$$

Here, $o$ stands for intensive property, * stands for extensive property.

Now each term on the right-hand side is a product of an intensive factor ($T$, $P$, $\gamma$, $\Delta\phi$, $\mu$) constant. Let the extensive factors be increased from their differential values to their absolute values for the system concerned $S$, $A$, $V$, $q'_m$, $n_i$. Therefore,

$$U = TS - PV - \gamma A - {}^m\Delta^s \phi q'_m - \sum_i \mu_i n_i \qquad (13.4)$$

On differentiating this equation, the result is

$$dU = \left[ TdS - PdV - \gamma dA - {}^{m_1}\Delta^s \phi dq'_m - \sum_i \mu_i dn_i \right]$$
$$+ \left[ SdT - VdP - Ad\gamma - q'_m d({}^{m_1}\Delta^s \phi) - \sum_i n_i d\mu_i \right] \qquad (13.5)$$

Equations (13.3) and (13.5) must be equal to each other

$$0 = SdT - VdP - Ad\gamma - q'_m d({}^{m_1}\Delta^s \phi) - \sum_i n_i d\mu_i \qquad (13.6)$$

Which, at constant temperature and pressure reduces to

$$0 = -Ad\gamma - q'_m d({}^{m_1}\Delta^s \phi) - \sum_i n_i d\mu_i \qquad (13.7)$$

$$d\gamma = -\frac{q'_m}{A} d({}^{m_1}\Delta^s \phi) - \sum_i \frac{n_i}{A} d\mu_i \qquad (13.8)$$

Here $q'_m$ is total excess charge on electrode. Thus, surface tension changes have been related to changes in the absolute potential differences across an electrode–electrolyte interface and to change in the chemical potential of all the species, i.e. to change in solution composition. Only one quantity is missing, the surface excess. But this can be easily introduced by recalling the definition of surface excess.

$$\Gamma_i = \frac{n_i}{A} - \frac{n_i^0}{A}$$

Here, $n_i$ is the actual number of moles of species $i$ in the interface region, $n_i^0$ is the number of moles in the absence of double layer. $A$ is the area of interface.

$$\frac{n_i}{A} = \Gamma_i + \frac{n_i^0}{A} \qquad (13.9)$$

$$\frac{n_i}{A} d\mu_i = \Gamma_i d\mu_i + \frac{n_i^0}{A} d\mu_i$$

$$\sum_i \frac{n_i}{A} d\mu_i = \sum_i \Gamma_i d\mu_i + \sum_i \frac{n_i^0}{A} d\mu_i \qquad (13.10)$$

It is known from Gibbs–Duhem equation that

$$\sum_i n_i^0 d\mu_i = 0$$

Therefore, Eq. (13.10) becomes

$$\sum_i \frac{n_i}{A} d\mu_i = \sum_i \Gamma_i d\mu_i \qquad (13.11)$$

Substituting the above result in Eq. (13.8), we get

$$d\gamma = -q_m d(^{m_1}\Delta^s\phi) - \sum \Gamma_i d\mu_i \qquad (13.12)$$

$q_m$ is excess charge density.

It may be recalled, however, that, though the absolute value of $^{m_1}\Delta^s\phi$ cannot be determined, a change in $^{m_1}\Delta^s\phi$, i.e. $d(^{m_1}\Delta^s\phi)$ can be measured provided: (1) the $m_1/s$ interface is a polarizable one and (2) the $m_1/s$ interface is linked to a nonpolarizable interface $m_2/s$ to form an electrochemical system or cell. If such a cell is connected to an external source of electricity. One has

$$V = {}^{m_1}\Delta^s\phi + {}^s\Delta^{m_2}\phi + {}^{m_2}\Delta^{m_1}\phi \qquad (13.13)$$

Since the sum of potential drops around a circuit must be zero. The inner potential difference $^{m_2}\Delta^{m_1'}\phi$ neither depends upon the potential $V$ supplied from the external source nor upon the solution composition. Hence, on differentiation.

$$-d(^{m_1}\Delta^s\phi) = -dV + d(^s\Delta^{m_2}\phi) \qquad (13.14)$$

Substituting this expression for $-d(^{m_1}\Delta^s\phi)$ in Eq. (13.13).

$$d\gamma = -q_m dV + q_m d(^s\Delta^{m_2}\phi) - \sum \Gamma_i d\mu_i \qquad (13.15)$$

The nonpolarizable characteristics of the second interface $m_2/s$ are now introduced. It is recalled that there is thermodynamic equilibrium at this interface, and thus

$$d(^s\Delta^{m_2}\phi) = -\frac{1}{Z_j F} d\mu_j \qquad (13.16)$$

By substituting Eq. (13.16) into Eq. (13.15), we obtain

$$d\gamma = -q_m dV - \frac{q_m}{Z_j F} d\mu_j - \sum \Gamma_i d\mu_i \qquad (13.17)$$

This is fundamental equation for thermodynamic treatment of polarizable interfaces. It is a relation among interface tension $\gamma$, surface excess $\Gamma_i$ applied potential $V$, charge density $q_m$ and solution composition $d\mu_i$.

## 13.7 DETERMINATION OF THE CHARGE DENSITY ON THE ELECTRODE

When an electro-capillary curve is obtained in the laboratory, a solution of a fixed composition is taken, i.e. $d\mu_i$ for all the species is zero. The condition of electro-capillary curve determinations corresponds, therefore to

$$\Sigma \Gamma_i d\mu_i = 0 \text{ and } d\mu_j = 0$$

So Eq. (13.17) becomes
$$\left(\frac{\partial \gamma}{\partial V}\right)_{\text{const. comp.}} = -q_m = \text{Slope} \qquad (13.18)$$

Above equation is known as *Lippmann equation*. The slope of the electro-capillary curve at any cell potential $V$ is equal to the charge density on the electrode. All that one has to do to know the excess electric charge on the electrode, one has to find the slope of the electro-capillary curve at the corresponding value of the cell potential.

The charge density on the electrode at a particular value of the potential difference $V$ is given by the slope of the electro-capillary curve at that potential. The curve shown is for mercury in contact with 1.0 N HCl (Figure 13.12).

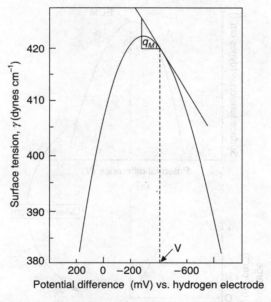

**FIGURE 13.12** The variation of surface tension with potential difference.

## 13.8 DETERMINATION OF THE ELECTRICAL CAPACITANCE OF THE INTERFACE

Differentiate the $\gamma$ versus $V$ electro-capillary curve at various values of cell potential and plot these values of the slope, i.e. charge density as a function of potential. If the electro-capillary curve were a perfect parabola then the charge (i.e. excess charge density) on the electrode would vary linearly with the cell potential. It can be considered a system capable of storing charge and hence act as electrical capacitor or a condenser.

Capacitance of a condenser is given by the total charge required to raise the potential difference across the condenser by 1 V.

$$K = \frac{q}{V}$$

This is integral capacitance for electrical capacitors, where the capacity is constant and independent of the potential. This constancy may not be the case with electrified interfaces and it is the best to define a differential capacity, $C$ thus

$$C = \left(\frac{\partial q_m}{\partial V}\right)_{\text{Const. Comp.}} = -\left(\frac{\partial^2 \gamma}{\partial V^2}\right)_{\text{Const. Comp.}}$$

It shows that slope of the curve of the electrode charge versus cell potential yields the value of the differential capacity of the double layer. In the case of an ideal parabolic $\gamma$ versus $V$ curve, which yields a linear charge ($q$) versus V curve, one obtains a constant capacitance.

By differentiating an electro-capillary curve, one obtains a curve for the variation of electrode charge density versus potential difference (Figure 13.13).

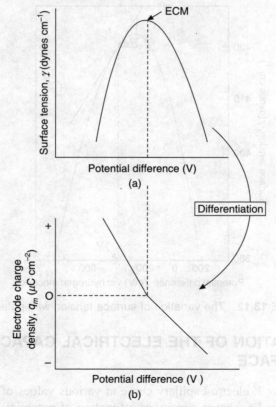

**FIGURE 13.13** (a) Surface tension versus potential difference, (b) Electrode charge density versus potential difference.

When the electro-capillary $\gamma$ versus $V$ curve is a perfect parabola the electrode charge density varies linearly with potential difference (Figure 13.14).

The straight line $q_m$ versus $V$ curve obtained by differentiating an ideally parabolic $\gamma$ versus $V$ electro-capillary curve yields on further differentiation a capacity which is potential independent (Figure 13.15).

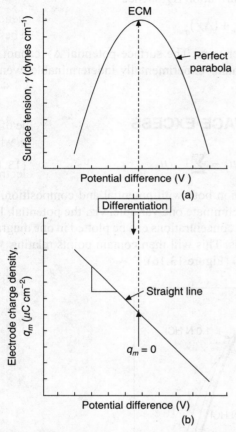

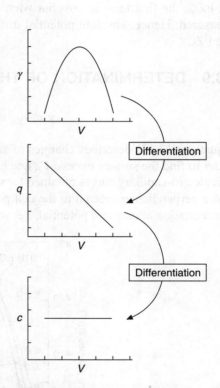

**FIGURE 13.14** (a) Perfect parabola. (b) Electrode charge density varies linearly with potential difference.

**FIGURE 13.15** Capacity which is potential independent.

## 13.8.1 The Potential at Which an Electrode has a Zero Charge

The potential difference across the system of cell at which the charge on the electrode is zero is known as *potential of zero charge* (PZC) and is given by the symbol $E_q = 0$ or $E_{PZC}$. Since $q_m$ is given by the slope of the curve, the PZC is defined by

$$q_m = -\left(\frac{\partial \gamma}{\partial V}\right)_{\text{const. comp.}} = 0$$

Hence, the PZC is the potential at which the ECM occurs. The $E_{PZC}$ is a fundamental reference potential in studies of electrified interfaces. This is because, one cannot measure the absolute potential difference $\Delta\phi$ across an electrode electrolyte interface. One can measure the cell potential at which the charge on the polarizable electrode is zero. Does this not mean that one has succeeded in knowing the absolute potential difference across the polarizable interface. No, because $\Delta\phi$ consists of charge contribution $\Delta\psi$ and a dipole contribution $\Delta\chi$. Hence

$$(\Delta\phi)_{q=0} = (\Delta\psi)_{q=0} + (\Delta\chi)_{q=0}$$

At PZC, the first term is zero but what of the other term. The surface potential $\Delta\chi$ cannot be measured. Hence, absolute potential difference remains experimentally indeterminable even at the PZC.

## 13.9 DETERMINATION OF THE SURFACE EXCESS

$$d\gamma = -q_m\, dV - \frac{q_m}{ZF} d\mu_j - \sum \Gamma_i d\mu_i \qquad (13.17)$$

Equation (13.17) describes changes of surface tension both with potential and composition. In order to find the surface excess, $\Gamma_i$, one has first to eliminate one variable, viz. the potential. For this electro-capillary curves obtained for various salt concentrations can be plotted in one diagram and a perpendicular erected to the cell potential axis. This will then contain points relating $\gamma$ to concentration at constant potential, i.e. when $dV = 0$ (Figure 13.16).

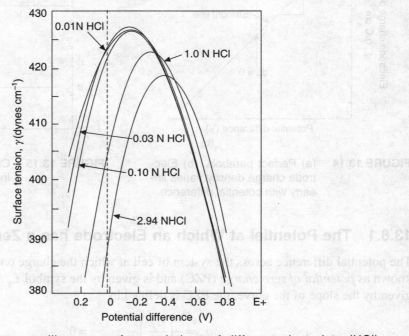

**FIGURE 13.16** Electro-capillary curve from solutions of different electrolyte (HCl) concentration.

Equation (13.15) contains the term $d(^s\Delta^{m_2}\phi)$, i.e. changes in the potential difference across the reference electrode solution interface, immersed in a solution of the same composition (concentration) as that of surrounding test electrode.

It is this value of $V$ which has to be kept constant. It is customary to denote these potential values referred not to the standard electrode but to reversible to ions of given concentration as $V_+$ or $V_-$ indicating at the same time whether the electrode is one at which cations or anions leak, respectively. Now a perpendicular erected on the axis of the cell potential $V_+$ or $V_-$ intersects the electro-capillary curves at points for which the condition $dV = 0$ is satisfied, whereupon

$$d\gamma = -\frac{q_m}{Z_j F} d\mu_j - \sum_i \Gamma_i d\mu_i \qquad (13.19)$$

This equation describes the changes of surface tension with composition at any particular cell potential $V_+$ or $V_-$.

Now consider a polarizable $(m_1/s)$ interface which consists of a metal electrode in contact with a solution of a 1:1 valent electrolyte, assembled into a cell along with nonpolarizable interface. Suppose that nonpolarizable interface is one at which negative ions interchange charge with the metal surface, i.e. $Z_j = -1$. Hence Eq. (13.19) for the polarizable interface becomes

$$d\gamma = +\frac{q_m}{Z_j F} d\mu_- - \Gamma_+ d\mu_+ - \Gamma_- d\mu_- \qquad (13.20)$$

where $\Sigma$ of Eq. (13.19) has been expanded.

The chemical potential $\mu$ of the electrolyte is the sum of the chemical potentials of the ions, i.e.

$$\mu = \mu_+ + \mu_-$$
$$d\mu = d\mu_+ + d\mu_-$$
$$d\mu_+ = d\mu - d\mu_- \qquad (13.21)$$

Using Eq. (13.21) one can substitute $d\mu_+$ in Eq. (13.20) to give

$$d\gamma = \frac{q_m}{F} d\mu_- - \Gamma_+ d\mu + \Gamma_+ d\mu_- - \Gamma_- d\mu_-$$

$$d\gamma = -\Gamma_+ d\mu + \left(\frac{q_m + F\Gamma_+ - F\Gamma_-}{F}\right) d\mu_- \qquad (13.22)$$

But before the formation of double layer, the metal is uncharged and in the solution, the charge per unit area of a Lamina due to positive and negative ions is zero, i.e.

$$F\left(\frac{n_+^0}{A}\right) - F\left(\frac{n_-^0}{A}\right) = 0 \qquad (13.23)$$

After the double layer is formed, electro-neutrality requires that

$$F\left(\frac{n_+}{A}\right) - F\left(\frac{n_-}{A}\right) + q_m = 0 \qquad (13.24)$$

Subtracting Eq. (13.23) from Eq. (13.24), one gets

$$q_m + F\left(\frac{n_+ - n_+^0}{A}\right) - F\left(\frac{n_- - n_-^0}{A}\right) = 0 \qquad (13.25)$$

But according to definition of surface excess

$$\Gamma_+ = \frac{n_+ - n_+^0}{A} \text{ and } \Gamma_- = \frac{n_- - n_-^0}{A} \qquad (13.26)$$

Hence, according to the electro-neutrality condition

$$q_m + F\Gamma_+ - F\Gamma_- = 0 \qquad (13.27)$$

and inserting this condition into Eq. (13.22), one finds that for a polarizable interface which is built into a cell with a nonpolarizable interface which leaks negative ions, the second term is zero, so that

$$d\gamma = -\Gamma_+ d\mu$$

$$\left(\frac{\partial \gamma}{\partial \mu}\right)_{\text{Const. } V_-} = -\Gamma_+ \qquad (13.28)$$

Now
$$\mu = (\mu_+ + \mu_-)$$
$$\mu = (\mu_+^0 + \mu_-^0) + (RT \ln a_+ + RT \ln a_-)$$
$$\mu = (\mu_+^0 + \mu_-^0) + (RT \ln a_+ a_-) \qquad (13.29)$$

Instead of $a_+ a_-$ one can introduce the mean ionic activity, $a_\pm$ obtained by multiplying mole fraction and fugacities. Thus,

$$a_\pm = (x_+ f_+)^{1/2} (x_- f_-)^{1/2} \Rightarrow (a_+ a_-)^{1/2}$$
$$= a_+ a_- = a_\pm^2 \qquad (13.30)$$

Hence, from Eqs. (13.29) and (13.30), we get

$$\mu = (\mu_+^0 + \mu_-^0) + (2RT \ln a_\pm)$$
$$d\mu = 2RT d \ln a_\pm \qquad (13.31)$$

Hence,
$$\left(\frac{\partial \gamma}{2RT d \ln a_\pm}\right)_{\text{Const. } V_-} = -\Gamma_+ \qquad (13.32)$$

Above equation shows that the slope of surface tension versus log $a_\pm$ curve at constant potential yields the surface excess. The activity of the electrolyte $a_\pm$ is obtained by taking the bulk concentration and multiplying it by the mean activity coefficient at that concentration. An interesting point to note is that the surface excess of the positive ion is determined by choosing a nonpolarizable interface which leaks negative ions and vice versa. Thus, in a study of Hg–HCl polarizable interface, the surface excess $\Gamma_-$ of $Cl^-$ ions may be obtained by coupling the Hg–HCl interface with a hydrogen refrence electrode which leaks hydrogen ions.

## 13.10 THE PARALLEL PLATE CONDENSER MODEL: HELMHOLTZ–PERRIN MODEL OF THE DOUBLE LAYER

As we know, Lippmann equation gives the relation between changes of surface tension and changes of cell potential.

$$\left(\frac{\partial \gamma}{\partial V}\right)_{\text{Const. Comp.}} = -q_m = \text{Slope} \qquad (13.33)$$

Experimentally, it is known that $\gamma$ versus $V$ curve is almost a parabola. Can this parabolic dependence of $\gamma$ on $V$ be explained in a thermodynamic framework? Yes by integrating above equation

$$\int d\gamma = -\int q_m dV \qquad (13.34)$$

But is $q_m$ a function of $V$. One cannot integrate without knowing this function.

In order to obtain the relation between $q_m$ and $V$, a model was given by Helmholtz and Perrin for the arrangement of charges at an electrified interface. The electrified interface was assumed of two sheets of charge. Hence, the term double layer was given to the interface. The charge densities on the two sheets are equal in magnitude but opposite in sign, exactly as in a parallel plate capacitor. The electrostatic theory of capacitors can be used for double layers. It is known, for example, that the potential difference $V$ across the condenser is

$$V = \frac{4\pi d}{\varepsilon} q \qquad (13.35)$$

where $d$ is the distance between the plates and $\varepsilon$ is the dielectric constant of the material between the plates.

$$dV = \frac{4\pi d}{\varepsilon} dq_m \qquad (13.36)$$

By putting this expression of $dV$ into Eq. (13.34), one has

$$\int d\gamma = -\frac{4\pi d}{\varepsilon} \int q_m dq_m$$

$$\gamma + \text{constant} = -\frac{4\pi d}{\varepsilon} \frac{1}{2} q_m^2 \qquad (13.37)$$

Now when $q_m = 0$, i.e at PZC, $\gamma$ is maximum at electro-capillary curve and therefore

$$\text{Constant} = -\gamma_{\text{max}}$$

Hence,

$$\gamma = \gamma_{\text{max}} - \frac{4\pi d}{\varepsilon} \frac{q_m^2}{2}$$

By putting value of $q_m$, i.e. $q_m^2 = \left(\frac{\varepsilon}{4\pi d}\right)^2 V^2$, we get $\gamma = \gamma_{\text{max}} - \frac{\varepsilon}{4\pi d} \frac{1}{2} V^2$

This is the equation of a parabola symmetrical about $\gamma_{\text{max}}$, i.e. about the ECM. It appears that the Helmholtz–Perrin model would be quite satisfactory for electro-capillary curves which are the perfect parabolas.

## 13.10.1 The Double Layer in Trouble: Neither Perfect Parabola Nor Constant Capacities

Now the questions is. Is the electro-capillary curve a perfect parabola? Almost, but not quite. There is always a slight asymmetry. The deviations from a parabolic shape are greater with some solutions than with others. Electro-capillary curves shows a marked sensitivity to the nature of the anions present in the electrolyte. In contrast, curves do not seem to be effected significantly by the cations present, unless they are large organic cations, e.g. tetraalkylammonium ions (Figure 13.17).

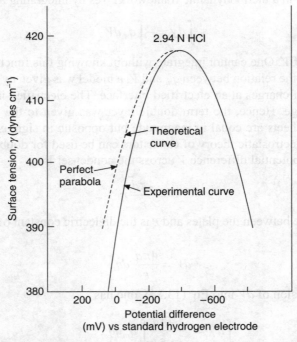

**FIGURE 13.17** Experimental and theoretical curves between $\gamma$ versus $V$.

The asymmetry of electro-capillary curves has important implications for the capacities of double layers. Consider a typical electro-capillary curve which is not exactly a parabola. It yields on differentiation a charge density versus cell potential plot which does not consist of one straight line but two straight lines meeting at $q = 0$. Now the differential capacity $C$ is given by

$$C = \frac{\partial q}{\partial V}$$

It turns out that, when one starts from an asymmetric electro-capillary curve, the interface displays a differential capacities which is not constant with cell potential. A parallel plate condenser has a specific capacity which is given by rearranging Eq. (13.36) in the form

$$\frac{dq}{dV} = \frac{\varepsilon}{4\pi d} = C$$

Thus, if $\varepsilon$ and $d$ are taken as constant. It means that parallel plate model predicts a constant capacity, i.e. one which does not change with potential. But, this is not what is observed. It appears that an electrified interface does not behave like a simple double layer.

## 13.11 IONIC CLOUD: THE GOUY–CHAPMAN DIFFUSE CHARGE MODEL OF THE DOUBLE LAYER

Helmholtz–Perrin model fixes charges onto a sheet parallel to the metal. But this model was too rigid to explain the asymmetry of electro-capillary curve. It was Gouy and Chapman, who thought of liberating the ions from a sheet parallel to the electrode. But once the ions are free, they become exposed to thermal buffeting and thermal jostling. Equilibrium between electric and thermal forces is attained and thus a time average ionic distribution takes place.

Therefore, it was assumed that charged electrode will be enveloped by an ionic cloud in the same way as anion in the bulk of the solution. This ionic atmosphere represents a falling off with distance from the electrode, of the net charge density in the lamina parallel to the electrode and at increasing distances out into the solution. The potential too will decay with distances asymptotically setting down to a constant value (taken as zero) in the bulk of the solution.

### 13.11.1 Ions Under Thermal and Electric Forces Near an Electrode or Mathematical Treatment of Diffuse Double Layer

Consider a lamina (in the electrolyte) parallel to the electrode and at a distance $x$ from it. The charge density in this lamina can be expressed in two ways, firstly in terms of the Poisson equation, which for $x$-dimension in rectangular co-ordinates reads (Figure 13.18):

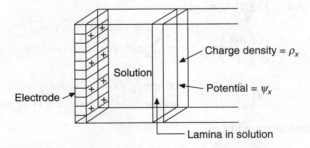

**FIGURE 13.18** A lamina in the solution parallel to a plane electrode.

$$\rho_x = -\frac{\varepsilon}{4\pi}\frac{d^2\psi_x}{dx^2} \quad \text{because} \quad V = \frac{4\pi}{\varepsilon}q_m x \quad \text{and} \quad q_m = \frac{\varepsilon}{4\pi}\frac{V}{x} \quad (13.38)$$

where $\psi_x$ is the outer potential difference between the lamina and the bulk of the solution (as $\psi_{x\to\infty} = 0$). According to Boltzmann equation, charge density is given by

$$\rho_x = \sum_i n_i z_i e_0 = \sum_i n_i^0 z_i e_0\, e^{-z_i e_0 \psi_x / kT} \quad (13.39)$$

where, $n_i$ and $n_i^0$ are the concentrations of the $i$th species in the lamina and in the bulk of the solution, respectively. $Z_i$ is the valency of the species $i$ and $e_0$ is the electronic charge. The factor $Z_i e_0 \psi_x / kT$ represents the ratio of the electrical and thermal energies of an ion at the distance $x$ from the electrode. From the two expressions for the charge density, one obtains Poisson–Boltzmann equation.

$$\frac{d^2\psi}{dx^2} = -\frac{4\pi}{\varepsilon} \sum_i n_i^0 z_i e_0 e^{-z_i e_0 \psi_x / kT} \tag{13.40}$$

In the Debye–Huckel theory of ion–ion interactions, it was considered that the potential was small enough to linearize the equation. This linearization would correspond in the present case to writing

$$e^{-z_i e_0 \psi_x / kT} = 1 - \frac{z_i e_0 \psi_x}{kT}$$

A simple transformation can now be used.

$$\frac{1}{2}\frac{d}{d\psi}\left(\frac{d\psi}{dx}\right)^2 = \frac{1}{2} 2\left(\frac{d}{d\psi}\frac{d\psi}{dx}\right)\frac{d\psi}{dx}$$

$$= \left(\frac{dx}{d\psi}\frac{d}{dx}\frac{d\psi}{dx}\right)\frac{d\psi}{dx}$$

$$\frac{1}{2}\frac{d}{d\psi}\left(\frac{d\psi}{dx}\right)^2 = \frac{d^2\psi}{dx^2} \tag{13.41}$$

By putting this value in Eq. (13.40), we get

$$\frac{d}{d\psi}\left(\frac{d\psi}{dx}\right)^2 = -\frac{8\pi}{\varepsilon} \sum_i n_i^0 z_i e_0 e^{-z_i e_0 \psi_x / kT} \tag{13.42}$$

$$d\left(\frac{d\psi}{dx}\right)^2 = -\frac{8\pi}{\varepsilon} \sum_i n_i^0 z_i e_0 e^{-z_i e_0 \psi_x / kT} d\psi \tag{13.43}$$

Above equation can be integrated to give

$$\left(\frac{d\psi}{dx}\right)^2 = \frac{8\pi}{\varepsilon} \int \sum_i n_i^0 z_i e_0 e^{-z_i e_0 \psi_x / kT} d\psi \tag{13.44}$$

$$= \frac{8\pi}{\varepsilon} \sum_i \frac{n_i^0 z_i e_0 e^{-z_i e_0 \psi_x / kT}}{(-z_i e_0 / kT)} + \text{constant}$$

$$= \frac{8\pi kT}{\varepsilon} \sum_i n_i^0 e^{-z_i e_0 \psi_x / kT} + \text{constant} \tag{13.45}$$

The integration constant can be evaluated by considering that deep in the solution, i.e. at $x \to \infty$ not only is the *Volta potential* zero $\psi_{x\to\infty} = 0$, but the field $\dfrac{\partial \psi}{\partial x}$ is also zero. Under these conditions

$$\text{Constant} = -\frac{8\pi kT}{\varepsilon} \sum_i n_i^0 \qquad (13.46)$$

By introducing this value of the integration constant into Eq. (13.45), we get

$$\left(\frac{\partial \psi}{\partial x}\right)^2 = \frac{8\pi kT}{\varepsilon} \sum_i n_i^0 (e^{-Z_i e_0 \psi_x / kT} - 1) \qquad (13.47)$$

Considering a Z:Z valent electrolyte, $|Z_+| = |Z_-| = Z$ and $n_+^0 = n_-^0 = n^0$ and

$$\left(\frac{d\psi}{dx}\right)^2 = \frac{8\pi kT}{\varepsilon} \sum_i n^0 (e^{-Z_i e_0 \psi_x / kT} - 1)$$

$$\left(\frac{d\psi}{dx}\right)^2 = \frac{8\pi kT}{\varepsilon} n^0 (e^{Ze_0 \psi_x / kT} - 1 + e^{-Ze_0 \psi_x / kT} - 1)$$

$$\left(\frac{d\psi}{dx}\right)^2 = \frac{8\pi kT}{\varepsilon} n^0 [e^{Ze_0 \psi_x / kT} - 2(e^{Ze_0 \psi_x / kT})(e^{-Ze_0 \psi_x / kT}) + e^{-Ze_0 \psi_x / kT}]$$

$$\left(\frac{d\psi}{dx}\right)^2 = \frac{8\pi kT}{\varepsilon} n^0 [e^{Ze_0 \psi_x / 2kT} - e^{-Ze_0 \psi_x / 2kT}]^2 \qquad (13.48)$$

But, $\qquad e^{+x} - e^{-x} = 2 \sinh x \qquad (13.49)$

And $\qquad (e^{+x} - e^{-x})^2 = 4 \sinh^2 x$

Hence, Eq. (13.48) becomes

$$\left(\frac{d\psi}{dx}\right)^2 = \frac{32\pi kT n^0}{\varepsilon} \sinh^2 \frac{Ze_0 \psi_x}{2kT} \qquad (13.50)$$

From this equation, one can get the field $\dfrac{d\psi}{dx}$ in the solution by taking square roots on both sides. There is, however, a positive and a negative square root. To decide which root is to be taken, one remembers that at positively charged electrode, $\psi > 0$ but $\dfrac{d\psi}{dx} < 0$, while at the negatively charged electrode $\psi < 0$ and $\dfrac{d\psi}{dx} > 0$. Hence, it is clear that only the negative root of Eq. (13.50) corresponds to the physical situation, i.e.

$$\frac{d\psi}{dx} = -\left(\frac{32\pi kT n^0}{\varepsilon}\right)^{1/2} \sinh \frac{Ze_0 \psi_x}{2kT} \qquad (13.51)$$

What is $\dfrac{d\psi}{dx}$? It is the field (or gradient of potential) at a distance $x$ from the electrode according to the diffuse charge model of Gouy and Chapmann. Hence, Eq. (13.51) tells out the relation between the electric field and the potential at any distance $x$ from the electrode. Instead of field it is preferable to have an expression for the total diffuse charge in the solution in terms of potential. According to Gaussian law of electrostatics, the charge contained in a closed surface or Gaussian box is equal to $\dfrac{\varepsilon}{4\pi}$ times the area of the closed surface (taken here as unity) times the component of the field normal to the surface.

$$q = \frac{\varepsilon}{4\pi}\frac{d\psi}{dx} \qquad (13.52)$$

Area is taken as unity. Hence, since $\dfrac{d\psi}{dx}$ is known from Eq. (13.51), the corresponding $q$ can be obtained. This $q$ is the charge enclosed within the closed volume at the surface of which the field is $\dfrac{d\psi}{dx}$.

To compute the total diffuse charge density $q_d$, the Gaussian box chosen is a rectangular box with one unit area side deep in the solution at $x \to \infty$, where $\psi_x$ and $\dfrac{d\psi}{dx} = 0$ with other side very close to the electrode, so as not to miss any diffuse charge. Hence, the total diffuse charge density scattered in the solution under the interplay of thermal and electrical forces $q_d$ is given by Eqs. (13.51) and (13.52) with $x = 0$, as

$$q_d = \sqrt{\frac{\varepsilon}{4\pi}}\sqrt{\frac{\varepsilon}{4\pi}} - \left(\frac{32\pi k T n^0}{\varepsilon}\right)^{1/2} \sinh\frac{Ze_0\psi_x}{2kT}$$

$$q_d = -2\left(\frac{\varepsilon k T n^0}{2\pi}\right)^{1/2} \sinh\frac{Ze_0\psi_x}{2kT} \qquad (13.53)$$

where $\psi_0$ is the potential at $x = 0$ relative to the bulk of the solution where the potential is taken as zero (i.e. $\psi_\infty = 0$). Equation (13.53) shows that the total diffuse charge varies with the total potential drop in the solution according to a hyperbolic sine relation.

### 13.11.2 A Picture of Potential Drop in the Diffuse Layer

The field $\dfrac{d\psi}{dx}$ is the gradient of the potential. By integrating the field, one obtains the variation of the potential with distance. Let this be done after assuming that

$$\sinh\frac{Ze_0\psi_x}{2kT} \approx \frac{Ze_0\psi_x}{2kT}$$

Thus, from Eq. (13.51), one has

$$\frac{d\psi_x}{dx} \approx -\left(\frac{32\pi kTn^0}{\varepsilon}\right)^{1/2} \frac{Ze_0\psi_x}{2kT}$$

$$\approx -\left(\frac{8\pi n^0 (Ze_0)^2}{\varepsilon kT}\right)^{1/2} \psi_x \qquad (13.54)$$

Examine the constants inside the brackets. It is the familiar, $\chi^2$ of the Debye–Huckel theory, where it was shown that $\chi^{-1}$ can be considered the effective thickness of the ionic cloud. In terms of $\chi$, Eq. (13.54) becomes

$$\frac{d\psi_x}{dx} = -\chi \psi_x \qquad (13.55)$$

By integration $\ln \psi_x = -\chi x + \text{constant}$

To evaluate the constant, the following boundary condition is used. At $x \to 0$, $\psi_x \to \psi_0$. It follows therefore that

$$\ln \psi_0 = 0 + \text{constant}$$
$$\psi_x = \psi_0 e^{-\chi x} \qquad (13.56)$$

That is potential decays exponentially into the solution, deep enough inside the solution, $x \to \infty$ the potential becomes zero. Further, as the solution concentration $n^0$ increases, $\chi$ increases and $\psi_x$ falls more and more sharply. This potential–distance relation is an important and simple result from the Gouy–Chapman model (Figure 13.19).

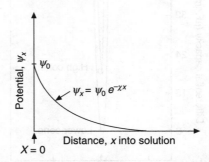

**FIGURE 13.19** Variation of potential with distance in diffuse charge region.

The potential difference across a parallel plate condenser is

$$dV = \frac{4\pi d}{\varepsilon} q \qquad (13.57)$$

where $q$ is the charge on the plates and $d$ their distance apart.

Rewriting Eq. (13.53) as $q_m$

$$q_m = -q_d = 2\left(\frac{\varepsilon kTn^0}{2\pi}\right)^{1/2} \sinh\frac{Ze_0\psi_x}{2kT}$$

and for small values of potential $\psi_0$ at the electrode sinh can be omitted so

$$q_m = -q_d = 2\left(\frac{\varepsilon k T n^0}{2\pi}\right)^{1/2} \frac{Ze_0 \psi_x}{2kT} \tag{13.58}$$

$$d = \chi^{-1} = \left(\frac{\varepsilon k T}{n^0 8\pi Z^2 e_0^2}\right)^{1/2} \quad \text{(from Debye–Huckel Onsager equation)} \tag{13.59}$$

On putting the value of $q_m$ and $d$ from Eqs. (13.58) and (13.59) into an Eq. (13.57), we get

$$\Delta V = \psi_0 \tag{13.60}$$

which means that the $x = 0$, plate is at a potential $\psi_0$ relative to the $x = \chi^{-1}$, plate at a potential zero. This is indeed the situation. Once the total diffuse charge $q_d$ is known and also the potential $\psi_0$ at $x = 0$, the differential capacity is easy to calculate. For getting capacity, differentiate $q_d$ with respect to $\psi_m - \psi_b \Rightarrow \psi_m$ because $\psi_b = 0$. Thus,

$$C = \frac{dq_m}{d\psi_m} = -\frac{dq_d}{d\psi_m} = \left(\frac{\varepsilon Z^2 e_0^2 n^0}{2\pi k T}\right)^{1/2} \cosh\frac{Ze_0 \psi_m}{2kT} \tag{13.61}$$

Now the cosh function gives inverted parabolas (Figure 13.20).

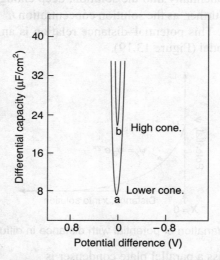

**FIGURE 13.20** cosh function gives inverted parabola dependence.

Hence, according to the simple diffuse charge theory, the differential capacity of an electrified interface should not be a constant. Rather, it should show an inverted parabola dependence on the potential across the interface. This, of course, a welcome result because the major weakness of Helmholtz–Perrin model is that it does not predict any variation of capacity with potential, although such a variation is found experimentally.

### 13.11.3 An Experimental Test of the Gouy–Chapman Model: Potential Dependence of the Capacitance

The experimental capacities potential curves are just not the inverted parabolas which the Gouy–Chapman diffuse model predicts (Figure 13.21). In a very dilute solution (<0.001 m) and at potentials near the PZC, there are portions of experimental curves, which suggest that the interface is behaving in a Gouy–Chapman way. But at potentials further away from that of zero charge and in concentrated solutions, the Gouy–Chapman model bears no relation to reality. Not only is the predicted shape of the capacity potential curves wrong, but there is also the matter of the concentration dependence. The Gouy–Chapman theory predicts that the capacity depends on the concentration to an extent that is not at all observed. At high concentration (e.g.1.0 m) the predicted capacity from Eq. (13.61) is almost an order of magnitude higher than the observed value (Table 13.1).

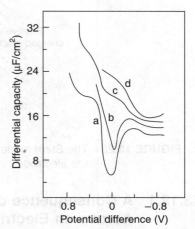

**FIGURE 13.21** Shape of capacity potential curve a = 0.001 m, b = 0.01 m, c = 0.1 m and d = 1.0 m.

**TABLE 13.1** Differential capacity of Hg–NaF interface at the potential of zero charge

| Concentration | Differential capacity ($\mu F/cm^2$) | |
|---|---|---|
| | Experimental | Calculated from Eq. (13.58) |
| 0.001 N | 6.0 | 7.2 |
| 0.01 N | 13.1 | 22.8 |
| 0.1 N | 20.7 | 72.2 |
| 1.0 N | 25.7 | 228.0 |

## 13.12 THE STERN MODEL

The simplest version of the Stern theory consists in eliminating the point charge approximation of the diffuse layer theory. This is done in exactly the same way as in the theory of ion–ion interactions, the ion-centres are taken as not coming closer than a certain critical distance $a$ from the electrode. This is tantamount to applying the Gouy–Chapman theory for ions which are of finite size, not point charges. In other words, the Gouy–Chapman treatment commences not with points on the electrode but from a layer at a certain distance $a$ from the electrode. This distance $a$ is the closest distance of approach of the ions to the electrode and will be specified more quantitatively as this account goes on (Figure 13.22).

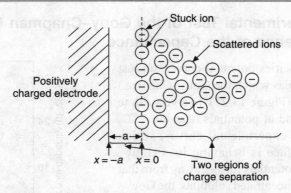

**FIGURE 13.22** The Stern model: the locus of centres of stuck ions is at a distance $a$ from the electrode.

## 13.12.1 A Consequence of the Stern Picture: Two Potential Drops Across an Electrified Interface

Under all conditions, the interface as a whole (the electrode side taken along with the electrolyte side) is electrically neutral. The net charge density $q_m$ on the electrode must be equal in magnitude and opposite in sign to the net charge density $q_s$ on the solution side, i.e. $-q_m = q_s$. But according to the Stern picture, the charge $q_s$ on the solution is partially stuck (Helmholtz–Perrin charge, $q_H$) to the electrode and the remainder $q_G$ is diffusely spread out (in Gouy–Chapman style) in the solution, i.e.

$$q_s = q_H + q_G$$

There are therefore two regions of charge separation. The first region is from the electrode to the Helmholtz plane (the plane defined by the locus of centres of the stuck ions and the second region is from this plane of fixed charge into the heart of the solution, where the net charge density is zero. When, however, charges are separated potential drops results. The Stern model implies, therefore, two potential drops, i.e,

$$\phi_m - \phi_b = (\phi_m - \phi_H) + (\phi_H - \phi_b) \qquad (13.62)$$

where $\phi_m$ and $\phi_H$ are the inner potential at the metal and the Helmholtz planes, and $\phi_b$ is the potential in the bulk of the solution.

Thus, the Stern model pictures the potential variation across an interface as consisting of two regions, a linear region corresponding to the ions stuck on the electrode and an exponential region corresponding to the ions which are under the combined influence of the ordering electrical and the disordering thermal forces.

## 13.12.2 Another Consequence of the Stern Model: An Electrified Interface is Equivalent to two Capacitors in Series

Now one can ask, how are the potential drops affected by small changes in the charge on the metal. In other words, what is the result of differentiating the expression for the potential difference across the interface with respect to charge on the metal. One obtains,

$$\frac{\partial(\phi_m - \phi_b)}{\partial q_m} = \frac{\partial(\phi_m - \phi_H)}{\partial q_m} + \frac{\partial(\phi_H - \phi_b)}{\partial q_m} \qquad (13.63)$$

In the denominator of the last term, one can replace $\partial q_m$ with $\partial q_d$ because the total charge on the electrode is equal to the total diffuse charge, i.e.

$$\frac{\partial(\phi_m - \phi_b)}{\partial q_m} = \frac{\partial(\phi_m - \phi_H)}{\partial q_d} + \frac{\partial(\phi_H - \phi_b)}{\partial q_d} \qquad (13.64)$$

Each term in above equation is the reciprocal of a quantity, which is of the form, i.e. it is the reciprocal of a differential capacity. Hence Eq. (13.64) can be rewritten, thus

$$\frac{1}{C} = \frac{1}{C_H} + \frac{1}{C_G} \qquad (13.65)$$

where $C$ is the total capacity of the interface, $C_H$ is the Helmholtz–Perrin capacity and $C_G$ is the Gouy–Chapman or diffuse charge capacity.

This result is formally identical to the expression for the total capacity displayed by two capacitors in series. The conclusion therefore is that an electrified interface has a total differential capacity, which is given, by Helmholtz–Gouy capacities in series. Capacitors in series imply that the region are consective in space each region accounting for only a part of the total potential difference. The Helmholtz and Gouy regions do store ions and are consective in a direction normal to the electrode and the total potential difference is across the whole interface.

## 13.13 THE RELATIVE CONTRIBUTIONS OF THE HELMHOLTZ–PERRIN AND GOUY–CHAPMAN CAPACITIES

Problem of diffuse charge layer theory, i.e. higher predicted capacity at higher concentration than experimental value can be solved by Stern theory quite easily.

Total capacities depends upon the two terms

$$\frac{1}{C} = \frac{1}{C_H} + \frac{1}{C_G}$$

What happens when the concentration $n^0$ of the electrolyte is large? From Eq. (13.61) it can be seen that $C_G$ becomes large while $C_H$ does not change. Hence, with increasing concentration the second term in Eq. (13.65), i.e., $\frac{1}{C_G}$ becomes small compared with the first $\frac{1}{C_H}$. Whereupon

$$\frac{1}{C_G} \ll \frac{1}{C_H}$$

For practical purpose, i.e. concentrated solution

$$\frac{1}{C} = \frac{1}{C_H}$$

or

$$C = C_H$$

That is in sufficiently concentrated solutions, the capacity of the interface is effectively equal to the capacity of the Helmholtz region, i.e. of the parallel plate model. The analogy between the electrified interface and electrical capacitors works well here. The total capacity of two capacitors in series is effectively equal to the smaller capacity when the other one is relatively large.

It means that if Helmholtz and Gouy regions are compared at sufficiently high concentration ($C_G$ high) most of the solution charge is squeezed onto the Helmholtz or confined in a region very near this plane. In other words, little charge is scattered diffusely into the solution in the Gouy–Chapman disarray. But what happens, when if $C_G$ is low, i.e. at sufficiently low concentration. Under these conditions

$$\frac{1}{C_H} \ll \frac{1}{C_G}$$

and
$$C = C_G$$

This means that the electrified interface has become in effect Gouy–Chapman like in structure with the solution charge scattered under the simultaneous influence of electrical and thermal forces.

It is to note that, older parallel plate condenser approach of Helmholtz–Perrin gives much more often the essential, if very rough, picture because, for most ordinary concentrations $C_G \gg C_H$, where upon the capacity of the double layer is virtually equal to the capacity derived with the Helmholtz–Perrin model.

But Helmhottz model has several difficulties, e.g. constant capacities and strictly parabolic electro-capillary curves.

## EXERCISES

1. What is an interface? How this interface becomes electrified?
2. Compare electrified interface with parallel plate capacitor.
3. What are various types of electrified interfaces?
4. Write down the importance of electrified interface.
5. Illustrate Helmholtz parallel plate model of electrified interface.
6. Draw a closer look diagram of electrified interface. Explains various terminology like OHP and IHP.
7. Discuss in detail Gouy–Chapman diffuse charge model of electrified interface.
8. Draw potential versus distance curve in case of super-equivalent adsorption at the interface.
9. How is Stern model different from Helmholtz parallel plate model and Gouy–Chapman diffuse charge model.
10. How we can measure surface tension at an electrified interface.
11. What is electro-capillary curve and electro-capillary maxima?
12. Derive fundamental equation which relates surface tension with electrode charge density, chemical potential and surface excess.
13. How we can determine charge density on the electrode?

14. How we can determine capacitance on the electrified interface?
15. How we can determine surface excess at the electrified interface?
16. Explain how this electrified interface is equivalent to two capacitors in series?

## SUGGESTED READINGS

Antropov, L.I., *Theoretical Electrochemistry*, 2nd ed., Mir Publishers, Moscow, 1972.

Bockris, J.O.M., *Modern Aspects of Electrochemistry*, No. 1, Butterworths, London, 1954.

Bockris, J.O.M., *Modern Aspects of Electrochemistry*, No. 2, Butterworths, London, 1959.

Bockris, J.O.M. and A.K.N. Reddy, *Modern Electrochemistry: An Introduction to an Interdisplinary Area*, Vols. 1 and 2, Plenum Press, New York, 1970.

Bokris, J.O.M., B.E. Conway, and E. Yeager, *The Double Layer Comprehensive Treatise of Electrochemistry*, Vol. 1, Plenum Press, New York, 1980.

Buttler, J.A.V., *Electrical Phenomenon at Interface in Chemistry, Physics and Biology*, Macmillan, New York, 1951.

Davies, C.W., *Electrochemistry*, Newnes, London, 1967.

Denaro, A.R., *Elementary Electrochemistry*, Butterworths, London, 1965.

Glasstone, S., *Fundamentals of Electrochemistry and Electrodeposition,* 2nd ed., Franklin, Palisade, New Jersey, 1960.

Hamer, W.J., *Structure of Electrolytic Solution*, Wiley, New York, 1959.

Kortum, G., *Treatise on Electrochemistry*, 2nd ed., Elsevier, New York, 1965.

Koryta, J., J. Dvorak, and V. Bohackova, *Electrochemistry*, Halsted, New York, 1973.

Mathewson, D.J., *Electrochemistry*, Macmillan, London, 1971.

Milazzo, G., *Electrochemistry, Theoretical, Principles and Practical Applications*, Elsevier, New York, 1963.

Rochow, E., *Electrochemistry*, Reinhold, New York, 1965.

Silk, T., *Electrochemistry*, Heinemann, London, 1968.

Sparnaay, M.J., *The Electrode Double Layer*, Pergamon Press, New York, 1972.

# 14

# Electrodics

## 14.1 INTRODUCTION

If the interface is ideally polarizable, a specific restriction is placed on the charge populating the interface, they must not cross it. The structure of the interface, i.e. how the ions distribute themselves, how dipoles are aligned in the interface region, etc. has nothing to say about how charges cross the interface.

But, once the transfer of charges across the interface is considered the perspective changes. The transfer of charge across electrically charged interface is a very significant part of nature. It compels electrochemistry to evolve new concepts and link itself with the everyday world, with batteries and brain cells, with energy producers and prevention of corrosion, with the avoidance of thrombosis and the synthesis of nylon.

Modern electrochemistry is the study of charge transfer across interfaces and ionics, the physical chemistry of the ionic side of the double layer a prerequisite.

### 14.1.1 Charge Transfer: Its Chemical and Electrical Implications

The transfer of charge across the electrified interface consists of essentially the exchange of electrons between the electrode and particles on the solution side of the interface.

Suppose that the particle is an ion positioned on the OHP and it accepts from or donates an electron to the electrode. Its valence state necessarily changes when such a transfer occurs. If for instance, the ion donates electrons to the electrode, i.e. the electron transfer reaction is given by

$$M^+ \xrightarrow[\text{to the electrode}]{\text{Electron donation}} M^{2+}$$

The valence state of the ion increase by 1, i.e. the ion is de-electronated or oxidized. An example of de-electronation or oxidation is the donation of an electron form a ferrous ion to the electrode, in which case, the divalent ferrous ion undergoes oxidation to a trivalent ferric ion.

$$Fe^{2+} \rightarrow F^{3+} + e^-$$

*A simple conclusion is:* charge transfer across an interface implies chemical transformations, i.e. transformation of substances into other substances. By controlling the direction, extent and rate of electron transfer across the interface, one can control the chemical reaction.

The transfer of electrons between the electrode and particles at the IHP or OHP can be considered to be directed normal to the plane of interface. It is the transport of charge, i.e. a current of electricity. Charge transfer implies an electric current across the interface.

If the electron transfer is from electrode to particle (electronation or reduction) and leads to the diminution of the positive charge on the particle, the current goes one way. If, however, the electron transfer is in the other direction, i.e. de-electronation, the current goes the other way. Whether, there is a net current flows across the interface or not, depends upon whether or not there are more electron transfer acts in one direction than the other (Figure 14.1).

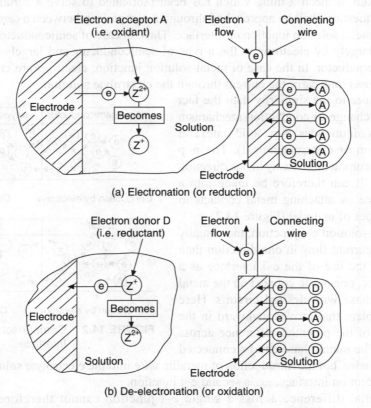

**FIGURE 14.1** Electronation and de-electronation reaction at the interface.

Now compare charge transfer at the interface and in the bulk of the solution. The de-electronation (oxidation) of ferrous to ferric ion can take place either at the electrode–electrolyte interface or in the bulk of the solution by ceric ion. In both cases, the reaction occurs by ferrous ions surrounding electrons, either to the electrode at the interface or to ceric ions in the bulk of the solution. Each act of electron transfer can be represented by a vector, showing the direction of the electron current. In the bulk of the solution, the vectors are all random in direction, hence, there is no current in solution. At the interface, however, one can consider all the vectors pointing towards the electrode. The interface situation is like the drift or flow of electrons, there is a net transport and a current.

In conclusion, electron transfer across charged interfaces has two implications which have for reaching importance in nature, a chemical implication and an electrical implication. Substance can be produced (chemistry) and currents can be generated (electricity). To realize these possibilities, however, one must understand more about the mechanics of charge transfer at interfaces.

### 14.1.2 Can an Insolated Electrode-Solution Interface be Used as a Device?

A device is taken to mean a thing which has been fabricated to serve a certain purpose. The answer to this question shall be approached through a comparison between a semiconductor n–p junction and a metal solution junction or interface. Thus, in case of semiconductor n–p junctions, conduction is largely by electrons in the n-type of semiconductor and largely by holes in the p-type of semiconductor. In the case of metal-solution junction, electrons are charge carriers in the metal, whereas ions carry the current through the electrolyte solution.

There is more to the similarity than the fact that, there is a change of conduction mechanism at the two types of junctions, n–p (solid state) and e–i (electron ion or electrochemical). The n–p junction passes current more easily in one direction than the other. It can therefore be made into a rectifying device by attaching metal contacts to the n- and p-types of material (Figure 14.2).

The metal–solution e–i junction, too, usually permits easier current flow in one direction than the other. But, the use of the e–i interface as a rectifying device, one has to connect up the metal and solution phase with desired circuits. Here comes the problem that proved awkward in the measurements of the potential difference across the interface. The metal phase can be connected

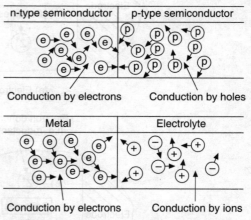

**FIGURE 14.2** Comparison of n-p and e-i interface.

with a metallic wire, but the introduction of metallic wire into the electrolyte solution produces a second metal solution interface, i.e. a second e–i junction.

The potential difference across a single e–i junction cannot therefore be measured. Correspondingly, the single metal solution interface cannot be used as a device. The isolated single interface holds out hypothetical possibilities of producing power and doing chemistry but these possibilities cannot be tapped for the only way of tapping them is with metallic contacts which, when dipped into the i-type of conductor (i.e. the solution) generate another interface.

## 14.1.3 Electrochemical System can be Used as Devices

It has been seen, however, that by assembling two metal–solution interfaces, i.e. forming an e–i–e junction, one has an electronic conductor at both ends of the assembly. The potential difference across the whole assembly, i.e. across the e–i–e junction can be measured.

## 14.1.4 An Electrochemical Device: The Substance Producer

Let the two interface assembly (the electrochemical system or cell) be connected to a source of direct current or a constant unidirectional flow of electrons. Electrons flow out of the DC source from its negative terminal, through the metallic leads and then up to the metal–solution interface (*I*) (e–i junction). At the junction, there is a change of charge carries from electrons to ions. The positive and negative ions carry the charge through the solution (according to their transport number) and at the second junction (*II*) there is another change of charge carries form ions to electrons.

The electron entry electrode (Faraday called such electrodes as *cathodes*) is the electron source; it gives electrons to particles in the solution. The electronation of electron accepting particles occurs at such electron sources. The electron exit electrode (Faraday's term was *anode*) is the electron sink, it takes electrons from particles in the solution. The de-electronation of electron donating particles occurs at such electron sinks.

All this can be exemplified by the following electrochemical system: two electronic conductors (e.g. two pieces of platinum) are dipped into water (made ionically conducting by the addition of some electrolyte) and then connected to an external electron source. At that conductor, which serves as an electron source, hydrogen ions, which may be present, if the electrolyte is an acid, are electronated and thus transform into hydrogen atoms, which combine to form hydrogen molecules. The transformation can be represented, thus

$$H^+ \xrightarrow{\text{Electron from source}} H_2$$

At the electron sink conductor, water molecules are de-electronated to produce oxygen molecules.

$$H_2O \xrightarrow{\text{Electron to sink}} O_2$$

If the external power source can keep pumping electrons through the system of cell, it can sustain the production of a gas of hydrogen molecules at the electron-source electrode and gas of oxygen molecules at the electron-sink electrode. An electrochemical device has been mentally fabricated a device which serves to produce the substances hydrogen and oxygen.

Thus, an externally driven electrochemical system is a device, which can be used as an electrochemical substance producer. This is electrochemistry.

## 14.2 THE INSTANT OF IMMERSION OF A METAL IN AN ELECTROLYTIC SOLUTION

At the time of immersion of metal electrode into the electrolyte, the metal is electro-neutral or uncharged, $q_m = 0$. Since the interface region as a whole must then be electroneutral, there must be zero excess charge on the solution side of the interface, i.e. $|q_m| = |q_s| = 0$. Hence, there is a zero potential difference and a zero field operating in the interface region. Thus, one has ignored the fact that, even when the metal charge is zero, there is a small net orientation of water molecules and hence, $H_2O$ will immediately start orienting itself in respect to the surface, which will create a dipole field.

Under these conditions of zero field, there are no electrical effects and one has pure chemistry, no electrochemistry. Let us assume that, electrode donate an electron to the electron acceptor ion $A^+$. After the receipt of the electron, the electron acceptor is transformed into a new substance $D$.

$$A^+ + e^- \to D$$

For example, electronation of ferric ions into ferrous ions. Now the question is: Will the electron transfer reaction occur of its own free will, or must it be driven? Such questions are answered by thermodynamics.

The precise criteria for equilibrium across an interface, is the equality of electrochemical potentials of the species which can leak across the interface.

Is $(\bar{\mu}_{A^+})_{\text{solution}}$ equal to $(\bar{\mu}_{A^+})_{\text{electrode}}$? In other words is $^m\Delta^s\bar{\mu}_{A^+} = 0$ But this criteria can be written in terms of the chemical and inner potentials, thus

$$^m\Delta^s\bar{\mu}_{A^+} = {}^m\Delta^s\mu_{A^+} + F\,{}^m\Delta^s\phi$$

Since there is zero field, the term $^m\Delta^s\phi$ can be set equal to zero. Hence, to know whether the interface is at equilibrium, one must check whether the chemical potentials of $A^+$ are same on both sides of the interface.

It is as in the process of diffusion. If there is a difference of the chemical potentials of a species in two regions, then the gradient of chemical potential act as the driving force for the diffusion. If therefore at $t = 0$, the chemical potentials of $A^+$ are not equal on both sides of the interface, there is no equilibrium across the interface. Thermodynamics, allows the electrode reaction to proceed spontaneously.

## 14.3 THE RATE OF CHARGE TRANSFER REACTIONS UNDER ZERO FIELD: THE CHEMICAL RATE CONSTANT

Consider the movement of the positive ions $A^+$ from the solution side of the interface to the metal surface. Somewhere along the way, the electron transfer occurs from electrode to ion. The progress of moving charge can be charted by specifying the values assumed by $x_1$ and $x_2$ in the movement. As the ion moves, its potential energy changes. Each point on this diagram, Figure 14.3 represents the energy corresponding to a certain location of the moving ion. The positive ion has to have a certain activation energy before the charge transfer reaction is accomplished. An example of actual diagram for such a situation is shown in Figure 14.4.

The Figure 14.4 shows potential–distance profile for the various successive steps of a metal ion undergoing electronation. This process of an ion's jumping from a solution side to the metal is

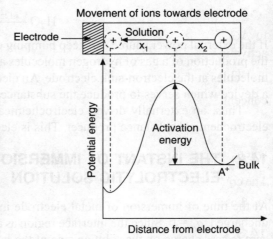

**FIGURE 14.3** Potential energy distance profile by consideration of the potential energy changes produced by varying $X_1$ and $X_2$.

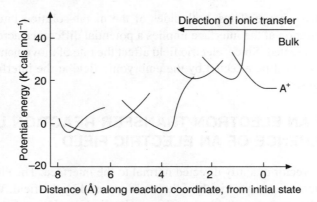

**FIGURE 14.4** Potential energy as a function of distance as the ion move towards electrode.

similar to the elementary act of diffusion, namely the jump of an ion from one site to another. In the case of diffusion too, there is a potential energy barrier. The frequency with which an ion successfully jumped the energy barrier for diffusion (i.e. jump frequency) has been shown to be

$$\text{Jump Frequency} = \vec{K} = \frac{kT}{h} e^{-\Delta \vec{G}^{0\#}/RT}$$

where $\Delta \vec{G}^{0\#}$ is the standard free energy of activation, the change in free energy required to climb to the top of the barrier when there is zero electric field acting on the ion in its motion.

When this jump frequency is multiplied by the concentration $C_{A^+}$ of electron acceptor ions $A^+$ on the solution side of the interface, one obtains the rate of the electronation reaction under zero electric field

$$\vec{V}_c = \frac{kT}{h} C_{A^+} e^{-\Delta \vec{G}_c^{0\#}/RT}$$

In $\Delta \vec{G}_c^{0\#}$ term $c$ stands for chemical or zero field rate, 0 stands for under zero field, ± stands for activation energy and → for electronation reaction.

This expression can be separated into a concentration-independent portion $\vec{K}_c$ and a concentration term $C_{A^+}$.

$$\vec{V}_c = \vec{k}_c C_{A^+}$$

where $\vec{k}_c = \frac{kT}{h} e^{-\Delta \vec{G}^{0\#}/RT}$

The concentration independent portion $\vec{k}_c$ is rate constant. It is the frequency of successful jumps, i.e. those in which the particle succeeds in passing over the barrier. The dimension of $\vec{k}_c$ is second$^{-1}$.

## 14.3.1 Some Consequences of Electron Transfer at an Interface

The emigration of the electron form the electrodes to the electron acceptor $A^+$ leaves the metal poorer by one negative charge. The metal has become charged positive. The solution side thus acquires a net negative charge. Thus, both sides of the metal–solution interface have become charged. Charge separation at the interface implies a potential difference across the interface. An electric field has been created. Since, electric field affect the rate of movement of charges, the rate of electrodic reactions must be affected by the embryonic field at the interface. Now chemistry becomes electrochemistry.

## 14.4 RATE OF AN ELECTRON TRANSFER REACTION UNDER THE INFLUENCE OF AN ELECTRIC FIELD

The electric field is a vector quantity directed normal to the interface. The electron transfer from the electrode to an electron acceptor in the solution is opposed by the field. When there is a field across the interface, the work done by the positive ion in climbing the potential energy barrier has to include the electrical work. In the presence of the field, the energies of all the charged particles will be altered. So the points and thus curve may shift up or down.

A highly simplified and approximate approach will be adopted. The electron will be forgotten and it will be assumed that the electronation reaction consists in moving the ion form its initial state right across the interface to its final position on the metal. On this basis, the electrical work of activating the ion, so that the energy representing it passes over the top of barrier is given by the charge ($e_0$) on the ion times the potential difference through the ion is moved to reach the top.

$$W = e_0 V = e_0 \Delta\phi = F\Delta\phi$$

As the positive ion begins to move across the double layer, it is the positive ion to do electrostatic work against the field in the double layer, i.e. the field does work on the ion. Let the total potential difference through which the ion passes be say $\Delta\phi$. However, with respect to the contribution of this electrostatic work to the standard free energy of activation, for the forward reaction (the ion from the solution to the electrode), only a part of the total $\Delta\phi$ is important, namely, that part through which the ion passes during passage to a point (perhaps somewhat halfway across the double layer) when the energy of the ion passes the summit of the energy barrier. The summit has been lowered by the electrical work done, i.e. potential difference passed through multiplied by the charge ($e_0$). Suppose that, instead of writing the important part of potential difference through which the ion moves as $\frac{1}{2}\Delta\phi$, one write it as $\beta\Delta\phi$, where $\beta$ is a factor greater than zero but less than unity. The title $\beta$ is unusually logical, it is called *symmetry factor*.

$$\beta = \frac{\text{Distance across the double layer to the summit}}{\text{Distance across the whole double layer}}$$

Then, $\beta\Delta\phi F$ is the amount by which the energy barrier for the ion to electrode transfer is lowered and hence $(1-\beta)\Delta\phi F$ is the amount it is raised for the metal to solution reaction (Figure 14.5).

Thus, once more, for the forward reaction, electrical contribution to the free energy of activation,

$$\Rightarrow +\beta F \Delta\phi$$

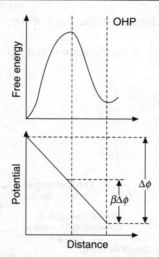

**FIGURE 14.5** (a) The electrical work of activating the ion is determined by the potential differences across which the ion has to be moved to reach the top of free energy–distance relation. (b) Variation of potential with distance across the interface.

In the presence of the field, therefore, the total free energy of activation for the electronation reaction is equal to the chemical free energy of activation $\Delta \vec{G}_c^{0\#}$ plus the electrical contribution $\beta F \Delta \phi^{\#}$.

$$\Delta \vec{G}_c^{0\#} = \Delta \vec{G}^{0\#} + \beta F \Delta \phi$$

Thus, the rate $\vec{V}_e$ of electronation reaction under the influence of the electric field can be written in any one of the following forms,

$$\vec{V}_e = \frac{kT}{h} C_{A^+} e^{-\Delta \vec{G}^{0\#}/RT}$$

$$= \frac{kT}{h} C_{A^+} e^{-\Delta \vec{G}_e^{0\#}/RT} e^{-\beta F \Delta \phi/RT}$$

$$= \vec{V}_c \, e^{-\beta F \Delta \phi/RT}$$

$$= \vec{K}_c \, C_{A^+} \, e^{-\beta F \Delta \phi/RT}$$

This rate $\vec{V}_e$ is the number of moles of positive ions reacting per second by crossing unit area of the interface. When this is multiplied by the charge per mole of positive charges, one obtains the electronation current density $\vec{i}$.

$$\vec{i} = F\vec{V}_e = F\vec{K}_c C_{A^+} \, e^{-\beta F \Delta \phi/RT}$$

$\vec{i}$ has dimensions of amperes per square centimetre. This exponential relationship, first established by Volmer and Erdey-Gruz but indicated earlier in a rough and inaccurate way by Butler, symbolizes the link between the electric field and the rate of electron transfer across the interface.

Figure 14.6 shows that small changes in the field at the electrified interface produce large change in current density $\left(\text{if } \beta = \dfrac{1}{2}\right)$. It can be shown that 120 mV change in $\Delta\phi$ produces a tenfold change in $i$.

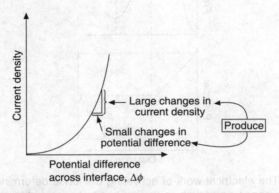

**FIGURE 14.6** The exponential nature of the current potential dependence.

If the metal is not connected to any other source of charge, every electronation of $A^+$ ions, charges the metal less negatively and the solution less positively, decreases the potential difference and field across the interface, increases the electrical work of activating the ion to the top of the barrier, decreases the electrical factor $e^{-\beta F \Delta\phi/RT}$ and reduces the rate of the electronation reaction. The larger the number of electrons transferred, the smaller is the attracting electric field and the smaller is the rate of the reaction. It looks, as if after a sufficient number of electron transfer, the electronation reaction should slow down or stop $\vec{V}_e = 0$.

## 14.5 THE EQUILIBRIUM EXCHANGE–CURRENT DENSITY ($i_0$)

At the metal–solution interface in its equilibrium state, there is no net current, no net electronation, and no net de-electronation, no substance produced and no change in potential difference or field across the interface. The current corresponding to these reactions are equal in magnitude and opposite in direction.

$$\vec{i} = F\vec{K}_c C_{A^+} e^{-\beta F \Delta\phi_e/RT} = \overleftarrow{i} = F\overleftarrow{K}_c C_D e^{(1-\beta)F\Delta\phi_e/RT} \tag{14.1}$$

where $\Delta\phi$ is absolute potential difference across the interface at equilibrium.

Significance of these equilibrium current is that it tells about quantitative measure of the rate of the reaction. It express in terms of numerical magnitudes, the rates of the two-way electron traffic between the electrode and particles in the electrolyte, when there is no net charge transport from one phase to the other. Hence, as suggested by Bulter, the individual electronation and de-electronation currents underlying the state of equilibrium can be designated by the same term. Such a magnitude is designated the equilibrium exchange current density ($i_0$).

$$i_0 = \vec{i} = F\vec{K}_c C_{A^+} e^{-\beta F \Delta\phi_e/RT} = \overleftarrow{i} = F\overleftarrow{K}_c C_D e^{(1-\beta)F\Delta\phi_e/RT} \tag{14.2}$$

Exchange current densities reflect the kinetic properties, i.e. rate at equilibrium of the particular interfacial systems concerned and thus can vary from one reaction to another and from one electrode material to another by many orders of magnitude. It was realized that, when equal numbers of identical particles moved in opposite directions between two parallel planes, their motions could not be detected. The equilibrium exchange current densities cannot therefore be directly and simply measured, because there is no net current at equilibrium.

## 14.6 THE NONEQUILIBRIUM DRIFT–CURRENT DENSITY ($i$)

The nonequilibrium drift–current densities $i$ is given by the difference between the de-electronation $\overleftarrow{i}$ and electronation $\overrightarrow{i}$ current.

$$i = \overleftarrow{i} - \overrightarrow{i} \qquad (14.3)$$

Here, putting the de-electronation current densities $\overleftarrow{i}$, first is meant to imply that when the magnitude of $\overleftarrow{i}$ is greater than the magnitude of $\overrightarrow{i}$, the net current $i$ is taken as positive. Hence, when there is net flow of electrons from solution to metal (i.e. net de-electronation) the net current is taken as positive.

$$i = \overleftarrow{i} - \overrightarrow{i} = F\bar{K}_c C_D e^{(1-\beta)F\Delta\phi/RT} - F\bar{K}_c C_{A^+} e^{-\beta F\Delta\phi/RT} \qquad (14.4)$$

where $\Delta\phi$ is the non equilibrium potential difference across the interface ($\Delta\phi \neq \Delta\phi_e$) corresponding to the current density $i$. One can split this nonequilibrium $\Delta\phi$ into the equilibrium potential difference $\Delta\phi_e$ and another portion, namely, the extra part $\eta$ by which the potential of the electrode departs from that at equilibrium, i.e. $(\Delta\phi - \Delta\phi_e) = \eta$ and write

$$\Delta\phi = \Delta\phi_e + (\Delta\phi - \Delta\phi_e) \Rightarrow \Delta\phi_e + \eta \qquad (14.5)$$

One can now write a net current density.

$$i = \overleftarrow{i} - \overrightarrow{i} = \{F\bar{K}_c C_D e^{(1-\beta)F\Delta\phi_e/RT}\} e^{(1-\beta)F\eta/RT} - \{F\bar{K}_c C_{A+} e^{-\beta F\Delta\phi_e/RT}\} e^{-\beta F\eta/RT} \qquad (14.6)$$

The two terms inside the brackets are simply the expressions for the equilibrium exchange current density ($i_0$).

Hence a convenient way of writing is

$$i = i_0 [e^{(1-\beta)F\eta/RT} - e^{-\beta F\eta/RT}] \qquad (14.7)$$

This is rather fundamental equation in electrodics. It may be termed the *Bulter–Volmer equation*. It shows how the current density across a metal solution interface depends on the difference $\eta$ between the actual nonequilibrium and equilibrium potential differences. Small changes in $\eta$ produce large changes in $i$.

## 14.7 THE OVERPOTENTIAL ($\eta$)

A potential difference in the case of a linear potential drop is equal to an electric field times distance. Thus,

$$\Delta\phi = \ell X \qquad (14.8)$$

where $X$ is the electric field and $\ell$ is distance between the metal surface and the locus of centres of the particles positioned for reaction on the solution side of the interface. Similarly, for the equilibrium potential.

$$\Delta\phi_e = \ell X_e \tag{14.9}$$

Thus,

$$\eta = \Delta\phi - \Delta\phi_e = \ell(X - X_e) = \ell\delta X \tag{14.10}$$

and

$$i = i_0[e^{(1-\beta)F\ell\delta X/RT} - e^{-\beta F\ell\delta X/RT}] \tag{14.11}$$

Note that as the current $I$ flows through the solution, there appears an additional potential difference $IR$ between the two electrodes of an actual cell. This acts as a driving force for the ionic motion required in the process, i.e. force required to give the ions in solution a preferential drift between electrodes. Equation (14.11) shows that the equilibrium field (corresponding to which $\delta X = 0$) cannot produce a net current, only an excess field ($\delta X \neq 0$) can drive a current $i$. We know that, for every flow or flux, there must be a driving force. The flux is the current density, the driving force is the excess field ($\delta X \neq 0$) acting on the charge concerned. Why should the driving force be an excess field and not just a field? In the expression for the conduction current through an electrolytic solution, one found that current density $i$ was driven by the field, not by the excess field. The reason is simple, in conduction by ionic migration, this entire field inside the electrolyte arises from the externally applied field. Switch off the externally applied field and the net potential drop inside the electrolyte collapses to zero and so does the ion migration or conduction current. This, however, is not the case at the interface. If one switches off the externally applied field the excess field $\delta X$ and, therefore, the current drops to zero but the field at the interface does not vanish. The equilibrium field still remains. It drives, the electronation and de-electronation current densities at an equal and opposite rate, i.e. gives no net current.

In case of interfaces, therefore, the net current is driven only by the excess field. This excess field $\delta X$ is best termed the current producing field.

When an externally driven electrochemical system or cell is considered, the excess potential $\eta$ (excess with reference to the equilibrium potential difference $\Delta\phi_e$) is the potential difference that drives the current, it is the current producing potential. On the other hand, if the system is a self-driving electrochemical system or cell then the current driven through the external load generates an excess potential $\eta$; this is a current produced potential. The term overpotential is used to refer both to the current producing potential $\eta$ in a dirven system and to the current produced potential $\eta$ in a self-driving cell.

## 14.8 SOME GENERAL AND SPECIAL CASES OF BUTLER-VOLMER EQUATION

All relations derived so far cover only one, the simplest, of all possible cases: a single step single electron exchange reaction. In more complex cases, important changes appear. Even from this simple case, much can still be learned. A better feel for the indications of the Bulter–Volmer relation is obtained by plotting $i$ against $\eta$. The $i$ versus $\eta$ curve so obtained looks much like the plot of a hyperbolic sine function (Figure 14.7).

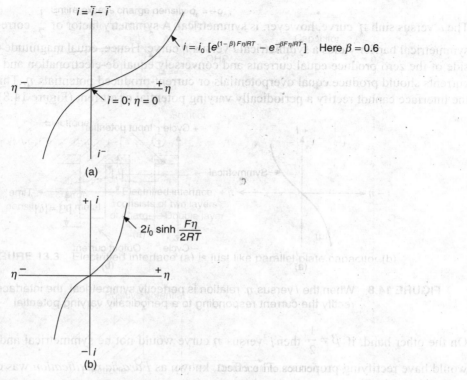

**FIGURE 14.7** (a) Dependence of current density on overpotential. (b) Shape of a hyperbolic sinh function.

*The hyperbolic sine function is explained as follows:* The symmetry factor is about $\frac{1}{2}$ then Eq. (14.7) becomes,

$$i = i_0[e^{(1-\beta)F\eta/RT} - e^{-\beta F\eta/RT}]$$
$$i = i_0[e^{+F\eta/2RT} - e^{-F\eta/2RT}] \tag{14.12}$$

On multiplying and dividing by 2, we get

$$i = 2i_0 \frac{(e^{+F\eta/2RT} - e^{-F\eta/2RT})}{2} \tag{14.13}$$

$$i = 2i_0 \frac{(e^{+x} - e^{-x})}{2}, \text{ where } x = \frac{\eta F}{2RT} \tag{14.14}$$

and since $\dfrac{e^{+x} - e^{-x}}{2} = \sinh x$

Therefore,
$$i = 2i_0 \sinh \frac{\eta F}{2RT} \tag{14.15}$$

The $i$ versus sinh $\eta$ curve, however, is symmetrical. A symmetry factor of $\frac{1}{2}$ corresponding to a symmetrical barrier yields a symmetrical $i$ versus $\eta$ curve. Hence, equal magnitude of $\eta$ on either side of the zero produce equal currents and conversely equal de-electronation and electronation currents should produce equal overpotentials or current-produced potentials $\eta$. This means that, the interface cannot rectify a periodically varying potential or current (Figure 14.8).

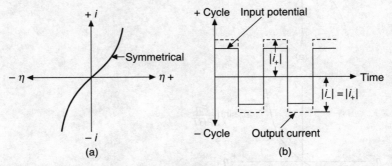

**FIGURE 14.8** When the $i$ versus $\eta$, relation is perfectly symmetrical, the interface cannot rectify the current responding to a periodically varying potential.

On the other hand, if $\beta \neq \frac{1}{2}$ then $i$ versus $\eta$ curve would not be symmetrical and the interface would have rectifying properties. The effect, known as *Faradic rectification* was discovered by Doss (Figure 14.9).

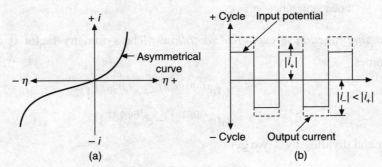

**FIGURE 14.9** If symmetry factor is different from $\frac{1}{2}$ the $i$ versus $\eta$, curve is asymmetrical and there is Faradic rectification effect or periodic varying potential.

The hyperbolic sine function has two interesting limiting cases. The first limiting case is when the overpotential $\eta$ or the excess field $\delta X$ numerically large. This is high overpotential or high field approximation. Under these conditions of large $\eta$ (if the example of a net de-electronation reaction is taken),

$$e^{F\eta/2RT} \gg e^{-F\eta/2RT} \tag{14.16}$$

and, since the $e^{-F\eta/2RT}$ terms tends to zero

$$2\sinh\frac{\eta F}{2RT} \approx e^{F\eta/2RT} \qquad (14.17)$$

Hence, under high fields, the Bulter–Volmer equation reduces to

$$i = i_0 e^{F\eta/2RT} \qquad (14.18)$$

that is the current density increases exponentially with the overpotential $\eta$ or with the driving force of the excess electric field across the double layer.

The second limiting case is when the overpotential $\eta$ or the excess field $\delta X$ is small. This is low overpotential or low field approximation. Under these conditions of small $\eta$ one can consider that $(F\eta/2RT) \ll 1$ and use the approximation.

$$\sinh\frac{\eta F}{2RT} \approx \frac{F\eta}{2RT} \qquad (14.19)$$

The low field approximation thus reduces the Bulter–Volmer equation to special case

$$i = i_0 \frac{F\eta}{RT} \qquad (14.20)$$

a linear relationship between the current density $i$ and the overpotential $\eta$ or driving force $\delta X$.

What excess fields or overpotentials, $\eta$ are low and what fields are high? What are quantitative criteria of low and high fields (or low and high overpotentials)? Consider the high field approximation. It is based on

$$e^{F\eta/2RT} \gg e^{-F\eta/2RT} \qquad (14.21)$$

Let it be assumed that right-hand side of Eq. (14.21) should be less than 1% of the left-hand side. Then, the condition for the high field approximation is

$$\frac{F\eta}{RT} > 2\ln_e 10 \qquad (14.22)$$

$$\eta = 0.12 \text{ V} \qquad (14.23)$$

where $2.303\, RT/F$ is equal to 0.058 at 298 K. Hence, when the interfacial potential difference $\Delta\phi$ exceeds the equilibrium potential $\Delta\phi_e$ by about 0.120 V for a one-electron transfer process, one can use the exponential $i$ versus $\eta$ high field law with an applicability of about 99%.

The condition for low field approximation is $F\eta/2RT \ll 1$. Let this be taken to be

$$\frac{F\eta}{RT} < \frac{1}{5} \qquad (14.24)$$

Then $\eta < 0.01$ V

Hence, when the overpotential $\eta$ is about 0.01 V or less for one-electron transfer reaction, the linear $i$ versus $\eta$ law can be used with good justification.

Hence, low field approximation implies that the interface is in a near equilibrium condition, and the high field approximation implies that the interface has been pushed far away from equilibrium.

## 14.8.1 The High-Field Approximation: The Exponential $i$ Versus $\eta$ Law

Bulter–Volmer equation contains two terms, one of them representing the de-electronation current density $\overleftarrow{i}$ and the other the electronation current density $\overrightarrow{i}$.

$$i = \overleftarrow{i} - \overrightarrow{i}$$

where, $\overleftarrow{i} = i_0 e^{(1-\beta)F\eta/RT}$ and $\overrightarrow{i} = i_0 e^{-\beta F\eta/RT}$ (14.25)

What happens when $\eta$ is increases? The electronation current density $\overrightarrow{i}$ decreases and deelectronation current-density $\overleftarrow{i}$, increases. When $\eta$ is large enough $\overleftarrow{i} \gg \overrightarrow{i}$ and the $\overrightarrow{i}$ becomes so small that it can be dropped out of the expression. Thus, the high field approximation of the Bulter–Volmer equation (valid at $\eta$ greater than about 0.10 V) yields

$$i = i_0 e^{(1-\beta)F\eta/RT}$$ (14.26)

The error involved in replacing the Bulter–Volmer equation with Eq. (14.26) as a function of potential is shown in Table 14.1.

**TABLE 14.1** High field approximation of the Bulter–Volmer equation $\beta = 0.5$, $i_0 = 1$ mA cm$^{-2}$ and $T = 25°C$

| $\eta$(V) | $F\eta/RT$ | $i_0 e^{(1-\beta)F\eta/RT}$ | $i$ | % age error in the high field approximation |
|---|---|---|---|---|
| 0 | 0 | 1.0 | 0 | Infinity |
| 0.001 | 0.039 | 1.02 | 0.04 | + 2450 |
| 0.005 | 0.195 | 1.10 | 0.195 | + 464 |
| 0.010 | 0.390 | 1.21 | 0.39 | + 210 |
| 0.020 | 0.780 | 1.48 | 0.80 | + 85 |
| 0.030 | 1.17 | 1.79 | 1.23 | + 45.5 |
| 0.050 | 1.95 | 2.65 | 2.27 | + 16.7 |
| 0.100 | 3.90 | 7.03 | 6.89 | + 2.0 |
| 0.20 | 7.8 | 49.4 | 49.38 | < + 0.1 |

For convenience of plotting, it is useful to put Eq. (14.26) into a logarithmic form by taking logarithms.

$$\log i = \log i_0 + \frac{F\eta}{RT}(1-\beta)\log_e e$$ (14.27)

$$\eta = -\frac{RT}{(1-\beta)F}\ln i_0 + \frac{RT}{(1-\beta)F}\ln i_i$$ (14.28)

$$\eta = -\frac{2.303\,RT}{(1-\beta)F}\log i_0 + \frac{2.303\,RT}{(1-\beta)F}\log i$$ (14.29)

Thus, when the electronation current $\overrightarrow{i}$ becomes too small for consideration, the current producing or the current produced potential (the overpotential) is a linear function of log $i$. If the potential difference across the interface is plotted against the logarithm of net current density, a straight line plot is obtained.

When $\eta$ values are obtained at various currents, it is possible to obtain $\eta$ versus log $i$ plots. Such $\eta$ versus log $i$ plots are known as *Tafel lines*, in recognition of Tafel who first published measurements, which showed the behaviour of Eq. (14.26). An example is shown in Figure 14.10.

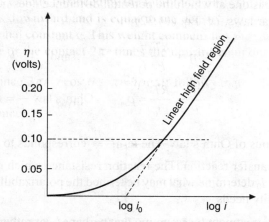

**FIGURE 14.10** A typical Tafel line for one electron transfer electrode reaction showing the experimental relationship at high over potentials which makes the relation between $\eta$ and log $i$ linear.

It will be noticed that the intercept $a$ in the Tafel plot permits a determination of the equilibrium exchange current density $i_0$. The slope $b$ has a meaning given by Eq. (14.29) only for the particular case of simple one-step electron transfer reactions.

## 14.8.2 The Low Field Approximation: The Linear $i$ Versus $\eta$ Law

The linear low field law is obtained by expanding the exponential and since $\eta$ is by definition small in this approximation by retaining only the first two terms of the expansion of each exponential term.

$$i = i_0 [e^{(1-\beta)F\eta/RT} - e^{-\beta F\eta/RT}] \tag{14.30}$$

$$i \approx i_0 \left[1 + \frac{(1-\beta)F\eta}{RT} - 1 + \frac{\beta F\eta}{RT}\right] \tag{14.31}$$

$$i = i_0 \frac{F\eta}{RT} \tag{14.32}$$

This special case of the electrodic reaction shows that electrodic reactions across interfaces exhibit ohmic behaviour under low field conditions. The current density is proportional to the current producing or current produced potential difference (the overpotential, $\eta$). One could write

$$i = \sigma_{m/s}(\eta) = \sigma_{m/s} \ell \delta X \tag{14.33}$$

where $\sigma_{m/s}$ is the conductivity of the metal-solution interface. Equation (14.33) is another example of the fact that near equilibrium, all flows can be taken to be proportional to their corresponding driving forces. This linear expression serves to emphasize that the potential producing the current in electrodic reactions driven from an outside source is the excess potential difference $\eta$.

## 14.9 NONPOLARIZABLE AND POLARIZABLE INTERFACE

Nonpolarizable interface is one at which the potential difference does not change easily with the consideration of the linear laws, i.e. Eq. (14.32)

$$\eta = \frac{RT}{F}\frac{i}{i_0} \qquad \text{and} \qquad i = i_0 \frac{\eta F}{RT}$$

$$\frac{\eta}{i} = \frac{RT}{Fi_0} = \rho_{m/s} \qquad \text{Ohm's law } \frac{V}{I} = R$$

The linear law is analogous of Ohm's law. The term $\dfrac{\eta}{i}$ corresponds to the resistance $\rho_{m/s}$ of the interface to the charge transfer reaction. The reaction resistance which mainly depends upon the exchange current density $i_0$ determine what may be termed the polarizability, i.e. what overpotential a particular current density needs.

One can improve the treatment by assuming that perhaps $i_0$ or rather the concentration inside $i_0$ may not be a constant with current whereupon it is better to use the differential resistance which for $\eta$ in the linear region is,

$$\left(\frac{\partial \eta}{\partial i}\right)_{C_A^+ CDT} = \frac{RT}{Fi_0} = \rho_{m/s}$$

Now observe what happens if the equilibrium exchange–current density $i_0$ tends to very high values, i.e. towards infinity. As $i_0 \to$ infinity $\rho_{m/s} = \left(\dfrac{\partial \eta}{\partial i}\right)_{C_A^+ CDT} \to 0$. Then, the slope of the $\eta$ versus $i$ curve is zero, i.e. despite the passage of a current density $i$ across the interface, the overpotential tends to be zero. The interface remains virtually at its equilibrium potential difference. This is precisely the behaviour of an ideally nonpolarizable interface. The higher the values of $i_0$ are, the less does the potential difference across an interface departs from the equilibrium value on the passage of a current. The (hypothetical) ideally nonpolarizable interface (with $i_0 \to \infty$), therefore is always at the equilibrium potential.

An exchange current density of infinity is of course an idealized case. All values of $i_0$ must be finite, which means that all interfaces show some degree of polarizability. The larger the values of $i_0$ are, the greater is the current density $i$ required to produce a given change of potential from the equilibrium value $\Delta \phi_e$ characteristic of the given reaction.

Similarly, one can conceive of the other extreme, the case of $i_0 \to 0$. Here $\dfrac{\partial \eta}{\partial i}$, the polarizability and the reaction resistance $\rho_{m/s}$ becomes infinite. The potential departs from the equilibrium values even with a very small current density leaking across the interface. This is, however, is precisely what could be expected for a highly polarizable interface, its potential is easily changeable.

Thus, the concepts of polarizable and nonpolarizable interface are quantified. The value $i_0 \to 0$ is the idealized extreme of a polarizable interface; $i_0 \to$ infinity is the idealized extreme of a nonpolarizable interface. The concept of the polarizability $\dfrac{\partial \eta}{\partial i}$ also shows, why instruments with high input impedance must be used by those making measurements of the potential differences across electrochemical systems or cells (Figure 14.11).

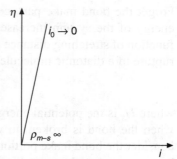

**FIGURE 14.11** At low $i_0$ values, the interface has high resistance or polarizability.

## 14.10 PHYSICAL MEANING OF SYMMETRY FACTOR $\beta$

The Bulter–Volmer equation (BVE) is

$$i = i_0 [e^{(1-\beta)F\eta/RT} - e^{-\beta F\eta/RT}]$$

BVE connects two basic aspects of charge transfer at an electrified interface, namely, equilibrium current, which represents the rate of transformation of substances at the interface without accelerating the effect of overpotential, and the electrical effects, that results from the applications of overpotential. Thus, equation embraces the chemical and electrical aspects of charge transfer.

Two quantities of interest in BVE are exponents $\beta$ and $\eta$. The quantity $\eta = \Delta\phi - \Delta\phi_e$, which determines how far the interfacial potential difference has departed from equilibrium, can be externally controlled. In a driven electrochemical system, it can be chosen at will by turning a knob on external power supply and in a self-driving cell, it depends on the current drawn from the cell and this current in turns depends on the resistance of the external load. Thus, $\eta$ is at the command of the experimenter.

The symmetry factor $\beta$, however, is an intrinsic characteristic of the given charge transfer reaction, at the given interface. It determines, what fraction of the electrical energy affects the rate of electrochemical transformation. In short, $\beta$ depends upon the interfacial reaction and not on experimenter's manipulation.

## 14.11 POTENTIAL–ENERGY DISTANCE RELATIONS OF PARTICLES UNDERGOING CHARGE TRANSFER

A simple transfer reaction will be considered, for example, the electronation of a hydronium ion, i.e. a proton hydrated by one water molecule $H^+ - H_2O$.

$$M(e) + H^+ - H_2O \to M - H + H_2O$$

The progress of the charge transfer reaction involves the breaking of the bond between $H^+$ and $H_2O$, and the making of a bond between the metal, M and the electronated electron acceptor, H.

Forget the bond make part, consider only the bond break portion. As this bond stretches, the energy of the system increases and finally the bond is broken. One can plot the energy as a function of stretching distance (Figures 14.12 and 14.13). A simple relation for the stretching and rupture of a diatomic molecule is known from spectroscopy. This is known as *Morse equation*.

$$U_x = D(1 - e^{-a\Delta x})^2 \tag{14.34}$$

where $U_x$ is the potential energy of the system. $\Delta x$ is stretching distance, which, goes to infinity when the bond is broken, in which case $e^{-a\Delta x} \to 0$ and $U = D$, the dissociation energy. Now consider the bond make portion. Again one has a Morse type of curve. Two Morse curve can now be associated as shown in Figure 14.14, so that one obtains a single diagram.

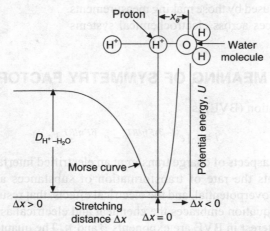

**FIGURE 14.12** A plot of potential energy with distance between atoms reflects the increase in energy as the bond is stretched.

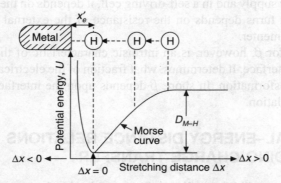

**FIGURE 14.13** Change in potential energy as hydrogen atom is approaching a metal surface.

The middle section of potential energy distance curve has the shape of a potential energy barrier. This barrier represents the energy of the system as the particle $H^+$ jumps from the OHP to the metal.

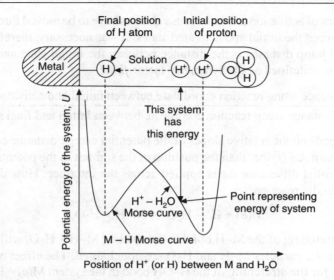

**FIGURE 14.14** Minima of two potential energy curve are positioned at proper place, the curve intersects, thus forming the potential energy barrier.

When the $H^+ - H_2O$ bond is stretching, the influence of the metal on the energy of the system has been ignored. Similarly, one has ignored the influence of the water molecules on the energy as the M–H bond is stretched. Thus, superposition of two Morse curve assumes that the energy of a three-body system (M, $H^+$, and $H_2O$) with three-body interaction can be obtained by properly superposing the energies from the two-body system each involving two-body interactions. If one takes into account the effects of the third atom on the energy in all configurations, particularly energy near the intersection of the Morse curves. This contributes (Figure 14.15) to the *rounded top* of an actual barrier obtained by sectioning the corresponding potential energy surface.

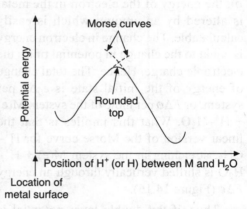

**FIGURE 14.15** Rounded top energy barrier as a result of three-body interaction.

## 14.12 A SIMPLE PICTURE OF THE SYMMETRY FACTOR

The charge transfer reaction was roughly pictured as the jump of an electron acceptor towards the electrode during which, somewhere in the route, an electron jumped to the particle and completed its job of electronation. Representing the energy of the system by a point on an energy–distance curve, once this point climbed to the top of the barrier (activated state), the rest of the jump was assured (automatic). But this climb to the peak requires some work to be done. The chemical activation work is altered by the presence of an electric field. More electrical work is done by or on the ion in climbing the barrier in the presence of the electric field at the interface than without it. The electrical work of activation arises, because charges have to be moved through the difference of the potential between the initial and activated states. It was necessary, therefore, to know what fraction of the total jump distance is the distance between the initial state and the barrier peak. This distance ratio was defined as the symmetry factor $\beta$, i.e.

$$\beta = \frac{\text{Distance along reaction coordinate between initial and activated state}}{\text{Distance along reaction coordinate between initial and final state}}$$

The value of $\beta$ depends on the relative slopes of the potential energy distance curves representing the energies of the particles (rather than the position of the summit of the potential energy barrier). Suppose that a potential difference $\Delta\phi$ is applied across the interface. How does this affect the barrier for the electrodic reaction:

$$M(e) + H^+ - H_2O \rightarrow M-H + H_2O$$

The curve for the stretching of the M–H bond in the system, M–H + $H_2O$, will not be influenced by the field because the particles M–H and $H_2O$ are not charged. The effect of the electric field on the linear curve for the stretching of the H–OH bond in the system $M(e) + H^+-OH_2$, has to be thought. When the potential difference across the interface is changed (from 0 to $\phi$), the energy of the electron in the metal is altered by an amount which is easily calculatable. The change in electron energy is equal to the change in potential times the electronic charge. Hence, The total change of energy of the initial state is $e_0\Delta\phi$ per system, or $F\Delta\phi$ per mole of the system $M(e) + H^+-H_2O$. What this implies is that the linear version of the Morse curve for $H^+-OH_2$, stretching in the system $M(e) + H^+-H_2O$ is shifted vertically through an energy $F\Delta\phi$ (Figure 14.16).

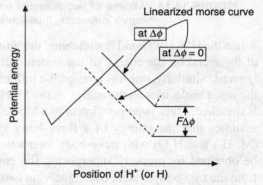

**FIGURE 14.16** When potential difference is changed from 0 to $\Delta\phi$, the Morse curve of the initial state is shifted vertically by the amount of electrical energy $F\Delta\phi$.

Thus, if the double layer potential is initially $\Delta\phi_e$ (i.e. interface is at equilibrium) and then the potential is changed to $\Delta\phi$, the Morse curve for the initial state is shifted vertically through an energy $F(\Delta\phi - \Delta\phi_e)$ or $F\eta$. The critical activation energy for the reaction is altered from $E_e^{\pm}$ at equilibrium to $E_\eta^{\pm}$ at the overpotential $\eta$. The difference $\Delta E^{\pm}$ between the two activation energies has resulted from the electrical energy $F\eta$ that has been introduced into the reaction. What is the relationship between $\Delta E^{\pm}$ and $F\eta$? The change $\Delta E^{\pm}$ in activation energy decides the net current output; the $F\eta$ is the input electrical energy channeled into the interface. Now the question is: How much did the activation energy decrease for the given energy input $F\eta$.

The question is answered by solving geometry in Figure 14.17.

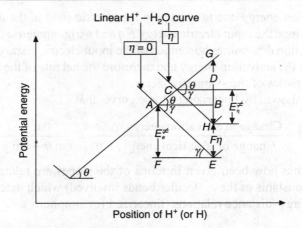

**FIGURE 14.17** The change in activation energy as a consequence of vertical shift of linear curve.

$$AB = FE = \frac{E_e^{\pm}}{\tan \gamma} \quad \text{(from } \triangle AEF\text{)}$$

and

$$AB = \frac{GE - E_e^{\pm}}{\tan \theta} \quad \text{(from } \triangle ABG\text{)}$$

and therefore,

$$E_e^{\pm} = \frac{\tan \gamma}{\tan \theta}(GE - E_e^{\pm}) \tag{14.35}$$

Further,

$$CD = \frac{E_\eta^{\pm}}{\tan \gamma} \quad \text{(from } \triangle CDH\text{) and } CD = \frac{GH - E_\eta^{\pm}}{\tan \theta} \text{ (from } \triangle CDG\text{)}$$

Hence,

$$E_\eta^{\pm} = \frac{\tan \gamma}{\tan \theta}(GH - E_\eta^{\pm}) \tag{14.36}$$

By making use of Eqs. (14.35) and (14.36), it follows that change in activation energy

$$\Delta E^{\pm} = E_e^{\pm} - E_\eta^{\pm}$$

$$\Delta E^{\pm} = \frac{\tan \gamma}{\tan \theta}(GE - GH) - (E_e^{\pm} - E_\eta^{\pm})$$

$$= \frac{\tan \gamma}{\tan \theta}(F\eta - \Delta E^{\pm})$$

$$\Delta E^{\pm} = \left(\frac{\tan \gamma}{\tan \theta + \tan \gamma}\right) F\eta \tag{14.37}$$

The change in activation energy due to change in the electric field in the double layer has been computed. It depends upon the input electric energy $F\eta$ and a trigonometric function which cannot exceed unity. This fraction determines how much of the input electric energy fed into the interface goes towards affecting the activation energy and therefore the net rate of the reaction. The fraction has the basic characteristics of the symmetry factor.

Thus, it has been shown, by linearizing Morse curve, that

$$\beta = \frac{\text{Change in activation energy, } \Delta E^{\pm}}{\text{Change of electrical energy, } F\eta} = \frac{\tan \gamma}{\tan \theta + \tan \gamma} \qquad (14.38)$$

The symmetry factor has now been given in terms of the slopes are related to those molecular quantities (i.e. force constants of the molecular bonds involved) which determines the shape and slopes of potential energy–distance relations (linearized for simplicity).

## 14.13 RELATIONSHIP BETWEEN $\beta$ AND OVER POTENTIAL $\eta$

It is clear that once a linear curve displaced vertically, it cannot, but yield a parallel shift of the curve and therefore a constant $\beta$. The apparent constancy of $\beta$ with potential is a result of the linearization of the potential energy–distance curves.

Figure 14.18 shows, however, that when the activation energy at equilibrium $E_e^{\pm}$ is large, i.e. for electrode reactions of low exchange-current density $i_0$ the slopes of the linear curves and hence, $\beta$ do not change significantly with overpotential. Such changes in $\beta$ become likely only for reactions which have very low equilibrium activation energies (i.e. very high $i_0$). One can take it, therefore, that for all but very fast electrode reactions, the symmetry factor $\beta$ will be independent of overpotential $\eta$ over a reasonably large range of potentials. At sufficiently high over potentials, the curves will be changed in relative position sufficiently that they begin to intersect at positions of differing curvature.

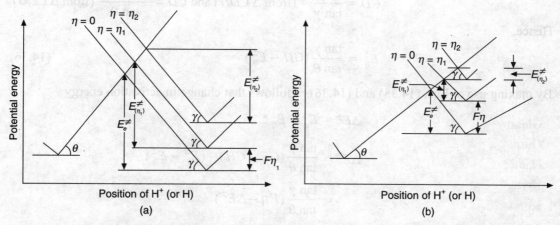

**FIGURE 14.18** The influence of over potential on the symmetry factor $\beta$: (a) reaction of low $i_0$: $\beta$ independent of $\eta$, (b) reaction of high $i_0$: $\beta$ dependent of $\eta$, tending to zero in the limit of high $\eta$.

## EXERCISES

1. Derive basic electrodic equation, i.e. Butler–Volmer equation.
2. Derive expression for rate of charge transfer under zero field.
3. What is the rate of electron transfer reaction under the influence of an electric field.
4. Derive equation for equilibrium exchange current density.
5. Derive equation for nonequilibrium exchange current density.
6. What is an overpotential? Can we measure overpotential?
7. Illustrate some general and special cases of Butler–Volmer equation.
8. Define high field and low field approximations of Butler–Volmer equation.
9. Define polarizable and nonpolarizable interface.
10. What is physical meaning of symmetry factor ($\beta$).
11. Draw potential energy distance relations of particles undergoing charge transfer.

## SUGGESTED READINGS

Allmand, A.J., *Principles of Applied Electrochemistry*, Longmans, Green, New York, 1912.

Balzani, V., *Electron Transfer in Chemistry*, Vols. 1–5, Wiley, New York, 2001.

Bockris, J.O.M. and A.K.N. Reddy, *Modern Electrochemistry: An Introduction to an Interdisplinary Area*, Vols. 1–2, Plenum Press, New York, 1970.

Bockris, J.O.M., A.K.N. Reddy, and M. Gamboa-Aldeco, *Modern Electrochemistry 2A: Fundamentals of Electrodics*, 2nd ed., Plenum Press, New York, 2000.

Bokris, J.O.M. and A.K.N Reddy, *Modern Electrochemistry 2B: Electrodics in Chemistry, Engineering, Biology and Environmental Science*, 2nd ed., Plenum Press, New York, 2000.

Bokris, J.O.M. and B.E. Conway, *Modern Aspects of Electrochemistry*, No. 3, Butterworths, London, 1964.

Brett, C.M.A. and A.M.O. Brett, *Electrochemistry: Principles, Methods and Application*, Oxford University Press, 1993.

Davis, C.W., *Electrochemistry*, Newnes, London, 1967.

Delahay, P., *Advances in Electrochemistry and Electrochemical Engineering*, Vol. 1, Wiley Interscience, New York, 1961.

Glasstone, S., *An Introduction to Electrochemistry*, East-West Press, New Delhi, 2003.

Gurney, R.W., *Ionic Process in Solution*, McGraw-Hill, New York, 1953.

Harned, H.S. and B.B. Owen, *Physical Chemistry of Electrolytic Solutions*, Reinhold, New York, 1950.

Kortum, G. and J.O.M. Bockris, *Textbook of Electrochemistry*, Vols. 1 and 2, Elsevier, Amsterdam, 1951.

Koryta, J. and J. Dvorak, *Principles of Electrochemistry*, Wiley, New York, 1987.

Kubasov, V. and A. Zaretsky, *Introduction to Electrochemistry*, Mir Publishers, Moscow, 1987.

Mac Innes, D.A., *Principles of Electrochemistry*, Dover, New York, 1961.

Potter, E.C., *Electrochemistry, Principles and Applications*, Macmillan, 1956.

Rochow, E., *Electrochemistry*, Reinhold, New York, 1965.

# 15
# Contact Adsorption on the Electrode

## 15.1 INTRODUCTION

Now the questions are: Are hydrated ions held on a hydrated electrode, i.e. an electrode covered with a sheet of water molecules shows arrangement O (Figure 15.1)? or the ions stripped of their solvent sheath in intimate contact with a bare electrode – arrangement I ? Are both these arrangements O and I realized in practice? Or does one ionic species, e.g. $K^+$ ion resort to arrangement O and another ionic species, e.g. chloride ion make use of arrangement I? If so, what determines which ion prefers one arrangement to the other? What are the forces, which influence the sticking of ions to electrode?

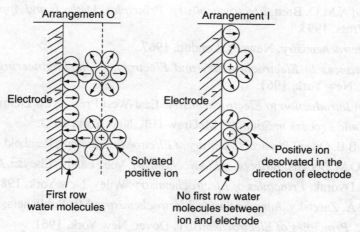

**FIGURE 15.1** Arrangements O and I.

## 15.2 SITUATION PRESENT AT METAL ELECTRODE

It is possible to show that in absence of directing forces, at least 70% of the metal surface must be covered with water molecules. All one has to do is to estimate the population density $n_w$ (no./cm²) of water molecules on a plane in bulk water and divide this by the number of sites $n_s$ per square centimetre on the metal surface. The quantity $n_w$ is roughly estimated by taking the two-third power of the number of water molecules per cubic centimetre, which is equal to Avogadro's number divided by the molar volume $V_m$. Thus,

$$n_w = \left(\frac{N_A}{V_m}\right)^{2/3} = \left(\frac{N_A \rho}{M}\right)^{2/3} \tag{15.1}$$

where $M$ and $\rho$ are the molecular weight and density of water, respectively. The number of sites per square centimetre can be taken as equal to the number of metal atoms per square centimetre. The fraction $\theta_w$ of the surface covered turns out (by inserting the appropriate numerical value into this zeroth-approximation approach) to be about 0.7.

## 15.3 METAL–WATER INTERACTIONS

There are, however, forces operating between water molecules and the metal electrode. The actual coverage with water molecules may therefore be even more than 70%. What are some of these forces? Firstly, there are the *image forces*. A water molecule is asymmetric, it is a dipole. Each of these charges induces a charge on the metal. But, since there are two charges in a dipole, one can think of an image dipole. Thus, the forces between a dipole and a metal can be computed by replacing the metal with an image dipole and considering the dipole–dipole interaction.

There is another type of the force between water molecules and the metal, *dispersion forces*. Dispersion force (London forces) can be seen classically as follows. Every atom has an instantaneous dipole moments, of course, the time average of all these dipole moments is zero. This instantaneous dipole will induce an instantaneous dipole in a contigous atom, and an instantaneous dipole–dipole force arises. When these forces are averaged over all the instantaneous electron configurations of the atoms, an attractive, nondirectional force arises, the *dispersion force* (Figure 15.2).

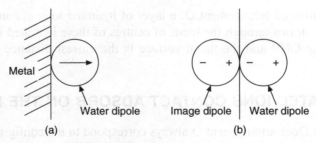

**FIGURE 15.2** (a) The interaction between a water molecule and a metal can be computed by (b) replacing the metal with an image dipole.

Dispersion forces are important at electrified interfaces. They contribute to adsorption at electrodes.

Apart from the image and dispersion forces, which attract water to the metal, there are also the forces, which induce chemical bonding. All this indicates that even an uncharged electrode has an attraction for water molecules and may overcome the forces, which bind water molecules into networks in a liquid phase.

## 15.4 THE ORIENTATION OF WATER MOLECULES ON CHARGED ELECTRODES

What happens to the water layer when the metal is charged? The charges on the metal will stimulate the water molecules to orient themselves (charge implies field, and dipoles tend to align with field). The net result of all these image, dispersion, chemical and orienting forces between an electrode and the water molecules adjacent to it is that the electrode is almost completely covered with a layer of oriented dipoles . However, it must not be imagined that the water molecules are unaffected by the presence of their neighbours.

After all, dipoles interact with dipoles. Hence, the oriented water molecules experience lateral interaction—a phenomenon which, affect the net number of water molecules oriented in one direction and therefore the value of the dipole potential $\Delta\chi$. Through, $\Delta\chi$ the potential difference across the interface and the structure of the interface will be affected.

## 15.5 CLOSEST APPROACH OF HYDRATED IONS TO A HYDRATED ELECTRODE

The ions in solution have been shown to solvate with a total number ($N_t$) of water molecules. Some of these water molecules are left behind when ions random walk and drift around. Others, however, the $n_1$ primary hydration molecules, resist the hydrogen bonding with the secondary water molecules and tolerate the lateral (dipole–dipole) repulsion of their like-oriented neighbours.

Thus, the ions wrapped in a primary hydration sheath, migrates upto the electrode. How close to the electrode can such a hydrated ion approach? In the first instance, it can proceed till the water molecules of the hydrated ion collide with the oriented water molecules of the hydrated electrode. The electron shells of the water molecule then start overlapping and repelling electron overlap interaction.

The ions have attained arrangement O, a layer of hydrated ions in contact with a solvated electrode. The plane drawn through the locus of centres of these hydrated ions is the OHP; the distance between the OHP and the metal surface is the closest distance of approach in this arrangement O.

## 15.6 DESOLVATED IONS CONTACT ADSORB ON THE ELECTRODE

Now the question is: Does arrangement O always correspond to the configuration of lowest free energy? Why ions cannot divest themselves (at least partly) of their primary water, nudge adsorbed water molecules away from electrode sites and come in contact with a bare electrode arrangement I.

Defining the locus of centres of these contact adsorbed ions as the IHP, the Question is: Do ions quit the OHP and populate the IHP. For this one must calculate the free energy change for (i.e. work done by) an ion to move from the OHP to IHP while at the same time displacing the appropriate number of adsorbed water molecules. If the free energy change is negative, ions will make this move.

## 15.7 THE FREE ENERGY CHANGE FOR CONTACT ADSORPTION

Approximate calculations of the free energy change involved in ions moving to the IHP have been made. The essence of these calculations consists in viewing the process of contact adsorption as a two-step process. In step I, a hole of area $\pi r_i^2$ ($r_i$ radius of contact adsorbing ion) must be swept free of water molecules in order to make room for the ion. Thus, a certain number of water molecules must be desorbed. In step 2, the ion strips itself a part of its solvent sheath and jumps into the hole.

In both the steps, the particles involve (water molecules in step 1 and the ion plus associated water in step 2) breaking off the old attachments and making new ones (change of enthalpy $\Delta H$) and also exchange old freedoms and restrictions for freedom and restrictions characteristic of the new neighbourhood (change of entropy, $\Delta S$).

These changes of interaction can be classified into three groups: (1) changes in water–electrode interactions, (2) changes in ion–electrode interactions (due to image, dispersion and electron overlap forces), and lastly (3) changes in ion–water interaction, i.e. changes in hydration. The enthalpy, entropy and free energy changes corresponding to these groups for various ions are shown in Table 15.1.

**TABLE 15.1** The enthalphy, entropy and free energy changes of three interacting groups

| Ion | Water–electrode interaction | | | Ion–electrode interaction | | | Ion-water interaction | | | Total |
|---|---|---|---|---|---|---|---|---|---|---|
| | $\Delta H$ | $\Delta S$ | $\Delta G$ | $\Delta H$ | $\Delta S$ | $\Delta G$ | $\Delta H$ | $\Delta S$ | $\Delta G$ | $\Delta G$ |
| $Na^+$ | 28.1 | 3.6 | 27.0 | −49.1 | 1.5 | −49.6 | 39.5 | 11.2 | 36.1 | +13.5 |
| $K^+$ | 21.1 | 6.7 | 19.1 | −48.2 | 2.0 | −48.8 | 33.9 | 5.2 | 32.3 | 2.6 |
| $Cs^+$ | 20.9 | 10.7 | 17.7 | −43.5 | 3.3 | −44.5 | 17.7 | −12.2 | 21.4 | −5.4 |
| $F^-$ | 38.1 | 6.7 | 36.1 | −47.1 | 1.4 | −47.5 | 35.0 | 6.4 | 33.1 | +21.7 |
| $Cl^-$ | 22.0 | 12.4 | 18.3 | −49.3 | 2.1 | −49.9 | 17.7 | −16.6 | 22.7 | −8.9 |
| $Br^-$ | 21.6 | 14.5 | 17.2 | −49.2 | 2.8 | −50.0 | 14.7 | −22.4 | 21.4 | −11.4 |
| $I^-$ | 21.9 | 18.0 | 16.5 | −49.2 | 3.6 | −50.3 | 10.9 | −32.6 | 20.7 | −13.1 |

It turns out that some ions loose free energy by moving to the IHP. These ions will therefore contact adsorb. Other ions would gain free energy by contact adsorbing, they will go no farther to the electrode than the OHP.

## 15.8 DEGREE OF CONTACT ADSORPTION

The changes of the enthalpy, entropy, etc. in table 15.1, reveal an important point about contact adsorption. When considered alone, the free energy change from water–electrode interactions is positive for all species. What tip over the total free energy change to negative values are the ion–electrode interactions (essentially dispersion and image interactions). But it is the group of ion⁻ water interactions, which gives no different ionic species, different signs of total free energy change and thus differing attitudes to contact adsorption—for the case of $Cs^+$, $Cl^-$, $Br^-$, and $I^-$ and against the case of $Na^+$, $K^+$ and $F^-$.

What are these ion–water interactions? They are hydration interactions. If the ions are strongly hydrated (with primary water), the changes in hydration free energy (required for contact adsorption) are too large, the transition to the IHP is not energetically worth while. Ions which are sufficiently and strongly hydrated do not contact adsorb. Now, hydration is much dependent on the radius of the ions, see for example, a plot of free energy of hydration versus ionic radius (Figure 15.3). The large ions ($Cs^+$, $Br^-$, $I^-$, $Cl^-$) are precisely those, which have negative free energy of contact adsorption. The smaller ions ($Na^+$, $K^+$ and $F^-$) tightly wrapped up in their solvent sheaths would not be expected to contact absorb.

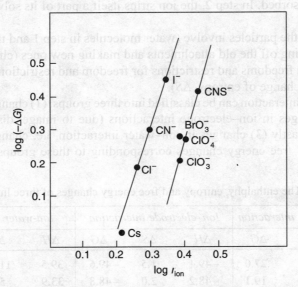

**FIGURE 15.3** The relation between contact adsorption and ionic radius.

The study of the forces operating in the inter-phase region has shown, therefore, that there are two versions of closest approach of ions to the electrode and thus two corresponding arrangements O and I. In one version, hydrated ions on the OHP touch the hydrated electrode, in the other, partially hydrated ions on the IHP are contact adsorbed onto a locally dehydrated electrode. Contact adsorption, however, is not for all ionic species, some do it and other do not. The main determining fact is: How easily may ions be partly dehydrated, i.e. what is the value of their free energy of hydration?

## 15.9 MEASUREMENT OF CONTACT ADSORPTION

The total surface excess $\Gamma_i$ of a particular species $i$ in an inter-phase can be obtained directly from electro-capillary measurement. If one wants, e.g. the surface excess of chloride ions, one obtains electro-capillary curve in solution of various hydrochloride acid concentration. A necessary condition, however, is to incorporate in the cell a non-polarizable electrode which allows, not chloride ions, but hydrogen ions to leak across its surface. Under these conditions, it has been shown that at constant cell potential $V$,

$$\left(\frac{\partial \gamma}{\partial \mu_{HCl}}\right)_{const.\ V} = \Gamma_i \tag{15.2}$$

Once the surface excess of the negative ions is obtained, it is easy to calculate the excess charge density $q^-$ due to the negative ions, from the relation

$$q^- = z_- F\Gamma_i = -F\Gamma_i \tag{15.3}$$

How is this $q^-$ distributed in the interface? Is it completely due to the contact adsorption of chloride ions? The answer to this question pertains to the location and distribution of charge, i.e. the structure of the interface. One approach is to take a close look at the table presenting the tendencies of various ionic species to contact adsorb (Table 15.1). It appears that the small cations do not contact adsorb. Thus, following structural assumption could be made in case of HCl solution: the $H_3O^+$ ions do not enter the IHP. But the OHP is the fence of the diffuse charge region, and therefore all the cationic excess charge $q^+$ in the inter-phase region must be in the Gouy-Chapman region.

One must get to know the magnitude of $q^+$. The slope of the electro-capillary curve furnishes the total value of the excess cationic and anionic charge $q_s$ on the solution side of the interface.

$$\left(\frac{\partial \gamma}{\partial V}\right)_{const.\ comp} = -q_m = q_s \tag{15.4}$$

This $q_s$ is composed of both positive and negative ions and have

$$q^+ = q_s - q^- \tag{15.5}$$

Having argued that $q^+$ is entirely in the diffuse layer, one can use the diffuse charge theory of Gouy–Chapman to get the potential $\psi_0$ at the OHP corresponding to the given potential difference at which the total surface excess of positive ions was determined. The following expression is used.

$$q_d^+ = \left(\frac{\varepsilon n^\circ kT}{2\pi}\right)^{1/2} (e^{-Z^+ e_o \psi_0 / kT} - 1) \tag{15.6}$$

Then, the calculated value of $\psi_0$ for the given applied potential can be used in a similar expression for the anionic diffuse charge, $q_d^-$.

$$q_d^- = \left(\frac{\varepsilon n^0 kT}{2\pi}\right)^{1/2} (e^{-Z^- e_o \psi_0 / kT} - 1)$$

The last step is to subtract this anionic charge in the diffuse region from the total surface excess of negative ions $\Gamma_i$ in the inter-phase expressed in the form of charge, i.e. $Z_- F \Gamma_i = q^-$. The result, of course, is the negative charge populating the IHP. This is the contact adsorbed charge $q^-_{CA}$ (CA means contact adsorbed).

By repeating the procedure (Table 15.2) at various cell potentials, i.e. various values of the charge on the electrode $q_m$, one can determine the variation in the amount of contact adsorbed charge $q^-_{CA}$ with $q_m$ (Table 15.2). A typical plot of $q^-_{CA}$ versus $q_m$ is shown in Figure 15.4. It shows unexpectedly varying behaviour of the capacity of the interface as a function of the potential difference across the interface or the charge on the metal.

**TABLE 15.2** Procedure for getting the contact adsorbed charge density $q^-_{CA}$.

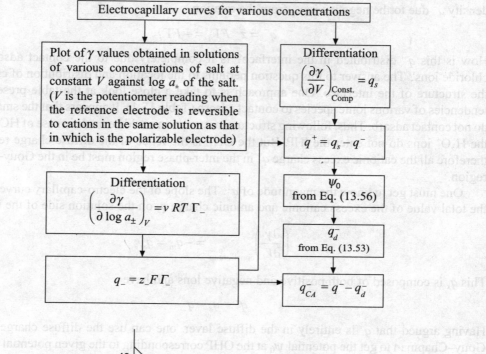

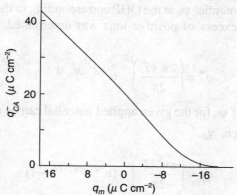

**FIGURE 15.4** A plot of contact adsorbed charge density $q^-_{CA}$ versus electrode charge density $q_m$.

## 15.10 CONTACT ADSORPTION, SPECIFIC ADSORPTION OR SUPER-EQUIVALENT ADSORPTION

The curve (Figure 15.4) shows contact adsorption of negative ions on a negatively charged metal surface. The term super-equivalent adsorption has been used based on consequence of contact adsorption. Since purely columbic forces do not operate, there need not to be an equality of charge density on the metal and IHP

$$|q_m| \neq |q_{CA}|$$

There is no equivalence, but super-equivalent of charge on the IHP. Since $|q_{CA}| > |q_m|$, it means that if for example, negative ions are contact adsorbed on a negatively charged electrode there must be a positively charged diffuse layer to maintain over all electro-neutrality across the interface, i.e. $q_m = q_{CA} + q_d$.

The term super-equivalent adsorption is thus only a formal description in contrast to the term contact adsorption, which suggests a model of the structure of the interface. Based on these views contact adsorption will be preferentially used in the following treatment.

## 15.11 CONTACT ADSORPTION: ITS INFLUENCE ON THE CAPACITY OF THE INTERFACE

In the presence of IHP, interface is triple layer and in absence of IHP interface is double layer. Thus, interface has two models. Model 1 (O type arrangement) is a double layer without contact adsorbing ions and model 2 (I type arrangement) is a triple layer with contact adsorbent ions. Now the type of model applicable to a given system, i.e. the presence or absence of contact adsorption, must influence the property of the interface for storing charge. To examine this influence, relationship between capacitance $C$, and $q_{CA}$ must be derived (Figure 15.5).

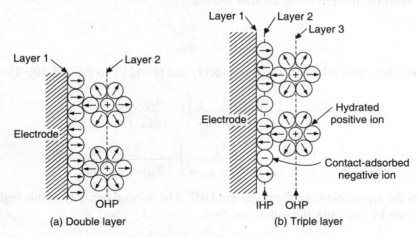

**FIGURE 15.5** Figure showing double layer and triple layer.

The procedure is same as in Stern demonstration, the total capacity of the interface being formally equivalent to two capacitors in series.

It will be assumed that all the diffuse charge is concentrated on the OHP, i.e. the solution is concentrated. Then the potential drop beyond the OHP into the solution tend to be negligible. The total potential difference across an interphase which includes contact adsorbed ions can be resolved into the component parts. Thus,

$$\phi_m - \phi_s = (\phi_m - \phi_{IHP}) + (\phi_{IHP} - \phi_{OHP}) \tag{15.7}$$

The potential drops across the two regions: (1) the metal surface to the IHP and (2) the IHP to OHP will now be expressed in terms of integral capacities of the two regions.

$$\phi_m - \phi_{IHP} = \frac{q_m}{K_{m \to IHP}} \text{ and } \phi_{IHP} - \phi_{OHP} = \frac{q_d}{K_{IHP \to OHP}} \tag{15.8}$$

Hence, Eq. (15.7) can be rewritten in the form

$$\phi_m - \phi_s = \frac{q_m}{K_{m \to IHP}} + \frac{q_d}{K_{IHP \to OHP}} \tag{15.9}$$

This expression is now differentiated with respect to $q_m$

$$\frac{d(\phi_m - \phi_s)}{dq_m} = \frac{1}{K_{m \to IHP}} + \frac{1}{K_{IHP \to OHP}} \frac{dq_d}{dq_m} \tag{15.10}$$

By definition, one has

$$\frac{d(\phi_m - \phi_s)}{dq_m} = \frac{1}{C} \tag{15.11}$$

and from the condition of electro-neutrality of the interface as a whole

$$q_m = q_{CA} + q_d$$

Thus, one obtains by differentiation another relation

$$\frac{dq_d}{dq_m} = 1 - \frac{dq_{CA}}{dq_m} \tag{15.12}$$

By substituting these two relations, i.e. Eqs. (15.11) and (15.12) into Eq. (15.10). The result is

$$\frac{1}{C} = \frac{1}{K_{m \to IHP}} + \frac{1}{K_{IHP \to OHP}} \left(1 - \frac{dq_{CA}}{dq_m}\right)$$

$$\frac{1}{C} = \left(\frac{1}{K_{m \to IHP}} + \frac{1}{K_{IHP \to OHP}}\right) - \frac{1}{K_{IHP \to OHP}} \frac{dq_{CA}}{dq_m} \tag{15.13}$$

Now consider the region between the metal and OHP. The integral capacity of this region may be considered given by two capacitors in series, thus.

$$\frac{1}{K_{m \to OHP}} = \frac{1}{K_{m \to IHP}} + \frac{1}{K_{IHP \to OHP}} \tag{15.14}$$

In terms of this idea, Eq. (15.13) can be rearranged to give.

$$\frac{1}{C} = \frac{1}{K_{m \to \text{OHP}}} - \left( \frac{1}{K_{m \to \text{OHP}}} - \frac{1}{K_{m \to \text{IHP}}} \right) \frac{dq_{\text{CA}}}{dq_m} \qquad (15.15)$$

This is the expression for the capacitance of an interface in the presence of contact adsorption. Now, how the differential capacity is affected by contact adsorbed ions populating the IHP. This situation is evidently a different from electrostatic theory of capacitors, where the charge on the opposite plates must be equal in magnitude and the capacity is purely dependent on the geometry and dielectric constant of the system, never on the charge. In the electrified interface, the double layer has become a triple layer, and the capacity varies with the contact adsorption through the quantity $dq_{CA}/dq_m$.

## 15.12 THE COMPLETE CAPACITY–POTENTIAL CURVE

If an electro-capillary $\gamma$ versus $V$ curve is twice differentiated, one obtains a curve of differential capacity versus potential (Figure 15.6). This curve consists of two sections, in which the capacity is apparently constant. Why apparently? Because, of the limitation of the whole procedure of deriving capacity from surface tension data. The $\gamma$ can be determined to an accuracy within ±0.2 dyne cm$^{-1}$ near the PZC and up to within ± 0.8 dyne cm$^{-1}$ far away from the PZC. The derived capacities, however, are only accurate to within ± 4 µF cm$^{-2}$. When sufficiently accurate electrical methods of determining the capacities are used, one finds that capacity potential curve breaks out into humps and flats. It has a complicated fine structure, which depends upon the ions, which populate the interphase region. Whereas there is a region of constant capacity at $V$ more negative than the ECM, there is also a hump in the capacity potential curve in a region positive to the ECM. At $V$, much more positive than the hump region, the capacity starts shooting up. Perhaps this is because the interface is on the verge of leaking and becoming nonpolarizable, in which case the $q$ in $C = dq/dv$ is contributed to by the transfer of charge across the double layer. The capacity potential curve presents two basic challenges, the challenge of interpreting the constant capacity region and that of interpreting the hump.

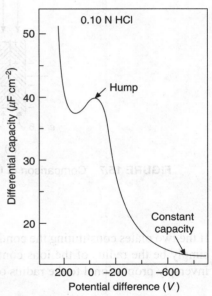

FIGURE 15.6 The experimental capacity–potential curve showing the constant capacity region and the hump.

## 15.13 THE CONSTANT CAPACITY REGION

One sees a relatively constant capacity region, e.g. on the negative side of $C$ versus $V$ curve. One's thought turns naturally to a simple parallel plate condenser model because such a model yields a potential independent capacity. One plate is located at the metal surface. But where is the second plate? On the IHP? or On the OHP?

One must think in terms of the capacity of a parallel plate condenser, i.e. capacity arising from the Helmholtz–Perrin model (Figure 15.7).

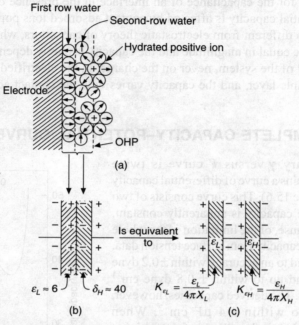

**FIGURE 15.7** Comparison of interface with parallel plate capacitor with two dielectric.

$$C = \frac{q}{\Delta\phi} = \frac{\varepsilon}{4\pi d} \quad (15.16)$$

If the two plates constituting the condenser are the metal and IHP, then $d$ for this condenser would simply be the radius of the ions contact adsorbed in the IHP. Hence, the capacitance should be inversely proportional to the radius of the ions

$$C = \frac{\varepsilon}{4\pi r_i} \quad (15.17)$$

Experiments do not show this radius dependence of the capacity. It seems therefore that there are no ions in the IHP. In other words, a constant capacity implies the absence of contact adsorption of ions.

The ions must therefore be in the OHP. Since, it is assumed that one is dealing with fairly concentrated solutions, the diffuse charge is effectively squeezed on to the OHP. The OHP has been pictured to be the locus of centres of hydrated ions in contact with a hydrated electrode. Will this model explain the constancy of capacity with the radius of ions? It will tend to do so because the $d$ will become $r_i$ plus other terms connected with the radii of water molecules separating the ions from the electrode.

## 15.14 THE POSITION OF THE OUTER HELMHOLTZ PLANE AND AN INTERPRETATION OF THE CONSTANT CAPACITY

A consideration of the dielectric constant of the material between the metal surface and the OHP suggests therefore, that even when there is no contact adsorption, there are two regions, one next to the electrode with a low dielectric constant $E_L \approx 6$ and another adjacent to the OHP with dielectric constant which changes as one proceed out to the solution, but can be represented by a mean value of about 40.

This is peculiar parallel plate capacitor with two dielectrics inside it. How does one calculate its capacity? One may imagine that charges arise at the boundary of the two dielectrics, charges equal in magnitude and opposite in sign to the plates facing them. Then it becomes obvious from previous argument that, in effect total capacity of the double layer is given by two capacitors in series, one across the first dielectric and the other across the second. Hence,

$$\frac{1}{C} = \frac{1}{K_{m \to OHP}} = 4\pi \frac{x_L}{\varepsilon_L} + 4\pi \frac{x_H}{\varepsilon_H} \tag{15.18}$$

$$\Rightarrow \quad 4\pi \frac{x_L}{\varepsilon_L} + 4\pi \frac{\delta - X_L}{\varepsilon_H} \tag{15.19}$$

It is obvious, however, from Figure 15.8 that $X_L$ is simply equal to the diameter $2r_w$ of a water molecule and that $\delta$ the distance between the metal and the OHP, is $2r_w + \sqrt{3}r_w + r_i$, where, $r_i$ is the radius of unhydrated ion at the OHP. Hence,

$$\frac{1}{C} = \frac{1}{K_{m \to OHP}} = \frac{4\pi 2r_w}{\varepsilon_L} + \frac{4\pi}{\varepsilon_H}\sqrt{3}r_w + \frac{4\pi}{\varepsilon_H} r_i \tag{15.20}$$

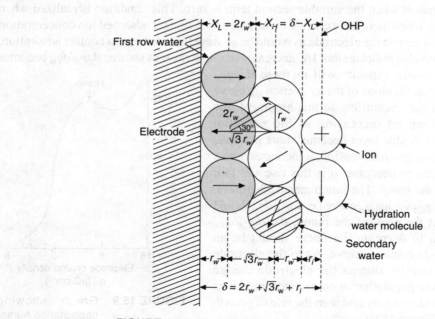

**FIGURE 15.8** A detailed look in to electrified interface.

the capacity can now be calculated. If $\varepsilon_L \approx 6$ and $\varepsilon_H \approx 40$, it turns out that the capacity is 16–17 µF cm$^2$ and fairly independent of the radius of the ion. Thus, the model has attained consistency with experiment for it predicts that in the absence of contact adsorption, the capacity of the interface does indeed become independent of the potential and of the radii of the ions in the OHP. Thus, the third term of Eq. (15.20) which contains the radii of ions, contribute little to the total capacity because the $r_i$ value is divided by the relatively large value of $\varepsilon_H \approx 40$ while the first term, which lacks $r_i$ is divided by the relatively small $\varepsilon_L \approx 6$, the second term is small and also independent of $r_i$. Hence, it is low dielectric constant region, which contributes most to the capacity. Thus, ionic radius scarcely attacks the double layer capacity. Simply stated, therefore, the double layer capacity in the region of constant capacity is determined by two capacitors in series, one capacitor immediately next to the electrode containing a (saturated) water dielectric and the second one bounded by the plane through the ion centres having a dielectric with a much higher value of dielectric constant than that of the first layer and containing an ionic radius term which however contributes little to the numerical value of capacitance.

## 15.15 THE CAPACITANCE HUMP

Now the question is: Why does the differential capacity of the interface increases when the electrode charge become positive with respect to the constant capacity region? Why does not the constant capacity maintain the constant value of 16–17 µF/cm$^2$ as the potential difference across the interface changes?

The general expression for the capacity of the electrode electrolyte interface is,

$$\frac{1}{C} = \frac{1}{K_{m \to \text{OHP}}} - \left( \frac{1}{K_{m \to \text{OHP}}} - \frac{1}{K_{m \to \text{IHP}}} \right) \frac{dq_{CA}}{dq_m} \qquad (15.21)$$

It says that $C$ is constant when the variable second term is zero. This condition is realized when $dq_{CA}/dq_m$ is zero, i.e. when there is no change in the amount of contact adsorbed ion concentration with change of the charge on the electrode as would be so were there to be zero contact adsorption. The above expression also indicates that $1/C$ decreases or $C$ increases as soon as $dq_{CA}/dq_m$ becomes nonzero. In other words, capacity will increase if $dq_{CA}$ increases with $dq_m$, i.e. the slope of the $q_{CA}$ versus $q_m$ curve is finite and increasing. According to this argument, the capacity should keep on increasing as the potential difference across the double layer becomes more positive. In practice, $C$ does not go on increasing. It increases upto a point and then begins to decrease. It is this rise and fall, which is known as *the hump*. The hump means, therefore, that the slope of the $q_{CA}$ versus $q_m$ curve i.e. $dq_{CA}/dq_m$ should at first increase. At the peak of the pump, the $dq_{CA}/dq_m$ slope should begin to decrease, i.e. there should be an inflection in the $q_{CA}$ versus $q_m$ curve. As to the electrode charge the hump implies that as the electrode charge becomes positive, the population of contact adsorbed ions increases more and more easily and then the rate of growth begins to decline (Figure 15.9).

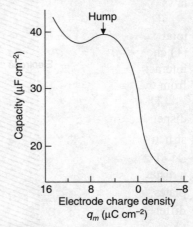

**FIGURE 15.9** Graph showing capacitance hump.

## 15.16 CONTACT ADSORBED IONS CHANGES WITH ELECTRODE CHARGE

Contact adsorption is formally similar to a chemical reaction

$$[\ ] + \Theta \leftrightarrow [\Theta]$$

where $\Theta$ is a contact adsorbing ion at the OHP, $[\ ]$ is an unoccupied site on the IHP, and $[\Theta]$ represents the result of a contact adsorbing ion jumping into a vacant site, i.e. $[\Theta]$ is a contact adsorbed ion.

From the law of mass action, it follows that

$$\frac{n_{CA}}{n_{[\ ]}} = n_\Theta \, e^{-\Delta G^0/RT} \tag{15.22}$$

where $n_{CA}$ is the population density (number per square centimetre) of contact adsorbed in the IHP, $n_{[\ ]}$ is a number of vacant sites per square centimetre of the IHP, $\Delta G^0$ is the standard free energy change accompanying the migration of a mole of ions from the OHP to the IHP, and $n_\Theta$ is concentration of contact adsorbing ions. The terms $\Delta G^0/RT$ and $e^{-\Delta G^0/RT}$ are dimensionless quantity so is $\frac{n_{CA}}{n}$, hence to ensure dimensionless equality of the two sides of the equation, $n_\Theta$ must not be expressed in the usual concentration unit of moles per litre. It is necessary to multiply the concentration or activity $a_i$ of ions $i$ in solution by $N_A 2r_i/1000$ to get the number of contact adsorbing ions per square centimetre and then divide by $N_T$, the total number of sites (free plus occupied) per square centimetre on IHP or parallel to it.

Thus, Eq. (15.22) becomes

$$n_{CA} = n_{[\ ]} N_A \, 2r_i a_i/1000 \, N_T \times e^{-\Delta G^0/RT}$$
$$n_{CA} = (1-\theta) N_A \, 2r_i a_i/1000 \times e^{-\Delta G^0/RT} \tag{15.23}$$

where $\theta$ is the fraction of the electrode surface occupied by adsorbed ions. If $\Delta G^0$ is evaluated, the laws of growth of the population of contact adsorbed ions have been formulated.

The problem, therefore, is to compute this work of transit, i.e. of contact adsorption. This total work when an ion is transported from the OHP to IHP can be divided into three contributions: (1) chemical work arising from forces between the metal and the ion, (2) work arising from the interaction of the ions with the electric field due to the charged electrode and (3) work arising from the lateral interaction of the ion with its surrounding contact adsorbed ions.

The nature of the chemical work has been already discussed. It concerns ion–electrode dispersion interaction and the changes in free energy resulting from the alternation of hydration structure. The chemical work per mole of ions may be left with the symbol $\Delta G_c^0$.

The electrical interaction with the field is a matter of the work of taking a charged ion through a distance $X_2 - X_1$ from the OHP to the IHP of the parallel plate condenser consisting of the two plates, namely the metal surface and the OHP. The field in such a condenser is $4\pi q_m/\varepsilon$ the potential difference of interest is $4\pi q_m (X_2 - X_1)/\varepsilon$ and the work of transporting one charge $e_0$ is $4\pi q_m e_0 (X_2 - X_1)/\varepsilon$. One must now compute the work of lateral interaction among the contact adsorbed ions. The first step is to visualize the spatial distribution of the ions populating the IHP. The simplest approach is to divide a unit area in the IHP into as many cells as there are contact adsorbed ions per unit area. The contact adsorbed ions are assumed to adopt time average positions corresponding to a hexagonal array.

Now consider any reference ion. The surrounding ion of the hexagonal array are effectively in circular rings about the central ion, the ring radii increasing thus $1r, 2r, 3r \ldots nr$. Further let the charges in a ring be smoothed out so that one can talk of a charge density and let this charge density be expressed in charge per unit angle (i.e. charge per unit radian). The number of charges in the first ring can be seen in Figure 15.10 to be 6, in the second 12 ..., and therefore in the $n$th ring $6n$. Hence the charge per unit charge density of $\sigma = 6ne_0/2\pi$ radian in the $n$th ring is

$$\sigma = \frac{6ne_0}{2\pi} \quad (15.24)$$

The contact adsorbed ions are effectively in circular rings so that there are $6n$ ions in the $n$th ring of radius $nr$. If the charge is smoothed out, there is a charge of $6ne_0$ in the $2\pi$ radians of the $n$th ring, i.e. a charge density of $\sigma = 6ne_0/2\pi$.

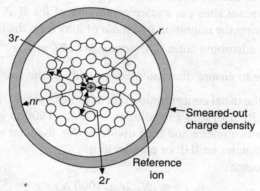

**FIGURE 15.10** The adsorbed ions are arranged in circular rings around the central ion.

The radius of the first ring is a known quantity. If $r$ is the mean distance between cell centres, i.e. between contact adsorbed ions, one cell has an area $\pi(r/2)^2$ and therefore there are $4/\pi r^2$ ions per unit area (Figure 15.11). But $n_{CA}$ is the number of contact adsorbed ions per unit area, hence

$$n_{CA} = \frac{4}{\pi r^2}; \quad r = \left(\frac{4}{\pi n_{CA}}\right)^{1/2} \quad (15.25)$$

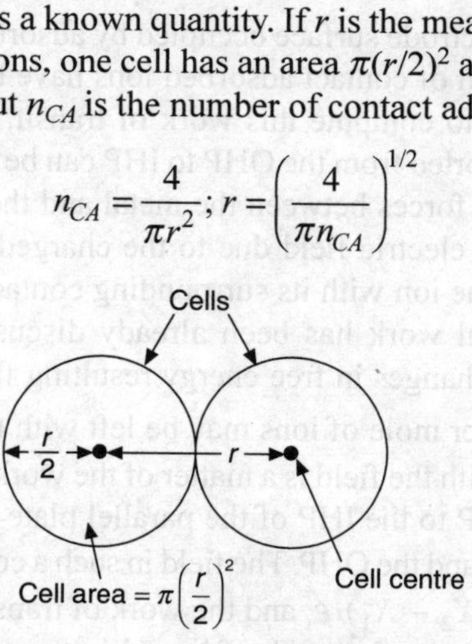

**FIGURE 15.11** Showing mean distance and area of cells.

The potential due to *n*th charged ring (on the IHP) at the site of reference ion is equal to (charge on the *n*th ring/$\varepsilon$ × distance of *n*th ring to ion) $2\pi\sigma/\varepsilon nr$ and the interaction energy (repulsive) between the reference ion (charge $e_0$) and the *n*th charge ring is $2\pi\sigma e_0/\varepsilon nr$.

But this calculation is for conditions near a metal, i.e. for charges on the metal surface and charges near a metal induced charges on the metal surface. This induced charge can be represented by image charges. So, every charged ring on the IHP has associated with it a ring of image charge located from the reference ion at a distance $[(nr^2) + (2r_i)^2]^{1/2}$. Hence, the attractive interaction between the reference ion and the *n*th ring of image charge is,

$$-\frac{2\pi\sigma e_0}{\varepsilon}\frac{1}{\left[(nr)^2 + (2r_i)^2\right]^{1/2}}$$

Thus, the interaction energy between the reference ion and the *n*th ring plus the interaction energy with the image charge due to the *n*th ring is

$$\frac{2\pi\sigma e_0}{\varepsilon nr}\left[1 - \frac{1}{\left[1 + (2r_i/nr)^2\right]^{1/2}}\right]$$

The total lateral interaction work $W_{L.I}$ is obtained by summing over all the ring as follows

$$W_{L.I} = \sum_{n=1}^{\infty} \frac{2\pi\sigma e_0}{\varepsilon nr}\left[1 - \frac{1}{\left[1 + (2r_i/nr)^2\right]^{1/2}}\right] \quad (15.26)$$

At this stage, the binomial expansion

$$(1+x)^{-1/2} = 1 - \frac{1}{2}x + \frac{3}{8}x^2 - \cdots \quad (15.27)$$

can be used, and only the first two terms taken. One has

$$W_{L.I} = \sum_{n=1}^{\infty} \frac{4\pi\sigma e_0 r_i^2}{\varepsilon n^3 r^3}\left[1 - \frac{3r_i^2}{n^2 r^2}\right] \quad (15.28)$$

The expression for $\sigma$ and for $r$, i.e Eqs. (15.24) and (15.25) are now inserted into Eq. (15.28).

$$W_{L.I} = \sum_{n=1}^{\infty} \frac{3\pi^{3/2} n_{CA}^{3/2} e_0^2 r_i^2}{2\varepsilon n^2}\left[1 - \frac{3\pi r_i^2 n_{CA}}{4}\frac{1}{n^2}\right]$$

$$\Rightarrow \frac{3\pi^{3/2} n_{CA}^{3/2} e_0^2 r_i^2}{2\varepsilon}\sum_{n=1}^{\infty}\frac{1}{n^2} - \frac{9\pi^{5/2} n_{CA}^{5/2} e_0^2 r_i^4}{8\varepsilon}\sum_{n=1}^{\infty}\frac{1}{n^4} \quad (15.29)$$

The series $\sum_{n=1}^{\infty}\frac{1}{n^2}$ and $\sum_{n=1}^{\infty}\frac{1}{n^4}$ have been evaluated to be $\frac{\pi^2}{6}$ and $\frac{\pi^4}{90}$, respectively. Hence,

$$W_{L.I} = \pi^2 \frac{\pi^{3/2} n_{CA}^{3/2} e_0^2 r_i^2}{4\varepsilon} - \pi^2 \frac{3\pi^{5/2} n_{CA}^{5/2} e_0^2 r_i^4}{16\varepsilon} \frac{\pi^2}{15} \qquad (15.30)$$

The chemical, electrical and interactional work term when summed together give $-\Delta G^0$, the free energy change associated with the equilibrium reaction of ions contact adsorbing on a charged electrode. When this sum of work terms is substituted in Eq. (15.23), One obtains

$$n_{CA} = (1-\theta) \frac{N_A 2 r_i}{1000} a_i \exp\left[ -\frac{\Delta G_c^0}{RT} + \frac{4\pi q_m e_0 (x_1 - x_2)}{\varepsilon kT} - \frac{\pi^2 n_{CA}^{3/2} e_0^2 r_i^2 \pi^{3/2}}{4\varepsilon kT} \left(1 - \frac{3}{4}\frac{\pi^2}{15} \pi r_i^2 n_{CA}\right) \right]$$

$$(15.31)$$

This is a relation which connects the population density of contact adsorbed ions on the IHP, i.e. number per unit area of contact adsorbed ions to the charge on the metal. It is seen that $n_{CA}$ depends on: (1) bulk activity $a_i$ of the adsorbing spices, (2) the radius $r_i$ of the contact adsorbing ion, (3) the chemical term is $\Delta G_c^0$ and (4) the electrode charge $q_m$.

Thus, for a particular contact adsorbing ionic species at a fixed bulk concentration, the population of contact adsorbing ion changes as the charge on the electrode changes. This is good.

If the population of contact adsorbed ions had remained a constant $\frac{dq_{CA}}{dq_m}$ would have been zero and there would have been no hope of explaining any increase of capacities as one approached the positive region by an equation such as Eq. (15.30) of contact adsorption for total capacitance.

## 15.17 TEST OF THE POPULATION LAW OF CONTACT ADSORBED IONS

A calculated $n_{CA}$ versus $q_m$ curve can be compared with a curve obtained from experimental electro-capillary curves. The charge $q_{CA}$ is of course simply $n_{CA} e_0$. Not only is there an initial increase of $\frac{dq_{CA}}{dq_m}$ from its zero value at $q_m = -12$ to $-13$ μC/cm$^2$, but more significantly there is an inflection in the theoretical $q_{CA}$ versus $q_m$ curve. Why this inflection is important? It is because that it has been argued that the hump in the $C$ versus $q_m$ curve is located at a value of the electrode charge corresponding to the inflection $\frac{d^2 q_{CA}}{dq_m^2} = 0$ in the $q_{CA}$ versus $q_m$ curve. A typical test therefore is to see whether the theory is able to predict at what value of $q_m$ the capacity hump will occur. This is easily done. All one has to do is to obtain an expression for $\frac{d^2 q_{CA}}{dq_m^2}$ and set is equal to zero, i.e. the procedure is to write down the mathematical condition of an inflection point (Figure 15.12).

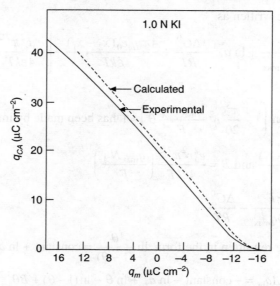

**FIGURE 15.12** Comparison of theoretical and experimental $q_{CA}$ versus $q_m$ curve.

It is convenient, however, to rewrite the expression for $n_{CA}$ Eq. (15.31), in a more convenient form. One makes use of the fact that the fraction $\theta$ of the IHP occupied by contact adsorbing ions is given by

$$\theta = \frac{n_{CA}}{n_T}$$

or

$$n_{CA} = \theta n_T = \frac{\theta n_T e_0}{e_0} = \frac{\theta q_{max}}{e_0}$$

$$n_{CA} = \frac{\theta q_{max} N_A}{F} \tag{15.32}$$

and

$$(n_{CA})^{3/2} = \left(\frac{q_{max} N_A}{F}\right)^{3/2} \theta^{3/2} \tag{15.33}$$

Relations (15.32) and (15.33) are inserted in Eq. (15.31), to give

$$n_{CA} = \frac{\theta q_{max} N_A}{F} = (1-\theta) \frac{N_A 2 r_i}{1000} a_i \exp\left[-\frac{\Delta G_c^0}{RT} + \frac{4\pi q_m e_0 (x_2 - x_1)}{\varepsilon k T}\right.$$
$$\left. - \frac{\pi^2 n_{CA}^{3/2} e_0^2 r_i^2 \pi^{3/2}}{4\varepsilon k T}\left(1 - \frac{3}{4}\frac{\pi^2}{15} \pi r_i^2 n_{CA}\right)\right]$$

$$\frac{\theta}{(1-\theta)} = \frac{2 r_i F}{1000 q_{max}} a_\pm \exp\left[-\frac{\Delta G_c^0}{RT} + \frac{4\pi q_m e_0 (x_2 - x_1)}{\varepsilon k T}\right.$$
$$\left. - \frac{e^2 r_i^2 \pi^{7/2}}{4\varepsilon k T}\left(\frac{q_{max} N_A}{F}\right)^{3/2} \theta^{3/2}\left(1 - \frac{\pi^3}{20} r_i^2 \frac{q_{max} N_A}{F} \theta\right)\right] \tag{15.34}$$

Equation (15.34) can be written as

$$\ln\frac{\theta}{(1-\theta)} = \ln\frac{2r_iF}{1000q_{max}} + \ln a_{\pm} - \frac{\Delta G_c^0}{RT} + \frac{4\pi q_m e_0(x_2-x_1)}{\varepsilon kT} - \left[\frac{e_0^2 r_i^2 \pi^{7/2}}{4\varepsilon kT}\left(\frac{q_{max}N_A}{F}\right)^{3/2}\right]\theta^{3/2} \quad (15.35)$$

where the approximation $\left(1-\frac{\pi^3}{20}r_i^2\frac{q_{max}N_A}{F}\theta\right) \approx 1$ has been made by introducing the following

notation $A = \dfrac{4\pi e_0(x_2-x_1)}{\varepsilon kT}$ and $B = \dfrac{e_0^2 r_i^2 \pi^{7/2}}{4\varepsilon kT}\left(\dfrac{q_{max}N_A}{F}\right)^{3/2}$ \quad (15.36 and 15.37)

And constant $= \ln\dfrac{2r_iF}{1000q_{max}} - \dfrac{\Delta G_c^0}{RT}$ \quad (15.38)

Equation (15.35) can be rewritten in the form $\ln\dfrac{\theta}{(1-\theta)} = \text{constant} + \ln a_{\pm} + Aq_m - B\theta^{3/2}$

$$Aq_m = -\text{constant} - \ln a_{\pm} + \ln\theta - \ln(1-\theta) + B\theta^{3/2} \quad (15.39)$$

One can now proceed rapidly to evaluate $\dfrac{d^2\theta_{CA}}{dq_m^2}$ and locate the inflection point in the $\theta$ versus $q_m$ curve and compare it with the position of the capacity hump, as experimentally observed by differentiating Eq. (15.39). Keeping the bulk concentration and therefore, the activity $a_{\pm}$ constant, one obtains

$$A\frac{dq_m}{d\theta} = \frac{1}{\theta} + \frac{1}{1-\theta} + \frac{3}{2}B\theta^{1/2}$$

$$\frac{d\theta}{dq_m} = \frac{A}{\left[\dfrac{1}{\theta}+\dfrac{1}{1-\theta}\right]+\dfrac{3}{2}B\theta^{1/2}} \quad (15.40)$$

$$\frac{d\theta}{dq_m} = \frac{A}{\rho} \quad (15.41)$$

where the symbol $\rho$ is used for the denominator of the right-hand side of Eq. (15.40). To proceed further equation used in the Gouy–Chapman theory can be resorted to

$$\frac{d^2\theta}{dq_m^2} = \frac{1}{2}\frac{d}{d\theta}\left(\frac{d\theta}{dq_m}\right)^2 = \frac{1}{2}\frac{d}{d\theta}\left(\frac{A}{\rho}\right)^2 = -\frac{A^2}{\rho^3}\left(\frac{d\rho}{d\theta}\right)$$

$$= -\frac{A^2}{\rho^3}\left(-\frac{1-2\theta}{\theta^2(1-\theta)^2} + \frac{3}{4}B\theta^{-1/2}\right)$$

$$= -\frac{A^2}{2\rho^3}\left(-\frac{2+4\theta+3/2B\theta^{3/2}-3B\theta^{5/2}}{\theta^2(1-\theta)^2}\right) \quad (15.42)$$

where $\theta^{7/2}$ term has been neglected.

The mathematical condition for an inflection point can now be used, i.e. $\dfrac{d^2\theta}{dq_m^2}$ is set equal to zero. Thus, from Eq. (15.42)

$$= -\frac{A^2}{2\rho^3} \frac{1}{\theta^2(1-\theta)^2}\left(-2 + 4\theta + \frac{3}{2}B\theta^{3/2} - 3B\theta^{5/2}\right) = 0 \quad (15.43)$$

If either $\rho$ or $\theta^2(1-\theta)^2$ is equal to zero, the left-hand side of Eq. (15.43) would tend to infinity rather than to zero. Hence, Eq. (15.43) or $\dfrac{d^2\theta}{dq_m^2}$ would be equal to zero only if

$$2 - 4\theta - \frac{3}{2}B\theta^{3/2} + 3B\theta^{5/2} = 0 \quad (15.44)$$

By solving this equation for $\theta$, one obtains the value of $\theta_{\text{inflec}}$, i.e. the value of $\theta$ at which an inflection is expected in the $\theta$ versus $q_m$ curve. This can be compared with $\theta_{\text{hump}}$, i.e. the value of $\theta$ at which the hump in the experiment $C$ versus $q_m$ curve is observed. Figure 15.13 shows good agreement between theory and experiment. A further check of the theory of the population growth of contact adsorbed ions can be made as:

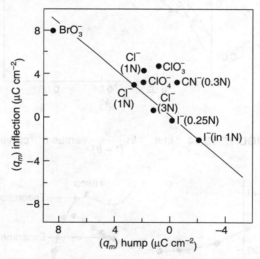

**FIGURE 15.13** The location of the hump on the $C$ versus $q_m$ curve, comparison between theory and experimental.

At constant $q_m$, Eq. (15.39) can be rearranged in the form

$$\ln a_\pm - \ln \frac{\theta}{(1-\theta)} = \text{constant} + B\theta^{3/2} \quad (15.45)$$

which shows that theory demands that if $\ln a_\pm - \ln \dfrac{\theta}{(1-\theta)}$ is plotted against $\theta^{3/2}$ at constant $q_m$, a straight line should be obtained with a slope $B$.

$$B = \frac{e_0^2 r_i^2 \pi^{7/2}}{4\varepsilon kT}\left(\frac{q_{max} N_A}{F}\right)^{3/2} \qquad (15.46)$$

According to theory, the hump should occur at the value of $q_m$ corresponding to the inflection in the $q_{CA}$ versus $q_m$ curve. Hence, when the ordinate of a point is equal to the abscissa, there is a perfect agreement between theory and experiment.

The theory is in quite agreement with experiment. The experimental plots are linear and there is reasonable agreement between the calculated and measured slopes (Figures 15.14 and 15.15).

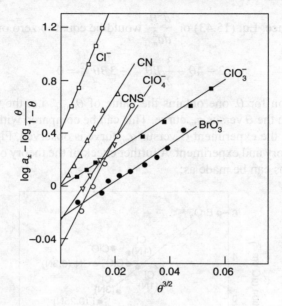

**FIGURE 15.14** $\ln a_\pm - \ln \dfrac{\theta}{(1-\theta)}$ versus $\theta^{3/2}$ curve.

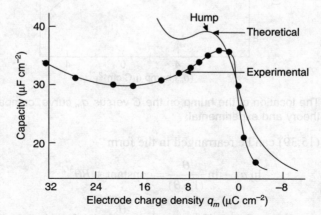

**FIGURE 15.15** Capacity versus electrode charge density curve.

## 15.18 LATERAL REPULSION MODEL FOR CONTACT ADSORPTION

The importance of $C$ versus $V$ or $C$ versus $q$ curve stems from the fact that the capacity of an electrified interface depends on the arrangement of charges in the interface region, i.e. molecular structure of the interface, i.e. manner in which, the IHP and OHP are populated with ions. The interpretation of the constant capacity region of the $C$ versus $V$ curve turns out to be a simple matter. If the IHP is denuded of contact adsorbing ions, the interface becomes literal double layer. Simple parallel plate condenser arguments permit the description of the structure of the interface in terms of two regions. One region has a low dielectric constant corresponding to oriented water and the second region has a high dielectric constant corresponding to partially oriented water.

The explanation of the hump was a little more sophisticated. As the electrode charge becomes positive with respect to the value ($-12$ to $13$ $\mu C\ cm^{-2}$) when there is simple double layer, ions starts to contact adsorb and populate the IHP. The dependence of differential capacity upon contact adsorption is contained in Eq. (15.15) which can be written in the form

$$\frac{1}{C} = \alpha - \beta \left( \frac{dq_{CA}}{dq_m} \right)$$

where $\alpha$ and $\beta$ are related to the constant integral capacities. Apparently, the capacity depends on the slope of the $q_{CA}$ versus $q_m$ curve. The variation of capacity with electrode charge must, therefore, be given by

$$\frac{d(1/C)}{dq_m} = -\beta \left( \frac{d^2 q_{CA}}{dq_m^2} \right)$$

If the capacity shows a hump or the reciprocal capacity $\frac{1}{C}$ shows an inverted hump, it means that, at the hump

$$\frac{d(1/C)}{dq_m} = 0$$

or

$$\left( \frac{d^2 q_{CA}}{d^2 q_m} \right) = 0$$

i.e. the inflection in the $q_{CA}$ versus $q_m$ curve locates the hump. The problem reduced, therefore, the understanding the rate of growth of the population of contact adsorbed ions and its change with the excess electric charge on the metal. A simple model was conceived. The migration of an ion from the OHP to the IHP (the contact-adsorption process) involves chemical interaction, interaction with the electrical field arising from the electrode charge $q_m$ and lateral interaction with already settled population of contact adsorbed ions. The final expression is of the form Eq. (15.31).

$$n_{CA} = \text{Constant } e^{Aq_m} e^{-\beta q_{CA}^{3/2}}$$

It is seen that the electrode charge encourages the growth of the population of contact adsorbed ions, $n_{CA}$ but this growth sets up and accentuates the lateral repulsion forces, which try to inhibit further growth. It is this example of negative feedback (electrical attracting forces giving rise to lateral repulsion forces) which generates the hump. At charges in the region of constant capacitance (16–17 µF cm$^{-2}$), $n_{CA} = 0$ and $e^{-\beta q_{CA}^{3/2}} \approx 1$ and at charges more positive than that of constant capacitance, $n_{CA}$ grows at first exponentially with increasing positivity of the electrode charge. Since $\dfrac{dn_{CA}}{dq_m}$ also increases exponentially with $q_m$ (around $n_{CA} = 0$) the capacity rises. But with the increasing departure of $n_{CA}$ form zero, the lateral repulsion term $e^{-\beta q_{CA}^{3/2}}$ increases in significance. It reduces the slope $\dfrac{dn_{CA}}{dq_m}$, i.e. it slows down the rate of growth of the population of contact adsorbed ions. At the inflection point $\dfrac{d^2 n_{CA}}{dq_m^2} = 0$, the capacity goes through the hump (Figure 15.16).

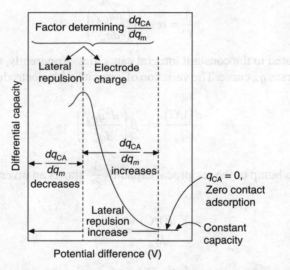

**FIGURE 15.16** Lateral repulsion model for the explanation of capacity potential curve.

The simple lateral repulsion model explained the capacitance hump in qualitative terms. The shape of the $q_{CA}$ versus $q_m$ curve with its inflection, the location of the hump in respect to charge on the electrode the magnitude of the capacity of the hump, the linearity of the $\ln a_{\pm} - \ln \dfrac{\theta}{(1-\theta)}$ versus $\theta^{3/2}$ plot and the magnitude of the slope of this relation, etc. all these can be reasonably well rationalized by the later repulsion model of contact adsorbing ions in the IHP.

## 15.19 FLIP–FLOPS WATER MOLECULES ON ELECTRODES

The highest observed values of the coverage of a mercury electrode with contact absorbed ions are less than 20%. Thus, even when contact adsorption is at a maximum, water molecules cover more than three quarters of the electrode surface and constitute the overwhelming majority of particles at the interface. It is surprising, therefore, that these adsorbed water dipoles have been ignored in all but very recent discussions of the structure of charged interfaces. The picture of the double layer has been ion-centric for too long.

In the present treatment too, the contribution of water to the picture of the electrified interface has been remembered only in one context, the dielectric constant of the water. Because the adsorbed water dipoles are largely oriented, there is dielectric saturation and the dielectric constant is not 78 (as it is in the bulk) but only about 6. This lowering of the dielectric constant was shown to explain the potential-independent part of a typical $C$ versus $V$ curve.

Forget, for a moment, the ions, and let the viewpoint become water-centric. Consider a number of water dipoles adsorbed on the electrode. Then there are two limiting conditions on the relation between the charge on the electrode and the orientation of the dipoles relative to the surface of the metal (Figure 15.17).

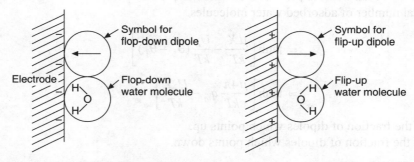

**FIGURE 15.17** The flop-up and flip-down orientation of water molecule, respectively.

One limiting condition arises on an electrode which has a high positive charge. The electric field vector is pointed from the metal into the solution. In a field, dipoles reduce their potential energy by aligning themselves, so that the dipole vector becomes parallel to the field. In other words, the water dipoles flips up so that the oxygen atoms are in contact with the electrode and the hydrogen end of water points into the solution. Let this orientation of a water molecule be called the flip-up state. The other limiting condition obtains, when electrons are pumped into the electrode to make it very negatively charged. What will the dipoles do? On the basis of a simple electrostatic argument, all that will happen is that the flipped up dipoles will turn around and flop-down. In the flop down state, the hydrogens are facing the electrode and the oxygen atom is towards the solution. This flip-flop model for water turn to be of consequence in the electrified interface and thus to electrochemistry. The equation for the flip-flop model are simple. They just tell one about the potential difference across a dipole layer and how it affects the kinetics of charge transfer process at electrodes. One implication of these equations will be pursued at first, that concerning the capacity of the interface.

## 15.20  THE CONTRIBUTION OF ADSORBED WATER DIPOLES TO THE CAPACITY OF THE INTERFACE

To expect from the dipole potential the contribution of water dipoles to the capacity, one has to differentiate $\Delta X$ potential with respect to electrode charge. The result is

$$\frac{1}{C_{\text{dipole}}} = \frac{\partial(\Delta X)}{\partial q_m} = \frac{4\pi\mu N_T}{\varepsilon} \frac{\partial Y}{\partial q_m} \tag{15.47}$$

What is $\dfrac{\partial Y}{\partial q_m}$ one writes for the field due to charge on the metal.

$$X = \frac{4\pi}{\varepsilon} q_m \tag{15.48}$$

And therefore $Y = \tanh x$ $\left[ \tanh x = \dfrac{\sinh x}{\cosh x} = \dfrac{N_\uparrow - N_\downarrow}{N_\uparrow + N_\downarrow} \right]$

Here $N_\uparrow$ represents the number per unit area of the flipped dipoles with hydrogen end towards the solution and $N_\downarrow$ be the number per unit area in the flop-down position.
$N_T$ is the total number of adsorbed water molecules.

$$= \tanh\left[ \frac{\mu X}{kT} - \frac{U_c}{kT}(\theta_\downarrow - \theta_\uparrow) \right] \tag{15.49}$$

$$= \tanh\left[ \frac{\mu 4\pi}{\varepsilon kT} q_m - \frac{U_c Y}{kT} \right] \tag{15.50}$$

where $\theta_\uparrow$ is the fraction of dipoles which points up.
$\theta_\downarrow$ is the fraction of dipoles which points down.

$$\theta_\uparrow = \frac{N_\uparrow}{N_T} \text{ and } \theta_\downarrow = \frac{N_\downarrow}{N_T}$$

$U$ is the interaction energy.
$X$ is the electric field arising from the charge $q_m$ on the electrode.
$c$ is the certain number of dipoles which surround a reference dipole with which it is coordinated.

$$\frac{dY}{dq_m} = \text{sech}^2 x \left[ \frac{\mu 4\pi}{\varepsilon kT} - \frac{U_c}{kT} \frac{dY}{dq_m} \right] \tag{15.51}$$

$$\frac{dY}{dq_m} = \frac{4\pi\mu}{\varepsilon kT} \frac{1-Y^2}{1 + \left(\dfrac{U_c}{kT}\right)(1-Y^2)} \tag{15.52}$$

The dipole capacity turns out therefore to be from Eq. (15.47).

$$\frac{1}{C_{\text{dipole}}} = \left[ \frac{16\pi^2 \mu^2 N_T}{\varepsilon^2 kT} \frac{1-Y^2}{1 + \left(\dfrac{U_c}{kT}\right)(1-Y^2)} \right] \tag{15.53}$$

Since, the parameter $Y$ contains the field $X$ which depends upon the electrode charge $q_m$. So, Eq. (15.53) predicts that dipole capacity should vary with electrode charge, when the calculated values of $C_{dipole}$ are plotted as a function of $q_m$. It turns out that the values of the dipole capacity are extremely large compared with the experimental values of the capacity (Figure 15.18). What does this imply? Consider the complete expression for the differential capacity.

$$\frac{1}{C} = \frac{\partial(\Delta\phi)}{\partial q_m} = \frac{\partial(\Delta\psi)}{\partial q_m} + \frac{\partial(\Delta\chi)}{\partial q_m}$$

$$\frac{1}{C} = \frac{1}{C_{charge}} + \frac{1}{C_{dipole}}$$

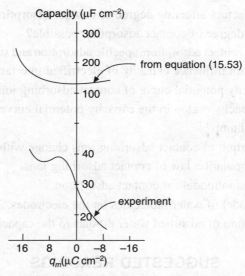

**FIGURE 15.18** Comparison between the calculated values of the dipole capacity and the experimental capacity.

The term $\dfrac{1}{C_{charge}}$ is given by Eq. (15.15) and has been the item of the interest in the discussion of constant capacity region and the hump. From the above equation, it is obvious that, whenever $C_{dipole}$ becomes too large, $\dfrac{1}{C_{dipole}}$ becomes too small to affect $1/C$ significantly. The situation is analogous to that in the Stern theory, where the Gouy capacity made a negligible contribution to the total capacity, whenever the magnitude of the Gouy capacity becomes large.

Hence,

$$\frac{1}{C} \approx \frac{1}{C_{charge}}$$

And it is quite justified to neglect the contribution of the water dipoles (constituting the hydration sheath of the electrode) to the differential capacity of an electrified interface.

The flip-flop water model will be now applied to a discussion of the model for the dependence of the adsorption of organic molecules on the charges on the metal electrodes, a subject which underlies, e.g. the use of organic inhibitors in metallic corrosion, and some aspect of whether a given electrode will act as a catalyst or not for a given fuel in electro-chemical energy converters.

## EXERCISES

1. Describes two types of arrangements O and I at an interface.
2. Draw a neat and clean diagram of ions contact adsorbing to metal electrode.
3. Illustrate free energy change involved in the process of contact adsorption.
4. Write down various factors affecting degree of contact adsorption.
5. Can measurement of degree of contact adsorption possible?
6. Differentiate between contact adsorption, specific adsorption and super-equivalent adsorption.
7. How contact adsorption influence capacity of electrified interface?
8. Draw complete capacity potential curve of contact adsorbing ions at the interface.
9. Explains constant capacity region in the capacity potential curve.
10. What is capacitance hump?
11. How does the population of contact adsorbing ions change with electrode charge?
12. Test the validity of population law of contact adsorbing ions.
13. Explain lateral repulsion model for contact adsorption.
14. Explains flip-flop model of water molecules on the electrodes.
15. Discuss the contribution of adsorbed water dipoles to the capacity of the interface.

## SUGGESTED READINGS

Balzani, V., *Electron Transfer in Chemistry*, Vols. 1–5, Wiley, New York, 2001.

Bockris, J.O.M., A.K.N. Reddy, and M. Gamboa-Aldeco, *Modern Electrochemistry 2A: Fundamentals of Electrodics*, 2nd ed., Plenum Press, New York, 2000.

Bockris, J.O.M. and A.K.N. Reddy, *Modern Electrochemistry: An Introduction to an Interdisplinary Area*, Vols. 1 and 2, Plenum Press, New York, 1970.

Bokris, J.O.M. and A.K.N Reddy, *Modern Electrochemistry 2B: Electrodics in Chemistry, Engineering, Biology and Environmental Science*, 2nd ed., Plenum Press, New York, 2000.

Glasstone, S., *An Introduction to Electrochemistry*, East-West Press, New Delhi, 2003.

Gurney, R.W., *Ionic Process in Solution*, McGraw-Hill, New York, 1953.

Kortum, G. and J.O.M. Bockris, *Textbook of Electrochemistry*, Vols. 1 and 2, Elsevier, Amsterdam, 1951.

Koryta, J. and J. Dvorak, *Principles of Electrochemistry*, Wiley, New York, 1987.

Kubasov V. and A. Zaretsky, *Introduction to Electrochemistry*, Mir Publishers, Moscow, 1987.

Potter, E.C., *Electrochemistry, Principles and Applications*, Macmillan, 1956.

# 16

# Structure of Semiconductor— Electrolyte Interface

## 16.1 INTRODUCTION

This aspect of electrodic's extends the scope of the subject beyond consideration of the metal–solution interface to all interfaces at which electrons are exchanged and thus opens up the prospect of understanding the electrochemistry of nonmetals, e.g. interfaces in biological systems and interfaces involving solid oxides.

Now we will study the distribution of excess charge inside the electrode. What is the situation inside the electrode? That depends upon whether the electrode is a metal or a semiconductor. What is most important difference between a metal or a semiconductor? Operationally speaking, it is the order of magnitude of conductivity. Metals have conductivities of the order of about $10^6$ $\Omega^{-1}$ cm$^{-1}$ and semiconductors about $10^2 - 10^{-9}$ $\Omega^{-1}$ cm$^{-1}$. These tremendous differences in conductivities reflect predominantly the concentration of free charge carrier. In crystalline solids, the atomic nuclei are relatively fixed and the charge carrier, which drift in response to electric fields are the electrons. So the question is what determines the concentration of mobile electrons? One has to take an inside look at electrons in crystalline solids.

## 16.2 THE BAND THEORY OF CRYSTALLINE SOLIDS

Consider a crystalline solid. The atoms are arranged according to a three-dimensional pattern in which they have equilibrium inter-atomic distances. A thought experiment is now performed. The lattice is expanded, i.e. inter-atomic distances are increased. Eventually, the atoms are so far apart that they can be considered isolated and independent atoms as in a gas. The purpose of the thought experiment is to discuss the electrons energy states in gaseous atoms and then to see how these energy states are modified as the atoms are brought closer and closer together until the lattice has contracted back to its original state. The electrons in a gaseous atom are arranged in shells. The shell structure is a result of the energy levels of the electrons having to follow quantum rules. The electrons in an incomplete outermost shell are known as *valence electrons* and these in filled inner shells are known as *core electrons*.

Suppose, now that two gaseous atoms are made to approach each other. As long as the electron clouds of the two atoms do not overlap, the electron energy states continue to follow the quantum rules for gaseous atoms. However, when the electron clouds begin to overlap and the electrons interact with both atoms, the rules for electron–energy states are upset and they start changing.

They change in an interesting way, each energy state from a gaseous atom splits into two states, one with a higher energy and the other with a lower energy. If three atoms are brought together, then each energy state of the gaseous atoms split into three energy states. In general, if there are $N$ atoms, each energy state of a gaseous atom splits into $N$ states. Some of these levels may be degenerate. The upper and lower level as shown in Figure 16.1, arise owing to the symmetrical and anti-symmetrical linear combination of atomic orbitals. The spacing between these $N$ energy states depends on the value of $N$, the larger the value of $N$ is, the closer together are the energy levels. In a bulky, solid electrode, where there may be some $10^{22}$ atoms cm$^{-3}$, the energy level are spaced so close and think of a continuous band of allowed energies.

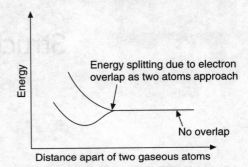

**FIGURE 16.1** Splitting of energy when two atoms approach each other.

This means that, as the expanded lattice is contracted in the thought experiment the discrete energy states of the atoms are replaced by energy bands.

Now, the splitting of the energy states of gaseous atoms occurs because of the overlap between electron clouds. Obviously, therefore, atoms must come much closer before the clouds of the core electrons begins to overlap compared with the distance at which the clouds of outer (or valence) electrons overlap.

## 16.3 CONDUCTORS, INSULATORS AND SEMICONDUCTORS

In solids, conduction requires the movement of electrons. But for an electron to move, there must be a partially vacant energy band. If all energy states in a band are completely filled, then an electron cannot move because the Pauli's principle says, it cannot go into a filled state. So, differences in conductivity between different substances must be a matter of vacant or partially filled bands. Consider the electron energy versus inter-atomic spacing diagram picturing the result of the thought experiment in which an expanded lattice is contracted. This contraction is stopped when the equilibrium inter-atomic spacing reached $d_m$ (Figures 16.2 and 16.3).

Then inside the crystal, the electron energy states would be grouped into bands. One can talk of lower or *valence band* which results from the overlap of filled valence orbitals of the individual atoms and upper or *conduction band* which results from the overlap of partially filled or empty higher orbitals of the atoms involved. In such a material, mobile electrons can arise in two ways. Either the valence band containing electrons is only partially filled and thus gives rise to electrons state to which electron can migrate, or even if this band (valence band) is completely filled, it can overlap an unfilled band (conduction band), where unoccupied energy states permits electron drift. Since, there are plenty of vacant states, the concentration of mobile charge carriers is high and so will be the conductivity which depends upon this concentration. The crystal will show metallic conduction.

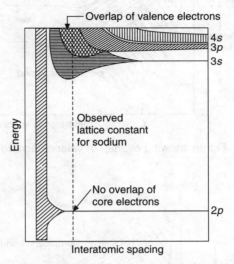

**FIGURE 16.2** The band picture of sodium metal.

If however, the equilibrium inter-atomic spacing in a certain solid is $d_I$. It is noticed that there exists a large range of forbidden energies between the first and second bands (Figures 16.3 and 16.4). Now, what happens if there are just enough available valence electrons to fill the first band, the valence band. Then, these electrons will not be able to find any easily accessible vacant energy states, in the valence band for them to move into. Further, if the energy gap $E_g$ is large compared with the thermal energy $kT$ of the electrons, the electrons cannot significantly be thermally excited into the conduction band. In effect, therefore there will be no mobile electrons in either band. The material will behave like an insulator (Figure 16.5).

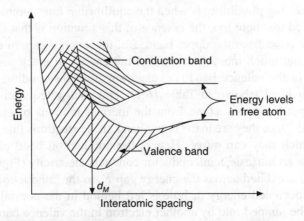

**FIGURE 16.3** Dependence of energy on inter-atomic spacing.

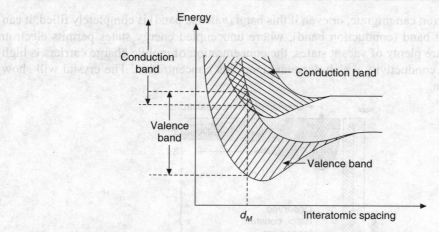

**FIGURE 16.4** Figure showing overlap of valence and conduction band.

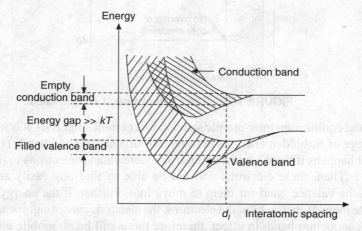

**FIGURE 16.5** Band picture of an insulator with an inter-atomic spacing $d_I$.

A third and most interesting possibility is when the equilibrium inter-atomic spacing in the solid is $d_{SC}$. It will be noticed that here too, the essence of this situation is that there is an energy gap separating the valence band from the upper band. But this energy gap, in contrast to that in the case of insulators, is not much more than the thermal energy of electrons and therefore small enough for electrons in the valence band (i.e. electrons used for bonding atom together) to be excited into the upper band (shown in Table 16.1). The energy required for the excitation of electrons into the upper band may come from the thermal motions of electrons or from light shining on the material. Once they are in the upper band, these electrons find plenty of unoccupied energy states into which they can move. Hence, the conduction band electrons can conduct electricity. This is how an intrinsic semiconductor conducts electricity (Figure 16.6).

When an electron is excited across the energy gap $E_g$ to the conduction band in an intrinsic semiconductor, an unoccupied energy or hole is left behind in the normally full valence band. This vacant state can be jumped into by another electron in the valence band, but this leads to a

**TABLE 16.1** Room temperature energy gap for some materials

| Substance | Energy gap (eV) |
|---|---|
| *Insulator* | |
| C, Diamond | 5.6 |
| *Semiconductor* | |
| Ge | 0.656 |
| Si | 1.089 |
| Te | 0.34 |
| Ga, As | 1.35 |
| In, Sb | 0.17 |

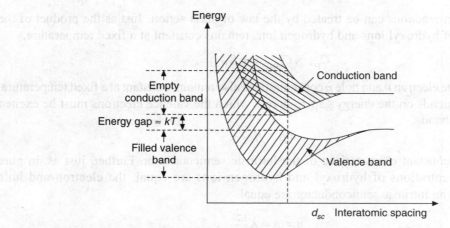

**FIGURE 16.6** The band picture of a semiconductor with an inter-atomic spacing $d_{SC}$.

vacant state in the place where the jumping electron was. The motion of electrons into unoccupied energy states or holes in the valence band is therefore equivalent to the movement of the vacant states or holes, in the opposite direction. Since, an electric field moves holes in an opposite direction to electrons the holes may be treated as if there were positively charged (Figures 16.7 and 16.8).

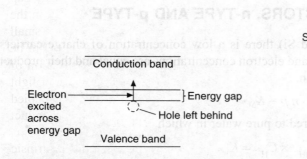

**FIGURE 16.7** Formation of a hole when an electron from the valence band is excited into conduction band.

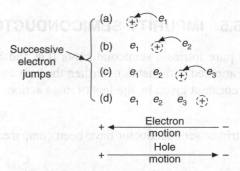

**FIGURE 16.8** Hole behaves as if it were positively charged.

## 16.4 COMPARISON OF SEMICONDUCTORS AND ELECTROLYTE SOLUTION

Since the electrons of the valence band are used for bonding together atoms, the removal of a valence electron by excitation into conduction band implies the rupture of a band in the lattice. The creation of an electron–hole pair may therefore be treated as an ionization reaction

$$\text{Lattice} \rightleftharpoons e^- + h$$

It turns out that there are remarkable parallels between the ionization of the lattice of an intrinsic semiconductor and the ionization of water.

$$\text{Water} \rightleftharpoons H^+ + OH^-$$

Both equilibrium reactions can be treated by the law of mass action. Just as the product of the concentrations of hydroxyl ions and hydrogen ions remains constant at a fixed temperature.

$$C_{H^+} \times C_{OH^-} \rightleftharpoons K_{H_2O}$$

The product of the electron $n$ and hole $p$ concentrations also remains constant at a fixed temperature. The constant depends on the energy gap $E_g$ across which the valence electrons must be excited into conduction band.

$$n\,p = K_{SC}$$

where $K_{SC}$ is a constant characteristic of the intrinsic semiconductor. Further, just as, in pure water, the concentrations of hydroxyl and hydrogen ions are equal, the electron and hole concentration in an intrinsic semiconductor are equal.

$$n = p = K_{SC}^{1/2}$$

When one examines the value of $n = p$, it turns out that the density of charge carriers in an intrinsic semiconductor at room temperature is in the range of $10^{13}$–$10^{16}$ cm$^{-3}$ compared with about $10^{22}$ cm$^{-3}$ in a metal. It is this relatively low concentration of charge carriers in intrinsic semiconductors which is responsible for the most important differences between semiconductor electrodes and metal electrodes.

## 16.5 IMPURITY SEMICONDUCTORS, n-TYPE AND p-TYPE

In pure intrinsic semiconductors (Ge and Si) there is a low concentration of charge carriers (compared with metals). Further, the hole and electron concentrations are equal and their product is constant given by the low of mass action.

$$n\,p = K_{SC}$$

Intrinsic semiconductor have been compared to pure water in which

$$C_{OH^-} \times C_{H^+} = K_w$$

In case of water, however, the concentration of either the OH$^-$ or H$^+$ ions can be decreased by adding proton donors (acids) or proton acceptors (bases). Thus, a proton donors releases hydrogen ions to the solution (i.e. $C_{H^+}$ increases) and the only way the product $K_w$ (i.e. $C_{OH^-}$ and $C_{H^+}$) remains constant is by a decrease in $C_{OH^-}$.

Is there an analogous situation in semiconductors? If one adds an electron donor (say arsenic) to an intrinsic semiconductor (say, Ge) than the ionization of arsenic ($As \rightleftharpoons As^+ + e^-$) releases electrons into the system and the hole concentration goes down to preserve the constancy of the product $np$. In this way, the electron concentration can be made so large compared with the hole concentration that the conduction is dominantly by electrons, the substance is known as an $n$-type of semiconductor (Figure 16.9).

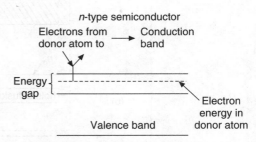

**FIGURE 16.9** Band picture of n-type semiconductor.

There is also a parallelism in semiconductor to the effect of adding a base or proton acceptor to water. This involves the addition of an electron acceptor (say Ga) to an intrinsic semiconductor. The electron acceptor ionize thus

$$Ga + e^- \rightleftharpoons Ga^-$$

and by accepting electrons, force up the hole concentration in the valence band. Such a doped semiconductor will conduct mainly by holes. It is known as a p-type of semiconductor (Figure 16.10).

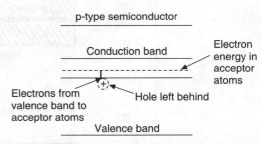

**FIGURE 16.10** Band picture of p-type semiconductor.

## 16.6 CURRENT POTENTIAL LAWS AT OTHER TYPES OF CHARGED INTERFACES: SEMICONDUCTOR N–P JUNCTIONS

Consider an interface formed by joining together an n-type of semiconductor (e.g. Ge-doped with arsenic atoms which donate electrons) and a p-type of semiconductor (e.g. Ge-doped with In atoms which accept electrons).

Before the junction is formed there is overall electro-neutrality in both types of material. Once the two types of material are brought face to face to form a junction, electrical contacts is established and a path is provided for electrons and holes to move from one side of the junction to the other (Figure 16.11).

Consider hole movement first. In the n-type of material, holes are generated when electrons from the valence band jump to acceptor atoms. These holes can random walk across the junction into n type of material. Conversely, holes from the p-side can random walk into the n-type of material where they are consumed in a hole–electron recombination process. Both electrons and holes have considerable mobility (Table 16.2).

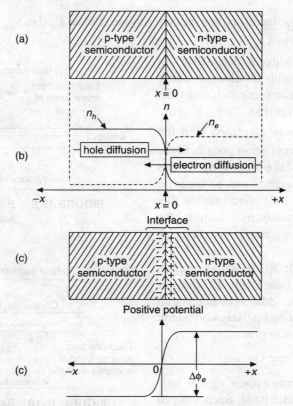

**FIGURE 16.11**  (a) Junction between p-type and n-type semiconductor (b) diffusion of hole and electrons in opposite directions, (c) results in separation of charge, (d) formation of electrical potential difference across the interface.

**TABLE 16.2**  Room temperature energy gap and electron and hole mobilities for some semiconductors

| Semiconductor | Energy gap (eV) | Electron mobility ($v_e$) (cm$^2$ V$^{-1}$ s$^{-1}$) | Hole mobility ($r_p$) (cm$^2$ V$^{-1}$ s$^1$) |
|---|---|---|---|
| Carbon (diamond) | 5.6 | 1800 | 1200 |
| Ge | 0.66 | 3900 | 1700 |
| Si | 1.09 | 1420 | 250 |
| Te | 0.34 | 300 | 200 |
| PbS | 0.37 | 500 | 150 |
| GaSb | 0.7 | 5000 | 1000 |

Since, one starts off with a far larger hole concentration in the p-type of material than the n-type of material, there will initially be more holes taking the p $\to$ n random walk. The net p $\to$ n transport of holes leaves a negative charge on the p material and confers a positive charge on the n material. A potential difference develops.

Further, this charging of the two sides of the interface and the resultant potential difference acts precisely in such a manner as to oppose further p → n hole diffusion.

Thus, equilibrium is reached when the driving force for the diffusion (concentration gradient) is just compensated for by the electric field (the potential gradient) (Figure 16.12). Under these conditions, there is an equilibrium net charge on each side of the junction and an equilibrium potential difference $\Delta\phi_e$. This whole process is analogous to the way charge transfer across a non-polarizable electrode–solution interface results in the establishment of an equilibrium potential difference $\Delta\phi_e$ across the interface (Figure 16.13).

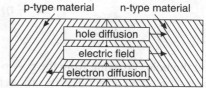

FIGURE 16.12 Established electric field opposes further diffusion of holes.

Since, there is no diffusion under equilibrium conditions, the n → p hole current is equal to the p → n hole current. These equilibrium currents are analogous to the equilibrium exchange currents at an electrode–solution interface. They represent the exchange of holes across the junction between the n-type and p-type of material and shall be designated by the symbol $i_0 h$. This $i_0$ shall now be examined more carefully.

Consider the holes which are making the p → n crossing. The number of holes approaching a unit area of the junction per second is proportional to the number per unit volume of holes on the p-side, i.e. $n_{np}$. Will they cross the junction? Each hole approaching the interface finds that it has to surmount the potential difference $\Delta\phi_e$ and the probability that it will climb the barrier is given by the Boltzmann term $e^{-e_0 \Delta\phi_e / kT}$. Hence, the p → n hole current density at equilibrium is

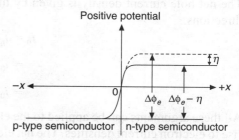

FIGURE 16.13 Potential difference across the interface is lowered by superimposing an external field.

$$\vec{i}_{h\Delta\phi e} \alpha \; n_{hp} e^{-e_0 \Delta\phi_e / kT}$$

$$\Rightarrow \qquad \vec{K} n_{hp} e^{-e_0 \Delta\phi_e / kT} \qquad (16.1)$$

where the arrow → over the $i$ represents p → n energy and $\vec{K}$ is the proportionality constant. Now think of the n → p hole current. When the holes from the n-type of medium reach the junction, they do not at all see any barrier due to an electrical potential drop. Hence, the n → p hole current density at equilibrium is controlled only by diffusion and is simply proportional to the number $n_{hn}$ (n-side) of holes in the n-type of material.

$$\overleftarrow{i}_{h\Delta\phi e} = \overleftarrow{K} n_{hn} \qquad (16.2)$$

where the arrow (←) represents p ← n crossings.

Hence, at equilibrium

$$i_{0h} = \vec{i}_{h\Delta\phi e} = \overleftarrow{i}_{h\Delta\phi e} \qquad (16.3)$$

Now, what happens if the potential difference across the junction is lowered by an amount $\eta$. The holes making the p $\rightarrow$ n crossing find a smaller barrier to climb, and hence, the p $\rightarrow$ n hole current density becomes

$$\vec{i}_h(\Delta\phi_e - \eta) = \bar{K} n_{hp} e^{-e_0(\Delta\phi_e - \eta)/kT}$$
$$= i_{0h} e^{e_0\eta/kT} \qquad (16.4)$$

But the holes crossing from the n $\rightarrow$ p type of material still have no barrier to climb at all. Hence, the n $\rightarrow$ p hole current density still depends only on the number of holes in the n-type of material and not on the potential difference across the junction, i.e. $i_h$ [n $\rightarrow$ p] is unaffected by the field.

$$\overleftarrow{i}_h(\Delta\phi_e - \eta) = i_{0h} \qquad (16.5)$$

The net hole current density is given by the difference in the hole current densities for the two directions.

$$i_h = \vec{i}_h(\Delta\phi_e - \eta) - \overleftarrow{i}_h(\Delta\phi_e - \eta)$$
$$= i_{0h} e^{e_0\eta/kT} - i_{0,h}$$
$$= i_{0h}(e^{e_0\eta/kT} - 1) \qquad (16.6)$$

All these arguments can be applied to the electrons making n $\rightarrow$ p and p $\rightarrow$ n crossings and giving rise to electron current densities. The net electron current density is given by an expression similar to Eq. (16.6) i.e.

$$i_e = i_{0e}(e^{e_0\eta/kT} - 1) \qquad (16.7)$$

The total current density across the junction is, therefore, equal to the sum of the electron and hole current densities, just as the total ionic migration current density in an electrolyte is equal to the sum of the current densities due to positive and negative ions.

Hence, the current potential laws for an n–p junction is

$$i = i'_0(e^{e_0\eta/kT} - 1) \qquad (16.8)$$

where

$$i'_0 = i_{0,h} + i_{0,e} \qquad (16.9)$$

Notice that, as $\eta$ (the departure from the equilibrium potential) increases, $e^{e_0\eta/kT}$ increases in comparison to unity until, when $e^{e_0\eta/kT} \gg 1$,

$$i = i'_0 e^{e_0\eta/kT} \qquad (16.10)$$

The exponential law for n–p junctions.

One is now in a position to compare the current potential law for an electrode–electrolyte interface (which has been referred to as e–i junction) with that for any n–p junction.

$$i = i_0 [e^{(1-\beta)e_0\eta/kT} - e^{-\beta e_0\eta/kT}] \qquad \text{e–i junction} \qquad (16.11)$$

$$i = i'_0 [e^{e_0\eta/kT} - 1] \qquad (16.12)$$

The current–potential relation at a p–n semiconductor junction differs from that at an electrode–solution interface by being totally asymmetrical (Figure 16.14).

For large departures from equilibrium, i.e. large $\eta$, both types of interface tend to give an exponential $i$ versus $\eta$ law. Thus,

$$i = i_0 e^{(1-\beta)e_0\eta/kT} \quad \text{[e–i junction]} \quad (16.13)$$

$$i = i_0' e^{e_0\eta/kT} \quad \text{[n–p junction]} \quad (16.14)$$

For small departures from equilibrium, i.e. small $\eta$ a linear $I$ versus $\eta$ law is obtained for both e–i and n–p junctions.

$$i = i_0 e_0 \eta/kT \quad \text{[e–i junction]} \quad (16.15)$$

$$i = i_0' e_0 \eta/kT \quad \text{[n–p junction]} \quad (16.16)$$

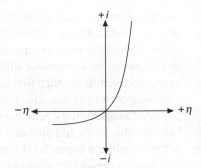

**FIGURE 16.14** Asymmetrical current potential relation at p–n semiconductor junction.

It is seen, therefore, that there are basic similarities in the $i$ versus $\eta$ laws for both types of interfaces, but there is an important difference. There is no symmetry factor $\beta$ in the exponential $i$ versus $\eta$ law for semiconductor n–p junction. In electrodics, the modification is such that only a fraction $(1-\beta)$ of the input electric energy $e_0\eta$ turns up in the change of activation energy and hence, in the rate expression. This is because the atom movements necessary for the system to reach the barrier peak are only a fraction of the total distance over which the potential difference extends. The situation in the case of the transfer of holes (or electrons) across n–p junctions is different. Since, the holes and electrons reach the barrier top only after transversing the whole distance over which the field extends, the entire $e_0\eta$ not a fraction $(1-\beta) e_0\eta$ affects the hole and electron movements.

## 16.7 THE CURRENT ACROSS BIOLOGICAL MEMBRANES

The exponential current–potential law has been shown to be obtained for charge transfer across solid–liquid (electrode–electrolyte) and solid–solid (semiconductor–junction) interfaces. The question is: Could liquid–liquid or perhaps gel–liquid interfaces display an exponential current–potential laws for charge transfer across their interfaces? Such gel–liquid interfaces are common in biological systems. The inside of a living cell is separated from the outside by a membrane which is a gel.

The axon membrane (~50–100 Å thick) separates the biological fluid inside and outside the cell. These fluids are aqueous electrolytic solutions of almost equal conductivity, but the chemical composition of the two solutions is different. Thus, $Na^+$ and $Cl^-$ ions constitute more than 90% of the charged species outside the cell. These ions, however, together account for less than 10% of the charged particles inside the cell. Here, the charged constituents are principally $K^+$ ions and a variety of negatively charged organic ions that are too large to move across the membrane, which is permeable only to $Na^+$, $K^+$ and $Cl^-$ ions. Measurement shows that the $Na^+$ concentration is normally 10 times higher outside the axon than inside it, whereas, $K^+$ concentration is 30 times higher inside the axon than outside it.

Further, the mobilities of these different ionic species in their passage through the membrane are not the same, $K^+$ and $Cl^-$ ions can permeate the membrane more easily than $Na^+$ ions and the inorganic ions, much more easily than the large organic ions. These differential mobilities or permeabilities across the membrane result in the development of a potential difference (~ 0.08 V) across the membrane such that the inside is negative with respect to the outside. Choosing the outside potential as an arbitrary zero, the potential difference between the inside and the outside of the membrane is the membrane potential (Figure 16.15).

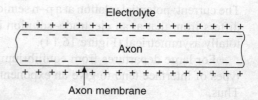

FIGURE 16.15 Different mobilities of ions across the membrane result in separation of charge, making inside negative.

Now suppose that a pulse of current is applied across the membrane with a polarity such that the net membrane potential is increased in the positive direction to a value above a certain threshold potential. Then by some, yet uncertain mechanism, the permeability of the membrane to different ions is found to alter and a flow of $Na^+$ ions occur from outside solution through the membrane to the solution inside it. It is said that membrane *opens up the sodium gates*. As a result of this transfer of positive charge into the cell, the inside of the axon becomes charged positively with respect to the outside and the potential difference across the nerve cell changes from about –0.08 V to about +0.04 V (Figure 16.16).

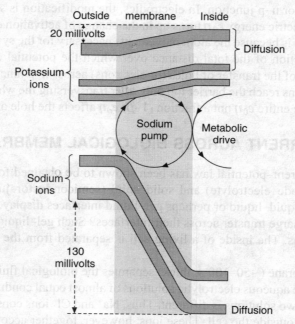

FIGURE 16.16 An unknown mechanism makes the membrane suddenly change in permeability to sodium ions.

This sudden inflow of $Na^+$ ions and the resulting change of potential difference across the membrane is termed the *action potential* or *nerve potential*. After this potential change has occurred, the original $Na^+$ and $K^+$ concentrations start recovering by some other, yet to be established, mechanism which makes the $Na^+$ and $K^+$ ions go back through the membrane against the concentration gradient, until the relative concentration inside and outside the cell assume the values they had before the potential difference was applied by the external source.

The action potential developed at a given point of the axon changes the membrane characteristics at an adjacent point and sets up the conditions for opening the Na$^+$ gates at this adjacent point, thus, a nerve impulse propagates along the axon. It is the propagation of such impulses which provides the basis for the communication between the various parts of the body and the brain (Figure 16.17).

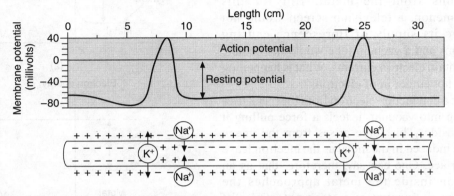

**FIGURE 16.17** The action potential, developed at one point, sets up the conditions for opening the sodium gates at the adjacent point, and thus, the nerve impulse propagates along the axon.

Now, the interesting point is that the transmembrane Na$^+$ current depends upon the potential difference across the membrane. What is the nature of variation of the Na$^+$ ion current with the membrane potential?

Experiment yields an interesting answer. The plot of $\Delta\phi$, the membrane potential, against log $i_{\text{ion}}$ is a straight line. In other words, the ion current varies with the membrane potential according to an exponential law (Figure 16.18)

$$i_{\text{ion}} = A e^{B\Delta\phi} \qquad (16.17)$$

where $A$ and $B$ are constants. How this exponential current–potential relationship comes about, will constitute exciting research for the immediate future. That the interpretation will be along electrochemical lines seems rather probable. For example, a number of examples shows that in $B = \dfrac{1}{2} F/RT$ as expected from Bulter–Volmer equation. That the research will have a far reaching impact is quite certain because membrane-liquid interfaces occur in all biological systems and the mechanism of their function is one of the most important aspects of molecular biology.

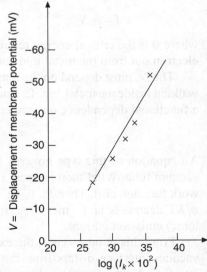

**FIGURE 16.18** The plot of membrane potential with log of the ionic flux is linear.

## 16.8 THE HOT EMISSION OF ELECTRONS FROM A METAL INTO VACUUM

Another interesting interface and one that has been studied for sometime is that between a metal and vacuum. By raising the temperature sufficiently, it is possible to evaporate the electrons from the metal. But for this phenomenon, a television screen would not receive its supply of florescence causing electrons and a vacuum tube would not be able to maintain electron currents. What is happening in this hot emission of electrons?

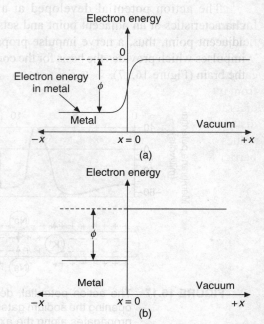

As a conduction electron from the metal tries to jump into vacuum, it feels a force pulling it back inside. This force arises from the image charge induced in the metal as the electron leave the surface. One can say, therefore, that as an electron inside the metal approaches the interface, it encounters a steep potential cliff (Figure 16.19). Hence, the emission current, i.e. the number of electrons jumping out of 1 cm² of the surface per second will be the electronic charge $e_0$ times the number $N_e$ striking the surface times a probability factor $e^{-\phi/KT}$

$$i = e_0 N_e e^{-\phi/KT} \quad (16.18)$$

**FIGURE 16.19** As an electron inside the metal approaches the surface, it encounters a steep potential cliff (a) which can be represented with sufficient accuracy as a potential step (b).

where $\phi$ is the critical energy required to lift an electron out from the metal into vacuum. The quantity $\phi$ is known as the *work function*.

The $N_e$ must depend on the temperature because it reflects how fast the electrons are random walking inside the metal, etc. To emphasize this point, one can write $N_e(T)$ which indicates formally a functional dependence on temperature. Thus,

$$i = e_0 N_e(T) e^{-\phi/KT} \quad (16.19)$$

An equation of this type governs the emission current from the electron source electrode in a vacuum tube. What then happen, if the majority of electrons inside the metal cannot climb the work function cliff. Then $i = 0$, i.e. hardly any emission. By raising the temperature of the metal, $\phi/KT$ decreases and $i$ increases. In such a situation one has a temperature-produced (thermo-ionic) emission current.

From this point of view, the exponential law for the charge transfer current across a metal–vacuum interface differs from the exponential law for the current at an electrode–electrolyte interface. In the thermo-ionic emission, one has to increase the temperature in order to increase the current. At a metal–solution interface $i = A\,e^{B\Delta\phi}$ and one can achieve the same result in a much easier way by changing the potential difference across the interface by turning a knob on an external power supply.

## 16.9 THE COLD EMISSION OF ELECTRONS FROM A METAL INTO VACUUM

Instead of boiling off the electrons from the metal into vacuum. Suppose that a strong electric field (~$10^6$ V cm$^{-1}$) is turned on to help the electrons jump out.

The potential barrier at the interface will be altered. The barrier in the presence of the field can be synthesized in a way similar to that for a charge transfer reaction can be approximated from two potential energy distance curves. In this case, however, one would combine into a single diagram the following curves: (i) an image energy distance curve, and (ii) an electric potential–distance curve, which will be a straight line, if the applied electric field is a constant. The calculated values of the image interaction energy as a function of distance is given in Table 16.3. The resulting barrier is shown in Figure 16.20.

**TABLE 16.3** The calculated values of the image interaction energy

| $x$ (cm) | Image energy (eV) |
|---|---|
| $10^{-8}$ | $-3.6$ |
| $10^{-7}$ | $-0.36$ |
| $10^{-6}$ | $-0.036$ |
| $10^{-5}$ | $-0.004$ |

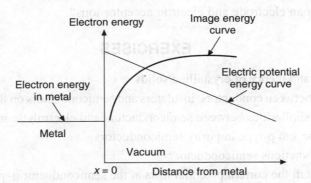

**FIGURE 16.20** A combination of electric potential and image interaction energy curve.

Common sense tells one that, if an electron had an energy less than the barrier peak it would be imprisoned for life inside the metal. Electrons, however, live by different laws, and these are not consistent with common sense notions because common sense has been built up from senses observing macroscopic objects very much larger than electrons. For example, electrons have the property of being able to penetrate into or leak through a barrier. They come out with the same energy as they had inside.

$$E_{inside} = E_{outside}$$

Thus, there can be a field induced cold emission of electrons and therefore a tunnelling current $I$, which, depends exponentially on the electric field at the interface–larger the field, the higher the tunnelling current (Figure 16.21).

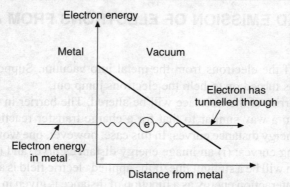

**FIGURE 16.21** Showing electro-tunnelling.

What has all this to do with charge transfer at an electrode–electrolyte interface? The picture of charge transfer (at electrode–solution interfaces) sketched so far has been conspicuously silent on the question of what the electron is doing. The picture has been completely ion centric, and the role of the electron has been dismissed with the statement that somewhere along an ion's movements to and from an electrode, electron transfer between the metal and the electrode occurs. What are the details of this electron transfer? Is there analogy to cold emission? There is an electric field at a metal–solution interface also and in place of a vacuum into which electrons can penetrate, there are electron acceptor ions or molecules to which electrons can tunnel. Is there electron tunnelling between an electrode and electron acceptor ions?

## EXERCISES

1. Discuss the band theory of crystalline solids.
2. Differentiate between conductors, insulators and semiconductors on the basis of band theory.
3. What are the similarities between semiconductors and electrolytic solutions.
4. Explain n-type and p-type impurity semiconductors.
5. What is n–p junctions semiconductor?
6. Discuss in detail the current potential laws at the semiconductor n–p junction.
7. Describes current across biological membranes.
8. Describe hot emission of electrons from a metal into vacuum.
9. Discuss cold emission of electrons from a metal into vacuum.

## SUGGESTED READINGS

Balzani, V., *Electron Transfer in Chemistry*, Vols. 1–5, Wiley, New York, 2001.

Bockris, J.O.M. and A.K.N. Reddy, *Modern Electrochemistry: An Introduction to an Interdisplinary Area*, Vols. 1 and 2, Plenum Press, New York, 1970.

Bokris, J.O.M. and A.K.N. Reddy, *Modern Electrochemistry 2B: Electrodics in Chemistry, Engineering, Biology and Environmental Science*, 2nd ed., Plenum Press, New York, 2000.

Bokris, J.O.M. and B.E. Conway, *Modern Aspects of Electrochemistry*, No. 3, Butterworths, London, 1964.

Harrison, W.A., *Electronic Structure and Properties of Solids: The Physics of Chemical Bond*, W.H. Freeman, 1980.

Holmes, P.J., *Electrochemistry of Semiconductors*, Academic Press, New York, 1962.

Kortum, G. and J.O.M. Bockris, *Textbook of Electrochemistry*, Vols. 1–2, Elsevier, Amsterdam, 1951.

Miamlin, V.A. and Y.V. Pleskov, *Electrochemistry of Semiconductors*, Plenum Press, New York, 1967.

Rickert, H., *Electrochemistry of Solids: An Introduction*, Springer-Verlag, 1982.

# 17

# Multistep Reactions

## 17.1 INTRODUCTION

The analysis of electrodic reactions has been restricted so far to one step, one electron charge transfer reactions. In practice, however, few reactions consist of only one step. For example, after electronation of $H_3O^+$ ions,

$$M(e) + H_3O^+ \longrightarrow MH + H_2O$$

One has a hydrogen atom adsorbed on the metal. The reaction does not terminate at this point. The adsorbed hydrogen atoms go on to combine and form molecular hydrogen and this hydrogen is evolved thus,

$$2MH \longrightarrow 2M + H_2 \uparrow$$

In fact, therefore the hydrogen evolution charge transfer reaction consists of several steps.

*Step* 1. $H_3O^+$ [in bulk of solution] $\xrightarrow{\text{Diffusion}}$ $H_3O^+$ [at OHP]

*Step* 2. $M(e) + H_3O^+$ [at OHP] $\xrightarrow[\text{transfer}]{\text{Charge}}$ $MH + H_2O$

*Step* 3. $2MH \xrightarrow{\text{Combination}} 2M + H_2$ [at electrode]

*Step* 4. $H_2$ [at electrode] $\xrightarrow[\text{formation}]{\text{Bubble}}$ $H_2$ [in atmosphere]

It has been assumed so far that the rate of the reaction is determined by the charge transfer, step 2. Is this always true? Should not one conceive of the possibility of the rates being controlled by steps other than the charge transfer step 2. In fact, how does one analyze the rates of a series of consecutive steps reactions, i.e. multistep reactions, which occur in a sequence of several steps?

## 17.2 QUEUES OR WAITING LINES

Queues or waiting lines can be understand by the following examples. Passengers arriving at the ticket counter in railway station. If the arrival rate of passengers at the ticket counter is greater than the servicing rate, then a queue or waiting lines of passengers builds up. Any service delay leads to the buildup of a queue. Since, the basic pattern in all these examples is the same, a general theory, known as *queuing* or *waiting times theory* has been developed.

Servicing centres may be generally quite complex. They essentially consist of subcentres. For example, at an airport there are several subcentres—landing, taxiing, unloading passengers, and freight, refueling, loading passengers, etc. Whenever there is a more than one subcentre, in the serving centre, a central question emerges in queuing theory: Which subcentre of the complex servicing centre is mainly responsible for the queue? For example, in an automobile factory what controls the overall rate of production, component manufacture, assembling the parts or the final finishing?

## 17.3 RELATION BETWEEN OVERPOTENTIAL, $\eta$ AND ELECTRON QUEUE

Think of the electrified metal–solution interface as a servicing centre for the electrons which flow into it from the metal to participate in the electronation reaction. The electrode reactions represent the servicing of the electrons.

Any servicing difficulties and delays, such as preconditions, which must be satisfied before electron tunnelling occurs, lead to queue of electrons on the electrode. In other words, the excess charge, $q_m$ on the electrode becomes more negative and thus the potential difference across the interface departs from the equilibrium value. The overpotential, therefore, is determined by the electron queue.

As in other complex servicing centres, the electrodic reaction may consist of a number of steps. For example, it has just been indicated that the overall electrodic reaction.

$$2H_3O^+ + 2M(e) \longrightarrow 2M + 2H_2O + H_2 \qquad (17.1)$$

This reaction includes the following steps

$$H_3O^+ + M(e) \longrightarrow MH + H_2O$$

and

$$2MH \longrightarrow 2M + H_2$$

Or as another example, the discharge of silver ions may consist of the transport of ions from the bulk of the solution

$$Ag^+ \text{[solution]} \longrightarrow Ag^+ \text{[OHP]}$$

and charge transfer

$$Ag^+ \text{[OHP]} + e^- \longrightarrow Ag \text{[metal]}$$

It can easily be seen that, whichever of the unit steps in the above reactions causes the hold up in the servicing centre, there will be a waiting line of electron forming. If for example, the recombination reaction of hydrogen evolution is slow, the electrons will accumulate waiting for the product of the transfer to pass through this bottleneck.

If on other side, the transport of Ag$^+$ ions is slow, the electrons will accumulate waiting for their partners like unfinished products on an assembly line waiting for parts to come to a particular place.

The above questions are obviously of crucial importance. Once there is an understanding of electron waiting lines, i.e. of the origin of the current produced potential $\eta$ then one can consider how to control the factor, that causes the electron waiting line and therefore how to control $\eta$ and perhaps significantly reduce it.

## 17.4 RELATION BETWEEN THE CURRENT DENSITY AND OVERPOTENTIAL

In case of multistep reactions, it is usually possible to single out one step and regard it as the essential cause of the overall electron queue and hence, the overpotential $\eta$.

Consider an overall electrodic reaction that takes place in $n$ steps (Figure 17.1).

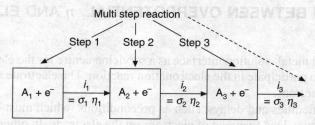

**FIGURE 17.1** Multistep electron exchange reaction.

In a multistep electron–exchange reaction, each step produces its individual current density. At a steady state, all these currents must be equal.

Let it be assumed, for convenience of exposition, that each step is a charge transfer reaction with an electron acceptor's receiving an electron.

Let it also be assumed that the $n$ individual electronation reactions are only slightly off equilibrium and that, therefore, for each reactions one can use the linear current density overpotential laws. The rate of any one step in such a case is proportional to its overpotential $\eta$.

$$i_j = \sigma_j \eta_j \qquad (17.2)$$

where in analogy to Ohm's law, $\sigma_j$ is the reciprocal resistance or the conductivity of the reaction step $j$.

Each of the $n$ electronation steps has associated with it an individual current density, which is produced by a corresponding overpotential at the interface. Thus, one can write

$$i_1 = \sigma_1 \eta_1, \quad i_2 = \sigma_2 \eta_2 \ldots i_n = \sigma_n \eta_n \qquad (17.3)$$

Now consider the situation where the overall reaction settles down and the intermediates do not change with time, i.e. when steady state conditions are reached. Since, consecutive currents are being considered and the current density from one reaction must be equal to the current density for the following reactions. Thus, the current densities of all the steps are equal to each other, i.e.

$$i_1 = i_2 = \cdots = i_n = i_j \, [j = 1, 2, 3, \ldots n]$$

$$= \sigma_1 \eta_1 = \sigma_2 \eta_2 = \cdots = \sigma_n \eta_n$$
$$= \eta_1/1/\sigma_1 = \eta_2/1/\sigma_2 = \cdots = \eta_n/1/\sigma_n \quad (17.4)$$

The equalities in Eq. (17.4) can also be written as

$$i_j \, 1/\sigma_1 = \eta_1, \quad i_j 1/\sigma_2 = \eta_2, \quad i_j \, 1/\sigma_n = \eta_n \quad (17.5)$$

and, summing all the equations, one obtains

$$i_j \left( \frac{1}{\sigma_1} + \frac{1}{\sigma_2} + \cdots + \frac{1}{\sigma_n} \right) = \eta_1 + \eta_2 + \cdots + \eta_n \quad (17.6)$$

$$i_j = \frac{\sum_{j=1}^{n} \eta_j}{\sum_{j=1}^{n} \frac{1}{\sigma_j}} \quad (17.7)$$

But the steps behave as parallel to each other as far as the electrons flow through the interface is concerned. Hence, the total current must also be equal to the sum of those individual currents, i.e.

$$i = i_1 + i_2 + \cdots + i_n = n\, i_j \quad (17.8)$$

Hence the total current flowing through the interface is from Eqs. (16.5) and (16.6),

$$i = \frac{\sum_{j=1}^{n} \eta_j}{\frac{1}{n}\sum_{j=1}^{n} \frac{1}{\sigma_j}} \quad (17.9)$$

The inverse of the conductivity of each reaction is its resistivity and the sum of all the resistivities divided by their number gives the average resistivity, $R_F$ of the reaction.

$$\frac{1}{n} \sum_{j=1}^{n} \frac{1}{\sigma_j} = R_F \quad (17.10)$$

This average resistivity $R_F$ shall be called the Faradaic resistance of the interface. The argument can be generalized without restricting it to near the equilibrium. The only difference is that, far from equilibrium, exponential current density potential relations are operative and the resistance of individual reactions as well as the Faradaic resistance are not constant any more, but dependent on overpotential. This is basic operational difference between the Faradaic and Ohmic resistances.

## 17.5  RATE-DETERMINING STEP

Observe what happens in Eqs. (17.7) and (17.9) if the conductivity $\sigma$ for one step $_r^R$ is much smaller than that for any other step $J \neq _r^R$, i.e.

$$\sigma_r^R \ll \sigma_j \; [j \neq _r^R] \quad (17.11)$$

In that case,

$$\sum_{j=1}^{n} \frac{1}{\sigma_j} = \frac{1}{\sigma_1} + \frac{1}{\sigma_2} + \cdots + \frac{1}{\sigma_r} + \cdots + \frac{1}{\sigma_n} \approx \frac{1}{\sigma_{R_r}} \quad (17.12)$$

because, all the terms $1/\sigma_j \, [j \neq {}^R_r]$ becomes insignificant in comparison with $\dfrac{1}{\sigma_{R_r}}$.

Same must apply to the overpotentials $\eta$. They must all become insignificant compared with $n_r$

$$n_r \gg n_j \, [j \neq r] \text{ or } n \approx n_r \quad (17.13)$$

Thus, Eq. (17.9) can be written as

$$i \approx \frac{\eta}{\dfrac{1}{n}\left(\dfrac{1}{\sigma_r}\right)} \quad (17.14)$$

Hence, a single step will control the overall rate of its conductivity is much smaller (or its resistivity is much larger) than that of any other step.

It was seen earlier that the conductivity $\sigma_j$ of any step is determined largely by its equilibrium exchange current densities, $i_{o,j}$. The smaller the $i_{o,j}$ is for the step, the lower is its conductivity. Thus, one can say that the step with the smallest $i_{o,j}$ generally determines the overall current.

In fact, one can imagine that the electrodic reaction is like a resistor and the Faradic resistance of the overall reaction is a series combination of resistors in an electrical circuit. Then, the overall conductance of the circuit is approximately given by the smallest conductance or largest resistance, so long as one of the resistors is significantly say 10 times larger than any of the other resistors.

One should note $i_{o,j}$ consists of two factors, the rate constant and the concentration of the substrate in the given step. Hence, either of those being small can be the cause of a slow rate-determining step (RDS).

If all the exchange current densities, except that for the RDS are very large, it means that the overpotential due to all other steps are negligibly small. Since, the magnitude of the overpotential for a step is a measure of how far the step is away from equilibrium, than if $\eta \to 0 \, [j \neq r]$. One concludes that the $j$th step is almost in equilibrium, i.e. it is in Quasi equilibrium. Hence, the existence of a unique RDS usually implies that other steps are virtually in equilibrium.

The electron waiting line problem is hence clear. In a particular multistep electron transfer reaction, the step with the lowest servicing rate or conductivity produces the largest queue at the RDS. In other words, in the steady state, all $n$ steps proceed at the rate of the rate-determining step $i_r$ and the total net current is

$$i = n i_r \quad (17.15)$$

where $n$ is the number of the single electron transfer steps in the overall reaction.

Since, $\quad i_r = \vec{i}_r - \overleftarrow{i}_r \quad$ then

$$i = n(\vec{i}_r - \overleftarrow{i}_r) \quad (17.15\text{a})$$

In order to develop the Bulter–Volmer equation for a multistep reaction, expressions for $\vec{i}_r$ and $\overleftarrow{i}_r$ must be found for this case. Consider a multistep reaction

$$\begin{array}{c} A + e^- \rightleftharpoons B\,[\text{Step 1}] \quad B + e^- \rightleftharpoons C\,[\text{Step 2}] \ldots P + e^- \rightleftharpoons R\,[\text{Step }\vec{\gamma}]\; R + e^- \rightarrow S[\text{RDS}] \\ S + e^- \rightleftharpoons T\,[\text{Step }\overleftarrow{\gamma} \equiv n - \vec{\gamma} - 1] \ldots Y + e^- \rightleftharpoons Z\,[\text{Step }n] \end{array} \quad (17.16)$$

In which $R + e^- \rightarrow S$ is the single electron transfer RDS preceded by $\vec{\gamma}$ other single electron transfer steps and followed by $\overleftarrow{\gamma}$ such steps.

The current $\vec{i}_r$ of the forward (electronation) reaction in the RDS is equal to the

$$\vec{i}_r = F\vec{K}_r C_R e^{-\beta F\Delta\phi/RT} \quad (17.17)$$

where, $C_R$ is the concentration of an intermediate. Species R was the result of a series of charge transfer mechanisms and thus its concentration is potential dependent. To unravel this dependence, it will be recalled that all steps preceding and following the RDS can often be assumed to be at equilibrium. Then, one can equate their forward and backward rates, e.g. for the first step

$$A + e^- \rightleftharpoons B$$

$$\vec{i}_1 \simeq \overleftarrow{i}_1 \quad (17.18)$$

Using the Eqs. (17.17) and (17.18)

$$F\vec{K}_1 C_A e^{-\beta\Delta\phi/RT} \simeq F\overleftarrow{K}_1 C_B e^{(1-\beta)F\Delta\phi/RT}$$

$$C_B = K_1 C_A e^{-F\Delta\phi/RT} \quad (17.19)$$

where $K_1 = \dfrac{\vec{K}_1}{\overleftarrow{K}_1}$

Similarly, we can derive

$$C_C = K_2 C_B e^{-F\Delta\phi/RT} = K_2 K_1 C_A e^{-2F\Delta\phi/RT}$$

$$C_D = K_3 C_C e^{-F\Delta\phi/RT} = K_3 K_2 K_1 C_A e^{-3F\Delta\phi/RT}$$

and finally

$$C_R = \left[\prod_{i=1}^{\vec{\gamma}} K_i\right] C_A e^{-\vec{\gamma} F\Delta\phi/RT} \quad (17.20)$$

By substituting Eq. (17.20) into Eq. (17.17), we get

$$\vec{i}_R = F\vec{K}_r \left[\prod_{i=1}^{\vec{\gamma}} K_i\right] C_A e^{-(\vec{\gamma}+\beta)F\Delta\phi/RT}$$

$$\vec{i}_R = i'_{0R} e^{-(\vec{\gamma}+\beta)F\eta/RT} \quad (17.21)$$

where

$$i'_{0R} = F\vec{K}_r \left[\prod_{i=1}^{\vec{\gamma}} K_i\right] C_A e^{-(\vec{\gamma}+\beta)F\Delta\phi_e/RT} \quad (17.22)$$

The prime at $i'_{0R}$ indicates that the rate is now related to the concentration of initial product A and not R. In complete analogy, the rate of the backward (de-electronation) reaction

$$S \rightarrow R + e^-$$

can be related to the concentration of the final product Z by the equation

$$\overleftarrow{i}_R = F\overleftarrow{K}_r \left[ \prod_{i=n-\overleftarrow{\gamma}-1}^{n} K_i \right] C_Z e^{(\overleftarrow{\gamma}+1-\beta)F\Delta\phi/RT}$$

$$\overleftarrow{i}_R = i'_{0R} = e^{(\overleftarrow{\gamma}+1-\beta)F\eta/RT} \tag{17.23}$$

where

$$i'_{0R} = F\overleftarrow{K}_r \left[ \prod_{i=n-\overleftarrow{\gamma}-1}^{n} K_i \right] C_Z e^{(\overleftarrow{\gamma}+1-\beta)F\Delta\phi_e/RT} \tag{17.24}$$

Thus, the Butler–Volmer equation for multistep reaction can be written as follows from equation

$$i = n(\overrightarrow{i}_R - \overleftarrow{i}_R) = ni'_{0R}[e^{(\overleftarrow{\gamma}+1-\beta)F\eta/RT} - e^{-(\overrightarrow{\gamma}+\beta)F\eta/RT}]$$

$$i = i_0 [e^{(\overleftarrow{\gamma}+1-\beta)F\eta/RT} - e^{-(\overrightarrow{\gamma}+\beta)F\eta/RT}]$$

$$i = i_0 [e^{(n-\overrightarrow{\gamma}-\beta)F\eta/RT} - e^{-(\overrightarrow{\gamma}+\beta)F\eta/RT}] \tag{17.25}$$

Since, $\overleftarrow{\gamma} = n - \overrightarrow{\gamma} - 1$ and where $i_0 = ni'_{0R}$ \hfill (17.26)

In high field approximation, the first exponential term can be neglected for $\eta \ll 0$, i.e. for net electronation, and the second exponential term for $\eta \gg 0$, i.e. for net de-electronation.

In the low field approximation, where both exponential terms in the Butler–Volmer equation can be linearized.

Equation (17.25) becomes,

$$i = ni'_{0R}\left(\frac{nF}{RT}\eta\right) = i_0\left(\frac{nF}{RT}\eta\right) \tag{17.27}$$

$$n = \overleftarrow{\gamma} + \overrightarrow{\gamma} + 1$$

This treatment remains valid for two other possible reaction sequences, these are sequences in which there are

(a) Chemical, i.e. noncharge-transfer steps before and after a charge transfer RDS, and

(b) Charge transfer steps before and after a chemical RDS. In the latter case, when no charge transfer occurs in the RDS, the number of electrons transferred after the RDS will be $n - \overrightarrow{\gamma}$. There will be no effect of potential on the rate of the RDS except from that arising from previous charge transfer steps, thus the Butler–Volmer equation for a chemical RDS is given as

$$i = i_0[e^{(n-\overrightarrow{\gamma})F\eta/RT} - e^{-\overrightarrow{\gamma}F\eta/RT}] \tag{17.28}$$

which, when applying the low field approximation, produces Eq. (17.27).

Equations (17.25) and (17.27) may be written in a general form by including a factor $\gamma^+$, e.g.

$$i = i_0 \left[ e^{(n-\bar{\gamma}-\beta r)F\eta/RT} - e^{-(\bar{\gamma}+\beta r)F\eta/RT} \right] \tag{17.29a}$$

Comparing Eqs. (17.14) and (17.27) allows the term $\sigma r$ to be identified as;

$$\sigma_r = i'_{0R}\left(\frac{nF}{RT}\right) \tag{17.29b}$$

## 17.6 RATE-DETERMINING STEPS AND ENERGY BARRIERS FOR MULTISTEP REACTIONS

Every reaction has an energy barrier associated with it. When, therefore, there are series of consecutive reactions, one has a series of consecutive barriers. The overall reaction corresponds to the passage in one direction of the point representing the system across all the barriers (Figures 17.2 and 17.3).

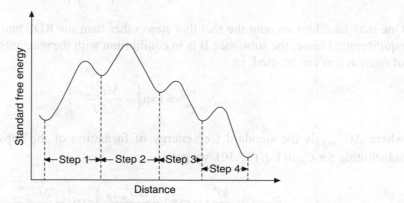

**FIGURE 17.2** Activation energy barrier for a multistep reaction.

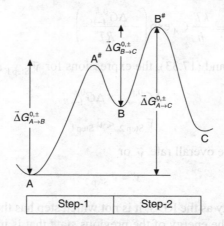

**FIGURE 17.3** Activation energy barrier is the total change in energy between the initial state and the activated state with the highest standard free energy.

It is noticed that step 1 has a larger standard free energy of activation than step, 2 i.e.

$$\Delta G^{0,\#}_{A \to B} > \Delta G^{0,\#}_{B \to C}$$

The activated state of step 2, $B^{\#}$ is higher with respect to the initial state $A$ than step 1 activated state $A^{\#}$.

The question is which step will determine the overall rate of the reaction?

Assume that step 1 determines the overall rate $\vec{v}$. One has

$$\vec{v}_{\text{Step1}} = \frac{kT}{h} C_A \exp\left(-\frac{\overline{\Delta G^{0,\#}_{A \to B}}}{RT}\right) \tag{17.30}$$

If, however, step 2 is controlling the overall rate, then

$$\vec{v}_{\text{Step2}} = \frac{kT}{h} C_B \exp\left(-\frac{\overline{\Delta G^{0,\#}_{B \to C}}}{RT}\right) \tag{17.31}$$

One may take into account the fact that steps other than the RDS can be considered in virtual equilibrium. Hence, the substance B is in equilibrium with the reactants of $A$. Therefore, the law of mass action can be used, i.e.

$$\frac{C_B}{C_A} = \exp\left(-\frac{\Delta G^0_{A \to B}}{RT}\right) \tag{17.32}$$

where $\Delta G^0_{A \to B}$ is the standard free energy of formation of the substances at $B$. Hence, by substituting for $C_B$ in Eq. (17.30), we get

$$\vec{v}_{\text{Step2}} = \frac{kT}{h} C_A \exp\left(-\frac{\Delta G^0_{A \to B}}{RT}\right) \exp\left(-\frac{\overline{\Delta G^{0,\#}_{B \to C}}}{RT}\right)$$

$$\vec{v}_{\text{Step2}} = \frac{kT}{h} C_A \exp\left(-\frac{\Delta G^{0,\#}_{A \to C}}{RT}\right) \tag{17.33}$$

On comparing Eqs. (17.30) and (17.33), the expressions for $\vec{v}_{\text{Step1}}$ and $\vec{v}_{\text{Step2}}$, it is clear that,

$$\Delta G^{0,\#}_{A \to C} > \Delta G^{0,\#}_{A \to B} \tag{17.34}$$

One has

$$\vec{v}_{\text{Step 2}} < \vec{v}_{\text{Step 1}} \tag{17.35}$$

That is, step 2 determines the overall rate $\vec{v}$ or

$$\vec{v} \approx \vec{v}_{\text{Step 2}} \tag{17.36}$$

One concludes that, to quality as the RDS, it is not which step has the highest activation standard free energy with respect to the energy of the previous state that is important, but which step has the highest standard free energy of the activated state compared with that of the initial state.

## 17.7 DETERMINATION OF ORDER

Consider an example:

$$-\frac{dC_A}{dt} = K C_A^a C_B^b \ldots C_N^n \qquad (17.37)$$

Each exponent is termed as the order of reaction in respect to the species concerned, while the sum of the exponents of the concentration terms define the overall order of a reaction.

Individual reaction orders are often expressed as derivative of the log of the rate in respect to the log of concentration of the particular species, at constant concentration of all other species, for it follows from Eq. (17.37),

$$\left(\frac{\partial \log \text{rate}}{\partial \log C_A}\right)_{C_B \ldots C_N} = a$$

or, in general case

$$\left(\frac{\partial \log \text{rate}}{\partial \log C_i}\right)_{C_{j \neq i}} = \rho_i \qquad (17.38)$$

In electrodics, the reaction rate is expressed in terms of current density $i$. Thus, one would expect, by analogy, the electrochemical order of the reaction to be given by an expression similar to Eq. (17.38) which should result from the Butler–Volmer equation,

$$i = n(\vec{i}_r - \vec{i}_r) = nF[K_r C_A^a C_A^a \ldots e^{\tilde{\alpha}F\Delta\phi/RT} - K_r C_A^{a'} C_A^{b'} \ldots e^{-\tilde{\alpha}F\Delta\phi/RT}] \qquad (17.39)$$

where $A'$, $B'$ … are the products of charge transfer reactions involving $A$, $B$, …, respectively. The exponents $a$, $b$, … and $a'$, $b'$, … in Eq. (17.39) which relates the rate of reaction (current density) to the concentration of various species, are termed as the electrochemical-reaction orders. It is stressed here that these electrochemical-reaction orders can only be related to equations such as Eq. (17.38) when $\Delta\phi$ is constant and hence, $\Delta\phi$ becomes an essential part of the definition of electrochemical reaction orders as given above.

It follows from the Eq. (17.39) that each reactant, $A$, $B$, … has a cathodic and an anodic reaction order, e.g. $a'$, $b'$, … $a$, $b$, …. At potential sufficiently anodic to neglect the cathodic reaction. Eq. (17.39) can be expressed in the form of Eq. (17.38), e.g.

$$\left(\frac{\partial \log i}{\partial \log C_A}\right)_{C_B, C_C \ldots \Delta\phi} = a \quad \text{and} \quad \left(\frac{\partial \log i}{\partial \log C_B}\right)_{C_A, C_C \ldots \Delta\phi} = b$$

At potential sufficiently cathodic to neglect the anodic reaction the electrochemical reaction order $a'$ and $b'$ are defined as

$$\left(\frac{\partial \log i}{\partial \log C_A'}\right)_{C_B', C_C' \ldots \Delta\phi} = a' \quad \text{and} \quad \left(\frac{\partial \log i}{\partial \log C_B'}\right)_{C_A', C_C' \ldots \Delta\phi} = b'$$

In a general form

$$\left(\frac{\partial \log \vec{i}}{\partial \log C_i}\right)_{C_{j \neq i}, \Delta\phi} = \rho_{i,\,\text{anodic}} \quad \text{and} \quad \left(\frac{\partial \log \vec{i}}{\partial \log C_i}\right)_{C_{i \neq j}, \Delta\phi} = \rho_{i,\,\text{cathodic}} \qquad (17.40)$$

Note that, in all these equations for electrochemical reaction orders, $\Delta\phi$ has been stipulated as a constant.

A study of the rates of reaction with respect to concentration of the $i$th species at constant overpotenital ($\eta$) produces equations which are related to Eq. (17.40), but are not obviously identical. Consider the relation between an anodic current density $\bar{i}$ and $\rho_i$, $\Delta\phi_e$ and $\eta$. Thus, one obtains,

$$\left(\frac{\partial \log \bar{i}}{\partial \log C_i}\right)_{C_{j\neq i},\eta} = \rho_{i,\,\text{anodic}} + \frac{\bar{\alpha} F}{RT}\left(\frac{\partial \Delta\phi_e}{\partial \log C_i}\right)_{C_{j\neq i}} \tag{17.41}$$

This shows that the dependence of the current density on the concentration at constant $\eta$ does not produce the electrochemical reaction order, since the expression involves $\Delta\phi_e$ which is itself a function of concentration $C_i$.

In order to obtain reaction order of say species $A$, one would measure current densities obtained of a certain potential $E$ referred to a standard electrode potential in solution containing various concentration of $A$ and constant concentration of all other reactants. The reaction order of the hydrogen evolution reaction (on mercury) in respect to hydrogen ion is equal to 1 as seen from the slope of the straight line (Figure 17.4).

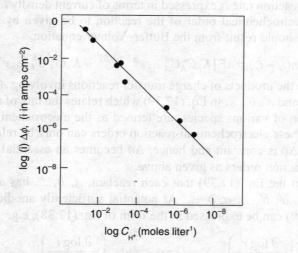

**FIGURE 17.4** The reaction order of hydrogen evolution reaction (on mercury) in respect to hydrogen ion is equal to 1.

Let us now consider a reaction whose order to be determined.

$$I_3^- + 2e \rightleftharpoons 3I^- \tag{17.42}$$

The reaction path, as shown by Vetter consists of three steps, with the third step being rate determining:

1. $I_3^- \rightleftharpoons I_2 + I^-$
2. $I_2 \rightleftharpoons 2I$
3. $2(I + e \rightarrow I^-)$

The rate equation is then written as

$$i = 2F(\vec{k}_3 C_{I^-} e^{(1-\beta)\Delta\phi F/RT} - \overleftarrow{k}_3 C_I e^{-\beta\Delta\phi F/RT}) \quad (17.43)$$

From the preceding equilibrium in steps 1 and 2,

$$C_I = (K_2 C_{I_2})^{1/2} = \left( K_1 K_2 \frac{C_{I_3^-}}{C_{I^-}} \right)^{1/2} \quad (17.44)$$

where, $K_1 = \dfrac{\vec{K}_1}{\overleftarrow{K}_1}$ and $K_2 = \dfrac{\vec{K}_2}{\overleftarrow{K}_2}$.

Substituting for $C_1$ in Eq. (17.43), produces

$$i = 2F(\vec{k}_3 C_{I^-} e^{F(1-\beta)\Delta\phi/RT} - \overleftarrow{k}_3 (k_1 k_2)^{1/2} C_{I_3^-}^{1/2} C_{I^-}^{1/2} e^{-\beta\Delta\phi F/RT}) \quad (17.45)$$

Thus, the electrochemical reaction order are

$$P_{\text{an}, I^-} = 1$$
$$P_{\text{cath}, I^-} = -\frac{1}{2}$$
$$P_{\text{cath}, I_3^-} = +\frac{1}{2}$$

## EXERCISES

1. What is a multistep reaction or process?
2. Explain the concept of queues or waiting line.
3. How overpotential is related to electron queue at an interface?
4. Establish a relation between current density and overpotential for a multistep reaction.
5. What is the rate determining step? How we can predict rate determining step in multistep reactions?
6. Explain energy barrier of rate determining reaction of multistep reactions.
7. Derive an equation for order of an electrode reaction in case of multistep process.

## SUGGESTED READINGS

Allmand, A.J., *Principles of Applied Electrochemistry*, Longmans, Green, New York, 1912.

Balzani, V., *Electron Transfer in Chemistry*, Vols. 1–5, Wiley, New York, 2001.

Bockris, J.O.M. and A.K.N. Reddy, *Modern Electrochemistry: An Introduction to an Interdisplinary Area*, Vols. 1–2, Plenum Press, New York, 1970.

Bockris, J.O.M., A.K.N. Reddy, and M. Gamboa-Aldeco, *Modern Electrochemistry 2A: Fundamentals of Electrodics*, 2nd ed., Plenum Press, New York, 2000.

Bokris, J.O.M. and A.K.N. Reddy, *Modern Electrochemistry 2B: Electrodics in Chemistry, Engineering, Biology and Environmental Science*, 2nd ed., Plenum Press, New York, 2000.

Bokris, J.O.M. and B.E. Conway, *Modern Aspects of Electrochemistry*, No. 3, Butterworths, London, 1964.

Brett, C.M.A. and A.M.O. Brett, *Electrochemistry: Principles, Methods and Application*, Oxford University Press, 1993.

Davis, C.W., *Electrochemistry*, Newnes, London, 1967.

Delahay, P., *Advances in Electrochemistry and Electrochemical Engineering*, Vol. 1, Wiley Interscience, New York, 1961.

Glasstone, S., *An Introduction to Electrochemistry*, East-West Press, New Delhi, 2003.

Gurney, R.W., *Ionic Process in Solution*, McGraw-Hill, New York, 1953.

Harned, H.S. and B.B. Owen, *Physical Chemistry of Electrolytic Solutions*, Reinhold, New York, 1950.

Kortum, G. and J.O.M. Bockris, *Textbook of Electrochemistry*, Vols. 1 and 2, Elsevier, Amsterdam, 1951.

Koryta, J. and J. Dvorak, *Principles of Electrochemistry*, Wiley, New York, 1987.

Kubasov, V. and A. Zaretsky, *Introduction to Electrochemistry*, Mir Publishers, Moscow, 1987.

Mac Innes, D.A., *Principles of Electrochemistry*, Dover, New York, 1961.

Potter, E.C., *Electrochemistry, Principles and Applications*, Macmillan, 1956.

Rochow, E., *Electrochemistry*, Reinhold, New York, 1965.

# 18

# Energy Conversion

## 18.1 ENERGY SOURCE

About 85% (coal 50%, oil and natural gas 35%) of the energy comes from the thermal combustion of coal, oil and natural gas. 15% comes from other sources such as 10% from waste, 4% from wood and 1% from hydro, wind, nuclear, etc. Therefore, the day must come when the reserves of fossil fuels are exhausted, particularly the easily transportable and usable oil and natural gas. Secondly, the efficiency of the process of the thermal combustion of these fossil fuel is very less.

## 18.2 HYDROCARBON FUEL

The feature common to both the external and internal combustion engine is that the chemical energy contained in the fuel is first converted into heat, and then this heat engine is used to produce mechanical power or its conversion by the generator to electricity.

## 18.3 ATMOSPHERIC POLLUTION FROM PRODUCTS OF INTERNAL COMBUSTION ENGINE

The organic compounds remaining incompletely burnt are several dozens in number in internal combustion engine and include particularly unsaturated compounds (Figure 18.1). $CO_2$, nitrogen oxides, sulphur oxides and lead containing compounds are also present in significant amounts. Certain organic compounds undergo a 3photochemical reaction of oxides of nitrogen to produce a complex addition compound which is the origin of *smog*. Smog is a mixture of PAN + FA + $O_3$ + LV + Soot + Ash (PAN is peroxyacylnitrate, FA is formaldehyde LV is liquid (water) vapour).

Effect of $CO_2$ in the atmosphere is *green house effect*. Temperature of earth's atmosphere will rise and leads to melting of glacier and rise in sea level.

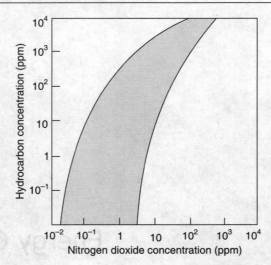

**FIGURE 18.1** An attempt to decrease the amount of hydrocarbon alone may increase the formation of harmful ozone.

## 18.4 WASTE OF CHEMICAL ENERGY AVAILABLE FROM BURNING HYDROCARBONS IN AIR

Carnot showed that all engines which convert heat to mechanical work operate by transferring heat from a source at a temperature $T_1$ to a sink at a lower temperature $T_2$ and that the efficiency $\varepsilon$ of such an engine is given by $\varepsilon = T_1 - T_2/T_1$, where temperature is in degree Kelvin. Since, $T_1 - T_2 < T_1$ then $\varepsilon < 1$, i.e. the efficiency is less than 100%. The maximum possible efficiency is prescribed is called *Carnot limitation*. This limitation is intrinsic. It cannot be avoided by improvement of engine design. A steam engine working between 356 and 100°C has a maximum efficiency of 41%.

In a real engine with moving parts, nonideal materials of construction, etc. extrinsic efficiency losses come in and reduce the efficiency. Thus, most mobile combustion engines have in practice percentage efficiency of 10–20%.

The intrinsic limitations of the heat engine method of obtaining mechanical energy from chemical reaction, is that some 60–90% of the energy is being wasted. The present method of providing available mechanical and electrical energy thus not only uses up the limited store of hydrocarbons, but wastes in doing so some two-third of the energy of the thermal combustion reaction concerned. At the same time, it pollutes the atmosphere and may even within decades, cause a significant reduction in land area by increasing sea level throughout the world.

## 18.5 DIRECT ENERGY CONVERSION

Methods discussed above, i.e. Carnot heat engine and usual method of producing electricity are indirect method.

Several methods called *direct energy conversion methods* have been known. In the thermo-ionic converter, an electronic conductor is heated till it emits electrons, these electrons are taken off by a counter electrode and a current is made to flow through an external circuit. In thermoelectric device two differing materials A and B are made to form two junctions each consisting of an A to B contact, one junction is maintained hot and other cold and electricity flows between them through a load.

In magneto-hydrodynamic converters, a hot plasma is circulated past the poles of a magnet. A fuel is used to heat the gas and ionize it. While passing between the magnets, the ions of opposite charge are attracted to the respective magnetic poles and produce a current which can then be lead through a load.

In these methods, heat is directly converted to electricity without an intermediate stage of mechanical work or a machine with moving parts. However, rather than *direct energy conversion methods* a better term might be *direct heat to electricity conversion*, for these methods, still share one very disadvantageous property. Energy in the form of heat is put in at a high temperature and comes out of the converter at a lower temperature. They are therefore still heat engine subject to the Carnot efficiency limitation, i.e. only a fraction of the energy of the chemical reaction can be turned into electricity.

What is needed is a method which avoids the loss of energy due to the intrinsic Carnot limitation.

## 18.5.1 Direct Energy Conversion by Electrochemical Means

The energy producer electrochemical system is a spontaneously working, self-driven electrochemical system. In it, there is the spontaneous occurrence of a de-electronation reaction at the electron-sink electrode and electronation reaction at the electron-source electrode. If an external load is connected to the two electrodes, a current of electrons flows in the external circuit. There is not an intermediate step in which the energy has to bring itself to power by expansion of a gas converting thereby only part of its thermal energy to mechanical work.

## 18.6 THE MAXIMUM INTRINSIC EFFICIENCY

As we know

$$-\Delta G = W_{rev} - P\Delta V \tag{18.1}$$

The change in free energy in a reaction is equal to the total reversible work obtainable from the reaction (this work to include all kinds of work i.e. gravitational, electrical surface, etc. and also the work of expansion) dimished by the work of expansion, $P\Delta V$. Hence,

$$-\Delta G = W'_{rev} \tag{18.2}$$

where $W'_{rev}$ is all the work obtainable from the reaction, exclusive of any work which can be obtained from a possible volume change in the system.

Chemical reaction in an electrochemical way

$$2H_2 \longrightarrow 4H^+ + 4e^-$$
$$O_2 + 4H^+ + 4e^- \longrightarrow 2H_2O$$
$$\overline{2H_2 + O_2 \longrightarrow 2H_2O} \qquad \text{Producer cell}$$

Hence, the normal change of free energy associated with this reaction at a given temperature and pressure has occurred in above reaction. But something else has occurred, namely transport of four electrical charges across a total potential difference of $V$. Based on assumption that the chemical change has been carried out near equilibrium, i.e. in the reversible way, the cell potential $V$ to which one here refers is the thermodynamic equilibrium potential $V = V_e$. Now, the electrical work of transporting such charges (4e⁻) is the total charge transported multiplied by the potential difference through which it is passed i.e. $4\,FV_e$. Thus, the general expression for the change of free energy in which the number of electrons transported externally $n$ is,

$$-\Delta G = n\,FV_e = W'_{rev} \tag{18.3}$$

It is in sense that, in an electrochemical energy converter the ideal maximum efficiency is 100%, i.e. the electrical energy one could draw from the reaction would be $n\,F\,V_e$ and this is all of the free energy change $\Delta G$, which is the maximum amount of useful work one can obtain from a chemical reaction (Figure 18.2).

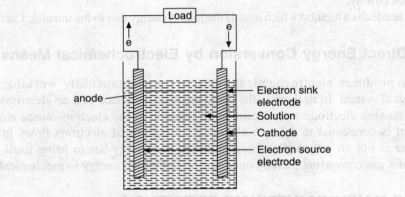

**FIGURE 18.2** Electrochemical energy producer.

Now the question is: In electrochemical methods of energy conversion, could intrinsic energy convert to electricity, all the energy which is intrinsically available as a result of a chemical reaction's occurring? Not quite all the energy difference between the reactants and products of an electrochemical reaction can be made available, however, even by the electrochemical method, some of it is wasted in very fundamental processes connected with ordering and disordering, i.e. entropy losses and gains which also occur in chemical reactions. It is the enthalpy change ($\Delta H$) which is equivalent to the total change in energy between the reactant and products of a reaction including the energy lost in entropy increases. It is more significant standard of comparison.

The $\Delta H$ is usually larger in magnitude than $\Delta G$ often by 10–20%. Hence, a second and better expression for the intrinsic maximum efficiency of an ideal electrochemical converter is

$$\varepsilon_{max} = \Delta G/\Delta H = -n\,FV_e/\Delta H \tag{18.4}$$

Thus, it can be seen from above equation that there is no general single number (e.g. 100%) which one can give for the maximum intrinsic efficiency of an electrochemical energy converter on heat content basis. Examples of values for typical overall reactions which are or might be used in fuel cell are given in Table 18.1.

**TABLE 18.1** Theoretical cell potential of various electrochemical reactions.

| Fuel | Reaction | $\Delta G^0$ | $V_e$ |
|---|---|---|---|
| Hydrogen | $H_2 + \frac{1}{2} O_2 \to H_2O$ | −56.69 | 1.229 |
|  | $H_2 + Cl_2 \to 2HCl$ | −62.70 | 1.370 |
| Propane | $C_3H_8 + 5O_2 \to 3CO_2 + 4H_2O$ | −503.9 | 1.093 |
| Methane | $CH_4 + 2O_2 \to CO_2 + 2H_2O$ | −195.5 | 1.060 |
| Formaldehyde | $CH_2O + O_2 \to CO_2 + H_2O$ | −68.2 | 1.480 |
| Formic acid | $HCOOH + \frac{1}{2} O_2 \to CO_2 + H_2O$ | −68.2 | 1.480 |
| Hydrazine | $N_2H_4 + O_2 \to N_2 + 2H_2O$ | −143.9 | 1.560 |
| Carbon | $C + O_2 \to CO_2$ | −94.26 | 1.020 |

The values can depend on whether the cell reaction is carried out in an acid or alkaline, since in the latter case, the reaction product is a carbonate with a somewhat different standard free energy than $CO_2$. Since the activity of the reactants and products depends on the concentration of the electrolyte, so does the potential of the cell reactions and the efficiency of conversion. Still one may say that the maximum intrinsic efficiency for electrochemical energy conversion on a heat content comparison basis is the region of 90% compared with heat engines which have a maximum intrinsic efficiency of 20–40%.

## 18.7 THE ACTUAL EFFICIENCY OF AN ELECTROCHEMICAL ENERGY CONVERTER

In an electrochemical energy converter, the maximum cell potential is the value $V_e$ obtainable, when the reaction in the cell is electrically balanced out to equilibrium, i.e. when no current is being drawn from the cell. As soon as, the cell drives a current through the external circuit, the cell potential falls from the equilibrium value $V_e$ to $V$. The value of actual potential $V$ at which the cell works, when delivering a current $i$ is always less than the equilibrium potential $V_e$. Hence, one has Eq. (18.4) modified as

$$\varepsilon_0 = -\frac{nFV_e}{\Delta H} \frac{V}{V_e} \tag{18.5}$$

$$\varepsilon_0 = \varepsilon_{max}\, \varepsilon_p \tag{18.6}$$

where $\varepsilon_{max}$ is maximum efficiency given by Eq. (18.4) and $\varepsilon_p$ is known as the *voltage efficiency* given by

$$\varepsilon_p = \frac{V}{V_e} \tag{18.7}$$

This picture is true if the reactants are completely converted to final reaction products, i.e. none of the electrons take part in some alternative reaction. We must consider the current or Faradic efficiency $\varepsilon_f$ to take into account the incomplete conversion of reactants into products. The overall efficiency will be

$$\varepsilon_0 = (\varepsilon_{max}\, \varepsilon_p)\, \varepsilon_f \tag{18.8}$$

In many reactions of interest $\varepsilon_f$ is virtually unity.

## 18.8 COLD COMBUSTION

Some reactions between some substances and oxygen that are carried out in an electrochemical way are called *cold combustion* (Justi and Winsel).

Thus, the net cell reaction $2H_2 + O_2 \longrightarrow 2H_2O$ is identical to the actual combustion reaction. It gives out heat which may be converted to mechanical work, with only an efficiency given by Carnot expression. The rest of heat is evolved as heat. But, in electrochemical reaction, the heat $\Delta H$ is not given out. What is given out is not hotter molecules, but a stream of electrons. The total energy of which is $\Delta G$. The combustion reaction has occurred, but cold.

It has been said that the same reaction has occurred as in combustion and the energy $\Delta G$ has been electrically drawn off. The total energy change in the reaction, however, is $\Delta H$. It is $\Delta H - \Delta G$ or $T\Delta S$ and is usually negative, i.e. heat is given out. The amounts are small, e.g. in the oxidation of 1 mol of propane into $CO_2$, $\Delta H = -530.6$ kcal/mol, but $T\Delta S$ is only $-32$ kcal/mol. Electrochemical combustion is almost cold.

## 18.9 MAKING $V$ NEAR $V_e$ IS THE CENTRAL PROBLEM OF ELECTROCHEMICAL ENERGY CONVERSION

One cannot change $I_{max}$ for a given reaction, and $\varepsilon_f$ is usually near unity. Consequently, the main efficiency determining quantity, which is subject to variation is $V$, the actual cell potential. Thus, the overall efficiency of electrochemical energy converters depends on how the overall cell potential varies with the current density, which the cell is producing.

For a very simple electrochemical energy converter having electrode of the same area $A$ and delivering a current $I$, one has

$$V = IR_e = (E_{e,so} - \eta_{a,so} - \eta_{c,so}) - (E_{e,si} + \eta_{a,si} + \eta_{c,si}) - IR_i \tag{18.9}$$

$$= (E_{e,so} - E_{e,si}) - \eta_{a,si} - \eta_{e,so} - \eta_{a,si} - \eta_{c,si} - IR_i \tag{18.10}$$

where, $E_e$ is equilibrium potential, $\eta$ is over potential, $R_e$ is the resistance external to the cell and $R_i$ is the resistance of the space between the electrodes, $\eta_{a,so}$ is overpotential at anode of source and $\eta_{c,si}$ is overpotential at cathode of sink. If both the deelectronation and the electronation reactions are running under high field or Tafel conditions, the high field approximation can be used to relate the two activation overpotentials to the current density $i_0$.

$$\eta_{a,so} = \frac{RT}{\alpha_{so}F} \ln \frac{(I/A_{so})}{i_{o,so}} \tag{18.11}$$

and

$$\eta_{a,si} = \frac{RT}{\alpha_{si}F} \ln \frac{(I/A_{si})}{i_{o,si}} \tag{18.12}$$

Further the expression for the concentration overpotential can be substituted for $\eta_{c,so}$ and $\eta_{c,si}$. They are:

$$\eta_{c,so} = \frac{RT}{nF} \ln \left(1 - \frac{I/A_{so}}{i_{L,so}}\right) \tag{18.13}$$

and
$$\eta_{c,si} = \frac{RT}{nF} \ln\left(1 - \frac{I/A_{si}}{i_{L,si}}\right) \quad (18.14)$$

Thus, the expression for the cell potential becomes,

$$V = IR_e = V_e - \left[\frac{RT}{\alpha_{so}F} \ln \frac{(I/A_{so})}{i_{o,so}} + \frac{RT}{nF} \ln\left(1 - \frac{I/A_{so}}{i_{L,so}}\right)\right]$$

$$- \left[\frac{RT}{\alpha_{si}F} \ln \frac{(I/A_{si})}{i_{o,si}} + \frac{RT}{nF} \ln\left(1 - \frac{I/A_{si}}{i_{L,si}}\right)\right] - IR_i \quad (18.15)$$

Above Eq. (15.15) represents a cell potential–cell current relation. Under current density conditions far below the limiting diffusion values and when the ohmic losses ($IR_i$) inside the cell are negligible, the activation–overpotential term dominates the expression for the relation of current to potential, i.e.

$$V \cong V_e - \frac{RT}{\alpha_{so}F} \ln \frac{(I/A_{so})}{i_{o,so}} - \frac{RT}{\alpha_{si}F} \ln \frac{(I/A_{si})}{i_{o,si}} \quad (18.16)$$

At higher current densities, the activation overpotential term in this equation change much less with current than the $IR_i$ drop term owing to the internal resistance of the cell. Under these conditions when $\frac{I/A}{i_L}$ continues to remain negligible and the variation of $V$ with $I$ is dominated by $IR_i$ term, one has

$$V \cong V_e - \text{constant} - IR_i \quad (18.17)$$

where the constant represents the activation overpotential which changes more slowly with current density than does the linear ohmic term (Figures 18.3 and 18.4).

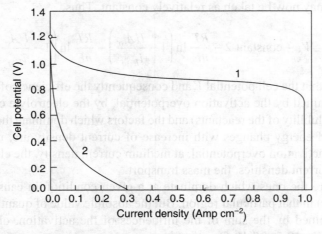

**FIGURE 18.3** Cell potential versus current density relation for an idealized electro chemical energy converter with planar, smooth electrodes.

*Curve 1.* $i_{o,si} > 1$ A cm$^{-2}$ $i_{o,so} = 10^{-3}$ A cm$^{-2}$ $i_{L,si} = 1$ A cm$^{-2}$ $i_{L,so} = 1$ A cm$^{-2}$, $\alpha_{si} = \infty$, $a_{so} = \dfrac{1}{2}$, $R_i = 10^{-2}$ Ω

*Curve 2.* $i_{o,si} = 10^{-3}$ A cm$^{-2}$, $i_{o,so} = 10^{-6}$ A cm$^{-2}$ $i_{L,si} = 1$ A cm$^{-2}$, $i_{L,so} = 1$ A cm$^{-2}$, $\alpha_{si} = \dfrac{1}{2}$, $\alpha_{so} = \dfrac{1}{2}$, $R_i = 1$ Ω.

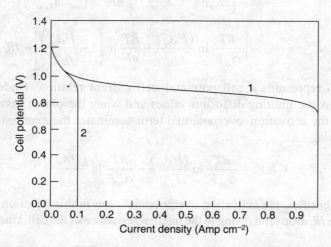

**FIGURE 18.4** Cell potential versus current density relation for an electrochemical energy converter for negligible values of *IR* showing the influence of limiting current. *Curve 1.* Same as Figure 18.3., *Curve 2.* Same as Figure 18.3 except $i_{L,si} = 10^{-1}$ A cm$^{-2}$ and $i_{L,so} = 10^{-1}$ A cm$^{-2}$.

At sufficiently high current densities the $I/A$ of Eq. (18.15) starts becoming comparable with $i_L$ and concentration overpotential starts to reduce the cell potential in a more significant way than the $IR_i$ term which may now be taken as relatively constant. Thus,

$$V \cong V_e - \text{constant } 2 - \frac{RT}{nF} \ln\left(1 - \frac{I/A_{so}}{i_{L,so}}\right) - \frac{RT}{nF} \ln\left(1 - \frac{I/A_{si}}{i_{L,si}}\right) \quad (18.18)$$

It is seen, therefore, that the cell potential $V$ and consequently the efficiency of an electrochemical converter are determined by the activation overpotential, by the electrolyte conductance and by mass transfer (i.e. solubility of the reactants) and the factors which dominate the way the efficiency of the conversion of energy changes with increase of current density are, respectively, at low current density, the activation overpotential, at medium current density, the electrolyte resistance and at the highest current densities, the mass transport.

These factors are the ones which dominate at a given condition in causing changes in the efficiency (or power) in that particular region. But the absolute value of quantities describing cell behaviour is determined by the sum of the influences of the activation, ohmic and diffusion overpotentials (Figures 18.5 and 18.6).

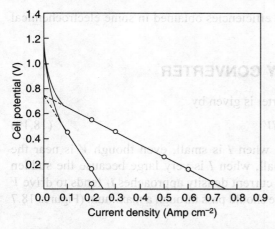

**FIGURE 18.5** Influence of the internal resistance of an electrochemical energy converter on the cell potential when mass transfer polarization is negligible.

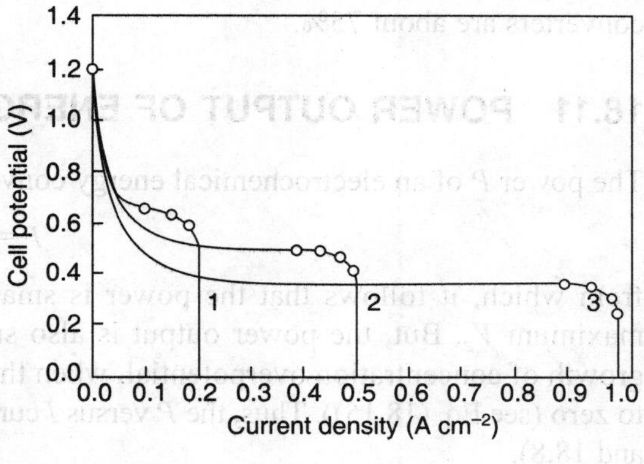

**FIGURE 18.6** The cell potential versus current density relation for an electrochemical energy converter. *Curve 1.* Shows influence of mass transfer limitations curve 1 is for $i_L = 0.2$ A cm$^{-2}$. *Curve 2.* For $i_L = 0.5$ A cm$^{-2}$. *Curve 3.* For $i_L = 1.0$ A cm$^{-2}$.

## 18.10 CONDITION FOR MAXIMUM EFFICIENCY

Inspection of Eq. (18.15) and Figures 18.2 to 18.6 shows that ideal reversible behaviour will be approached when $i_0$ and $i_L$ are very large and the internal electrolytic resistance of the cell is small. The maximumization of $i_L$ is a matter of designing and engineering cells to which the diffusion and convection of ions most easily occurs. To reduce $R_i$, electrolyte of as high a conductance as possible are used. The conductance of 1–5 M $H_2SO_4$ and KOH is in the range of 0.5 $\Omega^{-1}$cm$^{-1}$ while that of most other concentration aqueous solutions is much smaller. Hence, condition for maximum efficiency are:

1. $i_0$ can be increased by designing the electrochemical cell.
2. $i_L$ limiting current should be very large. In a manner cell is designed in which diffusion is maximum.
3. Electrolyte is made highly conducting to that $IR_i$ decreases so the $V \cong V_e$.

The main quantity is the exchange current density $i_0$. It is through this parameter that electrochemical energy conversion becomes linked up with electro-catalysis. The values of $i_0$ observed for some reactions vary over many order of magnitude (e.g. ~ $10^6$ for methanol as a fuel at different catalysts, and the aim of fundamental research on electro-catalysis is to understand the phenomenon of this considerable dependence of electrochemical reaction rates on the electronic structure of the substrate so that it is possible to make electrodes with high $i$ values (i.e. small polarization at high rates) for fuels which are cheap and which have, as do the hydrocarbons, a large amount of energy per gram.

In practice, the maximum energy conversion efficiencies obtained in some electrochemical converters are about 75%.

## 18.11 POWER OUTPUT OF ENERGY CONVERTER

The power $P$ of an electrochemical energy converter is given by

$$P = IV \tag{18.19}$$

from which, it follows that the power is small when $I$ is small, even though $V$ is near the maximum $V_e$. But, the power output is also small, when $I$ is very large because the sudden growth of concentration overpotential, when the current density approaches $i_L$ tends to drive $V$ to zero (see Eq. (18.15)). Thus, the $P$ versus $I$ curve should pass through a maximum (Figures 18.7 and 18.8).

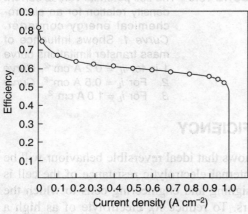

**FIGURE 18.7** The efficiency versus current density relation for an electrochemical energy converter.

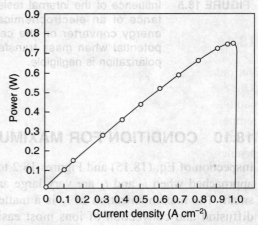

**FIGURE 18.8** The power versus current density relation for an electrochemical energy converter.

The distinction between the situation in which one needs predominantly high efficiency, and one in which one wants high power, becomes clear when one compares the $P$ versus $I$ and $\varepsilon$ versus $I$ curves. When its efficiency is at a maximum, the electrochemical energy converter is a less good power source. As the current density is increased, the power output increases, but the efficiency decreases. Of course, of the highest current drains both the power and efficiency fall towards zero.

## 18.12 WRONG STEP IN THE DEVELOPMENT OF POWER SOURCES

It is the overpotential associated with electrode reactions which is the general cause for the fall of the cell potential from the reversible value and the various types of overpotential have been related to cell potential and hence, to its efficiency and its power.

## 18.13 ELECTROCHEMICAL ELECTRICITY PRODUCERS: THE TWO TYPES

Electrochemical electricity producers are of two types, depending upon storage of the fuel is external or internal (open or closed). If the reactants (fuel) are stored outside the system, provision must be made for supply lines to feed the fuel to the electrolyte interfaces. The electrochemical system will then be an *open system*, exchanging matter with its environment. The reactants, however, can also be incorporated in the cell during manufacture and stored either as electrode material or as part of the electrolyte. In this case, one has a *closed electrochemical system*. Closed system has a limitation that when the stock of reactants is used up by the reactions, the cell's life is over.

The open electrochemical systems with external fuel storage are known as *fuel cells* and the closed *one shot* electrochemical systems with internal stocks of reactants are known as *primary batteries*. These two types of electricity producers are similar except in the location of fuel stocks.

## 18.14 FUEL CELL

Various fuel used are hydrogen, many hydrocarbons, several lower alcohol, hydrazine and ammonia. These fuels are generally used as anodically reacting materials in combination with an oxygen cathode. A rough division of present electrochemical generators may be made according to the temperature of operation. Thus, if this is below 150°C the device is called a *low temperature fuel cell* and above 500°C, a *high temperature fuel cell*. For the low temperature cells, the principal problem is that of electrocatalysis, how to raise the exchange current density for the oxidation of relatively cheap fuels by using electrode materials which do not make the initial cost of a device too high.

For the high temperature cells, the temperature alone causes the exchange current densities to be sufficiently high that there can be production of energy at a suitable rate without using expensive materials of catalysis. The counter problem then becomes the stability of the materials, the containers and electrode, under the highly corrosive action of the electrolytes.

### 18.14.1 The Hydrogen–Oxygen Fuel Cell

Discovered by Bacon in 1959 and operating at high gas pressure and somewhat elevated temperature up to 150°C. Fuel cells used in Gemini and Apollo systems of American Space program are outstanding. A schematic diagram of a single Gemini cell is shown in Figure 18.9. A unique feature of this cell is the use of a thin cation exchange membrane as electrolyte (polystyrene, sulphonic acid intimately mixed with a Kel-F spine [trifluorochloro ethylene risin, plasticizers, etc. are commercially known as *Kel-F spine*]. Each side of this rectangular membrane is covered by a titanium screen coated with a platinum catalyst. The thickness of entire cell is about 1/2 mm. Reaction taking place in the cell are:

At anode
$$2H_2 \longrightarrow 4H^+ + 4e^-$$
At cathode
$$O_2 + 4H^+ + 4e^- \longrightarrow 2H_2O$$

The performance of a single cell is shown graphically in Figure 18.10. The important overpotential loss is due to the membrane resistance and also as usual to the oxygen electrode. A storage system built of these cells having an average power of about 900 W and a maximum power of 2 kW was used in the first two man Gemini spacecraft.

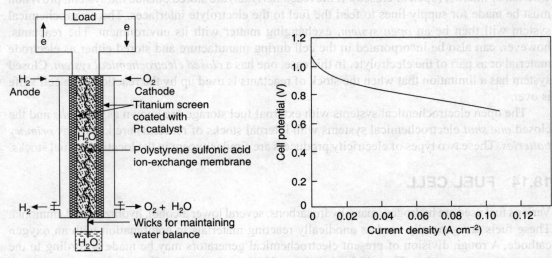

**FIGURE 18.9** Hydrogen–oxygen fuel cell.

**FIGURE 18.10** The cell potential versus current density relation for a Gemini $H_2$–$O_2$ fuel cell.

A noteworthy feature of this storage system is the self-contained provision for collecting water for drinking purpose in space. At present, some hydrogen oxygen fuel cells have attained a power level of 1 Wcm$^{-2}$.

### 18.14.2 Hydrogen–Air Fuel Cells

Pure hydrogen is expensive. Conversely, hydrogen is an excellent fuel because of its large $i_0$ value and the resulting possibility of catalyzing its dissolution well on cheap materials such as nickel. To meet this situation, a series of fuel cells utilize a system in which a cheap hydrocarbon fuel is the origin of the hydrogen, this being produced in an adjoining apparatus, separated from other gases and fed into the cell.

The steam hydrocarbon process is

$$C_nH_{2n+2} + n\,H_2O \longrightarrow (2n+1)\,H_2 + nCO$$
$$CO + H_2O \longrightarrow CO_2 + H_2$$

The $CO_2$ may be removed by absorption in ethylamine or by hydrogen separated by diffusion through palladium or silver–palladium membranes.

Ammonia or metal hydrides may also be used as the chemicals. The use of metal hydride is limited by poor economics, while, otherwise they are particularly suitable because of the ease of handling them.

The disadvantage of such a hydrogen supply is of course, in the weight and size added to the electrochemical device itself and also the cost, the start–stop difficulties, etc.

### 18.14.3 Hydrocarbon–Air Cells

Many hydrocarbons, including the main constituent of diesel oil have been oxidized electrochemically at levels of more than 99% completion. Platinum is the only suitable catalyst material at the present time.

The electrodes are constructed by depositing finely divided platinum on a porous Teflon substrate attached to a base of tantalum.

Unsaturated hydrocarbons can be oxidized at relatively low temperature. Saturated hydrocarbon, however, can be oxidized at practical rates if the temperature is in the range from 80 to 150°C. For example, oxidation of propane occurs by the following reaction:

$$C_3H_8 + 6\,H_2O \longrightarrow 3\,CO_2 + 2\,OH^+ + 20\,e^-$$

The supporting electrolyte is concentrated $H_3PO_4$. The power density of such cells is about $0.1\,W\,cm^{-2}$. The fact that the power density is about one order of magnitude less than that obtainable with cells which burn hydrogen directly, is of course, due to the much lower exchange current densities, which are exhibited during the oxidation of the hydrocarbon compared with those exhibited during the oxidation of $H_2$. This disadvantage must be compared with the lesser cost of the hydrocarbon fuel and the lesser weight and size which direct hydrocarbon converters have compared with reformer system.

Presently, the basic disadvantage of hydrocarbon burning systems are the cost of platinum catalyst and the low exchange current densities which cause less good conversion efficiency than with say hydrogen or even alcohol as fuels.

### 18.14.4 Natural Gas and CO–Air Cells

The analogy to the work of Bacon on $H_2$–$O_2$ cells in the low temperature range is the work of Broers in the high temperature range on cells with CO or natural gas (largely $CH_4$). Because methane is so unreactive, direct electrochemical oxidation of it is an inefficient process (high overpotential). Principally, there are two approaches to the use of natural gas as an electrochemical fuel. Mixed with steam, methane is reformed into $H_2$ and CO with a nickel catalyst either in situ (temperature 750°C) or externally (500–600°C) (Figures 18.11 and 18.12).

The electrolyte is the form of a paste with MgO and is a molten mixture of $LiCO_3$, $Na_2CO_3$, $K_2CO_3$. A thin layer of porous nickel is the anode while a finely divided silver constitutes the cathode.

The reactants at the cathode are $CO_2$ and air. The reactions occurring in this cell are
At cathode

$$2\,CO_2 + O_2 + 4\,e^- \longrightarrow 2\,CO_3^{2-}$$

At anode

$$CO + CO_3^{2-} \longrightarrow 2\,CO_2 + 2e^-$$
$$H_2 + CO_3^{2-} \longrightarrow CO_2 + H_2O + 2e^-$$

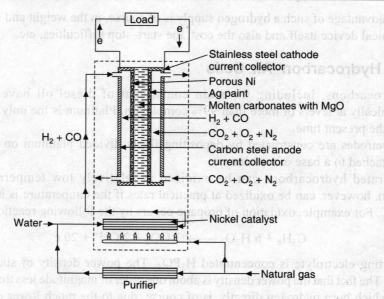

**FIGURE 18.11** High temperature (500°C) natural gas fuel cell.

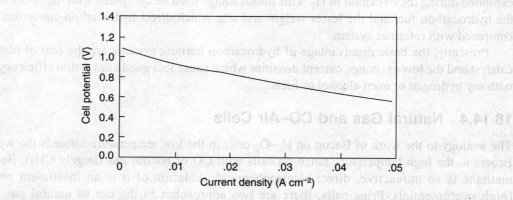

**FIGURE 18.12** The cell potential versus current density relation for natural gas fuel cell.

## 18.15 ELECTRICITY STORAGE

In an electrochemical converters described earlier, two reactants are stored outside the cells and are fed to each electrode, respectively. Now, suppose that, instead of feeding the reactants to the electrodes continually, one places the reactants on the electrodes.

On one inert electrode substrate, powdered Cd had been placed and on the other electrode, an oxide of nickel. When the external circuit is closed, the electrode on which the cadmium has been placed undergoes an anodic reaction, it de-electronates with the following overall charge transfer reaction (Figure 18.13).

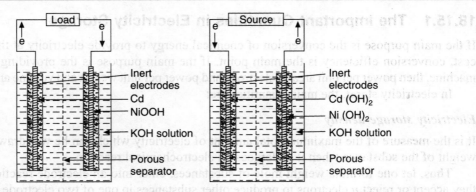

**FIGURE 18.13** A nickel–cadmium cell at the beginning of discharging and charging.

Reaction during charging are:
At the cathode
$$Cd(OH)_2 + 2\,e^- \longrightarrow Cd + 2\,OH^-$$
At anode
$$2\,Ni(OH)_2 + 2\,OH^- \longrightarrow 2\,NiOOH + 2\,H_2O + 2e^-$$
Reaction during discharging are:
At the cathode
$$2NiOOH + 2\,H_2O + 2e^- \longrightarrow 2\,Ni(OH)_2 + 2\,OH^-$$
At anode
$$Cd + 2\,OH^- \longrightarrow Cd(OH)_2 + 2e^-$$

Correspondingly, at the electrode on which the nickel has been placed, an electronation reaction occurs and can be represented as an overall reaction.
$$2\,NiOOH + 2\,H_2O + 2e^- \longrightarrow 2OH^- + 2\,Ni(OH)_2$$

But what happens, if one uses an external source of electric power to push electricity into the cell in the opposite way to that in which it passes, if this cell is allowed to function spontaneously. In an electrochemical converter with decane as the anodic fuel, the product of the overall cell reaction would be $CO_2$ and $H_3O^+$ and if the pH is acidic, $CO_2$ is evolved and expelled from the system. In an alkaline solution, the $CO_2$ would of course, form carbonate and eventually precipitate in the cell. When one tries with the aid of an external power source to push a current in a direction opposite to that in which the generator has been functioning spontaneously, there is no chance of reversing the overall reaction of decane oxidation, i.e. of reducing $CO_2$ to decane at the cathode, while evolving $O_2$ at the anode.

But in a device whose electrodes contain $Cd(OH)_2$ and $Ni(OH)_2$, the products of the spontaneous functioning of the cell are held on the electrodes. Nothing escaped from the system. Hence, on the Cd electrode, the reaction on the electrode plate covered with $Cd(OH)_2$ can be, when one puts electricity into the device, i.e. during charging:
$$Cd(OH)_2 + 2e^- \longrightarrow Cd + 2OH^-$$

and on the nickel electrode during charging:
$$2\,Ni(OH)_2 + 2\,OH \longrightarrow NiOOH + 2\,H_2O + 2e^-$$

One is back to the starting point again, charged up once more. Hence it is an electricity storer.

## 18.15.1 The Important Quantities in Electricity Storage

If the main purpose is the conversion of chemical energy to provide electricity at the minimum cost, conversion efficiency is the main point. If the main purpose is the providing power to a machine, then power per unit area of electrode and power per unit weight and volume are important.

In electricity storer, the main quantities are:

### Electricity storage density

It is the measure of the maximum total amount of electricity which can be withdrawn from unit weight of the substance when it undergoes an electrochemical reaction.

Thus, let one formula weight mole of substances, enter into an electrode reaction in which they accept or reject $n$ electrons to produce other substances in one of two electrode reactions in a cell. Then,

$$M \text{ (g of substance)} = n\, F \text{ (Coulombs of electricity)}$$

$$1 \text{ kg} = \frac{96{,}500 \times 1000\, n}{M} \text{ (C)}$$

Electricity storage density is $\dfrac{96.5 \times n \times 10^6}{M}$ $(\text{C kg}^{-1})$

The electricity storage density is the amount of electricity (coulomb) per unit weight, which the storer can hold. It states nothing concerning the energy (watts) which it can store per unit weight. Suppose the potential difference through which the amount of electricity stored passed was particularly small for a given system, then its electricity storage density could be high, but the energy it stored would be small.

Of course, for two substances to make a cell, the coulombs per kilogram of the cell can easily be calculated by using value of electricity storage density together with a knowledge of $n$ for the reaction concerned. Electricity storage density, say, little of actual behaviour of the substance at an electrode or the possibilities of making an electrode from it. Electricity storage density, therefore, only a paper parameter useful as an indication of what might be possible with certain substances if they are made to react reversibly at electrode which give the reactions large $i_0$. For example, large number of organic compounds have large $n$ values and hence, seems to indicate very attractive watt hours per kilogram. However, these substances often do not have enough electronic conductivity to form part of an electrode structure or the kinetics of their reactions at room temperature ($i_0$ factor) are so poor that cells comprising such reactions of organic substances deliver negligible power.

### Energy density

Energy which may be extracted from a given weight of a substance or a device per unit weight. Electric energy is measured by the quantity of electricity (current multiplied by the time) multiplied by the potential difference through which such a quantity passes. Thus,

$$\text{Energy density} = \text{Electricity storage density } (It) \times \text{Cell potential } (V)$$

So, maximum energy density $= \dfrac{9.65 \times 10^7\, n\, V_e}{\text{mole (kg)}}$ $(\text{C V kg}^{-1})$

where, $V_e$ is the reversible electrode potential in coulombs volts per kilogram.

Hence, maximum energy density is in kilowatt hours per kilogram

$$\frac{9.65 \times 10^7}{M} \frac{nV_e}{60 \times 60 \times 1000} = 26.8 \frac{nV_e}{\text{mole}} \text{ (kWh kg}^{-1}\text{)}$$

The real energy densities of cells are less than idealized energy density and sometimes of order of 1/5 (Table 18.2).

**TABLE 18.2** Energy storage density for some cell systems

| System density | Thermodynamic reversible pot. ($V_e$) | Realized energy density (kWh kg$^{-1}$) | Idealized energy (kWh kg$^{-1}$) |
|---|---|---|---|
| Single discharge | | | |
| M–O$_2$–KOH–Zn | 1.64 | 0.15 | 0.44 |
| Ag$_2$O$_2$–KOH–Zn | 1.81 | 0.11 | 0.22 |
| PbO$_2$–H$_2$SO$_4$–Pb | 2.04 | 0.02 | 0.18 |
| Multiple charge discharge | | | |
| Ag$_2$O$_2$–KOH–Zn | 1.70 | 0.11 | 0.44 |

Reason for this are obvious. The hypothetical maximum power density takes into account the weight of the active material with no accounting for grid, container, solution, connections, etc. In addition, $V_e$ value must of course, be reduced by the prevailing overpotential for the given conditions of the rate or current density at which, it is desired to supply electricity.

The hypothetical increase of current will simply reduce further the effective cell potential because of the increase in overpotential which an increase in current would bring about. The energy density of a storer is hence, firmly dependent by electrode kinetics on the current density or the rate at which the cell is discharged.

### Power

The question is: How much electricity can the device hold and deliver per unit weight? Another question is: rate at which the storer would be able to release the energy again? Rate in the form of current times potential is the rate of delivering energy, i.e. the power. One may imagine a device which has an excellent energy storage density, but can only release this energy at a slow rate. The disadvantage of such a device would be that, only a situation requiring low power could be coped with by it. Hence, the power density of a storer is also an important index of its merit as a device. Power is the product of the working potential of the cell multiplied by the current of the cell, which itself a function of overpotential. The power increases with the current density and passes through a maximum as shown in Figure 18.8.

## 18.16 DESIRABLE TREND IN STORER TO MAXIMIZE THE ENERGY DENSITY AND THE POWER OUTPUT

1. The cell reaction which would have a maximum thermodynamic reversible potential. $V_e$ = maximum reversible potential.
2. Cell reactions which involve in the overall reaction a maximum number of electrons ($n$ should be large).

3. Reactants of the lowest possible molecular weight.
4. Electrode reactions which have highest exchange current densities available, i.e. $i_0$ should be high.
5. A physical structure of the electrode which gives the highest possible limiting diffusion current densities.
6. Geometric structures, which gives the minimum distances between the electrodes so that energy-wasting ohmic drops caused by the passage of electricity between the electrodes are minimized.
7. The highest possible solution conductance and solubility of dissolved reaction products of discharge.
8. Highly insoluble reactants (to avoid diffusion onto the other electrode and self discharge when the cell is in the charged state.
9. The lightest possible materials of construction compatible with stability and mechanical strength.

## 18.17 LEAD ACID STORAGE BATTERY

In 1860, it was first discovered by Plante. It consists basically of two electrodes of lead sheet and the electrolytic solution $H_2SO_4$ saturated with $PbSO_4$ between them. Upon charging $Pb^{2+}$ is deposited on the lead sheet which is made cathodic by the application of a potential from an outside power source, i.e. electronation occurs. At other electrode, during charging, the $Pb^{2+}$ ions present in solution yield electrons to the anode lead plate and becomes $Pb^{4+}$ after which they undergo hydrolysis to form crystalline $PbO_2$, which deposits on the plate. The first attempt at formulating the electrode reaction during charging would therefore be

At cathode
$$Pb^{2+} + 2e^- \longrightarrow Pb$$

At anode
$$Pb^{2+} + 6\,H_2O \longrightarrow PbO_2 + 4H_3O^+ + 2e^-$$

During discharge, the reverse of the above reactions occur. There are two principal difficulties. The more important is the weight of the cell per unit of energy produced, i.e. power value of the real energy density. The theoretical value for this is 0.18 kWh kg$^{-1}$, but the practical value is 0.022 to 0.026 kWh kg$^{-1}$. The second difficulty is that plates undergo a process called *sulphation*. When the storer is not supplying electricity, spontaneous discharge takes place. The reaction occurring are:

At the lead electrode
$$Pb + SO_4^{2-} \longrightarrow PbSO_4 + 2e^-$$
$$2\,H_3O^+ + 2\,e^- \longrightarrow 2\,H_2O + H_2$$

At the $PbO_2$ electrode
$$PbO_2 + 2\,H_2SO_4 + 2e^- \longrightarrow PbSO_4 + 2\,H_2O + SO_4^{2-}$$
$$Pb + SO_4^{2-} \longrightarrow PbSO_4 + 2e^-$$

This gradual conversion of lead and lead dioxide to lead sulphate of larger particles size limit the life of the cell in practice.

Now the question is: Why such a low energy density cell as the lead acid battery used for so long? One reason is, there was not competitive product. Other reason is that the battery gives very high current densities per unit area. Its poor power weight is not so important because little total power is needed to start an automotive engine.

## 18.18 A DRY CELL

A typical dry cell is Leclanche cell (Figure 18.14). Reactions occurring in the cell during discharge are:

At anode
$$Zn \longrightarrow Zn^{2+} + 2\ e^-$$

At cathode
$$2\ MnO_2 + 2\ H_3O^+ + 2\ e^- \longrightarrow Mn_2O_3 + 3\ H_2O$$

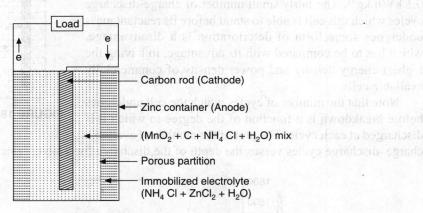

**FIGURE 18.14** A Lechlanche dry cell.

Since, hydroxide ions are produced during working (because $H_3O^+$ is consumed), the following irreversible side reaction occur.

$$OH^- + NH_4^+ \longrightarrow H_2O + NH_3$$
$$Zn^{2+} + 2\ NH_3 + 2\ Cl^- \longrightarrow Zn(NH_3)_2 Cl_2$$
$$Zn^{2+} + 2\ OH^- \longrightarrow ZnO + H_2O$$
$$ZnO + Mn_2O_3 \longrightarrow Mn_2O_3 \cdot ZnO$$

Due to above side reactions, the cell is only partially rechargeable and this to such a small extent that is never done in practice. In spite of this disadvantage, these cells are extensively used as primary batteries. They have energy density of about $0.14\ kWh\ kg^{-1}$ it discharged at low current densities.

### 18.18.1 Two More Electricity Storer

1. Silver–Zinc cell
2. Sodium Sulphur cell

## Silver–zinc cell

Discovered by Andre Yardney in 1950 (Figure 18.15). The materials used as fuel on the electrode substrates are zinc and AgO. Saturated KOH is the electrolyte. Reactions occurring in the cell during charging may be described as:

At anode

$$Ag + 2\ OH^- \longrightarrow AgO + H_2O + 2\ e^-$$

At cathode

$$ZnO + H_2O + 2e^- \longrightarrow Zn + 2\ OH^-$$

While discharging, the reverse of the above reactions takes place. The exchange current densities of the cell reactions are relatively large and the side reactions, comparatively slow in rate. These cells have energy density in the range of 0.1 kWh kg$^{-1}$. The fairly small number of charge-discharge cycles which this cell is able to stand before its reactant mass undergoes some form of deterioration is a disadvantage, which has to be compared with its advantage, in having the highest energy density and power density of commercially available cells.

Note that the number of cycles which the cell undergoes before breakdown is a function of the degree to which it is discharged at each cycle. The plot of the number of available charge–discharge cycles versus the depth of the discharge for a silver–zinc cell (Figure 18.16).

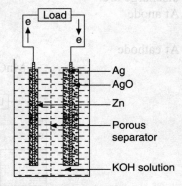

**FIGURE 18.15** A silver–zinc cell.

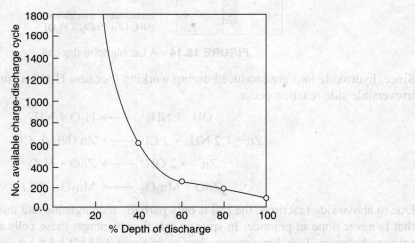

**FIGURE 18.16** Number of available charge–discharge cycles versus the depth of discharge for silver–zinc cell.

## Sodium sulphur cell

The hypothetical cell coupling on alkali metal with oxygen with the use of an aqueous solution would be an excellent storrer, were if not for the violent corrosion reaction with aqueous solution.

One might conceive of separating the alkali metal from the aqueous solution by a membrane through which only alkali metal ions would pass. The difficultly with this suggestion is that the conductance of known membrane does not attain a sufficiently high value until temperature above those at which aqueous solution boil. An analogous possibility is to use sulphur (melting point < 120°C; boiling point 444.6°C) in place of oxygen. The ions corresponding to $OH^-$ ions in aqueous electrolytes are now $S^{2-}$ ions. The advantage is that much higher temperature can be used than for the aqueous analogue and the conductance of certain membrane (in particular $Na_2O_{11}$ $Al_2O_3$ called $\beta$ Alumina) is now sufficient so that passage of current through them does not cause a large and wasteful loss in energy in the form of ohmic heat.

The reactions occurring during charging are:

At cathode
$$2 Na^+ + 2 e^- \longrightarrow 2 Na$$

At anode
$$S^{2-} \longrightarrow S + 2 e^-$$

while discharging, above reactions takes place in the reverse direction. The energy density of such a storrer is about 0.3 kWh kg$^{-1}$ and about half of this is attained in practice at 300°C.

Such a cell has the advantage of high current densities associated with the simple alkali metal reactions

$$Na \longrightarrow Na^+ + e^-$$

Conversely, the cell may suffer some long time corrosion problems.

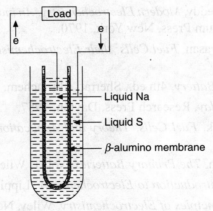

**FIGURE 18.17** A sodium sulphur cell.

## EXERCISES

1. Comment on the present situation on various types of fuel available for mankind.
2. Discuss the effect of thermal combustion engine waste on the environment.
3. Comment on direct energy conversion process by electrochemical means.
4. How can we find maximum intrinsic efficiency in electrochemical conversion of energy in a chemical reaction?

5. Find actual efficiency of an electrochemical energy converter.
6. What is cold combustion?
7. Write down relation between $V$ and $V_e$ and their effects on electrochemical energy conversion.
8. Write down the condition for maximum efficiency.
9. Write down equation for power output of an electrochemical energy converter.
10. What are the limitations of power source of electrochemical energy converter.
11. Explains two types of electrochemical electricity producers.
12. What are fuel cells. Explain with the help of few examples.
13. Explain construction, working and principle of hydrogen-oxygen fuel cell.
14. How hydrocarbon–air cell is different from natural gas and CO–air cells.
15. Explains construction, working and principle of lead acid electricity storer.
16. Explains various terms like: (i) electricity storage density, (ii) energy density, and (iii) power.
17. What are desirable trends to maximize the energy density and the power output from a storer?
18. Explains construction, working and principle of a dry cell.
19. Compare silver–zinc and sodium-sulphur cell.

## SUGGESTED READINGS

Bockris, J.O.M. and A.K.N. Reddy, *Modern Electrochemistry: An introduction to an Interdisplinary Area*, Vols. 1 and 2, Plenum Press, New York, 1970.

Bockris, J.O.M. and S. Srinivasan, *Fuel Cells: their Electrochemistry*, McGraw-Hill, New York, 1969.

Brown, H.G., *Lead Storage Battery*, 4th ed., Sherrat, Altronchem, UK, 1959.

Garret, A.B., *Batteries of Today*, Research Press, Dayton, 1957.

Hart, A.B. and G.H. Womack, *Fuel Cells: Theory and Application*, Chapman & Hall, London, 1967.

Heise, G.W. and N.C. Cahoon, *The Primary Batteries*, Vol. 1, Wiley, New York, 1971.

Klen, H.A., *Fuel Cells: An Introduction to Electrochemistry*, Lippincott, Philadelphia, 1966.

Koryta, J. and J. Dvorak, *Principles of Electrochemistry*, Wiley, New York, 1987.

Mitchell, W., *Fuel Cells*, Academic Press, New York, 1963.

Potter, E.C., *Electrochemistry, Principles and Applications*, Macmillan, 1956.

Vinal, G.W., *Primary Batteries*, Wiley, New York, 1950.

Vinal, G.W., *Storage Batteries*, 4th ed., Wiley, New York, 1955.

Williams, K.R., *An Introduction to Fuel Cells*, Elsevier, New York, 1966.

# 19 Corrosion

## 19.1 WHAT IS CORROSION

Corrosion is the loss of useful properties of a material as a result of chemical or electrochemical reaction with its environment. Almost all materials corrode, i.e. ceramics, plastics or concrete or noble metals.

The corrosion is unavoidable. However, losses due to corrosion can be considerably reduced.

## 19.2 CLASSIFICATION OF CORROSION

Based on the mechanisms, corrosion is of basically two types:

1. *Electrochemical corrosion* (*wet corrosion*): It involves an interface. It is further of three types:
   (a) Separable anode/cathode type
   (b) Interfacial anode/cathode type
   (c) Inseparable anode/cathode type. Here, the anodes and cathodes cannot be distinguished by experimental methods.

2. *Chemical corrosion* (*dry corrosion*): It involves direct chemical reaction of a metal with its environment. There is no transport of electric charge and the metal remains films free. It involves corrosion in gaseous environments, corrosion in liquid metals, fused halides and organic liquids.

## 19.3 EXPRESSIONS FOR CORROSION RATE

Weight loss measurements are most commonly used method to measure corrosion rate.

$$\text{Corrosion rate } (R) = \frac{KW}{DAT}$$

$K$ is a constant, $W$ is weight loss in mg, $A$ is area in square inch, $T$ is exposure time in hours, and $D$ is density in g cm$^{-3}$.

This method cannot be applied if the corrosion is highly selective such as intergranular or preferential and deep pitting corrosion.

Corrosion rate can be expressed in different units. For example:

1. mm/year
2. mm/diameter square/day
3. mils/year (mpy)
4. Inches/year
5. Inches/month
 (1 mils is equal to 0.0254 mm)

## 19.4 ELECTROCHEMICAL PRINCIPLES OF CORROSION

The basic difference between the chemical and electrochemical mechanism is in the overall reaction of metal with the environment. In the former case, it takes place due to direct chemical reaction of a metal with the environment. The metal remains film free and there is no transport of charge. In the latter case, it takes place by two different processes namely anodic and cathodic.

The electrochemical nature of corrosion can be illustrated by the attack of iron by HCl. When iron is placed in dilute HCl a vigorous reaction occurs, as a result of which $H_2$ gas is evolved and iron is dissolved. The reaction is

$$Fe + 2HCl \rightarrow FeCl_2 + H_2$$

Noting that the $Cl^-$ ion is not involved in the reaction, this equation can be rewritten in the simplified form

$$Fe + 2H^+ \rightarrow Fe^{2+} + H_2$$

Above reaction can be divided into two partial reactions
Oxidation (anodic reaction):

$$Fe \rightarrow Fe^{2+} + 2e^-$$

Reduction (cathodic reaction):

$$2H^+ + 2e^- \rightarrow H_2$$

The above anodic and cathodic reactions, known as *partial reactions* occur simultaneously and at the same rate on the metal surface. If this were not true the metal would spontaneously becomes electrically charged, which is clearly impossible.

However, the modern mixed potential theory based on electrode kinetics principle does not depend on assumptions regarding the distribution of local anodes and cathodes. In fact, these two theories are not conflicting—they merely represent two different approaches to the subject of corrosion.

## 19.5 TYPES OF ELECTROCHEMICAL CELL FORMATION

Three main types of electrochemical cell are:
(1) Galvanic cell, (2) Concentration cells, and (3) Electrolytic cells

1. *Galvanic cell:* Galvanic cells are formed when two dissimilar metals or the same metal consisting of dissimilar section joined together. This dissimilarity may be due to any of the following reasons acting separately or together:
    (a) Two dissimilar metals in contact.
    (b) Different heat treatment. For example, Tempered steel (cooled rapidly) is anodic to annealed steel (cooled slowly).
    (c) Scratches or abrasion, scratched or the abraded area will be anodic to the remaining portion.
    (d) *Differential strain*. A strained area is usually anodic to an unstrained area. The head and point of a nail are anodic to shank.
    (e) Differential grain and matrix composition. Many alloys may consist of heterogeneous phases which may have potential difference.
    (f) *Grain boundaries*. Grain boundary is anodic to the grain.
    (g) Different grain size, usually smaller grains are anodic to large grains.
    (h) Surface condition, a new section of pipe when welded in old pipe line becomes anodic and corrodes.
    (k) Differential curvature, more highly convex surface is anodic to the less convex surface.

2. *Concentration cell:* Difference in the electrolyte in contact with the metal can cause corrosion. Factors contributing to these difference are:
    (a) *Differential composition of the electrolyte:* Spilled battery acid seeping down to a underground pipe and other portion of pipe may be in contact with neutral ground water. This results in difference in potential and hence leads to corrosion. A section of pipe buried in deep soil is anode to pipe sections buried in loamy soil.
    (b) *Differential temperature of the electrolyte:* Two regions having temperature difference may have different potential.
    (c) *Differential motion of the electrolyte:* If a structure is subjected to varying velocities of the electrolyte, the area of higher velocity will be cathodic to area of lower velocity.
    (d) *Differential illumination:* The darker region is anodic to the brighter ones.
    (e) *Differential concentration:* The area of the metal in contact with the more dilute solution is usually anodic and corrodes.
    (f) *Differential oxygen concentration:* The electrode in deaerated solution (solution with less oxygen) becomes anode and corrodes. This type of cell account for pitting damage under rust or at the water line, less oxygen reaches the metal that is covered by rust and becomes anodic and corrodes (Figure 19.1).
    (g) *Differential temperature:* In $PbSO_4$ solution, Pb electrode at lower temperature is anode. Copper acts in the same way, but for silver the polarity is reversed.

For iron immersed in dilute aerated NaCl solutions, the hot electrode is anodic to cooler metal of same composition. But after some time, depending on aeration, stirring rate, the polarity and electrolyte composition reverse can happen.

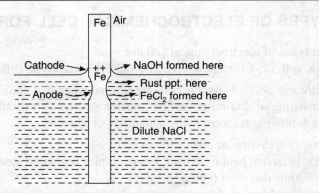

**FIGURE 19.1** Formation of differential oxygen concentration cell.

3. *Electrolytic cells:* When the surface of metal and electrolyte are homogeneous to such a degree that no cathodic and anodic areas are developed, even then the presence of energy from external source can develop cathodic and anodic areas. For example, if current enters a metallic structure at some point and leaves at other, then the area where the current leaves the metal surface becomes anodic to the area where the current enters and corrodes. Typical example of stray current corrosion of buried pipelines.

## 19.6 EXCHANGE CURRENT DENSITY ($i_0$)

On immersing a metal M in an electrolyte, an equilibrium is set up so that $M^{n+} + ne^- \rightarrow M$. The exchange current density ($i_0$) can be defined as the rate of oxidation or reduction at an equilibrium electrode expressed in terms of current density. At equilibrium there is no net current flow. $i_0$ is merely a convient way of representing the rates of oxidation and reduction of a given electrode at equilibrium. It cannot be determined theoretically. It can be determined experimentally only. Magnitude of exchange current density is a function of several variables given below:

(1) Electrode composition, (2) surface roughness (3) surface impurities (reduces by the presence of trace impurities such as arsenic, sulphur and antimony containing ions), and (4) temperature.

## 19.7 POLARIZATION OF THE ELECTRODE

When net oxidation and reduction processes occur over the electrodes, the potentials of these electrodes will be no longer be at their equilibrium value. This deviation from equilibrium potential is called *polarization*. Polarization can therefore be defined as the displacements of electrode potential resulting from a net current flow. Electrons can also be polarized by the application of external current.

The magnitude of polarization is usually measured in terms of overvoltage $\eta$ which is a measure of polarization with respect to the equilibrium potential of an electrode.
By anodic polarization of a specimen, we mean the acceleration of anodic processes on the electrode by changing the specimen potential in the cathodic direction.

At a metal electrode, many types of polarization can occur. Depending upon the cause, polarization of the following types can be easily recognized.

1. Concentration polarization ($\eta_c$)
2. Activation polarization ($\eta_a$), and
3. Resistance polarization ($\eta_r$)

1. *Concentration Polarization* ($\eta_c$): Consider the reduction of $M^{n+}$ in a solution at cathode.

$$M^{n+} + n\,e^- \rightarrow M$$

The electrode potential, according to Nernst equation is given by

$$E_1 = E_0 + \frac{0.0592}{n} \log(a_{M^{n+}})$$

where $a_{M^{n+}}$ is the activity of $M^{n+}$ ions.

If an external current is made to flow, so as to accelerate the reduction rate, then concentration of $M^{n+}$ ions in the solution near the electrode decreases. At higher currents, concentration of these ions changes to say $M_s^{n+}$. The potential of the electrode $E_2$ in this condition is given by

$$E_2 = E_0 + \frac{0.0592}{n} \log(a_{M_s^{n+}})$$

The difference in potential $E_2 - E_1$ is known as *concentration polarization* or *concentration overpotential* at the cathode ($\eta_{cc}$) and is given by

$$\eta_{cc} = E_1 - E_2 = \frac{0.0592}{n} \log \frac{a_{M^{n+}}}{a_{M_s^{n+}}}$$

Since $a_{M_s^{n+}}$ is less than $a_{M^{n+}}$, the potential of the polarized electrode $E_2$ is less noble than the unpolarized electrode $E_1$. If the external current is increased further, a limiting rate will be reached beyond which reduction cannot be increased any further. All the ions reaching the electrode are being reduced. This value of current is known as the *limiting current density* ($i_L$).

If $i_L$ is the limiting current density and $i$ the applied current density, the equation for concentration polarization for reduction process $\eta_{cc}$ is given by

$$\eta_{cc} = \frac{2.3RT}{F} \log\left(1 - \frac{i}{i_L}\right) \quad (19.1)$$

The limiting cathodic current density $i_L$ can be calculated from the following equation,

$$i_L = \frac{nDFC10^{-3}}{\delta t} \, \mu A/cm^2 \quad (19.2)$$

where $n$ is the number of $e^-$s, $D$ is diffusion coefficient of reacting ions, $F$ is Faraday's constant, $C$ is the concentration of diffusing ion (mole $dm^{-3}$), $\delta$ is the thickness of the stagnant layer (cm), $C$ is the concentration of diffusing layer (cm) and $t$ is the transport number of all ions in solution except the reacting ions and is equal to unity if many ions are present. Figure 19.2 shows concentration polarization curve for reduction process at the cathode in the absence of other types of polarization.

Concentration polarization for oxidation reactions at the anode is given by

$$\eta_{CA} = \frac{2.3RT}{F} \log\left(1 + \frac{i}{i_d}\right) \quad (19.3)$$

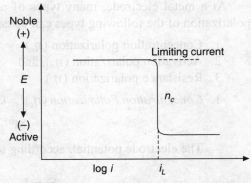

**FIGURE 19.2** Concentration polarization curve for reduction process.

where $i_d$ is the limiting anodic current density.

A comparison of Eqs. (19.1) and (19.3) shows that when the applied current density $i$ approaches limiting current density, concentration polarization at the cathode ($n_{cc}$), increases rapidly whereas concentration polarization at the anode is very small, i.e.

$$\eta_{CA} = \frac{2.3RT}{F} \log 2$$

According to Eq. (19.1), it appears that the concentration polarization at the cathode $n_{cc}$ approaches infinity as the applied current density approaches limiting current density. In practice, however, polarization never reaches infinity, instead another reaction establishes itself at a somewhat more negative potential. For example, electrodeposition of copper $E^0_{Cu^{2+}/Cu} = 0.345$ V as the applied current density approaches $i_L$, the potential moves to that of hydrogen evolution ($2H^+ + 2e^- \rightarrow H_2$) and hydrogen gas is evolved simultaneously with copper deposition. The concentration polarization decreases slowly with rise in temperature due to increased diffusion velocity. It is also reduced by stirring, since the diffusion layer $x$ then becomes thinner and the limiting current density is increased. Increased salt concentration also decreases concentration polarization at the cathode. Further, as concentration polarization is diffusion controlled, it disappears very slowly, several seconds after the current is shut off.

2. *Activation Polarization:* Activation polarization is caused by a slow electrode reaction. This is illustrated by considering hydrogen ion reduction at a cathode $2H^+ + 2e \rightarrow H_2$, the corresponding activation polarization term being called *hydrogen overvoltage* (Figure 19.3) shows the possible steps in hydrogen reduction on a platinum surface. The species must first be adsorbed on the electrode surface (step 1). Following this, electron transfer (step 2) resulting in the reduction of the species.

$$H^+ + e^- \rightarrow H_{ads}$$

As shown in step 3, two absorbed hydrogen atoms then combine to form a molecule of hydrogen,

$$2H_{ads} \rightarrow H_2$$

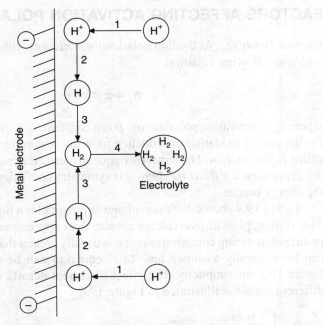

**FIGURE 19.3** Various possible steps in hydrogen reduction on a platinum surface.

These hydrogen molecules then combine to form a bubble of hydrogen gas (step 4). The rate of reduction of the hydrogen ions will be controlled by the slowest of the above steps. The controlling slow step is not always the same, but varies with metal, current density and environments. On platinum or palladium recombination of adsorbed hydrogen atoms, step 3 appears to be the rate controlling step and is the discharge of the hydrated hydrogen ions step 2.

Pronounced activation polarization also occurs with discharge of $OH^-$ ions at the anode in alkaline solutions accompanied by oxygen evolution:

$$4OH^- \rightarrow O_2 + 2H_2O + 4e^-$$

The activation polarization associated with the above reaction is known as *oxygen overvoltage*.

Overvoltage may also occur with $Cl^-$ or $Br^-$ discharge, but the values at a given current density are much smaller than for hydrogen or oxygen overvoltage.

Activation polarization also occurs during metal ion deposition or dissolution (corrosion). The value is large for transition metal like (Fe, Ni, Cr, Co, etc.), but it is small for the metals like Ag, Cu or Zn. The anion associated with the metal ion influenced metal overvoltage values more than in the case of hydrogen overvoltage. The controlling step is probably a slow rate of hydration of metal ion as it leaves the metal lattice or dehydration of the hydrated ions as it enters the lattice.

## 19.8 FACTORS AFFECTING ACTIVATION POLARIZATION

1. *Current Density:* Activation polarization increases with current density $i$ in accordance with the following equation.

$$\eta_a = \pm \beta \log \frac{i}{i_0}$$

where $\eta_a$ is activation polarization, $\beta$ is a constant, $i_0$ is the exchange current density and $i$ is the rate of oxidation or reduction in terms of current density. The above equation is called *Tafel equation*. The term $\beta$ is frequently termed *slope* or *Tafel constant* and represent the expression $2.3\ RT/\alpha\eta F$, here $\alpha$ is symmetry coefficient which describes the shape of the energy barrier.

Figure 19.4 shows that overvoltage or potential is a linear function of current density. This only applies at overvoltage greater than approximately $\pm 50$ mV. However, since polarization during corrosion reaction is usually greater than $\pm 50$ mV, a linear relationship can be generally assumed and Tafel equation can be graphically represented as in Figure 19.5 for simplicity. The effect of current density on hydrogen over potential on different metals is illustrated in Figure 19.6

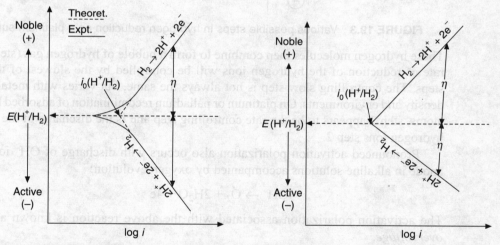

**FIGURE 19.4** Activation polarization as a function of current density for $H_2$ electrode.

**FIGURE 19.5** Activation polarization as a function of current density for hydrogen electrode.

2. *Cathode Material:* Activation polarization varies considerably from one cathode material to another. Figure 19.6 shows the effect of cathode material on hydrogen overvoltage. Higher value of exchange current density ($i_0$) and lower value of Tafel constant ($\beta$) results in lower activation polarization.
3. *Cathode Surface:* $H_2$ overvoltage is greater on smooth shining surface than on rough *unpolished* surface.
4. *Temperature:* Activation polarization decreases with increasing temperature since the exchange current density ($i_0$) increases rapidly with temperature.

5. *Pressure:* H$_2$ overvoltage on Cu, Ni and Hg cathode increases sharply on decreasing pressure whereas increasing pressure above atmospheric pressure has practically no effect.
6. *pH:* H$_2$ overvoltage first increases and then decreases with increasing pH (Figure 19.7).
7. *Impurity ions:* H$_2$ overvoltage is decreased by specific adsorption of anions and increased by specific adsorption of cations.
8. *Agitation:* Practically independent of degree of agitation of the electrolyte.

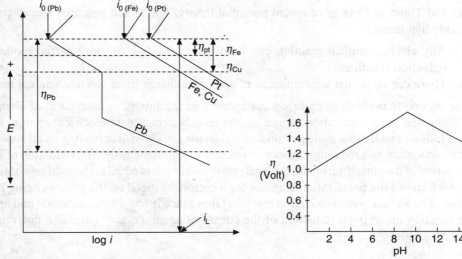

**FIGURE 19.6** Effect of electrode potential on hydrogen overvoltage.

**FIGURE 19.7** Effect of electrolyte pH on hydrogen overvoltage.

## 19.9 SIGNIFICANCE OF ACTIVATION POLARIZATION

Oxidation and reduction reactions corresponding to a H$_2$ electrode are plotted in Figure 19.5, with a $\beta$ value of 0.1 V. It is clear that the reaction rate represented by current density is very sensitive to small changes in electrode potentials. Further, it can be seen that at all potentials more active or more negative than the reversible potential, net reduction occurs and that at all potentials more noble or more positive than reversible potential a net oxidation occurs. At reversible potential, or at zero overvoltage, there is no net oxidation or reduction both being equal.

## 19.10 RESISTANCE POLARIZATION

Polarization measurements usually include an ohmic potential drop through either a portion of the electrolyte surrounding the electrode or through a metal reaction product film on the surface or both. Resistance polarization ($\eta_R$) may be written as

$$\eta_R = RI = \gamma_i$$

where $\gamma$ is film resistance for 1 cm$^2$ area in $\Omega$ cm$^{-2}$, and $i$ is current density in A cm$^{-2}$.

The passivity of metal surface caused by a film of an oxide may be considered as a special case of an anodic ohmic polarization.

Resistance polarization is practically independent of stirring rate and disappears very slowly after lapse of several seconds when the current is switched off.

## 19.11 MIXED POTENTIAL THEORY

Wagner and Traud in 1938 gave mixed potential theory. The mixed potential theory consists of two simple hypotheses.

1. Any electrochemical reaction can be divided into two or more partial oxidation and reduction reactions.
2. There can be no net accumulation of electrical charge in an electrochemical reaction.

The second hypothesis is merely a restatement of the law of conservation of charge, i.e. a metal immersed in an electrolyte cannot spontaneously accumulate electrical charge. From the above, it follows that the electrode potential of an electrically isolated corroding metal is determined by partial oxidation and reduction processes occurring simultaneously, viz. oxidation of the metal and reduction of the corroding metal, the total rate of oxidation is equal to the total rate of reduction. Figure 19.8 shows the polarization diagram for a corroding metal $m$, the process being hydrogen evolution. The surface where oxidation reaction takes place is considered as anode and as cathode for the reaction involving a reduction of the corrosive agent. ($E_{corr}$ = corrosion potential, $i_{corr}$ = corrosion current).

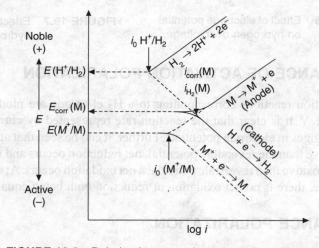

**FIGURE 19.8** Polarization curve for a corroding electrode.

If the metal was not corroding and was in reversible equilibrium with the solution it will have equilibrium potential $E(M^+/M)$ with exchange current density ($i_0$ $M^+/M$). If on the other hand, the noncorroding electrode were saturated with hydrogen gas at unit activity and pressure, it behaves as a hydrogen electrode and will assume the equilibrium electrode potential $E(H^+/H_2)$ with exchange current density $i_0$ ($H^+/H_2$).

During corrosion, these individual electrode potentials ($E(M^+/M)$ and $E(H^+/H_2)$) are displaced. The anode potential is displaced in a positive direction by anodic polarization ($\eta_{a\alpha}$) and the cathode potential in a negative direction by the cathode polarization ($\eta_{cc}$). The measurable electrode potential is a common, intermediate mixed potential the *corrosion potential* $E_{corr}(M)$ in Figure 19.8. The corrosion process is then going on with a current $i_{corr}(M)$ which in the absence of an outer current must be of equal size for the anodic and cathodic reaction. The corrosion potential and the corrosion current therefore appear at the point of intersection between two rectilinear polarization curves in the semi-logarithmic polarization diagram. The reducing species need not be hydrogen, but may be any species reducing at a more positive potential than $E_{M+/M}$. It is necessary to plot the logarithm of the total current on the abscissa rather than the logarithm of current density, as is customary for individual polarization curves. These diagrams are some times known as *Evan's diagram*.

Corrosion rate of a metal ($i_{corr}$) is affected by the polarization of the electrode. It is also affected by the resistance of the electrolyte. Resistance of the electrolyte is usually a secondary factor compared to the more important factor of polarization. When polarization occurs mostly at the cathode, it is said that the corrosion is under cathodic control (a). Typical polarization curves are shown in Figure 19.9.

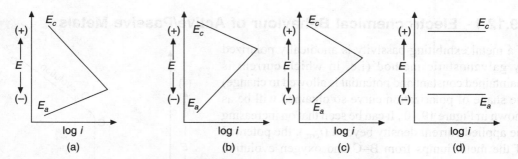

**FIGURE 19.9** Types of corrosion control.

When polarization occurs mostly at the anode the corrosion reaction is said to be under anodic control (b). When the resistance of the electrolyte is so high that the resultant current is insufficient to appreciably polarize anodes or cathodes, the corrosion is said to be under resistance control (d).

In majority of cases, polarization occurs in same degrees at both anodes and cathodes. The corrosion is then said to be under mixed control (c). The extent of polarization, depends not only on the nature of the metal and electrolyte, but also on the actual exposed area of the electrode. For example, if the cathdic area of a corroding metal be very small, there might be considerable cathodic polarization accompanying corrosion, through measurements may show that the unit area of the bare cathode polarizes only slightly at a given current density.

## 19.12 PASSIVITY

Certain metals including iron, aluminium, chromium and stainless steel almost loose their reactivity on anodic polarization under specific environmental conditions. This loss of chemical reactivity is known as *passivity*. Or

A metal or an alloy is passive if it subsequently resists corrosion in an environment where thermodynamically there is large free energy decrease associated with its passage from the metallic state to the appropriate corrosion product.

Nature of the passivity can be explained by the corrosion behaviour of iron in nitric acid. If small piece of iron or steel is immersed in concentrated (70%) nitric acid at room temperature, no reaction occurs. If the concentration of the acid is now reduced by adding water (1:1), still no change occurs and the sample remains inert as before. However, if the specimen is scratched with a glass rod, rapid reaction of the specimen occurs, whereas scratching has no effect on specimen immersed in concentrated acid. If the specimen is immersed in further diluted acid, vigorous reaction occurs without scratching. Thus, the same specimen is very reactive in dilute nitric acid and very inert in concentrated nitric acid, the difference in reaction rate being of the order of 100,000 to 1. The iron or steel specimen is considered passive in concentrated nitric acid and active in dilute acid or intermediate dilution with scratching. Examination of $E$/pH diagram for iron in aqueous solution shows that an iron specimen corroding at pH 7 can be passivated by either increasing the potential in positive direction (i.e. anodic polarization) or by increasing the solution pH. It can also be seen that in highly acidic solutions (lower pH), it may not be possible to achieve passivity even by anodic polarization.

### 19.12.1 Electrochemical Behaviour of Active/Passive Metals

If a metal exhibiting passivity is anodically polarized by galvanostatic method (i.e. in which current is maintained constant and potential is allowed to change) the shape of polarization curve so obtained will be as shown in Figure 19.10. It can be seen that on increasing the applied current density beyond ($i_{crit.}$), the potential of the metal jumps from B–C and oxygen evolution occurs. The metal at potentials between B and C is passivated. The behaviour of the metal between potential range B to C cannot be studied by galvano static methods.

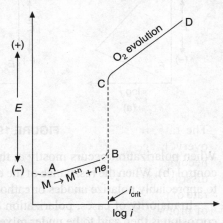

**FIGURE 19.10** Galvanostatic anodic polarization curve for metal exhibiting passivity.

However, if one uses a potentiostatic technique, the anodic polarization curve obtained is of the type shown in Figure 19.11. This potentiostatic curve clearly demonstrates that in the passive range the current suddenly drops to very low values, which is in accordance with low corrosion rates in the passive state. Initially, the metal demonstrates behaviour similar to nonpassivating metals. The point A corresponds to the equilibrium potential of the metal under the given environmental conditions and the curve AB corresponds to the anodic polarization behaviour of a normal corroding metal. As the electrode potential of the metal is raised in the positive direction, the dissolution rate increases. This range is called the *active range*. Point B corresponds to the initiation of passivation potential $E_{ip}$ at which electrochemical production of a protective oxide film is thermodynamically possible by a reaction of the type,

$$M + 2H_2O \rightarrow MO + 2H + 2e^-$$

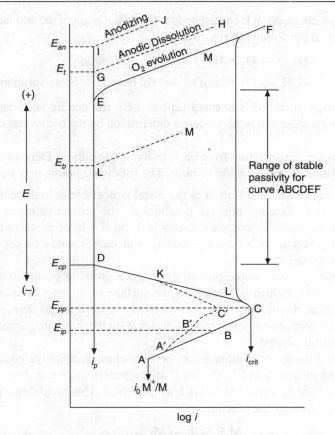

**FIGURE 19.11** Potentiostatic anodic polarization curve for metal exhibiting passivity.

The rate of anodic formation of a protective film at a potential $E_{ip}$ is small because of the overvoltage of the process. At point C, the potential $E_{pp}$ acceleration of the metal dissolution becomes exactly equal to the retardation of this process. A further increase in the rate of metal dissolution is not possible, thus a limiting passivation current ($i_{crit.}$) is reached. Beyond C, the rate of protective film growth already exceeds the rate of its chemical dissolution, and the process of protective film formation begins. This process is complete at D, at a potential of complete passivity $E_{cp}$, where entire surface of the electrode is covered with a continuous oxide film. Beginning from the point D, where formation of continous protective film is completed, the rate of anodic processes for the electrode is almost independent of potential and is largely determined by the rate of chemical dissolution of the protective film in the environment. The current drops to a very low value $i_p$ and the metal is said to be passivated. From this point D the anodic polarization curve becomes almost parallel to the potential axis indicating that the chemical process of the protective film dissolution is independent of potential. The range DE is known as passive range. Within the passive range D to E, the metal dissolution occurs at a constant rate through the passivating oxide film. This is because any increase in potential is accompanied by progressive thickening of the film so as to maintain the electrical field within it constant. This thickening usually proceeds by transport of cations $M^{+n}$ outwards and combination of these cations with $O^{-2}$ or $OH^-$ ions at the film/solution interface.

The end of the passive range, point E corresponds to the initial point of the anodic oxygen evolution (discharge) by either of the following reactions:

$$2H_2O \rightarrow O_2 + 4H^+ + 4e^- \text{ (in acid solution)}$$

$$4OH^- \rightarrow O_2 + 2H_2O + 4e^- \text{ (in neutral or basic solution)}$$

After oxygen discharge potential is reached (at point E), the anodic polarization curve on the segment EF will have a logarithmic dependence determined by the overvoltage of oxygen on the film surface.

There may, however, be deviation from basic polarization curves. Depending upon metal and environmental condition or combination of both. The following points may be observed:

1. In some cases, anodic dissolution of the metal is possible at high positive potentials. In such cases, at sufficiently positive potential $E_t$, the protective metal oxide changes to higher valence oxide or complex ions which do not have protective properties. This phenomenon is known as *transpassivity* and corresponds to segment GH on the polarization curve.
2. In some cases, at very noble potentials $E_{anode}$ a growing porous oxide film having a thickness of 200–300μm is formed on the surface of the thin nonporous passive film. This process of anodic oxidation is commonly known as *anodizing*. Aluminium and titanium are typical examples of metals showing such behaviour. Segment IJ corresponds to the anodizing process.
3. The passive film on iron, alluminium or iron–chromium alloys could be destroyed if certain active ions such as $Cl^-$, $Br^-$, or $I^-$ are present in the electrolyte. In Figure 19.11 at point L, potential $E_b$ known as *breakdown potential*. This breakdown potential is due to dissolution of film by the reaction,

$$M + 2Cl^- \rightarrow MCl_2 + 2e$$

    If this breakdown of passive film occurs locally, it results in *pitting corrosion*.
4. If the rate of the passive film dissolution is high, the anodic current in the passive rate could be quite high. This corresponds to segment CLMF in Figure 19.11. Typical example is stainless steel in a mixture of nitric acid and hydrofluoric acids. These conditions typically represent electropolishing.

### 19.12.2 Theories of Passivity

Two main theories are given below:

1. *Oxide-film theory:* According to this theory, the corrosion product, usually a metal oxide, acts as a diffusion barrier and slows down the corrosion rate by effectively separating the metal from the environment. This film is often very thin and invisible. Thus, when the passive state is established the physiochemical properties of the metal relative to the environment depend to a large extent on the properties of the protective film.
2. *Adsorption theory:* According to this theory, passivity is caused due to an adsorbed film of oxide or other passivating agents which decreases the exchange current density and increases the anode overvoltage for the anode oxidation reaction. Such a layer

displaces adsorbed water molecules and slows down the rate of metal hydration and thereby corrosion rate. This theory without rejecting the possibility of anodic film formation, responsible for passivity, describes anodic inhibition as an electrochemical mechanism.

## 19.13 FACTORS AFFECTING CORROSION RATE

Effects of various metallurgical and environmental factors on electrochemical behaviour and corrosion rate of metals exhibiting passivity is discussed here in detail:

1. *Effects of metallurgical factors:* Metallurgical factors include chemical composition, heat treatment and cold working. *Chemical composition* includes change of passivation characteristics of metals which are influenced by the presence of alloying elements. With increasing chromium content in iron, critical current density for passivation ($i_{crit}$) decreases and passivation potential ($E_{cp}$) becomes more negative. Introduction of more than 8% nickel into chromium steels promotes a decrease in the passivation current ($i_{crit}$), but shifts the passivation potential in the positive direction. *Differential heat treatment* could affect corrosion resistance of the steels. Annealed steels have increased the passive range and decreased $E_{pp}$.

   *Cold working* can also affect anodic polarization behaviour of metals. For commercial purity iron in hydrogen saturated 1N $H_2SO_4$ at 22°C, cold working reduced the passive potential range.

2. *Effects of environmental factors:* Environmental factors which effect anodic polarization behaviour and corrosion rate of metals are:

   (a) *Stirring rate of electrolyte:* Stirring of the electrolyte increases the limiting diffusion current and corrosion rate increases.

   (b) *Temperature and pH of the electrolyte:* An increase in temperature and pH of the electrolyte generally reduces the passive range, increases the critical current density for passivation ($i_{crit}$) and increases corrosion rate (i.e. corrosion current) $i_p$ in the passive range.

   (c) *Addition of oxidizers:* Increasing oxidizer concentration increases the corrosion rate of a normal corroding metal. Addition of increasing amount of oxidizer shifts the reversible electrode potential in accordance with Nerst equation.

$$E = E^0 + \frac{2.303RT}{nF} \log \frac{a_{oxy}}{a_{red}}$$

   (d) *Cathode overvoltage:* A decrease in cathode overvoltage caused by increase in exchange current density and/or increase in Tafel slope, increases the corrosion rate of a normal corroding metal. For a metal exhibiting passivity the behaviour is, however, different. Similar, effect is obtained when a metal exhibiting passivity is alloyed with small amounts of more noble constituents having low hydrogen overvoltage or low overvoltage for cathodic reduction of dissolved oxygen, for example, Pd, Pt or Cu. This explains why stainless steels, which may loose passivity in dilute sulphuric acid, retains it when alloyed with Pd, Pt or Cu.

(e) *Presence of halogen ions:* Presence of chloride ions prevents or destroys passivity in metals exhibiting passive behaviour, e.g. Fe, Cr, stainless steel, etc. The effect is similar to that of increasing temperature or hydrogen ion concentration. Breakdown of passivity due to chloride ions occurs locally, rather than generally over the entire passive surface, due to slight change in structure and thickness of the passive film. Resulting corrosion cell is known as passive–active cell. The effect of chloride and other halogen ions is less on Ti, Ta, Mo, etc. which may remain passive in solutions containing high halogen ion concentration, whereas Fe, Cr and Fe–Cr alloys loose there passivity readily in the presence of halogens.

(f) *Galvanic coupling:* In general, galvanic coupling with a more noble metal increases the corrosion rate of a corroding metal.

(g) *Cathode/anode area ratio:* Increase in cathode area or cathodic reaction favours passivity because of increased polarization of the anode area. For this reason, steel containers containing highest permissible amount of carbon from the stand point of mechanical properties are used for storing nitric acid and mixture of nitric acid and sulphuric acid.

(h) *Potential scan rate:* In some cases, the shape of anode polarization curve depends upon the rate at which the potential of the specimen is changed.

## 19.14 CORROSION PREVENTION METHODS

*Rusting* is the product of the reaction of iron or steel with atmospheric oxygen in the presence of water. The metal is converted back into its ore to form ferric oxide rust which is red-brown compound. Mechanism of this reaction is electrochemical oxidation of the iron metal, i.e. wet corrosion.

$$4Fe + 3O_2 + 2H_2O \rightarrow 4FeO.OH \tag{19.4}$$

Not only iron, nearly all metals including very noble metals like gold, platinum and silver, corrode in presence of an oxidizing (corrosive) environment forming compounds such as oxides, hydroxides and sulphides. The corrosion is an unavoidable phenomenon, but losses due to corrosion can be minimized by using various corrosion protection methods. The energy acts as driving force for the corrosion of metals as it spontaneously flows from higher value to a lower value. Similarly, metal like aluminium, which is used in aircraft, window frames and cooking utensils, is also attacked by oxygen to form the oxide as shown in Eq. (19.5) (dry corrosion):

$$4Al + 3O_2 \rightarrow 2Al_2O_3 \tag{19.5}$$

The reaction in Eq. (19.5) is highly exothermic, releasing about 1680 kJ of energy per mole of oxide formed. In fact the driving force of the reaction is so great that powdered aluminium will burn to produce a very high temperature, which is sufficient to melt steel.

Corrosion of metal is a surface phenomenon, hence any modification of the surface or its environment can change the rate of corrosion. Thus, engineers have suggested various modifications in designing methods to protect metals from corrosion so that losses due to corrosion can be minimized.

A number of methods have been developed to prevent metals and their alloys from corrosion. A few important are tabulated in Table 19.1.

**TABLE 19.1** Classification of different methods to prevent corrosion

| S. No. | Concept/Principle | Technique/process used |
|---|---|---|
| 1. | Avoiding contact of metal with the corrosive environment, i.e. removal of corrosive environment | (i) Barrier protection<br>(ii) Boiler water treatment<br>(iii) Cooling tower treatment |
| 2. | Prevention of reactions occurring at the surface of metals | (i) Cathodic protection either by sacrificial anode and impressed current method<br>(ii) Anodic protection |
| 3. | Inhibition of surface reaction | (i) By the use of chemical inhibitors like chromates, phosphates, phosphonates, quaternary ammonium salts, amines, surfactants, etc.<br>(ii) By controlling the pH of the medium as corrosion takes place in acidic medium and scaling occurs in basic medium |
| 4. | Protective coatings (barrier protection)<br>(a) Organic<br><br>(b) Metallic<br><br><br>(c) Non-metallic | By applying protective coating over the surface of metals, i.e.,<br>(i) Paint<br>(ii) Claddings<br>(i) Electroplating<br>(ii) Galvanizing<br>(iii) Metal spraying<br>(i) Anodizing<br>(ii) Conversion coatings |
| 5. | Modification of the metal by change in its chemical composition | Use of alloys, as pure metal is in high state of energy and prone to more corrosion than their alloys<br>– Stainless steel<br>– Cupronickel<br>– High temperature alloys |
| 6. | Modification of surface conditions as corrosion is the surface phenomenon. Corrosion first starts from surface than it penetrates deep inside the bulk | Surface conditions can be modified by:<br>(i) Maintenance to remove corrosive agents<br>(ii) Design to avoid crevices<br>(iii) Design to avoid reactive metal combinations as when two different metals come into contact, galvanic corrosion takes place |

Anodic and cathodic are the two types of electrode reactions occurring on the metal surface. Oxidation takes place at anode. Iron in presence of corrosive environment produce $Fe^{2+}$ ions (anodic reaction):

$$Fe \rightarrow Fe^{2+} + 2e^- \tag{19.6}$$

At a higher pH, the oxidation reaction at anode produces a surface film of ferric oxide as shown in Eq. (19.7).

$$2Fe + 3H_2O \rightarrow Fe_2O_3 + 6H^+ + 6e^- \tag{19.7}$$

Cathodic reactions involve reduction reaction and produce hydroxyl ions. An example of the most common cathodic reaction is the electrochemical reduction of dissolved oxygen with water by accepting four electrons according to Eq. (19.8).

$$O_2 + 2H_2O + 4e^- \rightarrow 4OH^- \qquad (19.8)$$

The cathodic reduction reaction of oxygen at an electrode leads to increase in pH of the medium due to the production of hydroxide ion. The potential difference $E$ is generated across the electrified interface between a metal and its solution which increase the rate of corrosion process. The potential difference itself is caused by layers of charges at the surface of two electrodes of the electrified interface. One electrode is of the electrons present at the surface layer of metal and other electrode is formed of excess anions or cations in the solution of the electrolyte, as shown in Figure 19.12.

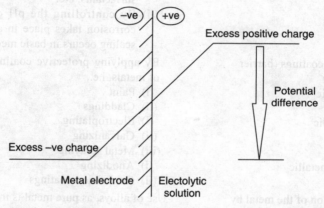

**FIGURE 19.12** Formation of electrified interface i.e. electric double layer at metal-electrolyte junction.

This separation of charges across the interface is also called as the *electrical double layer* or the *Helmholtz layer*. The magnitude of the potential $E$ is not fixed under non-equilibrium conditions and it can be further changed by either using an external electrical current or by changing speed of electrode reactions as given in Eqs. (19.6 to 19.8). Let us clear it by taking an example, at a high concentration of oxygen, the rate of cathodic reaction increases which will remove excess electrons from the metal surface, hence making the metal deficient of electrons which makes it more positively charged and thus increases its potential $E$. The effect of change in the electrode potential $E$ at the electrified interface, i.e. at the double layer on the products of corrosion reactions can be understand from Figure 19.13. At negative potentials, i.e. noble state, metallic iron is in the stable form hence, no corrosion is possible and it is termed as the *immunity condition* of the metal. At higher potentials and acidic pH values, metallic iron is in the active form, more and more ferrous ions are formed leading to active corrosion. At a still higher potential value, ferric ions are produced, i.e. at a potential value of greater than 0.7 V. If the pH lies in the range of 9–12, i.e. on the alkaline side, then insoluble surface oxides, i.e. $Fe_3O_4$ will be formed.

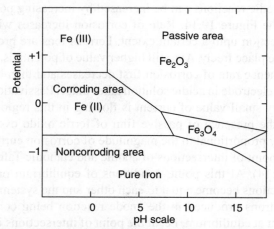

**FIGURE 19.13** Pourbaix diagram showing variation of potential versus pH of iron electrode.

The iron oxide, $Fe_3O_4$ (magnetite or black iron oxide), is formed at low electrode potentials. Iron oxide forms at the surface of the metal which blocks the surface reactions and hence corrosion rates are lowered. This is known as *passivity* and potential corresponding to the passivity is known as *passivation potential* and the oxide film formed on the surface is known as a *passive film*.

Corrosion process can be understood with the help of electrochemical polarization curves also known as *Evans diagram* (Figures 19.8 and 19.14) of the electrode reactions which take place on the metal surface. Figure 19.14 shows the anodic and cathodic polarization curve of iron in an acidic solution at room temperature. The rates of reaction occurring at the electrodes are dependent on the value of $E$. In a cathodic reaction in acidic solution, hydrogen gas is evolved by the reduction of hydrogen ions. The more negative the electrode potential $E$ the greater the surface concentration of electrons at the surface of metal and the faster will be the rate of evolution of hydrogen gas.

$$2H^+ + 2e^- \rightarrow H_2(g) \tag{19.9}$$

As we know the flow of electrons leads to production of current ($I$) the Figure 19.14 shows the variation of current $I$ as a function of potential $E$.

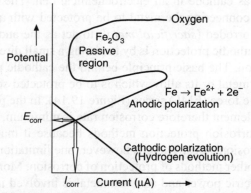

**FIGURE 19.14** Electrochemical anodic and cathodic polarization curves (Evans curves) of iron electrode dipped in acidic solution.

Similarly, rate of anodic reactions can be increased by increasing potential $E$ in the positive direction as shown in the Figure 19.14. Rate of corrosion increases with increase in value of potential in positive direction upto a certain extent. Ferrous ions are produced in this region in which corrosion will take place freely. At a still higher value of potentials, the reaction passes into the passivation region, hence rate of corrosion first decreases and then become almost constant and passivation of metal electrode in acidic solution takes place. Passivation of the metal electrode is observed due to a very small value of current is flowing in this region. The metal in passive region is protected by the presence of passive film of ferric oxide over the surface of metal. Corrosion rate is directly proportional to the magnitude of corrosion current density ($I_{corr}$) which can be found out at the point of intersections of anodic and cathodic Tafel polarization curves in the diagram (Figure 19.14). At this point, conditions of equilibrium prevails, the rates of the anodic and cathodic reactions become equal to each other and the system is behaving as a closed circuit with all the electrons produced in the anode reaction being consumed in the cathodic reaction. Value of current at equilibrium, i.e. at the point of intersections is known as *equilibrium exchange current density* ($i_0$). The potential corresponding to this equilibrium value of current is known as *equilibrium potential* and also known as *open circuit potential* (OCP). The polarization diagram of metal electrode dipped in acidic solutions can be helpful in predicting the rate of corrosion in mils per year (mpy) using Stern Gerry equation which is given as

$$\frac{\Delta E}{\Delta i} = \frac{\beta_a \times \beta_c}{2.303\, I_{corr}(\beta_a + \beta_c)} \tag{19.10}$$

where $\Delta E/\Delta i$ is the slope which is linear polarization resistance ($R_p$), $\beta_a$ and $\beta_c$ are anodic and cathodic Tafel slopes respectively and $I_{corr}$ is the corrosion current density in $\mu A/cm^2$.

$$\text{Corrosion rate} = \frac{0.1288 \times I_{Corr} \times \text{Eq. wt.}}{\text{Density}} \tag{19.11}$$

We shall discuss different methods of protecting metal and their alloys from corrosion one by one.

## 19.14.1 Cathodic Protection

Cathodic protection technique is used to minimize the corrosion of a metal by making the metal which is to be protected as cathode in an electrochemical cell. The simplest method to apply cathodic protection is by connecting the metal to be protected with another more electroactive metal which gets easily corroded (*sacrificial metal*) to act as the anode of the electrochemical cell. Another method of cathodic protection is by imposing a small direct current on a structure to be protected from corrosion. The basic principle behind the cathodic protection is to change the electrode potential of the metal or its alloy which is to be protected so that it lies in the passive region of the electromotive force (EMF) series (Figure 19.14). In the passive region, the metal is in the stable form of the element therefore corrosion rate is minimum. Cathodic protection is the most efficient form of corrosion protection methods because it makes the metal completely unreactive and hence corrosion stops. There is however, one limitation of cathodic protection as it is more expensive than other methods of protection of corrosion. More expensive in the sense is that it consumes extra electric power and an extra metal is involved in controlling the potential within this passive region. There are two ways of applying cathodic protection to a metal structure, i.e. *impressed current* and *sacrificial anode method*.

In the case of impressed current protection method, metal which is to be protected from corrosion which is in contact with an aqueous solution acting as electrolyte, the electrode potential should be fixed about −700 mV or greater than this with respect to standard hydrogen electrode scale so that the metal to be protected remains in the immunity region and hence no corrosion takes place. Cathodic protection systems are used to protect a wide range of metallic structures in different environments. Common applications of cathodic protection are steel water pipelines, fuel pipelines, metallic storage tanks, outer body of ships and boats, offshore oil platforms and onshore oil well casings and metal reinforcement bars in concrete buildings and structures.

### Cathodic protection by applying impressed current

This is more commonly used technique for the protection of buried pipelines and the hull of ships immersed in seawater. In impressed current technique, DC electric current is applied to the metallic structure to be protected from corrosion. The negative terminal of the battery is connected to the metal which is to be protected from corrosion. The positive terminal of the battery is connected to an auxiliary anode another metal electrode immersed in the same medium in order to complete the electrical circuit. The electric current charges the structure to be protected with excess of electrons and hence changes its electrode potential in the negative direction until the immunity region is reached.

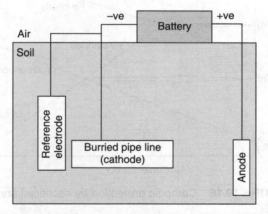

**FIGURE 19.15**  Schematic diagram of impressed current cathodic protection system.

Figure 19.15 shows the complete circuit diagram of a typical impressed cathodic current protection system. The another metal electrode which is connected to positive terminal of the battery acts as reference electrode (anode) which is designed to have a constant potential as no current passes through it. In the case of buried structures like buried pipe lines, the most commonly used reference electrodes are Cu/CuSO$_4$ (saturated), with a potential of +316 mV with respect to standard hydrogen scale (SHS), platinized titanium or lead alloys connected to an insulated cable. The buried anodes are fixed at a regular interval along the pipeline, normally few kilometres apart and several hundred metres from the nearest point of the pipeline.

Impressed current cathodic protection, if correctly designed and operated, is a very good technique which can be used very effectively in protecting different metallic structure. It is also used in defence systems. Royal New Zealand Navy has also applied impressed current systems technique for corrosion protection of their warships. Other examples include the natural gas pipelines which distribute methane and crude oil pipelines across the world.

## Cathodic Protection by Sacrificial Anode

Cathodic protection by sacrificial anode method is another efficient method of protecting metal structure from corrosion. In this technique, a more reactive metal which is placed higher in electrochemical series is connected with the metal structure to drive the potential in the negative direction until it reaches the passivity range. Figure 19.16 demonstrates the principle of cathodic protection by sacrificial anode. Metal with high oxidation potential like zinc is commonly used as the sacrificial anode. When zinc is not connected, the potential $E_{corr}$ is given by the point of intersection of the anodic and cathodic polarization curves. In the presence of zinc electrode, it produces an anodic dissolution current at a more negative potential as shown in Figure 19.16.

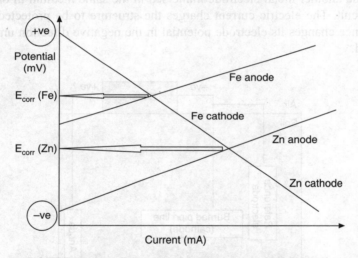

**FIGURE 19.16** Cathodic protection by sacrificial anode.

In order to protect steel structure in seawater, a potential of around −900 mV with respect to the Ag/AgCl reference electrode is applied. It is not practically feasible to apply surface coatings to the all metallic structure before installation. Hence, in order to protect around 6,000–10,000 tons of iron forming the different tower depends entirely on cathodic protection by sacrificial anodes made from the aluminium alloy (alanode). A small proportion of indium, about 0.1%, is included in the aluminium forming an alloy to provide efficient anodic action. Pure aluminium alone has such a resistant oxide film $Al_2O_3$ that its reactivity is insufficient to properly protect the steel structure. This technique is widely used for ships in seawater and for offshore oil and gas production platforms.

## 19.14.2 Anodic Protection Method

Anodic protection is simply opposite to the cathodic protection. In this technique, anodic current is applied to the structure to be protected from corrosion. It can be suitably applied to the metals that exhibit passivity (e.g., stainless steel) and suitably small passive current over a wide range of potentials. It can also be used in aggressive environments, for example, solutions of sulphuric acid. It is very less widely used method of protection of corrosion because it cannot be applied to all metals and their alloys. In addition to this, medium required is acidic and hence this technique cannot be applied in neutral and alkaline medium.

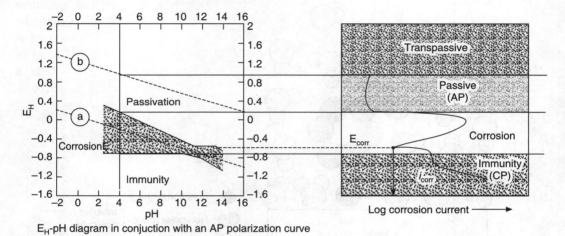

**FIGURE 19.17** Potential versus pH diagram and anodic polarization curve showing passive region and transpassive region.

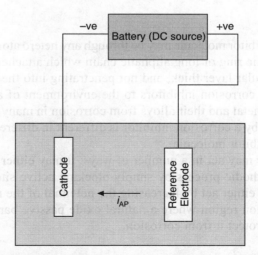

**FIGURE 19.18** Schematic diagram showing anodic protection method.

## 19.14.3 Corrosion Protection by the Use Corrosion Inhibitors

*Corrosion inhibitors* are the compounds (organic as well as inorganic) which are added deliberately into the system to minimize corrosion. Corrosion inhibitors act either by saturating the atmosphere by their suitable vapour pressure or by adsorbing on the reactive metal surface thus forming a protective film over the surface of metal which protects it from the attack of atmospheric gases.

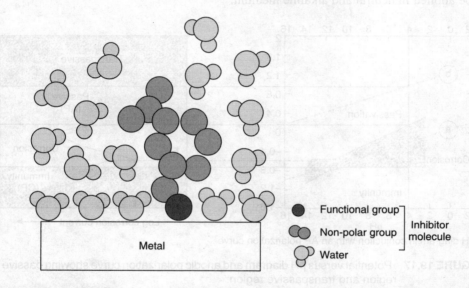

**FIGURE 19.19**  Adsorption of organic inhibitor onto a metal surface in aqueous environment.

The adsorption of inhibitor molecule may be through any hetero atom present or any electron donating group or aromatic ring or long aliphatic chain which attached directly to the surface, normally only one molecular layer thick, and not penetrating into the bulk of the metal itself. The technique of adding corrosion inhibitors to the environment of a metal is a well known method of protection of metal and their alloys from corrosion in many branches of technology. Mechanism of inhibition by a corrosion inhibitor is different in different conditions depending upon the structure of inhibitor molecule.

A corrosion inhibitor may act in a number of ways: it may either decrease the rate of the anodic process or the cathodic process by simply blocking active sites on the metal surface. Corrosion inhibitors may either act by increasing the potential of the metal surface so that the metal enters the passivation region where a natural oxide passive barrier film forms over the surface of metal which protect it from corrosion.

Table 19.2 shows classification of different corrosion inhibitor systems based on their modes of action with their suitable examples. Adsorption type corrosion inhibitors are widely used in different applications which are marketed by different chemical industries with different labels to control corrosion. For example, radiator fluids in the cooling circuits of automobile engines frequently contain amines such as hexylamine $C_6H_6NH_2$, or sodium benzoate which suppresses anodic reaction. Corrosion inhibitors also find its applications in the metal cleaning industry. For example, it is possible to clean steel articles by immersion in dilute sulphuric acid. Corrosive loss due to attack of acid can be minimized by adding antimony trichloride. $SbCl_3$ act as preferential scavenger which eats oxides and foreign metals such as zinc without affecting steel.

**TABLE 19.2** Classification of different corrosion inhibitors on the basis of their modes of action

| Mode of action of corrosion inhibitors | Compound or its functional group | Suitable examples of corrosion inhibitors |
|---|---|---|
| Adsorption type | Amines | $RNH_2$ |
| | Thiourea | $NH_2CSNH_2$ |
| | Antimony trichloride | $SbCl_3$ |
| | Benzoate | $C_6H_5COO^-$, $RCOO^-$, |
| Passivating type forming passive film over the surface | Nitrite, alkyl nitrites, aryl nitrites | $NO_2^-$, $CH_3NO_2^-$, $RNO_2^-$ |
| | Chromate | $CrO_4^{2-}$ |
| | Lead oxide (red lead) | $Pb_3O_4$ |
| | Calcium plumbate | $Ca_2PbO_4$ |
| Surface layer forming type | Phosphate, tri, tetra and polyphosphates | $H_2PO_4^-$, TSP, SPP |
| | Silicate | $H_2SiO_4^{2-}$ |
| | Hydroxide | $OH^-$ |
| | Bicarbonate | $HCO_3^-$ |
| | Hexametaphosphate or sodium hexametaphosphate (SHMP) | $Na_6(PO_3)_6$ |

Amine type corrosion inhibitors act as vapour phase corrosion inhibitors due to their high vapour pressure and are sometimes present as volatile corrosion inhibitors. These are used in packaging materials in airtight polythene packets in fixed concentration to prevent corrosion of metallic articles during storage and transport conditions. A good example is the wrapping used on automobile engines and other machinery equipment during their storage and shipment conditions.

The second classes of corrosion inhibitors are those which shift the potential of the metals in the passivation region. These include all oxidizing agents such as nitrites, plumbous plumbate $Pb_2(II)Pb(IV)O_4$ (a traditional pigment used in paints) and zinc chromate $ZnCrO_4$ containing elements in their higher oxidation states. The plumbate ion is an active oxidizing agent containing lead in the tetravalent state and serves to promote passivation of the underlying metal. All passivating inhibitors share the common property of corrosion protection to metal by using its own natural oxide film.

A few corrosion inhibitors like phosphates and chromates act by forming suitable surface film. Phosphates (tri, tetra and polyphosphates) are widely used as an additive in boiler water treatment or cooling tower treatment where corrosion, scale and sludge formation simultaneously occurs and in acid pickling baths for metals. Few commonly used phosphates are sodium tripolyphosphates (STPP) and trisodium phosphates (TSP). Chromate is also an extremely important surface film forming industrial corrosion inhibitor which is in practice from a very long time in spite of its toxicity and environmental problems. Now-a-days, few developed countries have totally banned the use of chromates as corrosion inhibitors due its environment problems.

The different corrosion inhibitors (Figure 19.20) are:

(a) Anodic
- Phosphates
- Silicate compounds

(b) Cathodic
- Poly-phosphates
- $Ca(HCO_3)_2$
- Methylamino-phosphate

(c) Mixed
- Amines
- Selenides

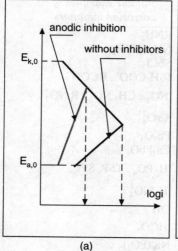
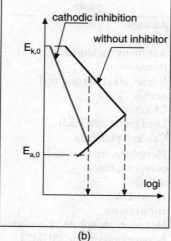
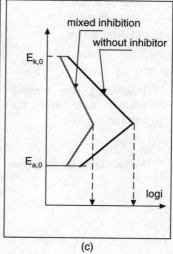

**FIGURE 19.20** Classification of different corrosion inhibitors.

Chromate inhibits the corrosion by two methods, one by its high oxidation state Cr (VI) which compels the metal to pass into the passivation region and second by the product of oxidation chromate changes into chromic oxide $Cr_2O_3$ which forms an inert, relatively insoluble surface film. The use of chromate as corrosion inhibitors started with its use in the thermal power station. It is also used in cooling tower where large quantities of water are circulated in the cooling tower plant through closed water pipe lines and sodium chromate, added at a level of about 400 mg/L, used as a corrosion and scale inhibitor. It proved to be very effective in protecting the steel from corrosion, but strict environmental regulations authority does not allow its use at a level above 5 µg/L. This use of sodium chromate as a corrosion inhibitor in cooling water is no longer in use and it was later replaced by a new corrosion inhibitor formulation involving the use of an organic zinc phosphate mixture. The well known commercial inhibitor formulation marketed in the name of Calgon is a mixed formulation of sodium hexametaphosphate (SHMP), a condensed phosphate and a polymer which contains a base unit of $(-PO_3-)_n$ which act as corrosion inhibitor and sludge control in cooling water treatment.

Scaling is another problem that is encountered in hard water or recirculating water where salts concentration goes on increasing with increase in cycle life of the recirculating water and complete water needs to be replaced after few cycles. SHMP also shows antiscaling properties because under some conditions, it dissolves substances like scales formed due to the presence of calcium and magnesium in water containing these cations and hence has a cleaning effect, which helps in the removal of scale deposits. Thus, SHMP can be used both as corrosion inhibitor and antiscalant. It is very pertinent here to mention that the simple hydroxide ion can also act as a corrosion inhibitor. In the presence of hydroxide, and hence at a high pH, metal oxides and hydroxides are insoluble, and these are effective in minimizing corrosion.

### 19.14.4 Corrosion Protection of the Metal by Protective Coating of the Surface (Surface Treatment)

Corrosion protection of the surface of metal can also be done by applying a protective coating over the surface of metal. Protective coatings of metallic, inorganic and organic materials can be applied satisfactorily which act as barrier between metal and its environment and can protect the metal surface against corrosion.

The following kinds of anticorrosion coatings are in common practice depending on the nature of surface to be protected:

(i) Metallic coatings for example, galvanising, electroplating, etc.
(ii) Inorganic coatings of phosphates, chromates, nitrites, sulphites, etc.
(iii) Organic coatings like amines, quaternary ammonium salts, alkyl chains etc.
(iv) Composite materials organosilicates, organochromates, organophospates, nanocomposites, polymeric nanocomposites, etc.
(v) Reactive coating like conducting polymers, ceramics, aluminiates, etc.
(vi) Anodizing coatings like coating with more reactive metal like zinc, magnesium, etc.
(vii) Bio-film coatings.

*Metallic coatings*

Metallic coating can be further classified into two categories:
(a) Cathodic and (b) anodic.

(a) *Cathodic:* Cathodic metallic coatings are composed from corrosion resistant metals, i.e. Ni, Cr, Cu and Co alloys, noble metals like Au, Ag and Pt protect underlying metal surface. Here metal used in cathodic coating acts as cathod in reference to the protected metal. It is based on principle of EMF series, a metal placed below in the EMF series than hydrogen acts as cathode and a metal placed above in the EMF series acts as anode. All noble metals like Cu, Ag and Au act as cathode and hence provide protection to other metallic surface and hence can be used in metallic coating. Any damage of the layer of coating like making a scratch, scar, etc. creates the Galvanic corrosion cell with the underlying metal to be act as anode and metal coating as a cathode.

(b) *Anodic (Zn, Cd and Al):* Anodic coating of metal is composed from metals which show more negative potential in corrosive medium which act as anod in reference to the underlying metal surface which is to be protected. These anodic coatings are highly efficient as it can prevent underlying metal surface even after the damage or perforation of the layer-cathodic protection. A metal placed above in the EMF series acts as anode and the metal placed below in the EMF series acts as sacrificial anode.

*Inorganic coatings*

A few examples of inorganic coating are:
 (i) Oxide films-anodization
 (ii) Chromates
 (iii) Phosphates
 (iv) Silica layers (Sol-Gel method)
 (v) Enamels

*Organic coatings*

A few examples of organic coating are:
 (i) Anticorrosion painting
 (ii) Polymer coatings

*Composite materials coating or applied coatings*

Plating, painting, and the application of enamel are the most common methods of protection of metals and their alloys from corrosive environments. Coating simply cut-off the metals and their alloys from their surrounding by a passive stable film which acts as barrier between the two, i.e. damaging environment and the structural material. Composite material coating includes either polymer nanocomposite or simply doped polymer coating which may be either conducting or non-conducting. Flexible polyurethane coating is an example of polymer coating that can provide an anti-corrosive seal with a highly durability even in acidic solutions. Electroplating with the help of electrolysis has two major disadvantages. One is, it usually fails in small sections, i.e. uniformity is lacking, and second is, it is costlier than others methods of protection of metal. In plating a more noble metal is used to coat the electroactive surface. For example, chromium is coated on the surface of steel.

*Reactive coatings*

In a closed system like re-circulating cooling water, barrier protection methods like painting and other organic coating methods cannot be applied and are not very effective. In a controlled system like this, corrosion inhibitors are deliberately added to the systems. These corrosion inhibitors simply adsorb on the surfaces of metals and their alloys thus act as an electrical insulator or provide chemically impermeable coating on exposed metal surfaces, to minimize corrosion. Some of the chemicals treatment formulations inhibit both corrosion and scaling include some of the salts of chromates, phosphates or polyaniline, other conducting polymers and a wide range of specially-designed chemicals that resemble surfactants added in recirculating water.

*Anodizing coatings*

Anodizing coating is very effective in weathering and etching of the surface, so it is commonly used for exposed areas of the surface that will come into regular contact with the aggressive environment. Anodizing coating involves coating the surface with more active material like zinc and magnesium which act as anode. A Galvanic couple is formed at the junction of two dissimilar metals leading to the formation of Galvanic cell resulting into the corrosion of anodic coating and below which metal surface act as cathode is protected.

*Biofilm coating*

Biofilm coating is a new form of bacterial corrosion protection method and has been developed by researchers by applying certain specific species of bacterial films like sulphate reducing bacteria (SRB) to the surface of metals in corrosive environments. Bacterial film produced by the activity of these specific bacteria increases the corrosion resistance characteristics substantially. Alternatively, antimicrobial-producing biofilms can be used to inhibit mild steel corrosion from sulphate-reducing bacteria (SRB). Film formed by the bacteria is stable, non-conductive and non-corrosive.

## 19.14.5 Volatile Corrosion Inhibitor (VCI) Method

Volatile corrosion inhibitors (VCI) or vapour phase corrosion inhibitors (VPCI) are those compounds which have high vapour pressure due to which they saturate the environment with their suitable vapour pressure which get adsorbed over the surface of metal in close environment thus protecting the metal from corrosion.

Their modes of action can be depicted as shown in Figure 19.21. Necessary condition for VCI is that they require close environment in order to produce suitable vapours and are deposited as a film onto the item to be protected (metal surfaces). There are different methods of application of VPCI, they may be applied either directly in solid or liquid form or by using suitable paper (Kraft paper), foam, cardboard or film supports or in a powder, spray or oil formulation. The advantage of VPCI is that they strongly adsorb to metal surfaces than that of water molecules, resulting in the formation of a stable and continuous protective film between the metal surface and the corrosive environment. So water vapour cannot come in direct contact with the metal surface, hence preventing every chances of corrosion. VPCI are placed in airtight plastic packing enclosing the whole metallic structure sufficiently close to it not more than 30 cm apart so that to produce

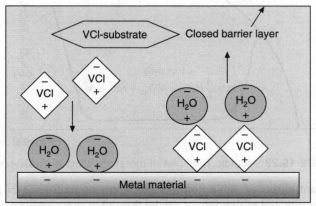

**FIGURE 19.21** Mechanism of action of VPCI.

sufficient vapour pressure. It should be kept in mind that there should be no leakage of VPCI from the plastic envelope. It should be properly sealed as any leakage leads to decrease in vapour pressure. The time of action of VPCI may vary depending upon type of system and environmental conditions.

Approximately, 40 g of VPCI are required per 1 m³ of air volume. The vapour phase corrosion inhibitor technique can be applied to carbon steel, stainless steel, cast iron, galvanized steel, nickel, chromium, aluminium and copper.

### 19.14.6 Corrosion Protection by Corrosion Resistant Alloys

As pure metals are in state of high energy, hence use of pure metals must be avoided. Instead of pure metal, corrosion resistant alloys with low energy and thermal stability can be produced which are less prone to corrosion. Noble metals such as Cu, Ag, Au, Pt, Pd etc. which are thermodynamically stable in aqueous environments may be used in making corrosion resistant alloys. Pourbaix diagrams solves the purpose to find thermal stability conditions of different metals like pH, redox potential and concentration of soluble species etc. and thus allow to predict stability conditions for pure metals in aqueous solutions. Metals and their alloys which are thermodynamically unstable cannot be used as corrosion resistant alloys. Corrosion resistance is connected with the passivation of metal surface (Figure 19.22). There are few metals which attain passivity in high potential range in an acidic environment, for eample, stainless steel, Ni and its alloys, high Si cast iron, valve metals, i.e. Ti, Ta, Al and Al alloys. The method of corrosion protection by use of corrosion resistant alloys is not economically favourable as noble metals are used in making corrosion resistant alloys which are very costlier. Hence, alternate method of protection from corrosion may be used which are ecofriendly, cost effective, easily available and easy to use.

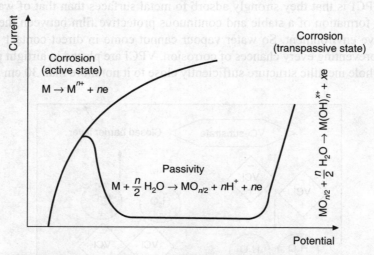

**FIGURE 19.22** Anodic Tafel plot of metal showing its passive range.

Corrosion resistant alloys of some non-metallic element such as P, N and Si and metallic element such as chromium and molybdenum can be formed at some active places of metal surface during corrosion process. These non-metallic and metallic elements react with the solvent and form non-soluble strongly adsorbed compounds on the joints, elbow, hinges and places where structural defects exist. These blocking of active areas of the alloy surface result in to the decrease of corrosion rate.

Figure 19.23 shows how the presence of small amounts of these metallic and non-metallic elements like Cr, Mo, Si, P in iron electrode can modify its chemical composition and thus making it resistant to corrosion.

The chemical and structural modification of corrosion resistant alloys in comparison to pure metal is proved by XRD and TEM techniques. The weathering steels are common example which shows protective properties of the films formed on its surface due to the presence of super paramagnetic goethite and maghemite. The lesser particle size of goethite and maghemite leads to increase in protective nature of the surface film in weathering steel.

Si and P element when present in the weathering steels promote the formation of super paramagnetic goethite and thus improve the corrosion resistance characteristics of the alloys.

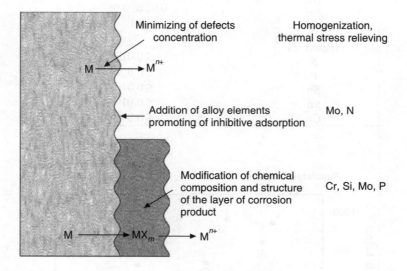

**FIGURE 19.23** Modification of chemical composition by the presence of non-metal and metallic elements in alloys.

Table 19.3 shows how the presence of impurities in trace amounts, different solid components, modifiers and carbide formers in metal can change the corrosion resistant characteristics of carbon steel towards hydrogen evolution reaction.

Figure 19.24 shows the effect of addition of passivity promoters and dissolution moderators in small amount in metal specimen, according to the synergy between the energy of the metal-metal bonds and heat of adsorption of oxygen. Metal aluminium, titanium and chromium show the best synergy between the energy of the metal-metal bonds and heat of adsorption of oxygen.

**TABLE 19.3** Optimal percentage composition of different element in carbon steel resistant to hydrogen evolution reaction.

| Name of element | Optimal percentage composition |
|---|---|
| Carbon | 0.2 to 0.3 |
| **Solid state components** | |
| Si | 0.4 to 0.7 |
| Mn | ≤ 1.2 |
| Ni | 0.5 to 1.0 |
| Co | ≤ 0.5 |
| Al | ≤ 0.25 |
| **Modifiers** | |
| Ce AlN, VN, NbN | 0.1 to 0.3 |
| | 0.2 |
| **Carbides formers** | |
| Cr | 1.0 to 1.5 |
| Mo | 0.4 to 0.5 |
| Ti | 0.05 |
| Nb | 0.02 to 0.06 |
| V | 0.1 |
| **Impurities** | |
| S | ≤ 0.01 |
| P | ≤ 0.015 |
| Sb | ≤ 0.01 |
| Sn | ≤ 0.01 |
| Cu | ≤ 0.05 |

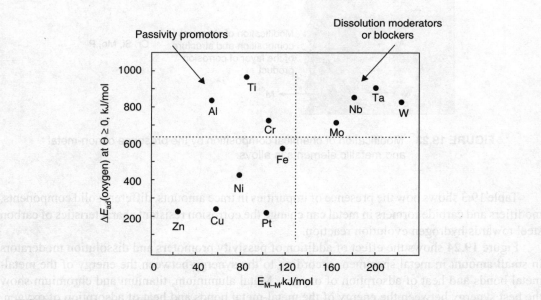

**FIGURE 19.24** Passivity promoters and dissolution moderators according to the synergy between the energy of the metal-metal bonds and heat of adsorption of oxygen.

## 19.14.7 Modification of Surface Conditions by Placement of Desiccant Bags

A desiccant bag is defined as a bag containing some amount of desiccants which has capacity to absorb certain amount of moisture present in the closed atmosphere. Few important characteristics of desiccant are, insoluble in water and must be chemically inert. Few examples of desiccant include silica gel, aluminium silicate, alumina, blue gel, bentonite, molecular sieves, etc. The desiccants are suspended from thread or strings in the upper part of the climate-controlled package to ensure good air circulation around them. Desiccant removes the humidity of the atmosphere due to the absorbency capacity, so it eliminates the risk of corrosion. Different manufacturers supply desiccants commercially in different trade names available in market as desiccant units. The number of desiccant units is a measure of the adsorption capacity of the desiccant bag. It is advisable to use large number of small desiccant bags in comparison to single large desiccant bag as this increases the available surface area of the desiccant.

## 19.15 DIFFERENT EXPERIMENTAL METHODS TO FIND CORROSION RATE

### 19.15.1 Weight Loss Method

For weight loss measurements, metal specimens of definite size were cut from the metal sheet. Mechanical polishing of the metal samples has to be carried out with the help of emery papers of different grades beginning from lower grade to higher one. Washing of metal specimen will be carried out with plenty of double distilled water and then with acetone. Then metal specimen will be dried and stored in desiccator.

The initial weights of metallic specimens can be recorded on a analytical balance, with a precision of 0.01 mg, they can be immersed in tilted position in 250 ml glass beakers (short form) each having 200 ml of corroding medium with or without the corrosion inhibitor. Experiment can be carried out in an electronically controlled air thermostat maintained at a constant temperature. After exposing the specimens for a definite time (hours), the specimens can be taken out from the beaker. Metal specimen will be washed initially under tap water. Help of rubber cork may be taken to remove loosely adhering corrosion products and then metal specimens will be washed with double distilled water and dried and then finally weighed again. Corrosion rate in mils per year (mpy) and percentage corrosion inhibition efficiency (PCIE) will be calculated using Eqs. (19.12) and (19.13) given below.

$$\text{Corrosion rate (mpy)} = \frac{534 \times W}{DAT} \qquad (19.12)$$

where, $W$ = weight loss (mg), $D$ = density of metal (gm/cm$^3$), $A$ = area of metal specimen exposed to corroding medium in sq. inch, $T$ = time of exposure of specimen to corroding medium in hours.

$$\text{PCIE} = \frac{CR_{(\text{Blank})} - CR_{(\text{Inhibitor})}}{CR_{(\text{Blank})}} \times 100 \qquad (19.13)$$

The degree of surface coverage ($\theta$) of the investigated surfactant compounds were calculated from the following equation:

$$\theta = \left[1 - \left(\frac{\Delta W_{\text{Inh}}}{\Delta W_{\text{Free}}}\right)\right] \quad (19.14)$$

where, $\Delta W_{\text{Free}}$ and $\Delta W_{\text{Inh}}$ are weight losses of metal per unit area in absence and presence of inhibitor at given time period and temperature, respectively.

The interaction of inhibitor molecules can be described by introducing of an parameter, $S_\theta$, obtained from the surface coverage values ($q$) of the anion, cation and both. Synergism parameter, $S_\theta$, was calculated using the following equation.

$$S_\theta = \frac{1 - \theta_{1+2}}{1 - \theta'_{1+2}} \quad (19.15)$$

where, $\theta_{1+2} = (\theta_1 + \theta_2) - (\theta_1 \theta_2)$, $\theta_1$ = surface coverage by anion, $\theta_2$ = surface coverage by cation and $\theta'_{1+2}$ = surface coverage by both, i.e. anion and the cation.

## 19.15.2 Electrochemical Method to Find Corrosion Rate

Electrochemical polarization experiments can be performed on electrochemical workstation, i.e. potentiostat/galvanostat. Linear polarization resistance measurements can be carried out potentiostatically by scanning through a potential range of 14 mV above and below the open circuit potential (OCP) also known as equilibrium potential, value in steps of 2.0 mV. Experiments were carried out in absence and presence of inhibitors at their different concentrations and at a constant desired temperature. The resulting current is plotted against the potential and slope of the line is measured.

After this, electrochemical polarization study can be performed in cathodic and anodic direction by scanning through a potential of 200 mV in cathodic direction and around 600 mV in anodic direction from OCP value. From the polarization curves anodic and cathodic Tafel slopes can be predicted. From the polarization data, corrosion current density can be estimated and from the value of corrosion current density, value of corrosion rate can be found out with the help of Stern-Geary equation Eq. (19.11) given below.

The corrosion current density, $I_{\text{corr}}$, is related to the corrosion rate by the equation,

$$\text{Corrosion rate (CR) (mpy)} = \frac{0.1288 \times I_{\text{Corr}} \times \text{Eq. wt.}}{D} \quad (19.11)$$

where, Eq. wt. = gram equivalent weight of metal/alloy, $D$ = density of metal (gm/cm$^3$), $I_{\text{corr}}$ = corrosion current density ($\mu$A/cm$^2$).

## 19.15.3 Electrochemical Impedance Spectroscopy to Measure Corrosion Rate

Electrochemical impedance spectroscopy (EIS) also known as *AC impedance spectroscopy* and also called as *dielectric spectroscopy* is based on the measurement of dielectric properties of a medium as a function of frequency. Measurement of dielectric properties are based on the interaction of an external field with the electric dipole moment of the sample often expressed in terms of its permittivity.

Impedance is a type of resistance that opposes the flow of alternating current (AC) in a complex system. Almost every physicochemical system, such as electrochemical cells including fuel cells and batteries and even biological tissue possesses provision of energy storage and its dissipation properties. EIS examines both of them.

It is an experimental electrochemical method performed by electrochemical workstation connected with electrochemical cells and their electrodes for characterizing electrochemical systems. In this technique, impedance of a system is measured over a wide range of frequencies, and therefore the frequency response of the system, including the energy storage, corrosion characteristics and dissipation properties are studies. Data obtained by EIS is expressed graphically in a Bode plot or a Nyquist plot.

The technique used in EIS is conceptually simpler than other techniques. In this technique, a low amplitude alternating potential (or current) wave is applied on top of a DC potential. The varying frequency is imposed from as high as $10^5$ to $10^{-3}$ Hertz in a set number (often between 5 and 10) steps per decade of frequency (Figure 19.25). The corrosion process compels the measured current to be out of phase with the input voltage. Ratio of the input voltage to the output current will give the value of impedance (Z). The variation in impedance (magnitude and phase angle) is used for the interpretation. EIS technique is based on the DC polarization resistance in which a direct current voltage or current ramp is applied.

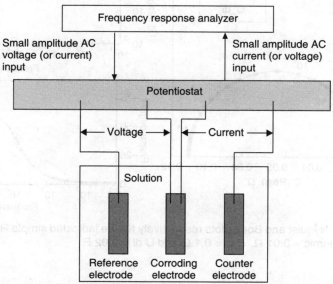

**FIGURE 19.25** A three-electrode based cell connected with the potentiostat to measure **FRA**.

## Interpretation of impedance data

Nyquist and Bode plots are the two types of plots which represent EIS data of electrochemical cells as shown in Figure 19.26. Bode plots refer to representation of the impedance magnitude (of the real or imaginary components of the impedance) and phase angle as a function of frequency. Both the impedance and the frequency often have large orders of magnitude, hence they are plotted on a logarithmic scale. Bode plots clearly show the dependence of the impedance on the frequency.

A complex plane or Nyquist plot depicts the imaginary impedance, which is indicative of the capacitive and inductive character of the electrochemical cell versus the real impedance. Nyquist plots have the advantage that activation-controlled processes with distinct time-constants show up as unique impedance arcs and the shape of the curve provides insight into possible mechanism or governing phenomena. However, this format of representing impedance data has the disadvantage that the frequency-dependence is not explained; therefore, the AC frequency of selected data points should be indicated. Both data formats, i.e. Nyquist and Bode plots have their own advantages; it is usually advisable to present the both Bode and Nyquist plots.

With the help of Nyquist diagram and an impedance plots, one can determine charge transfer resistance, double layer capacitance and ohmic resistance. The equilibrium exchange current density, ($I_0$) can also be easily calculated by measuring the impedance of a redox reaction at zero overpotential value. From the value of equilibrium exchange current density, one can find corrosion rate using Stern-Geary equation.

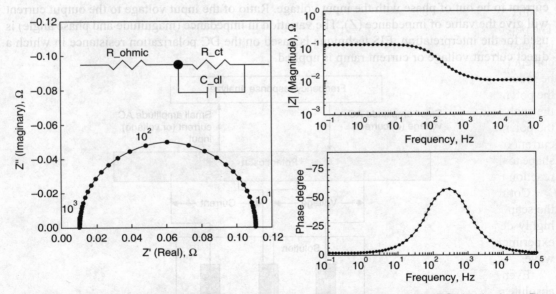

**FIGURE 19.26** Nyquist and Bode plots respectively for the indicated simple RC circuit where $R$ ohmic = 0.01 $\Omega$, $R$ ct = 0.1 $\Omega$ and C dl = 0.02 F.

## Advantages of electrochemical impedance spectroscopy (EIS)

(i) Major advantage of EIS is that measurements can be made under real situation, i.e. electrochemical cell operating conditions, i.e. in equilibrium condition (open circuit voltage) or under non-equilibrium condition (DC voltage or current).
(ii) More than one parameter can be simultaneously determined from a single experiment.
(iii) It gives information about Helmholtz–Perrin, Gouy–Chapman and Stern models, and provides information of bulk and interfacial properties of the system, e.g., membrane resistance and electrocatalysts.
(iv) Measurement carried out over the system does not disturb the system, i.e. it does not substantially remove or disturb the system from its operating condition.
(v) A high precision measurement—the data signal can be averaged over time to improve the signal-to-noise ratio.

### 19.15.4 Cyclic Voltammetry (CV) to Study Corrosion Rate

In a cyclic voltammetry experiment, simultaneous oxidation and reduction reaction are investigated by changing the potential of the working electrode linearly versus time like linear sweep voltammetry. CV takes the experiment a step further than linear sweep voltammetry which ends when it reaches a set potential. Whereas, in CV when it reaches a set potential, the working electrode's potential ramp is inverted. This inversion can happen multiple times during a single experiment. The current is plotted against the applied voltage to give the cyclic voltammogram trace (Figure 19.27(a)).

In CV, the electrode potential is investigated linearly versus time as shown in Figure (19.27(b)). The potential is applied between the reference electrode and the working electrode and the current is measured between the working electrode and the counter electrode. This data is then plotted as current ($I$) versus potential ($E$). The forward scan produces a current peak for any analyte that can be reduced (or oxidized) through the range of the potential scanned. The current will increase as the potential reaches the reduction potential of the analyte, but it decreases as the concentration of the analyte is reduced at the electrode surface. Then the applied potential is reversed, it will reach the potential that will re-oxidize the product formed in the first reduction reaction, and produce a current of reverse polarity from the forward scan. This oxidation peak will usually have a similar shape to the reduction peak. As a result, information about the redox potential and electrochemical reaction rates of the compounds can be obtained.

Cottrell equation (Eq. 19.16) gives the relationship between the current and square root of the scan rate and current is proportional to the square root of the scan rate. The utility of CV is highly dependent on the analyte being studied. The analyte has to be redox active within the experimental potential window. It is also highly desirable for the analyte to display a reversible wave.

Even reversible couples contain polarization overpotential and thus display a hysteresis between absolute potentials both in the reduction ($E_{pc}$) and oxidation peak ($E_{pa}$). This overpotential emerges from a combination of analyte diffusion rates and the intrinsic activation barrier of transferring electrons from an electrode to analyte. A theoretical description of polarization overpotential is in part described by the Butler–Volmer equation and Cottrell equation. Conveniently in an ideal system the relationships reduces to,

$$\left|E_{pc} - E_{pa}\right| = \frac{57 \text{ mV}}{n} \tag{19.16}$$

for an $n$ electron process.

Reversible couples will display a ratio of the peak currents passed at reduction ($I_{pc}$) and oxidation ($I_{pa}$) that is near unity ($I_{pa}/I_{pc}$) = 1. This ratio can be perturbed for reversible couples in the presence of chemical reaction, stripping wave, and nucleation event.

When such reversible peaks are observed, thermodynamic information in the form of half cell potential $E_{01/2}$ can be determined. When waves are semi-reversible such as when $I_{pa}/I_{pc}$ is less than or greater than 1, it can be possible to determine even more information especially kinetic processes.

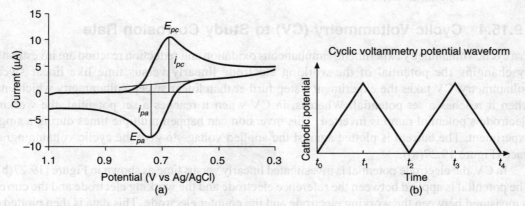

**FIGURE 19.27** (a) Cyclic voltammogram (b) Cyclic voltammetry potential waveform.

## EXERCISES

1. Define corrosion. What are various types of corrosion?
2. How can we express corrosion rate? Which is the most convenient method to express corrosion rate?
3. Write an equation to find corrosion rate. How can we measure corrosion rate.
4. Explains various types of electrochemical cell.
5. What are various possible ways for the formation of galvanic cell?
6. What are various possible ways for the formation of concentration cell?
7. Define exchange current density and polarization of metal electrode.
8. Define the following terms: (i) concentration polarization, (ii) activation polarization and (iii) resistance polarization.
9. What are the various factors affecting activation polarization.
10. Discuss in detail mixed potential theory.
11. What is Tafel solpe? Discuss anodic and cathodic polarization curves for a metal electrode.
12. What is passivity? Discuss electrochemical behaviour of active/passive material.

## SUGGESTED READINGS

Balzani, V., *Uhlig's Corrosion Handbook*, Vols. 1–5, Wiley, New York, 2001.

Fontana, M.G. and N.D. Greene, *Corrosion Engineering*, 2nd ed., McGraw-Hill.

Fontana, M.G., *Corrosion Engineering*, 3rd ed., McGraw-Hill, 1987.

Kaesche, H., *Metallic Corrosion*, 2nd ed., NACE Int. House, Houston Texas, 1985.

Molloy, E., *Electroplating and Corrosion Prevention*, Newnes, London, 1954.

Narain, R., *An Introduction to Metallic Corrosion*, Tata McGraw-Hill, New Delhi.

Piran, D.L., *Electrochemistry of Corrosion*, NACE Int. House, Houston Texas, 1991.

West, J.M. and Van Nostrand, *Electrodeposition and Corrosion Processes*, Reinhold, New York, 1965.

West, J.M. and Van Nostrand, *Electrodeposition and Corrosion Processes*, 2nd ed., Reinhold, New York, 1971.

# Part V
# Reaction Dynamics

# Chemical Kinetics

## 20.1 CHAIN REACTIONS

A reaction proceeding in a series of successive steps initiated by a suitable primary process is called a *chain reaction*.

Chain reactions proceed through a complex sequence of elementary steps. The various steps can be classified as follows:

1. *Chain initiating step:* The first step in which highly reactive species such as atoms and free radicals are produced is known as the *chain initiation step*.
2. *Chain propagation step:* In these steps, the highly reactive intermediate from the chain initiation steps reacts with one of the stable reactant molecules and there by produce a product molecule and another reactive intermediate. The produced reactive intermediate, is then reacts with another stable reactant molecule and produce a product molecule and another reactive intermediate when an old reactive intermediate is somehow destroyed.
3. *Chain inhibition step:* The reactive intermediate combines with a product molecule producing a reactant molecule and another reactive intermediate. Though a reactive intermediate is generated, the net effect of chain inhibition step is to decrease the rate of overall reaction.
4. *Chain terminating step:* The reactive intermediate is destroyed by combining with another reactive intermediate; certain substance, when added also help in terminating the reactive intermediate.

Kinetic study of chain reaction is based upon the principles of steady state. It is assumed that in a chain reaction, the intermediate such as atoms and free radicals which have low concentration acquire constant concentration during the course of the reaction.

581

## 20.2 FORMATION OF HBr FROM $H_2$ AND $Br_2$

$$H_2 + Br_2 \rightarrow 2\, HBr$$

To explain the above rate equation, the following chain mechanism was proposed and the steady state principles were applied for the first time to a complex reaction.

Chain initiation $\qquad\qquad Br_2 \xrightarrow{K_1} 2\,Br^{\bullet}$

Chain propagation $\qquad\quad Br^{\bullet} + H_2 \xrightarrow{K_2} HBr + H^{\bullet}$

$\qquad\qquad\qquad\qquad\qquad H^{\bullet} + Br_2 \xrightarrow{K_3} HBr + Br^{\bullet}$

Chain inhibition $\qquad\qquad H^{\bullet} + HBr \xrightarrow{K_4} H_2 + Br^{\bullet}$

Chain termination $\qquad\qquad 2\,Br^{\bullet} \xrightarrow{K_5} Br_2$

Evidently, the chain carriers are the $Br^{\bullet}$ and $H^{\bullet}$ atoms. They are present in low concentration. Hence, they can be assumed to be in steady state concentration during the chain reaction, which means that their concentrations do not change with time, i.e.

$$\frac{d[Br^{\bullet}]}{dt} = 0 \quad \text{and} \quad \frac{d[H^{\bullet}]}{dt} = 0 \qquad\qquad (20.1)$$

$$\frac{d[Br^{\bullet}]}{dt} = K_1[Br_2] - K_2[Br^{\bullet}][H_2] + K_3[H^{\bullet}][Br_2] + K_4[H^{\bullet}][HBr] - K_5[Br^{\bullet}]^2 = 0 \qquad (20.2)$$

$$\frac{d[H^{\bullet}]}{dt} = K_2[Br^{\bullet}][H_2] - K_3[H^{\bullet}][Br_2] - K_4[H^{\bullet}][HBr] = 0 \qquad (20.3)$$

Adding Eqs. (20.2) and (20.3), we get

$$K_1[Br_2] - K_5[Br^{\bullet}]^2 = 0$$

$$K_5[Br^{\bullet}]^2 = K_1[Br_2]$$

$$[Br^{\bullet}] = \left(\frac{K_1}{K_5}\right)^{1/2} [Br_2]^{1/2} \qquad\qquad (20.4)$$

Putting this value in Eq. (20.3), we get

$$K_2\left(\frac{K_1}{K_5}\right)^{1/2} [H_2][Br_2]^{1/2} - K_3[H^{\bullet}][Br_2] - K_4[H^{\bullet}][HBr] = 0$$

$$K_3[H^{\bullet}][Br_2] + K_4[H^{\bullet}][HBr] = K_2\left(\frac{K_1}{K_5}\right)^{1/2} [H_2][Br_2]^{1/2}$$

$$[H] = \frac{K_2\left(\dfrac{K_1}{K_5}\right)^{1/2} [H_2][Br_2]^{1/2}}{K_3[Br_2] + K_4[HBr]} \qquad\qquad (20.5)$$

The equation for the rate of formation of HBr is

$$\frac{d[\text{HBr}]}{dt} = K_2[\text{Br}^\bullet][\text{H}_2] + K_3[\text{H}^\bullet][\text{Br}_2] - K_4[\text{H}^\bullet][\text{HBr}]$$

Substituting the values of [Br] and [H] from Eqs. (20.4) and (20.5) and then rearranging, we get

$$\frac{d[\text{HBr}]}{dt} = \frac{2K_2(K_1/K_5)^{1/2}[\text{H}_2][\text{Br}_2]^{1/2}}{1 + K_4[\text{HBr}]/K_3[\text{Br}_2]}$$

## 20.3 DECOMPOSITION ACETALDEHYDES

The proposed mechanism of decomposition of acetaldehyde is as follows:

$$\text{CH}_3\text{CHO} \xrightarrow{K_1} {}^\bullet\text{CH}_3 + {}^\bullet\text{CHO}$$

$$\text{CH}_3\text{CHO} + {}^\bullet\text{CH}_3 \xrightarrow{K_2} \text{CH}_4 + \text{CH}_3\text{CO}^\bullet$$

$$\text{CH}_3\text{CO}^\bullet \xrightarrow{K_3} {}^\bullet\text{CH}_3 + \text{CO}$$

$$2\,{}^\bullet\text{CH}_3 \xrightarrow{K_4} \text{C}_2\text{H}_6$$

The steady state equation for methyl radicals will be:

$$\frac{d[{}^\bullet\text{CH}_3]}{dt} = K_1[\text{CH}_3\text{CHO}] - K_2[{}^\bullet\text{CH}_3][\text{CH}_3\text{CHO}] + K_3[\text{CH}_3\text{CO}^\bullet] - K_4[{}^\bullet\text{CH}_3]^2 = 0 \quad (20.6)$$

and for $\text{CH}_3\text{CO}^\bullet$ radicals:

$$\frac{d[\text{CH}_3\text{CO}^\bullet]}{dt} = K_2[{}^\bullet\text{CH}_3][\text{CH}_3\text{CHO}] - K_3[\text{CH}_3\text{CO}^\bullet] = 0 \quad (20.7)$$

Adding Eqs. (20.6) and (20.7), we get

$$K_1[\text{CH}_3\text{CHO}] - K_4[{}^\bullet\text{CH}_3]^2 = 0$$

$$[{}^\bullet\text{CH}_3] = \left(\frac{K_1}{K_4}\right)^{1/2}[\text{CH}_3\text{CHO}]^{1/2} \quad (20.8)$$

The rate of reaction which is based upon the rate of formation of methane, is given by

$$\frac{d[\text{CH}_4]}{dt} = K_2[{}^\bullet\text{CH}_3][\text{CH}_3\text{CHO}]$$

By putting value of [CH$_3$] from Eq. (20.8), we get

$$= K_2\left(\frac{K_1}{K_4}\right)^{1/2}[\text{CH}_3\text{CHO}]^{1/2}[\text{CH}_3\text{CHO}]$$

$$\frac{d[\text{CH}_4]}{dt} = K_2\left(\frac{K_1}{K_4}\right)^{1/2}[\text{CH}_3\text{CHO}]^{3/2}$$

## 20.4 DECOMPOSITION OF ETHANE

Proposed mechanism for ethane decomposition is:

Chain initiation $\quad C_2H_6 \xrightarrow{K_1} 2\,^{\bullet}CH_3$

Chain propagating $\quad ^{\bullet}CH_3 + C_2H_6 \xrightarrow{K_2} CH_4 + ^{\bullet}C_2H_5$

$$^{\bullet}C_2H_5 \xrightarrow{K_3} C_2H_4 + H^{\bullet}$$

$$H^{\bullet} + C_2H_6 \xrightarrow{K_4} H_2 + ^{\bullet}C_2H_5$$

Chain termination $\quad H^{\bullet} + ^{\bullet}C_2H_5 \xrightarrow{K_5} C_2H_6$

Steady state for methyl radical is:

$$K_1[C_2H_6] - K_2[^{\bullet}CH_3][C_2H_6] = 0 \qquad (20.9)$$

For the ethyl radicals:

$$K_2[^{\bullet}CH_3][C_2H_6] - K_3[^{\bullet}C_2H_5] + K_4[H^{\bullet}][C_2H_6] - K_5[H^{\bullet}][^{\bullet}C_2H_5] = 0 \qquad (20.10)$$

And for hydrogen atoms:

$$K_3[^{\bullet}C_2H_5] - K_4[H^{\bullet}][C_2H_6] - K_5[H^{\bullet}][^{\bullet}C_2H_5] = 0 \qquad (20.11)$$

Addition of Eqs. (20.9), (20.10) and (20.11) leads to

$$[H]^{\bullet} = \frac{K_1}{2K_5} \frac{[C_2H_6]}{[^{\bullet}C_2H_5]}$$

Putting the value of $[H^{\bullet}]$ in Eq. (20.11) after rearrangement, i.e. taking $[C_2H_5]^{\bullet}$ lowest common multiplier (LCM)

$$K_3K_5[^{\bullet}C_2H_5]^2 - K_1K_5[C_2H_6][^{\bullet}C_2H_5] - K_1K_4[C_2H_6]^2 \qquad (20.12)$$

The general solution of this quadratic equation is

$$[^{\bullet}C_2H_5] = \left[\frac{K_1}{2K_3} + \left\{\left(\frac{K_1}{2K_3}\right)^2 + \left(\frac{K_1K_4}{K_3K_5}\right)\right\}^{1/2}\right][C_2H_6] \qquad (20.13)$$

The constant $K_1$ is very small, since the initiating reaction has very high activation energy. The term involving $K_1/2K_3$ are therefore very small in comparison with $K_1K_4/K_3K_5$ and therefore

$$[^{\bullet}C_2H_5] = \left(\frac{K_1K_4}{K_3K_5}\right)^{1/2}[C_2H_6]$$

The rate of production of ethylene is:

$$\frac{d[C_2H_4]}{dt} = K_3[^{\bullet}C_2H_5]$$

$$= \left(\frac{K_1 K_3 K_4}{K_5}\right)^{1/2} [C_2H_6] \qquad (20.14)$$

The reaction is thus of first order.

## 20.5 PHOTOCHEMICAL HYDROGEN AND CHLORINE REACTIONS

Proposed mechanism is

$$Cl_2 + h\nu \xrightarrow{K_1} 2Cl^{\bullet}$$

$$Cl^{\bullet} + H_2 \xrightarrow{K_2} HCl + H^{\bullet}$$

$$H^{\bullet} + Cl_2 \xrightarrow{K_3} HCl + Cl^{\bullet}$$

$$H^{\bullet} + O_2 \xrightarrow{K_4} HO_2^{\bullet}$$

$$Cl^{\bullet} + O_2 \xrightarrow{K_5} ClO_2^{\bullet}$$

$$Cl^{\bullet} + X^{\bullet} \xrightarrow{K_6} ClX$$

Here $X^{\bullet}$ is any substance that can remove chlorine free radical. Let the rate of formation of Cl atoms in reaction (I) is $2I$ and the steady state equation for Cl atoms is,

$$\frac{d[Cl]}{dt} = 2I - K_2[Cl^{\bullet}][H_2] + K_3[H^{\bullet}][Cl_2] - K_5[Cl^{\bullet}][O_2] - K_6[Cl^{\bullet}][X^{\bullet}] = 0 \qquad (20.15)$$

The steady state equation for H atoms is

$$\frac{d[H^{\bullet}]}{dt} = K_2[Cl^{\bullet}][H_2] - K_3[H^{\bullet}][Cl_2] - K_4[H^{\bullet}][O_2] = 0 \qquad (20.16)$$

from which it follows that

$$[Cl^{\bullet}] = \frac{K_3[H^{\bullet}][Cl_2] + K_4[H^{\bullet}][O_2]}{K_2[H_2]} \qquad (20.17)$$

Putting this value in Eq. (20.15) with some rearrangement and neglecting small term involving $[O_2]^2$ gives

$$[H^{\bullet}] = \frac{2IK_2[H_2]}{K_3K_6[Cl_2][X^{\bullet}] + [O_2](K_2K_4[H_2] + K_3K_5[Cl_2] + K_4K_6[X])} \qquad (20.18)$$

Rate of formation of HCl is

$$V_{HCl} = K_2[Cl^{\bullet}][H_2] + K_3[H^{\bullet}][Cl_2] \qquad (20.19)$$

And subtracting Eq. (20.16) gives

$$V_{HCl} = 2K_3[H^{\bullet}][Cl_2] + K_4[H^{\bullet}][O_2]$$

At low concentration of $O_2$, the second term can be neglected, and then with the introduction of Eq. (20.18)

$$V_{HCl} = \frac{2K_2K_3[H_2][Cl_2]I}{K_3K_6[Cl_2][H^{\bullet}] + [O_2](K_2K_4[H_2] + K_3K_6[Cl_2] + K_4K_6[X^{\bullet}])} \quad (20.20)$$

$$= \frac{2K_3/K_4 I[H_2][Cl_2]}{\dfrac{K_3K_6}{K_2K_4}[Cl_2][X^{\bullet}] + [O_2]\left\{[H_2] + \left(\dfrac{K_3K_5}{K_2K_4}\right)[Cl_2] + K_6K_2[X^{\bullet}]\right\}}$$

Introducing new constants $K$, $M$ and $N$

$$V_{HCl} = \frac{KI[H_2][Cl_2]}{M[Cl_2] + [O_2]([H_2] + N[Cl_2])}$$

where $K = \dfrac{2K_3}{K_4}$, $M = \dfrac{[X^{\bullet}]K_3K_6}{K_2K_4}$ and $N = \dfrac{K_3K_5}{K_2K_4}$

## 20.6 GENERAL TREATMENT OF CHAIN REACTION

The derivation of expression for the overall reaction depends on the order of each of reaction step with respect to the chain carriers. The functional dependence of the rate of each step on the concentration of molecular species of the chain carriers can be indicated only after deriving general formulas. Let the mechanism be indicated as follows:

1. *Single chain carrier with second order breaking*

    Let the mechanism is

    Initiation ... $\longrightarrow$ MR + ... $mr_i$

    Propagation R + ... $\longrightarrow$ R + ... $r_p R$

    Breaking R + R + ... $\longrightarrow$ ... $r_b R^2$

    where, R represents a chain carrier, typically an atom or radical. The other molecules involved are not shown. Let the rate of various simple steps be designed by $mr_i$, $r_p R$ and $r_b R^2$.

    The steady state requires

    $$mr_i + r_p R - r_p R - 2 r_b R^2 = 0 \quad (20.21)$$

    $$mr_i = 2r_b R^2 \quad (20.22)$$

    $$R = \left[\frac{mr_i}{2r_b}\right]^{1/2} \quad (20.23)$$

    The result may then be used to eliminate the unknown $R$ in the expression for the rate of various steps

$$\text{Initiation} = r_i \quad (20.24)$$

$$\text{Propagation} = r_p \left(\frac{mr_i}{2r_b}\right)^{1/2} \quad (20.25)$$

$$\text{Breaking} = r_b \left(\frac{mr_i}{2r_b}\right) = \frac{mr_i}{2} \quad (20.26)$$

If the product is formed only in the propagation step, the overall rate is

$$\frac{dx}{dt} = r_p \left(\frac{mr_i}{2r_b}\right)^{1/2} \quad (20.27)$$

A specific example of this type of chain is the homogeneous *ortho–para* hydrogen conversion which goes by mechanism:

$$M + H_2 \,(o \text{ or } p) \longrightarrow 2\,H + M$$

$$H + H_2 \,(p) \longrightarrow H_2(o) + H$$

$$M + 2\,H \longrightarrow H_2 \,(ortho \text{ or } para) + M$$

In this case
$$r_i = K_i \,[H_2]\,[M]$$
$$r_p = K_p \,H_2(p)$$
$$r_b = K_b \,[M]$$

Substituting this in Eq. (20.27) for overall rate

$$\frac{dx}{dt} = K_p \left(\frac{K_i}{K_b}\right)^{1/2} [H_2]^{1/2}[H_2(p)]$$

and therefore, it is a three half order reaction, first order with respect to the variable *para*-hydrogen concentration.

2. *Single Carrier with first order breaking*

The chain breaking step is $R + M \longrightarrow \cdots$

And the steady state is:
$$mr_i - r_b R = 0$$

because
$$mr_i + r_p R - r_p R - r_b R = 0$$

$$mr_i = r_b R; \qquad R = \frac{mr_i}{r_b}$$

And chain propagation rate becomes:

$$r_p R = r_p \left(\frac{mr_i}{r_b}\right)$$

First order chain breaking may occur by absorption of chain carriers on the reaction vessel wall.

3. *Two chain carriers with second order Breaking*

Suppose the mechanism involving carrier R and S is of the form

Initiation ... $\longrightarrow mR + nS$     Rate $r_i$

Propagation    $R + ... \longrightarrow S$    $rp_1R$

                   $S + ... \longrightarrow R$    $rp_2R$

Breaking      $2R + ... \longrightarrow ...$    $2r_bR^2$

Steady state for R and S, respectively gives

$$mr_i - rp_1R + rp_2S - 2r_bR^2 = 0 \tag{20.28}$$

$$nr_i + rp_1R - rp_2S = 0 \tag{20.29}$$

$$nr_i + rp_1R = rp_2S \tag{20.30}$$

By placing Eq. (20.30), i.e. value of $rp_2S$ in Eq. (20.28)

$$mr_i - rp_1R + nr_i + rp_1R - 2r_bR^2 = 0$$

$$mr_i + nr_i = 2r_bR^2$$

$$R = \left[\frac{mr_i + nr_i}{2r_b}\right]^{1/2} \tag{20.31}$$

Place the value of $R$ in Eq. (20.30)

$$nr_i + rp_1\left(\frac{mr_i + nr_i}{2r_b}\right)^{1/2} = rp_2S$$

$$S = \frac{nr_i + rp_1(mr_i + nr_i)^{1/2}/2r_b^{1/2}}{rp_2} \tag{20.32}$$

The rate of two chain propagation are then

$$Rp_1R = rp_1 + \left[\frac{mr_i + nr_i}{2r_b}\right]^{1/2} \tag{20.33}$$

$$rp_2S = nr_i + \left[rp_1^1 \frac{(mr_i + nr_i)}{2r_b}\right]^{1/2} \tag{20.34}$$

Equations (20.33) and (20.34) become identical of either the second radical is not produced in the initial step ($n = 0$) or if chains are long.

An application of these general formulae to the hydrogen bromine reaction leads to the same result as that obtained earlier in this chapter. Here $R$ represents the Br atom, S the H atom, $m = 2$, and $n = 0$.

$$r_i = k_1[Br][M], \quad rp_1 = k_2[H_2],$$
$$rp_2 = k_3[Br_2] + k_4[HBr]$$
and
$$r_b = k_5[M].$$

The rate of formation of HBr is linear combination of the two chain propagation rate Eqs. (20.33) and (20.34). In particular, all the rate Eq. (20.33) but for Eq. (20.34) only fraction

$$\frac{K_3[Br_2] - K_4[HBr]}{K_3[Br_2] + K_4[HBr]} \text{ leads to net production of HBr, therefore.}$$

$$\frac{d[HBr]}{dt} = rp_1 R + \left(\frac{K_3[Br_2] - K_4[HBr]}{K_3[Br_2] + K_4[HBr]}\right) rp_2 S$$

$$= K_2[H_2]\left(\frac{K_1[Br_2]}{K_5}\right)^{1/2} + \left(\frac{K_3[Br_2] - K_4[HBr]}{K_3[Br_2] + K_4[HBr]}\right)^{1/2} \times K_2[H_2]\left(\frac{K_1[Br_2]}{K_5}\right)^{1/2}$$

$$\frac{d[HBr]}{dt} = \left(\frac{2K_2(K_1/K_5)^{1/2}[H_2][Br_2]}{1 + (K_4[HBr]/K_3[Br_2])}\right)^{1/2}$$

## 20.7 APPARENT ACTIVATION ENERGY OF CHAIN REACTION

Suppose each of the rate process has a simple Arrhenius temperature dependence

$$r_i \propto e^{-E_i/RT} \quad r_p \propto e^{-E_p/RT} \quad r_b \propto e^{-E_b/RT}$$

Substituting in Eq. (20.34)

Rate of propagation $\propto e^{-E_p/RT}\left(\dfrac{e^{-E_i/RT}}{e^{-E_b/RT}}\right)^{1/W}$ (20.35)

$$\propto e^{-[E_p + (1/W)(E_i - E_b)]/RT}$$

Therefore, apparent activation energy of chain propagation is

$$E_a = E_p + (1/w)(E_i - E_b) \quad (20.36)$$

where, $W$ is the order of chain breaking process with respect to chain carrier. It might have been expected that the apparent activation energy will be as great as that of the initiation process $E_i$. However, Eq. (20.36) shows that $E_a$ may be smaller either because of possible nonzero value for $E_b$ or more likely because of a second order chain breaking process where $W = 2$. This fact has already been illustrated in the hydrogen–bromine reaction. In the *ortho–para* hydrogen conversion mentioned above $E_a = 58.7$ kcal although $E_i$ is presumably about 103 kcal with $E_b = 0$, $W = 2$, $E_p$ would then be about 7 kcal which is in agreement with the directly measured value.

## 20.8 GENERAL MECHANISM TO STUDY DECOMPOSITION OF ORGANIC COMPOUND

$$M_1 \xrightarrow{K_1} R_1 + M_2 \qquad mr_i$$
$$R_1 + M_1 \xrightarrow{K_2} R_1H + R_2 \qquad rp_1R_1$$
$$M_1 + R_2 \xrightarrow{K_3} R_1 + M_3 \qquad rp_2R_2$$
$$R_1 + R_2 \xrightarrow{K_4} M_4$$
$$2R_1 \xrightarrow{K_5} M_5 \qquad r_bR_1^2$$
$$2R_2 \xrightarrow{K_6} M_6 \qquad r_bR_2^2$$

The rate can be expressed by

$$rp_1R_1 = rp_1\,[(m+n)r_i/2r_b]^{1/2}$$
$$rp_2R_2 = nr_i + rp_1\,[(m+n)\,r_i/2r_b]^{1/2}$$

An applying steady state approximation for $R_1$ and $R_2$ and then subtracting the two as we did earlier, we get

$$mr_i = 2r_b R_2^2; \quad R_2 = \left(\frac{mr_i}{2r_b}\right)^{1/2}$$

$$\frac{-d[M_1]}{dt} = K_1[M_1] + rp_1R_1 + rp_2\left(\frac{mr_i}{2r_b}\right)^{1/2}$$

$$= K_1[M_1] + K_1[M_1] + K_3\left(\frac{K_1[M_1]}{2K_6}\right)^{1/2}$$

$$= 2K_1[M_1] + K_3\left(\frac{K_1[M_1]}{2K_6}\right)^{1/2}$$

For the reaction having long chain propagation step, i.e. having long chain length:

$$K_3\left(\frac{K_1[M_1]}{2K_6}\right)^{1/2}$$

If second step is only propagation step and fifth is chain breaking step, then

$$\frac{-d[M_1]}{dt} = K_1[M_1] + K_2\left(\frac{K_1}{2K_5}\right)^{1/2}[M_1]^{3/2}$$

If third step is only propagating step and sixth is chain breaking step, then

$$\frac{-d[M_1]}{dt} = K_1[M_1] + K_3\left(\frac{K_1}{2K_6}\right)^{1/2}[M_1]^{1/2}$$

In long chain reactions the first term on the right become negligible and rates are three halves and one half order, respectively. In order to explain first order kinetics, Rice and Herzfeld showed that two unlike radical react to break the chain that is process (IV). This mechanism can be put in general term, similar to other general formulae as follows for two carriers with second order breaking involving different carriers.

$$\text{Initiation} \ldots \text{MR} + n\text{S} \qquad mr_i, nr_i$$
$$\text{Propagation R} + \ldots \longrightarrow \text{S} + \ldots \qquad rp_1 R$$
$$\text{Propagation S} + \ldots \longrightarrow \text{R} + \ldots \qquad rp_2 S$$
$$\text{Breaking R} + \text{S} \ldots \longrightarrow \ldots \qquad r_b RS$$

Application of steady state approximation gives for the first propagation step

$$mr_i - rp_1 R + Rp_2 S - r_b R\, S = 0 \qquad (20.37)$$
$$nr_i + rp_1 R - Rp_2 S - r_b R\, S = 0 \qquad (20.38)$$

On adding Eqs. (20.37) and (20.38), we get

$$mr_i + nr_i = 2\, r_b R\, S$$
$$S = (M + N)\frac{r_i}{2 r_b R}$$

On subtracting Eq. (20.38) from Eq. (20.37) and putting value of $S$, we have

$$(m-n) r_i - 2 rp_1 R + 2 rp_2 R\left(\frac{(m+n) r_i}{2 r_b R}\right) = 0$$

$$(m-n) r_i r_b R - 2\, rp_1 r_b R^2 + rp_2 r_i (m+n) = 0$$

By solving this quadric equation by the equation $x = \dfrac{-b \pm \sqrt{b^2 - 4ac}}{2a}$

$$R = \frac{(m-n) r_i r_b \pm \sqrt{(m-n)^2 r_i^2 r_b^2 + r(2 rp_1 r_b) \times rp_2 r_i (m+n)}}{2.2\, rp_1 r_b}$$

$$\text{Rate} = rp_1 R \Rightarrow \left(\frac{m-n}{4}\right) r_i + \sqrt{\left(\frac{m-n}{4} r_i\right)^2 + \frac{(m-n) r_i r_{p_1} r_{p_2}}{2 r_b}} \qquad (20.39)$$

A similar expression can be obtained for the second propagation step. This simplifies, if the chain are long, in which case the propagation and the rate of reaction becomes

$$rp_1 R = \sqrt{(m+n) r_i r_{p_1} r_{p_2} / 2 r_b} \qquad (20.40)$$

It is interesting to note that above equation is same as written before for the second order breaking with two carriers if $r_p$ is replaced by $(r_{p_1} r_{p_2})^{1/2}$ in above equation.

## 20.9 EXAMPLE OF RICE HERZFELD MECHANISM OR DECOMPOSITION OF ACETALDEHYDE MOLECULE

$$CH_3CHO \longrightarrow CH_4 + CO$$

The Rice Herzfeld mechanism to fit the data would involve following steps:

$$CH_3CHO \xrightarrow{K_1} {}^{\bullet}CH_3 + CHO^{\bullet} \qquad (I)$$

$$ {}^{\bullet}CHO \xrightarrow{K_2} {}^{\bullet}H + CO \qquad (II)$$

$$H^{\bullet} + CH_3CHO \xrightarrow{K_3} H_2 + CH_3CO^{\bullet} \qquad (III)$$

$$ {}^{\bullet}CH_3 + CH_3CHO \xrightarrow{K_4} CH_4 + CH_3CO^{\bullet} \qquad (IV)$$

$$CH_3CO^{\bullet} \xrightarrow{K_5} {}^{\bullet}CH_3 + CO \qquad (V)$$

$$2\,{}^{\bullet}CH_3 \xrightarrow{K_6} C_2H_6 \qquad (VI)$$

Applying steady state treatment to the radicals

$$\frac{d[{}^{\bullet}CH_3]}{dt} = K_1[CH_3CHO] - K_4[{}^{\bullet}CH_3][CH_3CHO] + K_5[CH_3CO^{\bullet}] - 2K_6[{}^{\bullet}CH_3]^2 \qquad (20.41)$$

$$\frac{d[CH_3CO^{\bullet}]}{dt} = K_3[H^{\bullet}][CH_3CHO] + K_4[{}^{\bullet}CH_3][CH_3CHO] - K_5[CH_3CO^{\bullet}] \qquad (20.42)$$

$$\frac{d[{}^{\bullet}CHO]}{dt} = K_1[CH_3CHO] - K_2[CHO^{\bullet}] \qquad (20.43)$$

$$\frac{d[H^{\bullet}]}{dt} = K_2[CHO^{\bullet}] - K_3[H^{\bullet}][CH_3CHO] \qquad (20.44)$$

By adding Eq. (20.43) and (20.44), we get

$$K_1[CH_3CHO] = K_3[H^{\bullet}][CH_3CHO]$$

$$K_3[H^{\bullet}] = K_1 \qquad (20.45)$$

$$[H^{\bullet}] = K_1/K_3$$

By adding Eqs. (20.41) and (20.42), we get

$$K_3[H^{\bullet}][CH_3CHO] + K_1[CH_3CHO] = 2K_6[{}^{\bullet}CH_3]^2 \qquad (20.46)$$

By putting the value of $K_1[CH_3CHO]$ from Eq. (20.45), we get

$$K_3[H^{\bullet}][CH_3CHO] + K_3[H^{\bullet}][CH_3CHO] = 2K_6[{}^{\bullet}CH_3]^2$$

$$2K_3[H^{\bullet}][CH_3CHO] = 2K_6[{}^{\bullet}CH_3]^2$$

$$[\text{}^\bullet CH_3]^2 = \frac{K_3}{K_6}[H^\bullet][CH_3CHO]$$

By putting the value of $[H^\bullet]$ from Eq. (20.45)

$$= \frac{K_3}{K_6} \times \frac{K_1}{K_3}[CH_3CHO]$$

$$[\text{}^\bullet CH_3] = \left[\frac{K_1}{K_6}\right]^{1/2}[CH_3CHO]^{1/2} \qquad (20.47)$$

When third step is slow then in Eq. (20.42)

$K_3[H^\bullet][CH_3CHO] \approx 0$. Then

$$\frac{d[CH_3CO^\bullet]}{dt} = K_4[\text{}^\bullet CH_3][CH_3CHO] - K_5[CH_3CO^\bullet] = 0$$

$$K_4[\text{}^\bullet CH_3][CH_3CHO] = K_5[CH_3CO^\bullet]$$

$$\frac{K_4}{K_5}[\text{}^\bullet CH_3][CH_3CHO] = [CH_3CO^\bullet] \qquad (20.48)$$

If Eq. (IV) is overall rate of reaction

$$\text{Rate} = K_4[\text{}^\bullet CH_3][CH_3CHO]$$

On putting value of $[\text{}^\bullet CH_3]$ from Eq. (20.47)

$$= K_4\left[\frac{K_1}{K_6}(CH_3CHO)\right]^{1/2}[CH_3CHO]$$

$$\text{Rate} = K_4 \times \left[\frac{K_1}{K_6}\right]^{1/2}[CH_3CHO]^{3/2}$$

If Eq. (V) is the overall rate of reaction

$$\text{Rate} = K_5[CH_3CO^\bullet]$$

$$= K_5\frac{K_4}{K_5} \times [\text{}^\bullet CH_3][CH_3CHO]$$

$$\text{Rate} = K_4 \times \left[\frac{K_1}{K_6}\right]^{1/2}[CH_3CHO]^{3/2}$$

## 20.10 STUDY OF FAST REACTIONS

Fast reactions are those where reaction times are less than a few seconds. Two general types of experimental difficulty arise in study of fast reactions: (1) defining initial time and (2) achieving suitable time resolution when measuring concentration. For kinetics experiments, reactants are mixed together to start the reaction, or a reactive species is suddenly formed in a system previously at equilibrium. In both the cases, the starting process takes a finite time, and for a satisfactory kinetic investigation, the mixing time or initiation time must be less than the half life of the reaction. After the initial time, the reaction is followed by measuring concentrations at a series of times. When these times are fraction of a second, kinetics enters a new technological world, one in which time resolution has gradually been refined from seconds to milliseconds to microseconds to nanoseconds and recently to picoseconds. Schemes for classifying these fast reactions technique have been devised, though demarcation is not meant to be rigid. One such scheme is to distinguish perturbation, competition and flow methods. In the first, a system originally at equilibrium is subjected to a sudden perturbation whereupon its rapid re-adjustment to a new equilibrium is followed. Large perturbations are provided by: intense pulses of light in flash photolysis and pulsed laser methods; by a pulse of high frequency electrons in pulse radiolysis; and by thermal heating with a supersonic shock wave in shock tubes. The large perturbations produce very high concentration of short-lived radicals or excited molecules which are usually observed by spectroscopic techniques. Small perturbations are employed in the wide range of chemical relaxation methods, where a small displacement is produced by temperature jump, pressure jump, electric field jump or ultrasonics and then relaxation to the new equilibrium position is traced. All the large perturbations methods are used in gas phase studies but small perturbations methods are more appropriate for solution kinetics, with the exceptions of ultrasonic dispersion which is a major source of data on vibrational–translational energy transfer in gases. Competition methods include observing line broadening in spectra (esp. NMR) and quenching of fluorescence. The latter refers to competition between the decay of molecular excited states by fluorescence, and delay by deactivation in molecular collisions. Flow methods represent an extension for methods in which reactants are mixed, since in flow systems mixing times can be generated by electrical discharge through a gas which is then mixed with other reactants in a fast flow systems. For solution kinetics, further refinements such as accelerated or stopped flow have been developed.

The reasons why conventional techniques lead to difficulties for very rapid reactions are:

1. The time that it takes to mix reactants or to bring them to a specified temperature may be significant in comparison to half life of the reaction. An appreciable error therefore will be made, since the initial time cannot be determined accurately.
2. The time that it takes to make a measurement to concentration may be significant compared to half life.

Here we will concentrate on techniques which have been fruitful in gas kinetics, or which seem to offer great future potential.

### 20.10.1 Flash Photolysis

This technique is developed by Norrish and Porter around 1950. They used an intense flash of light to produce transient species, radicals or excited molecules in concentrations that may be four or five order of magnitude greater than found in conventional systems. A flash of few microseconds duration allows kinetic studies in the microsecond–millisecond region which embraces a wide range of radical reactions and energy transfer process. A typical apparatus is represented schematically in Figure 20.1. The quartz reactions vessel is placed parallel to the initiating flash or photoflash, a quartz discharge tube containing an inert gas at low pressure. The condenser $C_1$ is charged to a high voltage and when a trigger pulse is applied the energy, hundreds or thousands of Joules discharge in a few microseconds producing an intense light flash which has the same duration. This radiation is emitted in a continuous range of wavelength from the lowest transmitted by quartz (~200 nm) throughout the ultraviolet and visible, thus covering much of the photochemically useful range. To increase the amount of radiation absorbed; photoflash and reaction vessel are enclosed in a cylindrical reflector. Radical concentration of $10^{-8} - 10^{-7}$ mol cm$^{-3}$, corresponding to pressure to about 1 Torr, 133 Nm$^{-2}$, can be attained.

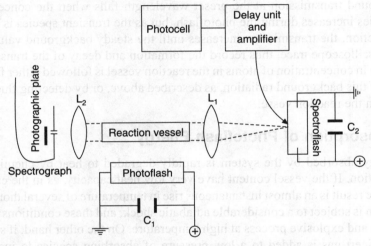

**FIGURE 20.1** Apparatus used in flash photolysis.

If transient species formed during the flash have suitable absorption spectra, the variation of spectral intensity with time can be followed. This intensity gives concentration directly if the absorption coefficient is known. The method is known as *kinetic spectroscopy* and can be employed in one of two modes, called by Porter, flash spectroscopy and kinetic spectrophotometry.

## 20.10.2 Flash Spectroscopy

The photoflash output is detected by a photocell and the resulting pulse is amplified to trigger a second flash lamp. The analyzing flash or spectroflash is similar to but smaller than the photoflash, with a discharge energy of around 100 J and a few microseconds duration. The spectroflash provides a continuum of background radiation against which the absorption spectra of species in the reaction vessel are photographed.

This is achieved by placing the spectroflash at the focus of lens ($L_1$) which gives an approximately parallel beam of background light. This passes through the reaction vessel and is then focused onto the slit of a spectrograph by a second lens ($L_2$). The photographic plate on the spectrograph records the absorption spectra at the time the spectroflash fired. This time is varied by the simple device of imposing a controlled and variable delay on the pulse that fires the spectroflash. The delay time thus corresponds to the reaction time after the initiating photoflash and a series of experiments at different delay time provides a photographic record of transient species spectra changes with time.

### 20.10.3 Kinetic Spectrophotometry

In this mode, if the spectrum of transient species is known, it is possible to obtain a complete kinetic run from one experiment. The spectroflash is replaced by a steady source of continuous background radiation extending over the appropriate wavelength range. The light passing through the reaction vessel is focused onto the slit of a monochromator fitted with a photoelectric detector set to the wavelength at which the transient absorb, and the photo cell output is fed to an oscilloscope. Thus, the instrument is essentially a scanning spectrophotometer. The originally steady, background transmission at the preset wavelength falls when the concentration of the absorbing species increases during the photoflash, but as the transient species is removed in the subsequent reaction, the transmission increases until the steady background value is eventually reached. The oscilloscope trace, thus record the formation and decay of the transient species.

The change in concentration of atoms in the reaction vessel is followed either from the change in absorption of this background radiation, as described above, or by detecting fluorescence from excited atoms in the reaction vessel.

### 20.10.4 Absorption of Photoflash Energy

The light energy absorbed by the system is rapidly degraded to heat by chemical reaction or collision relaxation. If the vessel content have very low heat capacity, as in the case for a gas at low pressure, the result is an almost instantaneous rise in temperature of several thousands degrees. Thus, the system is subject to a considerable adiabatic shock, and these conditions may be used to study pyrolytic and explosive process at high temperature. On the other hand, if several hundred fold excess of inert gas is added to a low pressure of absorbing species to increase the total thermal capacity the temperature rise can be kept below 10 K and reactions studied under reasonably isothermally conditions. These two general approach are called the *adiabatic* and *thermal methods*, and they will be illustrated with a few examples from the enormous number available.

### *The adiabatic method*

Among important reactions studied are the pyrolysis and combustion of hydrides. Such studies are relevant to the burning of fuels, as in internal combustion engine. With hydrogen or hydrocarbon, neither these fuel molecules nor the oxidant oxygen absorb radiation from the photoflash and a small concentration of photosensitizer must be added. This provides both adiabatic heating and radicals which initiate further reaction, and a good example is nitrogen dioxide ($NO_2 + H_2 \longrightarrow NO + O$). For hydrocarbons, flash photolysis clearly illustrates the effect of changing fuel: oxidant ratio. Knock and antiknock phenomenon have also been investigated. Adding lead tetraethyl to mixtures, lengthened the induction periods, reflecting the decreased tendency to premature ignition or knocking. The spectrum of gaseous lead oxide was seen during the induction period, but on ignition this was replaced by atomic lead. Such evidence shows that premature ignition is reduced by reactions of lead with hydroxyl and hydrocarbon radicals.

## The isothermal method

(1) Halogen molecules absorb light in the UV or visible range and dissociate. The intensity of the molecular spectrum being reduced as the concentration falls. Halogen atoms formed in the flash recombine over a period of a few milliseconds. The process is easily followed by observing the resultant increase in intensity of the halogen molecular spectrum:

$$X_2 + H_2 \longrightarrow 2X^{\cdot}$$
$$X^{\cdot} + X^{\cdot} + M \longrightarrow X_2 + M$$

Thus, the combination rate constant is easily measured and the activation energy can also be determined by enclosing the system in a suitable furnace to vary the temperature. This gave strange result that as temperature increases the rate decreases, the reaction has a negative activation energy.

## 20.10.5 Fast Flow Methods

Fast flow resembles conventional flow methods, but it uses a rapid flow of reactants of very low total pressure together with special mixing devices that restrict mixing times to under one millisecond. Reactions with half life of a few milliseconds and above can be studied. The apparatus is represented in Figure 20.2.

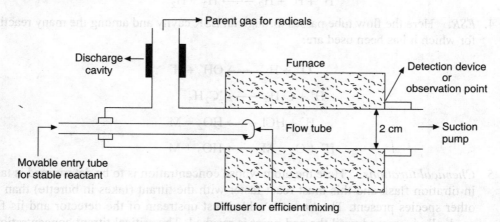

**FIGURE 20.2** Apparatus used in fast flow method.

It consists of discharge cavity for generating radicals from parent gas which are then mixed with stable reactant, flowing in a separate flow tube, by a special diffuser for efficient mixing of two reactants. This is the point where reaction starts. The mixture flows down the glass tube, perhaps 2 cm in diameter and 1 m long, past some detection device known as *observation point*. Gas flow is controlled by needle valves and the flow rate, usually a few hundred cm s$^{-1}$ is measured by a flow meter. Gas pressure in the flow tube is in the range 0.1–10 Torr, i.e. 13–1.3 × 10$^3$ Nm$^{-2}$. The concentration of reactants or products is determined at observation point down stream in the tube. Under steady flow conditions this corresponds to the reaction time $t = d/V$, where, $d$ is distance between mixing point and detector, and $V$ is linear velocity. The reaction time is varied by altering $d$, i.e. by placing reaction tube downstream the glass tube. Alternately, there may be a series of fixed reactant entry points at various positions along the glass tube or flow tube. To determine activation energy, the flow tube is enclosed within a thermostat bath or furnace.

Radicals are usually produced by flowing the parent gas through a microwave electrical discharge. Atoms such as hydrogen, oxygen, nitrogen and halogen are formed in high relative concentrations from the diatomic molecules some times with the inert gas present as carrier. Great care is necessary with very low pressure to prevent the removal of atoms at the wall of the flow tube and treatment of surfaces with phosphoric acid has proved successful in this respect.

1. *Absorption spectra*, e.g. for the reaction.

$$CN^{\bullet} + O_2 \longrightarrow CON^{\bullet} + O^{\bullet}$$

$$ClO^{\bullet} + H_2 \longrightarrow HOCl + H^{\bullet}$$

2. *Mass spectrometry*, e.g. for the reactions

$$OH^{\bullet} + OH^{\bullet} \longrightarrow H_2 + O^{\bullet\bullet}$$

$$H^{\bullet} + CH_3CHO \longrightarrow CH_4 + CHO^{\bullet}$$

3. *Calorimetric probes*, e.g. for the reaction

$$H^{\bullet} + H^{\bullet} + H_2 \longrightarrow H_2 + H_2$$

4. *ESR*: Here the flow tube passes through an ESR cavity and among the many reactions for which it has been used are:

$$O^{\bullet\bullet} + H_2 \longrightarrow OH^{\bullet} + H^{\bullet}$$

$$H^{\bullet} + C_2H_4 \longrightarrow {}^{\bullet}C_2H_5$$

$$H^{\bullet} + HCl \longrightarrow HO_2 + M$$

$$H^{\bullet} + O_2 + M \longrightarrow HO_2 + M$$

5. *Chemical titrations:* Here the atom whose concentration is to be determined is taken in titration flask and this must react faster with the titrant (takes in burette) than any other species present. The titrant is added just upstream of the detector and its flow gradually increased until the end point is reached. The critical titrant concentration is then equivalent to the atom concentration originally present. A classical example is the use of $NO_2$ as titrant to determine oxygen atom concentration. Nitrogen atoms may be titrated with NO as

$$N + NO \longrightarrow N_2 + O$$

$$N + O \longrightarrow NO^* \text{ (blue glow)}$$

The blue glow emitted from excited NO* molecules is extinguished when no nitrogen atoms remain. Some reactions that have been followed by the titration methods are

$$O + O + M \longrightarrow O_2 + M$$

$$N + N + M \longrightarrow N_2 + M$$
$$N + O_2 \longrightarrow NO + O$$

6. *Chemi luminescence:* Hydrogen atom concentration can be determined by adding NO just upstream to a photoelectric detector. HNO is formed in an electronically excited state:

$$H^{\cdot} + NO \longrightarrow HNO^* \text{ (red glow)}$$

and HNO* emits a red glow whose intensity is measured photoelectrically. Here, the concentration of atoms is calculated from the intensity of the chemiluminescene. The intensity is proportional to concentration.

$I = K$ [NO] [H] and $K$ is established by independent calibration. An example of reaction monitored in this way is

$$H^{\cdot} + O_2 + M \longrightarrow HO_2^{\cdot} + M$$

Another example where chemiluminescent detection has been employed is

$$S^{\cdot\cdot} + S^{\cdot\cdot} + M \longrightarrow S_2^* + M$$

7. *Atomic resonance fluorescence:* This method is a valuable extension to the fast flow method, due to the extra sensitivity of atomic resonance fluorescence.

## 20.10.6 Nuclear Magnetic Resonance (NMR) Methods

NMR and electron paramagnetic resonance (EPR) offer ways of studying very rapid exchange reactions in an equilibrium system. It is necessary that at least two different environments be accessible to the magnetically active nucleus or electron. These environments must be just that quite different spectra would be observed for a system in each environment in the absence of exchange or with slow exchange. Rapid exchange leads to an averaging of the spectra in a way which is predictable. A simple example is provided by the system hydrogen-peroxide–water, in which two major environment exists for proton. Figure 20.3 shows how the NMR spectrum of such a system, approximately equimolar, would appear for various environments. In this case the important variables are $\tau$, the mean lifetime of a proton in each environment, and d (chemical shift) the frequency difference in cycles per second between the two signals in the absence of exchange.

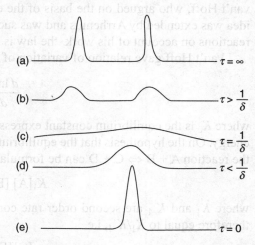

**FIGURE 20.3** The effect of chemical exchange on NMR line width.

Resonance frequency increasing form left to right. In general, value of $\tau$ of the order of $1/\delta$ can be found. For $\tau$ much larger than $1/\delta$, two separate signals are observed and for $\tau$ much less than $1/\delta$ one averaged signal is observed. Since $\delta$ for proton is generally of the order of 100 cycles per second. For every measurement, acid/base catalysis or change in temperature can be used to bring about change in $\tau$. Table 20.1 shows some proton transfer rate constants for amines obtained by NMR methods.

**TABLE 20.1**  Proton transfer rate constants for amines

| System | $K_1$ | $K_2$ |
|---|---|---|
| $NH_3 + NH_4^+$ | $10.6 \times 10^8$ | $0.9 \times 10^8$ |
| $CH_3NH_2 + CH_3NH_3^+$ | $2.5 \times 10^8$ | $3.4 \times 10^8$ |
| $(CH_3)_2NH + (CH_3)_2NH_2^+$ | $0.4 \times 10^8$ | $5.6 \times 10^8$ |
| $(CH_3)_3N + (CH_3)_3NH^+$ | $0.0 \times 10^8$ | $6.1 \times 10^8$ |

## 20.11 METHODS OF DETERMINING RATE LAWS: ARRHENIUS LAW

It is well known generalization that rate of reaction is doubled by a rise in temperature of 10°C. It was found empirically by J. Hood that rate constant $K$ of a reaction varies with the absolute temperature $T$ according to a law of the form

$$\log K = B - \frac{A^1}{T} \tag{20.49}$$

where $A^1$ and $B$ are constants. In 1884, some theoretical significance to this law was given by van't Hoff, who argued on the basis of the effect of temperature on equilibrium constants. This idea was extended by Arrhenius and was successfully applied by him to the data for a number of reactions on account of his work, the law is usually referred to as the *Arrhenius law*.

Van't Hoff gave relation of variation of equilibrium constant with temperature as

$$\frac{d \ln K_c}{dT} = \frac{\Delta E}{RT^2} \tag{20.50}$$

where $K_c$ is the equilibrium constant expressed in terms of concentration and $\Delta E$ is the change in energy. On the hypothesis that the equilibrium is dynamic in nature, the equilibrium condition for the reaction $A + B \Leftrightarrow C + D$ can be formulated by equating the rates of two opposing reactions.

$$K_1[A][B] = K_{-1}[C][D] \tag{20.51}$$

where $K_1$ and $K_{-1}$ are second order rate constants for the reaction. The equilibrium constant is therefore equal to $K_1/K_{-1}$, i.e.

$$\frac{[C][D]}{[A][B]} = \frac{K_1}{K_{-1}} = K_C \tag{20.52}$$

The reaction isochore may therefore be written as

$$\frac{d \ln K_1}{dT} - \frac{d \ln K_{-1}}{dT} = \frac{\Delta E}{RT^2} \tag{20.53}$$

which may be split into two equations:

$$\frac{d \ln K_1}{dT} = \frac{E_1}{RT^2} + I \quad \text{and} \quad \frac{d \ln K_{-1}}{dT} = \frac{E_{-1}}{RT^2} + I \quad \text{(20.54 and 20.55)}$$

where $E_1 - E_{-1} = \Delta E$ and $I$ is constant of integrations and found to be independent of temperature. Experimentally, it was found that $I$ can be set equal to zero, the rate constant therefore being related to the temperature by an equation of the form

$$\frac{d \ln K}{dT} = \frac{E}{RT^2} \quad (20.56)$$

There is therefore a close analogy between the equation for an equilibrium constant Eq. (20.50) and that for a rate constant Eq. (20.56).

The fact that the constant of integration $I$ turned out to be zero has a particular significance. If $I$ has not been zero, it would be necessary to conclude that the temperature dependencies of the rates in the forward and reverse directions were affected.

Arrhenius showed that reaction A + B to C + D may be divided into two parts, one of which is concerned with the rate from left to right, the other from right to left. As far as energy is concerned, one can say that between the initial and final states there is an intermediate state that has energy $E_1$ greater than that of A + B, this situation is represented in Figure 20.4.

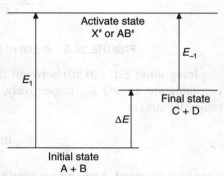

**FIGURE 20.4** Energy states of reactants, products and activated state.

When reaction occurs between A and B, there exists equilibrium between A + B on the one hand and a particular collision complex on the other. This complex is known as activated complex. The energy $E_1$ is the energy required for the system to pass from the state A + B to the activated complex state (AB$^\#$) and is known as energy of activation or activation energy.

The rate Eq. (20.56) integrates to

$$\ln K = -\frac{E}{RT} + \text{constant} \quad (20.57)$$

provided that, $E$ is independent of temperature. This equation may be written as

$$K = Ae^{-E/RT} \quad (20.58)$$

Above equation is known as *Arrhenius equation*. Here $A$ is a constant which is usually known as frequency factor for the reaction. The factor $e^{-E/RT}$ is recognized as the Boltzmann expression for the fraction of the system having energy in excess of the value $E$ so that, it may be identified with the fraction of reactant molecules that are activated complexes.

The Arrhenius law can be tested by plotting $\ln K$ against reciprocal of the absolute temperature. According to Eq. (20.57) a straight line should be obtained. Its slope is $-E/R$ ($=-E/4.57$ if common logarithms are used) so that the activation energy $E$ can be calculated. The law has been found to be obeyed with high degree of accuracy for all types of chemical reactions. Figure 20.5 shows a plot of log $K$ against $1/T$ for the thermal decomposition of HI, a reaction that was studied carefully by Bodenstein.

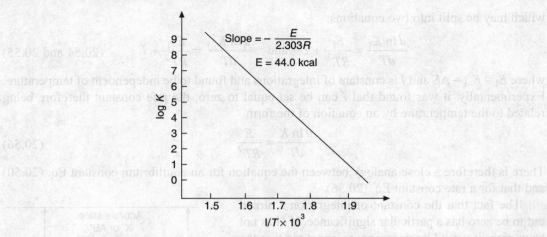

**FIGURE 20.5** A plot of log $K$ against $1/T$ for the thermal decomposition of HI.

Integrating Eq. (20.50) between the temperature $T_1$ and $T_2$ when the corresponding rate constants are $K_1$ and $K_2$, respectively, and assuming that $E_a$ is constant over this temperature range, we obtain

$$\ln \frac{K_2}{K_1} = \frac{E_a}{R}\left[\frac{T_2 - T_1}{T_1 T_2}\right] \tag{20.59}$$

This is integrated Arrhenius equation. Thus, knowing the rate constants at two different temperatures, the energy of activation $E_a$ can be readily determined.

## 20.12 COLLISION THEORY

In 1918, W.C. Lewis, identified frequency factor with the collision number and calculated its value using a simple version of the kinetic theory of gases in which the molecules are treated as hard spheres. However, for reactions between more complicated molecules it is necessary to treat collision in a more accurate and detailed manner. Such a treatment was first given in 1935, by Eyrring. Lewis was concerned with bimolecular reactions, and for a reaction between two identical gaseous molecules, he suggested that the rate in molecular units is

$$v = Z_{AA}\, e^{-E/RT} \quad \text{molecules/cc/s} \tag{20.60}$$

where $Z_{AA}$ is the number of collisions per second between two molecules of $A$ in 1 cc of gas. Application of simple kinetic theory to this problem leads to the result that the total number of collisions per second of all the $n$ molecules that are contained in 1 cc of gas is

$$Z_{AA} = 2\pi d^2\, \bar{c}\, n^2 \tag{20.61}$$

where $d$ is the distance between the centres of the spheres when the collision occurs and $\bar{c}$ is the average velocity of each molecule. This average speed $\bar{c}$ is given by kinetic theory

$$\bar{c} = \sqrt{\frac{8kT}{\pi m}} \tag{20.62}$$

The collision frequency $Z_{AA}$ is therefore

$$Z_{AA} = 2n^2 d^2 \sqrt{\frac{8\pi kT}{m}} \qquad (20.63)$$

The corresponding expression for the frequency $Z_{AB}$ between two unlike molecules $A$ and $B$ of masses $M_A$ and $M_B$, the concentration being $n_A$ and $n_B$ is

$$Z_{AB} = n_A n_B d_{AB}^2 \left[ 8\pi kT \frac{M_A + M_B}{M_A \times M_B} \right]^{1/2} \qquad (20.64)$$

Here $d_{AB}$ is the average of the diameters or sum of the radii. The quantity $d_{AB}^2$ is known as the collision cross-section. Let us calculate $Z_{AB}$ for the reaction between $H_2$ and $I_2$ at 700 K and 1 atm pressure, the quantities of the two gases being 1 mol each. Accordingly $n(H_2) = n(I_2) \approx 10^{19}$ molecules cm$^{-3}$. $d(H_2) = 2.2$Å $d(I_2) = 4.6$ Å so that $d_{av} = 3.4$ Å. Hence

$$Z_{AB} = (10^{19})^2 (3.4 \times 10^{-8})^2 \left[ \frac{(2+254) 8 \times 3.14 \times 8.314 \times 10^7 \times 700}{2 \times 254} \right]^{1/2}$$
$$= 10^{38} \times 1.16 \times 10^{-15} \times 8.58 \times 10^5 \approx 10^{29} \text{ collisions s}^{-1} \text{ cm}^{-3}$$

Since, there are approximately $10^{29}$ collisions/second for $10^{19}$ molecules of each species, each molecules makes about $10^{10}$ collisions/second with the molecules of other species. If each collision were to lead to a chemical reaction, then whole reaction would have been completed in about $10^{-10}$ s. However, this predicted rate of the reaction is in complete disagreement with the experimental rate. Hence, we conclude that all collisions do not result in chemical reaction. In order for a reaction to occur, the energy of collision must equal or exceed the threshold energy.

The rate of reaction between molecules $A$ and $B$ with an activation energy of $E$ is given by

$$v = n_A n_B d_{AB}^2 \left[ 8\pi kT \frac{M_A + M_B}{M_A M_B} \right]^{1/2} e^{-E/RT} \qquad (20.65)$$

This equation implies that the rate is the number of molecules colliding per second and having a joint energy $E$ in excess of the mean energy. If the concentrations are set equal to unity the resulting expression is

$$K^1 = d_{AB}^2 \left[ 8\pi kT \frac{M_A + M_B}{M_A M_B} \right]^{1/2} e^{-E/RT} \qquad (20.66)$$

This constant $K^1$, equal to $v/n_A n_B$ is a rate constant for the reaction in the molecular units, namely cc molecule$^{-1}$s$^{-1}$. It can be put into the units of cc mol$^{-1}$s$^{-1}$ by multiplication by $N$ the Avogadro number,

$$K = N d_{AB}^2 \left[ 8\pi kT \frac{M_A + M_B}{M_A M_B} \right]^{1/2} e^{-E/RT} \text{ (cc mol}^{-1}\text{s}^{-1}) \qquad (20.67)$$

Comparison with Eq. (20.58) shows that, according to this theory, the frequency factor $A$ is given by

$$A = N\, d_{AB}^2 \left[ 8\pi kT\, \frac{M_A + M_B}{M_A M_B} \right]^{1/2} \qquad (20.68)$$

The expression on the right-hand side is known as the collision number and is usually written as $Z$. The frequency factor of the reaction is therefore identified with the collision number. In the case of a reaction between two like molecules the expression for the frequency factor is

$$A = 2 N d_{AA}^2 \sqrt{\frac{8\pi kT}{m}} \qquad (20.69)$$

Collision number can be readily calculated using molecular diameter derived from viscosity data and $E$ can be determined from the experimental variation of the rate with the temperature. It is easy to test the theory. Calculation shows that $Z$ usually has a value lying between $4 \times 10^{11}$ and $4 \times 10^{12}$ cc mol$^{-1}$s$^{-1}$, depending upon molecule radii.

This is true for gas reactions involving relatively simple molecules and for many reactions in solutions, particularly those in which at least one of the reactants is a simple molecule or ion. However, a number of reactions especially those involving more complex molecules takes place at rates which are markedly different from those calculated on the basis of the collision theory. For example, certain gas phase reactions between radicals and molecules takes place much more slowly than expected, where as addition of tertiary amines to alkyl iodides in solution frequently proceeds at rates which are $10^{-5}$ to $10^{-8}$ of the calculated value. At one time, the anomalies in solution were thought to be due to solvent effects, but in some instances the rates were shown to be low in the gas phase also. Certain reactions were found to proceed more rapidly than calculated.

In order to account for deviations from the simple collision theory, it has been postulated that the number of effective collisions may be less than that given by kinetic theory, since for reaction to take place a critical orientation of the molecules before collision may be necessary. The rate constant was therefore written as

$$K = PZe^{-E/RT} \qquad (20.70)$$

where $P$ is referred to as a probability or steric factor. Several weaknesses of the collision treatment have become apparent in recent years. In the first place, attempts to correlate the value of $P$ with the structures and properties of the reacting molecules have not been very successful. Secondly, it is hardly possible to interpret the abnormally high rate values that are sometimes observed. Furthermore, logical weaknesses of the collision treatment apparent when reversible reactions are considered. Thus for reaction,

$$A_2 + B_2 \rightleftharpoons 2AB$$

The rate constant for the forward reaction would be given as

$$K_1 = P_1 Z_1 e^{-E_1/RT} \qquad (20.71)$$

While that for the reverse reaction is

$$K_{-1} = P_{-1} Z_{-1}\, e^{-E_{-1}/RT} \qquad (20.72)$$

The equilibrium constant $K$ is equal to $K_1/K_{-1}$ and is therefore given by

$$K = \frac{K_1}{K_{-1}} = \frac{P_1 Z_1}{P_{-1} Z_{-1}} e^{-(E_1 - E_{-1})/RT} \tag{20.73}$$

However, from the thermodynamics the equilibrium constant is equal to

$$K = e^{-\Delta G/RT} \tag{20.74}$$

$$K = e^{-\Delta S/R} e^{-\Delta H/RT} \tag{20.75}$$

If these expressions for $K$ are compared, it is clear that the terms $e^{-(E_1 - E_{-1})/RT}$ and $e^{-\Delta H/RT}$ correspond, consequently, the ratio $P_1 Z_1 / P_{-1} Z_{-1}$ must be equal to $e^{-\Delta S/R}$. If the molecules $A_2$, $B_2$ and $AB$ are of comparable dimensions, $Z_1$ will practically equal to $Z_{-1}$ so that the entropy term $e^{-\Delta S/R}$ must be approximately equal to the ratio of probability factors. It is therefore not sufficient to correlate the probability factors with the probability that certain reacting groups come together on collision, they should be interpreted in terms of entropy factors in a precise manner.

## 20.13 ACTIVATED COMPLEX THEORY OR ABSOLUTE REACTION RATE THEORY: OR TRANSITION STATE THEORY

In order for any chemical change to takes place, it is necessary for the atoms or molecules involved to come together to form an activated complex. This complex is regarded as being situated at the top of an energy barrier lying between the initial and final states, and the rate of reaction is controlled by the rate with which the complex travels over the top of the barrier. This type of formulation was first put forward by Marcelin, who considered the reacting molecules to cross a critical surface in phase space. The idea was somewhat further developed by Rodenbush and later by O.K. Rice and Gershinowitz. The first calculation of the rates of reaction in terms of specific potential energy surface was done by Pelzer and Wigner, who calculated the rate of reaction between hydrogen atoms and molecules. A clear formulation of the problem was made by Eyring who has applied his method with considerable success to a large number of physical and chemical processes. A somewhat similar formulation of the problem was made by Evans and Polanyi. The method of Eyring will be employed here.

Consider a reaction

$$A + B \Leftrightarrow X^{\#} \Leftrightarrow C + D \tag{20.76}$$

The theory involves the hypothesis that even when the reactants and products are not at equilibrium with each other, the activated complex is in equilibrium with the reactants. The equilibrium constant may be written as

$$K^{\#} = \frac{[X^{\#}]}{[A][B]} \tag{20.77}$$

In terms of partition function equilibrium constant may be written as

$$K^{\#} = \frac{F^{\#}}{F_A F_B} e^{-E_0/RT} \tag{20.78}$$

where $E_0$ is energy difference between the zero point energy per mole of the activated complex and that of the reactants. Since, this energy is amount of energy that the reactants must acquire at 0 K before they react, i.e. $E_0$ is the hypothetical energy of activation at this temperature.

This partition function can be factorized into contributions corresponding to translational, rotational, vibrational and electronic energy. If for example, the molecule A consists of $N_A$ atoms, there will be 3 $N_A$ such partition function, of which three are of translational motion, three for rotational motion (or two if the molecule is linear) and therefore $3N_A - 6$ for vibrational motion ($3N_A - 5$ for linear molecules). The same is true for activated complex which consists of $N_A + N_B$ atoms, giving $3(N_A + N_B) - 6$ vibrational terms if the molecule is nonlinear. One of these vibrational factors is of a different character from the rest, since it corresponds to a very loose vibration which allows the complex to dissociate into the products C and D.

From classical mechanics the energy of vibration is given by $RT/N_A$ (or $k_B T$ where $k_B$ is the Boltzmann constant) whereas from quantum mechanics, it is given by $h\nu$ so that $h\nu = \dfrac{RT}{N_A}$ or $\nu = RT/N_A h = k_B T/h$. The vibrational frequency $\nu$ is the rate at which the activated complex molecules move across the energy barrier.

For this one degree of freedom, one may therefore employ in place of the ordinary factor $(1 - e^{-h\nu/KT})^{-1}$, the value of this function calculated in the limit at which $\nu$ tends to zero. This is evaluated by expanding the exponential and taking only the first term.

$$\lim_{\nu \to 0} \frac{1}{1 - e^{-h\nu/kT}} = \frac{1}{1 - (1 - h\nu/kT)} = \frac{kT}{h\nu} \tag{20.79}$$

The equilibrium constant may therefore be expressed by including this term $kT/h\nu$ and replacing $F^{\#}$ by $F_{\#}$ which now refers only to $3(N_A + N_B) - 7$ degree of vibrational freedom [$3(N_A + N_B) - 6$ for linear complex] the resulting expression is

$$K^{\#} = \frac{[X^{\#}]}{[A][B]} = \frac{F_{\#}(kT/h\nu)}{F_A B_B} e^{-E_0/RT} \tag{20.80}$$

This expression rearranges to

$$\nu[X^{\#}] = [A][B] \frac{kT}{h} \frac{F_{\#}}{F_A F_B} e^{-E_0/RT} \tag{20.81}$$

The frequency $\nu$ is the frequency of vibration of the activated complexes in the degree of freedom corresponding to their decomposition. It is therefore the frequency of decomposition. The expression on the left hand side of equation (20.81) is therefore the product of concentration of complexes $X^{\#}$ and the frequency of their decomposition. It is, therefore, rate of reaction, i.e.

$$v = [A][B] \frac{kT}{h} \frac{F_{\#}}{F_A F_B} e^{-E_0/RT} \tag{20.82}$$

The molecularity of the reaction is in fact equal to the number of reactant molecules that exists in the activated complex.

The rate constant of the reaction, the rate of which is given by Eq. (20.82), is given by

$$k = \frac{[v]}{[A][B]} = \frac{kT}{h} \frac{F_{\#}}{F_A F_B} e^{-E_0/RT} \tag{20.83}$$

The quantity $kT/h$ which appears in these expressions is of great importance in rate theory, it has the dimensions of a frequency and its value is about $6 \times 10^{12}$ s$^{-1}$ at 300 K.

To allow for the possibility that not every activated complex reaching the top of the potential energy barrier is converted into reaction product. It is convenient to introduce a transmission coefficient ($K$), the rate constant being written as

$$k = K \frac{kT}{h} \frac{F_\#}{F_A F_B} e^{-E_0/RT} \tag{20.84}$$

The function $\frac{F_\#}{F_A F_B} e^{-E_0/RT}$ may be regarded as a modified equilibrium constant between the normal and activated states, the modification being the removal of the vibrational factor for the decomposition of the complex. If this modified equilibrium constant is denoted by $K^\#$ the rate expression becomes

$$k = K \frac{kT}{h} K^\# \tag{20.85}$$

This expression has been derived on the assumption that the rate constant is expressed in terms of concentration, under these conditions $K^\#$ is concentration equilibrium constant.

For a great many reactions the transmission coefficient is equal to unity, which means that every activated complex becomes a product. There are however two important classes of reactions for which the transmission coefficient may be considerably less than unity. The first comprises bimolecular atom recombination in the gas phase and the reverse decomposition of diatomic molecules. The second class of reactions in which the transmission coefficient may be appreciably less than unity comprises those in which there is a change from one type of electronic state to another.

## 20.14  FACTORS DETERMINING REACTION RATE IN SOLUTION

The reaction between two molecules in solution can be thought of as occurring in three well defined stages:

1. diffusion of the molecules to each other,
2. the actual chemical transformation, and
3. diffusion of the products away from each other.

Diffusion in liquid has an activation energy, but the magnitude of this is generally not greater than 5 kcal. Many chemical reactions have activation energies of more than this and thus cannot involve diffusion as the slow step, which must therefore be step 2, the purely chemical process. This conclusion is supported by the fact that the rates of these reactions do not depend much upon the viscosity of the solvent, as they would if diffusion were important, diffusion and viscosity being very closely related. On the other hand, a reaction occurring in a solvent, which solvates the reactants to a greater degree than the activated complex, will take place rapidly than it would in a solvent in which the reactants are not solvated. There are many reactions that do not occur at all in gas phase, an example is the formation of the quaternary ammonium salt $(C_2H_5)_4N^+I^-$ from ethyl iodide and triethyl amine. The activated complex for this reaction will be fairly polar. Ionizing solvents, which will stabilize the activated complex are therefore expected to favour the reaction. This explanation is supported by the fact that, rate is very much greater in an ionizing and polar solvents, such as nitrobenzene than it is in a nonpolar solvent such as hexane or benzene.

In this reaction and many other reactions in which the solvent has an important influence, the situation is dominated by the electrostatic forces between solvent and solute molecules.

## 20.15 REACTIONS BETWEEN IONS

The frequency factor of ionic reactions depends in a very simple way on the ionic charges. For reactions between ions of opposite signs the frequency factors are much higher than normal ($\sim 10^{10}$ l mol$^{-1}$ s$^{-1}$) whereas, if the ions are of the same sign, the frequency factors are abnormally low. The effects of electrostatic attraction or repulsion are obviously important. The situation can be considered in terms of collision theory, if the ions are of opposite signs, the frequency of collisions will be increased by the attractive forces. Scatchard and Moelwyn Hughes have modified the collision theory to allow for such electrostatic interactions. The problem of the influence of the solvent on the rates and frequency factors of reactions in solution will be considered first, afterward with the aid of Debye–Huckel theory, the influence of the ionic strength will be treated.

## 20.16 INFLUENCE OF SOLVENT: DOUBLE SPHERE MODEL

According to the theory of absolute reaction rate, the rate constant is related to the free energy of activation by the equation

$$k = \frac{kT}{h} e^{-\Delta G^{\pm}/RT} \tag{20.86}$$

In reactions between ions, the electrostatic interaction makes the most important contribution to the free energy of activation. According to simplest treatment of electrostatic interactions, the charged ions are considered to be conducting spheres and the solvent is regarded as a continuous dielectric having a fixed dielectric constant ($\varepsilon$).

This treatment, in which emphasis is placed on the increase in free energy when the system passes from the initial state to the activated state, is to be contrasted with the collision theory approach, according to the latter, the collision frequency is regarded as modified by the electrostatic interaction. The absolute rate method is more flexible than that the collision theory, since the structure of activated complex can be formulated in various ways. In an ideal case, when the reactant molecules simply approach one another and remain intact in the activated state, the two theories become equivalent.

A very simple model for a reaction between two ions in solution is represented in Figure 20.6. The reacting molecules are regarded as conducting spheres. Initially the ions are at infinite distance from each other,

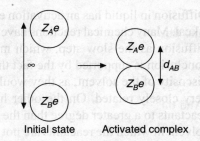

FIGURE 20.6 A double sphere model.

and in the activated state they are considered to be intact and they are at a distance $d_{AB}$ apart. This model is frequently referred to as the *double sphere model*. When the ions are at a distance $x$ apart the force acting between them is equal to

$$F = \frac{Z_A Z_B e^2}{\varepsilon x^2} \qquad (20.87)$$

The work that must be done in moving them together a distance $dx$ is

$$dw = -\frac{Z_A Z_B e^2}{\varepsilon x^2} dx \qquad (20.88)$$

The negative sign is used because $x$ decreases by $dx$ when the ions move together towards each other. The work that is done in moving the ions from an initial distance of infinity to a final distance of $d_{AB}$ is therefore.

$$W = -\int_{\infty}^{d_{AB}} \frac{Z_A Z_B e^2}{\varepsilon x^2} dx \qquad (20.89)$$

$$W = \frac{Z_A Z_B e^2}{\varepsilon d_{AB}} \qquad (20.90)$$

If the signs on the ions are the same, this work is positive, if they are different, it is negative. This work is equal to the electrostatic contribution to the free energy increase as the ions are moved up to each other. There is also a molar nonelectrostatic term $\Delta G_{nes}^{\pm}$. The free energy of activation per molecule may therefore be written as

$$\frac{\Delta G^{\pm}}{N} = \frac{\Delta G_{nes}^{\pm}}{N} + \frac{Z_A Z_B e^2}{\varepsilon d_{AB}} \qquad (20.91)$$

By multiplying both sides by $N$, we get $\Delta G^{\#} = \Delta G^{\#} + \dfrac{Z_A Z_B N e^2}{\varepsilon d_{AB}}$

And introduction of this into Eq. (20.86) gives

$$k = \frac{kT}{h} \exp\left(-\frac{\Delta G_{nes}^{\#}}{RT}\right) \exp\left(-\frac{Z_A Z_B e^2}{\varepsilon d_{AB} kT}\right) \qquad (20.92)$$

and this equation may be written in the logarithmic form

$$\ln k = \ln \frac{kT}{h} - \frac{\Delta G_{nes}^{\#}}{RT} - \frac{Z_A Z_B e^2}{\varepsilon d_{AB} kT} \qquad (20.93)$$

And this may be simplified to

$$\ln k = \ln K_0 - \frac{Z_A Z_B e^2}{\varepsilon d_{AB} kT} \qquad (20.94)$$

The rate constant $k_0$ is seen to be the value of $K$ in a medium of infinite dielectric constant, when the final term in Eq. (20.94) becomes zero (i.e. when there is no electrostatic forces). Equation (20.94) leads to the prediction that the logarithm of the rate constant of a reaction between ions should very linearly with the reciprocal of the dielectric constant.

The relationship is obeyed to a good approximation, although there are usually serious deviations at very low dielectric constants. An example of a test of Eq. (20.94) is shown in Figure 20.7. Deviations from linearity can be explained as due to failure of simple approximations involved in deriving Eq. (20.94) and in some cases to change in reaction mechanism as the solvent is varied.

The slope of the line obtained by plotting $\ln K$ against $1/\varepsilon$ is given by Eq. (20.94) as $Z_A Z_B e^2 / d_{AB} kT$. Since, every term in this expression is known except $d_{AB}$, it is possible to calculate $d_{AB}$ from the experimental slope. For the data shown in figure the value of $d_{AB}$ is 5.1 Å.

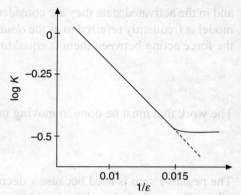

FIGURE 20.7 A plot of log $K$ with reciprocal of dielectric constant for the reaction between bromoacetate and thiosulphate ions in aqueous solution.

## 20.17 THE SINGLE-SPHERE ACTIVATED COMPLEX MODEL

In the double sphere model, the activated complex was regarded as having the form of a double sphere, the original changes residing at the centre of two spheres and being separated by a distance $d_{AB}$. The two spheres may alternatively be regarded as becoming merge into one single sphere which has a charge equal to the algebraic sum of the charges on the ions. It will be seen that the results to which these alternative treatments lead are quite similar to each other, and it is likely that the truth usually lies somewhere between the two.

The single sphere activated complex is represented schematically in Figure 20.8. The derivation of the rate equation for this case is based on an expression obtained by Born for the free energy of charging an ion in solution. Born expression is derived as follows. Consider the process of charging a conducting sphere of radius $r$ from an initial charge of zero to a final charge equal to $Ze$. This process will be carried out by transporting from an infinite distance small increments of charge equal to $ed\lambda$, where $\lambda$ is a parameter that varies from zero to $Z$. At any time, the charge on the sphere may be written as $\lambda e$ and if at a given instant the increment of charge is at a distance $x$ from the ion, the force acting on it is,

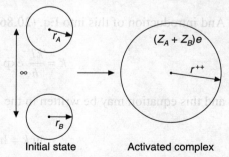

Initial state    Activated complex

FIGURE 20.8 Single sphere activated complex model.

$$df = \frac{\lambda e^2 d\lambda}{\varepsilon x^2} \qquad (20.95)$$

The work of moving the increment from $x$ to $x - dx$ is

$$dw = \frac{\lambda e^2 d\lambda}{\varepsilon x^2} dx \tag{20.96}$$

The total work of charging is obtained by carrying out the double integration $x$ being allowed to vary from infinity to $r$ and $\lambda$ from zero to $Z$. The total work is thus

$$W = \frac{e^2}{\varepsilon} \int_0^Z \int_\infty^r \frac{\lambda d\lambda dx}{x^2} \tag{20.97}$$

$$W = \frac{Z^2 e^2}{2\varepsilon r} \tag{20.98}$$

This work is the electrostatic contribution to the free energy of the ion

$$G_{es} = \frac{Z^2 e^2}{2\varepsilon r} \tag{20.99}$$

For the process represented in Figure 20.8, the electrostatic free energy of the reactant ions and the activated complex are given by

$$G_{es}(A) = \frac{Z_A^2 e^2}{2\varepsilon r_A} \tag{20.100}$$

$$G_{es}(B) = \frac{Z_B^2 e^2}{2\varepsilon r_B} \tag{20.101}$$

$$G_{es}(\neq) = \frac{(Z_A + Z_B)^2 e^2}{2\varepsilon r^{\neq}} \tag{20.102}$$

The increase in electrostatic free energy when the activated complex is formed is thus

$$\Delta G_{es}^{\#} = \frac{e^2}{2\varepsilon} \left[ \frac{(Z_A + A_B)^2}{r^{\neq}} - \frac{Z_A^2}{r_A} - \frac{Z_B^2}{r_B} \right] \tag{20.103}$$

Putting the value $\Delta G^{\#}$ in Eq. (20.86), we get

$$= \frac{kT}{h} \exp\left(-\frac{e^2}{2\varepsilon kT}\right) \left( \frac{(Z_A + Z_B)^2}{r^{\#}} - \frac{Z_A^2}{r_A} - \frac{Z_B^2}{r_B} \right) \tag{20.104}$$

Taking log on both sides

$$\ln K = \ln \frac{kT}{h} - \frac{e^2}{2\varepsilon kT} \left( \frac{(Z_A + Z_B)^2}{r^{\#}} - \frac{Z_A^2}{r_A} - \frac{Z_B^2}{r_B} \right) \tag{20.105}$$

$$\ln K = \ln K_0 - \frac{e^2}{2\varepsilon kT} \left( \frac{(Z_A + Z_B)^2}{r^{\#}} - \frac{Z_A^2}{r_A} - \frac{Z_B^2}{r_B} \right) \tag{20.106}$$

This equation is to be compared and contrasted with Eq. (20.94) which is based on the double-sphere model. Equation (20.106) reduces to Eq. (20.94) if one puts $r_A = r_B = r_{\neq}$. The experimental results may be fitted by Eq. (20.106) as readily by Eq. (20.94). There will be some merging of the electrical charges during the formation of the activated complex, at least in some cases, although perhaps not the complete merging that is assumed with the single sphere complex.

## 20.18 INFLUENCE OF IONIC STRENGTH

Theoretical treatments of the influence of ionic strength on the rates of reaction between ions were given by Bronsted, Bjerrum and Scatchard. Their discussion may be considered with reference to a reaction of the general type:

$$A + B \rightarrow X^{\neq} \rightarrow \text{Product}$$

The basis of the treatment is that the rate of a reaction will be proportional to the concentration of the activated complexes $X^{\pm}$ and not to their activity. The rate is therefore given by

$$v = K^1 [X^{\#}] \quad (20.107)$$

The equilibrium between the activated complexes and the reactants A and B may be expressed as

$$K = \frac{[X^{\neq}]}{[A][B]} \frac{f^{\neq}}{f_A f_B}; [X^{\#}] = k[A][B]\frac{f_A f_B}{f^{\#}} \quad (20.108)$$

where $f$s are activity coefficient. Introduction of Eq. (20.108) into Eq. (20.107) gives

$$v = K_0 [A][B] \frac{f_A f_B}{f^{\neq}} \quad (20.109)$$

Taking logarithms $k = \dfrac{v}{[A][B]} = k_0 \dfrac{f_A f_B}{f^{\#}}$

$$\log K = \log K_0 + \log \frac{f_A f_B}{f^{\neq}} \quad (20.110)$$

According to Debye–Hucked theory, the activity coefficient of an ion is related to its valency $Z$ and the ionic strength $\mu$ by the equation.

$$\log f = -QZ^2 \sqrt{\mu} \quad (20.111)$$

The coefficient $Q$ in this expression is given by

$$Q = \frac{N^2 e^3 (2\pi)^{1/2}}{2.303 \, (\varepsilon k T)^{3/2} (1000)^{1/2}} \quad (20.112)$$

and the ionic strength is defined, following G.N. Lewis by the equation:

$$02\,\mu = \frac{1}{2} \sum Z_i^2 C_i \quad (20.113)$$

Here $Z_i$ is the electrovalency of the ion and $C_i$ is the concentration. The introduction of Eq. (20.111) into the rate Eq. (20.110) gives

$$\log K = \log K_0 + \log f_A + \log f_B - \log f_{\#} \quad (20.114)$$

$$\Rightarrow \quad \log K_0 - Q\sqrt{\mu}\,[Z_A^2 + Z_B^2 - (Z_A + Z_B)^2] \quad (20.115)$$

$$\log K = \log K_0 + 2\,Q Z_A Z_B\,\sqrt{\mu} \quad (20.116)$$

The value of $Q$ is approx. 0.51 for aqueous solution at 25°C. Eq. (20.116) may therefore be written as

$$\log K = \log K_0 + 1.02\,Z_A Z_B\,\sqrt{\mu} \quad (20.117)$$

A plot of $\log K$ against $\sqrt{\mu}$ will give a straight line of slope $1.02\,Z_A Z_B$. Figure 20.9 shows a plot of results for reactions of various types. The line drawn are those with theoretical slopes. If one of the reactants is a neutral molecule, $Z_A Z_B$ is zero and the rate constant is expected to be independent of ionic strength. This is true for example, for the base catalyzed hydrolysis of ethyl acetate, shown in Figure 20.9. It will be seen later that a somewhat more elaborate treatment of the effects of the ionic strength on reactions between ions and neutral molecules indicates that there is a small ionic strength effect.

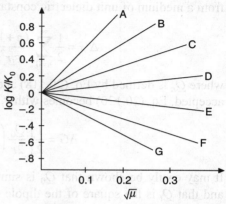

**FIGURE 20.9** Plot of $\log K$ against $\sqrt{\mu}$.

- A. $Co(NH_3)_5\,Br^2 + Hg^{2+}$    $(Z_A Z_B = 4)$
- B. $S_2O_3^{2-} + I^-$    $(Z_A Z_B = 2)$
- C. $Co(OC_2H_5)N{:}NO_2^- + OH^-$    $(Z_A Z_B = 1)$
- D. $[Cr(Urea)_6]^{3+} + H_2O$    $(Z_A Z_B = 0)$
  $CH_2\,COOC_2H_5 + OH^-$    $(Z_A Z_B = 0)$
- E. $H^+ + Br^- + H_2O_2$    $(Z_A Z_B = -1)$
- F. $Co(NH_3)_5\,Br^{2+} + OH^-$    $(Z_A Z_B = -2)$
- G. $Fe^{2+} + Co\,(C_2O_4)_3^{3-}$    $(Z_A Z_B = -6)$

## 20.19 REACTIONS INVOLVING DIPOLES: INFLUENCE OF SOLVENT

The treatment that will be given for reactions between dipolar molecules or ions is based on an expression derived by Kirkwood for the free energy of charging a sphere that has charges embedded in it in a number of specified position. The treatment will reduce to the equation derived above for the special case in which the sphere contains a single ion at its centre.

Kirkwood expression is for the free energy increase when a sphere containing a distribution of charges is transferred from a vacuum of unit dielectric constant to a medium where dielectric constant is $\varepsilon$. It involves the function $Q_n$ which relate to the charges on the sphere and their positions and are defined by

$$Q_n = \sum_{k=1}^{n}\sum_{l=1}^{n} e_k e_l\,r_k^n\,r_l^n\,P_n(\cos\theta_{kl}) \quad (20.118)$$

Here $e$s are the charges on the sphere, $r$s their distances from centre and $P_n (\cos \theta_{kl})$ is the Legendre polynomial. The summation in Eq. (20.118) converges rapidly and it is usually not necessary to go beyond $n = 1$. The two polynomial values that will be required for the present treatment are

$$P_0(x) = 1 \quad \text{and} \quad P_1(x) = x \tag{20.119}$$

Kirkwood's general expression for the free energy increase when the sphere of radius $r$ is transferred from a medium of unit dielectric constant to one of dielectric constant $\varepsilon$ is

$$\Delta G = \frac{1}{2} \sum_{n=0}^{\infty} \frac{(n+1)Q_n}{\varepsilon_i r^{2n+1}} \left[ \frac{\varepsilon_i - \varepsilon}{(n+1)\varepsilon + n\varepsilon_i} - \frac{\varepsilon_i - 1}{n+1+n\varepsilon_i} \right] \tag{20.120}$$

where $Q_n$ is defined by Eq. (20.118) and $\varepsilon_i$ is generally taken as 2. If only the first two terms are accepted, Eq. (20.120) becomes with $\varepsilon_i = 2$

$$\Delta G = \frac{1}{2} \left[ \frac{Q_0}{2r} \left( \frac{2-\varepsilon}{\varepsilon} - 1 \right) + \frac{Q_1}{r^2} \left( \frac{2-\varepsilon}{2\varepsilon+2} \right) - \frac{1}{4} \right] \tag{20.121}$$

It may easily be shown that $Q_0$ is simply the square of the total net charge $Ze$ on the sphere and that $Q_i$ is the square of the dipole moment $\mu$ of the molecule. Equation (20.121) therefore reduces to

$$\Delta G = \frac{z^2 e^2}{2r} \left( \frac{1}{\varepsilon} - 1 \right) + \frac{3\mu^2}{8r^3} \left( \frac{1-\varepsilon}{\varepsilon+1} \right) \tag{20.122}$$

The first term of this is simply Born expression for the free energy of transfer from a vacuum to a dielectric; the second represent the additional effect of the charge distribution on the sphere.

For the molecules A and B, and for the activated complex $X^{\neq}$ one may write the equation.

$$\Delta G_A = \frac{Z_A^2 e^2}{2r_A} \left( \frac{1}{\varepsilon} - 1 \right) + \frac{3\mu_A^2}{8r_A^3} \left( \frac{1-\varepsilon}{\varepsilon+1} \right) \tag{20.123}$$

$$\Delta G_B = \frac{Z_B^2 e^2}{2r_B} \left( \frac{1}{\varepsilon} - 1 \right) + \frac{3\mu_B^2}{8r_B^3} \left( \frac{1-\varepsilon}{\varepsilon+1} \right) \tag{20.124}$$

$$\Delta G_{\neq} = \frac{(Z_A + Z_B)^2 e^2}{2r_{\neq}} \left( \frac{1}{\varepsilon} - 1 \right) + \frac{3\mu_{\neq}^2}{8r_{\neq}^3} \left( \frac{1-\varepsilon}{\varepsilon+1} \right) \tag{20.125}$$

It follows from these expressions that the difference between the free energy of activation for the reaction in solution and that in the gas phase ($\varepsilon = 1$) is given by

$$\Delta G_{\text{sol}}^{\neq} - \Delta G_{\text{gas}}^{\neq} = \frac{e^2}{2} \left( \frac{1}{\varepsilon} - 1 \right) \left[ \frac{(Z_A + Z_B)^2}{r_{\neq}} - \frac{Z_A^2}{r_A} - \frac{Z_B^2}{r_B} \right] + \frac{3}{8} \left( \frac{1-\varepsilon}{\varepsilon+1} \right) \left[ \frac{\mu_{\neq}^2}{r_{\neq}^3} - \frac{\mu_A^2}{r_A^3} - \frac{\mu_B^2}{r_B^3} \right] \tag{20.126}$$

Since, the rate constant depends exponentially on the negative of the free energy of activation divided by $kT$, the relationship between the rate constant in solution and that in the gas phase is given by the expression.

$$\ln k = \ln k_g + \frac{e^2}{2kT}\left(\frac{1}{\varepsilon}-1\right)\left[\frac{Z_A^2}{r_A}+\frac{Z_B^2}{r_B}-\frac{(Z_A+Z_B)^2}{r_{\neq}}\right]+\frac{3}{8kT}\left(\frac{1-\varepsilon}{\varepsilon+1}\right)\left[\frac{\mu_A^2}{r_A^3}+\frac{\mu_B^2}{r_B^3}-\frac{\mu_{\neq}^2}{r_{\neq}^3}\right]$$

(20.127)

The second term of this equation is equivalent to the term obtained in Eq. (20.106) for single sphere model. For such reactions, the final term is less important than the second and can generally be neglected. For a reaction between two dipoles having no net charge, the second term disappears and the solvent effect is given entirely by the last term. For a reaction between an ion and a dipole or between two dipoles both terms must be included, however, the second term is then frequently small and the main effect of the dielectric is predicted by the last term.

Equation (20.127) may readily be put into a form, which is more convenient for the analysis of experiment data. The term $(1-\varepsilon)$ divide by $(1+\varepsilon)$ appearing in this equation may be shown to be approximately equal to 2 divide by $(\varepsilon-1)$ provided that $\varepsilon$ is sufficiently large. Equation (20.127) therefore becomes:

$$\ln k = \ln k_g + \frac{e^2}{2kT}\left(\frac{1}{\varepsilon}-1\right)\left[\frac{Z_A^2}{r_A}+\frac{Z_B^2}{r_B}-\frac{(Z_A+Z_B)^2}{r_{\neq}}\right]+\frac{3}{8kT}\left(\frac{2}{\varepsilon}-1\right)\left[\frac{\mu_A^2}{r_A^3}+\frac{\mu_B^2}{r_B^2}-\frac{\mu_{\neq}^2}{r_{\neq}^3}\right]$$

(20.128)

This equation predicts that the logarithm of $k$ will vary linearly with the reciprocal of the dielectric constant and gives an explicit expression for the slope in terms of the charge, radii and dipole moments.

## 20.20 ENZYME KINETICS

Enzymes are proteins that increase the rates of reactions, often by many orders of magnitude, without themselves being consumed in the reaction. Moreover, enzymes are highly selective in nature. The role of enzymes as unusual catalysts is well illustrated by the decomposition of hydrogen peroxide:

$$2H_2O_2 \rightarrow 2H_2O + O_2$$

This reaction occurs very slowly in pure aqueous solution, but its rate is greatly increased by a large variety of catalysis. The increase in rate is typically first order in concentration of catalyst. It may be approximately first order in concentration of $H_2O_2$ as well. However, with catalysis by the enzyme catalase the rate becomes zero order in $H_2O_2$ at higher initial concentrations of $H_2O_2$. This is common for enzyme catalyzed reactions. Inorganic catalyst such as iron salts or hydrogen halides increase the velocity of $H_2O_2$ decomposition by four to five orders of magnitude per mole of catalyst. The enzyme catalase, which occurs in blood and variety of tissues increases the velocity by more than 15 power of 10 over the uncatalyzed rate. For the maximum velocity, each molecule of catalase is able to decompose more than 10 million molecules of $H_2O_2$ per second.

Catalase is hemeoprotein with ferriprotoporphyrin (haematin) as a prosthetic group (the haematin group provides the catalytic function) at the active site. Alone in solution, haematin exhibits catalytic activity towards $H_2O_2$ decomposition that is two orders of magnitude higher than that of the inorganic catalysts but still more than a millionfold smaller than that of catalase.

Heme (like haematin, but with the iron in the ferrous state) is the prosthetic group present in haemoglobin, but there is no catalyst enhancement of $H_2O_2$ decomposition in comparison with free haematin. Wang in 1955, produced an iron complex $Fe(OH)_2TETA^+$ (TETA = triethylenetetraamine), which exhibited a velocity of $1.2 \times 10^3$ mol l$^{-1}$s$^{-1}$ for the decomposition of 1M $H_2O_2$.

The thermodynamics of the decomposition of hydrogen peroxide is also relevant to the function of catalysts. The reaction is exergonic, with $\Delta G^0_{298} = -24.64$ kcal mol$^{-1}$.

$$H_2O_2(aq) \rightarrow H_2O(l) + \frac{1}{2}O_2$$

The major contribution comes from enthalpy change, which is $\Delta H^0_{298} = -22.63$ kcal mol$^{-1}$. Thus, the reaction should go strongly to the right, if there is a suitable reaction path. The slowness of uncatalyzed reaction must be associated with the large activation barrier. The experimental activation energy is 17 kcal mol$^{-1}$. Using the observed rate of the uncatalyzed decomposition, $v < 4 \times 10^{-8}$ mol l$^{-1}$s$^{-1}$, we can calculate upper limit for the Arrhenius preexponential factor of $A = 1 \times 10^5$ s$^{-1}$. While the $A$ factor is significantly less than the diffusion limited rate, it is clear that the activation energy provides the larger contribution to the reaction barrier. This is illustrated in Figure 20.10.

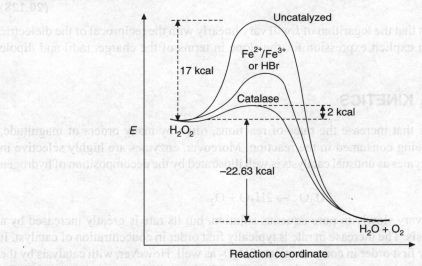

**FIGURE 20.10** Role of catalase in lowering the activation energy for the decomposition of $H_2O_2$.

By contrast with the uncatalyzed reaction, activation energy for the $H_2O_2$ reaction catalyzed by $Fe^{2+}/Fe^{3+}$ or HBr is only two-third as large. In the presence of catalase, the activation energy is only 2 kcal mol$^{-1}$. It appears that the enzyme succeeds in speeding up the reaction by providing a path with lower a substantially smaller energy of activation. The entropy barrier is also reduced, but its effect on the reaction velocity is less significant at room temperature. Note that decrease in barrier height for the forward reaction implies a decrease in activation energy for the reverse reaction as well. For this example, the reverse reaction is strongly endergonic and it is very slow for the reason.

For the most other biochemical reactions that we will be considering, the overall reaction has a standard free energy change closer to zero. In those cases, the enzyme pathway appreciably speeds both the forward and reverse reactions. A typical behaviour for enzyme catalyzed reactions is that the reaction is first order in substrate (reaction) at low substrate concentration and then becomes zero order in substrate at high substrate concentration. This means the reaction reaches a maximum velocity for a constant enzyme concentration. The turnover number is defined as the maximum velocity divided by the concentration of enzyme active sites. Its unit is obviously $s^{-1}$. Concentrations of active sites are used rather than concentration of enzyme so that a fair comparison can be made with enzyme that have more than one active site. Catalase has four active sites per molecule. Turn over number for this reaction is given as:

$$\text{TON} = \frac{V_{max}}{[\text{EAS}]}$$

where EAS is enzyme active sites.

The enzyme fumarase catalyzes the hydration of fumarate to $l$-malate with a turnover number of $2.5 \times 10^3$ $s^{-1}$ at 25° C. The rate constant for 1M fumarate hydrolysis catalyzed by acid (1M) or base (1M) are about $10^{-8}$ $s^{-1}$.

## 20.21 KINETICS OF ENZYMATIC REACTIONS: MICHAELIS MENTEN EQUATION

Enzymes can show their activity at exceedingly low concentration, typically $10^{-10}$ to $10^{-8}$ mole (Figure 20.11). Enzymology began therefore with a study of the kinetics of the disappearance of the substrate and formation of products, because their concentrations are typically $10^{-6}$ to $10^{-3}$ mol. Early observations resulted in the following generalizations about enzyme reactions:

1. The rate of substrate conversion increases linearly with increasing enzyme concentration.
2. For a fixed enzyme concentration, the velocity is linear in substrate concentration [S] at low value of [S].
3. The rate of enzymatic reactions approaches a maximum or saturation velocity at high substrate concentration.

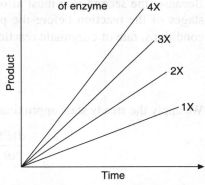

**FIGURE 20.11** Increase of product concentration with time for four different concentrations of enzyme.

At low substrate concentrations,

$$v_0 = k[\text{S}]$$

where $v_0$ is the initial velocity. At high substrate concentration,

$$v_0 = V \quad (\text{maximum velocity})$$

This behaviour is expected if the enzyme forms a complex with the substrate (Figure 20.12). At high substrate concentrations essentially all of the enzymes is tied up in the enzyme–substrate complex. Under these conditions the enzymes is working at full capacity as measured by its turnover number:

$$\text{Turnover number} = \frac{V}{[E]_0}$$

where $[E]_0$ is the total enzyme site concentration. At low substrate concentration the enzyme is not saturated and turnover is limited partly by the availability of substrate.

A kinetic formulation of these ideas was presented by Michaelis and Menten equation, with significant improvements by Briggs and Haldane in 1925. In the simplest picture, the enzyme and substrate binds reversibly to form a complex, followed by dissociation of the complex to form the products and regenerate the free enzyme.

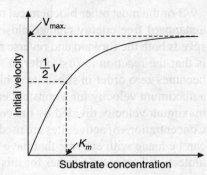

**FIGURE 20.12** Initial velocity of an enzyme catalyzed reaction as a function of substrate concentration [S] for a fixed amount of enzyme.

$$E + S \underset{k_{-1}}{\overset{k_1}{\rightleftharpoons}} ES \tag{20.129}$$

$$ES \xrightarrow{k_2} E + P \tag{20.130}$$

Because, the second step must also be reversible, this mechanism applies strictly to the initial stages of the reaction before the product concentration has become significant. Under these conditions, rate of enzymatic reaction is given as

$$v_0 = \left(\frac{d[P]}{dt}\right)_0 = k_2[ES] \tag{20.131}$$

We apply the steady state approximation for enzyme–substrate complex

$$\frac{d[ES]}{dt} = k_1[E][S] - k_{-1}[ES] - k_2[ES] \cong 0 \tag{20.132}$$

$$= k_1[E][S] - [k_{-1} + k_2][ES] \cong 0 \tag{20.133}$$

Therefore,

$$[ES] = \frac{k_1[E][S]}{k_{-1} + k_2} \tag{20.134}$$

[E] and [S] refer to the free concentrations of enzyme and substrate; however, these are difficult to measure, so we write the equation in terms of the measurable total enzyme $[E]_0$ and substrate $[S]_0$ concentration,

$$[E]_0 = [E] + [ES]; \quad [E] = [E]_0 - [ES] \tag{20.135}$$

$$[S]_0 = [S] + [ES]$$

$$[S]_0 \cong [S] \tag{20.136}$$

We can equate the total substrate concentration to the free substrate concentration, because the enzyme–substrate concentration is typically small compared to the substrate concentration. By substitution into Eq. (20.134) and rearranging terms, we obtain

$$[ES] = \frac{k_1 [[E]_0 - [ES]][S_0]}{k_{-1} + k_2} \quad (20.137)$$

$$[ES][k_{-1} + k_2] = [k_1[E]_0 - k_1[ES]][S_0]$$

$$[ES][(k_{-1} + k_2) + k_1[S]] = [k_1[E]_0][S]$$

$$[ES] = \frac{[k_1[E]_0][S]}{[(k_{-1} + k_2 + k_1[S])]} \quad (20.138)$$

By dividing numerator and denominator by $k_1[S]$, we get

$$[ES] = \frac{[E]_0}{\frac{(k_{-1} + k_2)}{k_1[S]} + 1}$$

$$[ES] = \frac{[E]_0}{1 + \frac{(k_{-1} + k_2)}{k_1[S]}} \quad (20.139)$$

By substituting the value of [ES] into Eq. (20.131), we get

$$v_0 = \frac{k_2[E]_0}{1 + \frac{(k_{-1} + k_2)}{k_1[S]}} \quad (20.140)$$

$$v_0 = \frac{V}{1 + \frac{K_m}{[S]}} \quad (20.141)$$

where $K_m = \frac{(k_{-1} + k_2)}{k_1}$ is the Michaelis constant for enzyme–substrate combination, and the substitution $k_2[E]_0 = V$ is the form of Eq. (20.131) appropriate to high substrate concentrations. That is at high substrate concentration all the enzyme will be present as [ES] enzyme substrate complex, i.e. $[E]_0 = [ES]$. The equation (20.141) is popularly known as *Michaelis Menten Equation*. The Michaelis constant is the ratio of rate constants for reactions involving dissociation of ES to those involving formation of ES. One interpretation is that $1/K_m$ is a measure of the affinity of an enzyme for its substrate. $K_m$ has the dimensions of concentration, and when $[S] = K_m$, then $v_0 = \frac{1}{2}V$.

In other words, the Michaelis constant is equal to the substrate concentration that is sufficient to give half the maximum velocity for the enzyme. One should get an intuitive feeling for the magnitude of $K_m$. A small value of $K_m$, means that the enzyme binds the substrate tightly and small concentrations of substrate are sufficient to saturate the enzyme and to reach the maximum catalytic efficiency of the enzyme.

## 20.22 KINETIC DATA ANALYSIS

For quantitative purposes it is useful to rewrite Eq. (20.141) in a form that suggests a straight line plot of the data. Several such approaches are used, the most popular was proposed by Lineweaver and Burk in 1934.

1. $\dfrac{1}{v_0}$ versus $1/[S]$ : Lineweaver Burk plot

   Taking the reciprocal of both sides of Eq. (20.141) and rearranging, we obtain

$$\frac{1}{v_0} = \frac{1}{V} + \frac{K_m}{V} \cdot \frac{1}{[S]} \qquad (20.142)$$

   Thus, a plot of the reciprocal of initial velocity versus the reciprocal of initial substrate concentration for experiments at fixed enzyme concentration should give a straight line. Furthermore, the intercept with the ordinate gives $1/V$, and the slope is $\dfrac{K_m}{V}$, from which these two constants can be determined. Alternately, $K_m$ can be determined by extrapolation to the abscissa intercept, to give $-1/K_m$ as shown in Figure 20.13.

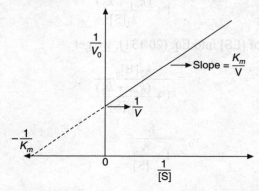

**FIGURE 20.13** Lineweaver Burk Plot of the reciprocal of initial reaction velocity versus reciprocal of initial substrate concentration for a series of experiments at fixed enzyme concentration.

2. $\dfrac{[S]}{v_0}$ versus $[S]$: Dixon plot

   Starting with Eq. (20.142), multiply both sides by $[S]$, to obtain

$$\frac{[S]}{v_0} = \frac{[S]}{V} + \frac{K_m}{V}$$

   A plot of $\dfrac{[S]}{v_0}$ versus $[S]$ for experiments at fixed enzyme concentration should give a straight line. From the slope $K_m$ can be determined.

3. $v_0$ versus $\dfrac{v_0}{[S]}$: Eadie–Hofstee plot (Figure 20.14)

On multiplying both sides of Eq. (20.142) by $v_0 \cdot V$, we get

$$\frac{v_0 \cdot V}{v_0} = \frac{v_0 \cdot V}{V} + \frac{v_0 \cdot V K_m}{V} \frac{1}{[S]}$$

$$V = v_0 + \frac{v_0 K_m}{[S]}$$

$$v_0 = V - \frac{v_0 K_m}{[S]}$$

$$v_0 = -K_m \frac{v_0}{[S]} + V$$

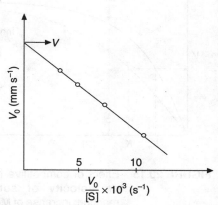

**FIGURE 20.14** Eadie-Hofstee plot of $v_0$ versus $V_0/[S]$ for experiments at fixed enzyme concentration.

A plot of $v_0$ versus $\dfrac{v_0}{[S]}$ for experiments at fixed enzyme concentration should give a straight line. From the slope $K_m$ can be determined.

Ideally with no experimental error each of these equations gives exactly the same desired information from a straight line plot. In practice, one or the other may be preferable because of the nature of the data involved. Qualitatively, the Eadie–Hofstee plot spreads the values at high substrate concentration (where $v_0 \to V$) whereas the Lineweaver–Burk plot compresses the points in this region. Quantitative analysis of the different plots may provide somewhat different values for $K_m$ and $V$.

## 20.23 COMPETITIVE INHIBITION

A molecule that resembles the substrate may be able to occupy the catalytic site because of its simplicity in structure, but may be nearly or completely unreactive. By occupying the active site, this molecule acts as a competitive inhibitor. In preventing normal substrate from being examined and catalyzed. Operationally competitive inhibitors are those that binds reversibly to the active sites (Figures 20.15 and 20.16). The inhibition can be reversed by:

1. diluting the inhibitor or
2. swamping the system with excess substrate.

Mechanistically, we can write a step

$$E + I \rightleftharpoons EI$$

Thus, in addition to usual Michaelis–Menton formulation. Thus,

$$E + S \underset{k_{-1}}{\overset{k_1}{\rightleftharpoons}} ES \underset{k_{-2}}{\overset{k_2}{\rightleftharpoons}} E + P \qquad (20.143 \text{ and } 20.144)$$

$$E + I \underset{k_{-3}}{\overset{k_3}{\rightleftharpoons}} EI \qquad (20.145)$$

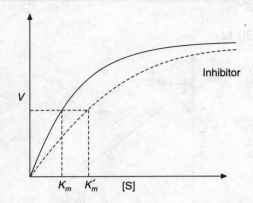 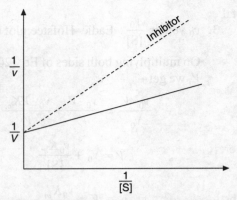

**FIGURE 20.15** Effect of competitive inhibitor on velocity of substrate reaction: increase of Michaelis constant, but no appreciable change in maximum velocity.

**FIGURE 20.16** Competitive inhibition viewed by a Lineweaver–Burk plot. The extrapolated maximum velocity is not changed by a competitive inhibitor.

where EI is an inactive form of the enzyme. Because the equilibria are reversible, the same maximum velocity is reached at sufficiently high substrate concentrations whether the inhibitor is present or not. But, the Michaelis constant is different in the presence of the inhibitor; that is, the concentration of the substrate required to reach one half $V$ is greater in the presence of competitive inhibitors than in their absence.

$$v_0 = \frac{d[\text{P}]}{dt} = k_2[\text{ES}] \qquad (20.146)$$

Steady equation for the formation of ES is

$$\frac{d[\text{ES}]}{dt} = k_1[\text{E}][\text{S}] - k_{-1}[\text{ES}] - k_2[\text{ES}] = 0$$

$$[\text{ES}] = \frac{k_1[\text{E}][\text{S}]}{k_{-1} + k_2} \qquad (20.147)$$

Steady equation for the formation of EI is

$$\frac{d[\text{EI}]}{dt} = k_3[\text{E}][\text{I}] - k_{-3}[\text{EI}] = 0 \qquad (20.148)$$

$$[\text{EI}] = \frac{k_3}{k_{-3}}[\text{E}][\text{I}]$$

$$[\text{EI}] = \frac{[\text{E}][\text{I}]}{k_I} \qquad (20.149)$$

where

$$K_I = \frac{k_3}{k_{-3}} \qquad (20.150)$$

Total enzyme concentration is given as

$$[E]_0 = [E] + [ES] + [EI]$$

$$= [E] + [ES] + \frac{[E][I]}{K_I}$$

$$[E]_0 = [E]\left[1 + \frac{[I]}{K_I}\right] + [ES]$$

Therefore,
$$[E] = \frac{[E]_0 - [ES]}{1 + \frac{[I]}{K_I}} \qquad (20.151)$$

Substituting the value of [E] into Eq. (20.147) for the concentration of [ES], we get

$$[ES] = \frac{k_1[S]}{k_{-1} + k_2} \cdot \frac{[E]_0 - [ES]}{1 + \frac{[I]}{K_I}} \qquad (20.152)$$

$$[ES] = \frac{k_1[S][E]_0}{(k_{-1} + k_2)\left(1 + \frac{[I]}{K_I}\right)} - \frac{k_1[S][ES]}{(k_{-1} + k_2)\left(1 + \frac{[I]}{K_I}\right)}$$

$$[ES]\left\{1 + \frac{k_1[S]}{(k_{-1} + k_2)\left(1 + \frac{[I]}{K_I}\right)}\right\} = \frac{k_1[S][E]_0}{(k_{-1} + k_2)\left(1 + \frac{[I]}{K_I}\right)}$$

$$[ES]\left\{\frac{(k_{-1} + k_2)\left(1 + \frac{[I]}{K_I}\right) + k_1[S]}{(k_{-1} + k_2)\left(1 + \frac{[I]}{K_I}\right)}\right\} = \frac{k_1[S][E]_0}{(k_{-1} + k_2)\left(1 + \frac{[I]}{K_I}\right)}$$

$$[ES] = \frac{k_1[S][E]_0}{(k_{-1} + k_2)\left(1 + \frac{[I]}{K_I}\right) + k_1[S]} \qquad (20.153)$$

According to rate Eq. (20.146), rate of reaction is given by

$$v_0 = \frac{d[P]}{dt} = k_2[ES]$$

On substituting the value of [ES] in above rate equation, we get

$$v_0 = \frac{k_2 k_1[S][E]_0}{k_1[S] + (k_{-1} + k_2)\left(1 + \frac{[I]}{K_I}\right)} \qquad (20.154)$$

By dividing numerator and denominator by $k_1[S]$, we get

$$= \frac{k_2[E]_0}{1 + \frac{(k_{-1} + k_2)\left(1 + \frac{[I]}{K_I}\right)}{k_1[S]}}$$

$$v_0 = \frac{k_2[E]_0}{1 + \frac{K_m}{[S]}\left(1 + \frac{[I]}{K_I}\right)}$$

$$v_0 = \frac{V_S}{1 + \frac{K'_m}{[S]}} \tag{20.155}$$

The new apparent Michaelis constant, $K'_m$ is given by $K'_m = K_m\left(1 + \frac{[I]}{K_I}\right)$, where [I] is the inhibitor concentration and $K_I$ is the dissociation constant for the enzyme inhibitor complex.

A classical example of competitive inhibition occurs in case of the enzyme succinic dehydrogenase. The normal substrate of this enzyme is succinic acid, and the enzyme catalazes the oxidation to fumaric acid. Malonic acid acts as a inhibitor for succinic dehydrogenase. There is a difference of only one methylene group between the structures of succinic acid and malonic acid; hence the two molecules resembles one another. In this case, it is clear that malonic acid cannot acts as an alternative substrate, because there is no oxidized form of the molecule analogous to fumaric acid. The value of $K_I$ for inhibition of yeast succinic dehydrgenase by malonate is $1 \times 10^{-5}$ M, which means that a concentration of $1 \times 10^{-5}$ M malonate decreases the apparent affinity of the enzyme for succinate by a factor of 2.

$$\begin{array}{ccc}
\text{COOH} & \text{COOH} & \text{COOH} \\
| & | & | \\
\text{CH}_2 & \text{CH} & \text{CH}_2 \\
| \quad \xrightarrow{\text{Succinic dehydrogenase}} & \| & | \\
\text{CH}_2 & \text{HC} & \text{COOH} \\
| & | & \\
\text{COOH} & \text{COOH} & \\
\text{Succinic acid} & \text{Fumaric acid} & \text{Malonic acid} \\
\text{(substrate)} & \text{(Product)} & \text{(Inhibitor)}
\end{array}$$

Inhibition by the product of the reaction is another example of competitive inhibition. Again a similarity in structure is implicit in the relation between a substrate and its product. Furthermore, inhibition by the product can serve the useful purpose of turning of the enzyme action when it has made sufficient product for the biochemical need of the cell.

## 20.24 NONCOMPETITIVE INHIBITION

Several types of inhibition occur that cannot be overcome by large amounts of substrate. These may occur as a consequence of

1. some permanent (irreversible) modification of the active sites,
2. reversible binding of the inhibitor to the enzyme, but not at the active site itself,
3. reversible binding to the enzyme–substrate complex.

The simplest form of noncompetitive inhibition occurs when only the value of $v$ is affected, but the affinity of the enzyme for substrate, as measured by $1/K'_m$ is not affected. The kinetic behaviour of noncompetitive inhibition is illustrated in Figures 20.17. and 20.18. As an example, consider the action of an irreversible modifier, such as the chemical alkylating agent iodoacetamide. This compound reacts with exposed sulfhydryl group, in particular with cysteine residues of the enzyme protein, to form a covalently modified.

$$\text{Enzyme—S—CH}_2\text{CONH}_2$$

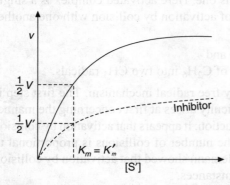
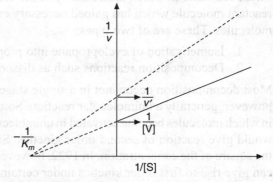

**FIGURE 20.17** Effect of noncompetitive inhibitor on velocity of substrate reaction: decrease of maximum velocity, but no change in Michealis constant.

**FIGURE 20.18** Noncompetitive inhibition viewed by a Lineweaver–Burk plot.

Others forms of noncompetitive inhibition occurs in which both $V$ and $K_m$ are affected by the inhibitor. A type designated uncompetitive results in Lineweaver–Burk plots that are parallel but displaced upward in the presence of inhibitors (Figure 20.18). In addition, where the enzyme has more than one substrate, the action of an inhibitor must be evaluated with respect to each of the substrates and to the order of addition of components.

An interesting form of inhibition is known as substrate inhibition. In this case the velocity of the reaction reaches a maximum and then declines at high substrate concentrations. The consequence in terms of the Lineweaver–Burk plot is shown in Figure 20.19. The origin of this behaviour can be that the substrate binds at a second site on the enzyme, and this site indirectly modifies the main catalytic site so as to render it less effective (Figure 20.20).

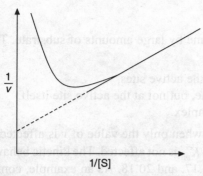
**FIGURE 20.19** Substrate inhibition viewed by a Lineweaver-Burk plot.

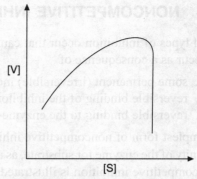
**FIGURE 20.20** [V] versus [S] plot of substrate inhibition.

## 20.25 UNIMOLECULAR REACTION RATE THEORY

Those reactions in which molecularity of the reaction is one. Here activated complex is a single reactant molecule which has gained necessary energy of activation by collision with one another molecules. These are of two types:

1. Isomerization of cyclopropane into propylene and
2. Decomposition reactions such as dissociation of $C_2H_6$ into two $CH_3$ radicals.

Most decomposition occurs not in a single stage but by free radical mechanism. The first step is however, generally a unimolecular reaction. Some difficulty arises at first concerning the manner in which molecules become activated in unimolecular reaction. It appears that activation by collision would give reaction of second order kinetics. Since, the number of collisions is proportional to the square of the concentration. In 1922, however, Lindemann showed that activation by collision can give rise to first order kinetics under certain circumstances.

### 20.25.1 Lindemann Theory

According to this theory, reactant molecule receive/gain energy by collision with another molecules, such molecules will be referred as *energized molecules*. If the energized molecule will be transferred into product at a rate, i.e. small compared to the rate with which they are de-energized by collision. A stationary concentration of them may be built up.

Let us clear it by taking an example of collision between like molecules,

$$A + A \underset{k_{-1}}{\overset{k_1}{\rightleftharpoons}} A^* + A \tag{20.156}$$

$$A^* \xrightarrow[\text{slow}]{k_2} \text{Product} \tag{20.157}$$

Since, these energized molecule are in equilibrium with normal molecule, their concentration is proportional to that of the normal molecule. Rate of the reaction is proportional to concentration of normal molecule. Reaction is therefore said to be of first order.

At low pressure, however, the collision cannot maintained a supply of energized molecule. Rate of the reaction then depends upon rate of energization, therefore, proportional to the square of concentration of reacting molecules.

A steady concentration of energized molecule will be established and net formation of energized molecule may be set equal to zero. That is steady state for the formation of energized molecule may be given as:

$$\frac{d[A^*]}{dt} = k_1[A]^2 - k_{-1}[A^*][A] - k_2[A^*] = 0 \qquad (20.158)$$

From this equation, it follows that concentration of energized molecule

$$[A^*] = \frac{k_1[A]^2}{k_{-1}[A] + k_2} \qquad (20.159)$$

Rate of the reaction is given by the second step as:

$$\text{Rate} = k_2[A^*] \qquad (20.160)$$

By putting the value of $[A^*]$ in above rate equation, we get

$$\text{Rate} = \frac{k_1 k_2 [A]^2}{k_{-1}[A] + k_2} \qquad (20.161)$$

At sufficient high pressure $k_{-1}[A] \ggg k_2$ so we can ignore $k_2$ from the denominator term of rate equation, hence

$$V = \frac{k_1 k_2 [A]^2}{k_{-1}[A]}$$

$$V = \frac{k_1 k_2 [A]}{k_{-1}}$$

$$V = K_\infty [A] \qquad (20.162)$$

where $K_\infty = \dfrac{k_1 k_2}{k_{-1}}$. The reaction is then first order in kinetics.

At low pressure, $k_2 \ggg k_{-1}[A]$, so Eq. (20.161) for the rate of reaction becomes,

$$\text{Rate} = \frac{k_1 k_2 [A]^2}{k_2}$$

$$V = k_1 [A]^2 \qquad (20.163)$$

So that reaction is now of second order. Lindemann theory therefore does give a satisfactory qualitative interpretation of unimolecular reaction. But quantitatively it is not completely satisfactory. This may be explained as for the first order rate coefficient $K'$.

$$V = K'[A] \qquad (20.164)$$

On comparing Eqs. (20.161) and (20.164), we will get

$$K' = \frac{k_1 k_2 [A]}{k_{-1}[A] + k_2}$$

On dividing both numerator and denominator by $k_{-1}[A]$, we get

$$K' = \frac{k_1 k_2 [A]/k_{-1}[A]}{k_{-1}[A]/k_{-1}[A] + k_2/k_{-1}[A]}$$

$$K' = \frac{K_\infty}{1 + k_2/k_{-1}[A]} \qquad (20.165)$$

A plot of $K'$ against [A] concentration is shown in Figure 20.21.

A coefficient $K'$ is constant in the high pressure range, but falls to zero at low pressure value. It follows from Eq. (20.165) that $K'$ become equal to 1/2 of $K_\infty$ when $k_2 = k_{-1}[A]$

$$K' = \frac{K_\infty}{2} \qquad (20.166)$$

Also, we know that $K_\infty = \dfrac{k_1 k_2}{k_{-1}}$. From $k_2 = k_{-1}[A]$, we get

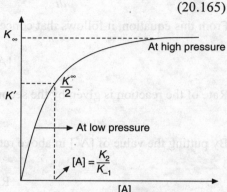

**FIGURE 20.21** A plot of $K'$ against [A].

$$[A] = \frac{k_2}{k_{-1}} \qquad (20.167)$$

So equation for

$$K_\infty = k_1 [A] \qquad (20.168)$$

$$[A] = \frac{K_\infty}{k_1} \qquad (20.169)$$

The value of $K_\infty$ the first order rate constant at high pressure, is found to be from experiment and according to simple collision theory

$$k_1 = Z\, e^{-E_a/RT}$$

Experimentally, however, it was found that the value of $[A]_{1/2}$ were always smaller than those estimated in this way. There can be no doubt about $K_\infty$ which is an experimental quantity therefore, the error must be in estimation of $k_1$.

It is thus necessary for the Lindemann theory to be modified in such a manner as to give larger value of $k_1$. Such a modification was made by Hinshelwood. Another difficulty with simple Lindemann theory for the study of unimolecular reaction become apparent when the Eq. (20.165) is considered from another point of view.

$$\frac{1}{K'} = \frac{K_{-1}}{k_1 k_2} + \frac{1}{k_1 [A]} \qquad (20.170)$$

A plot of $\dfrac{1}{K'}$ against the reciprocal of concentration should give a straight line (Figure 20.22). Deviations from linearity are found, these deviations are explained by theories of Kassel, Rice, and Rampsperger and Slater.

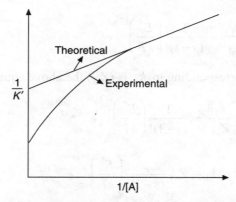

**FIGURE 20.22** A plot of $1/K'$ against the reciprocal of concentration.

## 20.25.2 Hinshelwood Treatment for Unimolecular Reaction

The basis of Hinshelwood treatment in the modification of Lindemann theory is that the rate constant for the energization process $k_1$ may be much greater for a complex molecule than for a simple molecule. This is because the energy possessed by complex molecule may be distributed among a considerable number of degree of vibrational freedoms, even single molecule having more than two atoms have large number of degree of vibrational freedom.

According to statistical mechanics, if a molecule has '$s$' degree of vibrational freedom the fraction of molecule having a total energy between $E$ and $E + dE$ distributed in such a manner that energy in the first degree of freedom is between $E_1 + dE_1$ and that in the second degree of freedom is $E_2$ and $E_2 + dE_2$ and so on …, and is given by

$$f = \frac{e^{-E/KT} dE_1 dE_2 \ldots dE_s}{(kT)^s} \qquad (20.171)$$

If on the other hand total energy $E$ is distributed between $s$ degree of freedom in any way.

$$f = \frac{1}{(s-1)!} \left(\frac{E}{kT}\right)^{s-1} \frac{1}{kT} e^{-E/kT} dE \qquad (20.172)$$

This function may be written as $\frac{dk_1}{k_{-1}}$, $dk_1$ represents the rate constant for the formation of molecule having energy lined between $E$ and $E + dE$. To obtain value of $\frac{k_1}{k_{-1}}$, the above expression must be integrated between $E^*$ and $\infty$. $E^*$ is the minimum energy that the molecule must have in order for it to decompose into products. Result of integration is

$$\frac{k_1}{k_{-1}} = \int_{E^*}^{\infty} \frac{1}{(s-1)!} \left(\frac{E}{kT}\right)^{s-1} \frac{1}{kT} e^{-E/kT} dE \qquad (20.173)$$

$$\frac{k_1}{k_{-1}} = \frac{1}{(s-1)!}\left(\frac{E^*}{kT}\right)^{s-1} e^{-E^*/kT} \tag{20.174}$$

If the collision frequency corresponding to $k_{-1}$ is ($Z_{-1}$) the above equation becomes

$$\frac{k_1}{Z_{-1}} = \frac{1}{(s-1)!}\left(\frac{E^*}{kT}\right)^{s-1} e^{-E^*/kT}$$

$$k_1 = Z_{-1} \frac{1}{(s-1)!}\left(\frac{E^*}{kT}\right)^{s-1} e^{-E^*/kT} \tag{20.175}$$

The above equation when compared with the expression for the rate constant for collision theory. According to collision theory

$$k_1 = Z_1 e^{-E^*/RT}$$

Compare the two equations. According to collision theory $E^*$ is the experimental energy of activation per molecule which is related to $E^*$ of Hinshelwood, according to relation

$$E^* = E_{\exp} + (s-1)kT \tag{20.176}$$

Thus, discrepancy (abnormal behaviour) arises may be due to temperature dependence of the frequency factor. It may be easily shown that the new theory gives rises to a much higher rate of energization and so to a much higher value of $\dfrac{k_1}{k_{-1}}$ than does older collision theory. If for example, the experimental energy of activation is taken as 40 kcal mol$^{-1}$ and $s = 12$ then Hinshelwood equation gives rise to a value of $k_1/k_{-1} = 9.9 \times 10^{-12}$. The simple collision theory gives $3.1 \times 10^{-18}$. The difference is of the order $10^6$. In actual practice $s$ is usually found to be equal to or less than the total number of normal modes of vibration in the molecule. The best agreement is usually obtained by taking $s$ as equal to about 1/2 of the total number of modes.

Hinshelwood treatment and other modification to Lindemann theory are to be considered in terms of following scheme of reactions,

$$A + A \underset{k_{-1}}{\overset{k_1}{\rightleftharpoons}} A^* + A \tag{20.177}$$

$$A^* \xrightarrow{k_2} A^{\#} \tag{20.178}$$

$$A^{\#} \xrightarrow{k^{\#}} \text{Products} \tag{20.179}$$

A distinction has been made between activated complex represented by symbol $A^{\#}$ and an energized molecule by the symbol $A^*$. An activated complex means one that is passing smoothly into final state. An energized molecule is one that has sufficient energy and can become an activated molecule without the attainment of further energy. It must, however, undergo vibrational changes before it can become an activated complex, in which energy has become localized in particular bond or bonds that are to be broken during the reaction. According to point of view of Hinshelwood treatment, the molecule may become energized much more readily than that had been considered possible on the basis of collision theory. A long period of time however elapsed before an energized molecule can become an activated molecule. Hinshelwood treatment predicts an abnormally large value of $k_1$ and corresponding low value of $k_2$. The theories that are now to be considered also postulate to a large value of $k_1$, but they considered that $k_2$ is larger, greater is the amount of energy residing in the energized molecule.

The theories of Kassel, Rice and Rampsperger (RRK theory) and Slater on the other hand represent two other alternatives ways of treatment to this problem of unimolecular reaction.

### Limitations of Hinshelwood treatment

1. The number of $s$ degree of vibrational freedom is about 1/2 of the total number of vibrational mode. There is no satisfactory explanation for this.
2. According to Hinshelwood theory $K'_\infty = \dfrac{k_1 k_2}{k_{-1}} = k_2 \dfrac{1}{(s-1)!} \left(\dfrac{E^*}{kT}\right)^{s-1} e^{-E^*/kT}$. Thus, one can expect a strong temperature dependence of the pre-exponential factor. No experimental evidence exists for this.
3. It cannot account for the lack of linearity found experimentally for the plot of $1/K'$ versus $1/[A]$.

## 20.25.3 RRK Theory

This theory is based on assumption that $k_2$ will be a function of energy possessed by the energized molecule $A^*$. The postulate made is that rate constant $k_2$ increases with the energy possessed by the molecule in its various degree of freedom at high pressure when there is a greater number of highly activated molecules. The rate will be higher than predicted by collision theory. It is supposed that on every vibration there is reshuffling of energy between the normal modes and after a number of vibrations, the critical energy ($E^*$) may be found in one particular normal mode and then reaction occurs, this mode may be referred as critical mode. Larger the energy $E$ possessed by the excited molecule, greater is the chance that the necessary amount $E^*$ can pass into the critical mode and greater is their rate of reaction.

In terms of quantum model RRK found statistical weight of a system of $s$ degree of vibrational freedom containing small $J$ quanta of vibrational energy = number of ways in which $J$ objects can be divided among $S$ boxes. Each of which contain any number of such ways is

$$\text{Statistical weight} = f = \dfrac{(J+S-1)!}{J!(S-1)!} \tag{20.180}$$

The statistical weight for the states in which $S$ oscillations have $J$ quanta among them and a particular one has $M$ critical quanta is given above

$$\text{Statistical weight} = f = \dfrac{(J-M+S-1)!}{(J-M)!(S-1)!} \tag{20.181}$$

The probability that a particular oscillations has $M$ quanta and all $S$ oscillations have $J$ quanta, then ratio of these can be obtained by dividing the two equations as

$$\dfrac{(J-M+S-1)!\,J!}{(J-M)!\,(J+S-1)!} \tag{20.182}$$

On solving above factorial provided that $J$ is very large, it results an approximately equal to

$$\left(\frac{J-M}{J}\right)^{S-1} \tag{20.183}$$

The total number of quanta $J$ may be taken as proportional to $E$ the total energy of the molecule while $M$ is proportional to $E^*$. $E^*$ is the minimum energy that a molecule must have for a reaction to takes place. So, above equation will be modified as

$$\left(\frac{E-E^*}{E}\right)^{S-1} \tag{20.184}$$

The rate with which acquired energy passes into particular degree of freedom is proportional to this quantity, so that

$$k_2 \propto \left(\frac{E-E^*}{E}\right)^{S-1}$$

$$k_2 = k^{\pm}\left(\frac{E-E^*}{E}\right)^{S-1} \tag{20.185}$$

The variation of $k_2$ with $E$ is shown in Figure 20.23. In order to obtain an expression for $K_\infty$ RRK employed the same expression as given by Hinshelwood for $dk_1/k_{-1}$ and used Eq. (20.185) for $k_2$. The expression for $K_\infty$ comes out to be $\left(K_\infty = \dfrac{k_1 k_2}{k_{-1}}\right)$ by inserting value of $k_1/k_{-1}$ from Eq. (20.174) and value of $k_2$ from Eq. (20.185), we get

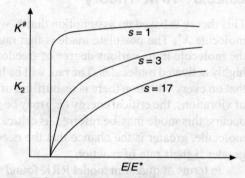

**FIGURE 20.23** Variation of $k_2$ with $E$.

$$K_\infty = \frac{k_1 k_2}{k_{-1}} = k^{\#} \int_{E^*}^{\infty} \left(\frac{E-E^*}{E}\right)^{S-1} \frac{1}{(s-1)!} \left(\frac{E^*}{kT}\right)^{s-1} e^{-E^*/kT} dE \tag{20.186}$$

The integration must be carried out between the limits and when this is done the result is simply Arrhenius equation,

$$K_\infty = k^{\#} e^{-E^*/KT} \tag{20.187}$$

Above equation would have been arrived at immediately using the method of absolute reaction rate. According to this theory

$$K^{\#} = \frac{kT}{h} \tag{20.188}$$

RRK theory would therefore predicts that frequency factor for the first order reaction should be the order $kT/h$. This is of the order $10^{13}$ at ordinary temperature. The conclusion however, requires some modification in RRK original formulation. In RRK theory, no particular significance was attached to the magnitude of $K^{\#}$.

According to absolute reaction rate theory

$$\text{Rate} = K^{\#} = \frac{kT}{h} e^{-E_a/RT}$$

But according to RRK theory $K_{\infty} = k^{\#} e^{-E^*/KT}$

### *Modification in RRK theory*

An important aspect of RRK theory is the manner in which it predicts the variation of $K'$ with the pressure of the reacting gas.

Equation (20.165) can be written as

$$K' = \frac{\dfrac{k_1 k_2}{k_{-1}}}{1 + \dfrac{k_2}{k_{-1}[A]}} \qquad (20.189)$$

Considering the rate constant associated with molecule having a energy between $E$ and $E + dE$.

So, we can write Eq. (20.189) as

$$dK' = \frac{(dk_1)\dfrac{k_2}{k_{-1}}}{1 + \dfrac{k_2}{k_{-1}[A]}} \qquad (20.190)$$

Making use of Eqs. (20.185) and (20.186), and integration between the limits $E^*$ and $\infty$ give rise to

$$K' = \int_{E^*}^{\infty} \frac{\dfrac{1}{(s-1)!}\left(\dfrac{E^*}{KT}\right)^{s-1} e^{-E^*/KT} K^{\pm}\left(\dfrac{E-E^*}{E}\right)^{S-1} dE}{1 + \dfrac{K^{\pm}}{k_{-1}[A]}\left(\dfrac{E-E^*}{E}\right)^{S-1}} \qquad (20.191)$$

This equation can be conveniently reduced by substitution

$$x = \frac{E - E^*}{kT} \quad \text{and} \quad b = \frac{E^*}{kT}$$

On inserting value of $x$ and $b$ in Eq. (20.191), we get

$$K' = \frac{K^{\#} e^{-b}}{(S-1)!} \int_0^{\infty} \frac{b^{s-1}\left(\dfrac{x}{b}\right)^{s-1} dx}{1 + \dfrac{K^{\#}}{k_{-1}[A]}\left(\dfrac{x}{b}\right)^{S-1}}$$

$$\frac{K'}{K^{\#}} = \frac{1}{(S-1)!} \int_0^\infty \frac{e^{-b}(x/b)^{S-1} b^{S-1} dx}{1 + \dfrac{K^{\#}}{k_{-1}[A]}\left(\dfrac{x}{b}\right)^{S-1}} \qquad (20.192)$$

The integration in the above equation for a fixed value of $S$ corresponding to variation of rate constant with concentration [A].

In order to test the theory, usually by trial and error method, what value of $S$ will predict the observed variation of $K'$ with the pressure. The application of this theory to a number of reaction has been extremely satisfactory. The value of $S$ that is required generally corresponds to about half the total number of normal modes in the molecule.

$$K' = \frac{K_\infty}{2}$$

It may be emphasized that in the treatment of RRK theory, the only condition for energization is that the molecule must acquire critical amount of energy $E^*$, any molecule that has acquired this energy will unless it is de-energized by collision, pass through the activated state into final state. This involves the assumption that the energy flows freely between the normal mode of vibration, the frequency $K^{\#}$ that appears in the above expression is really frequency of such energy redistribution. Marcus has developed what is exactly a quantum mechanical formulation of the Kassel, Rice, Rampsperger theory, zero point energy concept was taken into the account.

### 20.25.4 Rice Rampsberger-Kassel-Marcus (RRKM) Theory

During 1951–1952, R.A. Marcus extended the RRK treatment to bring it in line with the transition state theory. In the RRKM theory, the individual vibrational frequency of the energized species and activated complexes are considered axplicity. Account is taken of the way the various normal mode vibrations and rotation contribute to reaction and allowance is made for zero point energies.

The total energy contained in the energized molecule is classified as either active or inactive (also referred to as adiabatic). The inactive energy is the energy that remains in the same quantum state during the course of the reaction and that, therefore, cannot contribute to the breaking of bonds. The zero point energy is inactive, as is the energy of overall translational and rotation, since this energy is preserved as such when the activated molecule $A^{\#}$ is formed. Vibrational energy and the energy of the internal rotation are active. In the RRKM theory the distribution function (Figure 20.24),

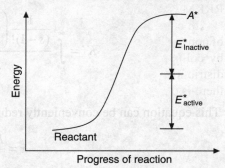

**FIGURE 20.24** Two different forms of energy given by Marcus.

$$f(E^*) dE^* = \frac{\rho(E^*) \exp(-E^*/kT) dE^*}{\int_0^\infty \rho(E^*) \exp(-E^*/kT) dE^*} \qquad (20.193)$$

where $\rho(E^*)$ is the density of states (DOS) having energy between $E^*$ and $E^* + dE^*$. The DOS is defined as the number of states per unit energy range. The denominator in above equation is partition function relating to the active energy contributions. According to the RRKM theory, the rate constant $k_2(E^*)$ is given by

$$k_2(E^*) = \frac{l^{\pm} \sum \rho(E^*_{\text{active}})}{l\rho(E^*)F_r} \qquad (20.194)$$

where $l^{\pm}$ is the statistical factor and $\sum \rho(E^*_{\text{active}})$ is the number of vibration rotation quantum states for the activated molecule corresponding to all energies up to and including $E^*_{\text{active}}$. The factor $F_r$ is introduced to correct for the fact that the rotation may not be same in the activated molecule as in the energized molecule. $\rho(E^*)$ is the density of the states.

A noteworthy feature of the RRKM theory is that it leads to the same expression for the limiting high pressure unimolecular rate constant that is given by transition state theory.

$$K'_{\infty} = \left(\frac{kT}{h}\right) \frac{q^{\#}}{q_i} \exp\left(-\frac{E_0^*}{kT}\right) \qquad (20.195)$$

where $q^{\#}$ and $q_i$ are the partition functions for the activated and initial states. Thus, the theory can explain the abnormally high pre-exponential factor that was sometimes observed. In order to use RRKM theory for detailed calculations, we must decide on models for the energized and activated molecules. Vibrational frequencies for the various normal mode must be estimated and decisions made as to which energies are active and which are inactive. Numerical methods are used to calculate the rate coefficients $K'$ at various concentrations. Reactions which have been successfully investigated include the isomerization of cyclopropoane, the isomerization of cyclobutane and the dissociation of cyclobutane into two ethylene molecules.

A major difficulty in applying the RRKM theory is that the vibrational frequencies of the activated complexes usually cannot estimated very reliably and there is evidence for the "non-RRKM behaviour".

In conclusion, none of the theories of unimolecular reactions have addressed the possibility of nonstatistical energy distribution in the original A molecule. We have concentrated on excitation by collisions, that is thermal excitation, which it seems should produce a statistical energy distribution. However, there are methods such as chemical and photochemical excitation wherein there is evidence for nonstatistical energy distributions.

# EXERCISES

1. What is meant by term rate constant and order of reaction?
2. State and explain the term temperature coefficient of a reaction. What is meant by energy of activation? Explain how energy of activation is determined with the help of Arrhenius equation.
3. Discuss in detail collision theory of bimolecular reactions. What are limitations of this theory?
4. Discuss in detail the activated complex theory (ACT) of bimolecular reactions. Explain how this theory helps in evaluating standard enthalpy of activation and standard entropy of activation. Show that for reactions involving simple molecules, the collision theory and ACT give identical results.

5. Discuss in detail the kinetics of: (i) reversible reactions, (ii) consective reactions and (iii) chain reactions.
6. Illustrate the influence of ionic strength and the nature of the solvent on the rate of ionic reactions.
7. Discuss the kinetics of reactions taking place in flow systems involving: (i) fast flow and (ii) stop flow methods.
8. Illustrate the technique used in studying kinetics of fast reactions. What is meant by the term relaxation time?
9. Describe the pulse method and flash photolysis for studying kinetics of fast reactions.
10. Discuss the mechanism of study of acetaldehyde molecule.
11. Kinetically study the mechanism of decomposition of ethane molecule.
12. Write down the kinetic mechanism of photochemical formation of HCl molecule.
13. How NMR technique is used to study fast reactions?
14. Discuss in detail double sphere model.
15. Discuss in detail single sphere activated complex model.
16. Derive Michaelis Menten equation for enzymatic reaction.
17. Differentiate between competive and non-competive inhibition.
18. Draw various plots to varify Michaelis Menten equation.
19. Write down the significance of Michaelis Menten constant.
20. Discuss Lindemann theory of unimolecular reaction in detail. Also write down its limitations.
21. Discuss Hinshelwood treatment for unimolecular reaction. Also write down its limitations.
22. Differentiate between RRK and RRKM theory.

## SUGGESTED READINGS

Amdur, I. and G. Hammes, *Chemical Kinetics: Principles and Selected Topics*, McGraw-Hill, 1966.

Baer, T. and W.L. Hase, *Unimolecular Reaction Dynamics—Theory and Experiments*, Oxford University Press, 1996.

Berry, RS., S.A. Rice, and J. Ross, *Physical and Chemical Kinetics*, Oxford University Press 2001.

Brouard, M., *Reaction Dynamics*, Oxford University Press, 1998.

Connors, K.A., *Chemical Kinetics: The Study of Reactions in Solutions*, VCH, 1990.

Hammes, G.G., *Principles of Chemical Kinetics*, Academic Press, 1978.

Houston, P.L., *Chemical Kinetics and Reaction Dynamics*, McGraw-Hill, 2001.

Laidler, K.J., *Chemical Kinetics*, 3rd ed., Pearson Education, 2004.

Laidler, K.J., *Theories of Chemical Reactions Rates*, 3rd ed., McGraw-Hill 1969.

Levine, R.D. and R.B. Bernstein, *Molecular Reaction Dynamics and Chemical Reactivity*, Oxford University Press, 1987.

Masel, R.I., *Chemical Kinetics and Catalysis*, Wiley Interscience, 2001.

Moore, J.W. and R. Pearson, *Chemical Kinetics*, 3rd ed., Wiley, 1983.

Piling, M.J. and P.W. Seakins, *Reaction Kinetics*, Oxford University Press, 1995.

Sharma, K.K. and L.K. Sharma, *A Textbook of Physical Chemistry*, 2nd Reprint, Vikas Publication House, New Delhi, 2006.

Steinfeld, J.I., J.S. Francisco, and W.L. Hase, *Chemical Kinetics and Dynamics*, Prentice Hall, 1989.

Weston, R.E. and H. Schwartz, *Chemical Kinetics*, Prentice Hall, 1972.

# Part VI
# Advanced Chemistry

# 21

# Chemistry of Nanomaterials

## 21.1 INTRODUCTION

Nanotechnology is miniaturization technology. Miniaturization is a general aim of the technology development that is taking place to produce smaller, faster, lighter and cheaper devices with greater functionality while using less raw materials and consuming less energy. Nanotechnology is design, fabrication and applications of nanostructures or nanomaterials, and the fundamental understanding of the relationships between physical properties or phenomenon and material dimensions. Nanotechnology deals with materials or structures in nanometre scales, typically ranging from subnanometres to several hundred nanometres. One nanometre is $10^{-9}$ metre. Similar to quantum mechanics, on nanometre scale, materials or structures may possess new physical properties or exhibits new physical phenomenon.

Properties of nanomaterials are different and often superior to their conventional counterparts available in polycrystalline form as they depend on the microstructure which is determined by the chemical composition, grain size, atomic structure, crystallographic orientation, co-ordination number and dimensionality.

## 21.2 CONSEQUENCE OF NANOSCALE

Nanoparticles due to their smaller size and larger surface to volume ratio, exhibit interesting novel properties which include nonlinear optical behaviour, increased mechanical strength, enhanced diffusitivity, high specific heat, magnetic behaviour and electrical resistivity, etc. Researchers have proposed a huge range of potential scientific applications of metal nanoparticles such as in the fields of biotechnology, sensors, medical diagnostics, catalysis, high-performance engineering materials, magnetic recording media, optics and conducting adhesives.

According to the finding of World Technology Evaluation Center (WTEC), nanotechnology has large potential to contribute to significant advances over a wide and diverse range of technological areas. A few potential application of nanotechnology includes producing stronger and lighter materials, it shorten the delivery time of nanostructured pharmaceuticals to the body's circulatory system, increasing the storage capacity of magnetic tapes and providing faster electronic switches to computers, laptops, tablet etc.

## 21.3 HISTORICAL PERSPECTIVE

In the fourth century A.D. Roman glassmakers fabricated soda lime glass containing silver and gold nanoparticles. An artifact from this period called the *Lycurgus Cup* resides in the British Museum in London.

Richard Feyman in one of his articles published in 1960, titles 'there is plenty of room at the bottom' discussed the idea of nanomaterials. He pointed out that if a bit of information required only 100 atoms, then all the books ever written could be stored in a cube with sides 0.02 inch long. The history of nanomaterials is not quite long. Major developments within nanoscience have taken place during the last two decades. The first form of nanomaterials, came into existence in the early 1990s by the material science community to represent particles that are composed of up to tens of thousand of atoms, but confined to size less than 100 nm Before the nanotechnology era, more general term like submicron and ultrafine particles were in use. The first nanoparticles based technology was heterogeneous catalysis, which is followed by the use of silver halide nanoparticles in photography.

Material scientists are conducting research to develop novel materials with better properties, more functionality and lower cost than the existing ones. Several physical, chemical and biological synthesis methods have been developed to enhance the performance of nanomaterials displaying improved properties with the aim to have a better control over the particle size, distribution and morphology. Synthesis of nanoparticles to have a better control over particles size, distribution, morphology, purity, quantity and quality by employing environment friendly economical processes has always been a challenge for the researchers.

## 21.4 NANOPARTICLE MORPHOLOGY AND ELECTRONIC STRUCTURE

To understand a nanomaterials, we must, first learn its structure, meaning that we must determine the type of atoms that constitute its building blocks and how these atoms are arranged relatives to each other. Most nanostructures are crystalline, meaning that their thousands of atoms have a regular arrangement in space called *crystal lattice*.

In two dimensions, there are 17 possible types of crystal structures called *space groups* and these are divided between the four crystal systems.

In three dimensions, there are three lattice constants $a$, $b$, and $c$ for the three dimensions $x$, $y$, and $z$ with respective angles, $\alpha$, $\beta$, and $\gamma$ between them. There are seven crystal systems in three dimensions with a total of 230 space groups.

The objective of a crystal structure analysis is to distinguish the symmetry and space groups, to determine the values of the lattice constants and angles to identify the positions of the atoms in the unit cell (Table 21.1).

**TABLE 21.1** Various possible dimensions, systems, conditions and space groups of crystal structure

| Dimension | System | Condition | Space groups |
|---|---|---|---|
| 2 | Oblique | $a \neq b, \gamma \neq 90°$ ($a \neq b, \gamma = 120°$) | 2 |
| 2 | Rectangular | $a \neq b, \gamma = 90°$ | 7 |
| 2 | Square | $a = b, \gamma = 90°$ | 3 |
| 2 | Hexagonal | $a = b, \gamma = 120°$ | 5 |
| 3 | Triclinic | $a \neq b \neq c, \alpha \neq \beta \neq \gamma$ | 2 |
| 3 | Monoclinic | $a \neq b \neq c, \alpha = \gamma = 90° \neq \beta$ | 13 |
| 3 | Orthorhombic | $a \neq b \neq c, \alpha = \beta = \gamma = 90°$ | 59 |
| 3 | Tetragonal | $a = b \neq c, \alpha = \beta = \gamma = 90°$ | 68 |
| 3 | Trigonal | $a = b = c, \alpha = \beta = \gamma < 120°$ | 25 |
| 3 | Hexagonal | $a = b \neq c, \alpha = \beta = 90°, \gamma = 120$ | 27 |
| 3 | Cubic | $a = b = c, \alpha = \beta = \gamma = 90°$ | 36 |

Certain special cases of crystal structures are important for nanocrystals, such as those involving simple cubic (SC), body-centred cubic (BCC) and face-centred cubic (FCC) unit cells. Another structural arrangement is formed by stacking planar hexagonal layers in the manner in which monoatomic (single atom) crystal provides the highest density or closed packed arrangement of identical spheres.

Some properties of nanomaterials depend on their crystal structure, while other properties such as catalytic reactivity and adsorption energies depend on the type of exposed surface. Epitaxial films prepared from FCC or HCP (Hexagonal close packed) crystal generally grow with planar close packed atomic arrangement just discussed. Face-centred cubic crystals tend to expose surfaces with this same hexagonal two-dimensional atomic array.

To determine the structure of a crystal, and thereby ascertain the positions of its atoms in the lattice, a collimated beam of X-rays, electrons or neutrons is directed at crystal, and the angles at which the beam is diffracted are measured. Various techniques to determine the structure of a crystal are: X-ray diffraction analysis, scanning electron microscopy (SEM) analysis, scanning tunnelling microscope (STM), atomic force microscopy (AFM), etc.

## 21.5 PROPERTIES OF NANOMATERIALS

Scientific interest of nanoparticles is due to their small size and they act as effective bridge between bulk materials and atomic or molecular structures. As compared with a bulk material which has a constant physical properties regardless of its size, but this is not the case at the nanoscale level. Nanoparticles are becoming interesting to scientist because of their size-dependent properties which have enormous applications such as quantum confinement in semiconductor particles, surface plasmon resonance in some metal particles and super-paramagnetism in magnetic materials.

The properties of materials change as their size approaches the nanoscale and as the percentage of atoms at the surface of a material become significant. For bulk materials larger than one micrometre, the percentage of atoms at the surface is minuscule relative to the total number of atoms of the material. The interesting and sometimes unexpected properties of nanoparticles are partly due to the aspects of the surface of the material dominating the properties in lieu of the bulk properties.

Nanoparticle also shows a number of special properties as compared to their bulk counterpart. Size dependent properties of metal nanoparticles are given here. Bending of bulk copper (in the form of wire or ribbon) takes place at a size of 50 nm. Whereas copper nanoparticles smaller than 50 nm are considered superhard materials. Ferroelectric materials of size less than 10 nm can change their magnetization direction using thermal energy at room temperature, thus making them useless for memory storage. Nanoparticles often have unexpected interesting optical properties because they are very small in size which helps to confine their electrons and produce quantum effects. Colour of metal nanoparticles are different from their bulk counterpart, for example, colour of Gold metal nanoparticles may appear from deep red to black colour in solution.

Sintering of nanoparticles can takes place at lower temperatures, over shorter time scales than for larger particles. The large surface to volume ratio also reduces the incipient melting temperature of nanoparticles.

Moreover, nanoparticles have been found to impart some extra properties to various day to day products. Like the presence of titanium dioxide nanoparticles impart what we call as the self-cleaning effect and the size being nanorange, the particles cannot be seen. Nanozinc oxide particles have been found to have superior UV blocking properties compared to its bulk substitute. This is one of the reason, why it is used in the sunscreen lotions. Nanoparticles have also been attached to textile fibres in order to create smart and functional design.

### 21.5.1 Mechanical Properties

The mechanical properties of materials increase with a decreasing size. Many studies have been focused on the mechanical properties of one-dimensional structure. It has been long known that the calculated strength of perfect crystals exceeds that of real ones by two or three orders of magnitude. It has also been found that the increase of mechanical strength becomes appreciable only when the diameter of a whisker is less than 10 microns. So, enhancement in mechanical strength starts in micrometre scale, which is noticeably different from other property (size dependence).

Two possible mechanisms have been proposed to explain the enhanced strength of nanowires or nanorods. One is to ascribe the increase of strength to the high internal perfection of the nanowires or whiskers. The smaller the crosssection of whisker or nanowires, the less is the probability of finding in it any imperfections such as dislocations, micro-twins, impurity precipitates, etc. Thermodynamically, imperfections in crystals are highly energetic and should be eliminated from perfect crystal structures. Small size makes such elimination of imperfection possible.

Another mechanism is the perfection of the side faces of whiskers or nanowires. In general, smaller structures have less surface defects. It is particularly true when the materials are made through a bottom-up approach.

Clearly two mechanisms are closely related. When a whisker is grown at low supersaturation, there is less growth fluctuation in the growth rate and both the internal and surface structures of the whiskers are more perfect.

Yield strength $\sigma_{TS}$ and hardness $H$ of polycrystalline materials are known to be dependent on the grain size on the micrometer scale, following the Hall–Petch relationship:

$$\sigma_{TS} = \sigma_0 + \frac{K_{TS}}{\sqrt{d}}$$

$$H = H_0 + \frac{K_H}{\sqrt{d}}$$

where $\sigma_0$ and $H_0$ are constants related to the lattice friction stress, $d$ the average grain size, and $K_{TS}$ and $K_H$ are material-dependent constants. The inverse square root dependence on the average grain size follows a scaling of the length of the pile up with the grain size. The Hall–Petch model treats grain boundaries as barriers to dislocation motion and thus dislocations pile up against boundary. Upon reaching a critical stress, the dislocations will crossover to the next grain and induce yielding.

### 21.5.2 Melting Point

Nanoparticles of metals, inert gases, semiconductors and molecular crystal are all found to have lower melting temperatures as compared with their bulk forms, when the particle size decreases below 100 nm. The lowering of the melting points is in general explained by the fact that the surface energy increases with decreasing size. The decrease in the phase transition temperature can be attributed to the changes in the ratio of surface energy to volume energy as a function of particle size. One can apply the known methods of phenomenological thermodynamics to systems of nanoparticles with finite size by introducing the Gibbs model to account for the existence of a surface. Some assumptions are applied to develop a model or approximation to predict a size dependence of melting temperature of nanoparticles.

The relationship between the melting points of a bulk material, $T_b$ and a particle $T_m$ is given by:

$$T_b - T_m = \left[\frac{2T_b}{\Delta H \rho_s r_s}\right]\left[\gamma_s - \gamma_l\left(\frac{\rho_s}{\rho_l}\right)^{2/3}\right]$$

where, $r_s$ is the radius of the particle, $\Delta H$ is the molar latent heat of fusion, and $\gamma$ and $\rho$ are surface energy and density, respectively.

It should be noted that the model is developed based on assumption that all nanoparticles have the equilibrium shape and are perfect crystals.

It is possible to make an experimental determination of the size dependence of melting temperature of nanoparticles. The following three different criteria have been explored for this determination:

1. the disappearance of the state of order in the solid,
2. the sharp variation of some physical properties such as evaporation rate, and
3. the sudden change in the particle shape.

The melting point of bulk gold is of 1337 K and decreases rapidly for nanoparticles with sizes below 5 nm. Such size dependence has also been found in other materials such as copper, tin, indium, lead and bismuth in the forms of particles and films. Size dependence is found not to be limited to the melting points of metallic nanoparticles. Similar, relationship has been reported in other materials including semiconductors and oxides.

## 21.6 SYNTHESIS OF NANOMATERIALS

Few very commonly used nanoparticles synthesis methods are listed below:

1. Physical vapour deposition
2. Chemical vapour deposition
3. Aerosol processing
4. Sol–gel process
5. Wet chemical synthesis
6. Mechanical alloying/milling
7. Inert gas condensation method
8. Micro–emulsion technique
9. Biological methods

### 21.6.1 Physical Vapour Deposition

Physical vapour deposition may be performed in many ways. Some of it very important are laser ablation, sputter deposition, electric arc deposition, ion implantation, pulse laser method, ball milling, etc. In ball milling, small balls are allowed to rotate around a drum and drop with gravity force on a solid enclosed in the drum. Ball milling is used to breakdown structures into nanocrystalline range. It is a preferred method for preparing nanometal oxide (Figure 21.1).

Pulsed laser method has been used in the synthesis of nanoparticles of metal. Metal salt solution and a reducing agent are allowed to flow through a blender like device. In the blender, a solid disk is present which rotates in the solution. The solid disk is subjected to pulses from a laser beam creating hot spots on the surface of disk. Metal salt and reducing agent react these hot spots, resulted in the formation of small metal nanoparticles which can be separated using a centrifuge. The size of particles is controlled by energy of laser and speed of disk (Figure 21.2).

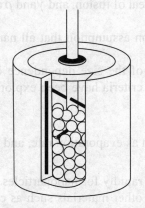

**FIGURE 21.1** A schematic diagram (sectional) of a ball mill vessel.

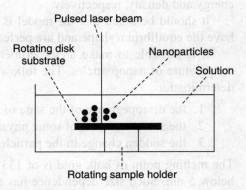

**FIGURE 21.2** Pulsed laser method.

### 21.6.2 Chemical Vapour Deposition

Chemical vapour deposition is a method of making nanoparticulate material from the gas phase. Material is heated to form gas and then allowed to deposit as a solid on a surface, usually under vacuum. There may be direct deposition or deposition by chemical reaction to form a new product that differs from material volatilized. It is a useful method for making aligned materials on surface.

### 21.6.3 Sol–Gel Method

Sol–gel is considered a low temperature synthesis method that gives pure, homogeneous nanoparticles with good size distribution. Chemical reactions can be tailored at the molecular level to have a better control over the process and hence the properties of product particles. Sol–gel synthesis has been used for the production of metal, metal oxide and ceramic nanoparticles with high purity and good homogeneity. Silica gels were synthesized in the year 1846, whereas alumina gels in 1870, since then, aerogels of zirconia, silazine, borate and other ceramics are being synthesized using sol–gel method. This process involves the evolution of inorganic networks through formation of a colloidal suspension (sol) and gelatin of the sol to form a network in a continuous liquid phase (gel). Three reactions generally describes the sol–gel process: hydrolysis, alcohol condensation and water condensation.

### 21.6.4 Inert Gas Condensation (IGC) Method

In IGC synthesis method, metal or a mixture of metals is evaporated inside the process chamber pumped down to ultrahigh vacuum (UHV) and subsequently filled with an inert gas such as helium or argon at a lower partial pressure. A boat or crucible is used to evaporate the materials. The temperature required for vaporization of metal has been achieved by resistive heating, although thermal plasmas, laser ablation, and spark ablation sources operate in the same manner. A convective flow of inert gas passes over the evaporation source and transports the nanoparticles formed above the source via thermophoresis towards a particle collection surface usually cooled by flowing liquid nitrogen or chilled water through it. The evaporated metal atoms loss kinetic energy due to inter-atomic collision with inert gas atoms and condense to form nanoparticles in gas phase. Nanoparticles are wiped off the collection surface by a scraper. In some cases, nanoparticles have been compacted into a pellet under UHV conditions. Historically, carbon nanoparticles have been produced by vapour phase synthesis route to reinforce the rubber to improve the wear properties of tyres.

### 21.6.5 Microemulsion Technique

Synthesis of nanoparticles in the cavity produced in microemulsion is a widely used method. A typical emulsion consists of a single phase having three components, oil, water, and surfactant. Normally water and oil are immiscible, but with the addition of surfactant, oil and water become miscible because the surfactant is able to bridge between two fluids. Surfactant consists of two main entities, a hydrophobic head group, and a hydrophilic tail group (Figure 21.3).

One class of emulsion is microemulsion. Microemulsions are distinct from other emulsions in that they are transparent, spontaneously self-assemble, more stable with respect phase separation and consist of aggregates on the scale of less than 100 nm. When oil, water and surfactant are mixed under some critical concentrations, *micelles* or *inverse micelles* are formed depending upon the concentration of water and oil. Micelles are formed with excess water and inverse micelles are formed in excess of oil. The ratio of water, oil and surfactant is important to decide which type of micelles will be formed.

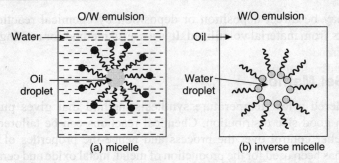

**FIGURE 21.3** Formation of micelles and inverse micelles.

### Synthesis by colloidal route

Colloidal nanoparticles are synthesized by reduction of metal salt or acid. For example, highly stable gold particles can be obtained by reducing chloroauric acid ($HauCl_4$) with tri-sodium citrate. The reaction is carried out in water using the set up as shown in Figure 21.4.

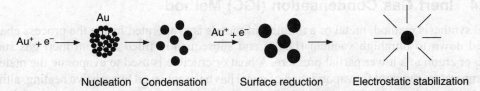

**FIGURE 21.4** Fomation of colloidal nanoparticles.

### 21.6.6 Biological Method

*Synthesis using plant extract*: Use of plants in synthesis of nanopoarticles is quite novel. Gold nanoparticles are produced from alfa–alfa plants. Leaves of germanium plant (*Pelargonium graveolens*) have been used to synthesize nanoparticles of gold.

## 21.7 CHARACTERIZATION TECHNIQUES

### 21.7.1 X-ray Diffraction Analysis

*X-ray diffraction analysis* is a widely used method to find out the mean size of nanoparticles. XRD data are used to examine the internal structure of crystal and to estimate the mean size variations with the changes in annealing temperature. X-ray diffraction study is a tool of determining the internal structure of crystals, i.e., arrangement of atoms within crystals, interplaner distance, size and shape by bombarding X-rays on the surface of crystals at different angles. Beam of X-rays causes spreading of beam of light in different specific directions from which a three-dimensional image of density of electrons can be generated which gives information about arrangement of atoms within crystals, nature of chemical bonds, irregularity in bonding and many more by using the mathematical model of Fourier transforms in combination with chemical information of the sample.

X-ray diffraction study can be applied to almost all types of materials that form crystals such as minerals, salts, semiconductors, inorganic, organic and biological molecules etc.

In X-ray diffraction study, a crystal is exposed to X-rays placed on a rotating table (goniometer) resulting into a diffraction pattern of regularly spaced spots (Figure 21.5). This refraction pattern provides information about internal structure of crystal. From XRD data and X-ray diffraction pattern (Figure 21.6) mean size of metal nanoparticles can be estimated by using Scherrer equation.

$$d = \frac{0.9\lambda}{\beta \cdot \cos\theta}$$

where $d$ is the mean diameter of the nanoparticles, $\lambda$ is the wavelength of X-ray radiation source, $\beta$ is the angular full width at half maximum of the X-ray diffraction peak at the diffraction angle $\theta$.

X-ray powder diffraction study is also a powerful tool for characterizing the products of solid state organic synthesis reactions. Products obtained at different steps in organic synthesis may be subjected to X-ray study for the phase determination, i.e., whether crystalline or amorphous or percentage crystallinity.

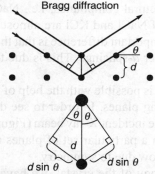

**FIGURE 21.5** Bragg's law of diffraction.

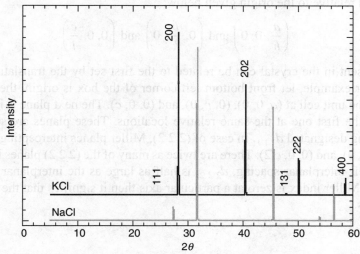

**FIGURE 21.6** Indexing of the peaks of XRD by powder method of NaCl (bottom) and KCl (top).

XRD diffraction pattern of the crystal gives qualitative information about internal structure of crystal as every crystalline substance has a characteristic diffraction pattern. There are number of shapes possible in a crystalline substance due to three different lengths and three different angles. Different shapes arise in crystalline substance depending on restrictions placed on the lengths of the three edges ($a$, $b$, and $c$) and the values of the three angles ($\alpha$, $\beta$ and $\gamma$). Higher is the symmetry present in a crystalline substance, lesser is the number of peaks observed in diffraction pattern and vice-versa. The $d$-spacings of the diffraction pattern peaks are functions of the repeating distances between planes of atoms in the crystal. The intensities of the peaks are related to the kinds of atoms present in the repeating planes of the crystal. The scattering intensities for X-rays, i.e., positions of the peaks are related to the number of electrons present in the atom of a crystalline substance. For example, lighter atoms having lesser number of electrons, scatter X-rays to small extent whereas heavier atoms with large number of electrons, scatter X-rays more strongly. Hence, three important features of a particular diffraction pattern, i.e., the *number*, *positions* and *intensities* of the peaks. XRD pattern are characteristic and act as fingerprint for every crystalline material which makes it an important characterization technique in crystal structure elucidation. For example, XRD pattern of the almost isostructural compounds, i.e., NaCl and KCl can be differentiated (Figure 21.6). The XRD patterns of NaCl and KCl are almost similar but differ in their relative intensities of the peaks. The other important difference is that the diffraction peaks for KCl appear at lower scattering angles and higher $d$-spacings. This is due to the larger size difference in the unit cell of KCl and NaCl.

Crystal structure determination is possible with the help of XRD because every peak in XRD arises from a unique set of repeating planes. In order to see diffraction from a specific set, the planes must be oriented relative to the incident X-ray beam (Figure 21.5). Hence, proper orientation is only one factor. Diffraction from a particular set of planes may not be observed at all or the peak intensity may be sufficiently low due to symmetry.

In order to know the morphology of the crystal, we have to assign a set of Miller indices ($h\ k\ l$) and are defined as the three integers that denote the orientation of the planes with respect to the unit cell. In a given set of planes, one plane in the set will intercept the unit cell at the points on $x$, $y$ and $z$ axes relative to the origin given below:

$$\left(\frac{a}{h}, 0, 0\right) \text{ and } \left(0, \frac{b}{k}, 0\right) \text{ and } \left(0, 0, \frac{c}{l}\right)$$

Other planes present in the crystal can be related to the first set by the translational symmetry (Figure 21.7). For example, let front bottom left corner of the box is origin, the (1 1 1) Miller plane intercepts the unit cell at $(a, 0, 0)$, $(0, b, 0)$, and $(0, 0, c)$. The next plane is intercepting the unit cell behind the first one at the same relative locations. These planes are separated by an interplanar spacing designated $d_{111}$. In case of (2 2 2), Miller planes intercepting the unit cell at $(a/2, 0, 0)$, $(0, b/2, 0)$, and $(0, 0, c/2)$. There are twice as many of the (2 2 2) planes in the repeating structure, and their interplanar spacing, $d_{222}$, is half as large as the interplanar spacing of the (1 1 1) planes. If Miller index is zero at a particular axis then it signifies that the plane does not intercept that axis.

Chemistry of Nanomaterials **651**

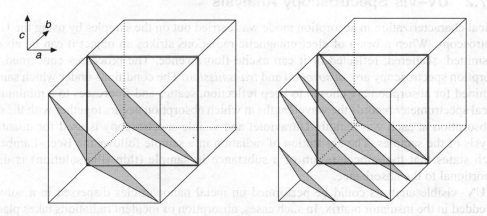

**FIGURE 21.7** (1 1 1) set of Miller planes (left) and the (2 2 2) set of Miller planes (right).

Indexing of the XRD pattern by assigning the Miller indices ($h\ k\ l$) to each diffraction peak is very essential in order to know crystal structure (Figures 21.6 and 21.8). This is done by comparing the data to that reported in the literature or in a database of diffraction patterns known as the *Joint Committee on Powder Diffraction Study* (JCPDS). The Miller indices relate the peak positions or $d$-spacings to the lattice parameters ($a$, $b$, $c$ and $\alpha$, $\beta$ and $\gamma$) by an equation specific to the crystal system.

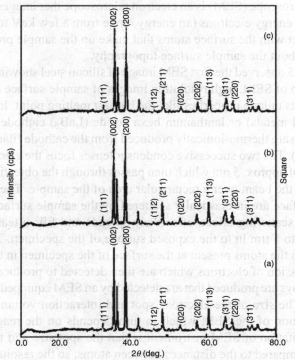

**FIGURE 21.8** XRD patterns of copper oxide nanoparticles in water to surfactant ratio (a) 5, (b) 10 and (c) 15.

## 21.7.2 UV–Vis Spectroscopy Analysis

Optical characterization in absorption mode was carried out on the samples by using the UV–vis spectroscopy. When a beam of electromagnetic radiations strikes an object it can be absorbed, transmitted, scattered, reflected or it can excite fluorescence. The processes concerned in the absorption spectroscopy are: *absorption* and *transmission*. The conditions under which sample is examined for absorption are chosen to keep reflection, scatter and fluoresces to a minimum. An optical spectrometer records the wavelengths at which absorption occurs together with the degree of absorption at each wavelength. Ultraviolet and visible spectroscopy is used for quantitative analysis of the samples. The absorption of radiation in a sample follows the Beer–Lambert law which states that the concentration of a substance in sample (thin film/solution) is directly proportional to the absorbance.

UV–visible analysis could be performed on metal nanoparticles dispersed in a solvent or embedded in the insulator matrix. In such cases, absorption of incident radiations takes place due to surface plasmon resonance (SPR) of the metal nanoparticles. Surface plasmons are essentially the light waves that are trapped on the surface because of their interaction with free electrons of the metals. When metal nanoparticles are embedded in dielectric media and specimen are exposed to the electromagnetic radiation, SPR absorption band is observed at a specific wavelength depending upon the nature of metal, matrix, size of the particles and the distribution.

## 21.7.3 Scanning Electron Microscopy Analysis

Scanning electron microscope (SEM) is an electron microscope that images a sample by scanning it with a beam of high energy electrons (an energy range from a few keV to 50 keV). Bombarded electron beams interact with the surface atoms that make up the sample producing signals which provide information about the sample surface topography.

Max Knoll, in 1935 observed the first SEM image of silicon steel showing electron channelling contrast. With the help of SEM, high resolution images of sample surface can be taken. In SEM, tungsten (because of its unique properties like the highest melting point, low cost and the lowest vapour pressure of all metals) or lanthanum hexa boride ($LaB_6$) cathode filament was used as electrons source which are thermo-ionically produced from the cathode filament towards an anode via field emission (FE). The two successive condenser lenses focus the electron beam into a beam of very fine spot size of approx. 5 nm which then passes through the objective lens, where pairs of scanning coils deflect the beam over a rectangular area of the sample. The high energy electrons striking the metal surface are in-elastically scattered by the sample surface atoms. The primary beam, resulting from scattering effect, effectively spreads and fills a tear-drop shaped volume extended about 1 μm to 5 μm in to the exposed surface of the specimen. The interactions of the beam of electrons with the atoms present at the surface of the specimen in the small volume leads to the subsequent emission of electrons which are then detected to produce an image. As a result of this interaction, X-rays are produced that are detected by an SEM equipped with energy dispersive X-ray spectroscopy. The size of the electrons spot and interaction volume limits the resolution power and magnification of the SEM which in turn depends on the magnetic electron-optical system (MEOS). Limitation of SEM instrument is that the spot size and the interaction volume are both very large compared to the distances between atoms, so the resolution of the SEM is not high enough to image down to the atomic scale but the advantage of SEM is that it has variety of analytical mode available for measuring the composition and the nature of the sample. SEM analysis on nanomaterials samples can be performed in order to investigate the morphology and distribution of nanoparticles. Figure 21.9 shows SEM image of silver oxide nanoparticles synthesized in author's nanotechnology lab.

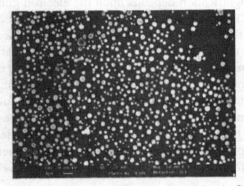

**FIGURE 21.9** SEM micrograph of silver nanoparticles in $SiO_2$ matrix.

## *Principle of SEM*

The signal results from interactions of the electron beam with surface atoms. The secondary electrons, back-scattered electrons (BSE), characteristic X-rays, specimen current and transmitted electrons affect the nature of signals produced by an SEM. The resolution power of the SEM can be as high as it can image a sample of less than 1.0 nm in size (Figure 21.10). The intensity of the BSE signal is strongly related to the atomic number ($Z$) of the specimen, and hence provides information about the distribution of different elements in the surface. High energy electron beam can remove an inner shell electron from the sample, X-rays are generated which are used to identify the composition and to measure the elemental density.

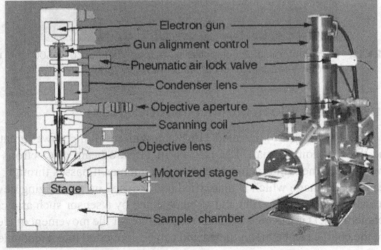

**Figure 21.10** An insight look of an SEM.

Resolution and magnification of a image in an SEM can be controlled over a range of up to 6 orders of magnitude from about 10 to 5,00,000 times of the optical microscope. In comparison to ordinary microscope, resolution and magnification of image in the SEM is not a function of the power of the objective and condenser lens. Here these lenes focus the beam to a spot and not to image the specimen. Magnification depends upon the electron gun, secondary electrons, back-scattered electrons (BSE), characteristic X-rays, specimen current and transmitted electrons.

Another limitation of the SEM is that the specimens must be electrically conductive and must be electrically grounded to prevent the accumulation of electrostatic charge at the surface. The electron beams tend to charge non-conductive specimens especially in secondary electron imaging mode, this causes scanning faults and other image artifacts. Therefore, it is usually coated with an ultrathin coating of electrically conducting materials like gold, gold/palladium alloy, platinum, osmium, iridium, tungsten, chromium, and graphite, deposited on the sample either by low-vacuum sputter coating or by high-vacuum evaporation. Advantage includes that it requires less special sample preparation except for cleaning and mounting on a specimen holder. Figure 21.11 shows SEM image of hair like follicles of toilet pages at a magnification of 500 X.

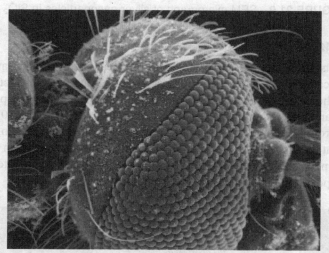

**FIGURE 21.11** SEM images of toilet paper at a magnification of X500 and mosquito head at a magnification of X200, respectively.

## 21.7.4 Transmission Electron Microscopy (TEM)

Max Knoll and Ernst Ruska (1931) invented TEM later with resolving power greater than that of light (1933) and the first commercial TEM (1939). In TEM, a beam of electrons is transmitted through an ultrathin specimen, interacting with the specimen as it passes through. Interaction of the electrons leads to an image which is magnified and focused on an imaging device, such as a fluorescent screen, or photographic film and to be detected by a sensor such as a charge coupled device (CCD) camera. CCD is a device which is responsible for the movement of electrical charge to an area where the charge can be manipulated and can be converted into a readable digital value.

Magnification and resolution power of TEM is even higher than SEM. TEM enables the user to examine fine structural detail at atomic scale. Other advantage of TEM over SEM is that it allows alternate modes of use to observe modulations in chemical identity, crystal orientation, electronic structure and sample induced electron phase shift as well as the regular absorption based imaging.

In TEM, electrons instead of light source travel through vacuum. Electromagnetic lenses are used in TEM whereas glass lenses are used in SEM. In TEM and STM, a field emission gun (connected to a high voltage source typically of ~100–300 kV) is used for the emission of electrons in vacuum which is smaller in diameter, more coherent and with upto three orders of magnitude greater current density or brightness than can be achieved with conventional thermionic emitters. Emitters used in TEM are either of cold-cathode type, or of the Schottky type. Cold cathode types are usually made of single crystal tungsten sharpened to a tip radius of about 100 nm. In Schottky type emitters are made by coating a tungsten tip with a layer of zirconium oxide, because ZrO has unusual property of decreasing electrical resistivity at high temperature (Figure 21.12).

The image produced in STM and TEM have significantly improved signal-to-noise ratio and spatial resolution, and greatly increased emitter life and reliability compared with other thermionic devices used in case of SEM.

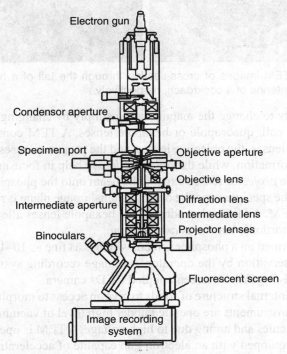

**FIGURE 21.12** An insight look and working principle of TEM.

The high energy electron beam is extracted from field emission gun is usually aided by the use of a Wehnelt cylinder. The upper lenses of the TEM allow for the formation of the electron probe to the desired size and location for later interaction with the sample.

The electron beam thus generated can be manipulated by two physical forces (effects), i.e., electrostatic fields and magnetic fields. The magnetic field will cause electrons to move according to the right hand rule. Electrostatic fields can cause the electrons to be deflected through a constant angle. Coupling of two deflections, i.e., electrical and magnetic in opposing directions with a small intermediate gap are used for beam shifting. Sufficient control over the beam path is possible in TEM and STM. Figure 21.13 shows TEM images of horizontal cross section through the tail of a human sperm and surface image of hair like antenna of a cockroach, respectively.

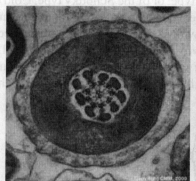 

**FIGURE 21.13** TEM images of cross-section through the tail of a human sperm and antenna of a cockroach, respectively.

TEM has the ability to change the magnification simply by changing the amount of current that flows through the coil, quadrupole or hexapole lenses. A TEM consists of three stages of lensing, i.e., condenser lenses, the objective lenses, and the projector lenses. The *condenser lenses* help in primary beam formation, while the *objective lenses* help in focusing the beam that comes through the sample. The *projector lenses* expand the beam onto the phosphor screen. The ratio of the distances between the specimen and the objective lens' image plane is responsible for the high magnification of the TEM. Additional quadrupole or hexapole lenses allow for the correction of asymmetrical beam distortions, known as *astigmatism*.

Sample image is formed on a phosphor screen, which is as fine as 10–100 μm particulate zinc sulphide, for direct observation by the operator. An image recording system like doped yttrium aluminum garnet, $Y_3Al_2O_{12}$ (YAG) screen coupled CCDs camera.

A TEM probes the internal structure of solids to give an access to morphological fine structural details. Electron beam instruments are operated under high level of vacuum to avoid scattering of electrons from air molecules and arcing due to high voltage. A TEM is operated under vacuum in the order of $10^{-7}$ torr, equipped with an electron gun capable of accelerating electrons through a potential difference in the range of 60 to 300 kV. A thin sample fixed on the grid is illuminated by an electron beam to perform analysis in the TEM. Electron beam is partially transmitted through the sample, while part of it is scattered either elastically or in-elastically. The elastically scattered electrons have undergone diffractions from the atomic planes of crystal whereas inelastic electrons arise from the phonon, plasmon and X-ray excitation. These electrons contribute towards formation of a diffuse halo around the transmitted beam, the diffraction beam and to the general background intensity between the diffraction spots. Most of the intensity at the exit surface of the crystal is in the transmitted and diffracted beam. In the bright field imaging mode, electrons transmitted through the samples are used to form image while intensity of scattered electrons is filtered out. TEMs find application in cancer research, virology, materials science as well as pollution, nanotechnology, and semiconductor research.

### Selected Area Diffraction Pattern (SADP) in TEM

A diffraction pattern is generated by adjusting the magnetic lenses in such a way that the back focal plane of the lens rather than the imaging plane is placed on the imaging apparatus. In the case of a single crystal, TEM produces an image that consists of a pattern of dots and in case of a polycrystalline or amorphous solid material, TEM image consists of series of rings. For the single crystal, the diffraction pattern is dependent upon the deflection angle $\alpha$, of the specimen and the structure of the crystal. The position of the diffraction spots provides information about the point group symmetries and the crystal's orientation to the beam path.

### Advantage of TEM

(i) It can measure size upto atomic scale, i.e., accessible size is < 1.0 nm.
(ii) It has very high resolution. It is a direct method of measurement.
(iii) No calibration is necessary before operating the instrument.
(iv) It can be directly combined with other analytical methods such as electron energy loss spectroscopy (EELS).
(v) Any particle shape is accessible with TEM.

### Disadvantage of TEM

(i) It is very expensive and complex equipment.
(ii) A very high vacuum is needed.
(iii) Sample preparation is required therefore it is slow and hence time-consuming.
(iv) It has poor statistics.
(v) It is an artifact instrument.
(vi) Influence of electron beam has great influence on instrument sensitivity.

## 21.7.5 Scanning Tunneling Microscope (STM)

Gerd Binnig and Heinrich Rohrer invented STM in the year 1981 and got Nobel Prize in Physics after five years, i.e.; 1986. In providing information, it is similar to SEM/TEM but it has certain advantage over others which includes good resolution, i.e., 0.1 nm lateral and 0.01 nm depth resolution. Hence, it can image and manipulate individual atoms. The other advantage is that it can be used in almost all types of atmospheric conditions like ultrahigh vacuum, air, water, and other liquids or gases. Further, it can be operated at a very wide range of temperature, i.e., from near 0 K to a few hundred °C. STM can image surfaces of conducting materials with atomic-scale resolution. There are few disadvantages of STM which includes: (a) it requires extremely clean and stable specimen surface, (b) sharp pointed tips, (c) vibration free mounting plane surface, and (d) sophisticated electronics (computer with printer).

The STM is based on the principle of the quantum tunnelling. In STM, a metallic conducting tip is brought very close to the surface of the specimen at a high voltage which allows electrons to tunnel through the vacuum/gap between tip and specimen. As a result tunnelling current is produced magnitude of which depends on: (i) position of the tip, (ii) potential applied, and (iii) local density of states (LDOS) of the sample.

The changes in surface height and density of states cause change in values of the current as the tip is moved across the surface. This change in values of current are recorded in the form of image. Image produced in STM are grayscale. In order to highlight the important characteristics features, colour is added in post-processing of the image.

## Instrumentation of STM

The components of an STM include scanning tip, piezoelectric controlled height and $X$ and $Y$ scanner, coarse sample-to-tip control, vibration isolation system, computer and printer (Figure 12.14). The radius of curvature of the scanning tips of STM limits the resolution quality of an image taken by an STM. The scanning tip of STM is usually made up of tungsten (W) or platinum–iridium, though gold (Au) may also be used (Figures 21.15 and 21.16). Recently, use of carbon nanotubes as scanning tips removes the problem of double tip imaging that are sometimes observed when tip has two tips at the end rather than single.

Vibration free mounting plane surface is an essential requirement of STM because value of tunnel current is affected by height, vibration, jerk etc. For this magnetic levitation was used by Binnig and Rohrer in their first STM. Recently, mechanical spring or gas spring systems are in common practice to reduce vibrations in STM. Few features of STM are computer controlled like movement of tip and maintaining tip position with respect to the sample, for acquiring the data and its enhancing, processing of the image formed and it quantitative measurements.

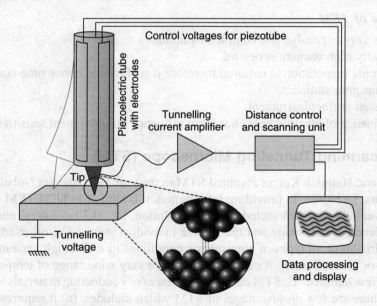

**FIGURE 21.14** Schematic view of STM instrument.

**Chemistry of Nanomaterials**     **659**

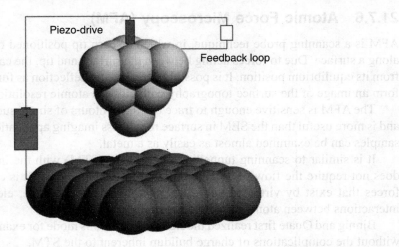

**FIGURE 21.15** Piezo-drive movement of tip of STM over sample surface.

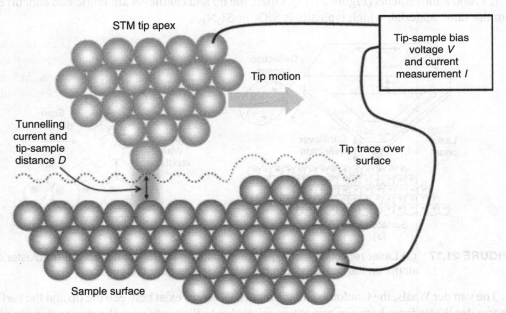

**FIGURE 21.16** Atomic scale diagram showing movement of STM tip over sample surface.

Although the STM itself does not need vacuum to operate (it works in air as well as under liquids), ultrahigh vacuum is required to avoid contamination of the samples from the surrounding medium.

A problem in investigating metal surfaces is the fact that these surfaces appear very flat to an STM, i.e., the apparent height of individual atoms (*corrugation*) is 1/100 to 1/10 of an atomic diameter.

## 21.7.6 Atomic Force Microscopy (AFM)

AFM is a scanning probe technique, in which a sharp tip positioned on a cantilever is scanned along a surface. Due to forces acting between the surface and tip, the cantilever is deflected away from its equilibrium position. It is possible to record the deflection as function of time and thereby form an image of the surface topography with close to atomic resolution.

The AFM is sensitive enough to trace out the contours of single surface atoms on the sample and is more useful than the SEM in surface roughness imaging application, because high-resistive samples can be examined almost as easily as a metal.

It is similar to scanning tunnelling microscopy (STM) with the important exception that it does not require the flow of a tunnelling current for the signal. It is capable of measuring tiny forces that exist by virtue of van der Waals, resonant exchange, electrostatic, and magnetic interactions between atoms of the tip and sample.

Binnig and Quate first realized the significance of this mode for examining insulating surfaces, without the complications of charge buildup inherent to the STM.

Here tip is fastened to a flexible cantilever, which in turn is mounted to piezoelectric drivers for $X$, $Y$, and $Z$ movements (Figure 21.17). Often, the tip and cantilever are fabricated concurrently from the same material, which typically is $SiO_2$ or $Si_3N_4$.

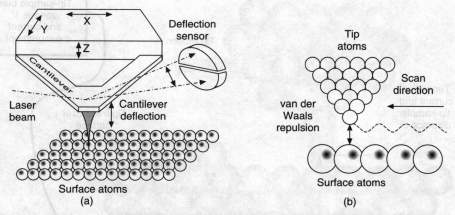

**FIGURE 21.17** (a) Laser reflection technique for measuring cantilever deflection, (b) cluster of atoms on the scanning van der Waals forces near the surface.

The van der Waals, the Coulomb and the meniscus forces exist between the tip and the surface. The van der Waals force between two atoms arises due to fluctuations in the charge-density of the atomic cores and the surrounding electrons and its value depends on the specific geometry of the objects (Figure 21.17b). The Coulomb force is between charged particles or objects. The meniscus force arises due to surface tension in a water-meniscus, which spans and connects tip and surface.

These three forces push the tip away or pull it towards (atomic level interactions may repel or attract atom clusters of the tip), provided it is mounted on a flexible, highly compliant cantilever. van der Waals forces are generally repulsive, provided the tip is not too close to the surface. The cantilever bends more over elevated points on the sample, and less over depressions. In this way, it is possible to map out roughness point-by-point across the surface without ever making direct contact. In practice, the tip and cantilever are usually raised and lowered to maintain a constant gap, or cantilever deflection.

## 21.8 APPLICATIONS OF NANOTECHNOLOGY

There are large numbers of potential applications of nanotechnology in daily life. Many everyday products like electronics, storing device, calculation device, chips, MEMs, quantum dots, etc. are the direct result of nanotechnology applications.

Nanotechnology involves the creation of small scale devices by the manipulation of particles at atomic level. Nanotechnology finds numerous applications in the fields of cosmetics, for example, nanoemulsion liquid products like cleaners and disinfectants for swimming pools that are not harmful to humans. Highly effective swimming pool antibacterial liquids are produced.

Now-a-days, silver nanoparticles are added to the bandages in order to create an inhospitable environment for bacteria. The silver literally kills or suffocates the bacteria. The infusion of metal nanoparticles with bandages is simple technology when compared to the technology that prevents infection around cuts and wounds.

Wires of tennis rackets are produced with the strength of steel buildings but weigh less. The tennis rackets created carbon nanotubes infused with graphite are stronger than the steel used to build weatherproof buildings. To make tennis ball bounce back strongly its core is modified by coating of nanoparticles of clay which means that air cannot escape from the core of the ball. Hence, it bounces back strongly (Figure 21.18).

**FIGURE 21.18** Polymer composite materials used in tennis rackets and balls.

Other applications of nanotechnology include the creation of nanobatteries, tiny capacitors, and microprocessors. As an extended application of nanotechnology, the production of a flexible high resolution screen which can be folded (Figure 21.19).

**FIGURE 21.19** Examples of high resolution screens.

There is a company using the application of nanotechnology to create something known as *self cleaning glass*. Self cleaning glass has the property that it does not allow the dust and water vapour to settle down on it.

**FIGURE 21.20** Photochromatic glass windows which change colour with sun light intensity.

In self cleaning glass, metal nanoparticles are used to make the glass hydrophilic, which means that the rainwater that touches the glass will spread out evenly. US army has designed military dress which is uses wrinkle free fabrics and self cleaning fabric. LED, Plasma and LCD screens which are used in display screen panel as well as TV have completely replaced hollow cathode ray discharge tube which is quite bulky and cumbersome. There is another example of nanotechnology in daily life is its use in photochromatic glass window which senses the change in outside temperature i.e. sunlight intensity and hence accordingly change its colour (Figure 21.20).

## 21.8.1  Use of Nanotechnology in Everyday Process and Technology

Nanotechnology has produced stronger, lighter, more durable, reactive, and flexible materials for a wide range of applications. Thousands of daily used commercial products are available which are based directly or indirectly on nanoscale materials.

For examples, nanoscale additives in polymer composite materials for tennis rackets, car bumpers, reinforced rubber tyres (Figure 21.21), luggage, etc. This makes them lighter, stronger, durable and advanced functional materials.

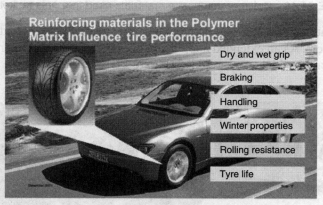

**FIGURE 21.21**  Toyota car manufacturer first time used reinforcing polymeric material in car tyre and bonnet.

Textile industries have modified the fabrics of cloth by nanoscale additives which incorporates few novel properties like wrinkle free, stain less, and bacterial growth resistant, and lightweight (Figure 21.22).

**FIGURE 21.22** Nanoscale additives fabric.

Nanoscale additives coating on display screens and spectacles, windows glass can make them water-repellent, scratch-resistant, anti-reflective, self-cleaning, UV resistant, anti-fogging and anti-microbial (Figure 21.23).

Phototchromic materials change reversibly color with changes in light intensity.

**FIGURE 21.23** Nanoscale additives in glasses and clothes.

Cosmetic industries have make significant advancement in synthesizing products like sunscreens lotions, creams and lotions, complexion treatments, shampoos and specialized makeup which novel properties like greater clarity or coverage, better cleansing power, greater absorption and anti-oxidant and anti-microbial (Figure 21.24).

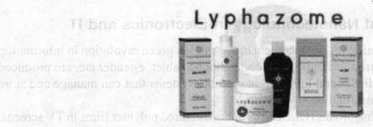

**FIGURE 21.24** Nanoscale materials in cosmetics.

Food industry have also made significant advancement by the use of nanoscale and nano-engineered materials like food containers to reduce carbon dioxide leakage, or reduce oxygen inflow in juice packaging, decrease in growth rate of bacteria in order to keep food longer fresher and safer.

Automobiles industry also use nano-engineered materials in their vehicles which includes:

(i) Rechargeable battery systems with high energy density and power output
(ii) Thermoelectric materials for the control of temperature in car cabin.
(iii) Reinforced tyres with lower-rolling-resistance and increased friction resistant.
(iv) Parking sensors and electronics.
(v) Thin-film smart solar panels
(vi) Fuel additives which increased octane and cetane value.

### Fuel cells

The main problem with the fuel cell is that it is very difficult to maintain an interface between solid, liquid and gaseous phase. By the use of nanotechnology this problem can be overcome (by the use of nanocatalysts). Catalyst is also used in fuel cells to produce hydrogen ions from fuel such as methanol and improve the efficiency of membranes used in fuel cells.

### Batteries

Nanocrystalline materials synthesized by sol-gel techniques are used as separator plates in new generation batteries because of their aerogel structure, which can hold more energy than conventional plates. Nickel metal hydride (Ni-MH) batteries made of nanocrystalline nickel and metal hydride can be recharged much faster and last for longer.

### Food

Nanotechnology is having an impact on several aspects of food science, from how food is grown to how it is packed. Nanomaterials will make a difference not only in the taste of food, but also in food safety, and health benefits that food delivers.

### Phosphors

The resolution of a television and monitor depends greatly on the size of pixel. These are made of material called *phosphors*, which glow when struck by stream of electrons inside the cathode ray tube (CRT). The resolution improves when the pixel or phosphors are smaller. The use of nanophosphors reduces the cost of making high resolution television.

## 21.8.2 Applications of Nanotechnology in Electronics and IT

Nanotechnology has made significant advancement or we can say green revolution in information and technology industry. Electronics items like mobile phones, tablet, e-reader etc. are produced which are faster, smaller, handheld, and lightweight portable systems that can manage and store larger and larger amounts of information.

Use of organic light-emitting diodes (OLEDs) as nanostructured polymer films in TV screens, laptop computers, cell phones, digital cameras, and other display devices. Brighter image, wider viewing angles, light weight, better picture density, energy efficient, and high lifetimes are possible in OLEDs as compared to LCD and plasma screens.

*Use of nanotechnology in electronics*

Nanotechnology gives answer of the question that how we might increase the capabilities of electronics devices while we reduce their weight and power consumption? In this context, we have replaces large and bulky cathode rays picture tube screens with liquid crystal display (LCD) screens. Later LCD screens are replaced by plasma screens and now plasma and LCD screens are replacing by light emitting diode (LED) screens. There is great revolution in the past few years in mobile technology. More sensitive capacitve screens smart phones are in the market with all features like e-reader, Google chrome, Google chat, email, camera, mp4 player, Wi-Fi, bluetooth, active synchronization, Facebook, voice recorder, games download, converter, memo, image editor, etc.

### 21.8.3 Use of Nanotechnology in Sustainable Energy Applications

The world's energy demand is increasing day by day and problem associated with over increasing energy demand is environmental pollution. Different research are carrying out in this field in order to develop different ways to create clean, affordable, and renewable energy sources along with different means to reduce energy consumption and lessen toxicity burdens on the environment.

In this context, prototype solar panels were developed incorporating nanotechnology which makes them more energy efficient than standard designs in converting sunlight to electricity. Nanostructured solar cells are already fabricated which are cheaper, easy to instal and flexible which can be rolled rather than discrete panels.

Efforts are consistently going on in the field of nano-bioengineering of enzymes which is aiming to enable conversion of cellulose into ethanol for fuel, from wood chips, corn stalks, unfertilized perennial grasses, etc.

Now-a-days research are going on to produce nanoelectrode for batteries that are less flammable, quicker-charging, more efficient, large charge-discharge cycle, high energy density, lighter weight, high power density and can hold loner electrical charge.

Research is being carried out in the field for converting waste heat generated in computers, automobiles, homes, power plants, etc., to usable electrical power by the use of nanotechnology.

The energy efficiency of windmill can be increased for sustainable energy production by fabricating an epoxy containing carbon nanotubes based windmill blade which are longer, stronger, and lighter-weight and hence result in an increase in the amount of electricity generation.

Energy efficiency products for sustainable energy production are increasing day by day in number and kinds of applications like more efficient lighting systems, lighter and stronger vehicle chassis and body, energy efficient advanced electronics, low-friction nano-engineered lubricants, energy efficient pumps, and fans, fast-recharging lanterns based on capacitors.

### 21.8.4 Environment Remedial Application of Nanotechnology

Nanotechnology have produced smart engines which are lighter, smaller and greater functionality that consumes less fuel or alternate fuel and emits less pollutant gas in atmosphere.

*Better air quality*

By the use nanotechnology, more advanced high performance catalyst can be fabricated which can convert exhaust of automobiles and industrial harmful gases into environment friendly gases.

This is possible because catalyst produced by the use of nanoparticles have greater surface area to interact with the harmful reacting chemicals present in exhaust gases.

*Cleaner water*

Underground water contains industrial waste as an impurity. Nanoparticles can be used to convert the contaminating chemicals into harmless chemicals. This treatment can be done when water is still present in earth at much lower cost than the method which requires pumping the whole water out of ground for treatment.

### 21.8.5 Nanotechnology in Health and Medical Applications

Now-a-days health and medical technology has advanced a lot by the use of nanotechnology. Nanotechnology has revolutionized a wide array of medical and biotechnology tools and procedures. Nanotechnology has given large number of medical apparatus that are more sophisticated, portable, cheaper, safer and easier to operate. Now-a-days quantum dots are used in biological imaging. *Quantum dots* are tiny nanoparticles especially semiconductors nanocrystals whose excitons are confined in all the three directions. It can be used to locate and identify specific kinds of target cells and specific biological activities.

Researchers have reported an imaging technology to measure the amount of an antibody-nanoparticle complex that accumulates specifically in plaque present in artery. Clinical scientists are able to monitor the development of plaque as well as its disappearance following its treatment. Others researchers have reported gold nanoparticles which can be used to detect early-stage Alzheimer's disease.

Different small pills are designed that are specifically fitted with biosensor (Figure 21.25), which senses the substance to be determined. Multifunctional therapeutics is also designed in which a nanoparticle serves as a stage to facilitate its specific targets to cancer cells and delivery of a potent treatment of medicine at a specific site, thus minimizing the risk of damage to normal tissues (Figure 21.26).

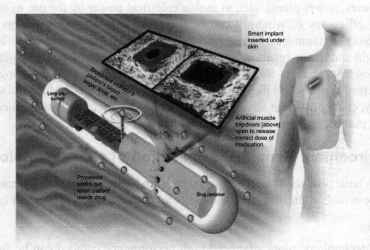

**FIGURE 21.25** A smart pill fitted with biosensor.

The biosensors sense the substance to be measured, say insulin. Once this quantity falls below a certain amount required by the body, the smart pill releases the drug.

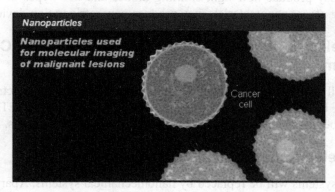

**FIGURE 21.26** Nanoscale devices have the potential to radically change cancer therapy for the better and to dramatically increase the number of highly effective therapeutic agents.

Researchers have designed microfluidic chip-based nanolabs capable of monitoring and manipulating individual cells and nanoscale probes which can track the movements of individual cells and molecules as they move about in their environments.

Research is also going in the field of nanotechnology to use nanomaterials to spur the growth of nerve cells, e.g., in damaged spinal cord or damaged brain cells. In one method, a nanostructured gel fills the space between existing cells and encourages new cells to grow.

### 21.8.6 Nanotechnology in Future Transportation

With the help of nanotechnology, different means of transport like vehicles, aircrafts and ships can be improved.

Nano-engineered steel, concrete, asphalt, and other cementitious materials, and their recycled forms, can be used in terms of improving the performance, and longevity of highway and transportation infrastructure components while reducing their cost.

#### *Motor vehicles*

Motor cars waste significant amount when thermal energy generated by engine is lost. This occurs in case of diesel engines. To prevent this waste, engine cylinders are coated with nanocrystalline ceramics such as zirconia and alumina.

## 21.9 FUTURE APPLICATIONS

### 21.9.1 Environmental Applications

Nanofilters will assist in cleaning up pollution and reducing wasteful by-products. New catalysts will create more efficient industrial process.

### 21.9.2 Optoelectronics Device

Nanotechnology may produce new light emitting diodes and new types of thermionic power. Electron tunnelling across nanomaterials may produce new refrigeration devices.

### 21.9.3 Nanoelectronic and Magnetic Devices and New Computing System

Nanoelectronic will replace microelectronics. The components of these electronics will include molecular mimics, biomolecules and DNA, single electron transistors (SETs) nuclei quantum dots. Energy may be transmitted in circuits via photons rather than electrons.

### 21.9.4 Microelectromechanical Systems

Micromechanical systems will be replaced by nanomechanical systems. Apart from nanometers based on biomotors, artificial muscles using nanoscale bundle of fibres will allow bending and gripping at the nanoscale 37–40.

## 21.10 SAFETY ISSUES OF NANOTECHNOLOGY

Nanoparticles present possible danger, both medically and environmentally. Most of these are due to high surface volume ratio, which can make the particle very reactive to catalytic. They are able to pass through cell membrane in organisms. However, free nanoparticles in the environment quickly tend to agglomerate and thus leave the nano–regime and nature itself presents many nanoparticles to which organisms on earth may have evolved immunity.

According to the "San Francisco Chronide" animal studies have shown that some nanoparticles can penetrate cells and tissues, more through the body and brain and cause biochemical damage. They also have shown to cause a risk factor in men for testicular cancer. But, whether cosmetics and sunscreens containing nanomaterials pose largely unknown pending completion of long-range studies recently begun by the FDA (Food and Drug Administration) and other agencies. Diesel nanoparticles have been found to damage the cardiovascular system in a mould mode.

## EXERCISES

1. Define the term nanotechnology. From where the term nanotechnology comes.
2. Comment upon the nanoparticle morphology and electronic structure.
3. Write down different physical properties of nanoparticles.
4. Write down different mechanical properties of nanoparticles.
5. Explain any three methods of synthesis of nanoparticles.
6. Discuss in detail any two characterization techniques of nanoparticles.
7. Write down various applications of nanoparticles.
8. What can be future applications of nanoparticles?
9. Are nanoparticles are safe from health points of view? Discuss various safety issues in nanotechnology.

10. Discuss in detail different methods to find size of nanoparticles.
11. Differentiate between physical method and chemical method of synthesis of nanoparticles.
12. How nanoparticles are different from bulk matter in physical and chemical properties.
13. Write down possible reason for different properties of different size of nanoparticles.

## SUGGESTED READINGS

Adair, J.H. and E. Suvaci, *Current Opinion in Colloid and Interface Science*, 5, 160–167, 2000.

Adair, J.H., T. Li. Kido, K. Havey, J. Moon, J. Mecholsky, A. Morrone, D.R. Talham, M.H. Ludwig, and L. Wang, *Materials Science and Engineering R-Reports*, 23, 139–242, 1948.

Drexler, K.E., C. Peterson, and G. Pergamit, *Unbounding the Future: The Nanotechnology Revolution*, Morrow, New York, 1991.

Drexler, K.E., *Engines of Creation*; Anchor Press/Doubleday, New York, 1986.

Drexler, K.E., *Nanosystems: Molecular Machinery, Manufacturing, and Computation*, Wiley, New York, 1992.

Hiemenz, P.C. and R. Rajagopalan, *Principles of Colloid and Surface Chemistry*, 3rd rev. and expanded ed., Marcel Dekker, New York, 1997.

Koch, C.C., *Nanostructured Materials: Processing, Properties, and Potential Applications*; Noyes Publications/William Andrew Pub., Norwich, New York, 2002.

Kruis, F.E. and H. Fissan, and A. Peled, *Journal of Aerosol. Science*, 29, 511–535, 1998.

Moriarty, P., *Reports on Progress in Physics*, 64, 297–381, 2001.

Nalwa, H.S., *Handbook of Nanostructured Materials and Nanotechnology*, Academic Press, San Diego, California, 2000.

Nalwa, H.S., *Nanostructured Materials and Nanotechnology*, Concise ed., Academic Press, San Diego, California, London, 2002.

National Research Council (U.S.) *Implications of Emerging Micro- and Nanotechnologies*, National Academies Press, Washington, D.C., 2002.

Newton, D.E., *Recent Advances and Issues in Molecular Nanotechnology*, Greenwood Press, Westport, Conn., 2002.

Shalaev, V.M. and M. Moskovits, American Chemical Society, Division of Physical Chemistry, American Chemical Society, *Meeting Nanostructured Materials:Clusters, Composites, and Thin Films*, American Chemical Society, Washington, DC, 1997.

Timp, G.L., *Nanotechnology*, Springer, New York, 1999.

Toshima, N. and T. Yonezawa, *New Journal of Chemistry*, 22, 1179–1201, 1998.

Yacaman, M.J., J.A. Ascencio and H.B. Liu, J., Gardea-Torresdey, *Journal of Vacuum Science and Technology B—an International Journal Devoted to Microelectronics and Nanometer Structures Processing Measurement and Phenomena*, 19, 1091–1103, 2001.

Zhang, J.Z., *Self-Assembled Nanostructures*, Kluwer Academic/Plenum Publishers, New York, 2003.

# Appendix

**TABLE A-1** Useful relations at 298.15 K

| | |
|---|---|
| $RT$ | 2.4790 kJ/mol |
| $RT/F$ | 25.693 mV |
| $RT \ln 10/F$ | 59.160 mV |
| $kT/hc$ | 207.23 cm$^{-1}$ |
| $kT/e$ | 25.693 meV |
| $V_m^\theta$ | $2.4790 \times 10^{-2}$ m$^3$/mol = 42.790 L mol$^{-1}$ |

**TABLE A-2** Conversion factors

| | |
|---|---|
| 1 eV | $1.60218 \times 10^{-19}$ J |
| | 96.485 kJ mol$^{-1}$ |
| | 8065.5 cm$^{-1}$ |
| 1 cal | 4.184 J |
| 1 atm | 101.325 kPa |
| | 760 Torr |
| 1 cm$^{-1}$ | $1.9864 \times 10^{-23}$ J |
| 1 D | $3.33564 \times 10^{-23}$ cm |
| 1 T | $10^4$ G |
| 1 $l$ atm | 101.325 J |
| $\theta$/°C | $T$/K $-273.15$ |

**TABLE A-3** Unit relations

| | |
|---|---|
| Energy | $1\,J = 1\,kg\,m^2 s^{-2}$ |
| | $= 1\,A\,V\,s$ |
| Force | $1\,N = 1\,kg\,m^2\,s^{-2}$ |
| Pressure | $1\,Pa = 1\,N\,m^{-2} = 1\,kg\,m^{-1}\,s^{-2}$ |
| | $= 1\,J\,m^{-3}$ |
| Charge | $1\,C = 1\,A\,s$ |
| Potential difference | $1\,V = 1\,J\,C^{-1} = 1\,kg\,m^2\,s^{-3}\,A^{-1}$ |

**TABLE A-4** General data and fundamental constant

| Quantity | Symbol | Value | Units |
|---|---|---|---|
| Speed of light | $c$ | $2.99 \times 10^8$ | $m\,s^{-1}$ |
| Elementary charge | $e$ | $1.602 \times 10^{-19}$ | C |
| Faradays constant | $F = N_A e$ | $9.64853 \times 10^4$ | $C\,mol^{-1}$ |
| Boltzmann Constant | $k$ | $1.38065 \times 10^{-23}$ | $J\,K^{-1}$ |
| Gas constant | $R = N_A \cdot k$ | $8.31447$ | $J\,K^{-1}\,mol^{-1}$ |
| | | $8.31447 \times 10^{-2}$ | $l\,bar\,K^{-1}\,mol^{-1}$ |
| | | $6.23637 \times 10^1$ | $l\,Torr\,K^{-1}\,mol^{-1}$ |
| Planck constant | $h$ | $6.62608 \times 10^{-34}$ | Js |
| Avogadros constant | $N_A$ | $6.02214 \times 10^{23}$ | $mol^{-1}$ |
| Atomic mass unit | $U$ | $9.10938 \times 10^{-31}$ | kg |
| Stefan-Boltzmann constant | $\sigma = 2\pi^5 k^4/15 h^3 c^2$ | $5.67051 \times 10^{-8}$ | $W\,m^{-2}\,K^{-4}$ |
| Rydberg constant | $R = m_e e^4 / 8 h^3 c \varepsilon_0^2$ | $1.09737 \times 10^5$ | $cm^{-1}$ |

**TABLE A-5** Surface tension of liquids at 293 K

| Molecule | Surface tension ($\gamma$ mNm$^{-1}$) |
|---|---|
| Benzene | 28.88 |
| Carbon tetrachloride | 27.0 |
| Ethanol | 22.8 |
| Hexane | 18.4 |
| Mercury | 472 |
| Methanol | 22.6 |
| Water | 72.75 |
| | 72.0 at 25°C |
| | 58.0 at 100°C |

**TABLE A-6** Refractive index relative to air at 20°C

| Molecule | 434 nm | 589 nm | 656 nm |
|---|---|---|---|
| Benzene | 1.5236 | 1.5012 | 1.4965 |
| Carbon tetrachloride | 1.4729 | 1.4676 | 1.4579 |
| Carbon disulphide | 1.6748 | 1.6276 | 1.6182 |
| Ethanol | 1.3700 | 1.3618 | 1.3605 |
| KCl | 1.5050 | 1.4904 | 1.4973 |
| KI | 1.7035 | 1.6664 | 1.6581 |
| Methanol | 1.3362 | 1.3290 | 1.3277 |
| Methyl benzene | 1.5170 | 1.4955 | 1.4911 |
| Water | 1.3404 | 1.3330 | 1.3312 |

**TABLE A-7** Viscosities of liquids at 298 K

| Liquid | Viscosity |
|---|---|
| Benzene | 0.601 |
| $CCl_4$ | 0.880 |
| Ethanol | 1.06 |
| Mercury | 1.55 |
| Methanol | 0.553 |
| Pentane | 0.224 |
| Sulphuric acid | 27 |
| Water | 0.891 |

**TABLE A-8** Dipole moments and polarizabilities

| | $\mu/(10^{-30} \text{ cm})$ | $\mu/D$ |
|---|---|---|
| Ar | 0 | 0 |
| $C_2H_5OH$ | 5.64 | 1.69 |
| $C_6H_5CH_3$ | 1.20 | 0.36 |
| $C_6H_6$ | 0 | 0 |
| $CCl_4$ | 0 | 0 |
| $CH_2Cl_2$ | 5.24 | 1.57 |
| $CH_3Cl$ | 6.24 | 1.87 |
| $CH_3OH$ | 5.70 | 1.71 |
| $CH_4$ | 0 | 0 |
| $CHCl_3$ | 3.37 | 1.01 |
| CO | 0.390 | 0.117 |
| $CO_2$ | 0 | 0 |
| $H_2$ | 0 | 0 |
| $H_2O$ | 6.17 | 1.85 |
| HBr | 2.67 | 0.80 |
| HCl | 3.60 | 1.08 |
| $N_2$ | 0 | 0 |
| $NH_3$ | 4.90 | 1.47 |

### TABLE A-6 Refractive index relative to air at 20°C

| Material | 434 nm | 589 nm | 650 nm |
|---|---|---|---|
| Benzene | 1.5236 | 1.5012 | 1.4965 |
| Carbon tetrachloride | 1.4729 | 1.4676 | 1.4579 |
| Carbon disulphide | 1.6748 | 1.6276 | 1.6182 |
| Ethanol | 1.3700 | 1.3618 | 1.3605 |
| KCl | 1.5050 | 1.4904 | 1.4973 |
| KI | 1.7035 | 1.6664 | 1.6581 |
| Methanol | 1.3362 | 1.3290 | 1.3277 |
| Methyl benzene | 1.5170 | 1.4955 | 1.4911 |
| Water | 1.3404 | 1.3330 | 1.3312 |

### TABLE A-7 Viscosities of liquids at 298 K

| Liquid | Viscosity |
|---|---|
| Benzene | 0.601 |
| CCl₄ | 0.880 |
| Ethanol | 1.06 |
| Mercury | 1.55 |
| Methanol | 0.553 |
| Pentane | 0.224 |
| Sulphuric acid | 27 |
| Water | 0.891 |

### TABLE A-8 Dipole moments and polarizabilities

| | $\mu/10^{-30}$ C m | $\mu/D$ |
|---|---|---|
| Ar | 0 | 0 |
| C₂H₅OH | 5.64 | 1.69 |
| C₆H₅CH₃ | 1.20 | 0.36 |
| C₆H₆ | 0 | 0 |
| CCl₄ | 0 | 0 |
| CH₃Cl | 5.24 | 1.57 |
| CH₂Cl₂ | 5.24 | 1.87 |
| CH₃OH | 5.70 | 1.71 |
| CH₄ | 0 | 0 |
| CHCl₃ | 3.37 | 1.01 |
| CO | 0.390 | 0.117 |
| CO₂ | 0 | 0 |
| H₂ | 0 | 0 |
| H₂O | 6.17 | 1.85 |
| HBr | 2.67 | 0.80 |
| HCl | 3.60 | 1.08 |
| N₂ | 0 | 0 |
| NH₃ | 4.90 | 1.47 |

# Logarithms Tables

**TABLE 1** Logarithms

|    | 0    | 1    | 2    | 3    | 4    | 5    | 6    | 7    | 8    | 9    | 1 | 2 | 3  | 4  | 5  | 6  | 7  | 8  | 9  |
|----|------|------|------|------|------|------|------|------|------|------|---|---|----|----|----|----|----|----|----|
| 10 | 0000 | 0043 | 0086 | 0128 | 0170 | 0212 | 0253 | 0294 | 0334 | 0374 | 5 | 9 | 13 | 17 | 21 | 26 | 30 | 34 | 38 |
|    |      |      |      |      |      |      |      |      |      |      | 4 | 8 | 12 | 16 | 20 | 24 | 28 | 32 | 36 |
| 11 | 0414 | 0453 | 0492 | 0531 | 0569 | 0607 | 0645 | 0682 | 0719 | 0755 | 4 | 8 | 12 | 16 | 20 | 23 | 27 | 31 | 35 |
|    |      |      |      |      |      |      |      |      |      |      | 4 | 7 | 11 | 15 | 18 | 22 | 26 | 29 | 33 |
| 12 | 0792 | 0828 | 0864 | 0899 | 0934 | 0969 | 1004 | 1038 | 1072 | 1106 | 3 | 7 | 11 | 14 | 18 | 21 | 25 | 28 | 32 |
|    |      |      |      |      |      |      |      |      |      |      | 3 | 7 | 10 | 14 | 17 | 20 | 24 | 27 | 32 |
| 13 | 1139 | 1173 | 1206 | 1239 | 1271 | 1303 | 1335 | 1367 | 1399 | 1430 | 3 | 6 | 10 | 13 | 16 | 19 | 23 | 26 | 29 |
|    |      |      |      |      |      |      |      |      |      |      | 3 | 7 | 10 | 13 | 16 | 19 | 22 | 25 | 29 |
| 14 | 1461 | 1492 | 1523 | 1553 | 1584 | 1614 | 1644 | 1673 | 1703 | 1732 | 3 | 6 | 9  | 12 | 15 | 19 | 22 | 25 | 28 |
|    |      |      |      |      |      |      |      |      |      |      | 3 | 6 | 9  | 12 | 14 | 17 | 20 | 23 | 26 |
| 15 | 1761 | 1790 | 1818 | 1847 | 1875 | 1903 | 1931 | 1959 | 1987 | 2014 | 3 | 6 | 9  | 11 | 14 | 17 | 20 | 23 | 26 |
|    |      |      |      |      |      |      |      |      |      |      | 3 | 6 | 8  | 11 | 14 | 17 | 19 | 22 | 25 |
| 16 | 2041 | 2068 | 2095 | 2122 | 2148 | 2175 | 2201 | 2227 | 2253 | 2279 | 3 | 6 | 8  | 11 | 14 | 16 | 19 | 22 | 24 |
|    |      |      |      |      |      |      |      |      |      |      | 3 | 5 | 8  | 10 | 13 | 16 | 18 | 21 | 23 |
| 17 | 2304 | 2330 | 2355 | 2380 | 2405 | 2430 | 2455 | 2480 | 2504 | 2529 | 3 | 5 | 8  | 10 | 13 | 15 | 18 | 20 | 23 |
|    |      |      |      |      |      |      |      |      |      |      | 3 | 5 | 8  | 10 | 12 | 15 | 17 | 20 | 22 |
| 18 | 2553 | 2577 | 2601 | 2625 | 2648 | 2672 | 2695 | 2718 | 2742 | 2765 | 2 | 5 | 7  | 9  | 12 | 14 | 17 | 19 | 21 |
|    |      |      |      |      |      |      |      |      |      |      | 2 | 4 | 7  | 9  | 11 | 14 | 16 | 18 | 21 |
| 19 | 2788 | 2810 | 2833 | 2856 | 2878 | 2900 | 2923 | 2945 | 2967 | 2989 | 2 | 4 | 7  | 9  | 11 | 13 | 16 | 18 | 20 |
|    |      |      |      |      |      |      |      |      |      |      | 2 | 4 | 6  | 8  | 11 | 13 | 15 | 17 | 19 |
| 20 | 3010 | 3032 | 3054 | 3075 | 3096 | 3118 | 3139 | 3160 | 3118 | 3201 | 2 | 4 | 6  | 8  | 11 | 13 | 13 | 17 | 19 |
| 21 | 3222 | 3243 | 3263 | 3284 | 3304 | 3324 | 3345 | 3365 | 3385 | 3404 | 2 | 4 | 6  | 8  | 10 | 12 | 14 | 16 | 18 |
| 22 | 3424 | 3444 | 3464 | 3483 | 3502 | 3522 | 3541 | 3560 | 3579 | 3598 | 2 | 4 | 6  | 8  | 10 | 12 | 14 | 15 | 17 |
| 23 | 3617 | 3636 | 3655 | 3674 | 3692 | 3711 | 3729 | 3747 | 3766 | 3784 | 2 | 4 | 6  | 7  | 9  | 11 | 13 | 15 | 17 |
| 24 | 3802 | 3820 | 3838 | 3856 | 3874 | 3892 | 3909 | 3927 | 3945 | 3962 | 2 | 4 | 5  | 7  | 9  | 11 | 12 | 14 | 16 |
| 25 | 3979 | 3997 | 4014 | 4031 | 4048 | 4065 | 4082 | 4099 | 4116 | 4133 | 2 | 3 | 5  | 7  | 9  | 10 | 12 | 14 | 15 |
| 26 | 4150 | 4166 | 4183 | 4200 | 4216 | 4232 | 4249 | 4265 | 4281 | 4298 | 2 | 3 | 5  | 7  | 8  | 10 | 11 | 13 | 15 |
| 27 | 4314 | 4330 | 4346 | 4362 | 4378 | 4393 | 4409 | 4425 | 4440 | 4456 | 2 | 3 | 5  | 6  | 8  | 9  | 11 | 13 | 14 |
| 28 | 4472 | 4487 | 4502 | 4518 | 4533 | 4548 | 4564 | 4579 | 4594 | 4609 | 2 | 3 | 5  | 6  | 8  | 9  | 11 | 12 | 14 |
| 29 | 4624 | 4639 | 4654 | 4669 | 4683 | 4698 | 4713 | 4728 | 4742 | 4757 | 1 | 3 | 4  | 6  | 7  | 9  | 10 | 12 | 13 |

*(Contd.)*

## TABLE 1  Logarithms (Contd.)

|    | 0 | 1 | 2 | 3 | 4 | 5 | 6 | 7 | 8 | 9 | 1 | 2 | 3 | 4 | 5 | 6 | 7 | 8 | 9 |
|----|---|---|---|---|---|---|---|---|---|---|---|---|---|---|---|---|---|---|---|
| 30 | 4771 | 4786 | 4800 | 4814 | 4829 | 4843 | 4857 | 4871 | 4886 | 4900 | 1 | 3 | 4 | 6 | 7 | 9 | 10 | 11 | 13 |
| 31 | 4914 | 4928 | 4942 | 4955 | 4969 | 4983 | 4997 | 5011 | 5024 | 5038 | 1 | 3 | 4 | 6 | 7 | 8 | 10 | 11 | 12 |
| 32 | 5051 | 5065 | 5079 | 5092 | 5105 | 5119 | 5132 | 5145 | 5159 | 5172 | 1 | 3 | 4 | 5 | 7 | 8 | 9 | 11 | 12 |
| 33 | 5185 | 5198 | 5211 | 5224 | 5237 | 5250 | 5263 | 5276 | 5289 | 5302 | 1 | 3 | 4 | 5 | 6 | 8 | 9 | 10 | 12 |
| 34 | 5315 | 5328 | 5340 | 5353 | 5366 | 5378 | 5391 | 5403 | 5416 | 5428 | 1 | 3 | 4 | 5 | 6 | 8 | 9 | 10 | 11 |
| 35 | 5441 | 5453 | 5465 | 5478 | 5490 | 5502 | 5514 | 5527 | 5539 | 5551 | 1 | 2 | 4 | 5 | 6 | 7 | 9 | 10 | 11 |
| 36 | 5563 | 5575 | 5587 | 5599 | 5611 | 5623 | 5635 | 5647 | 5658 | 5670 | 1 | 2 | 4 | 5 | 6 | 7 | 8 | 10 | 11 |
| 37 | 5682 | 5694 | 5705 | 5717 | 5729 | 5740 | 5752 | 5763 | 5775 | 5786 | 1 | 2 | 3 | 5 | 6 | 7 | 8 | 9 | 10 |
| 38 | 5798 | 5809 | 5821 | 5832 | 5843 | 5855 | 5866 | 5877 | 5888 | 5899 | 1 | 2 | 3 | 5 | 6 | 7 | 8 | 9 | 10 |
| 39 | 5911 | 5922 | 5933 | 5944 | 5955 | 5966 | 5977 | 5988 | 5999 | 6010 | 1 | 2 | 3 | 4 | 5 | 7 | 8 | 9 | 10 |
| 40 | 6021 | 6031 | 6042 | 6053 | 6064 | 6075 | 6085 | 6069 | 6107 | 6117 | 1 | 2 | 3 | 4 | 5 | 6 | 8 | 9 | 10 |
| 41 | 6128 | 6138 | 6149 | 6160 | 6170 | 6180 | 6191 | 6201 | 6212 | 6222 | 1 | 2 | 3 | 4 | 5 | 6 | 7 | 8 | 9 |
| 42 | 6232 | 6243 | 6253 | 6263 | 6274 | 6284 | 6294 | 6304 | 6014 | 6325 | 1 | 2 | 3 | 4 | 5 | 6 | 7 | 8 | 9 |
| 43 | 6335 | 6345 | 6355 | 6365 | 6375 | 6385 | 6395 | 6405 | 6415 | 6425 | 1 | 2 | 3 | 4 | 5 | 6 | 7 | 8 | 9 |
| 44 | 6435 | 6444 | 6454 | 6464 | 6474 | 6484 | 6493 | 6503 | 6513 | 6522 | 1 | 2 | 3 | 4 | 5 | 6 | 7 | 8 | 9 |
| 45 | 6532 | 6542 | 6551 | 6561 | 6571 | 6580 | 6590 | 6599 | 6609 | 6618 | 1 | 2 | 3 | 4 | 5 | 6 | 7 | 8 | 9 |
| 46 | 6628 | 6637 | 6646 | 6656 | 6665 | 6675 | 6684 | 6693 | 6702 | 6712 | 1 | 2 | 3 | 4 | 5 | 6 | 7 | 7 | 8 |
| 47 | 6721 | 6730 | 6739 | 6749 | 6758 | 6767 | 6776 | 6785 | 6794 | 6803 | 1 | 2 | 3 | 4 | 5 | 5 | 6 | 7 | 8 |
| 48 | 6812 | 6821 | 6830 | 6839 | 6848 | 6857 | 6866 | 6875 | 6884 | 6893 | 1 | 2 | 3 | 4 | 4 | 5 | 6 | 7 | 8 |
| 49 | 6902 | 6911 | 6920 | 6928 | 6937 | 6946 | 6955 | 6964 | 6972 | 6981 | 1 | 2 | 3 | 4 | 4 | 5 | 6 | 7 | 8 |
| 50 | 6990 | 6998 | 7007 | 7016 | 7024 | 7033 | 7042 | 7050 | 7059 | 7067 | 1 | 2 | 3 | 3 | 4 | 5 | 6 | 7 | 8 |
| 51 | 7076 | 7084 | 7093 | 7101 | 7110 | 7118 | 7126 | 7135 | 7143 | 7152 | 1 | 2 | 3 | 3 | 4 | 5 | 6 | 7 | 8 |
| 52 | 7160 | 7168 | 7177 | 7185 | 7193 | 7202 | 7210 | 7218 | 7226 | 7235 | 1 | 2 | 2 | 3 | 4 | 5 | 6 | 7 | 7 |
| 53 | 7243 | 7251 | 7259 | 7267 | 7275 | 7284 | 7292 | 7300 | 7308 | 7316 | 1 | 2 | 2 | 3 | 4 | 5 | 6 | 6 | 7 |
| 54 | 7324 | 7332 | 7340 | 7348 | 7356 | 7364 | 7372 | 7380 | 7388 | 7396 | 1 | 2 | 2 | 3 | 4 | 5 | 6 | 6 | 7 |
| 55 | 7404 | 7412 | 7419 | 7427 | 7435 | 7443 | 7451 | 7459 | 7466 | 7474 | 1 | 2 | 2 | 3 | 4 | 5 | 5 | 6 | 7 |
| 56 | 7482 | 7490 | 7497 | 7505 | 7513 | 7520 | 7528 | 7536 | 7543 | 7551 | 1 | 2 | 2 | 3 | 4 | 5 | 5 | 6 | 7 |
| 57 | 7559 | 7566 | 7574 | 7582 | 7589 | 7597 | 7604 | 7612 | 7619 | 7627 | 1 | 2 | 2 | 3 | 4 | 5 | 5 | 6 | 7 |
| 58 | 7634 | 7642 | 7649 | 7657 | 7664 | 7672 | 7679 | 7686 | 7694 | 7701 | 1 | 1 | 2 | 3 | 4 | 4 | 5 | 6 | 7 |
| 59 | 7709 | 7716 | 7723 | 7731 | 7738 | 7745 | 7752 | 7760 | 7767 | 7774 | 1 | 1 | 2 | 3 | 4 | 4 | 5 | 6 | 7 |
| 60 | 7782 | 7789 | 7796 | 7803 | 7810 | 7818 | 7825 | 7832 | 7839 | 7846 | 1 | 1 | 2 | 3 | 4 | 4 | 5 | 6 | 6 |
| 61 | 7853 | 7860 | 7868 | 7875 | 7882 | 7889 | 7896 | 7903 | 7910 | 7917 | 1 | 1 | 2 | 3 | 4 | 4 | 5 | 6 | 6 |
| 62 | 7924 | 7931 | 7938 | 7945 | 7952 | 7959 | 7966 | 7973 | 7980 | 7987 | 1 | 1 | 2 | 3 | 3 | 4 | 5 | 6 | 6 |
| 63 | 7993 | 8000 | 8007 | 8014 | 8021 | 8028 | 8035 | 8041 | 8048 | 8055 | 1 | 1 | 2 | 3 | 3 | 4 | 5 | 5 | 6 |
| 64 | 8062 | 8069 | 8075 | 8082 | 8089 | 8096 | 8102 | 8109 | 8116 | 8122 | 1 | 1 | 2 | 3 | 3 | 4 | 5 | 5 | 6 |
| 65 | 8129 | 8136 | 8142 | 8149 | 8156 | 8162 | 8169 | 8176 | 8182 | 8189 | 1 | 1 | 2 | 3 | 3 | 4 | 5 | 5 | 6 |
| 66 | 8195 | 8202 | 8209 | 8215 | 8222 | 8228 | 8235 | 8241 | 8248 | 8254 | 1 | 1 | 2 | 3 | 3 | 4 | 5 | 5 | 6 |
| 67 | 8261 | 8267 | 8274 | 8280 | 8287 | 8293 | 8299 | 8306 | 8312 | 8319 | 1 | 1 | 2 | 3 | 3 | 4 | 5 | 5 | 6 |
| 68 | 8325 | 8331 | 8338 | 8344 | 8351 | 8357 | 8363 | 8370 | 8376 | 8382 | 1 | 1 | 2 | 3 | 3 | 4 | 4 | 5 | 6 |
| 69 | 8388 | 8395 | 8401 | 8407 | 8414 | 8420 | 8426 | 8432 | 8439 | 8445 | 1 | 1 | 2 | 2 | 3 | 4 | 4 | 5 | 6 |
| 70 | 8451 | 8457 | 8463 | 8470 | 8476 | 8482 | 8488 | 8494 | 8500 | 8506 | 1 | 1 | 2 | 2 | 3 | 4 | 4 | 5 | 6 |
| 71 | 8513 | 8519 | 8525 | 8531 | 8537 | 8543 | 8549 | 8555 | 8561 | 8567 | 1 | 1 | 2 | 2 | 3 | 4 | 4 | 5 | 5 |
| 72 | 8573 | 8579 | 8585 | 8591 | 8597 | 8603 | 8609 | 8615 | 8621 | 8627 | 1 | 1 | 2 | 2 | 3 | 4 | 4 | 5 | 5 |
| 73 | 8633 | 8639 | 8645 | 8651 | 8657 | 8663 | 8669 | 8675 | 8681 | 8686 | 1 | 1 | 2 | 2 | 3 | 4 | 4 | 5 | 5 |
| 74 | 8692 | 8698 | 8704 | 8710 | 8716 | 8722 | 8727 | 8733 | 8739 | 8745 | 1 | 1 | 2 | 2 | 3 | 4 | 4 | 5 | 5 |
| 75 | 8751 | 8756 | 8762 | 8768 | 8774 | 8779 | 8785 | 8791 | 8797 | 8802 | 1 | 1 | 2 | 2 | 3 | 3 | 4 | 5 | 5 |
| 76 | 8808 | 8814 | 8820 | 8825 | 8831 | 8837 | 8842 | 8848 | 8854 | 8859 | 1 | 1 | 2 | 2 | 3 | 3 | 4 | 5 | 5 |
| 77 | 8865 | 8871 | 8876 | 8882 | 8887 | 8893 | 8899 | 8904 | 8910 | 8915 | 1 | 1 | 2 | 2 | 3 | 3 | 4 | 4 | 5 |
| 78 | 8921 | 8927 | 8932 | 8938 | 8943 | 8949 | 8954 | 8960 | 8965 | 8971 | 1 | 1 | 2 | 2 | 3 | 3 | 4 | 4 | 5 |
| 79 | 8976 | 8982 | 8987 | 8993 | 8998 | 9004 | 9009 | 9015 | 9020 | 9025 | 1 | 1 | 2 | 2 | 3 | 3 | 4 | 4 | 5 |
| 80 | 9031 | 9036 | 9042 | 9047 | 9053 | 9058 | 9063 | 9069 | 9074 | 9079 | 1 | 1 | 2 | 2 | 3 | 3 | 4 | 4 | 5 |
| 81 | 9085 | 9090 | 9096 | 9101 | 9106 | 9112 | 9117 | 9122 | 9128 | 9133 | 1 | 1 | 2 | 2 | 3 | 3 | 4 | 4 | 5 |
| 82 | 9138 | 9143 | 9149 | 9154 | 9159 | 9165 | 9170 | 9175 | 9180 | 9186 | 1 | 1 | 2 | 2 | 3 | 3 | 4 | 4 | 5 |
| 83 | 9191 | 9196 | 9201 | 9206 | 9212 | 9217 | 9222 | 9227 | 9232 | 9238 | 1 | 1 | 2 | 2 | 3 | 3 | 4 | 4 | 5 |
| 84 | 9243 | 9248 | 9253 | 9258 | 9263 | 9269 | 9274 | 9279 | 9284 | 9289 | 1 | 1 | 2 | 2 | 3 | 3 | 4 | 4 | 5 |
| 85 | 9294 | 9299 | 9304 | 9309 | 9315 | 9320 | 9325 | 9330 | 9335 | 9340 | 1 | 1 | 2 | 2 | 3 | 3 | 4 | 4 | 5 |
| 86 | 9345 | 9350 | 9355 | 9360 | 9365 | 9370 | 9375 | 9380 | 9385 | 9390 | 1 | 1 | 2 | 2 | 3 | 3 | 4 | 4 | 5 |
| 87 | 9395 | 9400 | 9405 | 9410 | 9415 | 9420 | 9425 | 9430 | 9435 | 9440 | 0 | 1 | 1 | 2 | 2 | 3 | 3 | 4 | 4 |
| 88 | 9445 | 9450 | 9455 | 9460 | 9465 | 9469 | 9474 | 9479 | 9484 | 9489 | 0 | 1 | 1 | 2 | 2 | 3 | 3 | 4 | 4 |
| 89 | 9494 | 9499 | 9504 | 9509 | 9513 | 9518 | 9523 | 9528 | 9533 | 9538 | 0 | 1 | 1 | 2 | 2 | 3 | 3 | 4 | 4 |

*(Contd.)*

**TABLE 1**  Logarithms (Contd.)

|    | 0    | 1    | 2    | 3    | 4    | 5    | 6    | 7    | 8    | 9    | 1 | 2 | 3 | 4 | 5 | 6 | 7 | 8 | 9 |
|----|------|------|------|------|------|------|------|------|------|------|---|---|---|---|---|---|---|---|---|
| 90 | 9542 | 9547 | 9552 | 9557 | 9562 | 9566 | 9571 | 9576 | 9581 | 9586 | 0 | 1 | 1 | 2 | 2 | 3 | 3 | 4 | 4 |
| 91 | 9590 | 9595 | 9600 | 9605 | 9609 | 9614 | 9619 | 9624 | 9628 | 9633 | 0 | 1 | 1 | 2 | 2 | 3 | 3 | 4 | 4 |
| 92 | 9638 | 9643 | 9647 | 9652 | 9657 | 9661 | 9666 | 9671 | 9675 | 9680 | 0 | 1 | 1 | 2 | 2 | 3 | 3 | 4 | 4 |
| 93 | 9685 | 9689 | 9694 | 9699 | 9703 | 9708 | 9713 | 9717 | 9722 | 9727 | 0 | 1 | 1 | 2 | 2 | 3 | 3 | 4 | 4 |
| 94 | 9731 | 9736 | 9741 | 9745 | 9750 | 9754 | 9759 | 9763 | 9768 | 9773 | 0 | 1 | 1 | 2 | 2 | 3 | 3 | 4 | 4 |
| 95 | 9777 | 9782 | 9786 | 9791 | 9795 | 9800 | 9805 | 9809 | 9814 | 9818 | 0 | 1 | 1 | 2 | 2 | 3 | 3 | 4 | 4 |
| 96 | 9823 | 9827 | 9832 | 9836 | 9841 | 9845 | 9850 | 9854 | 9859 | 9863 | 0 | 1 | 1 | 2 | 2 | 3 | 3 | 4 | 4 |
| 97 | 9868 | 9872 | 9877 | 9881 | 9886 | 9890 | 9894 | 9899 | 9903 | 9908 | 0 | 1 | 1 | 2 | 2 | 3 | 3 | 4 | 4 |
| 98 | 9912 | 9917 | 9921 | 9926 | 9930 | 9934 | 9939 | 9943 | 9948 | 9952 | 0 | 1 | 1 | 2 | 2 | 3 | 3 | 4 | 4 |
| 99 | 9956 | 9961 | 9965 | 9969 | 9974 | 9978 | 9983 | 9987 | 9991 | 9996 | 0 | 1 | 1 | 2 | 2 | 3 | 3 | 3 | 4 |

**TABLE 2**  Antilogarithms

|      | 0    | 1    | 2    | 3    | 4    | 5    | 6    | 7    | 8    | 9    | 1 | 2 | 3 | 4 | 5 | 6 | 7 | 8 | 9 |
|------|------|------|------|------|------|------|------|------|------|------|---|---|---|---|---|---|---|---|---|
| .00  | 1000 | 1002 | 1005 | 1007 | 1009 | 1012 | 1014 | 1016 | 1019 | 1021 | 0 | 0 | 1 | 1 | 1 | 1 | 2 | 2 | 2 |
| .01  | 1023 | 1026 | 1028 | 1030 | 1033 | 1035 | 1038 | 1040 | 1042 | 1045 | 0 | 0 | 1 | 1 | 1 | 1 | 2 | 2 | 2 |
| .02  | 1047 | 1050 | 1052 | 1054 | 1057 | 1059 | 1062 | 1064 | 1067 | 1069 | 0 | 0 | 1 | 1 | 1 | 1 | 2 | 2 | 2 |
| .03  | 1072 | 1074 | 1076 | 1079 | 1081 | 1084 | 1086 | 1089 | 1091 | 1094 | 0 | 0 | 1 | 1 | 1 | 1 | 2 | 2 | 2 |
| .04  | 1096 | 1099 | 1102 | 1104 | 1107 | 1109 | 1112 | 1114 | 1117 | 1119 | 0 | 1 | 1 | 1 | 1 | 2 | 2 | 2 | 2 |
| .05  | 1122 | 1125 | 1127 | 1130 | 1132 | 1135 | 1138 | 1140 | 1143 | 1146 | 0 | 1 | 1 | 1 | 1 | 2 | 2 | 2 | 2 |
| .06  | 1148 | 1151 | 1153 | 1156 | 1159 | 1161 | 1164 | 1167 | 1169 | 1172 | 0 | 1 | 1 | 1 | 1 | 2 | 2 | 2 | 2 |
| .07  | 1175 | 1178 | 1180 | 1183 | 1186 | 1189 | 1191 | 1194 | 1197 | 1199 | 0 | 1 | 1 | 1 | 1 | 2 | 2 | 2 | 2 |
| .08  | 1202 | 1205 | 1208 | 1211 | 1213 | 1216 | 1219 | 1222 | 1225 | 1227 | 0 | 1 | 1 | 1 | 1 | 2 | 2 | 2 | 3 |
| .09  | 1230 | 1233 | 1236 | 1239 | 1242 | 1245 | 1247 | 1250 | 1253 | 1256 | 0 | 1 | 1 | 1 | 1 | 2 | 2 | 2 | 3 |
| .10  | 1259 | 1262 | 1265 | 1268 | 1271 | 1274 | 1276 | 1279 | 1282 | 1285 | 0 | 1 | 1 | 1 | 1 | 2 | 2 | 2 | 3 |
| .11  | 1288 | 1291 | 1294 | 1297 | 1300 | 1303 | 1306 | 1309 | 1312 | 1315 | 0 | 1 | 1 | 1 | 2 | 2 | 2 | 2 | 3 |
| .12  | 1318 | 1321 | 1324 | 1327 | 1330 | 1334 | 1337 | 1340 | 1343 | 1346 | 0 | 1 | 1 | 1 | 2 | 2 | 2 | 2 | 3 |
| .13  | 1349 | 1352 | 1355 | 1358 | 1361 | 1365 | 1368 | 1371 | 1374 | 1377 | 0 | 1 | 1 | 1 | 2 | 2 | 2 | 3 | 3 |
| .14  | 1380 | 1384 | 1387 | 1390 | 1393 | 1396 | 1400 | 1403 | 1406 | 1409 | 0 | 1 | 1 | 1 | 2 | 2 | 2 | 3 | 3 |
| .15  | 1413 | 1416 | 1419 | 1422 | 1426 | 1429 | 1432 | 1435 | 1439 | 1442 | 0 | 1 | 1 | 1 | 2 | 2 | 2 | 3 | 3 |
| .16  | 1445 | 1449 | 1452 | 1455 | 1459 | 1462 | 1466 | 1469 | 1472 | 1476 | 0 | 1 | 1 | 1 | 2 | 2 | 2 | 3 | 3 |
| .17  | 1479 | 1483 | 1486 | 1489 | 1493 | 1496 | 1500 | 1503 | 1507 | 1510 | 0 | 1 | 1 | 1 | 2 | 2 | 2 | 3 | 3 |
| .18  | 1514 | 1517 | 1521 | 1524 | 1528 | 1531 | 1535 | 1538 | 1542 | 1545 | 0 | 1 | 1 | 1 | 2 | 2 | 2 | 3 | 3 |
| .19  | 1549 | 1552 | 1556 | 1560 | 1563 | 1567 | 1570 | 1574 | 1578 | 1581 | 0 | 1 | 1 | 1 | 2 | 2 | 3 | 3 | 3 |
| .20  | 1585 | 1589 | 1592 | 1596 | 1600 | 1603 | 1607 | 1611 | 1614 | 1618 | 0 | 1 | 1 | 1 | 2 | 2 | 3 | 3 | 3 |
| .21  | 1622 | 1626 | 1629 | 1633 | 1637 | 1641 | 1644 | 1648 | 1652 | 1656 | 0 | 1 | 1 | 2 | 2 | 2 | 3 | 3 | 3 |
| .22  | 1660 | 1663 | 1667 | 1671 | 1675 | 1679 | 1683 | 1687 | 1690 | 1694 | 0 | 1 | 1 | 2 | 2 | 2 | 3 | 3 | 3 |
| .23  | 1698 | 1702 | 1706 | 1710 | 1714 | 1718 | 1722 | 1726 | 1730 | 1734 | 0 | 1 | 1 | 2 | 2 | 2 | 3 | 3 | 4 |
| .24  | 1738 | 1742 | 1746 | 1750 | 1754 | 1758 | 1762 | 1766 | 1770 | 1774 | 0 | 1 | 1 | 2 | 2 | 2 | 3 | 3 | 4 |
| .25  | 1778 | 1782 | 1786 | 1791 | 1795 | 1799 | 1803 | 1807 | 1811 | 1816 | 0 | 1 | 1 | 2 | 2 | 2 | 3 | 3 | 4 |
| .26  | 1820 | 1824 | 1828 | 1832 | 1837 | 1841 | 1845 | 1849 | 1854 | 1858 | 0 | 1 | 1 | 2 | 2 | 3 | 3 | 3 | 4 |
| .27  | 1862 | 1866 | 1871 | 1875 | 1879 | 1884 | 1888 | 1892 | 1897 | 1901 | 0 | 1 | 1 | 2 | 2 | 3 | 3 | 3 | 4 |
| .28  | 1905 | 1910 | 1914 | 1919 | 1923 | 1928 | 1932 | 1936 | 1941 | 1945 | 0 | 1 | 1 | 2 | 2 | 3 | 3 | 4 | 4 |
| .29  | 1950 | 1954 | 1959 | 1963 | 1968 | 1972 | 1977 | 1982 | 1986 | 1991 | 0 | 1 | 1 | 2 | 2 | 3 | 3 | 4 | 4 |
| .30  | 1995 | 2000 | 2004 | 2009 | 2014 | 2018 | 2023 | 2028 | 2032 | 2037 | 0 | 1 | 1 | 2 | 2 | 3 | 3 | 4 | 4 |
| .31  | 2042 | 2046 | 2051 | 2056 | 2061 | 2065 | 2070 | 2075 | 2080 | 2084 | 0 | 1 | 1 | 2 | 2 | 3 | 3 | 4 | 4 |
| .32  | 2089 | 2094 | 2099 | 2104 | 2109 | 2113 | 2118 | 2123 | 2128 | 2133 | 0 | 1 | 1 | 2 | 2 | 3 | 3 | 4 | 4 |
| .33  | 2138 | 2143 | 2148 | 2153 | 2158 | 2163 | 2168 | 2173 | 2178 | 2183 | 0 | 1 | 1 | 2 | 2 | 3 | 3 | 4 | 4 |
| .34  | 2188 | 2193 | 2198 | 2203 | 2208 | 2213 | 2218 | 2223 | 2228 | 2234 | 1 | 1 | 2 | 2 | 3 | 3 | 4 | 4 | 5 |
| .35  | 2239 | 2244 | 2249 | 2254 | 2259 | 2265 | 2270 | 2275 | 2280 | 2286 | 1 | 1 | 2 | 2 | 3 | 3 | 4 | 4 | 5 |
| .36  | 2291 | 2296 | 2301 | 2307 | 2312 | 2317 | 2323 | 2328 | 2333 | 2339 | 1 | 1 | 2 | 2 | 3 | 3 | 4 | 4 | 5 |
| .37  | 2344 | 2350 | 2355 | 2360 | 2366 | 2371 | 2377 | 2382 | 2388 | 2393 | 1 | 1 | 2 | 2 | 3 | 3 | 4 | 4 | 5 |
| .38  | 2399 | 2404 | 2410 | 2415 | 2421 | 2427 | 2432 | 2438 | 2443 | 2449 | 1 | 1 | 2 | 2 | 3 | 3 | 4 | 4 | 5 |
| .39  | 2455 | 2460 | 2466 | 2472 | 2477 | 2483 | 2489 | 2495 | 2500 | 2506 | 1 | 1 | 2 | 2 | 3 | 3 | 4 | 5 | 5 |
| .40  | 2512 | 2518 | 2523 | 2529 | 2535 | 2541 | 2547 | 2553 | 2559 | 2564 | 1 | 1 | 2 | 2 | 3 | 4 | 4 | 5 | 5 |
| .41  | 2570 | 2576 | 2582 | 2588 | 2594 | 2600 | 2606 | 2612 | 2618 | 2624 | 1 | 1 | 2 | 2 | 3 | 4 | 4 | 5 | 5 |
| .42  | 2630 | 2636 | 2642 | 2649 | 2655 | 2661 | 2667 | 2673 | 2679 | 2685 | 1 | 1 | 2 | 2 | 3 | 4 | 4 | 5 | 6 |
| .43  | 2692 | 2698 | 2704 | 2710 | 2716 | 2723 | 2729 | 2735 | 2742 | 2748 | 1 | 1 | 2 | 3 | 3 | 4 | 4 | 5 | 6 |
| .44  | 2754 | 2761 | 2767 | 2773 | 2780 | 2786 | 2793 | 2799 | 2805 | 2812 | 1 | 1 | 2 | 3 | 3 | 4 | 5 | 5 | 6 |

(Contd.)

## TABLE 2  Antilogarithms

|      | 0 | 1 | 2 | 3 | 4 | 5 | 6 | 7 | 8 | 9 | 1 | 2 | 3 | 4 | 5 | 6 | 7 | 8 | 9 |
|------|---|---|---|---|---|---|---|---|---|---|---|---|---|---|---|---|---|---|---|
| .45 | 2818 | 2825 | 2831 | 2838 | 2844 | 2851 | 2858 | 2864 | 2871 | 2877 | 1 | 1 | 2 | 3 | 3 | 4 | 5 | 5 | 6 |
| .46 | 2884 | 2891 | 2897 | 2904 | 2911 | 2917 | 2924 | 2931 | 2938 | 2944 | 1 | 1 | 2 | 3 | 3 | 4 | 5 | 5 | 6 |
| .47 | 2951 | 2958 | 2965 | 2972 | 2979 | 2985 | 2992 | 2999 | 3006 | 3013 | 1 | 1 | 2 | 3 | 3 | 4 | 5 | 5 | 6 |
| .48 | 3020 | 3027 | 3034 | 3041 | 3048 | 3055 | 3062 | 3069 | 3076 | 3083 | 1 | 1 | 2 | 3 | 4 | 4 | 5 | 6 | 6 |
| .49 | 3090 | 3097 | 3105 | 3112 | 3119 | 3126 | 3133 | 3141 | 3148 | 3155 | 1 | 1 | 2 | 3 | 4 | 4 | 5 | 6 | 6 |
| .50 | 3162 | 3170 | 3177 | 3184 | 3192 | 3199 | 3206 | 3214 | 3221 | 3228 | 1 | 1 | 2 | 3 | 4 | 4 | 5 | 6 | 7 |
| .51 | 3236 | 3243 | 3251 | 3258 | 3266 | 3273 | 3281 | 3289 | 3296 | 3304 | 1 | 2 | 2 | 3 | 4 | 5 | 5 | 6 | 7 |
| .52 | 3311 | 3319 | 3327 | 3334 | 3342 | 3350 | 3357 | 3365 | 3373 | 3381 | 1 | 2 | 2 | 3 | 4 | 5 | 5 | 6 | 7 |
| .53 | 3388 | 3396 | 3404 | 3412 | 3420 | 3428 | 3436 | 3443 | 3451 | 3459 | 1 | 2 | 2 | 3 | 4 | 5 | 6 | 6 | 7 |
| .54 | 3467 | 3475 | 3483 | 3491 | 3499 | 3508 | 3516 | 3524 | 3532 | 3540 | 1 | 2 | 2 | 3 | 4 | 5 | 6 | 6 | 7 |
| .55 | 3548 | 3556 | 3565 | 3573 | 3581 | 3589 | 3597 | 3606 | 3614 | 3622 | 1 | 2 | 2 | 3 | 4 | 5 | 6 | 7 | 7 |
| .56 | 3631 | 3639 | 3648 | 3656 | 3664 | 3673 | 3681 | 3690 | 3698 | 3707 | 1 | 2 | 3 | 3 | 4 | 5 | 6 | 7 | 8 |
| .57 | 3715 | 3724 | 3733 | 3741 | 3750 | 3758 | 3767 | 3776 | 3784 | 3793 | 1 | 2 | 3 | 3 | 4 | 5 | 6 | 7 | 8 |
| .58 | 3802 | 3811 | 3819 | 3828 | 3837 | 3846 | 3855 | 3864 | 3873 | 3882 | 1 | 2 | 3 | 4 | 4 | 5 | 6 | 7 | 8 |
| .59 | 3890 | 3899 | 3908 | 3917 | 3926 | 3936 | 3945 | 3954 | 3963 | 3972 | 1 | 2 | 3 | 4 | 5 | 5 | 6 | 7 | 8 |
| .60 | 3981 | 3990 | 3999 | 4009 | 4018 | 4027 | 4036 | 4046 | 4055 | 4064 | 1 | 2 | 3 | 4 | 5 | 6 | 6 | 7 | 8 |
| .61 | 4074 | 4083 | 4093 | 4102 | 4111 | 4121 | 4130 | 4140 | 4150 | 4159 | 1 | 2 | 3 | 4 | 5 | 6 | 7 | 8 | 9 |
| .62 | 4169 | 4178 | 4188 | 4198 | 4207 | 4217 | 4227 | 4236 | 4246 | 4256 | 1 | 2 | 3 | 4 | 5 | 6 | 7 | 8 | 9 |
| .63 | 4266 | 4276 | 4285 | 4295 | 4305 | 4315 | 4325 | 4335 | 4345 | 4355 | 1 | 2 | 3 | 4 | 5 | 6 | 7 | 8 | 9 |
| .64 | 4365 | 4375 | 4385 | 4395 | 4406 | 4416 | 4426 | 4436 | 4446 | 4457 | 1 | 2 | 3 | 4 | 5 | 6 | 7 | 8 | 9 |
| .65 | 4467 | 4477 | 4487 | 4498 | 4508 | 4519 | 4529 | 4539 | 4550 | 4560 | 1 | 2 | 3 | 4 | 5 | 6 | 7 | 8 | 9 |
| .66 | 4571 | 4581 | 4592 | 4603 | 4613 | 4624 | 4634 | 4645 | 4656 | 4667 | 1 | 2 | 3 | 4 | 5 | 6 | 7 | 9 | 10 |
| .67 | 4677 | 4688 | 4699 | 4710 | 4721 | 4732 | 4742 | 4753 | 4764 | 4775 | 1 | 2 | 3 | 4 | 5 | 7 | 8 | 9 | 10 |
| .68 | 4786 | 4797 | 4808 | 4819 | 4831 | 4842 | 4853 | 4864 | 4875 | 4887 | 1 | 2 | 3 | 4 | 6 | 7 | 8 | 9 | 10 |
| .69 | 4898 | 4909 | 4920 | 4932 | 4943 | 4955 | 4966 | 4977 | 4989 | 5000 | 1 | 2 | 3 | 5 | 6 | 7 | 8 | 9 | 10 |
| .70 | 5012 | 5023 | 5035 | 5047 | 5058 | 5070 | 5082 | 5093 | 5105 | 5117 | 1 | 2 | 4 | 5 | 6 | 7 | 8 | 9 | 11 |
| .71 | 5129 | 5140 | 5152 | 5164 | 5176 | 5188 | 5200 | 5212 | 5224 | 5236 | 1 | 2 | 4 | 5 | 6 | 7 | 8 | 10 | 11 |
| .72 | 5248 | 5260 | 5272 | 5284 | 5297 | 5309 | 5321 | 5333 | 5346 | 5358 | 1 | 2 | 4 | 5 | 6 | 7 | 9 | 10 | 11 |
| .73 | 5370 | 5383 | 5395 | 5408 | 5420 | 5433 | 5445 | 5458 | 5470 | 5483 | 1 | 3 | 4 | 5 | 6 | 8 | 9 | 10 | 11 |
| .74 | 5495 | 5508 | 5521 | 5534 | 5546 | 5559 | 5572 | 5585 | 5598 | 5610 | 1 | 3 | 4 | 5 | 6 | 8 | 9 | 10 | 12 |
| .75 | 5623 | 5636 | 5649 | 5662 | 5675 | 5689 | 5702 | 5715 | 5728 | 5741 | 1 | 3 | 4 | 5 | 7 | 8 | 9 | 10 | 12 |
| .76 | 5754 | 5768 | 5781 | 5794 | 5808 | 5821 | 5834 | 5848 | 5861 | 5875 | 1 | 3 | 4 | 5 | 7 | 8 | 9 | 11 | 12 |
| .77 | 5888 | 5902 | 5916 | 5929 | 5943 | 5957 | 5970 | 5984 | 5998 | 6012 | 1 | 3 | 4 | 5 | 7 | 8 | 10 | 11 | 12 |
| .78 | 6026 | 6039 | 6053 | 6067 | 6081 | 6095 | 6109 | 6124 | 6138 | 6152 | 1 | 3 | 4 | 6 | 7 | 8 | 10 | 11 | 13 |
| .79 | 6166 | 6180 | 6194 | 6209 | 6223 | 6237 | 6252 | 6266 | 6281 | 6295 | 1 | 3 | 4 | 6 | 7 | 9 | 10 | 11 | 13 |
| .80 | 6310 | 6324 | 6339 | 6353 | 6368 | 6383 | 6397 | 6412 | 6427 | 6442 | 1 | 3 | 4 | 6 | 7 | 9 | 10 | 12 | 13 |
| .81 | 6457 | 6471 | 6486 | 6501 | 6516 | 6531 | 6546 | 6561 | 6577 | 6592 | 2 | 3 | 5 | 6 | 8 | 9 | 11 | 12 | 14 |
| .82 | 6607 | 6622 | 6637 | 6653 | 6668 | 6683 | 6699 | 6714 | 6730 | 6745 | 2 | 3 | 5 | 6 | 8 | 9 | 11 | 12 | 14 |
| .83 | 6761 | 6776 | 6792 | 6808 | 6823 | 6839 | 6855 | 6871 | 6887 | 6902 | 2 | 3 | 5 | 6 | 8 | 9 | 11 | 13 | 14 |
| .84 | 6918 | 6934 | 6950 | 6966 | 6982 | 6998 | 7015 | 7031 | 7047 | 7063 | 2 | 3 | 5 | 6 | 8 | 10 | 11 | 13 | 15 |
| .85 | 7079 | 7096 | 7112 | 7129 | 7145 | 7161 | 7178 | 7194 | 7211 | 7228 | 2 | 3 | 5 | 7 | 8 | 10 | 12 | 13 | 15 |
| .86 | 7244 | 7261 | 7278 | 7295 | 7311 | 7328 | 7345 | 7362 | 7379 | 7396 | 2 | 3 | 5 | 7 | 8 | 10 | 12 | 13 | 15 |
| .87 | 7413 | 7430 | 7447 | 7464 | 7482 | 7499 | 7516 | 7534 | 7551 | 7568 | 2 | 3 | 5 | 7 | 9 | 10 | 12 | 14 | 16 |
| .88 | 7586 | 7603 | 7621 | 7638 | 7656 | 7674 | 7691 | 7709 | 7727 | 7745 | 2 | 4 | 5 | 7 | 9 | 11 | 12 | 14 | 16 |
| .89 | 7762 | 7780 | 7798 | 7816 | 7834 | 7852 | 7870 | 7889 | 7907 | 7925 | 2 | 4 | 5 | 7 | 9 | 11 | 13 | 14 | 16 |
| .90 | 7943 | 7962 | 7980 | 7998 | 8017 | 8035 | 8054 | 8072 | 8091 | 8110 | 2 | 4 | 6 | 7 | 9 | 11 | 13 | 15 | 17 |
| .91 | 8128 | 8147 | 8166 | 8185 | 8204 | 8222 | 8241 | 8260 | 8279 | 8299 | 2 | 4 | 6 | 8 | 9 | 11 | 13 | 15 | 17 |
| .92 | 8318 | 8337 | 8356 | 8375 | 8395 | 8414 | 8433 | 8453 | 8472 | 8492 | 2 | 4 | 6 | 8 | 10 | 12 | 14 | 15 | 17 |
| .93 | 8511 | 8531 | 8551 | 8570 | 8590 | 8610 | 8630 | 8650 | 8670 | 8690 | 2 | 4 | 6 | 8 | 10 | 12 | 14 | 16 | 18 |
| .94 | 8710 | 8730 | 8750 | 8770 | 8790 | 8810 | 8831 | 8851 | 8872 | 8892 | 2 | 4 | 6 | 8 | 10 | 12 | 14 | 16 | 18 |
| .95 | 8913 | 8933 | 8954 | 8974 | 8995 | 9016 | 9036 | 9057 | 9078 | 9099 | 2 | 4 | 6 | 8 | 10 | 12 | 15 | 17 | 19 |
| .96 | 9120 | 9141 | 9162 | 9183 | 9204 | 9226 | 9247 | 9268 | 9290 | 9311 | 2 | 4 | 6 | 8 | 11 | 13 | 15 | 17 | 19 |
| .97 | 9333 | 9354 | 9376 | 9397 | 9419 | 9441 | 9462 | 9484 | 9506 | 9528 | 2 | 4 | 7 | 9 | 11 | 13 | 15 | 17 | 20 |
| .98 | 9550 | 9572 | 9594 | 9616 | 9638 | 9661 | 9683 | 9705 | 9727 | 9750 | 2 | 4 | 7 | 9 | 11 | 13 | 16 | 18 | 20 |
| .99 | 9772 | 9795 | 9817 | 9840 | 9863 | 9886 | 9908 | 9931 | 9954 | 9977 | 2 | 5 | 7 | 9 | 11 | 14 | 16 | 18 | 20 |

# Index

Absorption spectroscopy, 230
Acetaldehyde, 583, 592
Action potential, 498, 499
Activated complex, 605, 606, 607
    theory, 605
Activation energy, 457, 458, 511, 589, 616
    apparent, 589
    overpotential, 522, 523
    polarization, 543, 544, 545, 546, 547
Activity, 38, 543
    coefficient, 38, 422, 612
    mean ionic, 39, 422
Adiabatic process, 4
    method, 596
Adsorption theory, 552
Advantage voltage, 284
Aerosol, 646
Aluminiates, 565
Alzheimer's disease, 666
Amorphous solid, 321
Amperometric titrations, 361, 372, 374
Angular momentum, 213
    operator, 215
Angular velocity, 236
    variable, 177
Anodic current density, 544
Anodic polarization, 542
Anodic protection, 561
Anodizing, 552
Antimony trichloride, 563
Approximation method, 200, 208

Arrhenius law, 600
    equation, 601, 602
Asymmetric potential, 346
    tops, 235
Atomic force microscopy, 660
Avalanche voltage, 288
Azimuthal quantum number, 186

Back EMF, 353
Back-scattered electrons, 653
Ball milling, 646
Band theory, 270, 271, 487
Beer–Lambert law, 652
Biofilm coating, 567
Biosensor, 666
Biotechnology, 666
Blyholder model, 244
Bode plot, 573
Bohr theory, 184, 186, 195
    orbit, 184
    radius, 193, 194
Boltzmann distribution, 70
Bond energy, 11
Born–Lande's equation, 325, 326
    exponent, 324
    expression, 614
Bose–Einstein statistics, 124, 126, 139, 140
Bragg's law, 247, 251, 258, 313
Breakdown potential, 552
Butler–Volmer equation, 445, 446, 449, 450, 453,
    499, 509, 510, 513

Calcium plumbate, 563
Calgon, 564
Calomel electrode, 337, 338, 339
Calorific value, 7
Canonical ensemble, 112, 114
Carnot efficiency, 519
    heat engine, 518
    limitation, 518, 519
Cartesian coordinates, 198
Cathode ray tube (CRT), 664
Ceramics, 565
Chain
    initiation, 581
    propagation, 581
    reaction, 581, 586, 589
    termination, 581
Charge coupled device (CCD), 654
Charge transfer complex, 299
Chemical
    corrosion, 539
    free energy, 443
    potential, 22, 27, 412, 421, 440
    shift, 599
    work, 414, 473, 476
Chromate, 563
Clausius–Clapeyron equation, 25, 26
    inequility equation, 13
Cold emission, 501
    combustion, 522
Collision
    frequency, 603
    life time, 232
    line width, 232
    number, 604
    theory, 602, 608, 628, 629
Colour centre, 310
Commute, 147
Competition method, 594
    inhibition, 621, 624
Component, 40
Composite, 565
Concentration
    gradient, 495
    polarization, 354, 543
Condensation energy, 287
Condenser lenses, 656
Conductance, 382, 383
Conduction band, 488, 489, 490
Conductivity, 382
Conductometric titrations, 386
Congruent melting point, 45
Constant current coulometry, 395

Contact adsorption, 460, 463, 464, 466, 467, 473
Core electrons, 488
Corrosion, 563, 540, 539, 545
    current, 548, 553
    potential, 548, 549
    rate, 539, 540, 549, 553
Corrugation, 659
Cosmetics, 663
Cottrel equation, 376, 575
Coulomb forces, 467, 660
Coulometric analysis, 392
    method, 392
Coulometric titrations, 395, 396, 397
Coupled flow, 54
Critical effect, 286
Critical temperature, 285, 289
    field, 286
Cross phenomenon, 50
Crystal
    lattice, 247, 642
    system, 248, 642
Crystalline solids, 321
Crystallographic axis, 249, 312
Cubic crystal, 248
Current density, 506, 514, 546, 547, 549
    efficiency, 392
Current sampled polarography, 376


de Broglie, 150
    equation, 151
Debye
    $T^3$ law, 329
    temperature, 321, 331
    theory, 329
Debye–Huckel theory, 429, 608, 612
Decomposition potential, 355, 356, 362
De-electronation, 437, 445, 510
Defect, 299
Degree of freedom, 233
Depletion layer, 282, 436
Desiccant bag, 571
Dewar-Chatt model, 244
Diamagnetic, 286, 287
Dielectric constant, 471, 608, 610, 613
Dielectric properties, 573
Dielectric spectroscopy, 573
Differential strain, 541
    heat treatment, 553
Diffraction angle, 649
Diffraction method, 311
Diffusion current, 363, 367, 369
    coefficient, 543

Dipole moment, 243, 614
    potential, 462
Dipole–dipole interaction, 461
Discovery, 261
Dispersion force, 461, 462
Dissociation energy, 454
Dixon plot, 620
Doping, 278, 297
Doppler broadening, 232
Double layer, 405, 423, 425
Double sphere model, 608, 609, 612
Dropping mercury electrode, 362, 363, 366, 367, 373
Dry cell, 535
    corrosion, 539
Dulong–Petit, 329

Eadie-Hofstee plot, 621
Edge dislocations, 310
Efficiency, 526
Eigen function, 147, 148, 186, 188, 202
Eigenvalue, 202, 218
Einstein theory, 327
Electrical double layer, 405, 556
Electricity storage density, 532
Electrified interface, 405, 406, 409, 410, 414, 556
Electro-capillary maxima, 411, 423, 469
    curve, 420, 423, 464, 465, 466, 496
    measurement, 464
Electrochemical
    corrosion, 539
    impedance spectroscopy, 573
    polarization, 572
    potential, 412, 440
    series, 560
Electrolytic cell, 542
Electromagnetic radiation, 226, 229
    spectrum, 227, 228, 229
Electron energy loss spectroscopy (EELS), 657
Electronation, 437, 445
    current, 448
    current density, 443
Electronic partition function, 89
Electro-osmosis, 64
    flux, 50
    pressure, 64
Electroplating, 565
EMF series, 565
Emission spectroscopy, 230
Enamels, 566
Endergonic, 616
Energy
    bands, 273, 282

    conversion, 517, 522
    density, 532
    gap, 439, 490, 491, 494
    harmonic oscillator, 162
    hydrogen atom, 186
    particle in a one-dimensional box, 155
    rigid rotator, 176, 235
Ensemble, 69, 105, 111
Enthalpy, 5, 6, 27, 78
    of formation, 11
    of reaction, 11
Entropy, 13, 54, 73, 75
    change of surrounding, 16
    of allotrope, 17
    change of system, 16
    change total, 16
    change with temperature, 14, 15
    in a chemical reaction, 17
    of fusion, 17
    residual, 30
    of sublimation, 17
Entropy production due to
    chemical reaction, 59
    electrochemical reaction, 61
    heat flow, 56
    matter flow, 57
Enzyme active sites, 613, 617
Enzyme kinetics, 615
EPR, 599
Equation of state, 36
Equilibrium constant, 19, 78, 102, 600
    field, 446
    potential difference, 445, 446
Equilibrium exchange current density, 444, 452, 508, 558
Equivalent conductance, 383
Ergodic theorem, 52
ESR, 598
Ethane, 584
Eulerian angles, 97
Eutectic point, 44, 46
Evans diagram, 549
Exchange current density, 542, 558
Extreme gas degeneracy, 130, 131
Extrinsic semiconductor, 278

Faradic rectification, 448
    constant, 543
    efficiency, 521
Fast flow method, 597
Fast reactions, 594
Fermi–Dirac statistics, 129, 130, 139, 140

# Index

Fermi level, 275
Fick's law, 53
Field emission (FE), 652
Flash photolysis, 594
    spectroscopy, 594
Flip-Flop, 482, 483, 486
Flouresence, 594
Flow methods, 594, 597
Fluctuation, 112, 113, 115
Flux quantization, 289
Forward bias, 283
Fourier law, 53
Fourier transforms, 648
Frankel defect, 301, 304, 305, 307
Free energy, 17, 100, 463
Free energy of nucleation, 267
    activation, 609
    chemical, 443
Fuel cell, 527
Fugacity, 33, 34, 36, 37, 422
Fullerenes, 294

Galvanic cell, 541
Galvanic coupling, 554
Galvanising, 565
Galvanostat, 572
Galvanostatic anodic polarization, 550
Gas sensing electrode, 349
Geiger counter, 314
Gibbs–Duhem equation, 23, 24, 62, 416
Gibbs free energy, 18, 76, 94
Glass electrode, 343, 344, 345
Goniometer, 649
Gouy–Chapman model, 425, 429, 431
    theory, 478
Grain
    boundary, 541, 645
    size, 541, 641
Grand canonical ensemble, 112
Green house effect, 517

Hall–Petch relationship, 643, 644
Hamiltonian operator, 146, 152, 177, 202, 208
Hardness, 644
Harmonic oscillator, 158, 160
    energy level, 162
Hartree, 196, 200
Hauy's law, 249
Heat capacity, 6
    at constant pressure, 6
    at constant volume, 6, 73

    molar, 6
    specific, 6
Heat of combustion, 7
    formation, 7
    hydration, 7
    neutralization, 7
    solution, 7
Heisenberg uncertainty principle, 150
Helmholtz free energy, 77, 305
Helmholtz layer, 556
Helmholtz–Perrin model, 423, 430, 470
Hermite polynomial, 167, 169
Hess's law, 11
Hexapole lenses, 656
Hexylamine $C_6H_6NH_2$, 563
Hinselwood treatment, 629, 630
Hollow cathode ray discharge tube, 662
Hot emission, 500
Hump, 469, 472, 478, 479, 480
Hydrocarbon air cell, 529
Hydrogen overvoltage, 544, 545
Hydrogen oxygen fuel cell, 527, 528
    coulometer, 393
Hyperbolic sin function, 447
Hyperfine interaction, 240

Image force, 461
    energy, 501
    interaction energy, 501
Imperfect crystal, 299
Impressed current, 559
Incongruent melting point, 46
Inner Helmholtz plane, 406, 437, 464, 465, 467, 468
Inner potential, 412, 440
Insulator, 276, 488
Intensity of spectral line, 233
Intensive property, 4
Interaction energy, 484
Inter-atomic spacing, 489, 490
Interface, 265
Interfacial tension, 410, 414
Intermolecular effect, 295
Internal energy, 4, 5
Intramolecular effect, 295
Intramoleculer reactions, 295
Intrinsic semiconductor, 277
Iodine coulometer, 393
Ionic
    product, 385
    strength, 608, 612, 613
Ion–electrode interaction, 463, 464
Ion–water interaction, 463

Irreversible process, 50, 51
Irreversible thermodynamics, 50, 51
Isobaric process, 4
Isochoric process, 4
Isolated system, 4
Isothermal method, 597
        process, 4
Isotopic effect, 289

Joint Committee on Power Diffraction Study, 651
Jump frequency, 441

Kinetic
        spectrophotometry, 596
        theory, 602
Kirkwood expression, 613, 614
Kohlrausch law, 385

Ladder operator, 217
Lagendre's polynomial, 185, 186, 210
        equation, 174
Lagrange method, 110, 118
Laplacian operator, 146, 196, 198
Lateral interaction, 473
        repulsion model, 482
Lattice energy, 322
        planes, 255, 256
Lattice heat capacity, 327
Law of mass action, 305
Laws of thermodynamics, 9, 10, 11, 12
Lead acid storage battery, 534
Life time broadening, 232
Limiting potential, 363
        current, 363
        current density, 525, 543, 544
        diffusion current, 553
Lindemann theory, 625, 626
Line defect, 300, 310
Lineweaver–Burk plot, 620, 621
Liquid crystal display (LCD), 665
Liquid junction potential, 351
Liquid membrane electrode, 347
Local density of states (LDoS), 658
London force, 461

Macroscopic system, 04
Macrostate, 106
Madelung constant, 322, 323, 325
Magnetic electron-optical system (MEOS), 652

Magnetic quantum number, 186
Matrix mechanics, 145
Maxwell
        Boltzmann distribution law, 116, 119
        Boltzmann–statistics, 74, 139, 140
        distribution law of velocity, 121
Meissner effect, 286, 288, 291
MEMs, 661
Membrane electrode, 342
Meniscus forces, 660
Micelle, 647, 648
Michaelis–Menten equation, 617, 619
        constant, 619, 625
Microcanonical ensemble, 111
Micro-emulsion technique, 646, 647
Microstate, 106, 223
Microwave spectroscopy, 226, 227
Miller indices, 250, 251, 256, 650
Miller planes, 650
Mixed anodic cathodic wave, 368
Mixed potential theory, 540, 548
Modeling constant, 322, 325
Moment of inertia, 84, 171, 172, 235
Morse equation, 454
        curve, 454, 458
Most probable distribution, 110
Multiplication theorem, 71
Multistep reaction, 504, 511

$Na_6(PO_3)_6$, 563
Nanocomposites, 565
Nanorods, 644
Nanotechnology, 641, 661, 662
Nanowires, 644
Nernst equation, 354, 543, 553
Nernst heat theorem, 27
Nerve potential, 498
Newton's laws, 53, 145
Nickel–cadmium cell, 531, 594
Nickel metal hydride (Ni-MH), 664
Nitrites, 563
NMR, 594, 599, 600
Node, 155, 170
Non-competitive inhibition, 625, 626, 668
Non-equilibrium thermodynamics, 50, 51
        drift current density, 445
        potential difference, 445
Non-polarizable electrode, 465
        interface, 409, 414, 452
Non-stoichiometric defect, 300
n–p junction, 281, 438, 493, 496
n-type semiconductor, 278, 493

Nucleation, 265, 268, 269
Nyquist plot, 573

Objective lenses, 656
Ohm's law, 53, 452, 506
Ohmic potential, 352
Onsager's reciprocity relation, 51, 52
Open system, 527
Operator
    Hamiltonian, 146, 152
    Laplacian, 146
Optical characterization, 652
Organic light-emitting diodes (OLEDs), 664
Orthogonal, 147
Orthonormal, 147
Outer Helmholtz plane, 406, 436, 437, 454, 464, 465, 467, 468, 470
Outer potential, 412
Over potential, 445, 448, 449, 450, 458, 505, 506, 508
Over voltage, 354, 356, 542, 546, 547, 553
Oxide, 566
    film theory, 552
Oxygen overvoltage, 545
Oxygen wave, 368

Partial molar quantity, 22
Particle in a one-dimensional box, 152, 155
Partition coefficient, 75, 76, 77, 78, 91, 102, 105
Partition function, 69, 70, 71, 73, 605
Passivation potential, 553
Passivity, 549
Pauli's exclusion principle, 220
Pelargonium graveolens, 648
Perfect crystal, 299
pH, 547, 550, 553
Persistent current, 289
Perturbation theory, 201, 202
    method, 594
Phase diagram, 42
    one component system, 42
    two component system, 43
Phase rule, 3, 40, 41, 42
Phenomenological laws, 53
    coefficient, 54
    equations, 53
Photochemical, 585
Photochromatic glass, 662
Piezo-drive movement, 659
Pitting corrosion, 540, 552
Plane defect, 300
Plumbous plumbate $Pb_2(II)Pb(IV)O_4$, 563

Peasmon resonance, 643
p–n junction, 282, 283
Point defect, 300, 301
Polarizability, 452
Polarizable electrode, 466
    interface, 409, 452
Polarization, 353, 542
Polarogram, 361, 362, 368
Polarographic cell, 361
Polarography, 361, 369
Polyatomic molecules, 96, 97, 99
Polyurethane, 566
Population density, 233
Postulates of quantum mechanics, 147
Potential energy distance relations, 453
Potential gradient, 495
Potential of zero charge, 419, 420, 469
Potentiometric titrations, 340, 356
Potentiostat, 572
Potentiostatic anodic polarization, 551
Pourbaix diagrams, 568
Powder method, 313, 316
Power, 520, 533
Precipitate electrode, 349
Primary battery, 527
Primitive cubic lattice, 258
Primitive unit cell, 312
Probability, 69, 188
    factor, 500, 604
Producer cell, 519
Projector lenses, 656
Prosthetic group, 615
Protective coating, 565
p-type semiconductor, 278, 492
Pulse laser method, 646
Pulse polarography, 375

Quantized energy level, 155, 156
Quantum dots, 661, 668
Quantum mechanics, 68, 69, 79, 122, 145, 146, 147, 151
Quantum number, 186, 187
Quantum tunnelling, 658
Quaternary ammonium salts, 565

Rate determining step, 507, 508, 509, 510, 511, 512
Rault's law, 3
Reciprocity relation, 54
Recirculating water, 565
Recombination current, 282
Recursion formula, 161

# Index

Reduced mass, 172, 235
Reference electrode, 337
Relaxation method, 594
Repulsive potential exponent, 323, 326
Residual current, 363, 369
Residual entropy, 30
Resistance polarization, 543, 547, 548
Resolving power, 231
Reversed bias, 284
Rhydberg, 200
Rice–Herzfeld mechanism, 592
Rigid rotator, 170, 173, 174, 176, 179
Rotating platinum electrode, 373
Rotational energy, 86, 178, 236
Rotational partition function, 84, 97
Rotational quantum number, 177, 178, 235, 236
    constant, 84, 237
    spectra, 237
    spectroscopy, 234
RRK theory, 631, 633
RRKM theory, 634, 635
Russel–Saunder's coupling, 221
Rusting, 554

Sackur–Tetrode equation, 82
Sacrificial anode, 560
Saturated calomel electrode, 338
Saxen relation, 62, 64
$SbCl_3$, 563
Scale, 564
Scaling, 565
Scanning electron microscopy, 652, 653
    images, 652
Scanning tip, 658
Scanning tunnelling microscope (STM), 657
Schottky defect, 301, 302, 303, 305
Schrodinger wave equation, 146, 147, 151, 197
Screw dislocation, 311
Selected area diffraction pattern, 657
Selection rule, 236
Semiconductor, 270, 271, 273, 276, 487, 488, 492
Signal to noise ratio, 230, 655
Silver coulometer, 393
Silver zinc cell, 535, 536
Silver/silver chloride electrode, 339, 348
Single sphere model, 610, 615
Slight gas degeneracy, 130
Sludge, 564
Smog, 517
Sodium benzoate, 563
Sodium gate, 498

Sodium hexametaphosphate (SHMP), 564
Sodium sulphur cell, 535, 536, 537
Sol-Gel method, 566
Sol–gel process, 646, 647
Solid state chemistry, 264, 295
    electrode, 349
    reaction, 264, 266
Solubility, 385
Soot, 517
Space lattice, 247, 311
    groups, 642, 643
Spatial resolution, 655
Specific heat, 290
Spherical eigenfunction, 183
    coordinate, 198
Standard hydrogen scale (SHS), 559
State function, 4
Statistical mechanics, 105, 106, 111, 112, 116, 140
Statistical thermodynamics, 68, 69, 105, 305
Statistical weight factor, 69, 631
Steady state, 50, 582, 583, 584
    approximation, 618
Sterling approximation, 74, 82, 106, 117, 120, 305
Stern Gerry equation, 558
Stern model, 431, 432
    theory, 485
Stochiometric defect, 300
Streaming potential, 63
    current, 64
Striping analysis, 376
Sulphate reducing bacteria (SRB), 567
Super equivalent adsorption, 467, 468
Superconductivity, 285, 287, 292, 299
Superconductor, 293
    high temperature, 291
    low temperature, 291
    organic, 293
    type-I, 287
    type-II, 287
Surface plasmon resonance, 652
Surface potential, 412
Surface tension, 410, 411, 421, 422
    excess, 420, 422, 464
Surface treatment, 565
Surrounding, 3
Symmetric top, 235
Symmetry factor, 442, 448, 453, 455, 456, 458, 497
    coefficient, 546
Synergism parameter, 572
System, 3

$T^3$ law, 329

Tafel
 constant, 546
 equation, 546
 slope, 553
Thermal decomposition reaction, 266
Thermionic devices, 655
Thermodynamic probability, 107
Thermodynamics, 3
 first law, 9, 10, 11
 second law, 12
 third law, 29, 30
 zeroth law, 9
Thermo-ionic emission, 130, 135
Thermo-mechanical effect, 64, 65, 66
Thermo-molecular pressure difference, 64
Thermo-osmotic flow, 50
Thiourea, 563
Topography, 652, 660
Transition state theory, 605
Transition temperature, 285
Translational energy, 80, 81, 82
 enthalpy, 84
 entropy, 82
Translational partition function, 80, 81
 degree of freedom, 101
Transmission coefficient, 607
Transmission electron microscope, 654
Transpassivity, 552
Transport phenomenon, 51, 53
 coefficient, 53
Tripolyphosphates (STPP), 564
Trisodium phosphates (TSP), 564
Tunneling current, 501
Turn over number, 617
Type I superconductor, 287
Type II superconductor, 288

Unimolecular reaction rate theory, 626
Unit cell, 247, 312
 body centred cubic, 253, 254, 258, 312, 643
 face centred cubic, 253, 254, 258, 312, 643
 plane, 249
 simple cubic, 253, 254, 312, 643

Valence band, 488, 489, 490
 electrons, 488
van der Waals, 660
Vant Hoff, 3, 600

Vapour phase corrosion inhibitors (VPCI), 567
Variation method, 206, 207, 208, 210
Vibrational degree of freedom, 95, 96, 234, 606
 quantum number, 87, 167
Vibrational energy, 87
 entropy, 88
Vibrational partition function, 87
Volatile corrosion inhibitors, 567
Volta potential, 427
Voltage efficiency, 521
Voltametry, 361
Vortex state, 288

Wagner reaction mechanism, 265
Water electrode interaction, 463
Wave function, 155
 mechanics, 145
Wehnelt cylinder, 656
Weiss coefficient, 249, 250
Wet corrosion, 539
Width of spectral line, 231
Work
 adiabatic, 11
 electrical, 10, 412, 442, 443, 476
 electrochemical, 10
 expansion, 10, 519
 function, 17, 77, 94, 500
 gravitational, 10
 maximum, 10
 mechanical, 10
 reversible expansion, 10
 stretching, 10
 surface, 10
World Technology Evaluation Center (WTEC), 642

X-ray diffraction, 247, 251, 252, 260, 643, 648
 crystallography, 247, 313
 diffraction pattern, 257, 260
 spectrometer, 314

Yield strength, 644
Yttrium aluminium garnet, 656

Zenar voltage, 284
Zero point energy, 92, 93, 94, 243
Zeroth law of thermodynamics, 9
Zinc chromate $ZnCrO_4$, 563